FOLLOW THE
14-PART CASE STUDY ONLINE

How do the information systems concepts you're learning apply to the real world? Use the password card in this textbook to follow one running case study, made up of 14 parts, that provides up-to-date and relevant business examples online at **www.course.com/mis/stair**. Go to *Principles of Information Systems, Eighth Edition*, and then visit each chapter to see how principles are applied in the business setting of Whitmann Price Consulting. Corresponding discussion and critical thinking questions are included in the book.

Chapter	Running Case Topic
Chapter 1: An Introduction to Information Systems	A New Systems Initiative
Chapter 2: Information Systems in Organizations	Addressing the Needs of the Organization
Chapter 3: Hardware: Input, Processing and Output Devices	Choosing Hardware
Chapter 4: Software: Systems and Application Software	Software Considerations
Chapter 5: Organizing Data and Information	Database Considerations
Chapter 6: Telecommunications and Networks	Telecommunications and Network Considerations
Chapter 7: The Internet, Intranets, and Extranets	Telecommunications and Network Considerations
Chapter 8: Electronic and Mobile Commerce	E-Commerce Considerations
Chapter 9: Enterprise Systems	Enterprise Systems
Chapter 10: Information and Decision Support Systems	MIS and DSS Considerations
Chapter 11: Knowledge Management and Specialized Information Systems	Knowledge Management and Specialized Information Systems
Chapter 12: Systems Development: Investigation and Analysis	Systems Investigation and Analysis Considerations
Chapter 13: Systems Development: Design, Implementation, Maintenance, and Review	Systems Design, Implementation, and Review Considerations
Chapter 14: The Personal and Social Impact of Computers	Security, Privacy, and Ethical Considerations

LOOK FOR THESE OTHER POPULAR
THOMSON COURSE TECHNOLOGY
MIS TITLES

EIGHTH EDITION

Principles of Information Systems

A Managerial Approach

Eighth Edition

Ralph M. Stair
Florida State University

George W. Reynolds
The University of Cincinnati

THOMSON
COURSE TECHNOLOGY

Australia · Canada · Mexico · Singapore · Spain · United Kingdom · United States

THOMSON

COURSE TECHNOLOGY

Principles of Information Systems, A Managerial Approach, Eighth Edition
by Ralph M. Stair and George W. Reynolds

Senior Acquisitions Editor:
Maureen Martin

Product Manager:
Kate Hennessy

Development Editor:
Lisa Ruffolo, The Software Resource

Editorial Assistant:
Erin Kennedy

Content Product Manager:
Matthew Hutchinson

Marketing Manager:
Penelope Crosby

Text Designer:
Jennifer McCall

Cover Designer:
Elizabeth Paquin

Composition House:
Digital Publishing Solutions

Photo Researcher:
Abby Reip

Copyeditor:
Karen Annett

Proofreader:
Kathy Orrino

Indexer:
Liz Cunningham

Printed in Canada
1 2 3 4 5 6 7 8 9 QWV 09 08 07 06 05

Library of Congress Cataloging-in-Publication Data
Stair, Ralph M.
      Principles of information systems: a managerial approach / Ralph M. Stair, George W. Reynolds.
      p. cm.
      Includes bibliographical references and index.
      ISBN 1-4239-0115-0 (alk. paper)
      1. Management information systems. I. Reynolds, George Walter, 1944– II. Title.

      T58.6 .S72 2001
      658.4'038—dc21
      00-047561

For more information contact Thomson Course Technology, 25 Thomson Place, Boston, Massachusetts, 02210.

Or find us on the World Wide Web at: www.course.com

Disclaimer
Thomson Course Technology reserves the right to revise this publication and make changes from time to time in its content without notice.

ISBN: 1-4239-0115-0. International Student Edition ISBN: 1-4239-0119-3. Instructor Edition ISBN: 1-4239-0120-7.

BRIEF CONTENTS

PART 1   **An Overview**

     Chapter 1   An Introduction to Information Systems

     Chapter 2   Information Systems in Organizations

PART 2   **Information Technology Concepts**

     Chapter 3   Hardware: Input, Processing, and Output Devices

     Chapter 4   Software: Systems and Application Software

     Chapter 5   Organizing Data and Information

     Chapter 6   Telecommunications and Networks

     Chapter 7   The Internet, Intranets, and Extranets

     Part 2   World Views Case: Australian Film Studio Benefits from Broadband Networks...and Location

     Part 2   World Views Case: Virtual Learning Environment Provides Instruction Flexibility

PART 3   **Business Information Systems**

     Chapter 8   Electronic and Mobile Commerce

     Chapter 9   Enterprise Systems

     Chapter 10   Information and Decision Support Systems

     Chapter 11   Knowledge Management and Specialized Information Systems

     Part 3   World Views Case: Kulula.com:The Trials and Tribulations of a South African Online Airline

PART 4   **Systems Development**

     Chapter 12   Systems Development: Investigation and Analysis

     Chapter 13   Systems Design, Implementation, Maintenance, and Review

     Part 4   World Views Case: Brandon Trust Develops MIS Capability for Improved Operations and Services

     Part 4   World Views Case: Strategic Enterprise Management at International Manufacturing Corporation (IMC)

PART 5   **Information Systems in Business and Society**

     Chapter 14   The Personal and Social Impact of Computers

     Glossary

     Index

CONTENTS

Preface xxi

PART 1   An Overview   1

Chapter 1   An Introduction to Information Systems   2
Nissan Motor Company, Japan   3

Information Concepts 5

     Data, Information, and Knowledge   5

     The Characteristics of Valuable Information   6

     The Value of Information   7

System Concepts 8

     System Performance and Standards   8

Information Systems @ Work: Speeding Up Insurance Claims with New Information Systems 10

What is an Information System? 11

     Input, Processing, Output, Feedback   11

     Manual and Computerized Information Systems   12

     Computer-Based Information Systems   12

Business Information Systems 17

     Electronic and Mobile Commerce   17

     Enterprise Systems: Transaction Processing Systems and Enterprise Resource Planning   19

     Information and Decision Support Systems   20

     Specialized Business Information Systems: Knowledge Management, Artificial Intelligence, Expert Systems, and Virtual Reality   23

Systems Development 26

     Systems Investigation and Analysis   26

     Systems Design, Implementation, and Maintenance and Review   27

Information Systems in Society, Business, and Industry 27

     Security, Privacy, and Ethical Issues in Information Systems and the Internet   27

     Computer and Information Systems Literacy   28

Ethical and Societal Issues: Snipermail Executive Serves Hard Time 29

     Information Systems in the Functional Areas of Business   30

     Information Systems in Industry   30

Global Challenges in Information Systems 31

Case One: Shroff International Travel Care Opens Door to Philippines 37

Case Two: Discovery Communications Digs Out of Mountains of Documents 38

Chapter 2   Information Systems in Organizations   40
General Motors, United States   41

Organizations and Information Systems 42

     Organizational Structures   46

     Organizational Culture and Change   49

Reengineering and Continuous Improvement   50

User Satisfaction and Technology Acceptance   52

Quality   53

Outsourcing, On-Demand Computing, and Downsizing   54

Ethical and Societal Issues: Victims of Technology?   55

Competitive Advantage   56

Factors That Lead Firms to Seek Competitive Advantage   57

Strategic Planning for Competitive Advantage   58

Performance-Based Information Systems   59

Productivity   59

Return on Investment and the Value of Information Systems   60

**Information Systems @ Work: Wal-Mart Stores Manage Supply Chain
with RFID   62**

Risk   63

Careers in Information Systems   63

Roles, Functions, and Careers in the IS Department   64

Typical IS Titles and Functions   66

Other IS Careers   68

Case One: Itochu Kenzai Uses IS to Reduce Organizational Waste   74

Case Two: CIO Randy Mott Shakes Up Hewlett-Packard   75

PART 2   **Information Technology Concepts   79**

Chapter 3   Hardware: Input, Processing, and Output Devices   80
                Siebel Systems, United States   81

Computer Systems: Integrating the Power of Technology   82

Hardware Components   83

Hardware Components in Action   84

Processing and Memory Devices: Power, Speed, and Capacity   85

Processing Characteristics and Functions   85

Memory Characteristics and Functions   87

Multiprocessing   90

Parallel Computing   90

Secondary Storage   92

Access Methods   93

Devices   93

Ethical and Societal Issues: Backup Storage Media and Identity Theft   95

Enterprise Storage Options   98

Input and Output Devices: The Gateway to Computer Systems   99

Characteristics and Functionality   99

Input Devices   100

**Information Systems @ Work: Tablet PCs Serve Up Big Benefits for Holiday Inn
Restaurants   104**

Output Devices   106

Special-Purpose Input and Output Devices   109

Computer System Types, Selection, and Upgrading   **110**

    Computer System Types   110

    Selecting and Upgrading Computer Systems   114

Case One: Staples Unwires Point-of-Sale Terminals   **121**

Case Two: Bank of New York: Mainframes or Distributed Systems   **122**

Chapter 4   Software: Systems and Application Software   126

El-Al Airlines, Israel   127

An Overview of Software   **128**

    Systems Software   128

    Application Software   129

    Supporting Individual, Group, and Organizational Goals   129

    Installing and Removing Software for PCs   130

Systems Software   **131**

    Operating Systems   132

    Current Operating Systems   135

    Workgroup Operating Systems   139

    Enterprise Operating Systems   140

    Operating Systems for Small Computers, Embedded Computers, and

        Special-Purpose Devices   141

    Utility Programs   142

Information Systems @ Work: Gas Natural: Improving Customer Service with

    Mobile Software   **143**

    Middleware   145

Application Software   **146**

    Overview of Application Software   146

    Personal Application Software   147

    Workgroup Application Software   154

    Enterprise Application Software   154

    Application Software for Information, Decision Support, and Specialized

        Purposes   155

Programming Languages   **156**

    The Evolution of Programming Languages   156

Software Issues and Trends   **160**

    Software Bugs   160

    Copyrights and Licenses   160

    Open-Source Software   160

Ethical and Societal Issues: When Software Goes Wrong   **161**

    Shareware, Freeware, and Public Domain Software   162

    Software Upgrades   163

    Global Software Support   164

Case One: DreamWorks SKJ Goes Completely Open Source   **169**

Case Two: Select Comfort Finds Comfort in ERP   **170**

Chapter 5   Organizing Data and Information   174
Valero Energy, United States   175
Data Management   **177**
The Hierarchy of Data   177
Data Entities, Attributes, and Keys   178
Data Modeling and the Relational Database Model   **181**
Data Modeling   181
The Relational Database Model   182
Database Management Systems   **186**
Overview of Database Types   187
Providing a User View   188
Creating and Modifying the Database   188
Storing and Retrieving Data   190
Manipulating Data and Generating Reports   191
Database Administration   193
Popular Database Management Systems   194
Special-Purpose Database Systems   195
Selecting a Database Management System   195
Using Databases with Other Software   196
Database Applications   **196**
Linking Databases to the Internet   196
Data Warehouses, Data Marts, and Data Mining   197
Ethical and Societal Issues: Google: Protecting User Data from the
Government   **200**
Business Intelligence   202
Distributed Databases   202
Online Analytical Processing (OLAP)   204
Information Systems @ Work: Boston Celtics and the Business of Basketball   **205**
Object-Oriented and Object-Relational Database Management
Systems   206
Visual, Audio, and Other Database Systems   206
Case One: Motorola's Biometrics Solutions Require Large Self-Managing
Database   **212**
Case Two: Consolidating Database Management Systems in Georgia State
Courts   **213**

Chapter 6   Telecommunications and Networks   216
Virgin Entertainment Group, United States   217
An Overview of Telecommunications   **219**
Basic Communications Channel Characteristics   220
Channel Bandwidth   220
Communications Media   220
Guided Transmission Media Types   221
Telecommunications Hardware   225
Services   226
Digital Subscriber Line (DSL)   227

Networks and Distributed Processing 228

    Network Types 229

Ethical and Societal Issues: Sharing Workstations at Sisters of Mercy Health System 231

Information Systems @ Work: Wireless Networks at General Motors Connect People, Machines, and Parts 234

    Distributed Processing 235

    Client/Server Systems 235

    Communications Software and Protocols 236

    Wireless Communications Protocols 237

    Communications Software 240

Telecommunications and Network Applications 241

    Linking Personal Computers to Mainframes and Networks 241

    Voice Mail 241

    Home and Small Business Networks 242

    Electronic Document Distribution 242

    Call Centers 242

    Telecommuting and Virtual Workers and Workgroups 243

    Videoconferencing 244

    Electronic Data Interchange 245

    Public Network Services 246

    Electronic Funds Transfer 246

    Distance Learning 247

    Team Collaboration Systems 247

    Specialized Systems and Services 248

Case One: U.S. Census-Takers Go Wireless 253

Case Two: Delivering People More Efficiently Through Telecommunications 254

Chapter 7   The Internet, Intranets, and Extranets 256
    Firwood Paints, United Kingdom 257

Use and Functioning of the Internet 259

    How the Internet Works 261

    Internet Service Providers 263

Information Systems @ Work: Liz Claiborne Moves to IP Technologies 265

The World Wide Web 266

    Web Browsers 268

    Search Engines and Web Research 269

    Web Programming Languages 271

    Developing Web Content 272

    Web Services 273

Internet and Web Applications 274

    Business Uses of the Web 274

    E-Mail, Instant Messaging, and Push Technology 275

    Internet Cell Phones and Handheld Computers 276

    Career Information and Job Searching 277

    Telnet and FTP 278

Web Log (Blog), Video Log (Vlog), and Podcasting   278

Usenet and Newsgroups   279

Chat Rooms   279

Internet Phone and Videoconferencing Services   280

Content Streaming   281

Shopping on the Web   281

Web Auctions   281

Music, Radio, Video, and TV on the Internet   282

Office on the Web   284

Internet Sites in Three Dimensions   284

Other Internet Services and Applications   284

Intranets and Extranets   285

Net Issues   287

Ethical and Societal Issues: Surfing the Web in Anonymity   288

Case One: Web Portal for State of Michigan Provides Services and Savings   293

Case Two: FUJI Market Switches to Web-Based Ordering System   294

Part 2 World Views Case: Australian Film Studio Benefits from Broadband Networks...and Location   298

Part 2 World Views Case: Virtual Learning Environment Provides Instruction Flexibility   300

PART 3   Business Information Systems   303

Chapter 8   Electronic and Mobile Commerce   304

Nike, United States   305

An Introduction to Electronic Commerce   306

Business-to-Business (B2B) E-Commerce   306

Business-to-Consumer (B2C) E-Commerce   307

Consumer-to-Consumer (C2C) E-Commerce   307

eGovernment   307

Multistage Model for E-Commerce   308

E-Commerce Challenges   310

An Introduction to Mobile Commerce   312

Mobile Commerce in Perspective   312

Technology Needed for Mobile Commerce   312

Ethical and Societal Issues: The Power of the Cell Phone   313

Electronic and Mobile Commerce Applications   314

Retail and Wholesale   314

Manufacturing   315

Marketing   316

Investment and Finance   317

Information Systems @ Work: Acquiring a Mortgage Off and Online   319

Auctions   320

Anywhere, Anytime Applications of Mobile Commerce   320

Advantages of Electronic and Mobile Commerce   321

Threats to Electronic and Mobile Commerce   322

    Security   322

    Theft of Intellectual Property   322

    Fraud   323

    Invasion of Consumer Privacy   324

    Lack of Internet Access   324

    Return on Investment   324

    Legal Jurisdiction   324

    Taxation   325

Strategies for Successful E-Commerce   325

    Defining the Web Site Functions   325

    Establishing a Web Site   326

    Building Traffic to Your Web Site   326

    Maintaining and Improving Your Web Site   327

Technology Infrastructure required to Support E-Commerce and M-Commerce   327

    Hardware   328

    Web Server Software   328

    E-Commerce Software   329

    Electronic Payment Systems   331

Case One: Wells Fargo's Web-Enabled ATMs   338

Case Two: Japanese Mobile Music Service Provides Peek into M-Commerce Future   339

Chapter 9   Enterprise Systems   342

    Barilla Wasa, Germany   343

An Overview of Enterprise Systems: Transaction Processing Systems and Enterprise Resource Planning   345

    Traditional Transaction Processing Methods and Objectives   346

Transaction Processing Activities   349

    Data Collection   350

    Data Editing   351

Information Systems @ Work: Retailers Turn to Smart Carts   352

    Data Correction   353

    Data Manipulation   353

    Data Storage   353

    Document Production and Reports   353

Control and Management Issues   354

    Disaster Recovery Plan   354

    Transaction Processing System Audit   355

Ethical and Societal Issues: Battening Down the Systems for the Next Hurricane Season   356

Traditional Transaction Processing Applications   357

    Order Processing Systems   357

    Purchasing Systems   358

    Accounting Systems   359

Enterprise Resource Planning, Supply Chain Management, and Customer Relationship Management 362

An Overview of Enterprise Resource Planning 362

Advantages of ERP 363

Disadvantages of ERP Systems 364

Production and Supply Chain Management 366

Customer Relationship Management and Sales Ordering 367

Financial and Managerial Accounting 370

Hosted Software Model for Enterprise Software 371

International Issues Associated with Enterprise Systems 372

Different Languages and Cultures 372

Disparities in Information System Infrastructure 372

Varying Laws and Customs Rules 372

Multiple Currencies 373

Comparison of Several ERP Systems 373

Case One: Pico Electronics Develops TPS to Support Rapid-Fire Transactions 379

Case Two: Catholic Healthcare West Implements ERP 380

Chapter 10   Information and Decision Support Systems 382
Beauté Prestige International, France 383

Decision Making and Problem Solving 384

Decision Making as a Component of Problem Solving 384

Programmed versus Nonprogrammed Decisions 386

Optimization, Satisficing, and Heuristic Approaches 387

Sense and Respond 388

An Overview of Management Information Systems 388

Management Information Systems in Perspective 388

Inputs to a Management Information System 390

Outputs of a Management Information System 390

Characteristics of a Management Information System 392

Ethical and Societal Issues: Management Information Systems and National Security 394

Functional Aspects of the MIS 395

Financial Management Information Systems 396

Manufacturing Management Information Systems 397

Marketing Management Information Systems 400

Information Systems @ Work: Fully Integrated Decision Support 403

Human Resource Management Information Systems 404

Other Management Information Systems 405

An Overview of Decision Support Systems 406

Characteristics of a Decision Support System 407

Capabilities of a Decision Support System 409

A Comparison of DSS and MIS 410

Components of a Decision Support System 410

The Database 410

The Model Base   412

The User Interface or Dialogue Manager   412

Group Support Systems   413

Characteristics of a GSS That Enhance Decision Making   413

GSS Software   415

GSS Alternatives   417

Executive Support Systems   417

Executive Support Systems in Perspective   418

Capabilities of Executive Support Systems   419

Case One: American National Insurance Speeds Information to Customers with MIS and BPM   426

Case Two: Transit New Zealand Reroutes Information to Avoid Traffic Jams   427

Chapter 11   Knowledge Management and Specialized Information Systems   430

Lafarge, France   431

Knowledge Management Systems   432

Overview of Knowledge Management Systems   433

Data and Knowledge Management Workers and Communities of Practice   433

Obtaining, Storing, Sharing, and Using Knowledge   434

Technology to Support Knowledge Management   435

An Overview of Artificial Intelligence   437

Artificial Intelligence in Perspective   437

The Nature of Intelligence   438

The Difference Between Natural and Artificial Intelligence   440

The Major Branches of Artificial Intelligence   440

Expert Systems   440

Robotics   440

Vision Systems   443

Natural Language Processing and Voice Recognition   443

Learning Systems   444

Neural Networks   444

Other Artificial Intelligence Applications   445

Ethical and Societal Issues: Online Matchmaker eHarmony   446

An Overview of Expert Systems   447

When to Use Expert Systems   447

Components of Expert Systems   448

The Inference Engine   450

The Explanation Facility   450

The Knowledge Acquisition Facility   450

The User Interface   450

Expert Systems Development   451

Participants in Developing and Using Expert Systems   451

Expert Systems Development Tools and Techniques   452

Applications of Expert Systems and Artificial Intelligence   453

Information Systems @ Work: **Call Centers Use Artificial Intelligence to Improve Service to Customers 456**

Virtual Reality 457

    Interface Devices 457

    Forms of Virtual Reality 458

    Virtual Reality Applications 458

Other Specialized Systems 461

Case One: BMW Drives Virtual Prototypes 468

Case Two: Pacific Gas and Electric Corporation Increases Customer Satisfaction with Speech Recognition 469

Part 3 World Views Case: Kulula.com: The Trials and Tribulations of a South African Online Airline 472

PART 4 Systems Development 475

Chapter 12 Systems Development: Investigation and Analysis 476
Nordmilch, Germany 477

An Overview of Systems Development 478

    Participants in Systems Development 479

    Initiating Systems Development 480

    Information Systems Planning and Aligning Corporate and IS Goals 481

Information Systems @ Work: **Australia's Maroochy Shire Council Gains Control over Information Overload 484**

    Establishing Objectives for Systems Development 485

Systems Development Life Cycles 486

    The Traditional Systems Development Life Cycle 487

    Prototyping 488

    Rapid Application Development, Agile Development, Joint Application Development, and Other Systems Development Approaches 490

    The End-User Systems Development Life Cycle 491

    Outsourcing and On-Demand Computing 492

Factors Affecting Systems Development Success 494

    Degree of Change 494

Ethical and Societal Issues: **Sentinel: The FBI's New Case Management System 496**

    Quality and Standards 497

    Use of Project Management Tools 498

    Use of Computer-Aided Software Engineering (CASE) Tools 500

    Object-Oriented Systems Development 501

Systems Investigation 502

    Initiating Systems Investigation 502

    Participants in Systems Investigation 502

    Feasibility Analysis 503

    Object-Oriented Systems Investigation 504

    The Systems Investigation Report 504

Systems Analysis   505

General Considerations   505

Participants in Systems Analysis   505

Data Collection   506

Data Analysis   507

Requirements Analysis   511

Object-Oriented Systems Analysis   513

The Systems Analysis Report   514

Case One: Applebee's Simplifies Life for Managers while Improving Service to Customers   520

Case Two: Waeco Pacific Employees Work Together to Create New ERP System   521

Chapter 13   Systems Design, Implementation, Maintenance, and Review   524
Timex, United States   525

Systems Design   526

Logical and Physical Design   526

Object-Oriented Design   527

Information Systems @ Work: Installing Wi-Fi at Southern Wine's Mega Warehouse   528

Interface Design and Controls   529

Design of System Security and Controls   531

Generating Systems Design Alternatives   535

Evaluation Techniques   538

Freezing Design Specifications   539

The Contract   540

The Design Report   540

Systems Implementation   541

Acquiring Hardware from an IS Vendor   542

Acquiring Software: Make or Buy?   543

Acquiring Database and Telecommunications Systems   545

User Preparation   546

IS Personnel: Hiring and Training   547

Site Preparation   547

Data Preparation   547

Installation   547

Testing   548

Start-Up   549

User Acceptance   550

Systems Operation and Maintenance   550

Reasons for Maintenance   551

Ethical and Societal Issues: San Francisco's Bay Area Rapid Transit System Upgrade   552

Types of Maintenance   553

The Request for Maintenance Form   553

Performing Maintenance   553

The Financial Implications of Maintenance   553

The Relationship Between Maintenance and Design   555

Systems Review   555

Types of Review Procedures   556

Factors to Consider During Systems Review   557

System Performance Measurement   557

Case One: Port of Brisbane Develops System to Tighten Security   563

Case Two: The Christian Science Monitor Moves to Open Source for Content Management System   564

Part 4 World Views Case: Brandon Trust Develops MIS Capability for Improved Operations and Services   568

Part 4 World Views Case: Strategic Enterprise Management at International Manufacturing Corporation (IMC)   570

PART 5   Information Systems in Business and Society   573

Chapter 14   The Personal and Social Impact of Computers   574
Johnson & Johnson, United States   575

Computer Waste and Mistakes   576

Computer Waste   576

Computer-Related Mistakes   577

Preventing Computer-Related Waste and Mistakes   577

Establishing Policies and Procedures   577

Implementing Policies and Procedures   578

Monitoring Policies and Procedures   579

Reviewing Policies and Procedures   579

Computer Crime   580

The Computer as a Tool to Commit Crime   581

Cyberterrorism   582

Identity Theft   582

The Computer as the Object of Crime   583

Illegal Access and Use   583

Using Antivirus Programs   585

Information Systems @ Work: Westinghouse Finds New Tools to Guard Against Viruses   588

Information and Equipment Theft   589

Software and Internet Software Piracy   589

Computer-Related Scams   590

International Computer Crime   591

Preventing Computer-Related Crime   591

Crime Prevention by State and Federal Agencies   592

Crime Prevention by Corporations   592

Using Intrusion Detection Software   594

Using Managed Security Service Providers (MSSPs)   595

Internet Laws for Libel and Protection of Decency   595

Preventing Crime on the Internet   597

Privacy Issues   597

Privacy and the Federal Government   597

Privacy at Work   598

E-Mail Privacy   598

Privacy and the Internet   598

Fairness in Information Use   599

Corporate Privacy Policies   602

Ethical and Societal Issues: LexisNexis Gains Public Trust Through Disclosure   603

Individual Efforts to Protect Privacy   604

The Work Environment   604

Health Concerns   605

Avoiding Health and Environmental Problems   605

Ethical Issues in Information Systems   606

Case One: Indigo Books and Music Align Security Strategies with Corporate Goals   612

Case Two: Disgruntled Employee Plants Time Bomb in Get-Rich-Quick Scheme   613

Glossary   617

Index   633

As organizations continue to operate in an increasingly competitive and global marketplace, workers in all business areas including accounting, finance, human resources, marketing, operations management, and production must be well prepared to make significant contributions required for success. Regardless of your future role, you will need to understand what information systems can and cannot do and be able to use them to help you accomplish your work. You will be expected to discover opportunities to use information systems and to participate in the design of solutions to business problems employing information systems. You will be challenged to identify and evaluate information systems options. To be successful, you must be able to view information systems from the perspective of business and organizational needs. For your solutions to be accepted, you must recognize and address their impact on fellow workers, customers, suppliers, and other key business partners. For these reasons, a course in information systems is essential for students in today's high-tech world.

Principles of Information Systems: A Managerial Approach, Eighth Edition, continues the tradition and approach of the previous editions. Our primary objective is to provide the best information systems text and accompanying materials for the first information technology course required of all business students. We want you to learn to use information technology to ensure your personal success in your current or future job and to improve the success of your organization. Through surveys, questionnaires, focus groups, and feedback that we have received from current and past adopters, as well as others who teach in the field, we have been able to develop the highest-quality set of teaching materials available to help you achieve these goals.

Principles of Information Systems: A Managerial Approach, Eighth Edition, stands proudly at the beginning of the IS curriculum and remains unchallenged in its position as the only IS principles text offering the basic IS concepts that every business student must learn to be successful. In the past, instructors of the introductory course faced a dilemma. On one hand, experience in business organizations allows students to grasp the complexities underlying important IS concepts. For this reason, many schools delayed presenting these concepts until students completed a large portion of the core business requirements. On the other hand, delaying the presentation of IS concepts until students have matured within the business curriculum often forces the one or two required introductory IS courses to focus only on personal computing software tools and, at best, merely to introduce computer concepts.

This text has been written specifically for the introductory course in the IS curriculum. *Principles of Information Systems: A Managerial Approach, Eighth Edition*, treats the appropriate computer and IS concepts together with a strong managerial emphasis on meeting business and organizational needs.

APPROACH OF THE TEXT

Principles of Information Systems: A Managerial Approach, Eighth Edition, offers the traditional coverage of computer concepts, but it places the material within the context of meeting business and organizational needs. Placing IS concepts in this context and taking a general management perspective has always set the text apart from general computer books thus making it appealing not only to MIS majors but also to students from other fields of study. The text isn't overly technical, but rather deals with the role that information systems play in an organization and the key principles a manager needs to grasp to be successful. These principles of IS are brought together and presented in a way that is both understandable and relevant. In addition, this book offers an overview of the entire IS discipline, while giving students a solid foundation for further study in advanced IS courses as programming, systems analysis and design, project management, database management, data communications, Web site and systems development, electronic commerce and mobile commerce applications, and decision support. As such, it serves the needs of both general business students and those who will become IS professionals.

The overall vision, framework, and pedagogy that made the previous editions so popular have been retained in the eighth edition, offering a number of benefits to students. We continue to present IS concepts with a managerial emphasis. While the fundamental vision of this market-leading text remains unchanged, the eighth edition more clearly highlights established principles and draws out new ones that have emerged as a result of business, organizational, and technological change.

IS Principles First, Where They Belong

Exposing students to fundamental IS principles is an advantage for students who do not later return to the discipline for advanced courses. Since most functional areas in business rely on information systems, an understanding of IS principles helps students in other course work. In addition, introducing students to the principles of information systems helps future business function managers employ information systems successfully and avoid mishaps that often result in unfortunate consequences. Furthermore, presenting IS concepts at the introductory level creates interest among general business students who may later choose information systems as a field of concentration.

Author Team

Ralph Stair and George Reynolds have teamed up again for the eighth edition. Together, they have more than sixty years of academic and industrial experience. Ralph Stair brings years of writing, teaching, and academic experience to this text. He has written more than twenty-two books and a large number of articles while at Florida State University. George Reynolds brings a wealth of computer and industrial experience to the project, with more than thirty years of experience working in government, institutional, and commercial IS organizations. He has also authored seventeen texts and is an assistant professor at the University of Cincinnati, where he teaches the introductory IS course. The Stair and Reynolds team brings a solid conceptual foundation and practical IS experience to students.

GOALS OF THIS TEXT

Because *Principles of Information Systems: A Managerial Approach, Eighth Edition,* is written for all business majors, we believe it is important not only to present a realistic perspective on IS in business but also to provide students with the skills they can use to be effective business leaders in their organization. To that end, *Principles of Information Systems: A Managerial Approach, Eighth Edition,* has four main goals:

1. To provide a core of IS principles with which every business student should be familiar
2. To offer a survey of the IS discipline that will enable all business students to understand the relationship of IS courses to their curriculum as a whole
3. To present the changing role of the IS professional
4. To show the value of the discipline as an attractive field of specialization

By achieving these goals, *Principles of Information Systems, Eighth Edition,* will enable students, regardless of their major, to understand and use fundamental information systems principles so that they can function more efficiently and effectively as workers, managers, decision makers, and organizational leaders.

IS Principles

Principles of Information Systems: A Managerial Approach, Eighth Edition, although comprehensive, cannot cover every aspect of the rapidly changing IS discipline. The authors, having recognized this, provide students an essential core of guiding IS principles to use as they face the career challenges ahead. Think of principles as basic truths, rules, or assumptions that remain constant regardless of the situation. As such, they provide strong guidance in the face of tough decisions. A set of IS principles is highlighted at the beginning of each chapter. The application of these principles to solve real-world problems is driven home from the opening vignettes to the end-of-chapter material. The ultimate goal of *Principles of Information Systems* is to develop effective, thinking, action-oriented employees by instilling them with principles to help guide their decision making and actions.

Survey of the IS Discipline

This text not only offers the traditional coverage of computer concepts but also provides a broad framework to impart students with a solid grounding in the business uses of technology. In addition to serving general business students, this book offers an overview of the entire IS discipline and solidly prepares future IS professionals for advanced IS courses and their careers in the rapidly changing IS discipline.

Changing Role of the IS Professional

As business and the IS discipline have changed, so too has the role of the IS professional. Once considered a technical specialist, today the IS professional operates as an internal consultant to all functional areas of the organization, being knowledgeable about their needs and competent in bringing the power of information systems to bear throughout the organization. The IS professional views issues through a global perspective that encompasses the entire organization and the broader industry and business environment in which it operates.

The scope of responsibilities of an IS professional today is not confined to just his/her employer but encompasses the entire interconnected network of employees, suppliers, customers, competitors, regulatory agencies, and other entities, no matter where they are located. This broad scope of responsibilities creates a new challenge: how to help an organization survive in a highly interconnected, highly competitive global environment. In accepting that challenge, the IS professional plays a pivotal role in shaping the business itself and ensuring its success. To survive, businesses must now strive for the highest level of customer satisfaction and loyalty through competitive prices and ever-improving product and service quality. The IS professional assumes the critical responsibility of determining the organization's approach to both overall cost and quality performance and therefore plays an important role in the ongoing survival of the organization. This new duality in the role of the IS employee—a professional who exercises a specialist's skills with a generalist's perspective—is reflected throughout the book.

IS as a Field for Further Study

Despite the downturn in the economy at the start of the 21st century, especially in technology-related sectors, the outlook for computer and information systems managers is optimistic. Indeed, employment of computer and information systems managers is expected to grow much faster than the average for all occupations through the year 2012. Technological advancements will boost the employment of computer-related workers; in turn, this will boost the demand for managers to direct these workers. In addition, job openings will result from the need to replace managers who retire or move into other occupations.

A career in IS can be exciting, challenging, and rewarding! It is important to show the value of the discipline as an appealing field of study and that the IS graduate is no longer a technical recluse. Today, perhaps more than ever before, the IS professional must be able to align IS and organizational goals and to ensure that IS investments are justified from a business perspective. The need to draw bright and interested students into the IS discipline is part of our ongoing responsibility. Upon graduation, IS graduates at many schools are among the highest paid of all business graduates. Throughout this text, the many challenges and opportunities available to IS professionals are highlighted and emphasized.

CHANGES IN THE EIGHTH EDITION

We have implemented a number of exciting changes to the text based on user feedback on how the text can be aligned even more closely with how the IS principles and concepts course is now being taught. The following list summarizes these changes:

- **All new opening vignettes.** All of the chapter-opening vignettes are new, and continue to raise actual issues from foreign-based or multinational companies.

- **All new Information Systems @ Work special interest boxes.** Highlighting current topics and trends in today's headlines, these boxes show how information systems are used in a variety of business career areas.

- **All new Ethical and Societal Issues special interest boxes.** Focusing on ethical issues today's professionals face, these boxes illustrate how information systems professionals confront and react to ethical dilemmas.

- **New case studies.** Two new end-of-chapter cases provide a wealth of practical information for students and instructors. Each case explores a chapter concept or problem that a real-world company or organization has faced. The cases can be assigned as individual homework exercises or serve as a basis for class discussion.

- **Integrated, comprehensive, Web case.** New to this edition is an integrated and comprehensive case that runs throughout the text. The case follows the activities of two individuals employed at the fictitious Whitman Price Consulting firm as they are challenged to complete various IS-related projects. Each case has activities related to the material in the chapter in which it is presented.

Each chapter has been completely updated with the latest topics and examples. The following is a summary of the changes.

Chapter 1, An Introduction to Information Systems

Chapter 1 continues to create a framework for the entire book. Major sections in this chapter become entire parts. Any major changes to the book are directly reflected in Chapter 1. For example, several chapter titles and material have been changed in this edition and have been reflected in Chapter 1. The section on transaction processing and enterprise resource planning has been changed to enterprise systems to reflect the emphasis of importance of this material. Chapter 9 is now titled enterprise systems. We also have a new section that introduces knowledge management in Chapter 1. Knowledge management is now an important part of Chapter 11. The principles at the beginning of the chapter and all end-of-chapter material have been updated to reflect the changes in Chapter 1. This chapter continues to emphasize the benefits of an information system, including speed, accuracy, reduced costs, and increased functionality. New examples have been introduced in the section on computer-based information systems to give students a better understanding of these important components, and references of corporate IS usage have been stressed in the major section of business information systems. The section on Data Versus Information has been changed to Data, Information, and Knowledge to give knowledge and knowledge management more emphasis. Knowledge workers have been briefly defined and discussed. The best corporate users of IS, as reported by popular computer and business journals, are explored. As mentioned above, the section on transaction processing and enterprise resource planning will be called enterprise systems to reflect the nature of these systems. We continue stress that ERP systems can replace many applications with one unified set of programs. We have include a new section near the end of the chapter on global challenges in information systems that investigate cultural challenges, language challenges, time and distance challenges, infrastructure challenges, currency challenges, product and service challenges, technology transfer issues, and a variety of state and national laws. In addition, we have information on the eras of globalization as described in his book, The World Is Flat, by Thomas Friedman. Most sections in this chapter have been trimmed to reduce the overall book length. The section on system and modeling concepts, for example, has been reduced to cover only essential topics. The section on telecommunications, networks, and the Internet has been reduced slightly. The section on specialized business information systems has reduced to some extent, especially the section on virtual reality. The section on security, privacy, and ethical issues has also been trimmed slightly.

Chapter 2, Information Systems in Organizations

Chapter 2 gives an overview of business organizations and presents a foundation for the effective and efficient use of IS in a business environment. We have stressed that the traditional mission of the IS organization "to deliver the right information to the right person at the right time" has broadened to include how this information is used to improve overall performance and help people and organizations achieve their goals. New examples of how companies use information systems have been introduced throughout the chapter. In the section on virtual organizational structure, we discuss that virtual organizational structures allow work to be separated from location and time. Work can be done anywhere, anytime. The material on multidimensional structure has been deleted. The section on technology acceptance will now include information on user satisfaction. We have introduced the concept of business process reengineering (BPR) in the section on reengineering. We have stressed that a competitive advantage can result in higher quality products, better customer service, and lower costs. The section on strategic planning for competitive advantage has been modified to include cost leadership, differentiation, and niche strategies to deliver a competitive advantage. The material on total quality management and six sigma has been shortened and streamlined. The section on careers has been updated with a new list of top US employers, what makes a satisfied IS worker on the job, and new information on various visas and their impact on the workforce. As with previous editions, this chapter continues to stress performance-based management with new examples of how companies can use information systems to improve productivity and increase return on investment. The section on careers in information systems includes information on knowledge workers.

Chapter 3, Hardware: Input, Processing, and Output Devices

This chapter has been completely updated to reflect the latest equipment and computer capabilities. Both AMD and Intel computer chips are discussed including the use of Intel chips in Apple computers. Dual-core and multicore processors are fully covered. Parallel computing including SIMD and MIMD parallel processing and grid computing are covered. The use of flash memory (hard memory disks) to speed up hard drives is covered. The latest cost comparisons of various forms of data storage are presented. Holographic versatile disk technology is discussed. New examples of how companies use input and output devices to streamline operations and increase efficiency are presented. The discussion of RFID is expanded and includes new examples and cost data. The use of new cell phones with cameras and PDA capabilities is covered. Mobile computing devices including tablet PCs are covered.

As with all the chapters, all new examples are presented, and the opening vignette, special features, and cases are all new.

Chapter 4, Software: Systems and Application Software

In the introductory section that gives an overview of software, we have included new material on installing and removing software on PCs. This new material includes detailed steps that can be followed and new figures that show what is involved. We now include information on the use of end-user license agreements (EULAs). After reading the EULA material, people normally have to click on an "I agree" button before they can use the software. To reduce the length of the chapter, we have condensed the section on supporting individual, group, and enterprise goals. We mention that the enterprise sphere of influence includes business partners, such as suppliers that provide raw materials, retail companies that store and sell a company's products, and shipping companies that transport raw materials to the plant and finished goods to retail outlets. We also mention that both system and application software can be run from other devices, including CDs, DVDs, and small USB key chain or thumb storage devices.

We have included new material, references, and examples in the systems software section. The features and functions of Vista, Microsoft's new operating system, are fully discussed. We have reduced the material on earlier operating systems in the section on Microsoft PC Operating Systems. Tiger, the newest operating system from Apple, is discussed. We also cover Spotlight for Mac OS X Tiger, which is Apple's fast search technology that allows people to locate documents, music, images, e-mails, contacts, and other documents on their computers. Some of the historical information on Apple operating systems has been deleted. We have updated the section on utility programs. For example, we now cover virtualization software that can be used to make computers act like or simulate other computers. The result is often called a virtual machine. The section on application software has been completely updated. The section on the types and functions of application software has been renamed to Overview of Application Software. Figures 4.9 and 4.10 have been deleted as suggested by a reviewer.

The section on programming languages has also been fully updated. This section now includes material on Visual Studio.Net from Microsoft and includes information on Visual BASIC.NET, Visual C++.NET, Visual C#, and Visual J#. We have removed some of the historical information on visual languages from this section. We have also included more information about building object-oriented languages and the process that is used. We have included a new figure on the use of an object-oriented programming language. The section on selecting a programming language has been deleted to keep a reasonable chapter length. The software issues and trends section includes new examples and material. We have trimmed some of the material in the open-source software section and replaced it with new examples and references. This material also includes both compelling reasons for using open-source software and some of the disadvantages. In one case, it is estimated that open-source software may have saved an organization $10 million. On the other hand, getting good support using open-source software can be difficult in some cases.

Chapter 5, Organizing Data and Information

The chapter title has been simplified to database systems. We have included more information and examples on the ability of an organization to gather data, interpret it, and act on it quickly in highly competitive marketplaces. The brief paragraph on knowledge management has been expanded into a new section at the beginning of Chapter 11 on knowledge management. Relational database models continue to be emphasized. The role of the database administrator (DBA) to plan, design, create, operate, secure, monitor, and maintain databases is discussed. We have also stressed important database functions, including storing data, transforming data into information, providing security, and the ability to perform queries. The material on the database administrator (DBA) in the section on database management systems has been reorganized to stress that a DBA is expected to have a clear understanding of the fundamental business of the organization, be proficient in the use of selected database management systems, and stay abreast of emerging technologies and new design approaches. We have trimmed the material on open-source and special-purpose databases, as requested by a reviewer. As with all chapters, examples, boxes, cases, and references have redone to be fresh and new. A new section on data analysis shows students how "bad data" in data tables can be corrected by breaking a data table into two or more data tables. Database applications, including data warehouses and business intelligence, have stressed. We have reduced the total length of this chapter to some extent. The introductory paragraphs have been reduced. The material on the traditional approach to database management has also been trimmed. We have condensed the section on selecting a database as suggested by a reviewer. The material on data modeling has been cut to some extent. We have cut the material on subschemas and open database connectivity. We have also made other cuts to shorten the chapter.

Chapter 6 Telecommunications and Networks

An expanded section on voice and data convergence discusses the "big picture" of how we are changing from a telecommunications infrastructure based on POTS (plain old telephone service) using twisted pair copper wires intended for voice communications using analog signals, circuit switching, and "dumb" voice telephones. The new infrastructure is aimed at transmitting data, consists of fiber optic links used to send digital signals based on packet switching, and assume intelligent user devices that provide addressing information. The focus is on converting all signals to digital form and using a single digital network to carry all communications (including voice and data). The coverage of cellular transmission has been expanded and updated to include a discussion of the features and characteristics of 1G, 2G, 2.5G, 3G, and MMDS cellular communications. Personal Area Networks and Metropolitan Area Networks have been added under the discussion of Network Types. The discussion of communications protocols has been simplified and streamlined with additional protocols discussed including frame relay, FireWire, IEEE 802.11 g, IEEE 802.16 (WiMax), and IEEE 802.20 (MBWA). A discussion on the role of a hub has been added under the section on Network Switching Devices. The use of Call Centers and Electronic Funds Transfer were added to the discussion of telecommunications applications. Again, as with all the chapters, all new examples are presented, and the opening vignette, special features, and cases are all new.

Chapter 7, The Internet, Intranets, and Extranets

The updates to this chapter reflect all the changes in the Internet and its use. There is new information on the large investments that some countries are making in Internet infrastructure. There is also new information about how some countries, such as China, attempt to restrict Internet usage. In Figure 7.1, we have replaced the routers with a box that contains routers and gateways to include the importance of gateways. We have stressed the use of a Web portal as a starting point to enter the Internet and visit other Internet sites, including corporate Web portals. We have also included a brief history of the Internet in a new figure. The material on search engines includes new information on how search engines now allow searches for audio signals, video images, and other information. We also include more information on performing research on the Internet. Information on Wikipedia, an encyclopedia that has been created and edited by thousands of people, is also new to this edition. The section is now titled search engines and Web research. In the section on the business use of the Web, we have added some of the negative aspects of conducting business using the Internet. We also address the important issue of sales tax for commerce conducted over the Internet. Internet plug-ins, programs that enhance the capabilities of an Internet browser, to play audio, view video, and other tasks are covered. Browsers, including Internet Explorer for Windows and Mozilla Firefox and Safari for Apple computers are explored. The section on Internet and Web applications has been completely updated. We have included new information on VoIP, including companies such as Vonage, Verizon, and others. We also have new material and examples on e-mail, music, Web auctions, and text messaging. We have included new information on blogs and podcasts. The section on music, radio, and video on the Internet has been expanded to include Web TV. The section on other Internet services and applications has also been updated. We now cover how the Internet can be used to translate from one language to another. We also include information on a Google service that allows people to download popular TV programs. We continue to stress the section on Developing Web Content. This section includes new development tools that can make creating and posting Internet content easier. We have included information on Google's free Web site analytics that allow Web developers to monitor the number of Web page visits per day, product merchandising, and conversion—the number of Web site visits that are converted to buyers. This section also summarizes a study describing attractive Web page designs. New material on how some Web developers are creating programs and procedures to combine two or more Web sites into one Web site, called a mash-up, is also discussed. Getting TV over the Internet using IPTV has also been explored. We have reduced the chapter length in several sections. The section on the use and functioning of the Internet has been shortened.

The sections on search engines and Web programming languages have been trimmed. The material on VoIP has also been reduced to some extent. The section on Web services has also been reduced to some extent. The section on net issues has been reduced.

Chapter 8, Electronic and Mobile Commerce

The title of this chapter now includes mobile commerce to emphasize its importance in today's business. A brief summary of the status of e-commerce around the world has been added. M-commerce is introduced as a new business model. Many examples of M-commerce are provided, the technology required is discussed, and the anywhere/anytime capability of M-commerce is addressed. A new section exploring the advantages and disadvantages of e-commerce and m-commerce has been added. The section on threats to e-commerce and m-commerce has been expanded to include the new topics of lack of Internet access, return on investment, legal jurisdiction, and taxation. Web services are discussed including a definition, use of standards, examples, and a Gartner Group forecast of the growth of Web services. Phishing is discussed as a serious e-commerce issue. The role of the FTC in regulating Internet activities is addressed. The use of Web site customer experience technology to analyze the usability of a site is discussed. As with all the chapters, all new examples are presented, and the opening vignette, special features, and cases are all new.

Chapter 9, An Overview of Enterprise Systems: Transaction Processing and Enterprise Resource Planning Systems

The title of the chapter has been changed to reflect the increased emphasis on enterprise systems and ERP. The material on traditional transaction processing has been streamlined with the same sections included but with a reduced amount of detail. Sections are presented that cover the important management topics of disaster recovery planning and transaction processing system audit. The section on ERP has been greatly modified to focus more coverage on just what is an ERP and what business processes are involved and affected. The material on ERP has been reorganized to include an overview of ERP; advantages of ERP; disadvantages of ERP; and a description of how ERP is used in Production and Supply Chain Management, Customer Relationship Management, and Financial and Managerial Accounting. An interesting table identifying large, successful ERP implementations by company has been added. The hosted software model for enterprise software is also discussed and examples of companies using this approach are provided. The list of vendors providing CRM systems has been updated to be current. The list of vendors providing ERP systems has been updated to reflect recent acquisitions. The international issues associated with implementing and operating an ERP system are discussed.

Chapter 10, Information and Decision Support Systems

Like other chapters, all boxes, cases, and the opening vignette are new to this edition. We have included many new examples of how managers and decision makers can use information and decision support systems to achieve personal and corporate goals. The section on problem solving and decision making has new examples, including the use of MRI techniques to look inside the brain as it makes decisions and solves problems. A brief section on the sense and response (SaR) approach has been included with examples. We have emphasize that many of the information and decision support topics discussed in the chapter can be built into some ERP systems, discussed in Chapter 9. The use of digital dashboards and business activity monitoring has been introduced and explained in the section on decision support system, along with examples and references. The section on optimization includes new examples, showing the huge cost savings of the technique. The use of data warehousing, data marts, and data mining, first introduced in Chapter 5, has been stressed in this chapter. New corporate examples of the use of data warehousing, data marts, and data mining that can be used to provide information and decision support have been placed throughout the chapter. Figure 10.16 on a DSS now includes user interface and dialogue manager in the bottom box and the discussion in the text. This chapter continues to stress group support systems. We discuss some additional features of groupware, including group monitoring, idea collection, and idea organizing and voting features. New examples and approaches have been explored, including the use of the Web to deliver group support. We have trimmed several sections in this chapter. The section on decision making and problem solving has been shortened to some extent. The sections on overview of management information systems and the functional aspects of the MIS have also trimmed. Other cuts have also been made in other sections of the chapter.

Chapter 11, Specialized Information Systems

A new section on knowledge management has been added to the beginning of this chapter. This section is a natural extension of the material in Chapter 10 on information and decision support systems. It leads to a discussion of some of the special-purpose systems discussed in the chapter, including expert systems and knowledge bases. This section contains all new examples and references. The other topics discussed in this chapter also contains numerous new examples and references in robotics, vision systems, expert systems, virtual reality, and a variety of other special-purpose systems. We discuss embedded artificial intelligence, where artificial intelligence capabilities and applications are placed inside of products and services. Examples of embedded expert systems placed robots to allow them to make decisions on how to navigate difficult terrain or in a computer system or to diagnose a problem, such as a potential virus. These systems can take corrective action before the problem becomes serious and shuts down the operation of the computer. Material in this chapter has been cut to keep a reasonable length. The section on an overview of artificial intelligence has been reduced slightly. The section on robotics has also been reduced slightly, along with the section on natural language processing. More significant cuts have been made in the section on expert systems.

Chapter 12, Systems Investigation and Analysis

The section on performance objectives now includes information and examples on reducing project risk. We've included new examples on initiating systems development projects, such as the U.S. Department of Energy's (DOE) electrical infrastructure. Harrah's Casino that initiated a $130 million systems development project as a result of a merger with Caesars is another new example in this section. There are new examples in the section on quality and standards, including a British systems development project that was behind schedule and more than $400 million over budget. The impact of laws and regulations, such as Sarbanes-Oxley, has been included. Some of the disadvantages of new laws and regulations are also discussed, including the increased use of outsourcing by U.S. companies to reduce costs in complying with laws and regulations. The importance of user involvement and top management support has been included in the section on factors affecting systems development success. New examples of systems development failure have also been included, including United Airlines failed systems development project to implement a computerized baggage system at Denver International Airport (DIA). The material on systems development and the Internet and ERP have been integrated throughout the chapter. New examples of outsourcing, along with guidelines for outsourcing are new to this edition. New references on utility and on-demand computing have also been included. We have also included information on the scope of outsourcing, including hardware maintenance and management, software development, database systems, networks and telecommunications, Internet and Intranet operations, hiring and staffing, and the development of procedures and rules regarding the information system. In the section on systems development life cycles, we have inserted a new example of rapid application development (RAD). We briefly introduce the Java 2 Enterprise Edition (J2EE) specifications, which are a set of rules and procedures developed by Sun Microsystems to help reduce costs, minimize development complexities, increase security, and reduce development times in the section on quality and standards as recommended by a reviewer. We continue to stress the importance of software reuse, with new examples and references. We have trimmed many sections to reduce the length of the chapter. In addition, we have deleted the sections on the Internet, ERP, and the capability maturity model from the chapter. The end-of-chapter material has been completely updated with new questions and exercises. For example, we have included a new Web exercise on project management. As with other editions, all cases are new to this edition.

Chapter 13, Systems Design, Implementation, Maintenance, and Review

The section on systems design includes many new examples and references. It provides new information on the importance of enterprise-rights management. It also includes new references and examples on software reuse. In the section on the acquisition of software, we have also included new examples. We have included a new table that compares developing software internally versus purchasing it from outside vendors. There is new material on the use of Commercial off-the-shelf (COTS) development, which involves the use of commonly available products from software vendors in systems development projects. The COTS development process involves combining software from various vendors into a finished system. In the section on software development, we mention that Microsoft's Visual Studio includes several development tools, including Visual Basic, Visual C++, Visual J++, Visual FoxPro, and Visual InterDev. We also discuss that these stand-alone languages and products can access common databases and component libraries that are managed by Visual Studio. In the section on implementation, we mention that start up is also called cutover. In the systems operations section, we mention that the cost of operating a system can be far greater than the cost of developing it originally. We discuss that getting the most out of a new or modified system during its operation is the most important aspect of systems operations for many organizations. We also discuss that total maintenance costs can also be very high. New references for autonomic computing have been included. Trimming most sections has reduced the chapter length. For example, we have reduced the material on logical and physical design. We have also reduced the coverage of interface design and controls section to some extent along with the specific disaster recovery procedures. The section on software acquisition has been reduced in length.

Chapter 14, The Personal and Social Effects of Computers

As with all the chapters, all new examples are presented, and the opening vignette, special features, and cases are all new. All new examples of computer waste, mistakes, and fraud are provided including mention of the largest computer-related consumer fraud in the US by the Gambino crime family. Also added is a discussion by an FBI special agent citing how serious and widespread computer crime has become. The results of the latest available Computer Security Institute/FBI survey are presented showing the key issues. The most active computer viruses are identified. The serious issue of theft of laptops and their often more valuable data is discussed including the infamous Veterans Affairs case. Many new examples of computer scams are discussed. Additional, new privacy issues are discussed including the NSA capture of tens of millions of telephone records of US citizens. The potential privacy risks of children using social network services like MySpace are also discussed. The list of Web sites that provide health related info was eliminated. The idea of what is legal may not be ethical is raised to cause students to think about this important issue.

WHAT WE HAVE RETAINED FROM THE SEVENTH EDITION

The eighth edition builds on what has worked well in the past; it retains the focus on IS principles and strives to be the most current text on the market.

- **Overall principle.** This book continues to stress a single-all-encompassing theme: The right information, if it is delivered to the right person, in the right fashion, and at the right time, can improve and ensure organizational effectiveness and efficiency.

- **Information systems principles.** Information system principles summarize key concepts that every student should know. This important feature is a convenient summary of key ideas presented at the start of each chapter.

- **Global perspective.** We stress the global aspects of information systems as a major theme.

- **Learning objectives linked to principles.** Carefully crafted learning objectives are included with every chapter. The learning objectives are linked to thee Information Systems Principles and reflect what a student should be able to accomplish after completing a chapter.

- **Opening vignettes emphasize international aspects.** All of the chapter-opening vignettes raise actual issues from foreign-based or multinational companies.

- **Why Learn About features.** Each chapter has a "Why Learn About" section at the beginning of the chapter to pique student interest. The section sets the stage for students by briefly describing the importance of the chapter's material to the students-whatever their chosen field.

- **Information Systems @ Work special interest boxes.** Each chapter has an entirely new Information Systems @ Work box that shows how information systems are used in a variety of business career areas.

- **Ethical and Societal Issues special interest boxes.** Each chapter includes an "Ethical and Societal Issues" box that presents a timely look at the ethical challenges and the societal impact of information systems

- **Current examples, boxes, cases, and references.** As we have in each edition, we take great pride in presenting the most recent examples, boxes, cases, and references throughout the text. Some of these were developed at the last possible moment, literally weeks before the book went into publication. Information on new hardware and software, the latest operating systems, mobile commerce, the Internet, electronic commerce, ethical and societal issues, and many other current developments can be found throughout the text. Our adopters have come to expect the best and most recent material. We have done everything we can to meet or exceed these expectations.

- **Summary linked to principles.** Each chapter includes a detailed summary with each section of the summary tied to an associated information system principle.

- **Self-assessment tests.** This popular feature helps students review and test their understanding of key chapter concepts.

- **Career exercises.** End-of-chapter career exercises ask students to research how a topic discussed in the chapter relates to a business area of their choice. Students are encouraged to use the Internet, the college library, or interviews to collect information about business careers.

- **End-of-chapter cases.** Two end-of-chapter cases provide a wealth of practical information for students and instructors. Each case explores a chapter concept of problem that a real-world company or organization has faced. The cases can be assigned as individual homework exercises or serve as a basis for class discussion.

- **World Views cases.** While the text has always stressed the global factors affecting information systems, these factors are emphasized even more through the World View cases. These cases are written from people around the world.

STUDENT RESOURCES

MIS Companion CD

We are pleased to include in every textbook a free copy of Course Technology's MIS Companion CD, which is comprised of training lessons in Excel, Access, and MIS concepts. The Companion CD is integrated throughout the book. Wherever you see the CD icon in the chapter margins you will know that you can find additional related material on the CD.

Student Online Companion Web Site

We have created an exciting online companion, password protected for students to utilize as they work through the Eighth Edition. In the front of this text you will find a key code that provides full access to a robust Web site, located at *www.course.com/mis/stair*. This Web resource includes the following features:

- **PowerPoint slides**
  Direct access is offered to the book's PowerPoint presentations that cover the key points from each chapter. These presentations are a useful study tool.

- **Classic cases**
  A frequent request from adopters is that they'd like a broader selection of cases to choose from. To meet this need, a set of over 85 cases from the fifth, sixth, and seventh editions of the text are included here. These are the authors' choices of the "best cases" from these editions and span a broad range of companies and industries.

- **Links to useful Web sites**
  Chapters in *Principles of Information Systems, Eighth Edition* reference many interesting Web sites. This resource takes you to links you can follow directly to the home pages of those sites so that you can explore them. There are additional links to Web sites that the authors, Ralph Stair and George Reynolds, think you would be interested in checking out.

- **Hands-on activities**
  Use these hands-on activities to test your comprehension of IS topics and enhance your skills using Microsoft® Office applications and the Internet. Using these links, you can access three critical-thinking exercises per chapter; each activity asks you to work with an Office tool or do some research on the Internet.

- **Test yourself on IS**
  This tool allows you to access 20 multiple-choice questions for each chapter; test yourself and then submit your answers. You will immediately find out what questions you got right and what you got wrong. For each question that you answer incorrectly, you are given the correct answer and the page in your text where that information is covered. Special testing software randomly compiles 20 questions from a database of 50 questions, so you can quiz yourself multiple times on a given chapter and get some new questions each time.

- **Glossary of key terms**
  The glossary of key terms from the text is available to search.

- **Online readings**
  This feature provides you access to a computer database which contains articles relating to hot topics in Information Systems.

INSTRUCTOR RESOURCES

The teaching tools that accompany this text offer many options for enhancing a course. And, as always, we are committed to providing one of the best teaching resource packages available in this market.

Instructor's Manual

An all-new *Instructor's Manual* provides valuable chapter overviews; highlights key principles and critical concepts; offers sample syllabi, learning objectives, and discussion topics; and features possible essay topics, further readings and cases, and solutions to all of the end-of-chapter questions and problems, as well as suggestions for conducting the team activities. Additional end-of-chapter questions are also included. As always, we are committed to providing the best teaching resource packages available in this market.

Sample Syllabus

A sample syllabus with sample course outlines is provided to make planning your course that much easier.

Solutions

Solutions to all end-of-chapter material are provided in a separate document for your convenience.

Test Bank and Test Generator

ExamView® is a powerful objective-based test generator that enables instructors to create paper-, LAN- or Web-based tests from test banks designed specifically for their Course Technology text. Instructors can utilize the ultra-efficient QuickTest Wizard to create tests in less than five minutes by taking advantage of Course Technology's question banks or customizing their own exams from scratch. Page references for all questions are provided so you can cross-reference test results with the book.

PowerPoint Presentations

A set of impressive Microsoft PowerPoint slides is available for each chapter. These slides are included to serve as a teaching aid for classroom presentation, to make available to students on the network for chapter review, or to be printed for classroom distribution. Our presentations help students focus on the main topics of each chapter, take better notes, and prepare for examinations. Instructors can also add their own slides for additional topics they introduce to the class.

Figure Files

Figure files allow instructors to create their own presentations using figures taken directly from the text.

DISTANCE LEARNING

Course Technology, the premiere innovator in management information systems publishing, is proud to present online courses in WebCT and Blackboard.

- **Blackboard and WebCT Level 1 Online Content.** If you use Blackboard or WebCT, the test bank for this textbook is available at no cost in a simple, ready-to-use format. Go to *www.course.com* and search for this textbook to download the test bank.

- **Blackboard and WebCT Level 2 Online Content.** Blackboard 5.0 and 6.0 as well as Level 2 and WebCT Level 2 courses are also available for *Principles of Information Systems*, Eighth Edition. Level 2 offers course management and access to a Web site that is fully populated with content for this book.

For more information on how to bring distance learning to your course, instructors should contact their Course Technology sales representative.

ACKNOWLEDGMENTS

A book of this size and undertaking requires a strong team effort. We would like to thank all of our fellow teammates at Course Technology and the Software Resource for their dedication and hard work. Special thanks to Elizabeth Paquin, Alyssa Pratt, and Kate Hennessy, our Product Managers. Our appreciation goes out to all the many people who worked behind the scenes to bring this effort to fruition including Abigail Reip, our photo researcher, and Maureen Martin, our Senior Acquisitions Editor. We would like to acknowledge and thank Lisa Ruffolo, our development editor, deserves special recognition for her tireless effort and help in all stages of this project. Kelly Robinson and Matthew Hutchinson, our Content Product Managers, shepherded the book through the production process.

We are grateful to the sales force at Course Technology whose efforts make this all possible. You helped to get valuable feedback from current and future adopters. As Course Technology product users, we know how important you are.

We would especially like to thank Ken Baldauf for his excellent help in writing most of the boxes and cases for this edition. Ken also provided invaluable feedback for many topics discussed in the book.

Ralph Stair would like to thank the Department of Management Information Systems, College of Business Administration, at Florida State University for their support and encouragement. He would also like to thank his family, Lila and Leslie, for their support.

George Reynolds would like to thank the Department of Information Systems, College of Business, at the University of Cincinnati for their support and encouragement. He would also like to thank his family, Ginnie, Tammy, Kim, Kelly, and Kristy, for their patience and support in this major project. In addition, he would like to thank Kristen Duerr and Ralph Stair for asking him to join the writing team back in 1997.

To Our Previous Adopters and Potential New Users

We sincerely appreciate our loyal adopters of the previous editions and welcome new users of *Principles of Information Systems: A Managerial Approach, Eighth Edition*. As in the past, we truly value your needs and feedback. We can only hope the Eighth edition continues to meet your high expectations.

We would especially like to thank reviewers of the eighth edition; focus group members; and reviewers of previous editions

Reviewers for the Eighth Edition

We are indebted to the following reviewers for their perceptive feedback on early drafts of this text:

Kirk Atkinson, *Western Kentucky University*
Phillip D. Coleman, *Western Kentucky University*

Tom Johnston, *University of Delaware*
Efrem Mallach, *University of Massachusetts Dartmouth*
Zibusiso Ncube, *Concordia College*

Reviewers for the First, Second, Third, Fourth, Fifth, Sixth, and Seventh Editions

The following people shaped the book you hold in your hands by contributing to previous editions:

Jill Adams, *Navarro College*

Robert Aden, *Middle Tennessee State University*

A. K. Aggarwal, *University of Baltimore*

Sarah Alexander, *Western Illinois University*

Beverly Amer, *University of Florida*

Noushin Asharfi, *University of Massachusetts*

Yair Babad, *University of Illinois—Chicago*

Cynthia C. Barnes, *Lamar University*

Charles Bilbrey, *James Madison University*

Thomas Blaskovics, *West Virginia University*

John Bloom, *Miami University of Ohio*

Warren Boe, *University of Iowa*

Glen Boyer, *Brigham Young University*

Mary Brabston, *University of Tennessee*

Jerry Braun, *Xavier University*

Thomas A. Browdy, *Washington University*

Lisa Campbell, *Gulf Coast Community College*

Andy Chen, *Northeastern Illinois University*

David Cheslow, *University of Michigan—Flint*

Robert Chi, *California State University—Long Beach*

Carol Chrisman, *Illinois State University*

Miro Costa, *California State University—Chico*

Caroline Curtis, *Lorain County Community College*

Roy Dejoie, *USWeb Corporation*

Sasa Dekleva, *DePaul University*

Pi-Sheng Deng, *California State University—Stanislaus*

Roger Deveau, *University of Massachusetts—Dartmouth*

John Eatman, *University of North Carolina*

Gordon Everest, *University of Minnesota*

Juan Esteva, *Eastern Michigan University*

Badie Farah, *Eastern Michigan University*

Karen Forcht, *James Madison University*

Carroll Frenzel, *University of Colorado—Boulder*

John Gessford, *California State University—Long Beach*

Terry Beth Gordon, *University of Toledo*

Kevin Gorman, *University of North Carolina—Charlotte*

Costanza Hagmann, *Kansas State University*

Bill C. Hardgrave, *University of Arkansas*

Al Harris, *Appalachian State University*

William L. Harrison, *Oregon State University*

Dwight Haworth, *University of Nebraska—Omaha*

Jeff Hedrington, *University of Wisconsin—Eau Claire*

Donna Hilgenbrink, *Illinois State University*

Jack Hogue, *University of North Carolina*

Joan Hoopes, *Marist College*

Donald Huffman, *Lorain County Community College*

Patrick Jaska, *University of Texas at Arlington*

G. Vaughn Johnson, *University of Nebraska—Omaha*

Grover S. Kearns, *Morehead State University*

Robert Keim, *Arizona State University*

Karen Ketler, *Eastern Illinois University*

Mo Khan, *California State University—Long Beach*

Chang E. Koh, *University of North Texas*

Michael Lahey, *Kent State University*

Jan de Lassen, *Brigham Young University*

Robert E. Lee, *New Mexico State University—Carlstadt*

Joyce Little, *Towson State University*

Herbert Ludwig, *North Dakota State University*

Jane Mackay, *Texas Christian University*

Al Maimon, *University of Washington*

James R. Marsden, *University of Connecticut*

Roger W. McHaney, *Kansas State University*

Lynn J. McKell, *Brigham Young University*

John Melrose, *University of Wisconsin—Eau Claire*

Michael Michaelson, *Palomar College*

Ellen Monk, *University of Delaware*

Bertrad P. Mouqin, *University of Mary Hardin-Baylor*

Bijayananda Naik, *University of South Dakota*

Pamela Neely, *Marist College*

Leah R. Pietron, *University of Nebraska—Omaha*

John Powell, *University of South Dakota*

Maryann Pringle, *University of Houston*

John Quigley, *East Tennessee State University*
Mahesh S. Raisinghani, *University of Dallas*
Mary Rasley, *Lehigh-Carbon Community College*
Earl Robinson, *St. Joseph's University*
Scott Rupple, *Marquette University*
Dave Scanlon, *California State University—Sacramento*
Werner Schenk, *University of Rochester*
Larry Scheuermann, *University of Southwest Louisiana*
James Scott, *Central Michigan University*
Vikram Sethi, *Southwest Missouri State University*
Laurette Simmons, *Loyola College*
Janice Sipior, *Villanova University*
Anne Marie Smith, *LaSalle University*
Harold Smith, *Brigham Young University*
Patricia A. Smith, *Temple College*
Herb Snyder, *Fort Lewis College*
Alan Spira, *University of Arizona*
Tony Stylianou, *University of North Carolina*
Bruce Sun, *California State University—Long Beach*

Howard Sundwall, *West Chester University*
Hung-Lian Tang, *Bowling Green State University*
William Tastle, *Ithaca College*
Gerald Tillman, *Appalachian State University*
Duane Truex, *Georgia State University*
Jean Upson, *Lorain County Community College*
Misty Vermaat, *Purdue University—Calumet*
David Wallace, *Illinois State University*
Michael E. Whitman, *University of Nevada—Las Vegas*
David C. Whitney, *San Francisco State University*
Goodwin Wong, *University of California—Berkeley*
Amy Woszczynski, *Kennesaw State University*
Judy Wynekoop, *Florida Gulf Coast University*
Myung Yoon, *Northeastern Illinois University*

Focus Group Contributors for the Third Edition

Mary Brabston, *University of Tennessee*
Russell Ching, *California State University—Sacramento*
Virginia Gibson, *University of Maine*
Bill C. Hardgrave, *University of Arkansas*
Al Harris, *Appalachian State University*
Stephen Lunce, *Texas A & M International*
Merle Martin, *California State University—Sacramento*
Mark Serva, *Baylor University*
Paul van Vliet, *University of Nebraska—Omaha*

OUR COMMITMENT

We are committed to listening to our adopters and readers and to developing creative solutions to meet their needs. The field of IS continually evolves, and we strongly encourage your participation in helping us provide the freshest, most relevant information possible.

We welcome your input and feedback. If you have any questions or comments regarding *Principles of Information Systems: A Managerial Approach, Eighth Edition,* please contact us through Course Technology or your local representative, via e-mail at mis@course.com, via the Internet at *www.course.com,* or address your comments, criticisms, suggestions, and ideas to:

Ralph Stair
George Reynolds
Course Technology
25 Thomson Place
Boston, MA 02210

PART · 1 ·

An Overview

Chapter 1   An Introduction to Information Systems
Chapter 2   Information Systems in Organizations

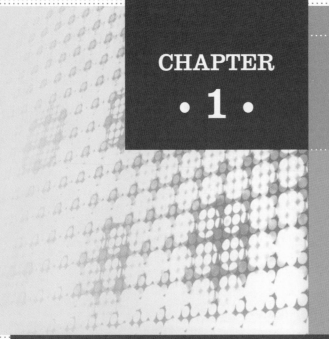

CHAPTER

• 1 •

An Introduction to Information Systems

PRINCIPLES	LEARNING OBJECTIVES
▪ The value of information is directly linked to how it helps decision makers achieve the organization's goals.	▪ Discuss why it is important to study and understand information systems. ▪ Distinguish data from information and describe the characteristics used to evaluate the quality of data.
▪ Computers and information systems are constantly making it possible for organizations to improve the way they conduct business.	▪ Name the components of an information system and describe several system characteristics.
▪ Knowing the potential impact of information systems and having the ability to put this knowledge to work can result in a successful personal career, organizations that reach their goals, and a society with a higher quality of life.	▪ List the components of a computer-based information system. ▪ Identify the basic types of business information systems and discuss who uses them, how they are used, and what kinds of benefits they deliver.
▪ System users, business managers, and information systems professionals must work together to build a successful information system.	▪ Identify the major steps of the systems development process and state the goal of each.
▪ Information systems must be applied thoughtfully and carefully so that society, business, and industry can reap their enormous benefits.	▪ Describe some of the threats to security and privacy that information systems and the Internet can pose. ▪ Discuss the expanding role and benefits of information systems in business and industry.

Information Systems in the Global Economy Nissan Motor Company, Japan
Nissan's 180-Degree Turn with Parts Distribution

In the early 1990s, Nissan Motor Company, Ltd. was highly respected as Japan's number two automotive company. In the mid 1990s, Nissan suffered an unfortunate series of events that overwhelmed the company with debt and nearly put it out of business. The company was revived thanks to a buy-in from Renault Motors and a new CEO named Carlos Ghosen. At the turn of the millennium, Nissan was working hard to reverse their downward spiral with a recovery program called "Nissan 180," the code name for the 180-degree turn that was required of the business.

One key to turning around Nissan's decline was the complete re-engineering of its North American parts distribution system—the pipeline that supplied parts to its 1,300 North American dealers. This information system was extremely complex, tracking 1.75 billion parts in transit at any point in time—approaching one trillion parts per year. Nissan knew that their distribution system was fraught with inefficiencies that detracted from their customer satisfaction ratings.

Nissan's "fill rate," its ability to fulfill their dealers' inventory needs, was measured at 95 percent while its key competitor's fill rate was at 99 percent. Although this might not seem a big difference, it indicates that 5 percent of Nissan customers who dropped off their vehicles for repair had to rent a car and wait several days for the repair to be completed due to the unavailability of parts.

To fix the system, Nissan reduced the number of parts suppliers they used from 700 to 350. Then it worked to improve the flow of parts from those suppliers. The focus was on delivering the right parts to the right dealers at the right time. This area of business is called supply chain management. Information systems are highly valued for their ability to streamline supply chain management. Nissan searched for the right information system to help with its supply chain. It ended up choosing a solution provided by two companies: IBM and Viewlocity Inc.

Viewlocity provided its Control Tower software, a popular supply chain visibility and event management system. IBM worked to integrate the software solution into Nissan's current information systems and into those of Nissan suppliers. The Control Tower software lets Nissan watch inventories more closely, order from suppliers earlier, and plan ahead. The results for Nissan have been significant: it has reduced procurement costs by 22 percent; increased on-time deliveries; for improved communication between supplier, carrier, and dealer; and improved customer satisfaction. It now projects a return on investment of up to 68 percent in three years.

Nissan's investment in solving one small problem in its business processes provides insight into the complexity of evaluating and developing information systems in today's businesses. The smooth flow of information in an organization has a dramatic effect on a business' ability to compete.

As you read this chapter, consider the following:

- How can improvements to information systems increase an organization's ability to fulfill its mission and compete in its industry?
- Consider the components of a computer-based information system. How does the quality of each component affect the effectiveness of a system?

Why Learn About Information Systems?

Information systems are used in almost every imaginable profession. Sales representatives use information systems to advertise products, communicate with customers, and analyze sales trends. Managers use them to make multimillion-dollar decisions, such as whether to build a manufacturing plant or research a cancer drug. Corporate lawyers use information systems to develop contracts and other legal documents. From a small music store to huge multinational companies, businesses of all sizes could not survive without information systems to perform accounting and finance operations. Regardless of your college major or chosen career, information systems are indispensable tools to help you achieve your career goals. Learning about information systems can help you land your first job, earn promotions, and advance your career.

This chapter presents an overview of information systems. The sections on hardware, software, databases, telecommunications, e-commerce and m-commerce, transaction processing and enterprise resource planning, information and decision support, special purpose systems, systems development, and ethical and societal issues are expanded to full chapters in the rest of the book. Let's get started by exploring the basics of information systems.

information system (IS)
A set of interrelated components that collect, manipulate, store, and disseminate data and information and provide a feedback mechanism to meet an objective.

People and organizations use information everyday. Many retail chains, for example, collect data from their stores to help them stock what customers want and to reduce costs. The collective components that perform these tasks are often called an information system. An **information system (IS)** is a set of interrelated components that collect, manipulate, store, and disseminate data and information and provide a feedback mechanism to meet an objective. It is the feedback mechanism that helps organizations achieve their goals, such as increasing profits or improving customer service. Businesses can use information systems to increase revenues and reduce costs. In Mantua, Italy, a consortium of cheese makers uses Radio Frequency Identification (RFID) chips and a computerized information system to track its cheese.[1] The chips contain information about where and when the cheese was produced, how the cheese looks and tastes, and what it will cost. The chip also helps prevent counterfeit cheese from entering the market. According to Carlo Buttasi, the technical director for the dairy consortium, "We calculate that [RFID] will eventually reduce our operating costs by up to 50%." This book emphasizes the benefits of an information system that the Italian dairy consortium is enjoying, including speed, accuracy, reduced costs, and increased functionality.

We interact with information systems every day, both personally and professionally. We use automated teller machines at banks, access information over the Internet, select information from kiosks with touch screens, and let checkout clerks scan our purchases using bar codes and scanners. Major *Fortune* 500 companies spend more than $1 billion per year on information systems. Knowing the potential of information systems and putting this knowledge to work can result in a successful career, organizations that reach their goals, and a society with a higher quality of life.

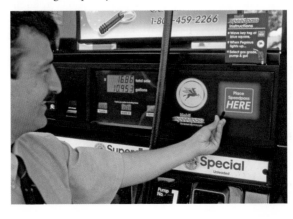

Information systems are everywhere. A customer at the gas pump waves a keychain tag at a reader that sends the information to a network to verify the customer's profile and credit information. The terminal processes the transaction, prints a receipt, and the customer's credit/check card is automatically billed.

(Source: © Michael Newman/Photo Edit.)

Information systems usually involve computers, and together, they are constantly changing the way organizations conduct business. Information systems that help organizations carry on daily operations are called operational systems. For example, Wal-Mart uses operational systems to transfer supplies from distribution centers and suppliers, and then stock store shelves with products and services that customers purchase. Computers and information systems also act as command-and-control systems that monitor processes and help supervisors control them. For example, air traffic control centers use these types of systems to monitor and direct planes in their air space. Also becoming popular are *embedded information systems*, which are placed within or as part of a product or service.

Today we live in an information economy. Information itself has value, and commerce often involves the exchange of information rather than tangible goods. Systems based on computers are increasingly being used to create, store, and transfer information. Using information systems, investors make multimillion-dollar decisions, financial institutions transfer billions of dollars around the world electronically, and manufacturers order supplies and distribute goods faster than ever before. Computers and information systems will continue to change businesses and the way we live. To prepare for these innovations, you need to be familiar with fundamental information concepts.

INFORMATION CONCEPTS

Information is a central concept of this book. The term is used in the title of the book, in this section, and in almost every chapter. To be an effective manager in any area of business, you need to understand that information is one of an organization's most valuable resources. This term, however, is often confused with *data*.

Data, Information, and Knowledge

Data consists of raw facts, such as an employee number, number of hours worked in a week, inventory part numbers, or sales orders. As shown in Table 1.1, several types of data can represent these facts. When facts are arranged in a meaningful manner, they become information. **Information** is a collection of facts organized so that they have additional value beyond the value of the facts themselves. For example, sales managers might find that knowing the total monthly sales suits their purpose more (i.e., is more valuable) than knowing the number of sales for each sales representative. Providing information to customers can also help companies increase revenues and profits.

data
Raw facts, such as an employee number, number of hours worked in a week, inventory part numbers, or sales orders.

information
A collection of facts organized in such a way that they have additional value beyond the value of the facts themselves.

Data	Represented by
Alphanumeric data	Numbers, letters, and other characters
Image data	Graphic images and pictures
Audio data	Sound, noise, or tones
Video data	Moving images or pictures

Table 1.1

Types of Data

Data represents real-world things. Hospitals and healthcare organizations, for example, maintain patient medical data, such as patient IDs and medical record numbers, which represents actual patients with specific health situations. In many cases, hospitals and healthcare organizations are converting data to electronic form.[2] Some organizations have developed electronic records management (ERM) systems to store, organize, and control important data.[3] However, data—simply raw facts—has little value beyond its existence. For example, consider data as pieces of railroad track in a model railroad kit. Each piece of track has little value beyond its inherent value as a single object. However, if you define a relationship among the pieces of the track, they will gain value. By arranging the pieces in a certain way, a railroad layout begins to emerge (see Figure 1.1a). Data and information work the same way. Rules and relationships can be set up to organize data into useful, valuable information.

The type of information created depends on the relationships defined among existing data. For example, you could rearrange the pieces of track to form different layouts. Adding new or different data means you can redefine relationships and create new information. For instance, adding new pieces to the track can greatly increase the value—in this case, variety and fun—of the final product. You can now create a more elaborate railroad layout (see Figure 1.1b). Likewise, a sales manager could add specific product data to his sales data to create monthly sales information organized by product line. The manager could use this information to determine which product lines are the most popular and profitable.

Figure 1.1

Defining and Organizing Relationships Among Data Creates Information

(a)

(b)

process
A set of logically related tasks performed to achieve a defined outcome.

knowledge
The awareness and understanding of a set of information and ways that information can be made useful to support a specific task or reach a decision.

Turning data into information is a **process,** or a set of logically related tasks performed to achieve a defined outcome. The process of defining relationships among data to create useful information requires knowledge. **Knowledge** is the awareness and understanding of a set of information and the ways that information can be made useful to support a specific task or reach a decision. Having knowledge means understanding relationships in information. Part of the knowledge you need to build a railroad layout, for instance, is understanding how much space you have for the layout, how many trains will run on the track, and how fast they will travel. Selecting or rejecting facts according to their relevancy to particular tasks is based on the knowledge used in the process of converting data into information. Therefore, you can also think of information as data made more useful through the application of knowledge. *Knowledge workers (KWs)* are people who create, use, and disseminate knowledge, and are usually professionals in science, engineering, business, and other areas.

In some cases, people organize or process data mentally or manually. In other cases, they use a computer. In the earlier example, the manager could have manually calculated the sum of the sales of each representative, or a computer could calculate this sum. Where the data comes from or how it is processed is less important than whether the data is transformed into results that are useful and valuable. This transformation process is shown in Figure 1.2.

Figure 1.2

The Process of Transforming Data into Information

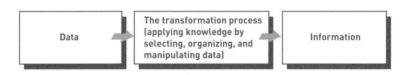

The Characteristics of Valuable Information

To be valuable to managers and decision makers, information should have the characteristics described in Table 1.2. These characteristics make the information more valuable to an organization. Many shipping companies, for example, can determine the exact location of inventory items and packages in their systems, and this information makes them responsive to their customers. In contrast, if an organization's information is not accurate or complete, people can make poor decisions, costing thousands, or even millions, of dollars. Many claim, for example, that the collapse and bankruptcy of some companies, such as drug companies and energy-trading firms, was a result of inaccurate accounting and reporting information, which led investors and employees alike to misjudge the actual state of the company's finances and suffer huge personal losses. As another example, if an inaccurate forecast of future demand indicates that sales will be very high when the opposite is true, an organization can invest millions of dollars in a new plant that is not needed. Furthermore, if information is not relevant, not delivered to decision makers in a timely fashion, or too complex to understand, it can be of little value to the organization.

Table 1.2

Characteristics of Valuable Information

Characteristics	Definitions
Accessible	Information should be easily accessible by authorized users so they can obtain it in the right format and at the right time to meet their needs.
Accurate	Accurate information is error free. In some cases, inaccurate information is generated because inaccurate data is fed into the transformation process. (This is commonly called garbage in, garbage out [GIGO].)
Complete	Complete information contains all the important facts. For example, an investment report that does not include all important costs is not complete.
Economical	Information should also be relatively economical to produce. Decision makers must always balance the value of information with the cost of producing it.
Flexible	Flexible information can be used for a variety of purposes. For example, information on how much inventory is on hand for a particular part can be used by a sales representative in closing a sale, by a production manager to determine whether more inventory is needed, and by a financial executive to determine the total value the company has invested in inventory.
Relevant	Relevant information is important to the decision maker. Information showing that lumber prices might drop might not be relevant to a computer chip manufacturer.
Reliable	Reliable information can be depended on. In many cases, the reliability of the information depends on the reliability of the data-collection method. In other instances, reliability depends on the source of the information. A rumor from an unknown source that oil prices might go up might not be reliable.
Secure	Information should be secure from access by unauthorized users.
Simple	Information should be simple, not overly complex. Sophisticated and detailed information might not be needed. In fact, too much information can cause information overload, whereby a decision maker has too much information and is unable to determine what is really important.
Timely	Timely information is delivered when it is needed. Knowing last week's weather conditions will not help when trying to decide what coat to wear today.
Verifiable	Information should be verifiable. This means that you can check it to make sure it is correct, perhaps by checking many sources for the same information.

Depending on the type of data you need, some quality attributes become more valuable than others. For example, with market-intelligence data, some inaccuracy and incompleteness is acceptable, but timeliness is essential. Market intelligence might alert you that competitors are about to make a major price cut. The exact details and timing of the price cut might not be as important as being warned far enough in advance to plan how to react. On the other hand, accuracy, verifiability, and completeness are critical for data used in accounting for the use of company assets such as cash, inventory, and equipment.

The Value of Information

The value of information is directly linked to how it helps decision makers achieve their organization's goals. For example, the value of information might be measured in the time required to make a decision or in increased profits to the company. Consider a market forecast that predicts a high demand for a new product. If you use this information to develop the new product and your company makes an additional profit of $10,000, the value of this information to the company is $10,000 minus the cost of the information. Valuable information can also help managers decide whether to invest in additional information systems and technology. A new computerized ordering system might cost $30,000, but generate an additional $50,000 in sales. The *value added* by the new system is the additional revenue from the increased sales of $20,000. Most corporations have cost reduction as a primary goal. Using information systems, some manufacturing companies have slashed inventory costs by millions of dollars.

SYSTEM CONCEPTS

system
A set of elements or components that interact to accomplish goals.

Figure 1.3

Components of a System

A system's four components consist of input, processing, output, and feedback.

Like information, another central concept of this book is that of a system. A **system** is a set of elements or components that interact to accomplish goals. The elements themselves and the relationships among them determine how the system works. Systems have inputs, processing mechanisms, outputs, and feedback (see Figure 1.3). For example, consider an automatic car wash. Tangible *inputs* for the process are a dirty car, water, and various cleaning ingredients. Time, energy, skill, and knowledge also serve as inputs to the system because they are needed to operate it. Skill is the ability to successfully operate the liquid sprayer, foaming brush, and air dryer devices. Knowledge is used to define the steps in the car wash operation and the order in which those steps are executed.

The *processing mechanisms* consist of first selecting which cleaning option you want (wash only, wash with wax, wash with wax and hand dry, etc.) and communicating that to the operator of the car wash. A *feedback mechanism* is your assessment of how clean the car is. Liquid sprayers shoot clear water, liquid soap, or car wax depending on where your car is in the process and which options you selected. The *output* is a clean car. As in all systems, independent elements or components (the liquid sprayer, foaming brush, and air dryer) interact to create a clean car.

System Performance and Standards

efficiency
A measure of what is produced divided by what is consumed.

System performance can be measured in various ways. **Efficiency** is a measure of what is produced divided by what is consumed. It can range from 0 to 100 percent. For example, the efficiency of a motor is the energy produced (in terms of work done) divided by the energy consumed (in terms of electricity or fuel). Some motors have an efficiency of 50 percent or less because of the energy lost to friction and heat generation.

Efficiency is a relative term used to compare systems. For example, a hybrid gasoline engine for an automobile or truck can be more efficient than a traditional gasoline engine because, for the equivalent amount of energy consumed, the hybrid gasoline engine produces more energy and gets better gas mileage.

Effectiveness is a measure of the extent to which a system achieves its goals. It can be computed by dividing the goals actually achieved by the total of the stated goals. For example, a company might want to achieve a net profit of $100 million for the year with a new information system. Actual profits, however, might only be $85 million for the year. In this case, the effectiveness is 85 percent (85/100 = 85%).

Evaluating system performance also calls for using performance standards. A **system performance standard** is a specific objective of the system. For example, a system performance standard for a marketing campaign might be to have each sales representative sell $100,000 of a certain type of product each year (see Figure 1.4a). A system performance standard for a manufacturing process might be to provide no more than 1 percent defective parts (see Figure 1.4b). After standards are established, system performance is measured and compared with the standard. Variances from the standard are determinants of system performance.

effectiveness
A measure of the extent to which a system achieves its goals; it can be computed by dividing the goals actually achieved by the total of the stated goals.

system performance standard
A specific objective of the system.

Figure 1.4

System Performance Standards

(a)

(b)

Speeding Up Insurance Claims with New Information Systems

Insurance companies employ claims adjusters to visit property damaged by disaster, analyze the damage, and estimate the dollar value required to repair the property. Each estimate can take considerable amount of time and effort. Adjusters might visit four or five properties in the morning and afternoon, then return to the office where they review notes taken at each property, consult reference charts to calculate repair costs, fill out paper forms, and enter data into the corporate information system. A check is then issued from the insurance company to the victim of the disaster. Multiply the complexity of this process by the thousands of claims that an insurance company processes each year, and the result is a lot of paperwork, wasted time, and mistakes.

The executives at Gore Mutual Insurance of Cambridge, Ontario, Canada, wanted to streamline the claims adjustment process, making it more effective and efficient by applying new information systems. They partnered with a start-up software company named Symbility to create a state-of-the-art solution. The goal of the new system was to process claims onsite during the damage inspection. Gore Mutual wanted to eliminate paper notes that adjusters carried back to the office for processing. Even processing on a notebook PC in a van onsite wasn't efficient enough. Symbility turned to tablet and handheld PCs with handwriting recognition to provide the ideal solution. Tablet and handheld PCs allow users to enter data by writing on the touch screen with a stylus.

Symbility designed pen-based software to run on Windows tablet PCs and handheld PCs. Adjusters can now take notes, sketch floor plans and diagrams, even transfer digital photos from cell phones. All forms and reference charts are accessible and easy to manipulate. The adjuster's PC is connected over a wireless network to a special information service provided by Symbility that supports all the calculations, data manipulation, and processing required to adjust claims. For a claims adjuster, it's like holding the power of the corporate server in his hand.

Gore Mutual found that the tablet PC solution let adjusters spend more time meeting with customers in the field. Estimates were calculated quickly and accurately. Symbility estimates that the system results in claims being settled up to six times faster and more accurately.

The pen-based solution was such a hit that Symbility now supports claim processing for many insurance companies at $20 per claim. In this way, Symbility plays the role of an application service provider (ASP). ASPs design and maintain software and systems and lease the use of the software to businesses. This benefits businesses in a number of ways. The business can focus on its primary goals rather than worrying about developing and maintaining the technology components of an information system. Because the ASP's primary purpose is to develop the best systems possible, the quality of the service is typically higher than if the business developed its own system.

As with most good information systems, the end result of the Symbility solution was a huge reduction of tedious procedures for claims adjusters and fewer opportunities for human error. More importantly, the system frees claim adjusters to do what they do best: evaluate damage and help people continue with their lives.

Considering recent devastation by hurricanes, mudslides, forest fires, and other natural disasters, the speed with which claims are processed affects more than an insurance company's bottom line. In catastrophic events, insurance companies are hard pressed to process claims efficiently. Imagine a technology that allows a company to process 6,000 claims in the amount of time that it typically processes 1,000. Such a system could mean the difference between life and death for homeless survivors. Symbility's pen-based claims processing system does just that.

Discussion Questions

1. Explain how the information system components hardware, software, databases, telecommunications, people, and procedures combine to provide a solution for insurance claims adjusters.
2. What benefits did Gore Mutual enjoy by working with Symbility rather than going it on their own?

Critical Thinking Questions

1. What technologies used in this system are more readily available today than they were five years ago? How long might it be before this system becomes outdated?
2. What other professions would benefit from a pen-based wireless system like Symbility's?

SOURCES: John Cox, "Insurance adjusters use pen-based GUI and wireless," *Network World*, October 5, 2005 *www.techworld.com*. "Symbility Solutions to Demonstrate the Power of 'Insurance Mobility,'" PRNewswire, January 17, 2006, *http://biz.yahoo.com/prnews/060117/nytu128.html?.v=41*. Symbility Web site, accessed February 22, 2006, *www.symbilitysolutions.com*.

WHAT IS AN INFORMATION SYSTEM?

As mentioned previously, an information system (IS) is a set of interrelated elements or components that collect (input), manipulate (process), store, and disseminate (output) data and information and provide a reaction (feedback mechanism) to meet an objective (see Figure 1.5). The feedback mechanism is the component that helps organizations achieve their goals, such as increasing profits or improving customer service.

Figure 1.5

The Components of an Information System

Feedback is critical to the successful operation of a system.

Input, Processing, Output, Feedback

Input

In information systems, **input** is the activity of gathering and capturing raw data. In producing paychecks, for example, the number of hours every employee works must be collected before paychecks can be calculated or printed. In a university grading system, instructors must submit student grades before a summary of grades for the semester or quarter can be compiled and sent to the students.

input
The activity of gathering and capturing raw data.

Processing

In information systems, **processing** means converting or transforming data into useful outputs. Processing can involve making calculations, comparing data and taking alternative actions, and storing data for future use. Processing data into useful information is critical in business settings.

Processing can be done manually or with computer assistance. In a payroll application, the number of hours each employee worked must be converted into net, or take-home, pay. Other inputs often include employee ID number and department. The required processing can first involve multiplying the number of hours worked by the employee's hourly pay rate to get gross pay. If weekly hours worked exceed 40 hours, overtime pay might also be included. Then deductions—for example, federal and state taxes, contributions to health and life insurance or savings plans—are subtracted from gross pay to get net pay.

After these calculations and comparisons are performed, the results are typically stored. *Storage* involves keeping data and information available for future use, including output, discussed next.

processing
Converting or transforming data into useful outputs.

Output

In information systems, **output** involves producing useful information, usually in the form of documents and reports. Outputs can include paychecks for employees, reports for managers, and information supplied to stockholders, banks, government agencies, and other groups. In some cases, output from one system can become input for another. For example, output from a system that processes sales orders can be used as input to a customer billing system. The design and manufacture of office furniture provides another example. The salesperson, customer, and furniture designer can go through several design iterations to meet the customer's needs. Special computer programs and equipment create the original design and allow the designer to rapidly revise it. After the customer approves the last design mock-up, the computer creates a bill of materials that goes to manufacturing to produce the order.

Computers typically produce output on printers and display screens. Output can also be handwritten or manually produced reports and documents.

output
Production of useful information, usually in the form of documents and reports.

feedback
Output that is used to make changes to input or processing activities.

Feedback

In information systems, **feedback** is information from the system that is used to make changes to input or processing activities. For example, errors or problems might make it necessary to correct input data or change a process. Consider a payroll example. Perhaps the number of hours an employee worked was entered as 400 instead of 40 hours. Fortunately, most information systems check to make sure that data falls within certain ranges. For number of hours worked, the range might be from 0 to 100 hours because it is unlikely that an employee would work more than 100 hours in a week. The information system would determine that 400 hours is out of range and provide feedback. The feedback is used to check and correct the input on the number of hours worked to 40. If undetected, this error would result in a very high net pay on the printed paycheck!

Feedback is also important for managers and decision makers. For example, a furniture maker could use a computerized feedback system to link its suppliers and plants. The output from an information system might indicate that inventory levels for mahogany and oak are getting low—a potential problem. A manager could use this feedback to decide to order more wood from a supplier. These new inventory orders then become input to the system. In addition to this reactive approach, a computer system can also be proactive—predicting future events to avoid problems. This concept, often called **forecasting**, can be used to estimate future sales and order more inventory before a shortage occurs. Forecasting is also used to predict the strength of hurricanes and possible landing sites, future stock-market values, and who will win a political election.

forecasting
Predicting future events to avoid problems.

Manual and Computerized Information Systems

As discussed earlier, an information system can be manual or computerized. For example, some investment analysts manually draw charts and trend lines to assist them in making investment decisions. Tracking data on stock prices (input) over the last few months or years, these analysts develop patterns on graph paper (processing) that help them determine what stock prices are likely to do in the next few days or weeks (output). Some investors have made millions of dollars using manual stock analysis information systems. Of course, today many excellent computerized information systems follow stock indexes and markets and suggest when large blocks of stocks should be purchased or sold (called *program trading*) to take advantage of market discrepancies.

Program trading systems allow traders to keep up with swift changes in stock prices and make better decisions for their investors.

(Source: © Yellow Dog Productions/ Getty Images.)

Computer-Based Information Systems

A **computer-based information system (CBIS)** is a single set of hardware, software, databases, telecommunications, people, and procedures that are configured to collect, manipulate, store, and process data into information. For example, a company's payroll, order entry, or inventory-control system is an example of a CBIS. CBISs can also be embedded into products. Some new cars and home appliances include computer hardware, software, databases, and even telecommunications to control their operations and make them more useful. This is often called embedded, pervasive, or ubiquitous computing. CBISs have evolved into sophisticated analysis tools. Because older CBISs didn't have sufficient power or speed, today's computers are just now analyzing decades-old data from the Apollo missions to the

computer-based information system (CBIS)
A single set of hardware, software, databases, telecommunications, people, and procedures that are configured to collect, manipulate, store, and process data into information.

moon.[4] The new data is revealing stunning information about the surface of the moon and ancient moon quakes. CBISs are also helping the military transform from "command and control" to "sense and respond" to meet tomorrow's military challenges.[5]

The components of a CBIS are illustrated in Figure 1.6. *Information technology (IT)* refers to hardware, software, databases, and telecommunications. A business's **technology infrastructure** includes all the hardware, software, databases, telecommunications, people, and procedures that are configured to collect, manipulate, store, and process data into information. The technology infrastructure is a set of shared IS resources that form the foundation of each computer-based information system.

technology infrastructure
All the hardware, software, databases, telecommunications, people, and procedures that are configured to collect, manipulate, store, and process data into information.

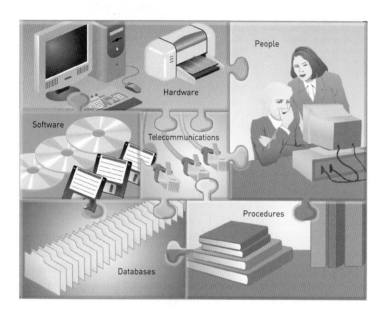

Figure 1.6

The Components of a Computer-Based Information System

Hardware

Hardware consists of computer equipment used to perform input, processing, and output activities. Input devices include keyboards, mice and other pointing devices, automatic scanning devices, and equipment that can read magnetic ink characters. Investment firms often use voice-response technology to allow customers to access their balances and other information with spoken commands. Processing devices include computer chips that contain the central processing unit and main memory. One processor chip, called the Bunny Chip by some, mimics living organisms and can be used by the drug industry to test drugs instead of using animals, such as rats or bunnies.[6] The experimental chip could save millions of dollars and months of time in drug research costs. Processor speed is also important. A large IBM computer used by U.S. Livermore National Laboratories to analyze nuclear explosions might be the fastest in the world (up to 300 teraflops—300 trillion operations per second).[7] The super fast computer, called Blue Gene, costs about $40 million. Mental Images of Germany and Pixar of the United States also need high processor speeds to produce award-winning images. The image technology is also used to help design cars, such as the sleek shapes of Mercedes Benz vehicles.

The many types of output devices include printers and computer screens. Bond traders, for example, often use an array of six or more computer screens to monitor bond prices and make split-second trades thousands of times each day. Another type of output device is a printer kiosk, which are located in some shopping malls.[8] After inserting a disk or a memory card from a computer or a camera, you can print photos and some documents. There are also many special-purpose hardware devices. Computerized event data recorders (EDRs) are now being placed into vehicles. Like an airplane's black box, EDRs record a vehicle's speed, possible engine problems, a driver's performance, and more. The technology is being used to monitor vehicle operation, determine the cause of accidents, and investigate whether truck drivers are taking required breaks. In one case, an EDR was used to help convict a driver of vehicular homicide.

hardware
Computer equipment used to perform input, processing, and output activities.

Hardware is a component of a computer-based information system and includes input, processing, and output.

(Source: Courtesy of Hewlett-Packard Company.)

Software

software
The computer programs that govern the operation of the computer.

Software consists of the computer programs that govern the operation of the computer. These programs allow a computer to process payroll, send bills to customers, and provide managers with information to increase profits, reduce costs, and provide better customer service. With software, people can work anytime at any place. Software along with manufacturing tools, for example, can be used to fabricate parts almost anywhere in the world.[9] Software, called Fab Lab, controls tools, such as cutters, milling machines, and other devices. A Fab Lab system, which costs about $20,000, has been used to make radio frequency tags to track animals in Norway, engine parts to allow tractors to run on processed castor beans in India, and many other fabrication applications.

The two types of software are system software, such as Microsoft Windows XP, which controls basic computer operations, including start-up and printing, and applications software, such as Microsoft Office 2003, which allows you to accomplish specific tasks, including word processing or tabulating numbers. Sophisticated application software, such as Adobe Creative Suite, can be used to design, develop, print, and place professional-quality advertising, brochures, posters, prints, and videos on the Internet. Artists are now developing digital art that can be viewed on TV screens, HDTV sets, and computer monitors when these devices are not used for their normal purposes.[10] This type of digital art is often purchased on DVDs or over the Internet.

Databases

database
An organized collection of facts and information.

A **database** is an organized collection of facts and information, typically consisting of two or more related data files. An organization's database can contain facts and information on customers, employees, inventory, competitors' sales, online purchases, and much more. Most managers and executives consider a database to be one of the most valuable parts of a computer-based information system. A California real estate development company uses databases to search for homes that are undervalued and purchase them at bargain prices.[11] It uses the database to analyze crime statistics, prices, local weather reports, school districts, and more to find homes whose values are likely to increase. The database has helped the company realize an average 50 percent return on investment. Increasingly, organizations are placing important databases on the Internet, which makes them accessible to many, including unauthorized users. One criminal illegally downloaded over 1.6 billion personal records from a database on the Internet.[12] The data contained names, phone numbers, and other personal information.

Telecommunications, Networks, and the Internet

telecommunications
The electronic transmission of signals for communications; enables organizations to carry out their processes and tasks through effective computer networks.

Telecommunications is the electronic transmission of signals for communications, which enables organizations to carry out their processes and tasks through effective computer networks. Large restaurant chains, for example, can use telecommunications systems and satellites to link hundreds of restaurants to plants and corporate headquarters to speed credit card authorization and report sales and payroll data. **Networks** connect computers and equipment in a building, around the country, or around the world to enable electronic communication. Investment firms can use wireless networks to connect thousands of investors with brokers or traders. Many hotels use wireless telecommunications to allow guests to connect to the Internet, retrieve voice messages, and exchange e-mail without plugging their computers or mobile devices into a phone jack. Wireless transmission also allows drones, such as Boeing's Scan Eagle, to fly using a remote control system and monitor buildings and other commercial establishments. The drones are smaller and less-expensive versions of the Predator and Global Hawk drones that the U.S. military used in the Afghanistan and Iraq conflicts.

The **Internet** is the world's largest computer network, actually consisting of thousands of interconnected networks, all freely exchanging information. Research firms, colleges, universities, high schools, and businesses are just a few examples of organizations using the Internet. Some towns and rural areas are investing in networks that provide wireless, high-speed access to the Internet for $20 per month or less.[13] Chaska, Minnesota, offers high-speed Internet access for about $16 per month to its 23,000 citizens.

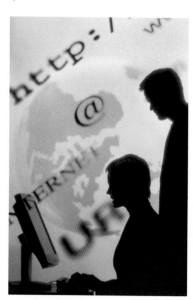

People use the Internet to research information, buy and sell products and services, make travel arrangements, conduct banking, and download music and videos, among other activities. After downloading music, you can use audio software to change a song's tempo, create mixes of your favorite tunes, and modify sound tracks to suit your personal taste. You can even play two or more songs simultaneously, called *mashing*. You can also use many of today's cell phones to connect to the Internet from around the world and at high speeds using broadband technology.[14] This not only speeds communications, but allows you to conduct business electronically. Some airline companies are providing Internet service on their flights so that travelers can send and receive e-mail, check investments, and browse the Internet. Internet users can create *Web logs (blogs)* to store and share their thoughts and ideas with others around the world.[15] Using *podcasting*, you can download audio programs or music from the Internet to play on computers or music players.[16] You can also record and store TV programs on computers or special viewing devices and watch them later.[17] Often called *place shifting*, this technology allows you to record TV programs at home and watch them at a different place when it's convenient. Table 1.3 lists a few companies that have used the Internet to their advantage.

The *World Wide Web (WWW)*, or the *Web*, is a network of links on the Internet to documents containing text, graphics, video, and sound. Information about the documents and access to them are controlled and provided by tens of thousands of special computers called Web servers. The Web is one of many services available over the Internet and provides access to millions of documents.

The technology used to create the Internet is also being applied within companies and organizations to create **intranets**, which allow people in an organization to exchange information and work on projects. One company, for example, uses an intranet to connect its 200 global operating companies and 20,000 employees. An **extranet** is a network based on Web technologies that allows selected outsiders, such as business partners and customers, to access authorized resources of a company's intranet. Companies can move all or most of their business activities to an extranet site for corporate customers. Many people use extranets every day without realizing it—to track shipped goods, order products from their suppliers, or access customer assistance from other companies. If you log on to the FedEx site (*www.fedex.com*) to check the status of a package, for example, you are using an extranet.

networks
Computers and equipment that are connected in a building, around the country, or around the world to enable electronic communications.

Internet
The world's largest computer network, actually consisting of thousands of interconnected networks, all freely exchanging information.

People often use the Internet to research information, buy and sell products and services, make travel arrangements, conduct banking, and download music and videos.

(Source: © Comstock Images / Alamy)

intranet
An internal network based on Web technologies that allows people within an organization to exchange information and work on projects.

extranet
A network based on Web technologies that allows selected outsiders, such as business partners and customers, to access authorized resources of a company's intranet.

Organization	Objective	Description of Internet Usage
Godiva Chocolatier	Increase sales and profits	The company developed a very profitable Internet site that allows customers to buy and ship chocolates. According to Kim Land, director of Godiva Direct, "This was set up from the beginning to make money." In two years, online sales have soared by more than 70% each year.
Environmental Defense	Alert the public to environmental concerns	The organization, formerly the Environmental Defense Fund, successfully used the Internet to alert people to the practice of catching sharks, removing their fins for soup, and returning them to the ocean to die. The Internet site also helped people fax almost 10,000 letters to members of Congress about the practice. According to Fred Krupp, the executive director of the Environmental Fund, "The Internet is the ultimate expression of 'think global, act local.'"
Buckman Laboratories	Improve employee training	The company used the Internet to train employees to sell specialty chemicals to paper companies, instead of bringing them to Memphis for training. According to one executive, "Our retention rate is much higher, and we removed a week [of training] in Memphis, which meant big savings." Using the Internet lowered the hourly cost of training an employee from $1,000 to only $40.
Siemens	Reduce costs	Using the Internet, the company, which builds and services power plants, reduced the cost of entering orders and serving customers. The Internet solution cost about $60,000, whereas a traditional solution would have cost $600,000.
Goldman Industrial Group	Save time	The company makes machine tools and was able to slash the time it takes to fill an order from three or four months to about a week using the Internet to help coordinate parts and manufacturing with its suppliers and at its plants.
Partnership America	Make better decisions	The company developed an Internet site for wholesalers of computer equipment and supplies. The wholesalers use the site to make better decisions about the equipment's features and prices. Wholesalers can also connect to Partnership America's site using cell phones. "When many of our customers need information, they're not at their desks," says one company representative.
Altra Energy Technologies	Get energy to companies that need it	The company developed an Internet site to help companies buy oil, gas, and wholesale power over the Internet.

Table 1.3

Uses of the Internet

procedures

The strategies, policies, methods, and rules for using a CBIS.

People

People are the most important element in most computer-based information systems. People make the difference between success and failure for most organizations. Information systems personnel include all the people who manage, run, program, and maintain the system. Large banks can hire IS personnel to speed the development of computer-related projects. Users are people who work with information systems to get results. Users include financial executives, marketing representatives, manufacturing operators, and many others. Certain computer users are also IS personnel.

Procedures

Procedures include the strategies, policies, methods, and rules for using the CBIS, including the operation, maintenance, and security of the computer. For example, some procedures describe when each program should be run. Others describe who can access facts in the database, or what to do if a disaster, such as a fire, earthquake, or hurricane, renders the CBIS unusable. Good procedures can help companies take advantage of new opportunities and avoid potential disasters. Poorly developed and inadequately implemented procedures, however, can cause people to waste their time on useless rules or result in inadequate responses to disasters, such as hurricanes or tornadoes.

Now that we have looked at computer-based information systems in general, we will briefly examine the most common types used in business today. These IS types are covered in more detail in Part 3.

BUSINESS INFORMATION SYSTEMS

The most common types of information systems used in business organizations are those designed for electronic and mobile commerce, transaction processing, management information, and decision support. In addition, some organizations employ special-purpose systems, such as virtual reality, that not every organization uses. Together, these systems help employees in organizations accomplish routine and special tasks—from recording sales, processing payrolls, and supporting decisions in various departments, to providing alternatives for large-scale projects and opportunities. Although these systems are discussed in separate sections in this chapter and explained in more detail later, they are often integrated in one product and delivered by the same software package. See Figure 1.7. For example, some enterprise resource planning packages process transactions, deliver information, and support decisions. Figure 1.8 shows a simple overview of the development of important business information systems discussed in this section.

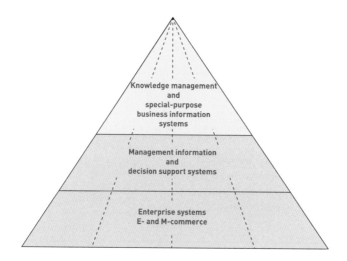

Figure 1.7

Business Information Systems

Business information systems are often integrated in one product and can be delivered by the same software package.

Figure 1.8

The Development of Important Business Information Systems

Electronic and Mobile Commerce

E-commerce involves any business transaction executed electronically between companies (business-to-business, B2B), companies and consumers (business-to-consumer, B2C), consumers and other consumers (consumer-to-consumer, C2C), business and the public sector, and consumers and the public sector. You might assume that e-commerce is reserved mainly for consumers visiting Web sites for online shopping. But Web shopping is only a small part of the e-commerce picture; the major volume of e-commerce—and its fastest-growing segment—is business-to-business (B2B) transactions that make purchasing easier for corporations. This growth is being stimulated by increased Internet access, growing user confidence, better payment systems, and rapidly improving Internet and Web security. E-commerce also offers opportunities for small businesses to market and sell at a low cost worldwide, allowing them to enter the global market. **Mobile commerce (m-commerce)** refers to transactions conducted anywhere, anytime. M-commerce relies on wireless

e-commerce
Any business transaction executed electronically between companies (business-to-business), companies and consumers (business-to-consumer), consumers and other consumers (consumer-to-consumer), business and the public sector, and consumers and the public sector.

mobile commerce (m-commerce)
Transactions conducted anywhere, anytime using wireless communications.

communications that managers and corporations use to place orders and conduct business with handheld computers, portable phones, laptop computers connected to a network, and other mobile devices.

IBM PartnerWorld® is an example of B2B (business-to-business) e-commerce that provides member companies with resources for product marketing, technical support, and training.

E-commerce offers many advantages for streamlining work activities. Figure 1.9 provides a brief example of how e-commerce can simplify the process of purchasing new office furniture from an office-supply company. In the manual system, a corporate office worker must get approval for a purchase that exceeds a certain amount. That request goes to the purchasing department, which generates a formal purchase order to procure the goods from the approved vendor. Business-to-business e-commerce automates the entire process. Employees go directly to the supplier's Web site, find the item in a catalog, and order what they need at a price set by their company. If approval is required, the approver is notified automatically. As the use of e-commerce systems grows, companies are phasing out their traditional systems. The resulting growth of e-commerce is creating many new business opportunities.

E-commerce can enhance a company's stock prices and market value. Today, several e-commerce firms have teamed up with more traditional brick-and-mortar businesses to draw from each other's strengths. For example, e-commerce customers can order products on a Web site and pick them up at a nearby store.

In addition to e-commerce, business information systems use telecommunications and the Internet to perform many related tasks. *Electronic procurement (e-procurement)*, for example, involves using information systems and the Internet to acquire parts and supplies. **Electronic business (e-business)** goes beyond e-commerce and e-procurement by using information systems and the Internet to perform all business-related tasks and functions, such as accounting, finance, marketing, manufacturing, and human resource activities. E-business also includes working with customers, suppliers, strategic partners, and stakeholders. Compared to traditional business strategy, e-business strategy is flexible and adaptable. See Figure 1.10.

electronic business (e-business)
Using information systems and the Internet to perform all business-related tasks and functions.

Figure 1.9

E-Commerce Greatly Simplifies Purchasing

Traditional process for placing a purchase order

E-commerce process for placing a purchase order

Figure 1.10

Electronic Business

E-business goes beyond e-commerce to include using information systems and the Internet to perform all business-related tasks and functions, such as accounting, finance, marketing, manufacturing, and human resources activities.

Enterprise Systems: Transaction Processing Systems and Enterprise Resource Planning

Transaction Processing Systems

Since the 1950s, computers have been used to perform common business applications. Many of these early systems were designed to reduce costs by automating routine, labor-intensive business transactions. A **transaction** is any business-related exchange such as payments to employees, sales to customers, or payments to suppliers. Thus, processing business transactions was the first computer application developed for most organizations. A **transaction processing system** (TPS) is an organized collection of people, procedures, software, databases, and devices used to record completed business transactions. If you understand a transaction processing system, you understand basic business operations and functions.

Enterprise systems help organizations perform and integrate important tasks, such as paying employees and suppliers, controlling inventory, sending out invoices, and ordering supplies. In the past, companies accomplished these tasks using traditional transaction processing

transaction

Any business-related exchange, such as payments to employees, sales to customers, and payments to suppliers.

transaction processing system (TPS)

An organized collection of people, procedures, software, databases, and devices used to record completed business transactions.

systems. Today, they are increasingly being performed by enterprise resource planning systems. For example, Whirlpool Corporation, the large appliance maker, used enterprise resource planning to reduce inventory levels by 20 percent and cut about 5 percent from its freight and warehousing costs by providing managers with information about inventory levels and costs.[18] The new system may have also helped the company increase its revenues by about $1 billion.

One of the first business systems to be computerized was the payroll system (see Figure 1.11). The primary inputs for a payroll TPS are the number of employee hours worked during the week and the pay rate. The primary output consists of paychecks. Early payroll systems produced employee paychecks and related reports required by state and federal agencies, such as the Internal Revenue Service. Other routine applications include sales ordering, customer billing and customer relationship management, and inventory control. Some automobile companies, for example, use their TPSs to buy billions of dollars of needed parts each year through Internet sites. Because these systems handle and process daily business exchanges, or transactions, they are all classified as TPSs.

Figure 1.11

A Payroll Transaction Processing System

The inputs (numbers of employee hours worked and pay rates) go through a transformation process to produce outputs (paychecks).

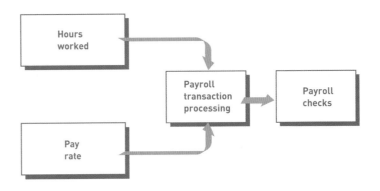

Enterprise Resource Planning

enterprise resource planning (ERP) system

A set of integrated programs capable of managing a company's vital business operations for an entire multisite, global organization.

An **enterprise resource planning (ERP) system** is a set of integrated programs that manages the vital business operations for an entire multisite, global organization. An ERP system can replace many applications with one unified set of programs, making the system easier to use and more effective. One healthcare company, with a large network of 33 hospitals and more than 4 million patients in northern California, uses an ERP system to process medical transactions and exchange information between hospitals, physicians, and employees. BASF, a large chemical company in Germany, is installing an ERP system that its employees can use every day to monitor the manufacture of chemicals.[19]

Although the scope of an ERP system might vary from company to company, most ERP systems provide integrated software to support manufacturing and finance. In such an environment, a forecast is prepared that estimates customer demand for several weeks. The ERP system checks what is already available in finished product inventory to meet the projected demand. Manufacturing must then produce inventory to eliminate any shortfalls. In developing the production schedule, the ERP system checks the raw materials and packing-materials inventories and determines what needs to be ordered to meet the schedule. Most ERP systems also have a purchasing subsystem that orders the needed items. In addition to these core business processes, some ERP systems can support functions such as human resources, sales, and distribution. The primary benefits of implementing an ERP system include easing adoption of improved work processes and increasing access to timely data for decision making.

Information and Decision Support Systems

The benefits provided by an effective TPS are tangible and justify their associated costs in computing equipment, computer programs, and specialized personnel and supplies. A TPS can speed business activities and reduce clerical costs. Although early accounting and financial TPSs were already valuable, companies soon realized that they could use the data stored in these systems to help managers make better decisions, whether in human resource management, marketing, or administration. Satisfying the needs of managers and decision makers continues to be a major factor in developing information systems.

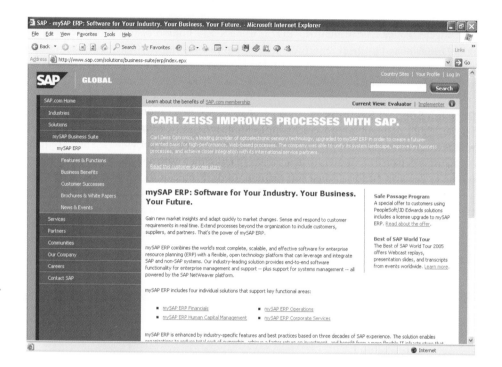

SAP AG, a German software company, is one of the leading suppliers of ERP software. The company employs more than 34,000 people in more than 50 countries.

Management Information Systems

A **management information system** (MIS) is an organized collection of people, procedures, software, databases, and devices that provides routine information to managers and decision makers. An MIS focuses on operational efficiency. Marketing, production, finance, and other functional areas are supported by MISs and linked through a common database. MISs typically provide standard reports generated with data and information from the TPS (see Figure 1.12). Producing a report that describes inventory that should be ordered is an example of an MIS.

management information system (MIS)
An organized collection of people, procedures, software, databases, and devices that provides routine information to managers and decision makers.

Figure 1.12

Management Information System

Functional management information systems draw data from the organization's transaction processing system.

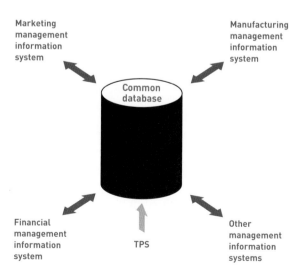

MISs were first developed in the 1960s and typically use information systems to produce managerial reports. In many cases, these early reports were produced periodically—daily, weekly, monthly, or yearly. Gambling casinos generate daily reports that tell their staff what specific customers like. Some casinos know whether a customer likes flowers in her room or a beverage in his hand and can accommodate individual needs or desires. Because of their value to managers, MISs have proliferated throughout the management ranks. For instance, the total payroll summary report produced initially for an accounting manager might also be useful to a production manager to help monitor and control labor and job costs. The Los Angeles Police Department (LAPD) is using a $35 million system to collect data on police behavior and produce reports on possible police misconduct.[20] The new system tracks complaints, lawsuits, uses of force, and other measures. According to the LAPD inspector general, "There definitely needs to be computerized management of officers."

Decision Support Systems

By the 1980s, dramatic improvements in technology resulted in information systems that were less expensive but more powerful than earlier systems. People at all levels of organizations began using personal computers to do a variety of tasks; they were no longer solely dependent on the IS department for all their information needs. People quickly recognized that computer systems could support additional decision-making activities. A **decision support system (DSS)** is an organized collection of people, procedures, software, databases, and devices that support problem-specific decision making. The focus of a DSS is on making effective decisions. Whereas an MIS helps an organization "do things right," a DSS helps a manager "do the right thing."

decision support system (DSS)
An organized collection of people, procedures, software, databases, and devices used to support problem-specific decision making.

Decisioneering provides decision support software called Crystal Ball, which helps business people of all types assess risks and make forecasts. Shown here is the Standard Edition being used for oil field development.

(Source: Crystal Ball screenshot courtesy of Decisioneering, Inc.)

In addition to assisting in all aspects of problem-specific decision making, a DSS can support customers by rapidly responding to their phone and e-mail inquiries. A DSS goes beyond a traditional MIS by providing immediate assistance in solving complex problems. Many of these problems are unique and complex, and information is often difficult to obtain. For instance, an auto manufacturer might try to determine the best location to build a new manufacturing facility. Traditional MISs are seldom used to solve these types of problems; a DSS can help by suggesting alternatives and assisting in final decision making.

Decision support systems are used when the problem is complex and the information needed to make the best decision is difficult to obtain and use. So, a DSS also involves managerial judgment and perspective. Managers often play an active role in developing and implementing the DSS. A DSS recognizes that different managerial styles and decision types require different systems. For example, two production managers in the same position trying to solve the same problem might require different information and support. The overall emphasis is to support, rather than replace, managerial decision making.

The essential elements of a DSS include a collection of models used to support a decision maker or user (model base), a collection of facts and information to assist in decision making (database), and systems and procedures (dialogue manager or user interface) that help decision makers and other users interact with the DSS (see Figure 1.13). Software is often used to manage the database—the database management system (DBMS)—and the model base—the model management system (MMS).

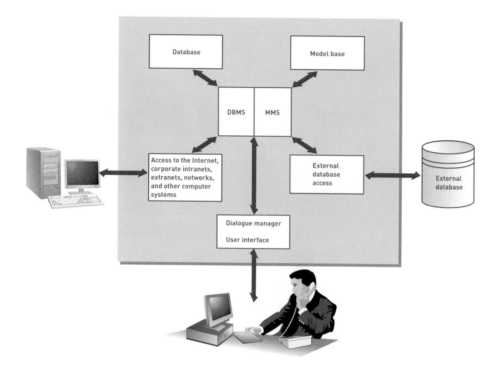

Figure 1.13

Essential DSS Elements

In addition to DSSs for managers, group decision support systems and executive support systems use the same approach to support groups and executives.[21] A group decision support system, also called a *group support system*, includes the DSS elements just described and software, called *groupware*, to help groups make effective decisions. An executive support system, also called an *executive information system*, helps top-level managers, including a firm's president, vice presidents, and members of the board of directors, make better decisions. An executive support system can assist with strategic planning, top-level organizing and staffing, strategic control, and crisis management.

Specialized Business Information Systems: Knowledge Management, Artificial Intelligence, Expert Systems, and Virtual Reality

In addition to TPSs, MISs, and DSSs, organizations often rely on specialized systems. Many use *knowledge management systems (KMSs)*, an organized collection of people, procedures, software, databases, and devices to create, store, share, and use the organization's knowledge and experience. Ryder Systems, for example, received help from the Accenture consulting company to streamline its transportation and logistics business.[22] According to Gene Tyndall, Ryder's executive vice president for Global Solutions and eCommerce, "People—the knowledge they have and the new knowledge they create—are the corporate assets that impact Ryder's performance more than any other form of capital." According to a survey of CEOs, firms that use KMSs are more likely to innovate and perform better.[23]

In addition to knowledge management, companies use other types of specialized systems. The Nissan Motor Company, for example, has developed a specialized system for automobiles and trucks called "Lane Departure Prevention" that nudges a car back into the correct lane if it veers off course.[24] The specialized system uses cameras and computers to adjust braking to get the vehicle back on course. The system switches off when the driver uses turn signals to change lanes. Other specialized systems are based on the notion of **artificial intelligence (AI)**, in which the computer system takes on the characteristics of human intelligence. The field of artificial intelligence includes several subfields (see Figure 1.14). Some people predict that in the future, we will have nanobots, small molecular-sized robots, traveling throughout our bodies and in our bloodstream, keeping us healthy.[25] Other nanobots will be embedded in products and services, making our lives easier and creating new business opportunities.

artificial intelligence (AI)
A field in which the computer system takes on the characteristics of human intelligence.

The Nissan Motor Company has developed a specialized system for automobiles and trucks called "Lane Departure Prevention" that nudges a car back into the correct lane if it veers off course.

(Source: © AP/Wide World Photos)

Figure 1.14

The Major Elements of Artificial Intelligence

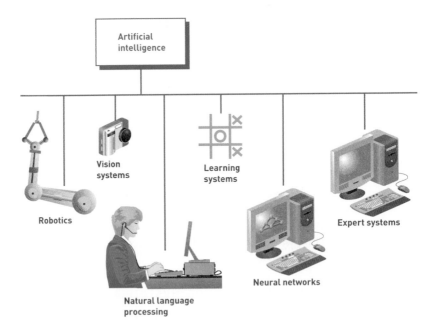

Artificial Intelligence

Robotics is an area of artificial intelligence in which machines take over complex, dangerous, routine, or boring tasks, such as welding car frames or assembling computer systems and components. Vision systems allow robots and other devices to "see," store, and process visual images. Natural language processing involves computers understanding and acting on verbal or written commands in English, Spanish, or other human languages. Learning systems allow computers to learn from past mistakes or experiences, such as playing games or making business decisions, and neural networks is a branch of artificial intelligence that allows computers to recognize and act on patterns or trends. Some successful stock, options, and futures traders use neural networks to spot trends and make them more profitable with their investments.

Expert Systems

Expert systems give the computer the ability to make suggestions and act like an expert in a particular field. It can help the novice user perform at the level of an expert. The unique value of expert systems is that they allow organizations to capture and use the wisdom of experts and specialists. Therefore, years of experience and specific skills are not completely lost when a human expert dies, retires, or leaves for another job. Expert systems can be applied to almost any field or discipline. They have been used to monitor nuclear reactors, perform medical diagnoses, locate possible repair problems, design and configure IS components, perform credit evaluations, and develop marketing plans for a new product or new investment strategy. The collection of data, rules, procedures, and relationships that must be followed to achieve value or the proper outcome is contained in the expert system's **knowledge base**.

Virtual Reality

Virtual reality is the simulation of a real or imagined environment that can be experienced visually in three dimensions. Originally, virtual reality referred to immersive virtual reality, which means the user becomes fully immersed in an artificial, computer-generated 3-D world. The virtual world is presented in full scale and relates properly to the human size. It can represent any 3-D setting, real or abstract, such as a building, an archaeological excavation site, the human anatomy, a sculpture, or a crime scene reconstruction. Virtual worlds can be animated, interactive, and shared. Through immersion, the user can gain a deeper understanding of the virtual world's behavior and functionality. Virtual reality can also refer to applications that are not fully immersive, such as mouse-controlled navigation through a 3-D environment on a graphics monitor, stereo viewing from the monitor via stereo glasses, stereo projection systems, and others.

A variety of input devices, such as head-mounted displays (see Figure 1.15), data gloves, joysticks, and handheld wands, allow the user to navigate through a virtual environment and to interact with virtual objects. Directional sound, tactile and force feedback devices, voice recognition, and other technologies enrich the immersive experience. Because several people can share and interact in the same environment, virtual reality can be a powerful medium for communication, entertainment, and learning.

It is difficult to predict where information systems and technology will be in 10 to 20 years. It seems, however, that we are just beginning to discover the full range of their usefulness. Technology has been improving and expanding at an increasing rate; dramatic growth and change are expected for years to come. Without question, a knowledge of the effective use of information systems will be critical for managers both now and in the long term. But how are these information systems created?

expert system
A system that gives a computer the ability to make suggestions and act like an expert in a particular field.

knowledge base
The collection of data, rules, procedures, and relationships that must be followed to achieve value or the proper outcome.

virtual reality
The simulation of a real or imagined environment that can be experienced visually in three dimensions.

Figure 1.15

A Head-Mounted Display

The head-mounted display (HMD) was the first device to provide the wearer with an immersive experience. A typical HMD houses two miniature display screens and an optical system that channels the images from the screens to the eyes, thereby presenting a stereo view of a virtual world. A motion tracker continuously measures the position and orientation of the user's head and allows the image-generating computer to adjust the scene representation to the current view. As a result, the viewer can look around and walk through the surrounding virtual environment.

(Source: Courtesy of 5DT, Inc. *www. 5dt.com.*)

SYSTEMS DEVELOPMENT

systems development
The activity of creating or modifying existing business systems.

Systems development is the activity of creating or modifying business systems. Systems development projects can range from small to very large in fields as diverse as stock analysis and video game development. The University of Denver, for example, offers a degree program in video game development.[26] According to the dean of the School of Engineering and Computer Science, "We wanted to create a niche area of specialty that people would identify us with."

People inside a company can develop systems, or companies can use *outsourcing*, hiring an outside company to perform some or all of a systems development project. Outsourcing allows a company to focus on what it does best and delegate other functions to companies with expertise in systems development. Outsourcing, however, is not the best alternative for all companies.

Developing information systems to meet business needs is highly complex and difficult—so much so that it is common for IS projects to overrun budgets and exceed scheduled completion dates. Her Majesty's Revenue & Customs (HMRC), which collects taxes in Britain, for example, might sue an outsourcing company to recover funds lost due to a tax-related mistake caused by a failed systems development project.[27] The failed project overpaid about $3.5 billion to some families with children or taxpayers in a low-income tax bracket. One strategy for improving the results of a systems development project is to divide it into several steps, each with a well-defined goal and set of tasks to accomplish (see Figure 1.16). These steps are summarized next.

Figure 1.16

An Overview of Systems Development

SYSTEMS INVESTIGATION
Understand problem

SYSTEMS ANALYSIS
Understand solutions

SYSTEMS DESIGN
Select and plan best solution

SYSTEMS IMPLEMENTATION
Place solution into effect

SYSTEMS MAINTENANCE
AND REVIEW
Evaluate results of solution

Systems Investigation and Analysis

The first two steps of systems development are systems investigation and analysis. The goal of the *systems investigation* is to gain a clear understanding of the problem to be solved or opportunity to be addressed. A cruise line company, for example, might launch a systems investigation to determine whether a development project is feasible to automate purchasing at ports around the world. After an organization understands the problem, the next question is, "Is the problem worth solving?" Given that organizations have limited resources—people and money—this question deserves careful consideration. If the decision is to continue with the solution, the next step, *systems analysis*, defines the problems and opportunities of the existing system. During systems investigation and analysis, as well as design maintenance and review, discussed next, the project must have the complete support of top-level managers and focus on developing systems that achieve business goals.[28] According to Mike McNall, director of application services for Vanguard Car Rental, "If we didn't have the backing of the business ... no matter what we built, it wouldn't be accepted."[29] Vanguard is a new car rental company that was formed by merging several other car rental businesses.

Systems Design, Implementation, and Maintenance and Review

Systems design determines how the new system will work to meet the business needs defined during systems analysis. *Systems implementation* involves creating or acquiring the various system components (hardware, software, databases, etc.) defined in the design step, assembling them, and putting the new system into operation. The purpose of *systems maintenance and review* is to check and modify the system so that it continues to meet changing business needs.

INFORMATION SYSTEMS IN SOCIETY, BUSINESS, AND INDUSTRY

Information systems have been developed to meet the needs of all types of organizations and people, and their use is spreading throughout the world to improve the lives and business activities of many citizens. To provide their enormous benefits, however, information systems must be implemented with thought and care. The speed and widespread use of information systems opens users to a variety of threats from unethical people.

Security, Privacy, and Ethical Issues in Information Systems and the Internet

Although information systems can provide enormous benefits, they do have some drawbacks.[30] Figure 1.17 reveals the annual cost of losses of about 250 organizations that responded to a survey. Computer-related mistakes are also a concern. In Japan, a financial services firm had trading losses of $335 million due to a typing mistake in entering a trade.

Increasingly, the ethical use of systems has been highlighted in the news. Ethical issues concern what is generally considered right or wrong. Some IS professionals believe that computers may create new opportunities for unethical behavior. For example, a faculty member of a medical school falsified computerized research results to get a promotion—and a higher salary. In another case, a company was charged with using a human resource information system to time employee layoffs and firings to avoid paying pensions. More and more, the Internet is also associated with unethical behavior. Unethical investors have placed false rumors or incorrect information about a company on the Internet and tried to influence its stock price to make money. Stolen property, such as Social Security numbers and credit card numbers, is another issue.

To protect against these threats, you can install security and control measures. For example, many software products can detect and remove viruses and spam, or unwanted e-mail, from computer systems. Information systems can help reduce other types of crime as well. In a New Zealand city, a free computer center has cut vandalism by keeping young people off the street and giving residents a sense of pride. When a pair of headphones

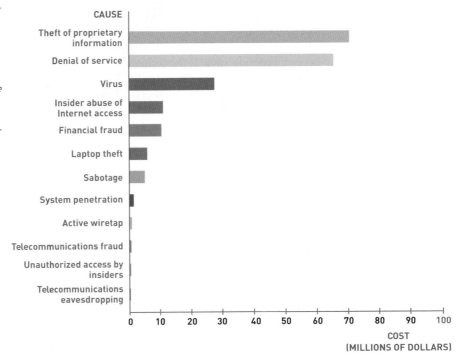

Figure 1.17

The Cost and Cause of
Computer Attacks

(Source: Data from Riva Richmond,
"How to Find Your Weak Spots," *The
Wall Street Journal*, September 29,
2003, p. R3.)

disappeared from the center, the community rallied to make sure that they were promptly returned.

You can install *firewalls* (software and hardware that protect a computer system or network from outside attacks) to avoid viruses and prevent unauthorized people from gaining access to your computer system. You can also use identification numbers and passwords. Some security experts propose installing Web cameras and hiring "citizen spotters" to monitor the Webcams. In response to possible abuses, a number of laws have been passed to protect people from invasion of their privacy, including The Privacy Act, enacted in the 1970s.

Use of information systems also raises work concerns, including job loss through increased efficiency and some potential health problems from making repetitive motions. *Ergonomics*, the study of designing and positioning workplace equipment, can help you avoid health-related problems of using computer systems.

Computer and Information Systems Literacy

In the twenty-first century, business survival and prosperity continue to become more difficult. For example, increased mergers among former competitors to create global conglomerates, continued downsizing of corporations to focus on their core businesses and to improve efficiencies, efforts to reduce trade barriers, and the globalization of capital all point to the increased internationalization of business organizations and markets. In addition, business issues and decisions are becoming more complex and must be made faster. Whatever career path you take, understanding information systems will help you cope, adapt, and prosper in this challenging environment.

A knowledge of information systems will help you make a significant contribution on the job. It will also help you advance in your chosen career or field. Managers are expected to identify opportunities to implement information systems to improve their business. They are also expected to lead IS projects in their areas of expertise. To meet these personal and organizational goals, you must acquire both computer literacy and information systems literacy. **Computer literacy** is a knowledge of computer systems and equipment and the ways they function. It stresses equipment and devices (hardware), programs and instructions (software), databases, and telecommunications.

computer literacy
Knowledge of computer systems
and equipment and the ways they
function; it stresses equipment and
devices (hardware), programs and
instructions (software), databases,
and telecommunications.

Snipermail Executive Serves Hard Time

When you picture a criminal hacker, you probably think of a young, socially-isolated deviant in a dark room tapping away at the keyboard in the middle of the night. You probably wouldn't picture a 46 year old Boca Raton executive hacking systems in his office in the middle of the afternoon. It's time to shatter that stereotype.

The ex-principle owner of Snipermail, Inc. is that executive. This chief executive officer (CEO) is not exactly what some would consider a reputable businessman, even when he's not hacking private networks. Snipermail is a spam company; it develops e-mail marketing strategies for businesses and matches its clients with e-mail lists of spam targets.

Snipermail contracted with a client that used Acxiom Corporation's services. Acxiom, with headquarters in Little Rock, Arkansas, manages personal information on consumers, along with financial and corporate data for a variety of companies, including Fortune 500 firms. To work with its client, Snipermail was provided with an Acxiom network account that allowed it to access certain limited database records at Acxiom.

Seizing the opportunity, associates at Snipermail used their limited access to the Acxiom system and sophisticated decryption software to illegally obtain passwords to access the entire Acxiom database. Over time, those associates downloaded more than 1 billion private data records from Acxiom containing names, e-mail addresses, and phone numbers of clients. The CEO planned to merge the data with the Snipermail database of spam targets and sell it to clients.

It isn't easy to download 8.2 GB of private valuable data without being noticed. Acxiom discovered the theft and it wasn't long before investigators traced the digital footprints to Snipermail.

On July 22, 2004, the CEO of Snipermail was charged with the largest computer crime indictment in U. S. history. On February 22, 2006 he was sentenced to eight years in prison for 120 counts of unauthorized access of a protected computer, two counts of access device fraud, and one count of obstruction of justice. The CEO's six associates and accomplices were of little help to him. They all struck deals with the Department of Justice and implicated the CEO in exchange for their own freedom.

The law enforcement agencies involved had strong words for anyone who might think stealing information over the Internet is harmless. "This sentence reflects the seriousness of these crimes," said U.S. Attorney Bud Cummins of the Eastern District of Arkansas. "At first blush, downloading computer files in the privacy of your office may not seem so terribly serious. But, if you are stealing propriety information worth tens of millions of dollars from a well-established and reputable company, you can expect to be punished accordingly."

"Neither the Internet nor cyberspace will ever be a safe haven for individuals who attempt this type of cyber crime. The Secret Service, along with our law enforcement partners, will hunt you down, keystroke by keystroke, until you face a jury of your peers," said Brian Marr, Special Agent in Charge of the Little Rock office of the U.S. Secret Service. "The investigation of cyber crime, particularly as it relates to computer intrusion, is one of the FBI's top priorities," said William C. Temple, Special Agent in Charge of the Little Rock office of the Federal Bureau of Investigation. "The success of this investigation should send a strong message to those who might consider becoming involved in similar criminal activity."

Discussion Questions

1. Who are the victims in the Snipermail case?
2. Do you think eight years is an appropriate sentence for this CEO? Why or why not?

Critical Thinking Questions

1. What actions could Acxiom have taken to prevent its customers and their associates from accessing sensitive information?
2. How has this incident affected Acxiom's reputation?

SOURCES: Grant Gross, "IT Exec Sentenced to Eight Years for Data Theft," IDG News Service, February 23, 2006, *www.idg.com.* "Scott Levine Gets 8 Years in Data Theft Case," 4Law Web site, accessed February 23, 2006, *www.4law.co.il/ arkan1.htm.*

information systems literacy
Knowledge of how data and information are used by individuals, groups, and organizations.

Information systems literacy goes beyond knowing the fundamentals of computer systems and equipment. **Information systems literacy** is the knowledge of how data and information are used by individuals, groups, and organizations. It includes knowledge of computer technology and the broader range of information systems. Most important, however, it encompasses *how* and *why* this technology is applied in business. Knowing about various types of hardware and software is an example of computer literacy. Knowing how to use hardware and software to increase profits, cut costs, improve productivity, and increase customer satisfaction is an example of information systems literacy. Information systems literacy can involve recognizing how and why people (managers, employees, stockholders, and others) use information systems; being familiar with organizations, decision-making approaches, management levels, and information needs; and understanding how organizations can use computers and information systems to achieve their goals. Knowing how to deploy transaction processing, management information, decision support, and expert systems to help an organization achieve its goals is a key aspect of information systems literacy.

Information Systems in the Functional Areas of Business

Information systems are used in all functional areas and operating divisions of business. In *finance* and *accounting*, information systems forecast revenues and business activity, determine the best sources and uses of funds, manage cash and other financial resources, analyze investments, and perform audits to make sure that the organization is financially sound and that all financial reports and documents are accurate. *Sales* and *marketing* use information systems to develop new goods and services (product analysis), select the best location for production and distribution facilities (place or site analysis), determine the best advertising and sales approaches (promotion analysis), and set product prices to get the highest total revenues (price analysis).

In *manufacturing*, information systems process customer orders, develop production schedules, control inventory levels, and monitor product quality. In addition, information systems help to design products (*computer-assisted design,* or *CAD*), manufacture items (*computer-assisted manufacturing,* or *CAM*), and integrate machines or pieces of equipment (*computer-integrated manufacturing,* or *CIM*). *Human resource management* uses information systems to screen applicants, administer performance tests to employees, monitor employee productivity, and more. *Legal information systems* analyze product liability and warranties and help to develop important legal documents and reports.

Information Systems in Industry

In addition to being used in every department in a company, information systems are used in almost every industry or field in business. The *airline industry* develops Internet auction sites to offer discount fares and increase revenue. *Investment firms* use information systems to analyze stocks, bonds, options, the futures market, and other financial instruments, and provide improved services to their customers. *Banks* use information systems to help make sound loans and good investments, as well as to provide online check payment for account holders. The *transportation industry* uses information systems to schedule trucks and trains to deliver goods and services at the lowest cost. *Publishing companies* use information systems to analyze markets and to develop and publish newspapers, magazines, and books. *Healthcare organizations* use information systems to diagnose illnesses, plan medical treatment, track patient records, and bill patients. Health maintenance organizations (HMOs) use Web technology to access patients' insurance eligibility and other information stored in databases to cut patient costs. *Retail companies* are using the Web to take orders and provide customer service support. Retail companies also use information systems to help market products and services, manage inventory levels, control the supply chain, and forecast demand. *Power management* and *utility companies* use information systems to monitor and control power generation and usage. *Professional services* firms employ information systems to improve the speed and quality of services they provide to customers. Management consulting firms use intranets and extranets to offer information on products, services, skill levels, and past engagements to their consultants. These industries are discussed in more detail as we continue through the book.

GLOBAL CHALLENGES IN INFORMATION SYSTEMS

Changes in society as a result of increased international trade and cultural exchange, often called globalization, has always had a big impact on organizations and their information systems. In his book, *The World Is Flat,* Thomas Friedman describes three eras of globalization.[31] See Table 1.4. According to Friedman, we have progressed from the globalization of countries to the globalization of multinational corporations and individuals. Today, people in remote areas can use the Internet to compete with and contribute to other people, the largest corporations, and entire countries. These workers are empowered by high-speed Internet access, making the world flatter. In the Globalization 3.0 era, designing a new airplane or computer can be separated into smaller subtasks and then completed by a person or small group that can do the best job. These workers can be located in India, China, Russia, Europe, and other areas of the world. The subtasks can then be combined or reassembled into the complete design. This approach can be used to prepare tax returns, diagnose a patient's medical condition, fix a broken computer, and many other tasks.

Era	Dates	Characterized by
Globalization 1.0	Late 1400–1800	Countries with the power to explore and influence the world
Globalization 2.0	1800–2000	Multinational corporations that have plants, warehouses, and offices around the world
Globalization 3.0	2000–today	Individuals from around the world who can compete and influence other people, corporations, and countries by using the Internet and powerful technology tools

Table 1.4

Eras of Globalization

Today's information systems have led to greater globalization. High-speed Internet access and networks that can connect individuals and organizations around the world create more international opportunities. Global markets have expanded. People and companies can get products and services from around the world, instead of around the corner or across town. These opportunities, however, introduce numerous obstacles and issues, including challenges involving culture, language, and many others.

- **Cultural challenges.** Countries and regional areas have their own cultures and customs that can significantly affect individuals and organizations involved in global trade.
- **Language challenges.** Language differences can make it difficult to translate exact meanings from one language to another.
- **Time and distance challenges.** Time and distance issues can be difficult to overcome for individuals and organizations involved with global trade in remote locations. Large time differences make it difficult to talk to people on the other side of the world. With long distance, it can take days to get a product, a critical part, or a piece of equipment from one location to another location.
- **Infrastructure challenges.** High-quality electricity and water might not be available in certain parts of the world. Telephone services, Internet connections, and skilled employees might be expensive or not readily available.

- **Currency challenges.** The value of different currencies can vary significantly over time, making international trade more difficult and complex.
- **Product and service challenges.** Traditional products that are physical or tangible, such as an automobile or bicycle, can be difficult to deliver to the global market. However, *electronic products (e-products)* and *electronic services (e-services)* can be delivered to customers electronically, over the phone, networks, through the Internet, or other electronic means. Software, music, books, manuals, and help and advice can all be delivered over the Internet.
- **Technology transfer issues.** Most governments don't allow certain military-related equipment and systems to be sold to some countries. Even so, some believe that foreign companies are stealing the intellectual property, trade secrets, copyrighted materials, and counterfeiting products and services.[32]
- **State, regional, and national laws.** Every state, region, and country have a set of laws that must be obeyed by citizens and organizations operating in the country. These laws can deal with a variety of issues, including trade secrets, patents, copyrights, protection of personal or financial data, privacy, and much more. Laws restricting how data enters or exits a country are often called *transborder data-flow* laws. Keeping track of these laws and incorporating them into the procedures and computer systems of multinational and transnational organizations can be very difficult and time consuming, requiring expert legal advice.
- **Trade agreements.** Countries often enter into trade agreements with each other. The North American Free Trade Agreement (NAFTA) and the Central American Free Trade Agreement (CAFTA) are examples.[33] The European Union (EU) is another example of a group of countries with an international trade agreement.[34] The EU is a collection of mostly European countries that have joined together for peace and prosperity. In addition to NAFTA, CAFTA, and the EU, there are many other trade agreements. The Australia-United States Free Trade Agreement (AUSFTA) was signed into law in 2005. There are free trade agreements between Bolivia and Mexico, Canada and Costa Rica, Canada and Israel, Chile and Korea, Mexico and Japan, the United States and Jordan, and many others.[35]

SUMMARY

Principle

The value of information is directly linked to how it helps decision makers achieve the organization's goals.

Information systems are used in almost every imaginable career area. Regardless of your college major or chosen career, you will find that information systems are indispensable tools to help you achieve your career goals. Learning about information systems can help you get your first job, earn promotions, and advance your career.

Data consists of raw facts; information is data transformed into a meaningful form. The process of defining relationships among data requires knowledge. Knowledge is an awareness and understanding of a set of information and the way that information can support a specific task. To be valuable, information must have several characteristics: It should be accurate, complete, economical to produce, flexible, reliable, relevant, simple to understand, timely, verifiable, accessible, and secure. The value of information is directly linked to how it helps people achieve their organization's goals.

Principle

Computers and information systems are constantly making it possible for organizations to improve the way they conduct business.

A system is a set of elements that interact to accomplish a goal or set of objectives. The components of a system include inputs, processing mechanisms, and outputs. A system uses feedback to monitor and control its operation to make sure that it continues to meet its goals and objectives.

System performance is measured by its efficiency and effectiveness. Efficiency is a measure of what is produced divided by what is consumed; effectiveness measures the extent to which a system achieves its goals. A systems performance standard is a specific objective.

Principle

Knowing the potential impact of information systems and having the ability to put this knowledge to work can result in a successful personal career, organizations that reach their goals, and a society with a higher quality of life.

Information systems are sets of interrelated elements that collect (input), manipulate and store (process), and disseminate (output) data and information. Input is the activity of capturing and gathering new data, processing involves converting or transforming data into useful outputs, and output involves producing useful information. Feedback is the output that is used to make adjustments or changes to input or processing activities.

The components of a computer-based information system (CBIS) include hardware, software, databases, telecommunications and the Internet, people, and procedures. The types of CBISs that organizations use can be classified into four basic groups: (1) e-commerce and m-commerce, (2) TPS and ERP systems, (3) MIS and DSS, and (4) specialized business information systems. The key to understanding these types of systems begins with learning their fundamentals.

E-commerce involves any business transaction executed electronically between parties such as companies (business to business), companies and consumers (business to consumer), business and the public sector, and consumers and the public sector. The major volume of e-commerce and its fastest-growing segment is business-to-business transactions that make purchasing easier for big corporations. E-commerce also offers opportunities for small businesses to market and sell at a low cost worldwide, thus allowing them to enter the global market right from start-up. M-commerce involves anytime, anywhere computing that relies on wireless networks and systems.

The most fundamental system is the transaction processing system (TPS). A transaction is any business-related exchange. The TPS handles the large volume of business transactions that occur daily within an organization. An enterprise resource planning (ERP) system is a set of integrated programs that can manage the vital business operations for an entire multisite, global organization. A management information system (MIS) uses the information from a TPS to generate information useful for management decision making.

A decision support system (DSS) is an organized collection of people, procedures, databases, and devices that help make problem-specific decisions. A DSS differs from an MIS in the support given to users, the emphasis on decisions, the development and approach, and the system components, speed, and output.

Specialized business information systems include knowledge management, artificial intelligence, expert, and virtual reality systems. Knowledge management systems are organized collections of people, procedures, software, databases, and devices used to create, store, share, and use the organization's knowledge and experience. Artificial intelligence (AI) includes a wide range of systems in which the computer takes on the characteristics of human intelligence. Robotics is an area of artificial intelligence in which machines perform complex, dangerous, routine, or boring tasks, such as welding car frames or assembling computer systems and components. Vision systems allow robots and other devices to have "sight" and to store and process visual images. Natural language processing involves computers interpreting and acting on verbal or written commands in English, Spanish, or other human languages. Learning systems let computers learn from past mistakes or experiences, such as playing games or making business decisions, while neural networks is a branch of artificial intelligence that allows computers to recognize and act on patterns or trends. An expert system (ES) is designed to act as an expert consultant to a user who is seeking advice about a specific situation. Originally, the term *virtual reality* referred to immersive virtual reality, in which the user becomes fully immersed in an artificial, computer-generated 3-D world. Virtual reality can also refer to applications that are not fully immersive, such as mouse-controlled navigation through a 3-D environment on a graphics monitor, stereo viewing from the monitor via stereo glasses, stereo projection systems, and others.

Principle

System users, business managers, and information systems professionals must work together to build a successful information system.

Systems development involves creating or modifying existing business systems. The major steps of this process and their goals include systems investigation (gain a clear understanding of what the problem is), systems analysis (define what the system must do to solve the problem), systems design (determine exactly how the system will work to meet the business needs), systems implementation (create or acquire the various system components defined in the design step), and systems maintenance and review (maintain and then modify the system so that it continues to meet changing business needs).

Principle

Information systems must be applied thoughtfully and carefully so that society, business, and industry can reap their enormous benefits.

Information systems play a fundamental and ever-expanding role in society, business, and industry. But their use can also raise serious security, privacy, and ethical issues. Effective information systems can have a major impact on corporate strategy and organizational success. Businesses around the globe are enjoying better safety and service, greater efficiency and effectiveness, reduced expenses, and improved decision making and control because of information systems. Individuals who can help their businesses realize these benefits will be in demand well into the future.

Computer and information systems literacy are prerequisites for numerous job opportunities, and not only in the IS field. Computer literacy is knowledge of computer systems and equipment, and information systems literacy is knowledge of how data and information are used by individuals, groups, and organizations. Today, information systems are used in all the functional areas of business, including accounting, finance, sales, marketing, manufacturing, human resource management, and legal information systems. Information systems are also used in every industry, such as airlines, investment firms, banks, transportation companies, publishing companies, healthcare, retail, power management, professional services, and more.

CHAPTER 1: SELF-ASSESSMENT TEST

The value of information is directly linked to how it helps decision makers achieve the organization's goals.

1. A(n) _____ is a set of interrelated components that collect, manipulate, and disseminate data and information and provide a feedback mechanism to meet an objective.

2. Numbers, letters, and other characters are represented by _____.
   a. image data
   b. numeric data
   c. alphanumeric data
   d. symmetric data

3. Knowledge is an awareness and understanding of a set of information. True or False?

Computers and information systems are constantly making it possible for organizations to improve the way they conduct business.

4. A(n) _____ is a set of elements or components that interact to accomplish a goal.

5. What is a measure of what is produced divided by what is consumed?
   a. efficiency
   b. effectiveness
   c. performance
   d. productivity

6. Graphs, charts, and figures are examples of physical models. True or False?

Knowing the potential impact of information systems and having the ability to put this knowledge to work can result in a successful personal career, organizations that reach their goals, and a society with a higher quality of life.

7. A(n) _____ consists of hardware, software, databases, telecommunications, people, and procedures.

8. Computer programs that govern the operation of a computer system are called _____.
    a. feedback
    b. feedforward
    c. software
    d. transaction processing systems

9. Payroll and order processing are examples of a computerized management information system. True or False?

10. What is an organized collection of people, procedures, software, databases, and devices used to create, store, share, and use the organization's experience and knowledge?
    a. TPS (transaction processing system)
    b. MIS (management information system)
    c. DSS (decision support system)
    d. KM (knowledge management)

11. _____ involves anytime, anywhere commerce that uses wireless communications.

System users, business managers, and information systems professionals must work together to build a successful information system.

12. What determines how a new system will work to meet the business needs defined during systems investigation?
    a. systems implementation
    b. systems review
    c. systems development
    d. systems design

Information systems must be applied thoughtfully and carefully so that society, business, and industry can reap their enormous benefits.

13. _____ literacy is a knowledge of how data and information are used by individuals, groups, and organizations.

CHAPTER 1: SELF-ASSESSMENT TEST ANSWERS

(1) information system (2) c (3) True (4) system (5) a (6) False (7) computer-based information system (CBIS) (8) c (9) False (10) d (11) Mobile commerce (m-commerce) (12) d (13) Information systems

REVIEW QUESTIONS

1. What is an information system? What are some of the ways information systems are changing our lives?
2. How would you distinguish data and information? Information and knowledge?
3. Identify at least six characteristics of valuable information.
4. Define the term *system*. Give several examples.
5. What are the components of any information system?
6. What is feedback? What are possible consequences of inadequate feedback?
7. How is system performance measured?
8. What is a knowledge management system? Give an example.
9. What is a computer-based information system? What are its components?

10. Identify three functions of a transaction processing system.
11. What is the difference between an intranet and an extranet?
12. What is m-commerce? Describe how it can be used.
13. What are the most common types of computer-based information systems used in business organizations today? Give an example of each.
14. Identify three elements of artificial intelligence.
15. What are computer literacy and information systems literacy? Why are they important?
16. What are some of the benefits organizations seek to achieve through using information systems?
17. Identify the steps in the systems development process and state the goal of each.

DISCUSSION QUESTIONS

1. Why is the study of information systems important to you? What do you hope to learn from this course to make it worthwhile?
2. Describe how information systems are used at school or work.

3. What is the value of software? Give several examples of software you use at school or home.
4. Why is a database an important part of a computer-based information system?

5. What is the difference between e-commerce and m-commerce?

6. What is the difference between an MIS and a DSS?

7. Suppose that you are a teacher assigned the task of describing the learning processes of preschool children. Why would you want to build a model of their learning processes? What kinds of models would you create? Why might you create more than one type of model?

8. Describe the "ideal" automated vehicle license plate renewal system for the drivers in your state. Describe the input, processing, output, and feedback associated with this system.

9. Discuss the potential use of virtual reality to enhance the learning experience for new automobile drivers. How might such a system operate? What are the benefits and potential disadvantages of such a system?

10. Discuss how information systems are linked to the business objectives of an organization.

11. What are your career goals and how can a computer-based information system be used to achieve them?

PROBLEM-SOLVING EXERCISES

1. Prepare a data disk and a backup disk for the problem-solving exercises and other computer-based assignments you will complete in this class. Create one directory for each chapter in the textbook (you should have 14 directories). As you work through the problem-solving exercises and complete other work using the computer, save your assignments for each chapter in the appropriate directory. On the label of each disk or USB flash drive, be sure to include your name, course, and section. On one disk write "Working Copy"; on the other write "Backup."

2. Search through several business magazines (*Business Week, Computerworld, PC Week,* etc.) for a recent article that discusses the use of information systems to deliver significant business benefits to an organization. Now use other resources to find additional information about the same organization (*Reader's Guide to Periodical Literature*, online search capabilities available at your school's library, the company's public relations department, Web pages on the Internet, etc.). Use word processing software to prepare a one-page summary of the different resources you tried and their ease of use and effectiveness.

3. Create a table that lists 10 or more possible career areas, annual salaries, and brief job descriptions, and rate how much you would like the career area on a scale from 1 (don't like) to 10 (like the most). Print the results. Sort the table according to annual salaries from high to low and print the resulting table. Sort the table from the most liked to least liked and print the results.

4. Do some research to obtain estimates of the rate of growth of e-commerce and m-commerce. Use the plotting capabilities of your spreadsheet or graphics software to produce a bar chart of that growth over a number of years. Share your findings with the class.

TEAM ACTIVITIES

1. Before you can do a team activity, you need a team! As a class member, you might create your own team, or your instructor might assign members to groups. After your group has been formed, meet and introduce yourselves to each other. Find out the first name, hometown, major, e-mail address, and phone number of each member. Find out one interesting fact about each member of your team as well. Brainstorm a name for your team. Put the information on each team member into a database and print enough copies for each team member and your instructor.

2. With the other members of your group, use word processing software to write a one-page summary of what your team hopes to gain from this course and what you are willing to do to accomplish these goals. Send the report to your instructor via e-mail.

WEB EXERCISES

1. Throughout this book, you will see how the Internet provides a vast amount of information to individuals and organizations. We will stress the World Wide Web, or simply the Web, which is an important part of the Internet. Most large universities and organizations have an address on the Internet, called a Web site or home page. The address of the Web site for this publisher is *www.course.com*. You can gain access to the Internet through a browser, such as Microsoft Internet Explorer or Netscape. Using an Internet browser, go to the Web site for this publisher. What did you find? Try to obtain information on this book. You might be asked to develop a report or send an e-mail message to your instructor about what you found.

2. Go to an Internet search engine, such as *www.yahoo.com*, and search for information about knowledge management. Write a brief report that summarizes what you found and the companies that provide knowledge management products.

3. Using the Internet, search for information on the use of information systems in a company or organization that interests you. How does the organization use technology to help it accomplish its goals?

CAREER EXERCISES

1. In the Career Exercises found at the end of every chapter, you will explore how material in the chapter can help you excel in your college major or chosen career. Write a brief report on the career that appeals to you the most. Do the same for two other careers that interest you.

2. Research careers in accounting, marketing, information systems, and two other career areas that interest you. Describe the job opportunities, job duties, and the possible starting salaries for each career area in a report.

CASE STUDIES

Case One

Shroff International Travel Care opens door to Philippines

In many markets, large superstores are seriously threatening the livelihood of small local businesses owners. This is true online as well. Imagine the concerns of the owner of a small local bookstore competing with huge online businesses such as Amazon.com and others who have both a Web and physical presence such as Barnes & Noble. The same can be said about the travel industry. How can a local travel agency compete with Travelocity, Expedia, and Priceline? One answer lies in finding your unique niche.

Shroff International Travel Care, Incorporated (SITCI) found its niche. SITCI is a small travel agency with two offices in and near Manila in the Philippines. SITCI prides itself on its extensive knowledge of travel in the region, and its high level of customer satisfaction. SITCI believes that it can provide customers with better deals, more effective service, and more options than the big online travel companies.

SITCI recently decided to automate their reservations system through a Web-based service. "If you take a look at the reservations process in the travel industry, most of them are excellent candidates for automation," states Arjun Shroff, CEO and managing director of the company. Taking the business online provides several advantages: 1) Shroff can present travel options to customers in a more organized manner to be viewed anytime, 2) the Web site provides self-service for customers to book their own flights, hotels, and ground transportation, and 3) the Web site transforms the business from a local entity to a global entity.

The Web site (*www.airlinecenter.info*) provides deals and information on tour packages, resorts and hotels, visa applications, airline reservations, embassy listings, and limousine services. Airline reservations are provided through the Amadeus global travel distribution system. Amadeus is a global provider of IT applications designed for the travel and tourism industry. Amadeus also provides the transaction processing system that allows customers to pay for flights and accommodations through SITCI's Web site.

The new system has freed up time for SITCI travel agents to work on the more complicated reservations and ticketing work. "Information technology allows our agency to enhance our product and service offerings, provide better and modern service to our existing customers, and even reach out to new customers. You simply cannot do without IT today," Shroff said.

Mr. Shroff takes his national responsibilities seriously and believes that taking his business online will help move the country forward. "We have to be very creative and innovative in attracting tourists to the Philippines; sincerity in dealings, continuous presence in all local and international travel trade-related shows will keep the country on the go," Mr. Shroff said. Arjun Shroff trained with the International Air Transport Associations and Universal Federation of Travel Agents Association (IATA/UFTA) in Switzerland and has spent 29 years in the travel industry in various countries.

Discussion Questions

1. Tour the *www.airlinecenter.info* Web site. Who do you think this Web site is primarily designed to assist—local customers or global customers?
2. How does *www.airlinecenter.info* empower SITCI travel agents to provide better personal service to customers?

Critical Thinking Questions

1. How might SITCI further develop its Web site to provide unique services to the global market that could not be provided by the big online companies?
2. If you were planning a trip to tour the Philippines, who would you rather work with, Expedia.com or SITCI? Why?

SOURCES: Jenalyn Rubio, "Local travel company invests in online reservation system," *Computerworld Philippines*, February 23, 2006, *www.itnetcentral.com/computerworld/default.asp.* Shroff International Travel Care Incorporated (SITCI) Web site, accessed February 23, 2006, *www.airlinecenter.info.*

Case Two

Discovery Communications Digs Out of Mountains of Documents

Discovery Communications, Inc. (DCI) is the leading global real-world media and entertainment company. DCI presents real-world content through documentaries and television programs over the Discovery Channel and many other network brands in 160 countries and 35 languages. DCI's unique brand of programming has been combining education with entertainment since 1985.

Like all global corporations, DCI works hard to distribute mission-critical information and materials to its 5,000-person global workforce. Unknown to most television viewers, each program produced involves a significant amount of legal and strategic paperwork; on average, this amounts to a six-inch stack of production documents for every program. The paperwork assists DCI in maintaining production lifecycles and articulating the legal rights of ownership. Creating and accessing these documents was a cumbersome and tedious chore for DCI personnel. The documents were stored at various locations, which made searches for documents time

consuming. Once located, it was difficult to tell if the document was current and up to date. DCI needed a system that would allow employees at any location to access up-to-date production documents for its programs without any time delay.

This type of business problem falls under the information system heading of knowledge management. Knowledge management, or KM, is a term used to identify systems that collect, transfer, secure, and manage knowledge in terms of resources, documents, and people skills within an organization. Successful knowledge management systems help an organization make the best use of that knowledge. DCI required a special type of KM system that focused on document management. Fortunately for them, KM is popular in industry today and many companies were eager to provide a solution for DCI's problem.

DCI worked with Carefree Technologies (an IBM partner company acquired by Integro, Inc.) for their document management system. Carefree Technologies turned to IBM's Lotus Domino Document Manager system, a document management solution that would centralize and streamline the process of document creation, filing, management, and retrieval. Carefree Technologies and DCI agreed on IBM's WebSphere Portal as the primary user interface for the document management system. As the name implies, WebSphere would act as a Web-based interface to the database of documents and allow Carefree Technology's development team to customize the system for DCI's needs. A portal is an application that provides access to a commonly used information system and communication tools from one central interface, typically a Web page.

Carefree Technology's developers found it easy to merge the document management system with other portal services such as news, information, and communications tools. The final product goes beyond the original hopes and expectations for the system. The portal helps employees track and manage the television production process and easily find the documents they need. In addition, employees can utilize links to external Web sites and a news service from LexisNexis to keep abreast of the latest trends in the television industry. Through the integration and customization of IBM's systems, Carefree Technologies and DCI's IT staff have enhanced its portal by integrating it with other business tools, instant messaging, and Web conferencing to further enhance productivity levels.

Discussion Questions

1. What companies were involved in developing DCI's new system? What role did each company play in the development process?
2. What is the purpose of a corporate portal? What convenience does DCI's new portal provide for its employees?

Critical Thinking Questions

1. Why do you think knowledge management is so popular today? What advantages can it provide a company?
2. What were the most important steps in organizing the millions of documents in DCI's systems? Why?

SOURCES: "Discovery Communications Unifies Working Environment with IBM Portal and Enterprise Document Management Solution," IBM Success Stories, August 24, 2005, *www.ibm.com*. "Volantis Chosen by Discovery to Deliver Global Mobile Portal," M2 Presswire, March 1, 2005, *www.lexis-nexis.com*. Discovery Communications, Inc. corporate home page, accessed February 23, 2006, *http://corporate.discovery.com/*.

Questions for Web Case

See the Web site for this book to read about the Whitmann Price Consulting case for this chapter. Following are questions concerning this Web case.

Whitmann Price Consulting: A New Systems Initiative

Discussion Questions

1. What advantages would the proposed Advanced Mobile Communications and Information System provide for Whitman Price Consulting? What problems might it assist in eliminating?
2. Why do you think Josh and Sandra have been asked to interview the managers of the six business units within WPC as a first step? As IT professionals, Josh, Sandra, and their boss Matt know much more about technology and information systems than the heads of the business units. Shouldn't they be able to design the system without suggestions from amateurs? Including more people in the planning stage is sure to complicate the process.

Critical Thinking Questions

1. If you were Josh or Sandra, what questions would you ask the heads of the six business units?
2. If you were Josh or Sandra, what additional research might you request of your IT staff at this point?

NOTES

Sources for the opening vignette: "Nissan North America scores impressive gains in getting parts to dealers", IBM Success Stories, October 7, 2005, *www.ibm.com/us/*. Nissan Corporate Information, accessed February 19, 2006, *www.nissan- global.com/EN/COMPANY*, Viewlocity on the Web, accessed February 19, 2006, *www.viewlocity.com*.

1   Kahn, Gabriel, "Who Made My Cheese?" *The Wall Street Journal*, July 7, 2005, p. B1.
2   Havenstein, Heather, "E-Health Records Slow to Catch On," *Computerworld*, February 21, 2005, p. 1.
3   Mitchell, Robert, "Businesses Are Using Technology to Gain Control Over Electronic Records," *Computerworld*, May, 30, 2005, p. 21.
4   Chang, Kenneth, "70's Apollo Data Yields New Information," *The Rocky Mountain News*, February 15, 2005, p. 34A.
5   Lin, Grace, et al., "New Model for Military Operations," *OR/MS Today*, December 2004, p. 26.
6   Schupak, Amanda, "The Bunny Chip," *Forbes*, August 15, 2005, p. 53.
7   Brennan, Peter, "IBM Claims Fastest Computer," *Rocky Mountain News*, June 14, 2005, p. 6B.
8   Tam, Pui-Wing, "Digital Snaps in a Snap," *The Wall Street Journal*, August 4, 2005, p. B1.
9   Port, Otis, "Desktop Factories," *Business Week*, May 2, 2005, p. 22.
10 Petersen, Andrea, "Art for When There's Nothing on TV," *The Wall Street Journal*, February 16, 2005, p. D1.
11 Barron, Kelly, "Hidden Value," *Fortune*, June 27, 2005, p. 184[B].
12 Carrns, Ann, "Trial Highlights Vulnerability of Databases," *The Wall Street Journal*, August 3, 2005, p. B1.
13 Woolley, Scott, "Backwater Broadband," *Forbes*, July 4, 2005, p. 64.
14 Yun, Samean, "New 3G Cell Phones Aim to Be Fast," *The Rocky Mountain News*, August 1, 2005, p. 1B.
15 Tynan, Dan, "Singing the Blog Electric," *PC World*, August 2005, p. 120.
16 Mossberg, Walter, "Podcasting Is Still Not Quite Ready for the Masses, *The Wall Street Journal*, July 6, 2005, D5.
17 Mossberg, Walter, "Device Lets You Watch Shows on a Home TV, TiVo from Elsewhere," *The Wall Street Journal*, June 30, 2005, p. B1.
18 Anthes, Gary, "Supply Chain Whirl," *Computerworld*, June 8, 2005, p. 27.
19 Sullivan, Laurie, "ERPzilla," *InformationWeek*, July 11, 2005, p. 30.
20 Marquez, Jeremiah, "Computer Policing L.A. Cops," *The Denver Post*, July 25, 2005, p. 10A.
21 Majchrzak, Ann, et al., "Perceived Individual Collaboration Know-How Development," *Information Systems Research*, March 2005, p. 9.
22 Staff, Ryder Systems, Inc., "Knowledge Management," *www.accenture.com/xd/xd.asp?it=enweb&xd=services%5Chp%5Ccase%5Chp_rydersystems.xml*, accessed on January 15, 2006.
23 Darroch, Jenny, "Knowledge Management, Innovation, and Firm Performance," *Journal of Knowledge Management*, Vol. 9, No. 3, March 2005, p. 101.
24 Staff, "Nissan Developing Smart Cars," *CNN Online*, March 1, 2005.
25 Kurzweil, Ray, "Long Live AI," *Forbes*, August 15, 2005, p. 30.
26 Johnson, Kimberly, "Video Gaming Serious Subject at DU," *The Denver Post*, June 12, 2005, p. 1K.
27 Songini, Marc, "U.K. Tax Agency Mulls Lawsuit Against EDS," *Computerworld*, July 4, 2005, p. 9.
28 Hess, H.M. "Aligning Technology and Business," *IBM Systems Journal*, Vol. 44, No. 1, 2005, p. 25.
29 Thibodeau, Patrick, "Mechanics of a Merger," *Computerworld*, February 14, 2005, p. 35.
30 Cavusoglu, Huseyin, et al., "The Value of Intrusion Detection Systems in Information Technology Security Architecture," *Information Systems Research*, March 2005, p. 28.
31 Friedman, Thomas, "The World Is Flat," *Farrar, Straus and Giroux*, 2005, p. 488.
32 Balfour, Frederik, "Invasion of the Brain Snatchers," *Business Week*, May 9, 2005, p. 24.
33 Smith, Geri, et al., "Central America Is Holding Its Breath," *Business Week*, June 20, 2005, p. 52.
34 *www.europa.eu.int*, accessed on January 15, 2006.
35 *www.sice.oas.org/tradee.asp*, accessed on January 15, 2006.

CHAPTER · 2 ·

Information Systems in Organizations

PRINCIPLES	LEARNING OBJECTIVES
▪ The use of information systems to add value to the organization is strongly influenced by organizational structure, culture, and change.	▪ Identify the value-added processes in the supply chain and describe the role of information systems within them. ▪ Provide a clear definition of the terms *organizational structure*, *culture*, and *change* and discuss how they affect the implementation of information systems.
▪ Because information systems are so important, businesses need to be sure that improvements or completely new systems help lower costs, increase profits, improve service, or achieve a competitive advantage.	▪ Identify some of the strategies employed to lower costs or improve service. ▪ Define the term *competitive advantage* and discuss how organizations are using information systems to gain such an advantage. ▪ Discuss how organizations justify the need for information systems.
▪ Cooperation between business managers and IS personnel is the key to unlocking the potential of any new or modified system.	▪ Define the types of roles, functions, and careers available in information systems.

Information Systems in the Global Economy ⟩⟩
General Motors, United States
Using the OnStar Information System for Strategic Purposes

U.S. auto manufacturers have always been challenged by low-priced, high-quality imports. In 1996, General Motors took a brave step and invested in a new information service that would dramatically alter their customers' driving experiences. They called the system OnStar and installed it in select models of their automobiles.

"We'll take this technology and make sure this technology has the ability to save lives or meaningfully impact somebody's peace of mind while they're driving" stated OnStar President Chet Huber. OnStar combines wireless communications with Global Positioning System (GPS) technologies to connect drivers to GM call centers in Michigan, North Carolina, and Ontario. With a press of a button on the OnStar control panel, drivers can ask an OnStar operator for driving directions, summon an ambulance, or make hands-free phone calls. Because the OnStar system tracks your vehicle's location using GPS, OnStar operators are aware of your location and are able to provide directions. Using a cell phone, a driver can have OnStar unlock a vehicle if locked out or track a stolen vehicle. OnStar automatically attempts to contact you and summons help if needed when an airbag is deployed. OnStar also provides vehicle diagnosis when you are having engine trouble and sends a tow truck when needed. A number of other OnStar services can be found listed at *www.onstar.com*.

The OnStar product has been a huge success with GM's customers, boosted their reputation in the marketplace, and provided them with a competitive advantage. The system required a huge investment on GM's part and took seven years before GM finally broke even on its investment and began making money. Around 4 million drivers now have OnStar, and GM expects that figure to rise as they work to make OnStar standard on all vehicles. OnStar services are provided for a monthly subscription fee: currently $16.95 per month for the basic OnStar safety features, and $34.95 for the navigation. Eighty-five percent of OnStar customers currently get only the safety package, while 15 percent get the navigation package.

Working to continuously improve their popular OnStar product, GM recently implemented a significant upgrade that allows customers to get real-time directions as they drive. Although the original navigation system allows the driver to record directions provided by an OnStar operator for playback throughout the trip, the new system is fully automated and talks drivers through each step of the route as they're driving. If the driver leaves the route, the system alerts the driver and recalculates the directions based on the new location. The new system is easy to use and safer than screen-based navigation because drivers never have to take their eyes off the road. Because it is automated and does not require the assistance of an operator, GM can provide the service at a lower fee than its original navigation system.

By harnessing information systems and using them as a unique service to customers, GM has been able to redefine itself as a high-tech company. The OnStar system has been unrivaled by competitors for over ten years now—surely a record in this tech-driven arena. By extending the capabilities of the OnStar system, GM continues to improve its market share while increasing its level of customer satisfaction.

As you read this chapter, consider the following:

- How have the organizations you encounter implemented information systems for strategic purposes?
- What types of changes within an organization are required to stay ahead of the competition?

Why Learn About Information Systems in Organizations?

Organizations of all types use information systems to cut costs and increase profits. After graduating, a management major might be hired by a shipping company to help design a computerized system to improve employee productivity. A marketing major might work for a national retailer using a network to analyze customer needs in different areas of the country. An accounting major might work for an accounting or consulting firm using a computer to audit other companies' financial records. A real estate major might use the Internet and work in a loose organizational structure with clients, builders, and a legal team located around the world. A biochemist might conduct research for a drug company and use a computer to evaluate the potential of a new cancer treatment.

Although your career might be different from your classmates', you will almost certainly work with computers and information systems to help your company or organization become more efficient, effective, productive, and competitive in its industry. In this chapter, you will see how information systems can help organizations produce higher-quality products and increase their return on investment. We begin by investigating organizations and information systems.

Technology has changed the way organizations work in recent years. Once used to automate manual processes, information technology has transformed the nature of work and the shape of organizations themselves. In this chapter and throughout the book, you will explore the benefits and drawbacks of information technology in today's organizations.

ORGANIZATIONS AND INFORMATION SYSTEMS

organization

A formal collection of people and other resources established to accomplish a set of goals.

An **organization** is a formal collection of people and other resources established to accomplish a set of goals. The primary goal of a for-profit organization is to maximize shareholder value, often measured by the price of the company stock. Nonprofit organizations include social groups, religious groups, universities, and other organizations that do not have profit as their goal.

An organization is a system, which means that it has inputs, processing mechanisms, outputs, and feedback. An organization constantly uses money, people, materials, machines and other equipment, data, information, and decisions. As shown in Figure 2.1, resources such as materials, people, and money serve as inputs to the organizational system from the environment, go through a transformation mechanism, and then are produced as outputs to the environment. The outputs from the transformation mechanism are usually goods or services, which are of higher relative value than the inputs alone. Through adding value or worth, organizations attempt to achieve their goals.

How does the organizational system increase the value of resources? In the transformation mechanism, subsystems contain processes that help turn inputs into goods or services of increasing value. These processes increase the relative worth of the combined inputs on their way to becoming final outputs. Let's reconsider the simple car wash example from Chapter 1 (see Figure 1.3). The first process is washing the car. The output of this system—a clean but wet car—is worth more than the mere collection of ingredients (soap and water), as evidenced by the popularity of automatic car washes. Consumers are willing to pay for the skill, knowledge, time, and energy required to wash their car. The second process is drying—transforming the wet car into a dry one with no water spotting. Again, consumers

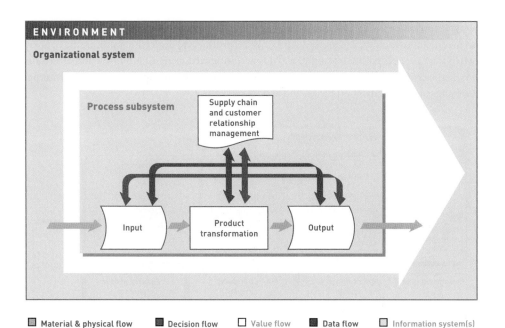

Figure 2.1

A General Model of an Organization

Information systems support and work within all parts of an organizational process. Although not shown in this simple model, input to the process subsystem can come from internal and external sources. Just prior to entering the subsystem, data is external. After it enters the subsystem, it becomes internal. Likewise, goods and services can be output to either internal or external systems.

are willing to pay for the additional skill, knowledge, time, and energy required to accomplish this transformation.

Another example of how inputs can be transformed into increased value for an organization is software that looks for new business opportunities. This type of software analyzes social networks by collecting data from an employee's computerized phone list, address book, e-mails, instant messages, and calendars. It then builds relationships among the data to identify potential new customers and business opportunities for increased sales, thereby increasing the value of the data. So, organizations establish processes to achieve their goals—to exploit opportunities and solve problems.

Providing value to a stakeholder—customer, supplier, manager, shareholder, or employee—is the primary goal of any organization. The value chain, first described by Michael Porter in a 1985 *Harvard Business Review* article, reveals how organizations can add value to their products and services. The **value chain** is a series (chain) of activities that includes inbound logistics, warehouse and storage, production, finished product storage, outbound logistics, marketing and sales, and customer service (see Figure 2.2).[1] You investigate each activity in the chain to determine how to increase the value perceived by a customer. An automotive parts supplier can use a value chain and just-in-time (JIT) inventory to deliver auto parts from a manufacturing plant in Mexico to a Mercedes automobile plant in the United States when they are needed.[2] "In the past, our primary focus was on our processes within the walls of a manufacturing plant," says Gary DeArment, manager for the supplier's North American JIT/JIS Centers. "Now, we have to manage the entire value stream." Depending on the customer, value might mean lower price, better service, higher quality, or uniqueness of product. The value comes from the skill, knowledge, time, and energy that the company invests in the product or activity. Analyzing value chains when developing e-business systems often results in efficient transaction processing systems, an expanding market, and the sharing of information.[3] The value chain is just as important to companies that don't manufacture products, but provide services, such as tax preparers and legal firms. By adding a significant amount of value to their products and services, companies ensure success.

Managing the supply chain and customer relationships are two key parts of managing the value chain. *Supply chain management (SCM)* helps determine what supplies are required for the value chain, what quantities are needed to meet customer demand, how the supplies should be processed (manufactured) into finished goods and services, and how the shipment of supplies and products to customers should be scheduled, monitored, and controlled.[4] For example, in an automotive company, SCM can identify key supplies and parts, negotiate

value chain
A series (chain) of activities that includes inbound logistics, warehouse and storage, production, finished product storage, outbound logistics, marketing and sales, and customer service.

with vendors for the best prices and support, make sure that all supplies and parts are available to manufacture cars and trucks, and send finished products to dealerships around the country when they are needed. Increasingly, SCM is accomplished using the Internet and electronic marketplaces (e-marketplaces.)[5] When an organization has many suppliers, it can use Internet exchanges to negotiate good prices and service.

Figure 2.2

The Value Chain of a Manufacturing Company

Managing raw materials, inbound logistics, and warehouse and storage facilities is called *upstream management,* and managing finished product storage, outbound logistics, marketing and sales, and customer service is called *downstream management.*

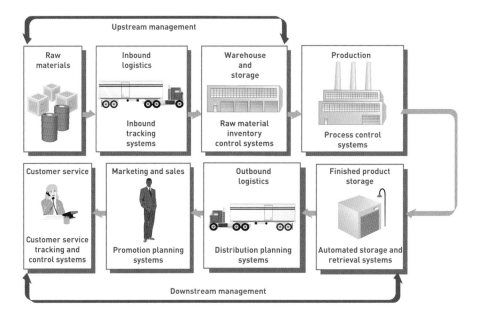

Information systems are an integral part of the UPS operation. UPS serves as a supply chain for many of its customers.

(Source: *www.ups.com.*)

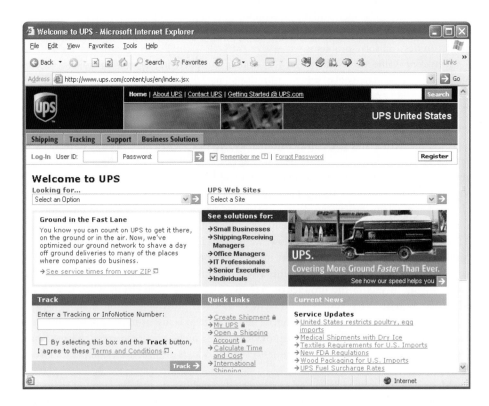

In addition to transporting packages around the world, shipping companies are increasingly serving as supply chains for their customers, and in some cases become part of the customer's information system as well. For example, a company such as UPS might pick up a broken laptop computer under warranty, ship it to a UPS repair facility where it is repaired by people certified by the laptop manufacturer, and then ship it back to the customer. The information system for the laptop company will be involved in confirming customer and warranty information. Shipping companies can also transport raw materials to their customer's factories and ship finished goods to warehouses, retail outlets, or consumers.

Customer relationship management (CRM) programs help a company manage all aspects of customer encounters, including marketing and advertising, sales, customer service after the sale, and programs to retain loyal customers. CRM can help a company collect customer data, contact customers, educate them about new products, and actively sell products to existing and new customers. Often, CRM software uses a variety of information sources, including sales from retail stores, surveys, e-mail, and Internet browsing habits, to compile comprehensive customer profiles. CRM systems can also get customer feedback to help design new products and services. Tesco, Britain's largest retail operation, uses a CRM Clubcard program to provide outstanding customer service and deliver loyalty rewards and perks to valued customers.[6] See Figure 2.3. Customers can earn services such as meals out, travel, dry cleaning, and car maintenance. The Clubcard loyalty program also extends to Tesco's business partners, introducing Tesco customers to other businesses.

Figure 2.3

Tesco uses its Web site to help with customer relationship management.

What role does an information system play in these processes? A traditional view of information systems holds that organizations use them to control and monitor processes and ensure effectiveness and efficiency. An information system can turn feedback from the subsystems into more meaningful information for employees. This information might summarize the performance of the subsystems and be used to change how the system operates. Such changes could involve using different raw materials (inputs), designing new assembly-line procedures (product transformation), or developing new products and services (outputs). In this view, the information system is external to the process and serves to monitor or control it.

A more contemporary view, however, holds that information systems are often so intimately involved that they are *part of* the process itself. From this perspective, the information system plays an integral role in the process, whether providing input, aiding product transformation, or producing output. Consider a phone directory business that creates phone books for international corporations. A corporate customer requests a phone directory listing all steel suppliers in Western Europe. Using its information system, the directory business can sort files to find the suppliers' names and phone numbers and organize them into an alphabetical list. The information system itself is an integral part of this process. It does not just monitor the process externally but works as part of the process to transform raw data into a product. In this example, the information system turns input (names and phone numbers) into a salable output (a phone directory). The same system might also provide the input (data files) and output (printed pages for the directory).

The latter view provides a new perspective on how and why businesses can use information systems. Rather than attempting to understand information systems independent of the organization, we consider the potential role of information systems within the process itself, often leading to the discovery of new and better ways to accomplish the process.

Organizational Structures

organizational structure
Organizational subunits and the way they relate to the overall organization.

Organizational structure refers to organizational subunits and the way they relate to the overall organization. An organization's structure depends on its goals and approach to management, and can affect how it views and uses information systems. The types of organizational structures typically include traditional, project, team, and virtual. Organizational structure can have a direct impact on the organization's information system.

Traditional Organizational Structure

traditional organizational structure
An organizational structure similar to a managerial pyramid, where the hierarchy of decision making and authority flows from strategic management at the top down to operational management and nonmanagement employees. Also called a hierarchical structure.

A **traditional organizational structure**, also called a *hierarchical structure,* is like a managerial pyramid where the hierarchy of decision making and authority flows from the strategic management at the top down to operational management and nonmanagement employees. Compared to lower levels, the strategic level, including the president of the company and vice presidents, has a higher degree of decision authority, more impact on corporate goals, and more unique problems to solve (see Figure 2.4). In most cases, major department heads report to a president or top-level manager. The major departments are usually divided according to function and can include marketing, production, information systems, finance and accounting, research and development, and so on (see Figure 2.5). The positions or departments that are directly associated with making, packing, or shipping goods are called *line positions*. A production supervisor who reports to a vice president of production is an example of a line position. Other positions might not be directly involved with the formal chain of command but instead assist a department or area. These are *staff positions,* such as a legal counsel reporting to the president.

Figure 2.4

A simplified model of the organization, showing the managerial pyramid from top-level managers to nonmanagement employees.

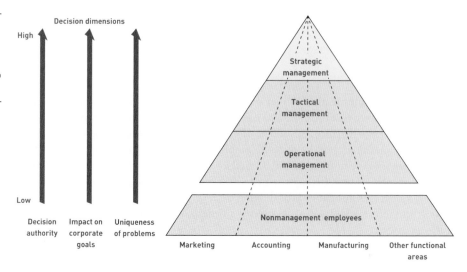

Decision dimensions

High

Low

Decision authority   Impact on corporate goals   Uniqueness of problems

Strategic management

Tactical management

Operational management

Nonmanagement employees

Marketing   Accounting   Manufacturing   Other functional areas

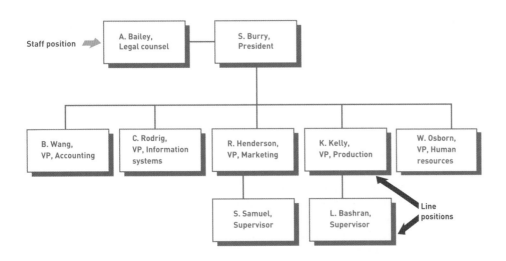

Today, the trend is to reduce the number of management levels, or layers, in the traditional organizational structure. This type of structure, often called a **flat organizational structure**, empowers employees at lower levels to make decisions and solve problems without needing permission from midlevel managers. **Empowerment** gives employees and their managers more responsibility and authority to make decisions, take action, and have more control over their jobs. For example, an empowered sales clerk could respond to certain customer requests or problems without needing permission from a supervisor. On a factory floor, empowerment might mean that an assembly-line worker can stop the production line to correct a problem or defect before the product is passed to the next station. Policies and programs that let employees share ownership in a company flattens the organizational structure. For example, the Royal Bank of Scotland uses a stock share ownership program to encourage employee ownership and empowerment.[7] The program resulted in about 100,000 employees receiving a benefit of 2,100 pounds, or about $4,000.

Information systems can be a key element in empowering employees because they provide the information employees need to make decisions. The employees might also be empowered to develop or use their own personal information systems, such as a simple forecasting model or spreadsheet.

flat organizational structure
An organizational structure with a reduced number of management layers.

empowerment
Giving employees and their managers more responsibility and authority to make decisions, take certain actions, and have more control over their jobs.

Project and Team Organizational Structures

A **project organizational structure** is centered on major products or services. For example, in a manufacturing firm that produces baby food and other baby products, each line is produced by a separate unit. Traditional functions such as marketing, finance, and production are positioned within these major units (see Figure 2.6). Many project teams are temporary—when the project is complete, the members go on to new teams formed for another project.

project organizational structure
A structure centered on major products or services.

The **team organizational structure** is centered on work teams or groups. In some cases, these teams are small; in others, they are very large. Typically, each team has a leader who reports to an upper-level manager. Depending on its tasks, the team can be temporary or permanent. A healthcare company, for example, can form small teams to organize its administrators, physicians, and others to work with individual patients.

Virtual Organizational Structure and Collaborative Work

A **virtual organizational structure** employs individuals, groups, or complete business units in geographically dispersed areas that can last for a few weeks or years, often requiring telecommunications or the Internet. These people might be in different countries, operating in different time zones. In other words, virtual organizational structures allow work to be separated from location and time. Work can be done anywhere, anytime. People might never meet physically, which explains the use of the word *virtual,* and highlights the difference between virtual organizations and traditional ones that have operations in more than one location—a virtual organization is geographically distributed, and uses information technology to communicate and coordinate their work. Despite their physical separation, members of a virtual organization can collaborate on any aspect of a project, such as supplying raw materials, producing goods and services, and delivering goods and services to the marketplace. In some cases, a virtual organization is temporary, lasting only a few weeks or months. In others, it can last for years or decades.

One way to use a virtual organizational structure is within a firm. A company can use a virtual organizational structure with dispersed workers who have distinct skills and abilities to save millions of dollars. PriceWaterhouseCoopers, a global accounting giant, uses virtual teams of 5 to 50 people in the learning and education department.[8] According to Peter Nicolas, the company's Learning Solutions manager, "Virtual teaming is the norm for us." The company takes advantage of software and technology, including Microsoft Live Meeting, Centra Software's Virtual-Classroom Application, and Lotus Notes from IBM, to help the teams work from distant locations. Often, however, virtual structures are formed with people or groups outside a company. For example, one company formed an organization called the Virtual Corporate Management System to assemble several small manufacturing businesses to compete with larger businesses.

In addition to reducing costs or increasing revenues, a virtual organizational structure can provide an extra level of security. Many companies are now dispersing employees and using a virtual structure in case of a terrorist attack or a disaster. If a disaster strikes at the primary location, the company still has sufficient employees at other locations to keep the company running. Today's workers are performing company work at home, at a customer's location, in coffee shops, on pleasure boats, and at convenient work centers in suburbia. People can also work at any time. Using the Internet and e-mail, workers can put the finishing touches on a new business proposal in Europe or Asia, while coworkers in North America are sleeping.

Successful virtual organizational structures share key characteristics. One strategy is to have in-house employees concentrate on the firm's core businesses and use virtual employees, groups, or businesses to do everything else. Using information systems to manage the activities of a virtual structure is essential, often requiring specialized software to coordinate joint work. Even with sophisticated IS tools, teams still need face-to-face meetings, especially at the beginning of new projects.

A virtual organizational structure allows *collaborative* work, in which managers and employees can effectively work in groups, even those composed of members from around the world. A management team, for example, can include executives from Australia and England. A programming team can consist of people in the United States and India. Collaborative work can also include all aspects of the supply chain and customer relationship management. An automotive design team, for example, can include critical parts suppliers, engineers from the company, and important customers.

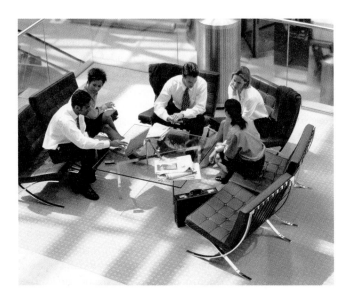

A virtual organizational structure allows collaborative work, in which managers and employees can effectively work in groups, even those composed of members from around the world.

(Source: Courtesy of ImageState/ Alamy.)

Organizational Culture and Change

Culture is a set of major understandings and assumptions shared by a group, such as within an ethnic group or a country. **Organizational culture** consists of the major understandings and assumptions for a business, corporation, or other organization. The understandings, which can include common beliefs, values, and approaches to decision making, are often not stated or documented as goals or formal policies. Employees might be expected to be clean-cut, wear conservative outfits, and be courteous in dealing with all customers. Sometimes organizational culture is formed over years. In other cases, it can be formed rapidly by top-level managers—for example, by starting a "casual Friday" dress policy.

Like organizational structure, organizational culture can significantly affect the development and operation of information systems. For example, a procedure associated with a newly designed information system might conflict with an informal procedural rule that is part of organizational culture. Furthermore, organizational culture can influence a decision maker's perception of the factors and priorities to consider in setting objectives. A company might have an unwritten understanding that all inventory reports must be prepared before 10:00 A.M. on Friday. Because of this understood deadline, the decision maker might reject a cost-reduction option that requires compiling the inventory report over the weekend.

Organizational change deals with how for-profit and nonprofit organizations plan for, implement, and handle change. Change can be caused by internal factors, such as those initiated by employees at all levels, or external factors, such as activities wrought by competitors, stockholders, federal and state laws, community regulations, natural occurrences (such as hurricanes), and general economic conditions. Many European countries, for example, adopted the euro, a single European currency, which changed how financial companies do business and use their information systems.

Change can be sustaining or disruptive.[9] *Sustaining change* can help an organization improve the supply of raw materials, the production process, and the products and services it offers. New manufacturing equipment to make disk drives is an example of a sustaining change for a computer manufacturer. The new equipment might reduce the costs of producing the disk drives and improve overall

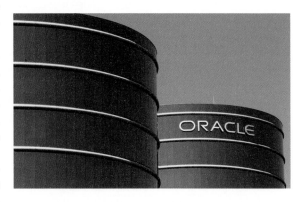

culture
A set of major understandings and assumptions shared by a group.

organizational culture
The major understandings and assumptions for a business, corporation, or other organization.

organizational change
The responses that are necessary so that for-profit and nonprofit organizations can plan for, implement, and handle change.

When Oracle recently acquired Siebel Systems, they combined their strengths to provide customers benefits neither company could have achieved on their own, especially in the area of customer relationship management (CRM) software, creating sustaining change.

(Source: Courtesy of Getty Images.)

performance. *Disruptive change,* on the other hand, often harms an organization's performance or even puts it out of business. In general, disruptive technologies might not originally have good performance, low cost, or even strong demand. Over time, however, they often replace existing technologies. They can cause good, stable companies to fail when they don't change or adopt the new technology.

Overcoming resistance to change, especially disruptive change, can be the hardest part of bringing information systems into a business. Occasionally, employees even attempt to sabotage a new information system because they do not want to learn the new procedures and commands. In most of these instances, the employees were not involved in the decision to implement the change, nor were they fully informed about the reasons for the change or the benefits they would enjoy.

The dynamics of change can be viewed in terms of a change model. A **change model** represents change theories by identifying the phases of change and the best way to implement them. Kurt Lewin and Edgar Schein propose a three-stage approach for change (see Figure 2.7). *Unfreezing* is ceasing old habits and creating a climate receptive to change. *Moving* is learning new work methods, behaviors, and systems. *Refreezing* involves reinforcing changes to make the new process second nature, accepted, and part of the job.[10] When a company introduces a new information system, a few members of the organization must become agents of change—champions of the new system and its benefits. Understanding the dynamics of change can help them confront and overcome resistance so that the new system can be used to maximum efficiency and effectiveness.

change model
A representation of change theories that identifies the phases of change and the best way to implement them.

Figure 2.7

A Change Model

Unfreezing

Moving

Refreezing

organizational learning
The adaptations to new conditions or alterations of organizational practices over time.

Organizational learning is closely related to organizational change. According to the concept of **organizational learning**, organizations adapt to new conditions or alter their practices over time. Assembly-line workers, secretaries, clerks, managers, and executives learn better ways of doing business and incorporate them into their day-to-day activities. Collectively, these adjustments based on experience and ideas are called organizational learning. In some cases, the adjustments can be a radical redesign of business processes, often called *reengineering*. In other cases, these adjustments can be more incremental, a concept called *continuous improvement*. Both adjustments reflect an organization's *strategy*, the long-term plan of action for achieving their goals.

Reengineering and Continuous Improvement

To stay competitive, organizations must occasionally make fundamental changes in the way they do business. In other words, they must change the activities, tasks, or processes they use to achieve their goals. **Reengineering**, also called **process redesign** and *business process reengineering (BPR),* involves the radical redesign of business processes, organizational structures, information systems, and values of the organization to achieve a breakthrough in business results. Reengineering can cause changes of an organization's values and information systems (see Figure 2.8).[11] Reengineering can reduce delivery time, increase product and service quality, enhance customer satisfaction, and increase revenues and profitability. For example, as a result of increased electronic trading, the Securities Industry Association is radically reengineering the communications systems used for stock and securities trading. The reengineering also calls for a new settlement system, called Straight-Through Processing.

reengineering (process redesign)

The radical redesign of business processes, organizational structures, information systems, and values of the organization to achieve a breakthrough in business results.

Figure 2.8

Reengineering

Reengineering involves the radical redesign of business processes, organizational structure, information systems, and values of the organization to achieve a breakthrough in business results.

With the increased volume of securities trading in the global marketplace, the Securities Industry Association has released a new model to reengineer the process for settling and clearing transactions that reduces costs and saves time.

(Source: Courtesy of Getty Images.)

In contrast to simply automating the existing work process, reengineering challenges the fundamental assumptions governing their design. It requires finding and vigorously challenging old rules blocking major business process changes. These rules are like anchors weighing down a firm and keeping it from competing effectively. Table 2.1 provides some examples of such rules.

Some organizations are using reengineering to improve processes and reduce errors. For example, to reengineer their inventory systems, hospitals can attach radio frequency chips to

Rule	Original Rationale	Potential Problem
Hold small orders until full-truckload shipments can be assembled.	Reduce delivery costs.	Customer delivery is slow—lost sales.
Do not accept an order until customer credit is approved.	Reduce potential for bad debt.	Customer service is poor—lost sales.
Let headquarters make all merchandising decisions.	Reduce number of items carried in inventory.	Customers perceive organization has limited product selection—lost sales.

Table 2.1

Selected Business Rules That Affect Business Processes

continuous improvement
Constantly seeking ways to improve business processes to add value to products and services.

drugs. In addition to cutting costs and reducing errors, the chips can curb drug counterfeiting. Other companies use reengineering to increase their competitive position in the marketplace. However, although reengineering can introduce breakthrough innovations, it can also be disruptive, expensive, and time consuming.

In contrast to reengineering, the idea of **continuous improvement** is to constantly seek ways to improve business processes and add value to products and services. This continual change will increase customer satisfaction and loyalty and ensure long-term profitability. Manufacturing companies make continual product changes and improvements. Service organizations regularly find ways to provide faster and more effective assistance to customers. By doing so, these companies increase customer loyalty, minimize the chance of customer dissatisfaction, and diminish the opportunity for competitive inroads. Table 2.2 compares these two strategies.

Business Process Reengineering	Continuous Improvement
Strong action taken to solve serious problem	Routine action taken to make minor improvements
Top-down change driven by senior executives	Bottom-up change driven by workers
Broad in scope; cuts across departments	Narrow in scope; focus is on tasks in a given area
Goal is to achieve a major breakthrough	Goal is continuous, gradual improvements
Often led by outsiders	Usually led by workers close to the business
Information system integral to the solution	Information systems provide data to guide the improvement team

Table 2.2

Comparing Business Process Reengineering and Continuous Improvement

technology acceptance model (TAM)
A model that describes the factors that lead to higher levels of acceptance and usage of technology.

User Satisfaction and Technology Acceptance

To be effective, reengineering and continuous improvement efforts must result in satisfied users and be accepted and used throughout the organization. Over the years, IS researchers have studied user satisfaction and technology acceptance as they relate to IS attitudes and usage. Although user satisfaction and technology acceptance started as two separate theories, some believe that they can be integrated into one.[12]

User satisfaction with a computer system and the information it generates often depends on the quality of the system and the information.[13] A quality information system is usually flexible, efficient, accessible, and timely. Recall that quality information is accurate, reliable, current, complete, and delivered in the proper format.[14]

The **technology acceptance model** (TAM) specifies the factors that can lead to better attitudes about the information system, along with higher acceptance and usage of the system in an organization. These factors include the perceived usefulness of the technology, the ease of its use, the quality of the information system, and the degree to which the organization supports its use.[15]

You can determine the actual usage of an information system by the amount of technology diffusion and infusion. **Technology diffusion** is a measure of how widely technology is spread throughout an organization. An organization in which computers and information systems are located in most departments and areas has a high level of technology diffusion.[16] Some online merchants, such as Amazon.com, have a high diffusion and use computer systems to perform most of their business functions, including marketing, purchasing, and billing. **Technology infusion**, on the other hand, is the extent to which technology permeates an area or department. In other words, it is a measure of how deeply embedded technology is in an area of the organization. Some architectural firms, for example, use computers in all aspects of designing a building from drafting to final blueprints. The design area, thus, has a high level of infusion. Of course, a firm can have a high level of infusion in one part of its operations and a low level of diffusion overall. The architectural firm might use computers in all aspects of design (high infusion in the design area), but not to perform other business functions, including billing, purchasing, and marketing (low diffusion). Diffusion and infusion often depend on the technology available now and in the future, the size and type of the organization, and the environmental factors that include the competition, government regulations, suppliers, and so on. This is often called the technology, organization, and environment (TOE) framework.[17]

Although an organization might have a high level of diffusion and infusion, with computers throughout the organization, this does not necessarily mean that information systems are being used to their full potential. In fact, the assimilation and use of expensive computer technology throughout organizations varies greatly.[18] Companies hope that a high level of diffusion, infusion, satisfaction, and acceptance will lead to greater performance and profitability.[19]

technology diffusion
A measure of how widely technology is spread throughout the organization.

technology infusion
The extent to which technology is deeply integrated into an area or department.

Quality

The definition of the term *quality* has evolved over the years. In the early years of quality control, firms were concerned with meeting design specifications—that is, conforming to standards. If a product performed as designed, it was considered a high-quality product. A product can perform its intended function, however, and still not satisfy customer needs. Today, **quality** means the ability of a product (including services) to meet or exceed customer expectations. For example, a computer that not only performs well but is easy to maintain and repair would be considered a high-quality product. Increasingly, customers expect helpful support after the sale. This view of quality is completely customer oriented. A high-quality product satisfies customers by functioning correctly and reliably, meeting needs and expectations, and being delivered on time with courtesy and respect. Organizations now use techniques to ensure quality, including total quality management and six sigma. See Table 2.3.

quality
The ability of a product (including services) to meet or exceed customer expectations.

Table 2.3

Total Quality Management and Six Sigma

Technique	Description	Examples
Total quality management (TQM)	Involves developing a keen awareness of customer needs, adopting a strategic vision for quality, empowering employees, and rewarding employees and managers for producing high-quality products.	The U.S. Postal Service uses the Mail Preparation Total Quality Management (MPTQM) program to certify leading mail sorting and servicing companies. MAA Bozell, a communications company in India, used TQM to improve the quality for all of its business processes.[20] According to the chairman of the MAA group, "We are looking at continuous improvement in three areas with the help of TQM: quality, accuracy, and faster delivery to our clients. We do this by empowering all employees to take ownership of existing processes and continuously find ways to improve/change these processes."
Six sigma	A statistical term that means products and services will meet quality standards 99.9997% of the time. In a normal distribution curve used in statistics, six standard deviations (six sigma) is 99.9997% of the area under the curve.	Transplace, a $57 million trucking and logistics company, uses six sigma to improve quality by eliminating waste and unneeded steps.[21] According to CEO Tom Sanderson, "If you drive out waste, you have a better chance to make the process reliable. If you study the process and eliminate extra steps, you have a better chance to make the process reliable. If you study the process and eliminate extra steps, you are more apt to improve it. Why bother improving a process that doesn't need to be done?"

Outsourcing, On-Demand Computing, and Downsizing

A significant portion of an organization's expenses are used to hire, train, and compensate talented staff. So organizations try to control costs by determining the number of employees they need to maintain high-quality goods and services. Strategies to contain costs are outsourcing, on-demand computing, and downsizing.

outsourcing
Contracting with outside professional services to meet specific business needs.

Outsourcing involves contracting with outside professional services to meet specific business needs. Often, companies outsource a specific business process, such as recruiting and hiring employees, developing advertising materials, promoting product sales, or setting up a global telecommunications network. Organizations often outsource a process to focus more closely on their core business—and target limited resources to meet strategic goals. Boeing, for example, has outsourced some design work for most of its aircraft.[22] "The most interesting development of recent years has been the participation of Russian engineers and scientists in the development of the Boeing 787, 50 percent of which will be manufactured from composite materials, therefore representing a new stride in the development of the world aircraft industry," said Sergev Kravchenko, a Russian engineer in Moscow. About a third of the design work on a new aircraft model, the Boeing 777-300, has been done in Russia.

Other reasons for outsourcing are to trim expenses or benefit from the expertise of a service provider. A computer company, for example, can outsource the manufacturing of its personal computers and save hundreds of millions of dollars while its in-house staff concentrates on developing new designs and features. Companies can save millions or billions of dollars by transferring some work from the United States to lower-wage countries such as Costa Rica and the Philippines. Companies that are considering outsourcing to cut the cost of their IS operations need to review this decision carefully, however. A growing number of organizations are finding that outsourcing does not necessarily lead to reduced costs. One of the primary reasons for cost increases is poorly written contracts that tack on charges from the outsourcing vendor for each additional task. Other potential drawbacks of outsourcing include loss of control and flexibility, overlooked opportunities to strengthen core competency, and low employee morale.

on-demand computing
Contracting for computer resources to rapidly respond to an organization's varying workflow. Also called on-demand business and utility computing.

On-demand computing is an extension of the outsourcing approach, and many companies offer on-demand computing to business clients and customers. **On-demand computing**, also called *on-demand business* and *utility computing,* involves rapidly responding to the organization's flow of work as the need for computer resources varies. It is often called *utility computing* because the organization pays for computing resources from a computer or consulting company just as it pays for electricity from a utility company. This approach treats the information system—including hardware, software, databases, telecommunications, personnel, and other components—more as a service than as separate products. In other words, instead of purchasing hardware, software, and database systems, the organization only pays a fee for the systems it needs at peak times. The approach can save money because the organization does not pay for systems that it doesn't routinely need. It also allows the organization's IS staff to concentrate on more-strategic issues. Organic grocer Wild Oats Market, Inc., for example, uses utility computing for its processing and storage needs.[23] According to the chief information officer of the grocer, "You schedule it, and they make it happen." Wild Oats uses VeriCenter for its utility computing needs.

downsizing
Reducing the number of employees to cut costs.

Downsizing involves reducing the number of employees to cut costs. The term *rightsizing* is also used. Rather than pick a specific business process to downsize, companies usually look to downsize across the entire company. Downsizing clearly reduces total payroll costs, though employee morale can suffer.

Employers need to be open to alternatives for reducing the number of employees but use layoffs as the last resort. It's simpler to encourage people to leave voluntarily through early retirement or other incentives. Voluntary downsizing programs often include a buyout package offered to certain classes of employees (for example, those over 50 years old). The buyout package offers employees certain benefits and cash incentives if they voluntarily retire from the company. Other options are job sharing and transfers.

Victims of Technology?

New technologies often replace older, less-efficient ways of doing things. In this manner, technological innovation pushes people to progress toward a more efficient and, some would say, more fulfilling existence. People often feel intense pressure to use the latest and greatest technologies. Those that cling to old technology find themselves out of sync with the rest of society.

Although a person can ignore technological advances and still survive, the same is not true of businesses in competitive markets. New businesses with new innovations regularly eclipse old businesses based on old technologies. A business must continuously evaluate its mission and practices in light of current day trends.

Consider Kodak, the film company. When digital photography became the norm for personal and professional photographers, some thought that a company that made its money by developing film would go belly up. Kodak surprised many by redefining itself as a digital photography company rather than a film company. Kodak invested heavily in its Web presence and now provides its customers with valuable digital photography services rather than film developing. Kodak has done considerably more than just survive. There are more photographers now than ever before, and Kodak continues to profit from selling digital cameras, prints, and other photograph services.

Almost all large traditional businesses have repositioned themselves as technology companies. Another example is Time, Inc., the magazine giant. Time recently cut hundreds of jobs to reinvest and build their Web business. Time is transforming from a magazine publishing company to a multiplatform media company that will have a presence on paper, on TV, and on the Web.

When most people think of Western Union, they think of telegrams. Western Union was established in 1856 to exploit the hot technology of the day, the telegraph, to send cross-country messages in less than a day. Prior to this, messages were sent via pony express. In the early days of electronic communication, Western Union telegrams were an important part of person-to-person communications across America. Telegrams reached their peak in the 1920s and 30s when it was cheaper to send a telegram than it was to make a long-distance phone call. In 1933, telegrams became much more amusing with the invention of the singing telegram. Early in 2006, the last telegram was delivered as Western Union discontinued its telegram service.

What has happened to Western Union? Western Union foresaw the end of its telegram service long ago. The invention of the telephone and fax machine reduced the need for telegrams. The Internet and e-mail was the nail that sealed the coffin. Western Union continued providing important telecommunications services through the 1980s. It was the first to launch a domestic communications satellite into orbit. It also was the first to use transcontinental microwaves for communications rather than poles and wires. In the 1990s, the company began focusing on financial transactions. Today, Western Union's core business is in helping consumers and businesses transfer money or make payments.

These three cases—Kodak, Time, and Western Union—show how new technologies can cause businesses to redefine themselves. These businesses implemented sustaining change by changing their organizations to move with new trends. Some might consider businesses that refuse to change, and consequently lose out to others that embrace new technologies, as the victims of technology—as though technology itself has callously plowed them under. Kodak, Time, and Western Union show that even the oldest of traditional businesses can harness technology for themselves. This is true on the individual employee level as well. For each employee who loses a job because of outdated skills or talents, another employee is hired for his technical prowess. Employees, like the companies that hire them, must adapt to the needs of the market by learning new and desirable skills.

Technology motivates change at all levels, personal, business, and social. It is important to consider the changes brought about by new technologies and gauge their value in terms of quality of life. The most successful people, businesses, and societies are those that are able to identify and adopt technologies that are truly beneficial while sorting out and discarding technologies that are trendy, superfluous, or even dangerous.

Discussion Questions

1. What, if anything, has the world lost in terms of social value as the result of the retirement of Western Union telegrams? How does this balance against the value of telephones, faxes, e-mail, and text messaging—those technologies that displaced the telegram?

2. How have online retailers such as Amazon.com affected change in traditional brick-and-mortar businesses such as Sears?

Critical Thinking Questions

1. It is often difficult for old, established industries to change business practices. How can this be observed in the impact of the digital music boom (MP3 file sharing) on record companies and their traditional methods of music distribution?

2. Consider the uses of technology at your school. How are educational institutions taking advantage of technology to compete and win students?

SOURCES: Seelye, Katharine Q., "Time Inc. to Cut 100 More Jobs as It Focuses on Web Business," *The New York Times,* January 31, 2006, *www.nytimes.com.* Weiss, Todd R., "Technology Victim: Western Union Sends Its Last Telegram," *Computerworld,* February 3, 2006, *www.computerworld.com.* Associated Press, "STOP — Telegram Era Over, Western Union Says," MSNBC, February 2, 2006, *www.msnbc.msn.com/id/11147506/.*

Accenture provides consulting, technology, and outsourcing services.

(Source: *www.accenture.com*.)

COMPETITIVE ADVANTAGE

competitive advantage
A significant and (ideally) long-term benefit to a company over its competition.

A **competitive advantage** is a significant and (ideally) long-term benefit to a company over its competition, and can result in higher-quality products, better customer service, and lower costs. Establishing and maintaining a competitive advantage is complex, but a company's survival and prosperity depend on its success in doing so. An organization often uses its information system to help achieve a competitive advantage. In his book *Good to Great,* Jim Collins outlines how technology can be used to accelerate companies from good to great.[24] Table 2.4 shows how a few companies accomplished this move. Ultimately, it is not how much a company spends on information systems but how it makes and manages investments in technology. Companies can spend less and get more value.

Table 2.4

How Some Companies Used Technologies to Move from Good to Great

(Source: Data from Jim Collins, *Good to Great*, Harper Collins Books, 2001, p. 300.)

Company	Business	Competitive Use of Information Systems
Circuit City	Consumer electronics	Developed sophisticated sales and inventory-control systems to deliver a consistent experience to customers
Gillette	Shaving products	Developed advanced computerized manufacturing systems to produce high-quality products at low cost
Walgreens	Drug and convenience stores	Developed satellite communications systems to link local stores to centralized computer systems
Wells Fargo	Financial services	Developed 24-hour banking, ATMs, investments, and increased customer service using information systems

Factors That Lead Firms to Seek Competitive Advantage

A number of factors can lead to attaining a competitive advantage. Michael Porter, a prominent management theorist, suggested a now widely accepted competitive forces model, also called the **five-forces model**. The five forces include (1) the rivalry among existing competitors, (2) the threat of new entrants, (3) the threat of substitute products and services, (4) the bargaining power of buyers, and (5) the bargaining power of suppliers. The more these forces combine in any instance, the more likely firms will seek competitive advantage and the more dramatic the results of such an advantage will be.

Rivalry Among Existing Competitors

Typically, highly competitive industries are characterized by high fixed costs of entering or leaving the industry, low degrees of product differentiation, and many competitors. Although all firms are rivals with their competitors, industries with stronger rivalries tend to have more firms seeking competitive advantage. To gain an advantage over competitors, companies constantly analyze how they use their resources and assets. This *resource-based view* is an approach to acquiring and controlling assets or resources that can help the company achieve a competitive advantage. For example, a transportation company might decide to invest in radio-frequency technology to tag and trace products as they move from one location to another.

Threat of New Entrants

A threat appears when entry and exit costs to an industry are low and the technology needed to start and maintain a business is commonly available. For example, a small restaurant is threatened by new competitors in the restaurant industry. Owners of small restaurants do not require millions of dollars to start the business, food costs do not decline substantially for large volumes, and food processing and preparation equipment is easily available. When the threat of new market entrants is high, the desire to seek and maintain competitive advantage to dissuade new market entrants is also usually high.

five-forces model
A widely accepted model that identifies five key factors that can lead to attainment of competitive advantage, including (1) the rivalry among existing competitors, (2) the threat of new entrants, (3) the threat of substitute products and services, (4) the bargaining power of buyers, and (5) the bargaining power of suppliers.

In the restaurant industry, competition is fierce because entry costs are low. Therefore, a small restaurant that enters the market can be a threat to existing restaurants.

(Source: Courtesy of Medioimages/ Alamy.)

Threat of Substitute Products and Services

Companies that offer one type of goods or services are threatened by other companies that offer similar goods or services. The more consumers can obtain similar products and services that satisfy their needs, the more likely firms are to try to establish competitive advantage. For example, consider the photographic industry. When digital cameras became popular, traditional film companies had to respond to stay competitive and profitable. Traditional film companies, such as Kodak and others, started to offer additional products and enhanced services, including digital cameras, the ability to produce digital images from traditional film cameras, and Web sites that could be used to store and view pictures.

Bargaining Power of Customers and Suppliers

Large customers tend to influence a firm, and this influence can increase significantly if the customers can threaten to switch to rival companies. When customers have a lot of bargaining power, companies increase their competitive advantage to retain their customers. Similarly, when the bargaining power of suppliers is strong, companies need to improve their competitive advantage to maintain their bargaining position. Suppliers can also help an organization gain a competitive advantage. Some suppliers enter into strategic alliances with firms and eventually act as a part of the company. Suppliers and companies can use telecommunications to link their computers and personnel to react quickly and provide parts or supplies as necessary to satisfy customers. Government agencies are also using strategic alliances. The investigative units of the U.S. Customs and Immigration and Naturalization Service entered into a strategic alliance to streamline investigations.

Strategic Planning for Competitive Advantage

To be competitive, a company must be fast, nimble, flexible, innovative, productive, economical, and customer oriented. It must also align its IS strategy with general business strategies and objectives.[25] Given the five market forces just mentioned, Porter and others have proposed a number of strategies to attain competitive advantage, including cost leadership, differentiation, niche strategy, altering the industry structure, creating new products and services, and improving existing product lines and services.[26] In some cases, one of these strategies becomes dominant. For example, with a cost leadership strategy, cost can be the key consideration, at the expense of other factors if need be.

- **Cost leadership.** Deliver the lowest possible products and services. Wal-Mart and other discount retailers have used this strategy for years. Cost leadership is often achieved by reducing the costs of raw materials through aggressive negotiations with suppliers, becoming more efficient with production and manufacturing processes, and reducing warehousing and shipping costs. Some companies use outsourcing to cut costs when making products or completing services.
- **Differentiation.** Deliver different products and services. This strategy can involve producing a variety of products, giving customers more choices, or delivering higher-quality products and services. Many car companies make different models that use the same basic parts and components, giving customers more options. Other car companies attempt to increase perceived quality and safety to differentiate their products. Some consumers are willing to pay higher prices for these vehicles that differentiate on higher quality or better safety.
- **Niche strategy.** Deliver to only a small, niche market. Porsche, for example, doesn't produce inexpensive station wagons or large sedans. It makes high-performance sports cars and SUVs. Rolex only makes high-quality, expensive watches. It doesn't make inexpensive, plastic watches that can be purchased for $20 or less.
- **Altering the industry structure.** Change the industry to become more favorable to the company or organization. The introduction of low-fare airline carriers, such as Southwest Airlines, has forever changed the airline industry, making it difficult for traditional airlines to make high profit margins. To fight back, airlines such as Delta are launching their own low-fare flights. Creating strategic alliances can also alter the industry structure. A **strategic alliance**, also called a **strategic partnership**, is an agreement between two or more companies that involves the joint production and distribution of goods and services.
- **Creating new products and services.** Introduce new products and services periodically or frequently. This strategy always helps a firm gain a competitive advantage, especially for the computer industry and other high-tech businesses. If an organization does not introduce new products and services every few months, the company can quickly stagnate, lose market share, and decline. Companies that stay on top are constantly developing new products and services. A large U.S. credit-reporting agency, for example, can use its information system to help it explore new products and services in different markets.

strategic alliance (strategic partnership)
An agreement between two or more companies that involves the joint production and distribution of goods and services.

- **Improving existing product lines and service.** Make real or perceived improvements to existing product lines and services. Manufacturers of household products are always advertising new and improved products. In some cases, the improvements are more perceived than real refinements; usually, only minor changes are made to the existing product, such as to reduce the amount of sugar in breakfast cereal. Some direct-order companies are improving their service by using Radio Frequency Identification (RFID) tags to identify and track the location of their products as they are shipped from one location to another. Customers and managers can instantly locate products as they are shipped from suppliers to the company, to warehouses, and finally to customers.
- **Other strategies.** Some companies seek strong *growth* in sales, hoping that it can increase profits in the long run due to increased sales. Being the *first to market* is another competitive strategy. Apple Computer was one of the first companies to offer complete and ready-to-use personal computers. Some companies offer *customized* products and services to achieve a competitive advantage. Dell, for example, builds custom PCs for consumers. *Hire the best people* is another example of a competitive strategy. The assumption is that the best people will determine the best products and services to deliver to the market and the best approach to deliver these products and services. Companies can also combine one or more of these strategies. In addition to customization, Dell attempts to offer low-cost computers (cost leadership) and top-notch service (differentiation).

> To maintain a competitive advantage, companies introduce innovative products. The DocuPen RC800 is a pen-sized portable color scanner that lets you scan printed text and then transfer the text to a smartphone, PDA, or other computer.
>
> (Source: Copyright © 2005 Planon System Solutions Inc.)

PERFORMANCE-BASED INFORMATION SYSTEMS

Businesses have passed through at least three major stages in their use of information systems. In the first stage, organizations focused on using information systems to reduce costs and improve productivity. Companies generally ignored the revenue potential, not looking for opportunities to use information systems to increase sales. The second stage was defined by Porter and others. It was oriented toward gaining a competitive advantage. In many cases, companies spent large amounts on information systems and downplayed the costs. Today, companies are shifting from strategic management to performance-based management of their information systems. In this third stage, companies carefully consider both strategic advantage and costs. They use productivity, return on investment (ROI), net present value, and other measures of performance to evaluate the contributions their information systems make to their businesses. Figure 2.9 illustrates these stages. This balanced approach attempts to reduce costs and increase revenues.

Productivity

Developing information systems that measure and control productivity is a key element for most organizations. **Productivity** is a measure of the output achieved divided by the input required. A higher level of output for a given level of input means greater productivity; a lower level of output for a given level of input means lower productivity. The numbers assigned to productivity levels are not always based on labor hours—productivity can be based on factors such as the amount of raw materials used, resulting quality, or time to produce the goods or service. The value of the productivity number is not as significant as how it compares with other time periods, settings, and organizations.

productivity
A measure of the output achieved divided by the input required.

Productivity = (Output / Input) × 100%

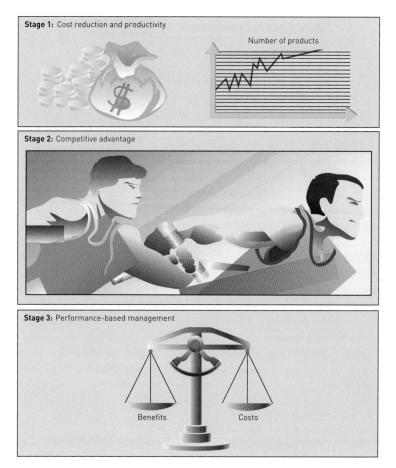

Stage 1: Cost reduction and productivity

Number of products

Stage 2: Competitive advantage

Stage 3: Performance-based management

Benefits   Costs

After a basic level of productivity is measured, an information system can monitor and compare it over time to see whether productivity is increasing. Then, a company can take corrective action if productivity drops below certain levels. In addition to measuring productivity, an information system can be used within a process to significantly increase productivity. Thus, improved productivity can result in faster customer response, lower costs, and increased customer satisfaction.

In the late 1980s and early 1990s, overall productivity did not seem to improve as a company increased its investments in information systems. Often called the *productivity paradox,* this situation troubled many economists who were expecting to see dramatic productivity gains.[27] In the early 2000s, however, productivity again seemed on the rise.

Return on Investment and the Value of Information Systems

return on investment (ROI)
One measure of IS value that
investigates the additional profits or
benefits that are generated as a
percentage of the investment in IS
technology.

One measure of IS value is **return on investment (ROI)**. This measure investigates the additional profits or benefits that are generated as a percentage of the investment in IS technology. A small business that generates an additional profit of $20,000 for the year as a result of an investment of $100,000 for additional computer equipment and software would have a return on investment of 20 percent ($20,000/$100,000). In many cases, however, it can be difficult to accurately measure ROI.[28] According to the Butler Group based in England, "Nothing inhibits IT investment more than an inability to measure benefits and keep control of costs." Because of the importance of ROI, many computer companies provide ROI calculators to potential customers. ROI calculators are typically found on a vendor's Web site and can be used to estimate returns.

Earnings Growth

Another measure of IS value is the increase in profit, or earnings growth, it brings. For instance, a mail-order company might install an order-processing system that generates a 7 percent earnings growth compared with the previous year.

Market Share

Market share is the percentage of sales that a product or service has in relation to the total market. If installing a new online catalog increases sales, it might help a company increase its market share by 20 percent.

Customer Awareness and Satisfaction

Although customer satisfaction can be difficult to quantify, about half of today's best global companies measure the performance of their information systems based on feedback from internal and external users. Some companies use surveys and questionnaires to determine whether the IS investment has increased customer awareness and satisfaction.

Total Cost of Ownership

Another way to measure the value of information systems was developed by the Gartner Group and is called the **total cost of ownership** (TCO). This approach breaks total costs into areas such as the cost to acquire the technology, technical support, administrative costs, and end-user operations. Other costs in TCO include retooling and training costs. TCO can help to develop a more accurate estimate of the total costs for systems that range from small PCs to large mainframe systems. Market research groups often use TCO to compare products and services. For example, a survey of large global enterprises ranked messaging and collaboration software products using the TCO model.[29] The survey analyzed acquisition, maintenance, administration, upgrading, downtime, and training costs. In this survey, Oracle had the lowest TCO of about $65 per user for messaging and collaboration. Union Bank of the Philippines also used reductions in TCO to justify an agile new system from Sun Microsystems.[30]

Return on investment, earnings growth, market share, customer satisfaction, and TCO are only a few measures that companies use to plan for and maximize the value of their IS investments. Regardless of the difficulties, organizations must attempt to evaluate the contributions that information systems make to assess their progress and plan for the future. Information technology and personnel are too important to leave to chance.

total cost of ownership (TCO)
The measurement of the total cost of owning computer equipment, including desktop computers, networks, and large computers.

Wal-Mart Stores Manage Supply Chain with RFID

Wal-Mart has a reputation for squeezing as much as possible out of every dollar. Information management and efficient management of its supply chain have been key to the company's success. When Wal-Mart management learned about Radio Frequency Identification (RFID) technology, they jumped at the chance to automate the process of tracking merchandise through its supply chain. RFID technology involves attaching tiny RFID chips to merchandise, which can then be scanned with handheld or stationary scanning devices to divulge information about that particular product.

Wal-Mart's RFID Tag Initiative requires that its suppliers attach RFID tags to all shipped cartons of products. Wal-Mart employees can then scan a sealed crate to determine what it contains, how long it has been in transit, what its expiration date is, and a host of other useful information. Wal-Mart suppliers have had to quickly learn about this new technology and integrate it into their packaging systems at no small cost; however, when Wal-Mart speaks, suppliers jump.

After piloting the RFID system at distribution centers in Dallas, Wal-Mart went live with RFID in January 2005. One hundred suppliers were ready on that first day and delivered RFID tagged cartons. The packages are now scanned and tracked as they leave the manufacturing facility, and as they enter and leave trucks and warehouses. Each time they are scanned, their related records are updated in Wal-Mart's database.

At the end of its first year, 300 suppliers were feeding RFID-tagged goods to 500 Wal-Mart facilities through five distribution centers. This number is expected to double over the next year.

Wal-Mart is experiencing impressive results with its new system. Items using RFID tags that are marked as out of stock are replenished three times faster than items from companies not using RFID tags. The amount of out-of-stock items that have to be manually filled has been cut by 10 percent.

Wal-Mart is experimenting with new and valuable applications for RFID technology. They are working to tag perishable items. The tags would inform Wal-Mart grocery and produce managers of how long a crate of bananas, for example, has been in transit. The manager can then determine the ripeness of the fruit providing more control over choices of which fruits and vegetables to hold on to and which to place on sale or throw out. RFID will become very useful to businesses that sell perishable foods because RFID chips can provide precise information about when the product was packaged and how long it has been in transit—all without needing to open a crate.

RFID tags can also be useful in determining the order for unloading crates from trucks. Each Wal-Mart truck carries about 7,000 boxes that have to be organized and unpacked. Associates will be equipped with wearable devices that will detect a high-priority box of goods that can be unloaded immediately to be stocked on store shelves.

Tracking crates of products through the supply chain provides managers with valuable information regarding the sale of products. Wal-Mart can examine the sales of a given item, store by store, and determine whether a product didn't sell well because it wasn't on the floor on the best day of the week or timed with an advertising campaign.

Through the use of RFID technology, Wal-Mart is automating many of the processes involved in traditional supply chain management. Low-inventory items are automatically reordered from the supplier, and packed on trucks without any human involvement. In Wal-Mart distribution centers, pallets of cartons are shuffled from one location to another and items are automatically sorted on conveyor belts without the need for a human inspection of the cartons. All of this automation means money saved for Wal-Mart, which then passes on the savings to its customers to maintain its competitive advantage over competing companies that employ older technologies.

Discussion Questions

1. What benefits has RFID technology provided or is anticipated to provide for Wal-Mart in the management of its supply chain?
2. How do you think information systems are used to automate communications between links in Wal-Mart's supply chain?

Critical Thinking Questions

1. How does Wal-Mart's decision to use RFID technology and require suppliers to adopt RFID affect other retailers and the future of RFID technology itself?
2. What effect has Wal-Mart's RFID Tag Initiative had on Wal-Mart as an organization? How are managers and floor workers affected?

SOURCES: Songini, Marc, "Wal-Mart Details Its RFID Journey," *Computerworld,* March 2, 2006, *www.computerworld.com.* "RFID Progress at Wal-Mart," *IDTechEx,* Oct 1, 2005, *www.idtechex.com/products/en/articles/ 00000161.asp.* Sullivan, Laurie, "Wal-Mart Assesses New Uses for RFID," *Information Week,* March 28, 2005, *www.informationweek.com.*

Risk

In addition to the return-on-investment measures of a new or modified system discussed in Chapter 1 and this chapter, managers must also consider the risks of designing, developing, and implementing these systems. Information systems can sometimes be costly failures. Some companies, for example, have attempted to implement ERP systems and failed, costing them millions of dollars. In other cases, e-commerce applications have been implemented with little success. The costs of development and implementation can be greater than the returns from the new system. The risks of designing, developing, and implementing new or modified systems are covered in more detail in Chapters 12 and 13, which discuss systems development.

CAREERS IN INFORMATION SYSTEMS

Realizing the benefits of any information system requires competent and motivated IS personnel, and many companies offer excellent job opportunities. As mentioned in Chapter 1, *knowledge workers (KWs)* are people who create, use, and disseminate knowledge. They are usually professionals in science, engineering, business, and other areas who specialize in information systems. Numerous schools have degree programs with such titles as information systems, computer information systems, and management information systems. These programs are typically offered by information schools, business schools, and within computer science departments. Graduating students with degrees in information systems have attracted high starting salaries. In addition, students are increasingly completing business degrees with a global or international orientation. Table 2.5 lists the top places to work in information systems, according to a survey reported by *Computerworld.* IS employees seek jobs that are interesting and challenging, that have access to new technologies, and that give IS workers the opportunity to work on interesting projects.[31]

Company Name	Business	Number of IS Employees
American Fidelity Assurance Co.	Insurance	130
Network Appliance	Telecommunications	148
Quicken Loans	Financial services	175
University of Pennsylvania	Education	275
The Mitre Corporation	Nonprofit consulting for government agencies	275
University of Miami	Education	284
American Century Investments	Financial services	308
Universal Health Services	Medical	334
Minnesota Life Insurance	Insurance	382
Qualcomm	Chips and software for telecommunications companies	751

Table 2.5

The Top Places to Work in Information Systems

(Source: Hoffman, Thomas, "First-Rate Five," *Computerworld*, June 27, 2005, p. 34.)

The job market in the early 2000s has been tight. Many jobs have been lost in U.S. companies as firms merged, outsourced certain jobs overseas, or went bankrupt. Today, demand for IS personnel is on the rise, along with salaries.[32] About 69 percent of IS personnel salaries increased in 2005, 22 percent had no change, and 9 percent decreased. The U.S. Department of Labor's Bureau of Labor Statistics predicts that software publishing will increase about 68 percent from 2002 to 2012, making it one of the fastest growing industries in the U.S. economy.[33] Software publishers produce and distribute computer software, such as designing, providing documentation, assisting in installation, and providing support services to software purchasers. The Bureau of Labor Statistics also projects network systems and data communications analysts, computer software engineers, and database administrators to be among the fastest growing occupations in the next decade.

On the job, computer systems are making IS professionals' work easier. Called *autonomics* by some, the use of advanced computer systems can help IS professionals spend less time maintaining existing systems and more time solving problems or looking for new opportunities. Colgate-Palmolive Co., for example, uses autonomics to keep its computer systems running better in more than 50 countries.

Opportunities in information systems are also available to people from foreign countries, including Russia and India. The U.S. H-1B and L-1 visa programs seek to allow skilled employees from foreign lands into the United States. In 2005, the U.S. Congress increased the H-1B program by 20,000 foreign workers.[34] According to Jesus Arriaga, chief information officer of Keystone Automotive Industries, which distributes auto parts, "It's just like offshoring. You're probably going to get similar skills at a lesser cost." Arriaga tries to hire U.S. workers when possible.[35] The L-1 visa program is often used for intracompany transfers for multinational companies.

Roles, Functions, and Careers in the IS Department

Professionals with careers in information systems typically work in an IS department as Web developers, computer programmers, systems analysts, computer operators, or other positions. They might also support other departments or areas such as finance or logistics. In addition to technical skills, they need skills in written and verbal communication, an understanding of organizations and the way they operate, and the ability to work with people and in groups. Today, many good information, business, and computer science schools require these business and communications skills of their graduates.

In general, IS professionals are charged with maintaining the broadest perspective on organizational goals. Most medium to large organizations manage information resources through an IS department. In smaller businesses, one or more people might manage information resources, with support from outsourced services. (Recall that outsourcing is also popular with larger organizations.) As shown in Figure 2.10, the IS organization has three primary responsibilities: operations, systems development, and support.

Operations

People in the operations component of a typical IS department work with information systems in corporate or business unit computer facilities. They tend to focus more on the *efficiency* of IS functions rather than their effectiveness.

System operators focus on the efficiency of IS functions, rather than their effectiveness. Their primary responsibility is to run and maintain IS equipment.

(Source: Courtesy of BananaStock/ Alamy.)

System operators primarily run and maintain IS equipment, and are typically trained at technical schools or through on-the-job experience. They are responsible for starting, stopping, and correctly operating mainframe systems, networks, tape drives, disk devices, printers, and so on. Other operations include scheduling, hardware maintenance, and preparing input and output. Data-entry operators convert data into a form the computer system can use. They can use terminals or other devices to enter business transactions, such as sales orders and payroll data. Increasingly, data entry is being automated—captured at the source of the transaction rather than entered later. In addition, companies might have local area network and Web operators who run the local network and any Web sites the company has.

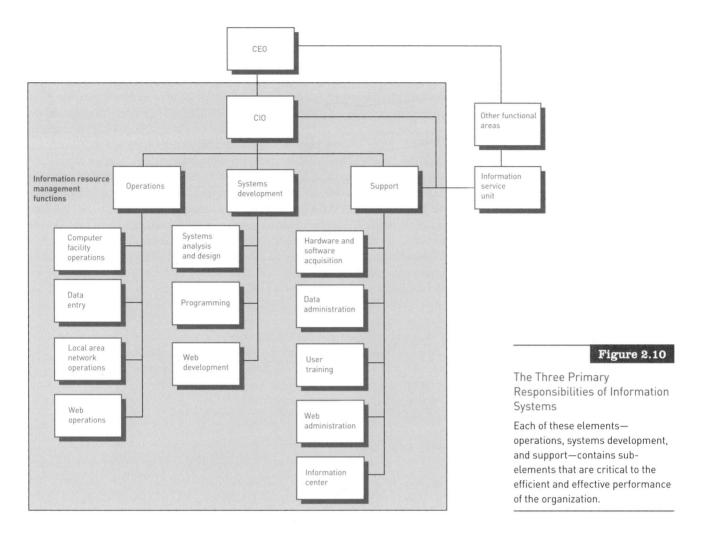

Figure 2.10

The Three Primary Responsibilities of Information Systems

Each of these elements— operations, systems development, and support—contains sub- elements that are critical to the efficient and effective performance of the organization.

Systems Development

The systems development component of a typical IS department focuses on specific development projects and ongoing maintenance and review. Systems analysts and programmers, for example, address these concerns to achieve and maintain IS effectiveness. The role of a systems analyst is multifaceted. Systems analysts help users determine what outputs they need from the system and construct plans for developing the necessary programs that produce these outputs. Systems analysts then work with one or more programmers to make sure that the appropriate programs are purchased, modified from existing programs, or developed. A computer programmer uses the plans the systems analyst created to develop or adapt one or more computer programs that produce the desired outputs. To help businesses select the best analysts and programmers, companies such as TopCoder offer tests to evaluate the proficiency and competence of current IS employees or job candidates. TopCoder Collegiate Challenge allows programming students to compete with other programmers around the world. Some companies, however, are skeptical of the usefulness of these types of tests.

With the dramatic increase in the use of the Internet, intranets, and extranets, many companies have Web or Internet developers who create effective and attractive Web sites for customers, internal personnel, suppliers, stockholders, and others who have a business relationship with the company.

Support

The support component of a typical IS department provides user assistance in hardware and software acquisition and use, data administration, user training and assistance, and Web administration. In many cases, support is delivered through an information center.

Because IS hardware and software are costly, a specialized support group often manages computer hardware and software acquisitions. This group sets guidelines and standards for the rest of the organization to follow in making purchases. They must gain and maintain an understanding of available technology and develop good relationships with vendors.

A database administrator focuses on planning, policies, and procedures regarding the use of corporate data and information. For example, database administrators develop and disseminate information about the corporate databases for developers of IS applications. In addition, the database administrator monitors and controls database use.

User training is a key to get the most from any information system, and the support area ensures that appropriate training is available. Training can be provided by internal staff or from external sources. For example, internal support staff can train managers and employees in the best way to enter sales orders, to receive computerized inventory reports, and to submit expense reports electronically. Companies also hire outside firms to help train users in other areas, including the use of word processing, spreadsheets, and database programs.

Web administration is another key area for support staff. With the increased use of the Internet and corporate Web sites, Web administrators are sometimes asked to regulate and monitor Internet use by employees and managers to make sure that it is authorized and appropriate. Web administrators also maintain the corporate Web site to keep it accurate and current, which can require substantial resources.

The support component typically operates the information center. An **information center** provides users with assistance, training, application development, documentation, equipment selection and setup, standards, technical assistance, and troubleshooting. Although many firms have attempted to phase out information centers, others have changed their focus from technical training to helping users find ways to maximize the benefits of the information resource.

information center
A support function that provides users with assistance, training, application development, documentation, equipment selection and setup, standards, technical assistance, and troubleshooting.

information service unit
A miniature IS department.

Information Service Units

An **information service unit** is basically a miniature IS department attached and directly reporting to a functional area in a large organization. Notice the information service unit shown in Figure 2.10. Even though this unit is usually staffed by IS professionals, the project assignments and the resources necessary to accomplish these projects are provided by the functional area to which it reports. Depending on the policies of the organization, the salaries of IS professionals staffing the information service unit might be budgeted to either the IS department or the functional area.

Typical IS Titles and Functions

The organizational chart shown in Figure 2.10 is a simplified model of an IS department in a typical medium or large organization. Many organizations have even larger departments, with increasingly specialized positions such as librarian or quality assurance manager. Smaller firms often combine the roles shown in Figure 2.10 into fewer formal positions.

The Chief Information Officer

The role of the chief information officer (CIO) is to employ an IS department's equipment and personnel to help the organization attain its goals. The CIO is usually a vice president concerned with the overall needs of the organization, and sets corporate-wide policies, and plans, manages, and acquires information systems. Some of the CIO's top concerns include integrating IS operations with corporate strategies, keeping up with the rapid pace of technology, and defining and assessing the value of systems development projects. The high level of the CIO position reflects that information is one of the organization's most important resources. A CIO works with other high-level officers of an organization, including the chief financial officer (CFO) and the chief executive officer (CEO), in managing and controlling total corporate resources. CIOs must also work closely with advisory committees, stressing effectiveness and teamwork and viewing information systems as an integral part of the organization's business processes—not an adjunct to the organization. Thus, CIOs need both technical and business skills. For federal agencies, the Clinger-Cohen Act of 1996 requires that a CIO coordinate the purchase and management of information systems.[36] According to Dianah Neff, CIO for the city of Philadelphia, "Street smarts are needed. You need to be more politically astute, absolutely. Politics was never in any of our training agendas to become CIOs, but being politically savvy is more of our job today."[37] Neff was involved with delivering wireless Internet access points throughout the city of Philadelphia. The service competed with private communications companies.

A company's CIO is usually a vice president who plans, manages, and acquires information systems and sets corporate-wide policies.

(Source: Courtesy of Leland Bobbe/ Getty Images.)

Depending on the size of the IS department, several people might work in senior IS managerial levels. Some job titles associated with IS management are the CIO, vice president of information systems, manager of information systems, and chief technology officer (CTO). A central role of all these people is to communicate with other areas of the organization to determine changing needs. Often these employees are part of an advisory or steering committee that helps the CIO and other IS managers make decisions about the use of information systems. Together they can best decide what information systems will support corporate goals. The CTO, for example, typically works under a CIO and specializes in networks and related equipment and technology.

LAN Administrators

Local area network (LAN) administrators set up and manage the network hardware, software, and security processes. They manage the addition of new users, software, and devices to the network. They also isolate and fix operations problems. LAN administrators are in high demand and often solve both technical and nontechnical problems.

Internet Careers

The bankruptcy of some Internet start-up companies in the early 2000s, called the *dot-gone era* by some, has resulted in layoffs for some firms. Executives of these bankrupt start-up Internet companies lost hundreds of millions of dollars in a few months. Yet, the use of the Internet to conduct business continues to grow and has stimulated a steady need for skilled personnel to develop and coordinate Internet usage. As shown in Figure 2.10, these careers are in the areas of Web operations, Web development, and Web administration. As with other areas in IS, many top-level administrative jobs are related to the Internet. These career opportunities are found in both traditional companies and those that specialize in the Internet.

Internet jobs within a traditional company include Internet strategists and administrators, Internet systems developers, Internet programmers, and Internet or Web site operators. Some companies suggest a new position, chief Internet officer, with responsibilities and a salary similar to the CIO's.

In addition to traditional companies, Internet companies offer exciting career opportunities. These companies include Amazon.com, Yahoo!, eBay, and many others. Systest, for example, specializes in finding and eliminating digital bugs that could halt the operation of a computer system.

Many Web sites, such as Monster.com, post job opportunities for Internet careers and more traditional careers. Most large companies list job opportunities on their Internet sites. These sites allow prospective job hunters to browse job opportunities, locations, salaries, benefits, and other factors. In addition, some sites allow job hunters to post their résumés.

Internet job sites such as Monster.com allow job hunters to browse job opportunities and post their résumés.

(Source: *www.monster.com.*)

certification

A process for testing skills and knowledge, which results in a statement by the certifying authority that states an individual is capable of performing a particular kind of job.

Often, the people filling IS roles have completed some form of certification. **Certification** is a process for testing skills and knowledge resulting in an endorsement by the certifying authority that an individual is capable of performing a particular job. Certification frequently involves specific, vendor-provided or vendor-endorsed coursework. Popular certification programs include Microsoft Certified Systems Engineer, Certified Information Systems Security Professional (CISSP), Oracle Certified Professional, and many others.

Other IS Careers

To respond to the increase in attacks on computers, new and exciting careers have developed in security and fraud detection and prevention. A recent survey reported that about 22 percent of responding companies have a chief information security officer, 18 percent have a chief security officer, and 8 percent have a chief privacy officer.[38] The University of Denver, for example, offers a masters program in cybersecurity. The National Insurance Crime Bureau, a nonprofit organization supported by roughly 1,000 property and casualty insurance companies, uses computers to join forces with special investigation units and law enforcement agencies, as well as to conduct online fraud-fighting training to investigate and prevent these types of crimes. The University of Denver also offers a program in video-game development.[39] According to one student, "This is every kid's dream in my age group. I know very few people who wouldn't want to get into the video-game industry." It is even possible to work from home in an IS field. Pat Misterovich, for example, is a stay-at-home dad developing an MP3 music player the size of a Pez dispenser.[40] The likely slogan for the new music player will be "Candy for your ears." Programmers, systems developers, and others are also working from home in developing new information systems.

In addition to working for an IS department in an organization, IS personnel can work for a large consulting firm, such as Accenture, IBM, EDS, and others. These jobs often entail frequent travel because consultants are assigned to work on various projects wherever the client is. Such roles require excellent people and project management skills in addition to IS technical skills.

Other IS career opportunities include being employed by a hardware or software vendor developing or selling products. Such a role enables an individual to work on the cutting edge of technology, which can be extremely challenging and exciting. As some computer companies cut their services to customers, new companies are being formed to fill the need. With names such as Speak With a Geek and the Geek Squad, these companies are helping people and organizations with their computer-related problems that computer vendors are no longer solving.

SUMMARY

Principle

The use of information systems to add value to the organization is strongly influenced by organizational structure, culture, and change.

Organizations use information systems to support their goals. Because information systems typically are designed to improve productivity, organizations should devise methods for measuring the system's impact on productivity.

An organization is a formal collection of people and other resources established to accomplish a set of goals. The primary goal of a for-profit organization is to maximize shareholder value. Nonprofit organizations include social groups, religious groups, universities, and other organizations that do not have profit as the primary goal.

Organizations are systems with inputs, transformation mechanisms, and outputs. Value-added processes increase the relative worth of the combined inputs on their way to becoming final outputs of the organization. The value chain is a series (chain) of activities that includes (1) inbound logistics, (2) warehouse and storage, (3) production, (4) finished product storage, (5) outbound logistics, (6) marketing and sales, and (7) customer service.

Organizational structure refers to how organizational subunits relate to the overall organization. Several basic organizational structures include traditional, project, team, and virtual. A virtual organizational structure employs individuals, groups, or complete business units in geographically dispersed areas. These can involve people in different countries operating in different time zones and different cultures.

Organizational culture consists of the major understandings and assumptions for a business, corporation, or organization. Organizational change deals with how profit and nonprofit organizations plan for, implement, and handle change. Change can be caused by internal or external factors. The change model consists of these stages: unfreezing, moving, and refreezing. According to the concept of organizational learning, organizations adapt to new conditions or alter practices over time.

Principle

Because information systems are so important, businesses need to be sure that improvements or completely new systems help lower costs, increase profits, improve service, or achieve a competitive advantage.

Business process reengineering involves the radical redesign of business processes, organizational structures, information systems, and values of the organization to achieve a breakthrough in results. Continuous improvement to business processes can add value to products and services.

The extent to which technology is used throughout an organization can be a function of technology diffusion, infusion, and acceptance. Technology diffusion is a measure of how widely technology is in place throughout an organization. Technology infusion is the extent to which technology permeates an area or department. User satisfaction with a computer system and the information it generates depends on the quality of the system and the resulting information. The technology acceptance model (TAM) investigates factors, such as the perceived usefulness of the technology, the ease of use of the technology, the quality of the information system, and the degree to which the organization supports the use of the information system, to predict IS usage and performance.

Total quality management consists of a collection of approaches, tools, and techniques that fosters a commitment to quality throughout the organization. Six sigma is often used in quality control. It is a statistical term that means products and services will meet quality standards 99.9997 percent of the time.

Outsourcing involves contracting with outside professional services to meet specific business needs. This approach allows the company to focus more closely on its core business and to target its limited resources to meet strategic goals. Downsizing involves reducing the number of employees to reduce payroll costs; however, it can lead to unwanted side effects.

Competitive advantage is usually embodied in either a product or service that has the most added value to consumers and that is unavailable from the competition or in an internal system that delivers benefits to a firm not enjoyed by its competition. A five-forces model covers factors that lead firms to seek competitive advantage: the rivalry among existing competitors, the threat of new market entrants, the threat of substitute products and services, the bargaining power of buyers, and the bargaining power of suppliers. Strategies to address these factors and to attain competitive advantage include cost leadership, differentiation, niche strategy, altering the industry structure, creating new products and services, improving existing product lines and services, and other strategies.

Developing information systems that measure and control productivity is a key element for most organizations. A useful measure of the value of an IS project is return on investment (ROI). This measure investigates the additional profits or benefits that are generated as a percentage of the investment in IS technology. Total cost of ownership (TCO) can also be a useful measure.

Principle

Cooperation between business managers and IS personnel is the key to unlocking the potential of any new or modified system.

Information systems personnel typically work in an IS department that employs a chief information officer, chief technology officer, systems analysts, computer programmers, computer operators, and other personnel. The chief information officer (CIO) employs an IS department's equipment and personnel to help the organization attain its goals. The chief technology officer (CTO) typically works under a CIO and specializes in hardware and related equipment and technology. Systems analysts help users determine what outputs they need from the system and construct the plans needed to develop the necessary programs that produce these outputs. Systems analysts then work with one or more programmers to make sure that the appropriate programs are purchased, modified from existing programs, or developed. The major responsibility of a computer programmer is to use the plans developed by the systems analyst to build or adapt one or more computer programs that produce the desired outputs.

Computer operators are responsible for starting, stopping, and correctly operating mainframe systems, networks, tape drives, disk devices, printers, and so on. LAN administrators set up and manage the network hardware, software, and security processes. Trained personnel are also needed to set up and manage a company's Internet site, including Internet strategists, Internet systems developers, Internet programmers, and Web site operators. Information systems personnel can also support other functional departments or areas.

In addition to technical skills, IS personnel need skills in written and verbal communication, an understanding of organizations and the way they operate, and the ability to work with people (users). In general, IS personnel are charged with maintaining the broadest enterprise-wide perspective.

Besides working for an IS department in an organization, IS personnel can work for a large consulting firm, such as Accenture, IBM, EDS, and others. Another IS career opportunity is to be employed by a hardware or software vendor developing or selling products.

CHAPTER 2: SELF-ASSESSMENT TEST

The use of information systems to add value to the organization is strongly influenced by organizational structure, culture, and change.

1. The value chain is a series of activities that includes inbound logistics, warehouse and storage, production, finished product storage, outbound logistics, marketing and sales, and customer service. True or False?

2. A(n) _____ is a formal collection of people and other resources established to accomplish a set of goals.

3. User satisfaction with a computer system and the information it generates often depends on the quality of the system and the resulting information. True or False?

4. The concept in which organizations adapt to new conditions or alter their practices over time is called _____.
   a. organizational learning
   b. organizational change
   c. continuous improvement
   d. reengineering

Because information systems are so important, businesses need to be sure that improvements or completely new systems help lower costs, increase profits, improve service, or achieve a competitive advantage.

5. _____ involves contracting with outside professional services to meet specific business needs.

6. Today, quality means _____.
   a. achieving production standards
   b. meeting or exceeding customer expectations
   c. maximizing total profits
   d. meeting or achieving design specifications

7. Technology infusion is a measure of how widely technology is spread throughout an organization. True or False?

8. Reengineering is also called _____.

9. What is a measure of the output achieved divided by the input required?
   a. efficiency
   b. effectiveness
   c. productivity
   d. return on investment

10. _____ is a measure of the additional profits or benefits generated as a percentage of the investment in IS technology.

Cooperation between business managers and IS personnel is the key to unlocking the potential of any new or modified system.

11. Who is involved in helping users determine what outputs they need and constructing the plans needed to produce these outputs?
   a. CIO
   b. applications programmer

c.   systems programmer
d.   systems analyst

12. The systems development component of a typical IS department focuses on specific development projects and ongoing maintenance and review. True or False?

13. The _____ is typically in charge of the IS department or area in a company.

REVIEW QUESTIONS

1. What is the value chain?
2. What is the difference between a virtual organizational structure and a traditional organizational structure?
3. What role does an information system play in today's organizations?
4. What is reengineering? What are the potential benefits of performing a process redesign?
5. What is user satisfaction?
6. What is the difference between reengineering and continuous improvement?
7. What is the difference between technology infusion and technology diffusion?
8. What is quality? What is total quality management (TQM)?
9. What are organizational change and organizational learning?
10. List and define the basic organizational structures.
11. Sketch and briefly describe the three-stage organizational change model.
12. What is downsizing? How is it different from outsourcing?
13. What are some general strategies employed by organizations to achieve competitive advantage?
14. What are several common justifications for implementing an information system?
15. Define the term *productivity*. How can a company best use productivity measurements?
16. What is on-demand computing? What two advantages does it offer to a company?
17. What is the total cost of ownership?
18. What is the role of a systems analyst? What is the role of a programmer?

DISCUSSION QUESTIONS

1. You have been hired to work in the IS area of a manufacturing company that is starting to use the Internet to order parts from its suppliers and offer sales and support to its customers. What types of Internet positions would you expect to see at the company?
2. You have decided to open an Internet site to buy and sell used music CDs to other students. Describe the value chain for your new business.
3. What sort of IS career would be most appealing to you—working as a member of an IS organization, consulting, or working for an IT hardware or software vendor? Why?
4. What are the advantages of using a virtual organizational structure? What are the disadvantages?
5. How would you measure user satisfaction with a registration program at a college or university? What are the important features that would make students and faculty satisfied with the system?
6. You have been asked to participate in preparing your company's strategic plan. Specifically, your task is to analyze the competitive marketplace using Porter's five-forces model. Prepare your analysis, using your knowledge of a business you have worked for or have an interest in working for.
7. Based on the analysis you performed in Discussion Question 6, what possible strategies could your organization adopt to address these challenges? What role could information systems play in these strategies? Use Porter's strategies as a guide.
8. There are many ways to evaluate the effectiveness of an information system. Discuss each method and describe when one method would be preferred over another method.
9. Assume you are the manager of a retail store and need to hire a CIO to run your new computer system. What characteristics would you want in a new CIO?

PROBLEM-SOLVING EXERCISES

1. For an industry of your choice, find the number of employees, total sales, total profits, and earnings growth rate of 15 firms. Using a database program, enter this information for the last year. Use the database to generate a report of the three companies with the highest earnings growth rate. Use your word processor to create a document that describes the 15 firms. What other measures would you use to determine which is the best company in terms of future profit potential?

2. A new IS project has been proposed that will produce not only cost savings but also an increase in revenue. The initial costs to establish the system are estimated to be $500,000. The rest of the cash flow data is presented in the following table.

	Year 1	Year 2	Year 3	Year 4	Year 5
Increased Revenue	$0	$100	$150	$200	$250
Cost Savings	$0	$ 50	$ 50	$ 50	$ 50
Depreciation	$0	$ 75	$ 75	$ 75	$ 75
Initial Expense	$500				

Note: All amounts are in 000s.

a. Using a spreadsheet program, calculate the return on investment (ROI) for this project. Assume that the cost of capital is 7 percent.
b. How would the rate of return change if the project delivered $50,000 in additional revenue and generated cost savings of $25,000 in the first year?

3. Using a word processing program, write a detailed job description of a CIO for a medium-sized manufacturing company. Use a graphics program to make a presentation on the requirements for the new CIO.

TEAM ACTIVITIES

1. With your team, interview one or more managers at a local company. Find out how they define and measure quality.

2. With your team, research a firm that has achieved a competitive advantage. Write a brief report that describes how the company was able to achieve its competitive advantage.

WEB EXERCISES

1. This book emphasizes the importance of information. You can get information from the Internet by going to a specific address, such as *www.ibm.com*, *www.whitehouse.gov*, or *www.fsu.edu*. This will give you access to the home page of the IBM corporation, the White House, or Florida State University, respectively. Note that "com" is used for businesses or commercial operations, "gov" is used for governmental offices, and "edu" is used for educational institutions. Another approach is to use a search engine, which is a Web site that allows you to enter key words or phrases to find information. Yahoo!, developed by two Tulane University students, was one of the first search engines on the Internet. You can also locate information through lists or menus. The search engine will return other Web sites (hits) that correspond to a search request. Using Yahoo! at *www.yahoo.com*, search for information about a company or topic discussed in Chapter 1 or 2. You might be asked to develop a report or send an e-mail message to your instructor about what you found.

2. Use the Internet to search for information about user satisfaction. You can use a search engine, such as Yahoo!, or a database at your college or university. Write a brief report describing what you found.

CAREER EXERCISES

1. Organizations can use traditional, project, virtual, and other organizational structures. For the career of your choice, describe which organizational structure or structures are likely to be used and how computers and information systems can help you communicate and work with others in this structure.

2. Pick the five best companies for your career. Describe the quality of the products and services offered by each company. Are the best companies always the ones with the highest quality products and services?

CASE STUDIES

Case One

Itochu Kenzai Uses IS to Reduce Organizational Waste

An environmental management system (EMS) is a set of cohesive elements that an organization can use to minimize its impact on the environment. The International Organization for Standardization (ISO) is the world's largest developer of standards. ISO standards contribute to making the development, manufacturing, and supply of products and services more efficient, safer, and cleaner. The ISO 14001 is an internationally accepted standard for an EMS.

When a company is certified as complying with these standards, it implies that the company is making an effort to reduce its effect on the natural environment. Recently, companies have found that having a positive environmental reputation makes for good business. Consider GE and its move over the last five years to promote most of its technologies as environmental solutions (see *http://ge.ecomagination.com/*) and its success with this platform. GE has been ranked as the most respected company in the world by several sources.

A company can prove that it has respectable environmental intentions by becoming certified as ISO 14001-compliant. Such a company would find numerous doors open leading to new customers and partners, global affiliations, government support, and other opportunities. Itochu Kenzai Corporation, a building materials company in Tokyo, Japan, sought to enjoy this distinction.

Itochu Kenzai was certified ISO 14001-compliant, but only under the condition that it would take steps to reduce its environmental waste. One way in which it was to do so was to dramatically reduce its use of paper. This, of course, required the use of electronic documents and information systems that supported their management.

"We started out with the desire to eliminate paperwork for large quantities of forms that had a low frequency of use among the various forms generated," relates Mr. Yoshihiro Sakamoto, team manager of systems operations at Itochu Kenzai. Before the introduction of an electronic document management system, "...the output of accounting-related legal forms alone was responsible for about 80,000 pages per month, and on the fourth and fifth of each month, printers were operating night and day for monthly processing. Employees had to attend to the printers to keep them from running out of paper. Because 30,000 items of building materials are necessary to build one house, building materials extend to a wide variety of areas, and the number of items handled reached as high as 700,000 per month. All those items require documentation."

Itochu Kenzai chose IBM DB2 Content Manager OnDemand as the electronic document solution and the IBM eServer iSeries system as the server. The IBM iSeries system was installed at the head office, collected accounting data, and performed electronic data interchange (EDI) with several dozen suppliers. The company also installed servers at 25 of its business centers throughout Japan to run business systems for the issuance of delivery slips, invoices, and other forms of content. After four years of effort, the company is entirely operating with digitally converted business-related documents, and retrieving data and storage on compact discs.

The company has eliminated the printing of approximately 190,000 pages (equivalent to 100 boxes of paper) per month of legal documents and business-related documents. Itochu Kenzai is storing all these documents in electronic form. If these documents were stored for 10 years, the savings obtained would be equivalent to about 12,000 boxes of paper. Not only is the company saving in paper, there is also substantial savings in personnel costs and warehouse storage expenses. The savings amounted to more than ¥700,000 ($6,000) per month, or about ¥8 million ($69,000) per year. In addition, staff has been freed up to concentrate on core operations. The solution has become an extremely important part of the company system.

Before adopting the IBM content management solution, printed documents were archived in cardboard boxes stored in warehouses and typically never used. With DB2 Content Management OnDemand, that information is now easily retrieved and utilized. Itochu Kenzai management is discovering that the information that used to be discarded to warehouses has value.

Discussion Questions

1. How has Itochu Kenzai's move to become ISO 14001-compliant helped the environment?
2. How has Itochu Kenzai's move to become ISO 14001-compliant helped Itochu Kenzai?

Critical Thinking Questions

1. Who is ultimately responsible for the pressure placed on companies to become environmentally friendly? Businesses? Governments? Citizens?
2. Often, refitting a company with new environmentally friendly technologies ends up saving the company money in the long run. What reasons might a company have for resisting such change?

SOURCES: ISO14001 "Itochu Kenzai Reduces Printing by 190,000 Pages per Month with Electronic Storage Solution from IBM," IBM Case Studies, November 21, 2005, *www.ibm.com.* ISO Web site, accessed March 8, 2006, *www.iso14000.com.* Itochu Kenzai Web site, accessed March 8, 2006, *www.ick.co.jp/english/default.htm.*

Case Two

CIO Randy Mott Shakes Up Hewlett-Packard

Chapters 1 and 2 illustrate how important the technological infrastructure is to an organization's ability to function and compete. The person who controls the development and growth of an organization's technological infrastructure, the CIO, carries a lot of responsibility. Good CIOs with experience in running large enterprise systems are among the highest valued members of an organization. Recruiting a top-notch CIO is similar to a baseball team that recruits the league's best pitcher; in both cases, the organization increases its value and has a better chance of beating the competition. Such is the case with Hewlett-Packard (HP) and their new CIO Randall "Randy" Mott.

Randy Mott built a reputation for himself as a top-notch CIO during his 22 years at Wal-Mart where he was a key figure in automating Wal-Mart's supply-chain systems, arguably one of the biggest and most successful information system projects in the retail industry. After this project was completed, in search of another challenge, Mr. Mott accepted the CIO position at computer giant, Dell, where he was embraced by Dell management's inner circle. Five years later in 2005, Dell's archrival Hewlett-Packard hired Mr. Mott away from Dell.

Winning Mr. Mott was a triumph for HP considering Mott's experience in the field and his intimate knowledge of Dell's strategic plans. Acquiring Randy Mott was a key piece in HP's restructuring puzzle. HP's previous CIO took a lateral move to executive vice president of HP's global operations division. The shifting of CIOs represented a larger reengineering of

HP's organizational structure. HP was working to reverse some of the corporate consolidation imposed by former CEO Carly Fiorina. For an example of HP's reverse consolidation, consider how it recently separated its PC and handheld business from its imaging and printing group. Todd Bradley, previously president and CEO of PalmOne, was named to lead the PC division.

Now that the reengineering effort is complete, it is time to trim the fat. Randy Mott's initial analysis of HP's IT department determined that it spends far more than its competitors in its IT staff and projects. Randy Mott's first obligation was to cut costs and improve operations. Some serious changes were called for.

Mr. Mott was not only out to save HP substantial sums of money, but also to drum up some business. Because HP sells and manages information systems for other corporations, Mr. Mott sees this as an opportunity to show clients and future clients the types of savings that HP can make possible.

Mott set challenging and specific goals for the company: to reduce the company's IT spending from about 4 percent of its annual revenue to about 1.5 percent by the end of 2008 or 2009. A considerable amount of consolidating, simplifying, and downsizing would be required to meet this goal. Mr. Mott plans to consolidate 85 data centers down to six. He is also working to centralize all HP's data from dozens of locations into one large database. He hopes that consolidating all information into one location will help decision makers. HP wants a "single version of the truth"—consistent information—to "take people out of the role of gathering information and instead focus on decision making," said Mott.

Mott is also working to simplify processes by using standardized system configurations. He hopes to cut the number of internal IT projects from about 1,200 to 500. This will mean a reduction in the workforce. Some analysts anticipate that HP will cut 15,000 jobs, a bit less than 10 percent of HP's workforce, in the near future.

Such sweeping changes in an organization as large as HP are not implemented without a considerable amount of planning and courage. It takes CIOs with a lot of experience to exert the type of leadership required to turn such initiatives into successes. Randy Mott is under additional pressure because HP is giving his work a high profile in the media. With his extensive experience and the support of the HP leadership, his initiatives are bound to be a success.

Discussion Questions

1. What advantages over its competition did HP achieve by taking on Randy Mott as CIO?
2. What types of professional skills do you think are required of a good CIO?

Critical Thinking Questions

1. Who is most likely to blame for the bloat that existed in HP's IT department prior to Mott's arrival?
2. How do you think the sweeping changes at HP will impact HP employees, stockholders, and customers?

SOURCES: Thibodeau, Patrick, "Ex-Dell CIO Seeks to Reshape HP's IT," *Computerworld,* March 6, 2006, *www.computerworld.com.* Singer, Michael, "HP Bags Randy Mott as New CIO," July 12, 2005, Silicon.com, *www.silicon.com.*

Questions for Web Case

See the Web site for this book to read about the Whitmann Price Consulting case for this chapter. Following are questions concerning this Web case.

Whitmann Price Consulting: Addressing the Needs of the Organization

Discussion Questions

1. Compare and contrast the benefits and shortcomings of notebook PCs and handheld PCs in accessing information on the road.

2. If Josh and Sandra are able to provide a solution that meets all of the requirements, how might it impact the effectiveness and efficiency of Whitmann Price?

Critical Thinking Questions

1. The six business units of Whitmann Price have some needs for this new system in common and others that are unique. Do you think it will be possible to satisfy all of them with one solution? What are the possibilities for customizing a solution that meets all the requirements?
2. Why might Whitmann Price prefer one common device for all business units rather than unique devices for each unit?

NOTES

Sources for the opening vignette: "OnStar Debuts Updated Navigation System at Chicago Auto Show," Associated Press and Motortrend, February 8, 2006, *www.motortrend.com/features/news/2006/112_news39/index1.html.* OnStar Web site, accessed March 6, 2006, *www.onstar.com.* Koslowski, Thilo, "OnStar's New Navigation Offering Will Disrupt the Industry," *Gartner,* 14 February 2006, *www.gartner.com/DisplayDocument?doc_cd=137880.*

1   Staff, "Market Orientation of Value Chains," *European Journal of Marketing,* May 1, 2005, p. 428.
2   Trebilcock, Bob, "From Juarez to Cottondale," *Modern Materials Handling,* May 1, 2005, p. 39.
3   Zhu, K.; Kraemer, K., "Post-Adoption Variations in Usage and Value of E-Business by Organizations," *Information Systems Research,* March 2005, p. 61.
4   McDougall, Paul, "Tools to Tune Supply Chains," *Information Week,* January 9, 2006, p. 62.
5   Grey, W., et al., "The Role of E-Marketplaces in Relationship-based Supply Chains," *IBM Systems Journal,* Vol. 44, No. 1, 2005, p. 109.
6   Rowley, Jennifer, "Customer Relationship Management Through the Tesco Clubcard Loyalty Scheme," *International Journal of Retail & Distribution Management,* March 1, 2005, p. 194.
7   Blackman, Trevor, "Employee Empowerment," *Employee Benefits,* April 11, 2005, p. 82.
8   Gordon, Jack, "Do Your Virtual Teams Deliver Only Virtual Performance?" *Training Magazine,* June 1, 2005.
9   Christensen, Clayton, *The Innovator's Dilemma,* Harvard Business School Press, 1997, p. 225 and *The Inventor's Solution,* Harvard Business School Press, 2003.
10 Schein, E. H., *Process Consultation: Its Role in Organizational Development,* (Reading, MA: Addison-Wesley, 1969). See also Peter G. W. Keen, "Information Systems and Organizational Change," *Communications of the ACM,* Vol. 24, No. 1, January 1981, pp. 24–33.
11 Staff, "SBI Implements Business Process Re-Engineering," *Business Line,* August 16, 2005.
12 Wixom, Barbara; Todd, Peter, "A Theoretical Integration of User Satisfaction and Technology Acceptance," *Information Systems Research,* March 2005, p. 85.

13 Bailey, J.; Pearson, W., "Development of a Tool for Measuring and Analyzing Computer User Satisfaction," *Management Science,* 29(5), 1983, p. 530.
14 Chaparro, Barbara, et al., "Using the End-User Computing Satisfaction Instrument to Measure Satisfaction with a Web Site," *Decision Sciences,* May 2005, p. 341.
15 Davis, F., "Perceived Usefulness, Perceived Ease of Use, and User Acceptance of Information Technology," *MIS Quarterly,* 13(3) 1989, p. 319. Kwon et al., "A Test of the Technology Acceptance Model," *Proceedings of the Hawaii International Conference on System Sciences,* January 4–7, 2000.
16 Loch, Christoph; Huberman, Bernardo, "A Punctuated-Equilibrium Model of Technology Diffusion," *Management Science,* February 1999, p. 160.
17 Tornatzky, L.; Fleischer, M., "The Process of Technological Innovation," *Lexington Books,* Lexington, MA, 1990, and Zhu, K.; Kraemer, K., "Post-Adoption Variations in Usage and Value of E-Business by Organizations," *Information Systems Research,* March 2005, p. 61.
18 Armstrong, Curtis; Sambamurthy, V., "Information Technology Assimilation in Firms," *Information Systems Research,* April 1999, p. 304.
19 Agarwal, Ritu; Prasad, Jayesh, "Are Individual Differences Germane to the Acceptance of New Information Technology?" *Decision Sciences,* Spring 1999, p. 361.
20 Staff, "MAA Bozell Implements Total Quality Management System," *Business Line,* August 4, 2005.
21 Staff, "Forward Thinking 3PL Folds in Lean Six Sigma," *Manufacturing Business Technology,* July 1, 2005, p. 60.
22 Staff, "Boeing Official Hails Outsourcing of Design Work to Russian Engineers," *BBC Monitoring International Reports,* August 18, 2005.
23 Cone, Edwards, "Organic Grocer Embraces Utility Computing," *CIO Insight,* June 5, 2005, p. 18.
24 Collins, Jim, *Good to Great,* Harper Collins Books, 2001, p. 300.
25 Porter, M. E., *Competitive Advantage: Creating and Sustaining Superior Performance,* New York: Free Press, 1985; *Competitive Strategy: Techniques for Analyzing Industries and Competitors,* The Free Press, 1980; and *Competitive Advantage of Nations,* The Free Press, 1990.

26 Porter, M. E.; Millar, V., "How Information Systems Give You Competitive Advantage," *Journal of Business Strategy*, Winter 1985. *See also* Porter, M. E., *Competitive Advantage* (New York: Free Press, 1985).

27 Staff, "Paradox Lost," *The Economist*, September 11, 2003.

28 Huber, Nick, "Return on Investment: Analysts to Offer Tips on Measuring the Value of IT," *Computer Weekly*, April 26, 2005, p. 20.

29 Staff, "Detailed Comparison of a Range of Factors Affecting Total Cost of Ownership (TCO) for Messaging and Collaboration," *M2 Presswire*, September 5, 2003.

30 Staff, "Union Bank Replaces Mainframe Legacy System," *The Asian Banker Journal*, July 15, 2005.

31 Brandel, Mary, "What's the Secret to Building a Strong and Satisfied IT Workforce?" *Computerworld*, June 27, 2005, p. 25.

32 Staff, "Salary Survey 2005," *Computerworld*, September 24, 2005, p. 41.

33 "Software Publishers," *U.S. Department of Labor Bureau of Labor Statistics, http://stats.bls.gov/oco/cg/cgs051.htm*, accessed on November 28, 2005.

34 Thibodeau, Patrick, "Government to Add 20,000 H-1B Visas," *Computerworld*, May 9, 2005, p. 8.

35 Thibodeau, Patrick, "The H-1B Equation," *Computerworld*, February 28, 2005, p. 4.

36 Staff, "The Clinger-Cohen Act of 1996," *The Governments Accounts Journal*, Winter 1997, p. 8.

37 Hamblen, Matt, "City CIOs Are Using Hot New Technologies," *Computerworld*, February 28, 2005, p. 35.

38 Staff, "Security's Shaky State," *Information Week*, December 5, 2005, p. 49.

39 Johnson, Kimberly, "Video Gaming Serious Subject at DU," *The Rocky Mountain News*, June 12, 2005, p. K1.

40 Roth, Daniel, "The Amazing Rise of the Do-It-Yourself Economy," *Fortune*, May 30, 2005, p. 45.

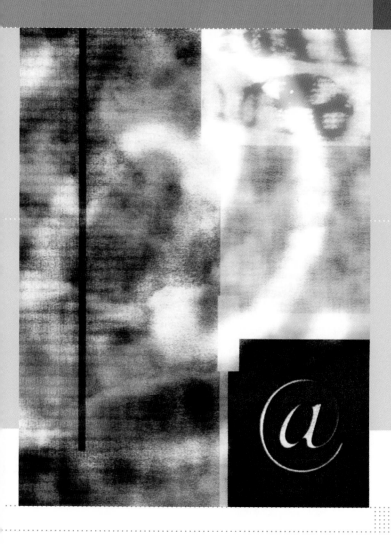

PART · 2 ·

Information Technology Concepts

Chapter 3 Hardware: Input, Processing, and Output Devices

Chapter 4 Software: Systems and Application Software

Chapter 5 Organizing Data and Information

Chapter 6 Telecommunications and Networks

Chapter 7 The Internet, Intranets, and Extranets

CHAPTER · 3 ·

Hardware: Input, Processing, and Output Devices

PRINCIPLES

- Assembling an effective, efficient set of computer hardware devices requires understanding their role in supporting the underlying information systems and the needs of the organization. The computer hardware objectives are subordinate to, but supportive of, the information systems and the needs of the organization.

- When selecting computer hardware, you must consider the current and future needs of the information systems and the organization. Your choice of a hardware device should always allow for later improvements to meet evolving organizational needs.

LEARNING OBJECTIVES

- Describe how to select and organize computer hardware components to support information system (IS) objectives and business needs.

- Describe the power, speed, and capacity of central processing and memory devices.
- Describe the access methods, capacity, and portability of secondary storage devices.
- Discuss the speed, functionality, and importance of input and output devices.
- Identify popular classes of computer systems and discuss the role of each.

Information Systems in the Global Economy ▶
Siebel Systems, United States
Information Systems Provider Powers Up with Blade Servers

Siebel Systems, based in San Mateo, California, and recently acquired by Oracle, is a leading provider of customer relationship management (CRM) and business intelligence systems for businesses around the world. More than 3.7 million people working for over 4,000 businesses and organizations in 20 industries worldwide trust Siebel to provide systems for improving customer service, increasing profitability, and reducing costs.

The information systems and services that Siebel provides run on hundreds of large server computers owned and operated by Siebel. Using these resources, Siebel offers hosted software solutions so that its clients do not need to purchase and maintain the supportive hardware. Most of Siebel's servers are three to four years old and manufactured by a variety of vendors. Each server is maintained individually, requiring vendor-specific tools and software.

Siebel is a well-established company, and, like many technology companies, is facing increased competition from smaller start-up businesses looking to win a piece of the CRM market. To compete, Siebel recently needed to cut costs and boost the capabilities of its hardware systems. Siebel reduced costs by moving some server operations overseas where labor, taxes, and other expenses were greatly reduced—a strategy known as offshoring. The distribution of its server centers around the world also made it possible to better serve its global customers. However, offshoring provided an additional challenge of managing these servers remotely from headquarters in the United States.

In its restructuring efforts, Siebel reviewed many options for upgrades to its server hardware. Siebel decided on a solution from IBM. It purchased over 700 blade servers to replace its aging servers. Blade servers are smaller than traditional servers, house many computer motherboards that include one or more processors, computer memory, computer storage, and computer network connections, and can be combined for powerful processing. By standardizing on one type of server from one manufacturer and by using the fastest processors, Siebel gained serious advantages. The information systems staff now spends 67 percent less time on managing its worldwide servers. The new hardware is more powerful and reliable than the old servers it replaced and requires much less space and cabling. Through remote systems management, the company can resolve every potential system problem without having to travel to the server location.

Like many companies, Siebel has learned that using the most recent innovations in hardware technologies can provide significant operational and competitive advantages. By overhauling its hardware, Siebel increased the productivity of its employees and positioned itself for growth.

As you read this chapter, consider the following:

- How do companies use computer hardware to compete and meet their business's objectives?
- What are the important considerations for selecting processing, storage, and input and output hardware?

Why Learn About Hardware?

Organizations invest in computer hardware to improve worker productivity, increase revenue, reduce costs, and provide better customer service. Those that don't might be stuck with outdated hardware that often fails and cannot take advantage of the latest software advances. As a result, obsolete hardware can place an organization at a competitive disadvantage. Managers, no matter what their career field and educational background, are expected to know enough about hardware to help define the business needs that the hardware must support. In addition, managers must be able to ask good questions and evaluate options when considering hardware investments for their area of the business. Managers in marketing, sales, and human resources often help IS specialists assess opportunities to apply computer hardware and evaluate the options and features specified for the hardware. Managers in finance and accounting especially must keep an eye on the bottom line, guarding against overspending, yet be willing to invest in computer hardware when and where business conditions warrant it.

hardware
Any machinery (most of which uses digital circuits) that assists in the input, processing, storage, and output activities of an information system.

Today's use of technology is practical—it's intended to yield real business benefits, as demonstrated by Siebel. Employing information technology and providing additional processing capabilities can increase employee productivity, expand business opportunities, and allow for more flexibility. This chapter concentrates on the hardware component of a computer-based information system (CBIS). **Hardware** consists of any machinery (most of which uses digital circuits) that assists in the input, processing, storage, and output activities of an information system. When making hardware decisions, the overriding consideration of a business should be how hardware can support the objectives of the information system and the goals of the organization.

COMPUTER SYSTEMS: INTEGRATING THE POWER OF TECHNOLOGY

To assemble an effective and efficient system, you should select and organize components while understanding the trade-offs between overall system performance and cost, control, and complexity. For instance, in building a car, manufacturers try to match the intended use of the vehicle to its components. Racecars, for example, require special types of engines, transmissions, and tires. Selecting a transmission for a racecar requires balancing how much engine power can be delivered to the wheels (efficiency and effectiveness) with how expensive the transmission is (cost), how reliable it is (control), and how many gears it has (complexity). Similarly, organizations assemble computer systems so that they are effective, efficient, and well suited to the tasks that need to be performed.

People involved in selecting their organization's computer hardware must clearly understand current and future business requirements so they can make informed acquisition decisions. Consider the following examples of applying business knowledge to reach critical hardware decisions.

- Sailors from allied nations travel on U.S. Navy ships, and, until now, used separate personal computers and onboard networks to communicate with their own military branches. Recognizing that this system was unwieldy and knowing that such multinational missions will continue, the Navy is upgrading the hardware on its 160-ship surface fleet by replacing bulky personal computers with thin-client Sun Ray systems from Sun Microsystems, which are simpler, more secure, easier to administer, and can be used by multiple sailors. This will decrease the number of computers each ship needs, cut costs, save space, and reduce power requirements.[1]

- Like most airlines, American Airlines is under intense pressure to reduce costs and provide better customer service. As a result, management is replacing 35,000 computers in airports, in back offices, and at corporate headquarters with more powerful Hewlett-Packard desktop computers so employees can run new applications that help them serve customers and receive training more efficiently. American is also working with Panasonic Corporation to provide rugged laptop computers that can hold all the information maintenance workers need to work on a plane to increase worker productivity and improve on-time departures.[2]

As these examples demonstrate, choosing the right computer hardware requires understanding its relationship to the information system and the needs of the organization. Furthermore, hardware objectives are subordinate to, but supportive of, the information system and the current and future needs of the organization.

Hardware Components

Computer system hardware components include devices that perform input, processing, data storage, and output (see Figure 3.1). To understand how these hardware devices work together, consider an analogy from a paper-based office. Imagine a one-room office occupied by a single person named George. George (the processing device) can organize and manipulate data. George's mind (register storage) and his desk (primary storage) are places to temporarily store data. Filing cabinets fill the need for more permanent storage (secondary storage). In this analogy, the incoming and outgoing mail trays are sources of new data (input) or places to put the processed paperwork (output).

Recall that any system must be able to process (organize and manipulate) data, and a computer system does so through an interplay between one or more central processing units and primary storage. Each **central processing unit (CPU)** consists of three associated elements: the arithmetic/logic unit, the control unit, and the register areas. The **arithmetic/logic unit (ALU)** performs mathematical calculations and makes logical comparisons. The **control unit** sequentially accesses program instructions, decodes them, and coordinates the flow of data in and out of the ALU, registers, primary storage, and even secondary storage and various output devices. **Registers** are high-speed storage areas used to temporarily hold small units of program instructions and data immediately before, during, and after execution by the CPU.

Primary storage, also called **main memory** or **memory**, is closely associated with the CPU. Memory holds program instructions and data immediately before or after the registers. To understand the function of processing and the interplay between the CPU and memory, let's examine the way a typical computer executes a program instruction.

central processing unit (CPU)
The part of the computer that consists of three associated elements: the arithmetic/logic unit, the control unit, and the register areas.

arithmetic/logic unit (ALU)
The part of the CPU that performs mathematical calculations and makes logical comparisons.

control unit
The part of the CPU that sequentially accesses program instructions, decodes them, and coordinates the flow of data in and out of the ALU, registers, primary storage, and even secondary storage and various output devices.

register
A high-speed storage area in the CPU used to temporarily hold small units of program instructions and data immediately before, during, and after execution by the CPU.

primary storage (main memory; memory)
The part of the computer that holds program instructions and data.

Figure 3.1

Hardware Components

These components include the input devices, output devices, communications devices, primary and secondary storage devices, and the central processing unit (CPU). The control unit, the arithmetic/logic unit (ALU), and the register storage areas constitute the CPU.

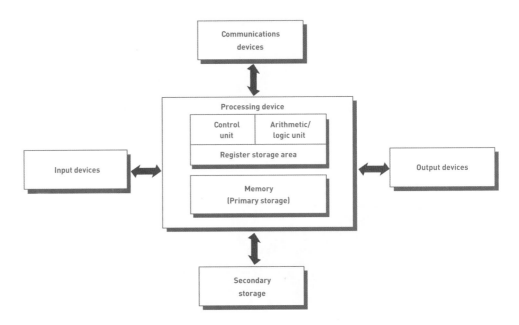

Hardware Components in Action

Executing any machine-level instruction involves two phases: instruction and execution. During the instruction phase, a computer performs the following steps:

- **Step 1: Fetch instruction.** The computer reads the next program instruction to be executed and any necessary data into the processor.
- **Step 2: Decode instruction.** The instruction is decoded and passed to the appropriate processor execution unit. Each execution unit plays a different role: The arithmetic/logic unit performs all arithmetic operations, the floating-point unit deals with noninteger operations, the load/store unit manages the instructions that read or write to memory, the branch processing unit predicts the outcome of a branch instruction in an attempt to reduce disruptions in the flow of instructions and data into the processor, the memory-management unit translates an application's addresses into physical memory addresses, and the vector-processing unit handles vector-based instructions that accelerate graphics operations.

instruction time (I-time)
The time it takes to perform the fetch-instruction and decode-instruction steps of the instruction phase.

The time it takes to perform the instruction phase (Steps 1 and 2) is called the **instruction time (I-time)**.

The second phase is execution. During the execution phase, a computer performs the following steps:

- **Step 3: Execute instruction.** The hardware element, now freshly fed with an instruction and data, carries out the instruction. This could involve making an arithmetic computation, logical comparison, bit shift, or vector operation.
- **Step 4: Store results.** The results are stored in registers or memory.

execution time (E-time)
The time it takes to execute an instruction and store the results.

The time it takes to complete the execution phase (Steps 3 and 4) is called the **execution time (E-time)**.

machine cycle
The instruction phase followed by the execution phase.

After both phases have been completed for one instruction, they are performed again for the second instruction, and so on. Completing the instruction phase followed by the execution phase is called a **machine cycle** (see Figure 3.2). Some processing units can speed processing by using **pipelining**, whereby the processing unit gets one instruction, decodes another, and executes a third at the same time. The Pentium 4 processor, for example, uses two execution unit pipelines. This means the processing unit can execute two instructions in a single machine cycle.

pipelining
A form of CPU operation in which multiple execution phases are performed in a single machine cycle.

Figure 3.2

Execution of an Instruction

In the instruction phase, a program's instructions and any necessary data are read into the processor (1). Then the instruction is decoded so the central processor can understand what to do (2). In the execution phase, the ALU does what it is instructed to do, making either an arithmetic computation or a logical comparison (3). Then the results are stored in the registers or in memory (4). The instruction and execution phases together make up one machine cycle.

PROCESSING AND MEMORY DEVICES: POWER, SPEED, AND CAPACITY

The components responsible for processing—the CPU and memory—are housed together in the same box or cabinet, called the *system unit*. All other computer system devices, such as the monitor, secondary storage, and keyboard, are linked directly or indirectly into the system unit housing. In this section, we investigate the characteristics of these important devices.

Processing Characteristics and Functions

Because organizations want efficient processing and timely output, they use a variety of measures to gauge processing speed. These measures include the time it takes to complete a machine cycle and clock speed.

Machine Cycle Time

As you've seen, a computer executes an instruction during a machine cycle. The time in which a machine cycle occurs is measured in *nanoseconds* (one-billionth of one second) and *picoseconds* (one-trillionth of one second). Machine cycle time also can be measured by how many instructions are executed in one second. This measure, called **MIPS**, stands for millions of instructions per second. MIPS is another measure of speed for computer systems of all sizes.

MIPS
Millions of instructions per second.

Clock Speed

Each CPU produces a series of electronic pulses at a predetermined rate, called the **clock speed**, which affects machine cycle time. The control unit in the CPU manages the stages of the machine cycle by following predetermined internal instructions, known as **microcode**. You can think of microcode as predefined, elementary circuits and logical operations that the processor performs when it executes an instruction. The control unit executes the microcode in accordance with the electronic cycle, or pulses of the CPU "clock." Each microcode instruction takes at least the same amount of time as the interval between pulses. The shorter the interval between pulses, the faster each microcode instruction can be executed.

Because the number of microcode instructions needed to execute a single program instruction—such as performing a calculation or printing results—can vary, the clock speed is not directly related to the true processing speed of the computer.

clock speed
A series of electronic pulses produced at a predetermined rate that affects machine cycle time.

microcode
Predefined, elementary circuits and logical operations that the processor performs when it executes an instruction.

megahertz (MHz)
Millions of cycles per second.

gigahertz (GHz)
Billions of cycles per second.

Clock speed is often measured in **megahertz** (MHz, millions of cycles per second) or **gigahertz** (GHz, billions of cycles per second). Unfortunately, as the clock speed of the CPU increases, additional heat is generated that can corrupt the data and instructions the computer is trying to process. Because this can lead to errors that cause a program to behave erratically, chip and computer manufacturers must be wary of potential heat problems in their new designs.

In past years, organizations typically bought the fastest chips with no concern for the chip's power requirements. Now the chip's power requirement has become an important consideration. For example, IT services provider Fiducia AG processes some 20 million transactions per day for 900 German banks. The firm plans to upgrade to more powerful and more energy-efficient UltraSPARC T1-microprocessor-based servers from Sun Microsystems that will replace four old servers with one server and cut power costs by nearly $1 million per year.[3]

Physical Characteristics of the CPU

Most CPUs are collections of digital circuits imprinted on silicon wafers, or chips, each no bigger than the tip of a pencil eraser. To turn a digital circuit on or off within the CPU, electrical current must flow through a medium (usually silicon) from point A to point B. The speed the current travels between points can be increased by either reducing the distance between the points or reducing the resistance of the medium to the electrical current.

Moore's Law
A hypothesis that states that transistor densities on a single chip double every 18 months.

Reducing the distance between points has resulted in ever-smaller chips, with the circuits packed closer together. In the 1960s, shortly after patenting the integrated circuit, Gordon Moore, former chairman of the board of Intel (the largest microprocessor chip maker), hypothesized that progress in chip manufacturing ought to make it possible to double the number of transistors (the microscopic on/off switches) on a chip roughly every 18 months. When actual results bore out his idea, the doubling of transistor densities on a single chip every 18 months became known as **Moore's Law**, and this "rule of thumb" has become a goal that chip manufacturers have met for over four decades. As shown in Figure 3.3, the number of transistors on a chip continues to climb. The minimum width of basic circuit features of a chip today is 90 nanometers (one billionth of a meter, 10^{-9} meter). Intel has already announced its Pentium Extreme Edition 955 desktop processor based on a chip with 65-nanometer processing technology.[4] Intel is also constructing a $3.5 billion advanced semiconductor manufacturing facility in Israel capable of high-volume production of 45-nanometer chips.[5] In addition to increased processing speeds, Moore's Law means decreasing costs. As silicon-based components and computers gain in performance, they become cheaper to produce, and therefore more plentiful, more powerful, and more a part of our everyday lives.

Figure 3.3

Moore's Law

Transistor densities on a single chip double every 18 months.

(Source: Data from "Moore's Law: Overview," Intel Web site at *www.intel.com/technology/silicon/mooreslaw/index.htm*, accessed January 21, 2006.)

Researchers are taking many approaches to continue to improve the performance of computers:

- Chips built from superconductive metals such as carbon that allow current to flow with minimal electrical resistance
- Optical processors that use light waves instead of electrical current to represent bits
- Strained silicon, a technique that boosts performance and lowers power consumption by stretching silicon molecules farther apart, thus allowing electrons to meet less resistance and flow faster
- Forming the tiny circuits for a computer from tiny carbon nanotubes only 1 to 3 nanometers in diameter
- Extreme miniaturization using radio waves to manipulate atoms into executing a simple computer program

The Aeroflex MIP7000 family of MIPS processors provides high-performance and low-power solutions for microprocessor products.

(Source: Courtesy of Aeroflex Incorporated.)

When selecting a CPU, organizations must balance the benefits of speed with cost. CPUs with faster clock speeds and machine cycle times are usually more expensive than slower ones. This expense, however, is a necessary part of the overall computer system cost, for the CPU is typically the single largest determinant of the price of many computer systems.

Memory Characteristics and Functions

Main memory is located physically close to the CPU, but not on the CPU chip itself. It provides the CPU with a working storage area for program instructions and data. The chief feature of memory is that it rapidly provides the data and instructions to the CPU.

Storage Capacity

Like the CPU, memory devices contain thousands of circuits imprinted on a silicon chip. Each circuit is either conducting electrical current (on) or not (off). Data is stored in memory as a combination of on or off circuit states. Usually 8 bits are used to represent a character, such as the letter *A*. Eight bits together form a **byte (B)**. In most cases, storage capacity is measured in bytes, with 1 byte equivalent to one character of data. The contents of the Library of Congress, with over 126 million items and 530 miles of bookshelves, would require about 20 petabytes of digital storage. Table 3.1 lists units for measuring computer storage.

byte (B)
Eight bits that together represent a single character of data.

Types of Memory

There are several forms of memory, as shown in Figure 3.4. Instructions or data can be temporarily stored in and read from **random access memory (RAM)**. With the current design of RAM chips, they are volatile storage devices, meaning they lose their contents if the current is turned off or disrupted (as in a power surge, brownout, or electrical noise generated by lightning or nearby machines). RAM chips are mounted directly on the computer's main circuit board or in other chips mounted on peripheral cards that plug into the computer's main circuit board. These RAM chips consist of millions of switches that are sensitive to changes in electric current.

random access memory (RAM)
A form of memory in which instructions or data can be temporarily stored.

Table 3.1

Computer Storage Units

Name	Abbreviation	Number of Bytes
Byte	B	1
Kilobyte	KB	2^{10} or approximately 1,024 bytes
Megabyte	MB	2^{20} or 1,024 kilobytes (about 1 million)
Gigabyte	GB	2^{30} or 1,024 megabytes (about 1 billion)
Terabyte	TB	2^{40} or 1,024 gigabytes (about 1 trillion)
Petabyte	PB	2^{50} or 1,024 terabytes (about 1 quadrillion)
Exabyte	EB	2^{60} or 1,024 petabytes (about 1 billion billion, or 1 quintillion)

Figure 3.4

Basic Types of Memory Chips

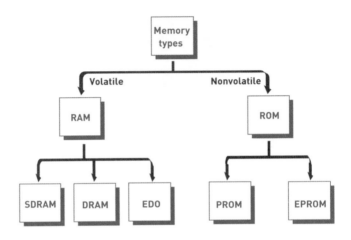

RAM comes in many varieties. The mainstream type of RAM is extended data out, or EDO RAM, which is faster than older types of RAM. Another kind of RAM is called dynamic RAM (DRAM) and is based on single-transistor memory cells. SDRAM, or synchronous DRAM, employs a minimum of four transistors per memory cell and needs high or low voltages at regular intervals—every two milliseconds (two one-thousands of a second)—to retain its information. Compared with EDO RAM, SDRAM provides a faster transfer speed between the microprocessor and the memory.

read-only memory (ROM)
A nonvolatile form of memory.

Read-only memory (ROM), another type of memory, is usually nonvolatile. In ROM, the combination of circuit states is fixed, and therefore its contents are not lost if the power is removed. ROM provides permanent storage for data and instructions that do not change, such as programs and data from the computer manufacturer, including the instructions that tell the computer how to start up when power is turned on.

There are other types of nonvolatile memory as well. In programmable read-only memory (PROM), the desired data and instructions—and hence the desired circuit state combination—must first be programmed into the memory chip. Thereafter, PROM behaves like ROM. A common use of PROM chips is for storing the instructions to popular video games, such as those for GameBoy and Xbox. Game instructions are programmed onto the PROM chips by the game manufacturer. Instructions and data can be programmed onto a PROM chip only once.

Erasable programmable read-only memory (EPROM) is similar to PROM except, as the name implies, the memory chip can be erased and reprogrammed. A form of EPROM memory is used to store information on RFID chips that identify and track the items to which they are attached. The contents of EPROM can be changed each time the item is moved throughout the supply chain.

Although microprocessor speed has doubled every 18 months over the past decade, memory performance has not kept pace. In effect, memory has become the principal bottleneck to system performance. The use of **cache memory**, a type of high-speed memory that a processor can access more rapidly than main memory (see Figure 3.5), helps to ease this bottleneck. Frequently used data is stored in easily accessible cache memory instead of slower memory such as RAM. Because cache memory holds less data, the CPU can access the desired data and instructions more quickly than selecting from the larger set in main memory. Thus, the CPU can execute instructions faster, improving the overall performance of the computer system. There are three types of cache memory. The Level 1 (L1) cache is on the CPU chip. The Level 2 (L2) cache memory can be accessed by the CPU over a high-speed dedicated interface. The latest processors go a step further and place the L2 cache directly on the CPU chip itself and provide high-speed support for a tertiary Level 3 (L3) external cache. Deerfield, the low-power version of Intel's 64-bit servers called the Itanium 2 chip, was introduced with 1.5 MB of Level 3 cache.

cache memory
A type of high-speed memory that a processor can access more rapidly than main memory.

Figure 3.5

Cache Memory

Processors can access this type of high-speed memory faster than main memory. Located on or near the CPU chip, cache memory works with main memory. A cache controller determines how often the data is used, transfers frequently used data to cache memory, and then deletes the data when it goes out of use.

When the processor needs to execute an instruction, the instruction's operation code indicates whether the data will be in a register or in memory. If the operation code specifies a register as the source, it is taken from there. Otherwise, the processor looks for the data in the L1 cache, then the L2 cache, and then the L3 cache. If the data is not in any cache, the CPU requests the data from main memory. If the data is not even stored in main memory, the system has to retrieve the data from secondary storage. It can take from one to three clock cycles to fetch information from the L1 cache, while the CPU waits and does nothing. It takes 6 to 12 cycles to get data from an L2 cache on the processor chip. It can take dozens of cycles to fetch data from an L3 cache and hundreds of cycles to fetch data from secondary storage. Because this hierarchical arrangement of memory helps the CPU find data faster, it bridges a widening gap between processor speeds, which are increasing at roughly 50 percent per year, and DRAM access rates, which are climbing at only 5 percent per year.

Costs for memory capacity continue to decline. However, when considered on a megabyte-to-megabyte basis, memory is still considerably more expensive than most forms of secondary storage.

Memory capacity contributes to the effectiveness of a CBIS. The specific applications of a CBIS determine the amount of memory required for a computer system. For example, complex processing problems, such as computer-assisted product design, require more memory than simpler tasks such as word processing. Also, because computer systems have different types of memory, they might need other programs to control how memory is accessed and used. In other cases, the computer system can be configured to maximize memory usage. Before purchasing additional memory, an organization should address all these considerations.

Multiprocessing

multiprocessing
The simultaneous execution of two or more instructions at the same time.

coprocessor
The part of the computer that speeds processing by executing specific types of instructions while the CPU works on another processing activity.

multicore microprocessor
A microprocessor that combines two or more independent processors into a single computer so they can share the workload and deliver a big boost in processing capacity.

parallel computing
The simultaneous execution of the same task on multiple processors to obtain results faster.

massively parallel processing systems
A form of multiprocessing that speeds processing by linking hundreds or thousands of processors to operate at the same time, or in parallel, with each processor having its own bus, memory, disks, copy of the operating system, and applications.

single instruction/multiple data (SIMD)
A form of parallel computing in which the processors all execute the same instruction on many data values simultaneously.

multiple instruction/multiple data (MIMD)
A form of parallel computing in which the processors all execute different instructions.

grid computing
The use of a collection of computers, often owned by multiple individuals or organizations, to work in a coordinated manner to solve a common problem.

Generally, **multiprocessing** involves the simultaneous execution of two or more instructions at the same time. One form of multiprocessing uses coprocessors. A **coprocessor** speeds processing by executing specific types of instructions while the CPU works on another processing activity. Coprocessors can be internal or external to the CPU and can have different clock speeds than the CPU. Each type of coprocessor performs a specific function. For example, a math coprocessor chip speeds mathematical calculations, and a graphics coprocessor chip decreases the time it takes to manipulate graphics.

A **multicore microprocessor** combines two or more independent processors into a single computer so that they can share the workload and boost processing capacity. "A dual-core processor is like a four-lane highway—it can handle up to twice as many cars as its two-lane predecessor without making each car drive twice as fast."[6] In addition, a dual-core processor enables people to perform multiple tasks simultaneously such as playing a game and burning a CD. AMD and Intel are battling for leadership in the multicore processor marketplace. In late 2005, Intel announced that it would replace a planned multicore Xeon processor (Whitefield) with a new chip (Tigerton) designed to reduce the current performance gap between Intel's server multicore chips and AMD Opteron processors.[7]

Examples of processes that require a multicore microprocessor include a medical CAT scan, which generates a three-dimensional image of the internals of an object based on many two-dimensional x-ray images taken around the same axis of rotation. In a similiar process, Chevron uses seismic waves (rather than x-rays) that travel through rock in search of energy reserves. The prodigious amount of data generated is processed by over 700 dual-core processor servers from Advanced Micro Devices to yield a three-dimensional image of the Earth's subsurface. The increased memory capabilities and processing speed of the systems yield an improved image resolution and in less time than was previously required.[8]

In January 2006, Apple unveiled a new laptop computer called the MacBook Pro as well as new 17-inch and 20-inch iMac computers based on Intel's Core Duo processor. The new computers provide roughly five times as much performance per watt of power as the comparable computers that use the IBM PC G4 processor. In a statement at the Macworld Conference and Expo, Apple CEO Steve Jobs said, "It's not a secret we've been trying to shoehorn a G5 [processor] into a notebook, and have been unable to do so because of its power consumption."[9]

Parallel Computing

Parallel computing is the simultaneous execution of the same task on multiple processors to obtain results faster. Systems with thousands of such processors are known as **massively parallel processing systems**. There are different approaches to achieving parallel computing. **Single instruction/multiple data (SIMD)** parallel processors all execute the same instruction on many data values simultaneously. **Multiple instruction/multiple data (MIMD)** parallel processors all execute different instructions. The processors might communicate with one another to coordinate when executing a computer program or they might run independently of one another but under the direction of another processor that distributes the work to the other processors and collects their processing results. The dual-core processors mentioned earlier are a simple form of parallel computing.

Grid computing is the use of a collection of computers, often owned by multiple individuals or organizations, to work in a coordinated manner to solve a common problem. Grid computing is a low-cost approach to parallel computing. The grid can include dozens, hundreds, or even thousands of computers that run collectively to solve extremely large processing problems. Key to the success of grid computing is a central server that acts as the grid leader and traffic monitor. This controlling server divides the computing task into subtasks and assigns the work to computers on the grid that have (at least temporarily) surplus processing power. The central server also monitors the processing, and if a member of the grid fails to complete a subtask, it restarts or reassigns the task. When all the subtasks are completed, the controlling server combines the results and advances to the next task until the whole job is completed.

Through the World Community Grid, more than 100,000 people donate unused time from about 170,000 computers. The grid invites volunteers to download Rosetta software, which runs during idle time, and enables computers on the grid to solve scientific problems and create public databases for scientific research. A project must hold potential for contributing to the greater good to be eligible for support. For example, the World Community Grid is currently using its massive computational powers to test thousands of human immunodeficiency virus (HIV) mutations against tens of thousands of chemical compounds. The goal is to help scientists design effective therapies to stop potential drug-resistant viral strains from causing AIDS.[10]

By installing the SETI@home screen-saver program on their personal computers, millions of people worldwide contribute their idle CPU time to analyzing radio data from space for signs of intelligent life. With SETI, a central server doles out the work by giving each computer on the grid data from a tiny slice of the sky to analyze.

(Source: Copyright © 2006 SETI@home.)

The most frequent business uses for parallel computing include modeling, simulating, and analyzing large amounts of data. In today's challenging marketplace, consumers are demanding increased product features and a whole array of new services. These consumer demands have forced companies to find more effective and insightful ways of gathering and analyzing information about existing and potential customers. Collecting and organizing this enormous amount of data is difficult. Acxiom is an IT services provider that processes billions of records each month for its clients. Its specialty is helping clients build marketing programs to boost sales, reduce customer turnover, and improve customer relationships. For example, a credit card issuer might ask Acxiom for consumer information to target new customers or to cross-sell to existing ones. The client sends Acxiom millions of its own records, which are then processed in its massive database of 20 billion consumer records of various types to produce the detailed consumer files the credit card company needs. During the 1990s, Acxiom ran its internal processes and client applications on parallel processing systems. It cost more than $150 million per year for equipment to maintain this capability and still the firm experienced lengthy processing times. Acxiom recently converted from parallel computing to grid computing. The result was lower hardware costs and a dramatic reduction in the time required to process customer requests. For example, one common client request would take nearly 30 days to run. Now it takes less than a day on the grid system consisting of 4,000 dual-processor computers. Converting to grid computing has enabled Acxiom to accept more client requests and reduce hardware spending.[11]

SECONDARY STORAGE

Driven by such factors as needing to retain more data longer to meet government regulatory concerns, store new forms of digital data such as audio and video data, and keep systems running under the onslaught of ever-increasing volumes of e-mail, the amount of data that companies store digitally is increasing at a rate of close to 100 percent per year! Organizations need a way to store large amounts of data and instructions more permanently than main memory allows. **Secondary storage**, also called *permanent storage*, serves this purpose.

Compared with memory, secondary storage offers the advantages of nonvolatility, greater capacity, and greater economy. On a cost-per-megabyte basis, most forms of secondary storage are considerably less expensive than primary memory (see Table 3.2). The selection of secondary storage media and devices requires understanding their primary characteristics—access method, capacity, and portability.

Table 3.2

Cost Comparison for Various Forms of Data Storage

All forms of secondary storage cost considerably less per megabyte of capacity than SDRAM, although they have slower access times. A CD-R spindle costs about $.07 per gigabyte, while SDRAM can cost around $138.60 per gigabyte—nearly 200 times more expensive. (Source: Office Depot Web site, *www.officedepot.com*, February 5, 2006.)

Description	Storage Capacity (GB)	Cost Per GB
Office Depot CD-R Spindle	0.7	$0.07
3½-inch bulk diskette, IBM Format, DS/HD	.14	$0.10
HP DDS-3 Tape Cartridge	24.0	$0.22
Maxell Data Tape, 4MM	24.0	$0.30
HP AIT-2 Data Cartridge	100.0	$0.77
Maxell CD-RW Discs with Jewel Cases	0.7	$1.86
Office Depot DVD-RW Rewriteable Media Spindle	4.7	$5.32
HP 9.1 GB Rewriteable Optical Disk	9.1	$10.51
Office Depot DVD-R Recordable Media Spindle	4.7	$12.76
SanDisk Compact Flash Memory Card	1.0	$99.99
SanDisk Memory Stick Flash Memory Card	0.256	$136.88
PNY Optima DDR SDRAM Memory Upgrade	0.512	$138.46

As with other computer system components, the access methods, storage capacities, and portability required of secondary storage media are determined by the information system's objectives. An objective of a credit card company's information system, for example, might be to rapidly retrieve stored customer data to approve customer purchases. In this case, a fast access method is critical. In other cases, such as equipping a sales force with laptop computers, portability and storage capacity might be major considerations in selecting and using secondary storage media and devices.

Storage media that allow faster access are generally more expensive than slower media. The cost of additional storage capacity and portability vary widely, but they are also factors to consider. In addition to cost and portability, organizations must address security issues to allow only authorized people to access sensitive data and critical programs. Because the data and programs kept in secondary storage devices are so critical to most organizations, all of these issues merit careful consideration.

Access Methods

Data and information access can be either sequential or direct. **Sequential access** means that data must be accessed in the order in which it is stored. For example, inventory data might be stored sequentially by part number, such as 100, 101, 102, and so on. If you want to retrieve information on part number 125, you must read and discard all the data relating to parts 001 through 124.

Direct access means that data can be retrieved directly, without the need to pass by other data in sequence. With direct access, it is possible to go directly to and access the needed data—for example, part number 125—without having to read through parts 001 through 124. For this reason, direct access is usually faster than sequential access. The devices used only to access secondary storage data sequentially are simply called **sequential access storage devices (SASDs)**; those used for direct access are called **direct access storage devices (DASDs)**.

Devices

The most common forms of secondary storage include magnetic tapes, magnetic disks, virtual tapes, and optical discs. In general, magnetic tapes are the oldest storage medium, while optical discs are the most recent. Some of these media (magnetic tape) allow only sequential access, while others (magnetic and optical discs) provide direct and sequential access. Figure 3.6 shows some different secondary storage media.

Magnetic Tape

Magnetic tape is a type of sequential secondary storage medium, now used primarily for storing backups. Similar to the tape found in audio- and videocassettes, magnetic tape is a Mylar film coated with iron oxide. Portions of the tape are magnetized to represent bits. If the computer needs to read data from the middle of a reel of tape, it must first pass all the tape before the desired piece of data—one disadvantage of magnetic tape. When information is needed, it can take time to retrieve the proper tape and mount it on the tape reader to get the relevant data into the computer. Despite the falling prices of hard drives, tape storage is still a popular choice for low-cost data backup for off-site storage in the event of a disaster.

Technology is improving to provide tape storage devices with greater capacities and faster transfer speeds. In addition, the bulky tape drives used to read and write on large reels of tapes in the early days of computing have been replaced with much smaller tape cartridge devices measuring a few millimeters in diameter that take up much less floor space and allow hundreds of tape cartridges to be stored in a small area.

Magnetic Disks

Magnetic disks are also coated with iron oxide; they can be thin metallic platters (hard disks, see Figure 3.7) or Mylar film (diskettes). As with magnetic tape, magnetic disks represent bits using small magnetized areas. When reading from or writing data onto a disk, the disk's read/write head can go directly to the desired piece of data. Thus, the disk is a direct-access storage medium. Because direct access allows fast data retrieval, this type of storage is ideal for companies that need to respond quickly to customer requests, such as airlines and credit card firms. For example, if a manager needs information on the credit history of a customer

Figure 3.6

Types of Secondary Storage

Secondary storage devices such as magnetic tapes and disks, optical discs, CD-ROMs, and DVDs are used to store data for easy retrieval at a later date.

(Source: Courtesy of Imation Corp.)

magnetic tape
A secondary storage medium; Mylar film coated with iron oxide with portions of the tape magnetized to represent bits.

magnetic disk
A common secondary storage medium, with bits represented by magnetized areas.

Hard Disk

Hard disks give direct access to stored data. The read/write head can move directly to the location of a desired piece of data, dramatically reducing access times, as compared with magnetic tape.

(Source: Courtesy of Seagate Technology.)

or the seat availability on a particular flight, the information can be obtained in seconds if the data is stored on a direct access storage device.

Magnetic disk storage varies widely in capacity and portability. Removable magnetic disks, such as diskettes or Zip disks, are nearly obsolete. Hard disks, though more costly and less portable, are more popular because of their greater storage capacity and quicker access time.

Making magnetic media hold more data has traditionally been a matter of making the grains in the recording medium, and the spots that hold the recorded bits, ever smaller and closer together.

RAID

Putting an organization's data online involves a serious business risk—the loss of critical business data can put a corporation out of business. The concern is that the most critical mechanical components inside a disk storage device—the disk drives, the fans, and other input/output devices—can break (like most things that move).

Organizations now require that their data-storage devices be fault tolerant—can continue with little or no loss of performance if one or more key components fails. A **redundant array of independent/inexpensive disks (RAID)** is a method of storing data that generates extra bits of data from existing data, allowing the system to create a "reconstruction map" so that if a hard drive fails, it can rebuild lost data. With this approach, data is split and stored on different physical disk drives using a technique called *striping* to evenly distribute the data. RAID technology has been applied to storage systems to improve system performance and reliability.

RAID can be implemented in several ways. In the simplest form, RAID subsystems duplicate data on drives. This process, called **disk mirroring**, provides an exact copy that protects users fully in the event of data loss. However, to keep complete duplicates of current backups, organizations need to double the amount of their storage capacity. Thus, disk mirroring is expensive. Other RAID methods are less expensive because they only partly duplicate the data, allowing storage managers to minimize the amount of extra disk space (or overhead) they must purchase to protect data. Dell now offers an optional second drive for some personal computers for users who need to mirror critical data for less than $100.[12]

Virtual Tape

Virtual tape is a storage technology that manages less frequently needed data so that it appears to be stored entirely on tape cartridges, although some parts might actually be located on faster hard disks. The software associated with a virtual tape system is sometimes called a *virtual tape server*. Virtual tape can be used with a sophisticated storage-management system that moves data to slower but less costly forms of storage media as people use the data less often. Virtual tape technology can decrease data access time, lower the total cost of ownership, and reduce the amount of floor space consumed by tape operations. IBM and Storage Technology are well-established vendors of virtual tape systems.

redundant array of independent/inexpensive disks (RAID)

A method of storing data that generates extra bits of data from existing data, allowing the system to create a "reconstruction map" so that if a hard drive fails, the system can rebuild lost data.

disk mirroring

A process of storing data that provides an exact copy that protects users fully in the event of data loss.

virtual tape

A storage device that manages less frequently needed data so that it appears to be stored entirely on tape cartridges, although some parts of it might actually be located on faster hard disks.

ETHICAL AND SOCIETAL ISSUES

Backup Storage Media and Identity Theft

After two weeks of internal investigations, Marriott International reported that backup computer tapes had gone missing from its time-share division offices. The tapes contained credit card account information and the Social Security numbers of about 206,000 time-share owners and Marriott customers and employees. Company officials stated that it was not clear whether the tapes were stolen or lost. All involved parties were notified in case identity thieves gained access to the information.

Identity theft is the crime in which an imposter uses stolen personal identification information to obtain credit, merchandise, or services in the name of the victim. Companies assume the responsibility of keeping such information secure.

Marriott is not alone in neglecting to secure customer records. Two weeks later, a computer tape from a Connecticut bank containing personal data on 90,000 customers, including names, addresses, Social Security numbers, and checking account numbers, was lost in transit. In 2005, at least 134 such data breaches affected 57 million people. Many of these data breaches involved the loss of data on backup tapes. So far, no direct correlation has been proven between such data breaches and cases of identity theft. However, there are about 10 million cases of identity theft each year with a total loss of about $53 billion. The sources of the stolen data are often unknown. It is assumed that much of the information is stolen from businesses.

Costly loss due to identity theft from businesses has caused U.S. state and federal governments to fight for tighter security measures. In 2003, California was the first state to pass a rigorous disclosure law requiring that organizations inform people when their personal information is compromised. Since then, many states have followed suit. The U.S. Congress is considering similar laws.

Besides the danger to personal privacy, the companies responsible also suffer. The public embarrassment resulting from data theft detracts from a company's reputation and ability to win customers. Substantial costs are also involved in notifying victims of the possible theft. Companies could avoid this trouble and expense by using inexpensive methods of protecting data.

Most companies employ courier services to transport backup tapes to and from storage. However, the backups are not always safely delivered to the storage facility. In addition, accounting and inventory of backup tapes in storage can be slack. When tapes are lost, the courier or storage facility is usually to blame, leading to suspicions of theft.

To better protect the data stored for back up, it can be encrypted so that if stolen it remains unreadable. Encryption is a technique that converts data into a secret code. In-line appliances can encrypt backup data prior to being written to tape with little effect on performance. Several tape drive manufacturers have introduced data encryption into their products. Services such as U.S. DataTrust (*www.usdatatrust.com*) also encrypt and back up data to secure locations over the Internet, eliminating the need for portable storage media all together. Because the data is encrypted prior to transport, it is secure in transit and in storage.

Some feel that for businesses to take this problem seriously, government needs to apply pressure through additional legislation. Others argue that data reported as lost typically cannot be used on its own by thieves and is deactivated or changed when the theft is realized anyway, reducing the need for legislation. Improvements in networking and storage technologies will eventually solve this dilemma for those concerned about privacy. Until then, thousands of database records will likely be accessed without authorization due to the loss or theft of corporate storage media.

Discussion Questions

1. How can the theft of a credit card number, Social Security number, or bank account number be a danger to the victim?
2. Why is tape the most common medium used for backing up data?

Critical Thinking Questions

1. To what extent should laws hold companies responsible for the security of the data they keep? Should encryption be required?
2. If you were responsible for the safety of Marriott's backup data, what security measures would you take?

SOURCES: Rosenwald, Michael S., "Marriott Discloses Missing Data Files," *Washington Post*, December 28, 2005, *www.washingtonpost.com*. Damoulakis, Jim, "Do We Really Care About Storage Security?" *Computerworld*, February 7, 2006, *www.computerworld.com*. Lawson, Stephen, "Bank Tape Lost with Data on 90,000 Customers," *Computerworld*, January 11, 2006, *www.computerworld.com*.

One organization using virtual tape technology is the Chicago Mercantile Exchange, the Merc for short, which supports electronic trading for futures and options contracts. Its computer storage system holds over 180 terabytes of data managed across four tiers of storage devices. Tier 1 storage devices hold the critical application data and databases required to support the high-performance trading systems. Tier 2 devices hold noncritical production databases. Tier 3 devices archive data required by government regulation. At one time, tier 4 storage was a tape-based system to support the backup and recovery process. This setup was slow and resulted in a high rate of backup-and-restore failures. The Merc replaced the tier 4 setup with a special disk system called Maid (massive array of idle disks). Maid disks only spin when a piece of data that resides on it is requested; the rest of the time, the disks are idle. This reduces the risk of disk failure and extends the useful life of the devices. The Maid system backs up and restores data three times faster and with no failed backups to date. To NetBackup, Merc's backup software, the Maid system appears as a "virtual tape" library because the data appears to be stored on tape cartridges but is actually being stored on faster hard disks.[13]

Optical Discs

optical disc

A rigid disc of plastic onto which data is recorded by special lasers that physically burn pits in the disc.

Another type of secondary storage medium is the **optical disc**. Like a ROM chip, an optical disc is simply a rigid disk of plastic onto which data is recorded by special lasers that physically burn pits in the disk. Data is directly accessed from the disc by an optical disc device, which operates much like a stereo's compact disc player. This optical disc device uses a low-power laser that measures the difference in reflected light caused by a pit (or lack thereof) on the disc.

compact disc read-only memory (CD-ROM)

A common form of optical disc on which data, once it has been recorded, cannot be modified.

A common optical disc is the **compact disc read-only memory (CD-ROM)** with a storage capacity of 740 MB of data. After data is recorded on a CD-ROM, it cannot be modified— the disc is "read-only." A CD burner, the informal name for a CD recorder, is a device that can record data to a compact disc. *CD-recordable (CD-R)* and *CD-rewritable (CD-RW)* are the two most common types of drives that can write CDs, either once (in the case of CD-R) or repeatedly (in the case of CD-RW). CD-rewritable (CD-RW) technology allows PC users to back up data on CDs.

Digital Video Disc

digital video disc (DVD)

A storage medium used to store digital video or computer data.

A **digital video disc (DVD)** looks like a CD but can store about 135 minutes of digital video or several gigabytes of data (see Figure 3.8). Software, video games, and movies are often stored or distributed on DVDs. At a data transfer rate of 1.352 MB/second, the access speed of a DVD drive is faster than that of the typical CD-ROM drive.

Figure 3.8

Digital Video Disc and Player

DVDs look like CDs but have a greater storage capacity and can transfer data at a faster rate.

(Source: Courtesy of Toshiba America Information Systems.)

DVDs have replaced recordable and rewritable CD discs (CD-R and CD-RW) as the preferred format for sharing movies and photos. Whereas a CD can hold about 740 MB of data, a single-sided DVD can hold 4.7 GB, with double-sided DVDs having a capacity of 9.4 GB. Unfortunately, DVD manufacturers haven't agreed on a recording standard, so there are several types of recorders and discs. Recordings can be made on record-once discs (DVD-R and DVD+R) or on rewritable discs (DVD-RW, DVD+RW, and DVD-RAM). Not all types of rewritable DVDs are compatible with other types. Dell and Hewlett-Packard use DVD+RW; Apple, Gateway, and IBM offer DVD-R. Dell and other manufacturers use DVD +/- RW.

The two types of competing high-definition video-disc formats are called *HD-DVD* and *Blu-ray Disc*. Both formats were originally based on blue-laser technology that stores at least three times as much data as a DVD now holds.[14] Traditional magneto-optical (MO), CD, and DVD formats all use red lasers. Because the wavelength of blue light is shorter than that of red light, the beam from a blue laser makes a much smaller spot on the recording layer of a disc. A smaller spot means less space is needed to record one bit of data, so more data can be stored on a disc. The primary use for these new formats is in home entertainment equipment to store high definition video, though these formats can also store computer data.

Holographic Disc

Holographic Versatile Disc (HVD) is an advanced optical disc technology still in the research stage that would store more data than even the Blu-ray and HD DVD optical disc systems. One approach to HVD records data through the depth of the storage media in three dimensions by splitting a laser beam in two—the signal beam carries the data, and the reference beam positions where the data is written and reads it. This approach can record and read 1 billion bits of data with each flash of light. A holographic disc is expected to hold about 200 GB of data—50 times as much as a DVD. HVD will also write data as much as ten times faster.[15]

Memory Cards

A group of computer manufacturers formed the Personal Computer Memory Card International Association (PCMCIA) to create standards for a peripheral device known as a *PC memory card*. These PC memory cards are credit card-sized devices that can be installed in an adapter or slot in many personal computers. To the rest of the system, the PC memory card functions as a hard disk drive. Although the cost per megabyte of storage is greater than for traditional hard disk storage, these cards are less prone to fail than hard disks, are portable, and are relatively easy to use. Software manufacturers often store program instructions on a memory card for use with laptop computers.

Flash Memory

Flash memory is a silicon computer chip that, unlike RAM, is nonvolatile and keeps its memory when the power is shut off. It gets its name from the fact that the microchip is organized so that a section of memory cells (called a *block*) is erased or reprogrammed in a single action, or "flash." Solid-state disks (SSDs) that use flash memory are supplementing or replacing traditional hard drives that employ power-hungry spinning platters with mobile read/write heads near data surfaces.[16] The result is longer laptop battery life and more protection for data. Another advantage is that a flash memory system reboots faster than hard disks.[17] Digital music players and cameras use flash memory to hold music and photos. Compared with other types of secondary storage, flash memory can be accessed more quickly and consumes less power and storage space. The primary disadvantage is cost. Flash memory chips cost much more per megabyte than a traditional hard disk.

A flash memory device called Migo is a keychain device that comes with software that captures files and settings from your e-mail, word processing, Web browser, and presentation programs. These preferences are stored in flash memory (up to 2 GB), and are transferred when you plug the Migo device into a computer so that you can work as if you were at your own PC. When you unplug the Migo, the software wipes all traces of the files and settings from the PC. Then, when you plug the Migo into your own PC, it synchronizes the data and updates all changes to your desktop. The device means people do not have to lug their laptop computers with them everywhere they go. Course Technology sometimes packages USB flash drives with textbooks or makes them available to attendees of various conferences.

flash memory
A silicon computer chip that, unlike RAM, is nonvolatile and keeps its memory when the power is shut off.

The Migo keychain device comes with software that captures files and settings from your e-mail, word processing, Web browser, and presentation programs.

(Source: Courtesy of Migo Software, Inc.)

Enterprise Storage Options

Businesses increasingly need to store large amounts of data created throughout the organization. Such large secondary storage is called *enterprise storage*, and comes in three forms: attached storage, network-attached storage (NAS), and storage area networks (SANs).

Attached Storage

Attached storage methods include the tape, hard disks, and optical devices discussed previously, which are connected directly to a single computer. Attached storage methods, though simple and cost effective for single users and small groups, do not allow systems to share storage, and they make it difficult to back up data.

Because of the limitations of attached storage, firms are turning to network-attached storage (NAS) and storage area networks (SANs). These alternative forms of enterprise data storage enable an organization to share data-storage resources among a much larger number of computers and users, resulting in improved storage efficiency and greater cost-effectiveness. In addition, they simplify data backup and reduce the risk of downtime. Nearly one-third of system downtime is a direct result of data-storage failures, so eliminating storage problems as a cause of downtime is a major advantage.

Network-Attached Storage

network-attached storage (NAS)

Storage devices that attach to a network instead of to a single computer.

Network-attached storage (NAS) employs storage devices that attach to a network instead of to a single computer. NAS includes software to manage storage access and file management and relieve the users' computers of those tasks. The result is that both application software and files can be served faster because they are not competing for the same processor resources. Computer users can share and access the same information, even if they are using different types of computers. Common applications for NAS include consolidated storage, Internet and e-commerce applications, and digital media. St. Joseph's Hospital Health Center in Syracuse, New York, implemented a NAS storage system to hold more than 2 terabytes of patient data, and will grow to 10 terabytes of data per year to include digital versions of patient x-rays, CT scans, and MRIs.[18]

Storage Area Network

storage area network (SAN)

The technology that provides high-speed connections between data-storage devices and computers over a network.

A **storage area network (SAN)** is a special-purpose, high-speed network that provides direct connections between data-storage devices and computers across the enterprise (see Figure 3.9). A SAN also integrates different types of storage subsystems, such as multiple RAID storage devices and magnetic tape backup systems, into a single storage system. Use of a SAN off-loads the network traffic associated with storage onto a separate network. The data can then be copied to a remote location, making it easier for companies to create backups and implement disaster recovery policies.

Figure 3.9

Storage Area Network

A SAN provides high-speed connections between data-storage devices and computers over a network.

Using a SAN, an organization can centralize the people, policies, procedures, and practices for managing storage, and a data-storage manager can apply the data consistently across an enterprise. This centralization eliminates inconsistent treatment of data by different system administrators and users, providing efficient and cost-effective data-storage practices. For example, the U.S. Navy linked data and software at the Navy's Surface Combat Systems Center in Wallops Island, VA, and the Naval Sea Systems Command in nearby Dahlgren, VA, using SAN technology. The Navy could then accelerate the testing and certification of the weapons developed at these two locations.[19]

A fundamental difference between NAS and SAN is that NAS uses file input/output, which defines data as complete containers of information, while SAN deals with block input/output, which is based on subsets of data smaller than a file. SAN manufacturers include EMC, Hitachi Data Systems Corporation, Xiotech, and IBM.

As organizations set up large-scale SANs, they use more computers and network connections, which become difficult to manage. In response, software tools designed to automate storage using previously defined policies are finding a place in the enterprise. Known as **policy-based storage management**, the software products from industry leaders such as Veritas Software Corporation, Legato Systems, Inc., EMC, and IBM automatically allocate storage space to users, balance the loads on servers and disks, and reroute networks when systems go down—all based on policies set up by system administrators.

policy-based storage management
Automation of storage using previously defined policies.

The trend in secondary storage is toward higher capacity, increased portability, and automated storage management. Organizations should select a type of storage based on their needs and resources. In general, storing large amounts of data and information and letting users access it quickly makes an organization more efficient. Businesses can also choose pay-per-use services, where they rent space on massive storage devices housed either at a service provider (e.g., Hewlett-Packard or IBM) or on the customer's premises, paying only for the amount of storage they use. This approach is sensible for organizations with wildly fluctuating storage needs, such as those involved in the testing of new drugs or developing software.

INPUT AND OUTPUT DEVICES: THE GATEWAY TO COMPUTER SYSTEMS

Your first experience with computers is usually through input and output devices. These devices are the gateways to the computer system—you use them to provide data and instructions to the computer and receive results from it. Input and output devices are part of a computer's user interface, which includes other hardware devices and software that allow you to interact with a computer system.

As with other computer system components, an organization should keep their business goals in mind when selecting input and output devices. For example, many restaurant chains use handheld input devices or computerized terminals that let waiters enter orders efficiently and accurately. These systems have also cut costs by helping to track inventory and market to customers.

Characteristics and Functionality

In general, businesses want input devices that let them rapidly enter data into a computer system, and they want output devices that let them produce timely results. When selecting input and output devices, businesses also need to consider the form of the output they want, the nature of the data required to generate this output, and the speed and accuracy they need for both. Some organizations have very specific needs for output and input, requiring devices that perform specific functions. The more specialized the application, the more specialized the associated system input and output devices.

The speed and functions of input and output devices should be balanced with their cost, control, and complexity. More specialized devices might make it easier to enter data or output information, but they are generally more costly, less flexible, and more susceptible to malfunction.

The Nature of Data

Getting data into the computer—input—often requires transferring human-readable data, such as a sales order, into the computer system. "Human-readable" means data that people can read and understand. A sheet of paper containing inventory adjustments is an example of human-readable data. In contrast, machine-readable data can be understood and read by computer devices (e.g., the universal bar code that grocery scanners read) and is typically stored as bits or bytes. Inventory changes stored on a disk is an example of machine-readable data.

Some data can be read by people and machines, such as magnetic ink on bank checks. Usually, people begin the input process by organizing human-readable data and transforming it into machine-readable data. Every keystroke on a keyboard, for example, turns a letter symbol of a human language into a digital code that the machine can understand.

Data Entry and Input

Getting data into the computer system is a two-stage process. First, the human-readable data is converted into a machine-readable form through **data entry**. The second stage involves transferring the machine-readable data into the system. This is **data input**.

Today, many companies are using online data entry and input—they communicate and transfer data to computer devices directly connected to the computer system. Online data entry and input places data into the computer system in a matter of seconds. Organizations in many industries require the instantaneous updating offered by this approach. For example, when airline clerks need to enter a last-minute reservation, they can use online data entry and input to record the reservation as soon as it is made. Reservation agents at other terminals can then access this data to make a seating check before they make another reservation.

Source Data Automation

Regardless of how data gets into the computer, it should be captured and edited at its source. **Source data automation** involves capturing and editing data where the data is originally created and in a form that can be directly input to a computer, thus ensuring accuracy and timeliness. For example, using source data automation, salespeople enter sales orders into the computer at the time and place they take the order. Any errors can be detected and corrected immediately. If any item is temporarily out of stock, the salesperson can discuss options with the customer. Prior to source data automation, orders were written on paper and entered into the computer later (usually by a clerk, not the person who took the order). Often the handwritten information wasn't legible or, worse yet, got lost. If problems occurred during data entry, the clerk had to contact the salesperson or the customer to "recapture" the data needed for order entry, leading to further delays and customer dissatisfaction.

Input Devices

You can use hundreds of devices for data entry and input. They range from special-purpose devices that capture specific types of data to more general-purpose input devices. Some of the special-purpose data entry and input devices are discussed later in this chapter. First, we focus on devices used to enter and input general types of data, including text, audio, images, and video for personal computers.

Personal Computer Input Devices

A keyboard and a computer mouse are the most common devices used for entry and input of data such as characters, text, and basic commands. Some companies are developing keyboards that are more comfortable, more easily adjusted, and faster to use than standard keyboards. These ergonomic keyboards, such as the split keyboard by Microsoft and others, are designed to avoid wrist and hand injuries caused by hours of typing. Other keyboards include touchpads that let you enter sketches on the touchpad and text using the keys. Another innovation is wireless mice and keyboards, which keep a physical desktop free from clutter.

data entry
Converting human-readable data into a machine-readable form.

data input
Transferring machine-readable data into the system.

source data automation
Capturing and editing data where the data is initially created and in a form that can be directly input to a computer, thus ensuring accuracy and timeliness.

You use a computer mouse to point to and click symbols, icons, menus, and commands on the screen. The computer takes a number of actions in response, such as placing data into the computer system.

A keyboard and mouse are two of the most common devices for computer input. Wireless mice and keyboards are now readily available.

(Source: Courtesy of Hewlett-Packard Company.)

Speech-Recognition Technology

Speech-recognition technology enables a computer equipped with a source of speech input such as a microphone to interpret human speech as an alternative means of providing data or instructions to the computer. The most basic systems require you to train the system to recognize your speech patterns or are limited to a small vocabulary of words. More advanced systems can recognize continuous speech without requiring you to break up your speech into discrete words. Very advanced systems used by the government and military can interpret a voice it has never heard and understand a rich vocabulary.

Companies that must constantly interact with customers are eager to reduce their customer support costs while improving the quality of their service. Pacific Gas and Electric (PG&E) serves 5 million electric customers and 4 million natural gas customers. The firm implemented speech-recognition technology to automate the account identification process and to provide other customer self-service functions. Some 38 percent of customer service calls are satisfied using the system without speaking directly to a customer service representative. This yields a savings of nearly $3 million per year based on the average cost of $7 for a PG&E customer service representative to handle a call.[20]

speech-recognition technology
Input devices that recognize human speech.

Digital Cameras

Digital cameras record and store images or video in digital form (see Figure 3.10). When you take pictures, the images are electronically stored in the camera. You can download the images to a computer either directly or by using a flash memory card. After you store the images on the computer's hard disk, you can edit them, send them to another location, paste them into another application, or print them. For example, you can download a photo of your project team captured by a digital camera and then post it on a Web site or paste it into a project status report. Digital cameras have eclipsed film cameras used by professional photographers for photo quality and features such as zoom, flash, exposure controls, special effects, and even video-capture capabilities. With the right software, you can add sound and handwriting to the photo.

More than two dozen camera manufacturers offer at least one digital camera model for under $225 with sufficient resolution to produce high-quality 5×7-inch photos.[21] Some manufacturers offer a video camera that records full-motion video.

The primary advantage of digital cameras is saving time and money by eliminating the need to process film. In fact, digital cameras that can easily transfer images to CDs have made the consumer film business of Kodak and Fujitsu nearly obsolete. Until film-camera users switch to digital cameras, Kodak is allowing photographers to have it both ways. When you want to develop print film, Kodak offers the option of placing pictures on a CD in addition

digital camera
An input device used with a PC to record and store images and video in digital form.

to the traditional prints. After the photos are stored on the CD, they can be edited, placed on a Web site, or sent electronically to business associates or friends around the world.

Organizations are now using digital cameras for security and other business purposes. The New York Metropolitan Transport Authority awarded a $212 million contract to Lockheed Martin to develop and deploy thousands of cameras and intelligent video and sensor technologies to monitor New York's subway system, commuter rail platforms, and highway bridges and tunnels. The goal is to protect the transit system from future terrorist attacks.[22] In this case, the cameras provide input to a computer system that provides access control, alarm monitoring, and digital video surveillance.

Terminals

Inexpensive and easy to use, terminals are input and display devices that perform data entry and input at the same time. A terminal is connected to a complete computer system, including a processor, memory, and secondary storage. After you enter general commands, text, and other data via a keyboard or mouse, it is converted into machine-readable form and transferred to the processing portion of the computer system. Terminals, normally connected directly to the computer system by telephone lines or cables, can be placed in offices, in warehouses, and on the factory floor.

Scanning Devices

You can input image and character data using a scanning device. A page scanner is like a copy machine. You typically insert a page you want to input into the scanner or place it face down on the glass plate of the scanner, cover it, and then scan it. With a handheld scanner, you manually move or roll the scanning device over the image you want to scan. Both page and handheld scanners can convert monochrome or color pictures, forms, text, and other images into machine-readable digits. Considering that U.S. enterprises generate an estimated 1 billion pieces of paper daily, many companies are looking to scanning devices to help them manage their documents and reduce the high cost of using and processing paper.

Optical Data Readers

You can also use a special scanning device called an *optical data reader* to scan documents. The two categories of optical data readers are for optical mark recognition (OMR) and optical character recognition (OCR). You use OMR readers for test scoring and other purposes when test takers use pencils to fill in boxes on OMR paper, which is also called a "mark sense form." OMR systems are used in standardized tests, including the SAT and GMAT tests, and are being considered as a means to capture voters' choices on Election Day. In comparison, most OCR readers use reflected light to recognize and scan various characters. With special software, OCR readers can convert handwritten or typed documents into digital data. After being entered, this data can be shared, modified, and distributed over computer networks to hundreds or thousands of people.

Magnetic Ink Character Recognition (MICR) Devices

In the 1950s, the banking industry became swamped with paper checks, loan applications, bank statements, and so on. The result was the development of magnetic ink character recognition (MICR), a system for reading banking data quickly. With MICR, data is placed on the bottom of a check or other form using a special magnetic ink. Using a special character set, data printed with this ink is readable by people and computers (see Figure 3.11).

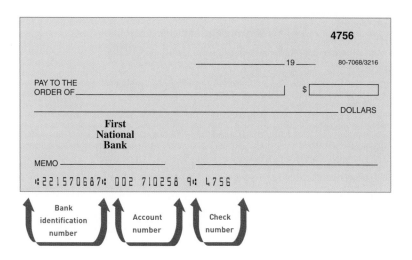

Figure 3.11

MICR Device

Magnetic ink character recognition technology codes data on the bottom of a check or other form using special magnetic ink, which is readable by people and computers. For an example, look at the bottom of a bank check.

(Source: Courtesy of NCR Corporation.)

Magnetic Stripe Card

A **magnetic stripe card** stores limited amounts of data by modifying the magnetism of tiny iron-based particles contained in a band on the card. The magnetic stripe is read by physically swiping the card past a reading head. Magnetic stripe cards are commonly used in credit cards, transportation tickets, and driver's licenses. The Los Angeles police department (LAPD) purchased 1,200 Symbol Technologies handheld computers for police officers to record information about each interaction they have with the public. They expanded the use of this hardware to reduce the time-consuming and error-prone tasks of writing and processing traffic citations. Officers capture identification data encoded on the magnetic stripe of a driver's license by swiping the card on their handheld computer. They then enter information about the violation. The driver signs the citation displayed on the handheld computer, and the citation is printed on the spot by a wireless printer. All citations issued for the day are transmitted to the Los Angeles municipal court system for processing and billing.[23]

magnetic stripe card

A type of card that stores limited amounts of data by modifying the magnetism of tiny iron-based particles contained in a band on the card.

Point-of-Sale Devices

Point-of-sale (POS) devices are terminals used in retail operations to enter sales information into the computer system. The POS device then computes the total charges, including tax. Many POS devices also use other types of input and output devices, such as keyboards, bar-code readers, printers, and screens. Much of the money that businesses spend on computer technology involves POS devices.

point-of-sale (POS) device

A terminal used in retail operations to enter sales information into the computer system.

Automated Teller Machine (ATM) Devices

Another type of special-purpose input/output device, the automated teller machine (ATM) is a terminal that bank customers use to perform withdrawals and other transactions with their bank accounts. The ATM, however, is no longer used only for cash and bank receipts. Companies use various ATM devices, sometimes called *kiosks,* to support their business processes. Some can dispense tickets, such as for airlines, concerts, and soccer games. Some colleges use them to produce transcripts. For this reason, the input and output capabilities of ATMs are varied. Like POS devices, ATMs can work with other types of input and output devices. Unisys, for example, has developed an ATM kiosk that allows bank customers to make cash withdrawals, pay bills, and receive advice on investments and retirement planning.

Tablet PCs Serve Up Big Benefits for Holiday Inn Restaurants

Kemmons Wilson opened the first Holiday Inn in Memphis, Tennessee, in 1952. Kemmons was tired of the low quality and high prices of hotels of the day and thought that he could do better. Kemmons' first Holiday Inn offered telephones, air-conditioning, a swimming pool, and a restaurant. Kids stayed free and there was no charge for ice or parking. Such amenities were unheard of in 1952, and Kemmons roadside inn quickly multiplied along U.S. highways. Today, there are more than 1,500 Holiday Inns worldwide.

The pioneering spirit of Kemmons Wilson lives on in the management of today's Holiday Inn Hotels and Resorts. Recently, Holiday Inn has been experimenting with new technologies to improve their amenities. One technology that has captured the imagination of Holiday Inn management is the Tablet PC.

At a Holiday Inn prototype hotel near Atlanta, Georgia, the company is experimenting with new technology services. The designers of this hotel are providing more than the standard wireless Internet services in an attempt to enhance the guest experience and increase hotel productivity. To create an electronic menu in the hotel restaurant, Kem's café, they are using a Tablet PC. Guests who eat at Kem's can use a paper or electronic menu. The e-menu consists of a wireless Tablet PC with touch screen and stylus. Guests can browse through the menu and select items in the same way they navigate a Web site.

The e-menu offers several benefits over a paper menu. For example, the e-menu includes high-resolution images of all items on the menu, as well as ingredients and nutritional information. The system can estimate the total calories and fat of the selected menu items and suggest choices for customers following popular diets such as Weight Watchers, South Beach, and Atkins. Other benefits include the following:

- A running total of all items selected that includes tax and gratuity
- Support for multiple languages and currencies
- Adjustable font size, which is especially valued by seniors

The e-menu system is also a benefit to managers. Because it is electronic and wirelessly connected to the restaurant network, the contents can be updated at a moment's notice. For instance, if the kitchen runs out of ingredients for Flaming Duck, that dish can be removed from the menu. The restaurant staff does not have to disappoint customers by reporting that the item they requested is no longer available. Likewise, management might find that they have too much of an item and update the menu to display a special sale price.

The Tablet PC was the ideal hardware device for an e-menu for the following reasons:

- The high-resolution display supports the visual experience that diners expect from a menu in a fine restaurant.
- The mobility of the device makes it easy to pass around.
- The stylus and touch screen provide an intuitive and easy-to-use interface.
- The wireless networking capabilities provide two-way communications for placing orders and updating menu information.

As it introduced the e-menu system, Holiday Inn saw additional value opportunities in their Tablet PCs. After customers place their food order, they can continue using the Tablet PC to read the daily news, exchange e-mail, check on flight status, reserve a room at another Holiday Inn, access street maps, and send documents to the printer in the lobby.

The e-menu pilot project has been a complete success, earning soaring customer satisfaction ratings. The message that e-menus send to guests is that Holiday Inn understands and is committed to meeting their needs, the same principle on which Holiday Inn was founded. In 1952, it was swimming pools and AC. Today, businesses must impress customers using the latest that technology has to offer.

Discussion Questions

1. Why does Holiday Inn feel it needs to adopt cutting-edge technologies? What does it hope to gain?
2. For restaurant menus, would you prefer paper or Tablet PC? Which e-menu services would you find most useful? Can you foresee a day when paper menus are completely retired?

Critical Thinking Questions

1. What are the drawbacks or risks involved in implementing the e-menu system?
2. What other services might Holiday Inn offer using Tablet PCs outside of Kem's Café?

SOURCES: Microsoft Customer Solution Case Study: "Wireless Solution Improves Guest Experience, Drives Increased Revenues for Worldwide Hotel Chain," Microsoft Web site, accessed February 12, 2006, *http://members.microsoft.com/CustomerEvidence.* "Intercontinental Hotels Breaks Ground in Southern Michigan," HotelInteractive.com, January 31, 2006, *www.hotelinteractive.com/.* Holiday Inn Web site, accessed February 12, 2006, *www.holiday-inn.com.*

Pen Input Devices

By touching the screen with a pen input device, you can activate a command or cause the computer to perform a task, enter handwritten notes, and draw objects and figures. Pen input requires special software and hardware. Handwriting recognition software can convert handwriting on the screen into text. The Tablet PC from Microsoft and its hardware partners can transform handwriting into typed text and store the "digital ink" just the way a person writes it. Users can use a pen to write and send e-mail, add comments to Word documents, mark up PowerPoint presentations, and even hand-draw charts in a document. That data can then be moved, highlighted, searched, and converted into text. If perfected, this interface is likely to become widely used. Pen input is especially attractive if you are uncomfortable using a keyboard. The success of pen input depends on how accurately handwriting can be read and translated into digital form and at what cost.

Touch-Sensitive Screens

Advances in screen technology allow display screens to function as input as well as output devices. By touching certain parts of a touch-sensitive screen, you can start a program or trigger other types of action. Touch-sensitive screens are popular input devices for some small computers because they do not require a keyboard, which conserves space and increases portability. Touch screens are frequently used at gas stations for customers to select grades of gas and request a receipt, on photocopy machines to enable users to select various options, at fast-food restaurants for order clerks to enter customer choices, at information centers in hotels to allow guests to request facts about local eating and drinking establishments, and at amusement parks to provide directions to patrons. They also are used in kiosks at airports and department stores. Touch-sensitive screens are also being considered as a technology to use in capturing voter choices.

Bar-Code Scanners

A bar-code scanner employs a laser scanner to read a bar-coded label. This form of input is used widely in grocery store checkouts and warehouse inventory control. Often, bar-code technology is combined with other forms of technology to create innovative ways for capturing data.

Radio Frequency Identification

The purpose of a **Radio Frequency Identification (RFID)** system is to transmit data by a mobile device, called a tag, which is read by an RFID reader and processed according to the needs of an IS program. One popular application of RFID is to place a microchip on retail items and install in-store readers that track the inventory on the shelves to determine when shelves should be restocked. Recall that the RFID tag chip includes a special form of EPROM memory that holds data about the item to which the tag is attached. A radio frequency signal can update this memory as the status of the item changes. The data transmitted by the tag might provide identification, location information, or details about the product tagged, such as date manufactured, retail price, color, or date of purchase.

The two basic types of RFID tags are passive and active. Passive tags have no internal power source. An incoming radio frequency signal triggers a tiny electrical current in the antenna of the RFID tag, which powers the integrated circuit in the tag and transmits a response communicating the information it has been programmed to deliver. Passive RFID tags are small (see Figure 3.12), do not require batteries, have an unlimited life span, and send signals that can be heard from a distance of a few yards. Active RFID tags have an internal power source, which powers the tag so that it can generate an outgoing signal. Active tags can be heard over a longer distance and have more memory than passive tags. Currently, the smallest active tags are about the size of a coin. Because passive tags are less expensive to manufacture and require no battery, most RFID tags today are of the passive type. As of this writing, you can purchase passive tags for less than $.07 each in volumes of 10 million units or more.

The popularity of RFID tags continues to grow. Toyota was one of the first automobile manufacturers to employ a smart key with an active RFID tag that allows a car to acknowledge

Radio Frequency Identification (RFID)

A technology that employs a microchip with an antenna that broadcasts its unique identifier and location to receivers.

RFID tag

the key's presence within approximately three feet of the sensor. The driver can open the doors and start the car while the key remains in a purse or pocket. The U.S. Food and Drug Administration (FDA) is considering the use of RFID tags to help fight counterfeit prescription drugs by tracking shipping containers and crates of medicine. Privacy advocates fear that the FDA could also use the tags to track individual medicine bottles or even individual tablets. That, privacy advocates said, would be pervasive.[24]

Output Devices

Computer systems provide output to decision makers at all levels of an organization so they can solve a business problem or capitalize on a competitive opportunity. In addition, output from one computer system can provide input into another computer system. The desired form of this output might be visual, audio, or even digital. Whatever the output's content or form, output devices are designed to provide the right information to the right person in the right format at the right time.

Display Monitors

The display monitor is a device similar to a TV screen that displays output from the computer. Because some monitors use a cathode-ray tube to display images, they are sometimes called *CRTs*. Such a monitor works much the same way a traditional TV screen does—the cathode-ray tubes generate one or more electron beams. As the beams strike a phosphorescent compound (phosphor) coated on the inside of the screen, a dot on the screen called a **pixel** lights up. A pixel is a dot of color on a photo image or a point of light on a display screen. It appears in one of two modes: on or off. The electron beam sweeps across the screen so that as the phosphor starts to fade, it is struck and lights up again.

pixel
A dot of color on a photo image or a point of light on a display screen.

With today's wide selection of monitors, price and overall quality can vary tremendously. The quality of a screen image is often measured by the number of horizontal and vertical pixels used to create it. The more pixels per square inch, the higher the resolution, or clarity and sharpness, of the image. For example, a screen with a 1,024×768 resolution (786,432 pixels) has a higher sharpness than one with a resolution of 800×600 (480,000 pixels). Another way to measure image quality is the distance between one pixel on the screen and the next nearest pixel, which is known as *dot pitch*. The common range of dot pitch is from .25 mm to .31 mm. The smaller the dot pitch, the better the picture. A dot pitch of .28 mm or smaller is considered good. Greater pixel densities and smaller dot pitches yield sharper images of higher resolution.

The characteristics of screen color depend on the quality of the monitor, the amount of RAM in the computer system, and the monitor's graphics adapter card. The Color/Graphics Adapter (CGA) was one of the first technologies to display color images on the screen. Today, Super Video Graphics Array (SVGA) displays are standard, providing vivid colors and high resolution. Digital Video Interface (DVI) is designed to maximize the visual quality of digital display devices such as flat-panel LCD computer displays.

Liquid Crystal Displays (LCDs)

LCD displays are flat displays that use liquid crystals—organic, oil-like material placed between two polarizers—to form characters and graphic images on a backlit screen. These displays are easier on your eyes than CRTs because they are flicker-free, brighter, and don't emit the type of radiation that makes some CRT users worry. In addition, LCD monitors take up less space and use less than half of the electricity required to operate a comparably sized CRT monitor.[25] *Thin-film transistor (TFT) LCDs* are a type of liquid crystal display that assigns a transistor to control each pixel, resulting in higher resolution and quicker response to changes on the screen. TFT LCD monitors are rapidly displacing competing CRT technology, and are commonly available in sizes from 12 to 30 inches. LCD monitors with exciting new features include the following:

- The Dell UltraSharp Widescreen 3007WFP is a high-end, 30-inch display with stunning resolution of 2,560×1,600 pixels suitable for graphics professionals and very serious gamers. It competes directly with the Apple 30-inch HD Display.[26]
- Sharp Corporation has developed an LCD that can simultaneously display different information and image content in right and left views. Called a controlled or two-way viewing-angle LCD, the screen can provide information and content tailored to specific users depending on the angle at which they view the screen.[27]
- The Sharp PC-XG50H and PC-XG70H are two multimedia notebook PCs with 15-inch screens as bright as the company's LCD TVs and a resolution of 1,024×768 pixels. The PCs are designed to be used as TVs as much as they are PCs.[28]
- Sharp has also developed a 3-D LCD that users can use to view realistic three-dimensional images without having to wear special goggles. The 3-D LCD can switch between flat display and three-dimensional display, so it can show word processor and spreadsheet files in the 2-D mode while using the 3-D mode for computer graphics and games.

LCD displays

Flat displays that use liquid crystals—organic, oil-like material placed between two polarizers—to form characters and graphic images on a backlit screen.

The Dell UltraSharp Widescreen 3007WFP is a high-end, 30-inch display with stunning resolution of 2,560×1,600 pixels suitable for graphics professionals and very serious gamers.

(Source: Courtesy of Dell Inc.)

Organic Light-Emitting Diodes

Organic light-emitting diode (OLED) technology is based on research by Eastman Kodak Company and is appearing on the market in small electronic devices. OLEDs use the same base technology as LCDs, with one key difference: Whereas LCD screens contain a fluorescent backlight and the LCD acts as a shutter to selectively block that light, OLEDs directly emit light. OLEDs can provide sharper and brighter colors than LCDs and CRTs, and because they don't require a backlight, the displays can be half as thick as LCDs and used in flexible displays. Another big advantage is that OLEDs don't break when dropped. OLEDs are currently limited to use in cell phones, car radios, and digital cameras, but might be used in computer displays—if the average display lifetime can be extended.[29]

Printers and Plotters

One of the most useful and popular forms of output is called *hard copy*, which is simply paper output from a printer. The two main types of printers are laser printers and inkjet printers, and they are available with different speeds, features, and capabilities. Some can be set up to accommodate paper forms, such as blank check forms and invoice forms. Newer printers allow businesses to create customized printed output for each customer from standard paper and data input using full color. Ticket-receipt printers such as those used in restaurants, ATMs, and point-of-sale systems are in wide-scale use.

The speed of the printer is typically measured by the number of pages printed per minute (ppm). Like a display screen, the quality, or resolution, of a printer's output depends on the number of dots printed per inch (dpi). A 600-dpi printer prints more clearly than a 300-dpi printer. A recurring cost of using a printer is the inkjet or laser cartridge that must be replaced periodically—every few thousand pages for laser printers and every 500 to 900 pages for inkjet printers. Figure 3.13 shows a laser printer. A later section discusses considerations for selecting a printer.

Figure 3.13

Laser Printer

Laser printers, available in a wide variety of speeds and price ranges, have many features, including color capabilities. They are a popular solution for printing hard copies of information.

(Source: Courtesy of Lexmark International.)

Laser printers are generally faster than inkjet printers and can handle more volume than inkjet printers. Laser printers print 15 to 50 pages per minute (ppm) for black and white and 4 to 20 ppm for color. Inkjet printers print 10 to 30 ppm for black and white and 2 to 10 ppm for color.

For color printing, inkjet printers print vivid hues and with an initial cost much less than color laser printers. Inkjet printers can produce high-quality banners, graphics, greeting cards, letters, text, and prints of photos.

Plotters are a type of hard-copy output device used for general design work. Businesses typically use plotters to generate paper or acetate blueprints, schematics, and drawings of buildings or new products. Standard plot widths are 24 inches and 36 inches, and the length can be whatever meets the need—from a few inches to many feet.

Digital Audio Player

digital audio player
A device that can store, organize, and play digital music files.

MP3
A standard format for compressing a sound sequence into a small file.

A **digital audio player** is a device that can store, organize, and play digital music files. **MP3** (MPEG-1 Audio Layer-3) is a popular format for compressing a sound sequence into a very small file while preserving the original level of sound quality when it is played. By compressing the sound file, it requires less time to download the file and less storage space on a hard drive.

You can use many different music devices about the size of a cigarette pack to download music from the Internet and other sources. These devices have no moving parts and can store hours of music. Apple expanded into the digital music market with an MP3 player (the iPod) and the iTunes Music Store, which allows you to find music online, preview it, and download it in a way that is safe, legal, and affordable. In October 2005, Apple unveiled a new iPod with a 2.5-inch screen that can play video, including selected TV shows you can download from the iTunes Music Store.[30] There are dozens of MP3 manufacturers including Dell, Sony, Samsung, Iomega, and Motorola, whose Rokr product is the first iTunes-compatible phone.

Apple's iPod, a digital music player for Mac and Windows, has a 2.5-inch screen and can play video.

(Source: Courtesy of Apple Computer.)

Special-Purpose Input and Output Devices

Many additional input and output devices are used for specialized or unique applications. Some examples of such devices are discussed in the following sections.

Computer-Based Navigation Systems

Computer-based navigation systems are global positioning systems (GPSs), satellite-based radio navigation systems that can guide you to specified destinations. These systems come in all shapes and sizes and with varying capabilities—from PC-based systems factory-installed in automobiles for guiding you across the country to handheld units you carry while hiking through a national forest. All systems need a GPS antenna to receive satellite signals pin-pointing your location. High-end systems for automobiles include a gyroscope and a connection to the auto's speed sensor to enable accurate positioning even if the GPS signal is blocked by tunnels or tall buildings. Your location is superimposed on a map stored on CDs or a DVD. The latter allows for more detailed maps and eliminates the need to shuffle CDs. Some systems are voice activated so you don't have to take your eyes off the road to enter a destination. Portable systems can be moved from one car to another or carried in your knapsack. Some systems come with dynamic rerouting capability where the path recom-mended depends on weather and road conditions continually transmitted to a receiver in your car connected to a satellite radio system.

Multiple Function Printers

A number of manufacturers offer multiple-function printers that can copy, print (in color or black and white), fax, and scan. Such multifunctional devices are often used when people need to do a relatively low volume of copying, printing, faxing, and scanning. The typical price of multifunction printers ranges from $150 to $500, depending on features and capa-bilities. Because these devices take the place of one or more pieces of equipment, they are less expensive to acquire and maintain than a stand-alone fax, plus a stand-alone printer, plus a stand-alone copier, and so on. Also, eliminating equipment that was once located on a coun-tertop or desktop clears a significant amount of workspace for other work-related activities. As a result, such devices are popular in homes and small office settings.

Eyebud Screens

Eyebud screens are portable media devices that display video in front of one eye. They employ optical technology that provides very high resolution and "enlarges" the video or images. With the proximity of the screen to the eye and the magnifying effect of the optical technology, using an eyebud screen is like watching a 105-inch display from 12 feet away.

Such devices enable users of portable media devices to capture the big-screen, movie-screen, or home-theater experience, wherever they are.

The eyebud screen displays "enlarged" and high-resolution video.

(Source: Courtesy of eMagin Corp.)

COMPUTER SYSTEM TYPES, SELECTION, AND UPGRADING

In general, computers can be classified as either special purpose or general purpose. *Special-purpose computers* are used for limited applications by military and scientific research groups such as the CIA and NASA. Other applications include specialized processors found in appliances, cars, and other products. For example, automobile repair shops connect special-purpose computers to your car's engine to identify specific performance problems.

General-purpose computers are used for a variety of applications and to execute the business applications discussed in this text. General-purpose computer systems can be divided into two major groups: systems used by one user at a time and systems used by multiple concurrent users. Table 3.3 shows the general ranges of capabilities for various types of computer systems.

Computer System Types

Computer systems can range from small handheld computers to massive supercomputers that fill an entire room. We start first with the smallest computers.

Handheld Computers

handheld computer
A single-user computer that provides ease of portability because of its small size.

Handheld computers are single-user computers that provide ease of portability because of their small size—some are as small as a credit card. These systems often include a variety of software and communications capabilities. Most are compatible with and can communicate with desktop computers over wireless networks. Some even add a built-in GPS receiver with software that can integrate the location data into the application. For example, if you click an entry in an electronic address book, the device displays a map and directions from your current location. Such a computer can also be mounted in your car and serve as a navigation system. One of the shortcomings of handheld computers is that they require lots of power relative to their size.

PalmOne is the company that invented the Palm Pilot organizer in 1996. The Palm personal digital assistant (PDA) lets you track appointments, addresses, and tasks. PalmOne has now signed licensing agreements with Handspring, IBM, Sony, and many other

Factor	Single-User Systems					Multiuser Systems (Servers)		
	Handheld	Portable	Thin Client	Desktop	Workstation	Server	Mainframe	Supercomputer
Cost Range	$200 to $1,500	$1,000 to $3,500	$250 to $1,000	$600 to $3,500	$4,000 to $40,000	$500 to $50,000	>$100,000	>$250,000
Weight	<24 oz.	<7 lbs.	<15 lbs.	<25 lbs.	<25 lbs.	>25 lbs.	>200 lbs.	>200 lbs.
Typical Size	Palm size	Size of a notebook	Fits on desktop	Fits on desktop	Fits on desktop	Three-drawer filing cabinet	Refrigerator	Refrigerator and larger
CPU Speed	>200 MHz	>2 GHz	>200 MHz	>3 GHz	>3 GHz	>2 GHz	>300 MIPS	>2 teraflops
Typical Use	Organize personal data	Improve worker productivity	Enter data and access the Internet	Improve worker product-ivity	Perform engineering, CAD, and software development	Perform network and Internet applications	Perform computing tasks for large organ-izations and provide massive data storage	Run scientific applications; perform intensive number crunching
Example	Handspring Treo 600 smart phone	Motion Computing M1300 Mainstream Tablet PC	Max-speed Max-Term 8400	iMac Power PC G4	Sun Microsystems Sun Blade 2500 Workstation	Hewlett-Packard HP ProLiant BL	Unisys ES5000	IBMs RS/6000 SP

manufacturers, permitting them to make what amounts to Palm clones. As a result of the popularity of the Palm PDA, all handheld computers are often referred to as PDAs.

A **smartphone** combines the functionality of a mobile phone, personal digital assistant, camera, Web browser, e-mail tool, MP3 player, and other devices into a single handheld device. Smartphones will continue to evolve as new applications are defined and installed on the device. The applications might be developed by the manufacturer of the handheld device, by the operator of the communications network on which it operates, or by any other third-party software developer.

Portable Computers

Many computer manufacturers offer a variety of **portable computers**, those that can be carried easily—from laptops, to notebooks, to subnotebooks, to tablet computers. A *laptop computer* is a small, lightweight PC about the size of a three-ring notebook. The even smaller and lighter *notebook* and *subnotebook* computers offer similar computing power. Some notebook and subnotebook computers fit into docking stations of desktop computers to provide additional storage and processing capabilities.

Table 3.3

Types of Computer Systems

smartphone
A phone that combines the functionality of a mobile phone, personal digital assistant, camera, Web browser, e-mail tool, and other devices into a single handheld device.

portable computer
A computer small enough to be carried easily.

The Nokia 6101 smartphone can function as a phone, camera, FM radio, or e-mail device, and allows you to send text messages with audio and video clips.

(Source: Courtesy of Nokia.)

Tablet PCs (introduced earlier) are portable, lightweight computers that allow you to roam the office, home, or factory floor carrying the device like a clipboard. Recall that you can enter text with a writing stylus directly on the screen thanks to built-in handwriting recognition software. Other input methods might include an on-screen (virtual) keyboard, speech recognition, or a physical keyboard. Tablet PCs that only support input via a writing stylus are called *slates*. The *convertible tablet PC* comes with a swivel screen and can be used as a traditional notebook or as a pen-based tablet PC.

Portable computers are especially popular and useful in the healthcare industry. For example, New York Presbyterian Healthcare System has implemented an information system that uses data from patient histories and artificial intelligence software to create suggested treatment plans for patients in the cardiac intensive care unit. The system continuously receives data from patient monitoring equipment (such as heart rate and blood pressure) and alerts physicians via Tablet PCs if patients' conditions vary from projected outcomes.[31] At some Stop & Shop stores, shoppers can place an order through a Tablet PC mounted on their carts and connected to a wireless network. A message is sent to the customers when their orders are ready.[32]

Fujitsu Computer Systems Corporation makes a LifeBook P1500 notebook PC that weighs only 2.2 pounds with a nine-inch screen and can be flipped open and swiveled to convert into a tablet device with touch-screen and handwriting recognition capabilities.

(Source: © Fujitsu-Siemens Computers.)

Thin Client

thin client

A low-cost, centrally managed computer with essential but limited capabilities and no extra drives, such as a CD or DVD drive, or expansion slots.

A **thin client** is a low-cost, centrally managed computer with no extra drives, such as a CD or DVD drive, or expansion slots. These computers have limited capabilities and perform only essential applications, so they remain "thin" in terms of the client applications they include. These stripped-down versions of desktop computers do not have the storage capacity or computing power of typical desktop computers, nor do they need it for the role they play. With no hard disk, they never pick up viruses or suffer a hard disk crash. Unlike personal computers, thin clients download software from a network when needed, making support, distribution, and updating of software applications much easier and less expensive. Some organizations are able to take advantage of thin clients, including Fireman's Fund Insurance Company, which is providing employees in the field with thin clients to let them make billing inquiries and view policy data.[33] Other organizations find it difficult to have users accept thin clients, who prefer the features of a full-fledged computer.

Desktop Computers

desktop computer

A relatively small, inexpensive, single-user computer that is highly versatile.

Desktop computers are relatively small, inexpensive single-user computer systems that are highly versatile. Named for their size—the parts are small enough to fit on or beside an office desk—*desktop computers* can provide sufficient memory and storage for most business computing tasks.

Workstations

Workstations are more powerful than personal computers but still small enough to fit on a desktop. They are used to support engineering and technical users who perform heavy mathematical computing, computer-aided design (CAD), and other applications requiring a high-end processor. Such users need very powerful CPUs, large amounts of main memory, and extremely high-resolution graphic displays.

Servers

A **server** is a computer used by many users to perform a specific task, such as running network or Internet applications. Servers typically have large memory and storage capacities, along with fast and efficient communications abilities. A Web server handles Internet traffic and communications. An Internet caching server stores Web sites that a company uses frequently. An enterprise server stores and provides access to programs that meet the needs of an entire organization. A file server stores and coordinates program and data files. A transaction server processes business transactions. Server systems consist of multiuser computers, including supercomputers, mainframes, and other servers. Often an organization will house a large number of servers in the same room where the access to the machines can be controlled and authorized support personnel can more easily manage and maintain them from this single location. Such a facility is called a *server farm*.

Servers offer great **scalability**, the ability to increase the processing capability of a computer system so that it can handle more users, more data, or more transactions in a given period. Scalability is increased by adding more, or more powerful, processors. *Scaling up* adds more powerful processors, and *scaling out* adds many more equal (or even less powerful) processors to increase the total data-processing capacity.

A **blade server** houses many computer motherboards that include one or more processors, computer memory, computer storage, and computer network connections. These all share a common power supply and air-cooling source within a single chassis. By placing many blades into a single chassis, and then mounting multiple chassis in a single rack, the blade server is more powerful, but less expensive than traditional systems based on mainframes or server farms of individual computers. In addition, the blade server approach takes much less physcial space than traditional server farms.[34] The New York-based IT unit of Sumitomo Mitsui Banking Corporation consolidated 149 of its physical servers into 14 blade servers to save the company millions of dollars over three years by saving floor space, reducing the labor required to operate and maintain its servers, and cutting hardware costs.[35]

workstation
A more powerful personal computer that is used for technical computing, such as engineering, but still fits on a desktop.

server
A computer designed for a specific task, such as network or Internet applications.

scalability
The ability to increase the capability of a computer system to process more transactions in a given period by adding more, or more powerful, processors.

blade server
A server that houses many individual computer motherboards that include one or more processors, computer memory, computer storage, and computer network connections.

The Dell Power Edge 1855 Chassis can hold up to 10 blade servers.

(Source: Courtesy of Dell Inc.)

Mainframe Computers

A **mainframe computer** is a large, powerful computer shared by dozens or even hundreds of concurrent users connected to the machine over a network. The mainframe computer must reside in a data center with special heating, ventilating, and air-conditioning (HVAC) equipment to control temperature, humidity, and dust levels. In addition, most mainframes are kept in a secure data center with limited access to the room. The construction and maintenance of a controlled-access room with HVAC can add hundreds of thousands of dollars to the cost of owning and operating a mainframe computer.

The role of the mainframe is undergoing some remarkable changes as lower-cost, single-user computers become increasingly powerful. Many computer jobs that used to run on

mainframe computer
A large, powerful computer often shared by hundreds of concurrent users connected to the machine via terminals.

Mainframe computers have been the workhorses of corporate computing for more than 50 years. They can support hundreds of users simultaneously and handle all of the core functions of a corporation.

(Source: Courtesy of IBM Corporation.)

mainframe computers have migrated onto these smaller, less-expensive computers. This information-processing migration is called *computer downsizing*. One company that is using computer downsizing to its advantage is Starwood Hotels & Resorts Worldwide, which owns the Sheraton, Westin, W, Le Meredien, St. Regis, and Four Points hotel chains. Starwood is downsizing its reservation system from an aging mainframe to a number of servers. The firm can now improve the performance of the reservation system and add new features as necessary.[36]

The new role of the mainframe is as a large information-processing and data-storage utility for a corporation—running jobs too large for other computers, storing files and databases too large to be stored elsewhere, and storing backups of files and databases created elsewhere. The mainframe can handle the millions of daily transactions associated with airline, automobile, and hotel/motel reservation systems. It can process the tens of thousands of daily queries necessary to provide data to decision support systems. Its massive storage and input/output capabilities enable it to play the role of a video computer, providing full-motion video to multiple, concurrent users.

Supercomputers

supercomputers
The most powerful computer systems with the fastest processing speeds.

Supercomputers are the most powerful computers with the fastest processing speed and highest performance. They are *special-purpose machines* designed for applications that require extensive and rapid computational capabilities. Originally, supercomputers were used primarily by government agencies to perform the high-speed number crunching needed in weather forecasting and military applications. With recent reductions in the cost of these machines, they are now used more broadly for commercial purposes. For example, golf club maker Ping, Inc., uses a Cray supercomputer to run simulations of golf club designs and reduce the development time from weeks to days.[37]

Selecting and Upgrading Computer Systems

Computer systems are often upgraded by installing additional memory, faster processors, more hard disk storage, or various input and output devices to support new or changing business needs.

Given typical failure rates, most organizations can safely plan to keep a desktop computer four or more years. However, if the system fails after the standard three-year warranty period, it might be more cost effective to replace it with an upgraded model rather than fixing it. Companies plan to replace notebooks, which face rougher handling by users, every three or

IBM's Blue Gene/L System at the Lawrence Livermore National Laboratory is the fastest supercomputer in the world and can perform 136.8 trillion floating-point operations per second.

(Source: Courtesy of IBM Corporation.)

four years. As they replace workstations and notebooks, organizations should consider rolling over the most capable, but older personal computers to less-demanding users. Companies are sometimes forced to upgrade otherwise usable machines because of the memory and CPU requirements of newer versions of resource-intensive software.

Many large corporations set their own internal computer standards by selecting specific computer configurations from a small set of manufacturers. The goal is to reduce hardware support costs and increase the organization's flexibility. Business units that adopt different hardware complicate future corporate IS projects. For example, installing software is much easier if similar equipment from the same manufacturer is involved, rather than new and different systems at each installation site.

When upgrading to new computer hardware, whether it be for a single individual or for a large, multinational corporation, you must dispose of the old equipment properly. The National Recycling Coalition estimates nearly 500 million personal computers will have become obsolete in the period from 1997 to 2007. As a result, the EPA has tagged e-waste as the fastest-growing stream of waste in the United States. A typical CRT monitor contains 3 to 9 pounds of lead. Printed circuit boards contain beryllium, cadmium, flame retardants, and other compounds that can contaminate the air and groundwater, exposing people to carcinogens and other toxins. Unfortunately, some people and organizations do not realize the potential for harm from improper disposal of their computer hardware, nor have they budgeted for the costs associated with proper disposal. Three major computer hardware vendors—Dell, IBM, and Hewlett-Packard—offer disposal programs for any computer hardware, regardless of brand. They follow environment-friendly recycling practices and do not export the waste. Some reputable vendors offer e-waste disposal and recycling services. If the computer hardware still has a useful life, it can be donated to schools or charitable organizations.

Hard Drive Considerations

The optimal hard drive for a computer depends on several considerations. Because its main role is to provide long-term data storage, capacity, speed, and media capabilities are key features. Today's business software and large video, audio, and graphics files require lots of storage, so a hard drive capacity of at least 100 GB is recommended. Other considerations are access speed (look for 10 milliseconds or less) and hard drive cache size. Many single-user computers employ fans to help cool the internal components. These fans can cause the case to resonate and make the system seem loud. It is wise to select a system with fans, hard drive, and power supply that are classified as "quiet."

Main Memory Considerations

Main memory stores software code, while the processor reads and executes the code. Having more RAM means you can run software faster. Systems with 1 GB of RAM are well suited to take advantage of today's advanced personal productivity software (word processing, spreadsheet, graphics, and database) and multimedia programs.

As discussed earlier, your system's processor, main memory, and cache memory depend heavily on each other to achieve optimal system functionality. The original manufacturer of your computer considers this interdependency when designing and choosing the parts for the system. If you plan to upgrade your system's main memory beyond 512 MB, you should consult your supplier to understand the trade-off between the amount of main memory, the size of cache memory, and system performance.

Printer Considerations

As mentioned earlier, laser printers and inkjet printers are the two primary types of printers, and the differences between the two are diminishing. Although most inkjet printers are color and laser printers are monochrome, there are also color laser printers. All produce sharp images, with resolutions of 600×600 dots per inch (dpi) to 1,800×1,800 dpi now common. The major differences are in price, color, and speed.

You should consider two cost factors when purchasing a printer. The first is the cost of the printer. Prices for laser printers range from $200 to over $2,000, while prices for inkjet printers range from $50 to $400. Also consider operating cost. Laser printers can handle larger volumes of printing with a longer life for the ink/toner products, giving them a much lower operating cost than inkjet printers. Laser printers typically have operating costs from $.01 to $.04 per page for black-and-white pages and $.08 to $.18 for color pages. Inkjet printers can have operating costs from $0.03 to $0.08 per page for black-and-white pages and $0.10 to $0.20 for color pages. The cost for printing photos on special paper can exceed $.50 per page for either type of printer because of the high cost of the photo-quality paper.

DVD Burners

CD burners have been supplanted by DVD burners that can burn both DVDs and CDs. A DVD burner that mounts inside a computer costs under $200. Making sure that your DVD player will handle the discs you create is the most difficult part of choosing a DVD writer. Most players can handle DVD-R and DVD+R, but not all work with the various rewritable formats. Check the DVD player manufacturer's Web site to be sure. When comparing DVD writers, you'll see numbers such as 16X/8X/16X. These are the numbers for one-time recordable (-R or +R), rewritable (-RW or +RW), and read (ROM) formats. 1X translates to about 1.3 MB per second.

As mentioned throughout this chapter, you should select a computer system's components and architecture to support fundamental objectives, current business processes, and future needs of the organization and information system. Each computer system component—processing, memory, storage, input, and output devices—has a critical role in the successful operation of the computer system, the information system, and the organization. A thorough understanding of the broader system goals and the characteristics of the hardware as they relate to these goals will be an important guide for the future IS professional or business user.

SUMMARY

Principle

Assembling an effective, efficient set of computer hardware devices requires understanding their role in supporting the underlying information systems and the needs of the organization. The computer hardware objectives are subordinate to, but supportive of, the information systems and the needs of the organization.

Hardware includes any machinery (often using digital circuitry) that assists with the input, processing, and output activities of a computer-based information system (CBIS). A computer system is an integrated assembly of physical devices with at least one central processing mechanism; it inputs, processes, stores, and outputs data and information.

You should select and organize hardware to effectively and efficiently attain computer system objectives. These objectives should, in turn, support IS objectives and organizational goals. Balancing specific computer system objectives in terms of cost, control, and complexity should guide selection.

Processing is performed by cooperation between the central processing unit (CPU) and memory. The CPU has three main components: the arithmetic/logic unit (ALU), the control unit, and the register areas. The ALU performs calculations and logical comparisons. The control unit accesses and decodes instructions and coordinates data flow. The registers are temporary holding areas for instructions to be executed by the CPU.

Principle

When selecting computer hardware, you must consider the current and future needs of the information systems and the organization. Your choice of a hardware device should always allow for later improvements to meet evolving organizational needs.

Instructions are executed in a two-phase process that includes the instruction phase and the execution phase. The completion of this two-phase process is a machine cycle.

Computer system processing speed is also affected by clock speed, which is measured in megahertz (MHz). As the clock speed of the CPU increases, heat is generated that can corrupt the data and instructions the computer is trying to process.

Moore's Law is a hypothesis stating that the number of transistors on a single chip will double every 18 months. This hypothesis has been accurate since it was introduced in 1970.

Processing speed is also limited by physical constraints, such as the distance between circuitry points and circuitry materials. Advances in superconductive metals, optical processors, strained silicon, carbon nanotubes, and extreme miniaturization will result in faster CPUs.

Primary storage, or memory, provides working storage for program instructions and data to be processed and provides them to the CPU. Storage capacity is measured in bytes. A common form of memory is random access memory (RAM). RAM is volatile—loss of power to the computer erases its contents and comes in many different varieties, including EDO, dynamic RAM (DRAM), and synchronous DRAM (SDRAM).

Read-only memory (ROM) is nonvolatile and contains permanent program instructions for execution by the CPU. Other nonvolatile memory types include programmable read-only memory (PROM) and erasable programmable read-only memory (EPROM). Cache memory is a type of high-speed memory that CPUs can access more rapidly than RAM.

Processing that uses several processing units is called *multiprocessing*, of which there are many forms. A coprocessor executes one type of instruction while the CPU works on others. A multicore microprocessor is one that combines two or more independent processors into a single computer so they can share the workload. Parallel computing is the simultaneous execution of the same task on multiple processors to obtain results faster. Massively parallel processing involves linking many processors to work together to solve complex problems. Single instruction/multiple data (SIMD) parallel processors all execute the same instruction on many data values simultaneously. Multiple instruction/multiple data (MIMD) parallel processors all execute different instructions. Grid computing is the use of a collection of computers, often owned by multiple individuals or organizations, to work in a coordinated manner to solve a common problem.

Computer systems can store larger amounts of data and instructions in secondary storage, which is less volatile and has greater capacity than memory. The primary characteristics of secondary storage media and devices include access method, capacity, portability, and cost. Storage media can implement either sequential access or direct access. Sequential access requires data to be read or written in sequence. Direct access means that data can be located and retrieved directly from any location on the media.

Common forms of secondary storage include magnetic tape, magnetic disk, virtual tape, optical disc, digital video disc (DVD), holographic versatile disc (HVD), memory cards, and flash memory. Redundant array of independent/inexpensive disks (RAID) is a method of storing data that generates extra bits of data from existing data, allowing the system to more easily recover data in the event of a hardware failure. Network-attached storage (NAS) and storage area networks (SAN) are alternative forms of data storage that enable an organization to share data resources among a much larger number of computers and users for improved storage efficiency and greater cost-effectiveness. Software tools known as policy-based storage management automate storage using previously defined policies. The overall trend in secondary storage is toward direct access methods, higher capacity, increased portability, and automated storage management. Interest in renting space on massive storage devices is increasing.

Input and output devices allow users to provide data and instructions to the computer for processing and allow subsequent storage and output. These devices are part of a user interface through which human beings interact with computer systems. Input and output devices vary widely, but they share common characteristics of speed and functionality.

Data is thus placed in a computer system in a two-stage process: Data entry converts human-readable data into machine-readable form; data input then transfers it to the computer. Online data entry and input immediately convert and transfer data from devices to the computer system. Source data automation involves automating data entry and input so that data is captured close to its source and in a form that can be input directly to the computer.

Computer input devices include a keyboard, a mouse, speech recognition, digital cameras, terminals, scanners, optical data readers, magnetic ink character recognition devices, magnetic stripe cards, point-of-sale devices, automated teller machines, pen input devices, touch-sensitive screens, bar-code scanners, and Radio Frequency Identification tags.

Output devices provide information in different forms, from hard copy to sound to digital format. Display monitors are standard output devices; monitor quality is determined by size, color, and resolution. Liquid crystal display and organic light-emitting diode technology is enabling improvements in the resolution and size of computer monitors. Other output devices include printers, plotters, and computer output microfilm. Printers are popular hard-copy output devices whose quality is measured by speed and resolution. Plotters output hard copy for general design work. MP3 is a standard format for compressing the amount of data it takes to accurately represent sound. Numerous music devices based on the compressed music are available.

Computer systems are generally divided into two categories: single user and multiple users. Single-user systems include handheld, portable, thin client, desktop, and workstation computers. Handheld computers provide ease of portability because of their small size. They include palmtop computers, personal digital assistants, and smartphones. A variety of portable computers are available, including the laptop, notebook, subnotebook, and tablet computer. The thin client is a diskless, inexpensive computer used for accessing server-based applications and the Internet. Desktop computers are relatively small, inexpensive computer systems that are highly versatile. Workstations are advanced PCs with greater memory, processing, and graphics abilities.

Multiuser systems include servers, blade servers, mainframes, and supercomputers. A computer server is a computer designed for a specific task, such as network or Internet applications. Servers typically have large memory and storage capacities, along with fast and efficient communications abilities. A blade server houses a number of individual computer motherboards that include one or more processors, computer memory, computer storage, and computer network connections. These all share a common power supply and air-cooling source within a single chassis. Mainframe computers have greater processing capabilities, while supercomputers are extremely fast computers used to solve the most intensive computing problems.

Computer systems can be upgraded by changing hard drives, memory, printers, DVD burners, and other devices. Care must be taken in the disposal of obsolete systems because computer hardware contains many potentially harmful metals and chemicals.

CHAPTER 3: SELF-ASSESSMENT TEST

Assembling an effective, efficient computer system requires understanding its relationship to the information system and the organization. The computer system objectives are subordinate to, but supportive of, the information system and the needs of the organization.

1. Organizations that don't invest in computer hardware might find they are at a competitive disadvantage. True or False?

2. The computer hardware that most nonprofit organizations choose is virtually identical. True or False?

3. The overriding consideration for a business making hardware decisions should be how the hardware can support the objectives of the information system and _____.

4. Which represents a larger amount of data—a terabyte or a gigabyte?

5. Which of the following components performs mathematical calculations and makes logical comparisons?
    a. control unit
    b. register
    c. ALU
    d. main memory

6. Executing an instruction by the CPU involves two phases: the instruction phase and the _____ phase.

7. _____ involves capturing and editing data when the data is originally created and in a form that can be directly input to a computer, thus ensuring accuracy and timeliness.

When selecting computer hardware, you must consider the current and future needs of the information systems and the organization. Your choice of a hardware device should always allow for later improvements to meet evolving organizational needs.

8. A form of parallel processing in which a number of computers, often owned by multiple individuals or organizations, work in a coordinated manner to solve a common problem is _____.
   a. massively parallel processing
   b. multicore processing
   c. grid computing
   d. coprocessing

9. Three fundamental strategies for providing data storage are _____.

   a. expandable, nonexpandable, and static
   b. attached storage, network-attached storage, and storage area networks
   c. sequential, direct, indirect
   d. hard drive, CD-ROM, DVD

10. As the clock speed of the CPU increases, what problem occurs?

11. Which of the following technologies employs a microchip to broadcast its unique identifier and location to receivers?
   a. Radio Frequency Identification
   b. bar-code scanning
   c. biometrics
   d. GPS

CHAPTER 3: SELF-ASSESSMENT TEST ANSWERS

(1) True (2) False (3) the goals of the organization (4) terabyte (5) c (6) execution (7) Source data automation (8) c (9) b (10) Heat is generated that can corrupt data and instructions and cause the computer to behave erratically. (11) a

REVIEW QUESTIONS

1. When determining the appropriate hardware components of a new information system, what role must the user play?
2. What is a blade server? What advantages does it offer over an ordinary server?
3. Identify three basic characteristics of RAM and ROM.
4. What is RFID technology? Identify three practical uses for this technology.
5. Identify at least two ways businesses can take advantage of digital photography technology.
6. Give three practical examples of using speech-recognition devices.
7. What is the difference between data entry and data input?
8. What is RAID storage technology?
9. What is the difference between SIMD and MIMD parallel processing?
10. Explain the two-phase process for executing instructions.

11. Why are the components of all information systems described as interdependent?
12. Identify the three components of the CPU and explain the role of each.
13. What is the difference between sequential and direct access of data?
14. Identify several types of secondary storage media in terms of access method, capacity, portability, and cost per GB of storage.
15. Identify and briefly describe the various classes of personal computers.
16. What is the difference between cache memory and main memory?
17. What is source data automation?
18. What is the overall trend in secondary storage devices?

DISCUSSION QUESTIONS

1. Briefly describe how RFID technology works. Why is RFID technology being used heavily in the pharmaceutical industry?

2. Briefly outline some issues and associated solutions for discarding old computer hardware.

3. Imagine that you are the business manager for your university. What type of computer would you recommend for broad deployment in the university's computer labs—a standard desktop personal computer or a thin client? Why?

4. What is a dual-core processor? What advantages does it offer its user?

5. Assuming that cost was no object, identify and briefly describe the features of your ideal computer monitor.

6. Briefly describe Moore's Law. What are the implications of this? Are there any practical limitations to Moore's Law?

7. Identify and briefly describe the three fundamental approaches to data storage.

8. Describe how you would select the best DVD burner to meet your needs.

9. Briefly discuss several data-storage issues that face the modern organization.

10. Briefly explain how computer-based navigation systems work. What are some of the options and features of such systems? Would you like to have such a system in your car? Why or why not?

11. If cost were not an issue, describe the characteristics of your ideal computer. What would you use it for? Would you choose a handheld, portable, desktop, or workstation computer? Why?

PROBLEM-SOLVING EXERCISES

1. Use word processing software to document what your needs are as a computer user and your justification for selecting either a desktop or laptop computer. Find a Web site that allows you to order and customize a computer and select those options that meet your needs in a cost-effective manner. Assume that you have a budget of $1500. Enter the computer specifications into an Excel spreadsheet that you cut and paste into the document defining your needs. E-mail the document to your instructor.

2. Research the latest color laser printers. Develop a spreadsheet that compares the features, the initial purchase price, and the ongoing operating costs for three printers. Now do the same for three ink-jet printers. Write a brief memo on which printer you would choose and why. Cut and paste the spreadsheet into a document.

3. Enter data from Figure 3.3 (Moore's Law) into a spreadsheet program. Use the forecasting capabilities of the program to estimate the number of transistors on a chip for the next five years and draw a chart depicting this. Are there basic limitations that can keep this forecast from being met? If so, what are they?

TEAM ACTIVITIES

1. With one or two of your classmates, visit a retail store that employs Radio Frequency Identification chips to track inventory. Interview an employee involved in inventory control and document the advantages and disadvantages they see in this technology.

2. With two or three of your classmates, visit the computer operations site of your college or university. Interview members of the operations group and identify the primary processing and data-storage devices used. Define the role that each plays.

WEB EXERCISES

1. Do research on the Web to document the current state of computer hardware disposal. What are some of the issues? What solutions are there to this problem? Write a brief report summarizing your findings.

2. Do research on the Web to identify the current status of the use of Radio Frequency Identification chips in the pharmaceutical industry. Write a brief report summarizing your findings.

CAREER EXERCISES

1. Imagine that you are going to buy a single handheld device to improve your communications and organizational abilities. What tasks do you need it to perform? What features would you look for in this device? Visit a computer store or a consumer electronics store and see whether you can purchase such a device for under $400.

2. Your company's marketing department plans to acquire 150 new computers and monitors, plus several new printers. The marketing vice president has asked you to lead a three-person project team assigned to define user computer hardware needs. Who else (role, department) would you select to be a member of the team? How would you go about defining users' needs? Do you think that one hardware configuration (computer, monitor, printer) will meet everyone's needs? Should you define multiple configurations based on the needs of various classes of end user? What business justification can you define to substantiate this expenditure of roughly $300,000?

CASE STUDIES

Case One

Staples Unwires Point-of-Sale Terminals

Staples is the world's leading seller of office products with over $14.4 billion in sales annually. Throughout the United States and Canada, 1,491 Staples retail stores offer more than 7,000 office products. In its role as office product advisor, Staples feels obligated to implement best practices in its own business—to practice what it preaches. To do so, Staples must use the latest and greatest technology tools available to offer customers every convenience.

Staples analyzed the shopping experience in their stores to determine how it could be improved. One apparent problem was the bottleneck at checkout. Stores such as Wal-Mart and others have worked to address this problem with self-serve checkout lanes. Staples decided to address it in another way. Staples recognized that checkout clerks stay busy while store associates working the floor often had free time. Staples' goal was to use all store personnel to their fullest potential while eliminating the bottleneck at checkout.

Staples decided to use a hardware device from Fujitsu called the iPAD as a solution. The iPAD is a powerful handheld computer that uses an Intel Pentium M processor to provide mobile point-of-sale (POS) transaction processing. Weighing only 10 ounces, the iPAD includes wireless networking capabilities, a touch screen, a built in scanner for reading inventory data, and a magnetic stripe reader for processing credit cards. Using the iPAD, store associates can check out customers from any location in the store. In other words, the iPAD acts as an instant checkout station.

Besides checking out customers from store aisles, store associates can access a wealth of store inventory data from the corporate database. Associates can retrieve up-to-date corporate data without leaving the floor, providing them with more time for selling and contributing to higher revenues. For example, if an item isn't stocked on the shelf, the iPAD guides the associate through a standard procedure to satisfy the customer, saying "This item is out of stock but we have another item that can serve the same purpose, or we can order your item and have it delivered to your home within two days." The associate no longer has to hunt down a manager for solutions.

"The POS and handhelds are the tip of the iceberg," says George Lamson, director of application development for Staples. "Their performance brings the information and applications to the store associate and the customers. The iceberg itself is all the work we're doing to streamline the backend and the supply chain. It's the combination that's so powerful."

Staples contracted with Dallas-based Fujitsu Transaction Solutions, Inc., to handle the bulk of its in-store IT needs and life-cycle management. Fujitsu performs in-store POS conversions, placing iPADs in the hands of store associates and setting up PCs as POS terminals at the front of the store. Managers and store personnel can use the PCs to check out customers or access other store systems. Fujitsu also integrates Staples' in-store servers, and prepares and executes rapid-deployment store conversions and installations. In addition, Fujitsu offers a dispatch center that replaces or fixes system components within precise time allotments—four to eight hours in many cases—and a warehouse operation that helps Staples collect, sort, and dispose of obsolete IT equipment.

By turning over IT hardware management to Fujitsu, Staples can focus on business drivers rather than on in-store IT needs. By integrating hardware components throughout the organization, the flow of information through the supply chain—from suppliers to store associates—is becoming seamless.

Discussion Questions

1. What new hardware technologies made it possible for Staples to bring the cash register to the customer?
2. What benefits has Staples reaped by being one of the first to incorporate a mobile POS device?

Critical Thinking Questions

1. What benefits and risks do you think are involved in outsourcing hardware procurement and maintenance as Staples has done?
2. What considerations regarding processing, storage, and input/output specifications do you think were involved in Staples' choice of the iPAD device?

SOURCES: Intel Case Studies: "Staples Makes Buying Office Product Easy," Intel Web site, accessed February 12, 2006, *www.intel.com/business/casestudies/staples.pdf*. Staples Corporate Overview, Staples Web site, accessed February 12, 2006, *www.staples.com/sbd/content/about/media/overview.html*. Fujitsu Case Studies, "Fujitsu Provides Staples with Complete IT Lifecycle Management," Fujitsu Web site, accessed February 12, 2006, *www.fujitsu.com/us/casestudies/case_Staples.html*.

Case Two

Bank of New York: Mainframes or Distributed Systems

When organizations need to provide information services to thousands of employees, they traditionally have turned to mainframe computers—referred to as "big iron" by those in the business. The most popular mainframes are those manufactured by IBM such as the z series mainframes (*www.ibm.com/servers/eserver/zseries*). Over the past dozen years or so, this tradition has been challenged by the increasing use of smaller, less-powerful servers such as midrange servers and blade servers working together over a network to provide the services traditionally handled by mainframes. A system that divides business processes among network servers is referred to as a distributed system because the work is distributed over multiple servers.

Mainframes and distributed systems can handle many of the same organizational demands, though each has its strengths and weaknesses. Information systems specialists in many businesses must determine which is best for the varying needs of their organization.

At the Bank of New York, mainframes play a role in every transaction. Three-quarters of the transactions run on the mainframe and the rest depends on the mainframe in some way. Edward Mulligan, managing director of the technology services division of the bank, believes that the mainframe is the perfect solution for the bank's needs. However, he notes that the tide may be turning. "Most of the server solutions available today are morphing to become more like a mainframe." Many utilities that were once exclusive to mainframes are now provided by smaller servers. Increasing amounts of work at the bank is being done on Windows and UNIX servers.

The bank is discovering that distributed systems are more open and agile than mainframes. Application services can be deployed more rapidly on the distributed system than on the mainframe. Smaller servers are also less expensive to purchase. Another plus is that they are scalable: They can grow and shrink with the needs of the business by adding or removing servers.

At the same time, mainframe technologies are evolving to support many of the new techniques that have been designed for distributed systems. They support many of the most popular programming languages and can be partitioned into logical units to handle a variety of services. As mainframes become more like distributed systems, and distributed systems more like mainframes, it becomes difficult to determine which is better.

Mainframes provide the benefit of ease of management. Rather than maintaining dozens or hundreds of individual servers, mainframe operators have the entire system contained in one cabinet. Some system administrators find this approach to be more secure and stable than distributed systems.

Vendors promote mainframes as a useful means of consolidation, and distributed systems as conveniently scalable. An important consideration is the amount of requests the server can handle per second. This boils down to processor power and the ability to manage input and output. Most experts agree that mainframes will continue to be the preferred system for very large complex systems. However, the less-demanding environments—applications of up to 1,000 MIPS, where most mainframe applications fall, are slowly giving way to smaller servers. Dale Vecchio, an analyst at Garter, Inc., believes that cost is a big factor. "IBM mainframes are going to become marginalized to the high end if IBM can't significantly reduce the cost," he stated. IBM is well aware of this and through innovative technologies and improved manufacturing techniques has been able to reduce its mainframe prices by 20 percent each year. Still, many believe that mainframes will continue to decline in use.

Mainframe use will probably continue to decline gradually because it is expensive to switch from mainframe to distributed system. Considering that many mainframe systems have been around for over 40 years, they have a deep and complex legacy. An estimated trillion dollars worth of legacy code is tightly tied to the mainframe's hardware and operating system. Edward Mulligan at the Bank of New York estimates that he is responsible for tens of millions of lines of mainframe programming code. He believes that it would be impossible to transfer all of that functionality to a distributed system in his lifetime.

Considering the pros and cons of both mainframes and distributed systems, you can see why many businesses today maintain information systems that employ both technologies. Using mixed systems allows IS professionals to take advantage of the best of all worlds while eliminating many of the drawbacks.

Discussion Questions

1. List the strengths and weaknesses of distributed systems and mainframes.
2. Why would it be easier and less expensive for a new company to base its systems on a distributed system, compared to an established company that has been using a mainframe?

Critical Thinking Questions

1. How might a 40-year-old legacy mainframe system encumber an established company that is trying to compete with new competitors who use the latest technologies? If you were the CEO of the established company, what types of strategies might you employ?
2. What benefits would a distributed system provide for a multinational company?

SOURCES: Mitchell, Robert L., "Morphing the Mainframe," *Computerworld*, January 30, 2006, *www.computerworld.com*. Thibodeau, Patrick, "Mixed Systems Remain Common in Data Centers," *Computerworld*, December 12, 2005, *www.computerworld.com*. IBM Servers Web site, accessed February 16, 2006, *www.ibm.com/products/shop/servers/us*. Bank of New York Web site, accessed February 16, 2006, *www.bankofny.com*.

Questions for Web Case

See the Web site for this book to read about the Whitmann Price Consulting case for this chapter. Following are questions concerning this Web case.

Whitmann Price Consulting: Choosing Hardware

Discussion Questions

1. What considerations led Josh and Sandy to lean toward a BlackBerry as the handheld device on which to run the AMCIS?
2. How did Josh and Sandy organize their hardware considerations for the new system?

Critical Thinking Questions

1. What role does system compatibility play in Josh and Sandy's decision?
2. What device(s) might have been chosen if the system requirements called for a 12-inch display, the ability to take handwritten notes, and the ability to communicate through voice, text, and video?

NOTES

Sources for the opening vignette: "Siebel Systems Speeds Performance and Enhances Productivity with IBM eServer BladeCenter," IBM Success Stories, February 7, 2006, *www.ibm.com/us/*. "New IBM Blade Computers Speed Business Data Up to Ten Times Faster," Market Wire, February 8, 2006, *www.lexis-nexis.com*. "Siebel Systems at a Glance," Siebel Web site, accessed February 11, 2006, *www.siebel.com/crm-company/index.shtm*.

1 Weiss, Todd R., "U.S. Navy Gets Shipshape with Sun Ray Thin Clients," *Computerworld*, December 16, 2005.
2 Rosencrance, Linda, "IT Gives Airlines a Lift," *Computerworld*, September 5, 2005.
3 Thibodeau, Patrick, "Cool Chips Offer Some Help to Data Centers," *Computerworld*, December 12, 2005.
4 Krazit, Tom, "Intel Readying New Extreme Edition Processor," *Computerworld*, November 17, 2005.
5 Kirk, Jeremy, "Intel to Build $3.5B Chip Plant in Israel," *Computerworld*, December 1, 2005.
6 "Intel Dual-Core Processors," Intel Web site, *www.intel.com/technology/computing/dual-core/*, accessed January 13, 2006.
7 Krazit, Tom, "Intel Alters Server Processor Road Map," *Computerworld*, October 25, 2005.
8 Thibodeau, Patrick, "Chevron Adopts Opteron for High Performance Computing," *Computerworld*, December 13, 2005, *www.computerworld.com*.
9 Krazit, Tom, "Update: Apple Unveils Intel-Based Laptop, iMac," *Computerworld*, January 10, 2006.
10 Jones, K.C., "Supercomputer Fights AIDS," *InformationWeek*, November 21, 2005.
11 Pratt, Mary K., "Case Study: Acxiom Corporation's Homegrown Grid," *Computerworld*, August 8, 2005.
12 Keizer, Gregg, "Dell Adds Mirroring to Desktops," *InformationWeek*, October 26, 2005.

13 Marlin, Steve, "MAID Service at the Merc," *InformationWeek*, October 17, 2005.
14 Williams, Martyn, "Analysis: Fuzzy HD Picture to Come into Focus at CES," *Computerworld*, January 3, 2006.
15 Mearian, Lucas, "Japanese Holographic Storage Firm to Ship 200GB Drives in '06," *Computerworld*, October 24, 2005.
16 Krazit, Tom, "Intel's Latest Flash Chip Runs Faster with More Storage," *Computerworld*, November 17, 2005.
17 Jacobi, Jon L., "Flash Memory to Speed Up Hard Drives," *PCWorld Magazine*, September 2005, *www.pcmagazine.com*.
18 Marlin, Steve, "IBM Storage Supports Hospital's Move to Digital Health Records," *InformationWeek*, September 19, 2005.
19 Greenemeier, Larry, "SANs Are in the Navy Now," *InformationWeek*, June 13, 2005.
20 Hoffman, Thomas, "Speech Recognition Powers Utility's Customer Service," *Computerworld*, September 12, 2005.
21 Information from Deal Time Web site, *www.dealtime.com*, accessed February 4, 2006.
22 Vijayan, Jaikumar, "New York Signs Pact to Boost Transit Security," *Computerworld*, August 29, 2005.
23 Wailgum, Thomas, "The Fast Lane for Speeding Tickets," *CIO Magazine*, January 15, 2006.
24 Weiss, Todd R., "Privacy Groups Question RFID Use in Medicine Tracking," *Computerworld*, October 14, 2005.
25 Marrin, John, "Review: Six 19-Inch LCD Monitors," *InformationWeek*, June 8, 2005.
26 "Dell UltraSharp Widescreen 3007WFP," *PC Magazine*, January 5, 2006, *www.pcmag.com*.
27 "Sharp to Mass Produce World's First LCD to Simultaneously Display Different Information in Right and Left Viewing Directions," *Sharp Press Release*, July 14, 2005, *http://sharp-world.com/corporate/news*.

28 Kallender, Paul, "Sharp Claims World's Brightest Screen for New Notebooks," *Computerworld*, January 18, 2005.

29 Nystedt, Dan, "Top LCD Maker Bets Research on LCD Backlights," *Computerworld*, September 21, 2005.

30 Krazit, Tom, "Apple Unveils Video iPod, Strikes Deal with ABC," *Computerworld*, October 12, 2005.

31 Havenstein, Heather, "Health Care System Turns to IT for Patient Care Plans," *Computerworld*, August 8, 2005.

32 Sliwa, Carol, "Retailers Make Trial Offers on Shopping Technologies," *Computerworld*, January 17, 2005.

33 Fontana, John, "Fireman's Fund Rescues Remote Users," *Computerworld*, December 19, 2005.

34 Dunn, Darrell, "Businesses Are Switching to Blade Servers," *InformationWeek*, January 13, 2006.

35 Mearian, Lucas, "Japanese Bank Expects Server Consolidation to Save It Millions," *Computerworld*, October 10, 2005.

36 Lai, Eric, "Starwood Checks in with Object Database for Reservations," *Computerworld*, January 16, 2006.

37 Thibodeau, Patrick, "Smaller Companies Eye Supercomputing," *Computerworld*, August 2, 2006.

CHAPTER
• 4 •

Software: Systems and Application Software

PRINCIPLES	LEARNING OBJECTIVES
▪ Systems and application software are critical in helping individuals and organizations achieve their goals.	▪ Identify and briefly describe the functions of the two basic kinds of software. ▪ Outline the role of the operating system and identify the features of several popular operating systems.
▪ Do not develop proprietary application software unless doing so will meet a compelling business need that can provide a competitive advantage.	▪ Discuss how application software can support personal, workgroup, and enterprise business objectives. ▪ Identify three basic approaches to developing application software and discuss the pros and cons of each.
▪ Choose a programming language whose functional characteristics are appropriate for the task at hand, considering the skills and experience of the programming staff.	▪ Outline the overall evolution and importance of programming languages and clearly differentiate among the generations of programming languages.
▪ The software industry continues to undergo constant change; users need to be aware of recent trends and issues to be effective in their business and personal life.	▪ Identify several key software issues and trends that have an impact on organizations and individuals.

Information Systems in the Global Economy »
El-Al Airlines, Israel

El-Al Adopts Software Solution for Scheduling Complexity

El-Al is Israel's national airline and has grown into a prestigious international carrier with 77 sales offices worldwide. As with most airlines, El-Al's second largest operating expense is the cost of flight crews. An airline can cut expenses by efficiently scheduling flight crews.

Although scheduling flights and crews might seem fairly straightforward, it is not an easy task. The challenge is in the amount of variables to consider. The efficient scheduling of flights involves much more than planning which planes fly where and when. Round trips for each of the crew members must also be considered. This involves many considerations, including legal requirements for rest and duty, equipment ratings, home bases, overtime, and crew preferences. Many of these variables directly affect cost.

El-Al required an efficient way to build schedules that would reduce the variable costs of flight crews as much as possible. The complexity of the variables made it impossible to create the schedules with simple software tools. El-Al turned to professionals at IBM for a software solution.

The software solution had to satisfy all legal criteria, completely cover the flight schedules, reduce costs, optimize resources, and be easy to use. Working with IBM technicians, El-Al divided the problem into its three primary challenges:

- **Pairing.** Create the best pairing sequences of duty and rest periods for crew members.
- **Assignment.** Assign crew members to the pairings.
- **Data management.** Manage the rules and data required to solve the problem.

The programming team then created code modules, or objects, to address each of the three challenges.

The software they created uses advanced mathematical models and algorithms that work through the billions of possible scheduling combinations to find a solution that meets all the criteria. The processing is so intensive that it requires a parallel-processing computer system to run quickly and efficiently.

El-Al has seen impressive results since implementing the new scheduling system. They noted a 5–8 percent reduction in variable crew-related costs and an increase of 4–7 percent in crew usage. Also, scheduling itself takes much less time with the new easy-to-use software.

The positive results have been noticeable to others in the industry as well. El Al was recently ranked as one of the world's three most efficient air carriers by the International Air Transport Association (IATA).

As you read this chapter, consider the following:

- In what ways can software assist an organization to run more smoothly and efficiently and develop new and valuable consumer services and products?
- When it comes to acquiring new software, what options are available to businesses?

Why Learn About Software?

Software is indispensable for any computer system and the people using it. In this chapter, you will learn about systems and application software. Without systems software, computers would not be able to input data from a keyboard, make calculations, or print results. Application software is the key to helping you achieve your career goals. Sales representatives use software to enter sales orders and help their customers get what they want. Stock and bond traders use software to make split-second decisions involving millions of dollars. Scientists use software to analyze the threat of global warming. Regardless of your job, you most likely will use software to help you advance in your career and earn higher wages. Today, most organizations could not function without accounting software to print payroll checks, enter sales orders, and send out bills. You can also use software to help you prepare your personal income taxes, keep a budget, and play amazing games. Software can truly advance your career and enrich your life. We begin with an overview of software.

Software has a profound impact on individuals and organizations. It can make the difference between profits and losses, survival and bankruptcy. As Figure 4.1 shows, companies recognize this impact, and spend more on software than on computer hardware.

Figure 4.1

The Importance of Software in Business

Since the 1950s, businesses have greatly increased their expenditures on software compared with hardware.

Hardware

Software

AN OVERVIEW OF SOFTWARE

computer programs
Sequences of instructions for the computer.

documentation
The text that describes the program functions to help the user operate the computer system.

As you learned in Chapter 1, software consists of computer programs that control the workings of computer hardware. **Computer programs** are sequences of instructions for the computer. **Documentation** describes the program functions to help the user operate the computer system. The program displays some documentation on screen, while other forms appear in external resources, such as printed manuals. People using commercially available software are usually asked to read and agree to End-User License Agreements (EULAs).[1] After reading the EULA, you normally have to click an "I agree" button before you can use the software, which can be one of two basic types: systems software and application software.

Systems Software

Systems software is the set of programs that coordinates the activities and functions of the hardware and other programs throughout the computer system. Each type of systems software is designed for a specific CPU and class of hardware. The combination of a hardware configuration and systems software is known as a computer system platform.

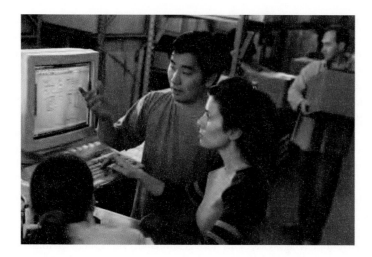

Application software has the greatest potential to affect processes that add value to a business because it is designed for specific organizational activities and functions.

(Source: © Walter Hodges/CORBIS.)

Application Software

Application software consists of programs that help users solve particular computing problems. In most cases, application software resides on the computer's hard disk before it is brought into the computer's memory and run. Application software can also be stored on CDs, DVDs, and even flash or keychain storage devices that plug into a USB port.[2] Before a person, group, or enterprise decides on the best approach for acquiring application software, they should analyze their goals and needs carefully.

Lotus Notes is an application that enables a workgroup to schedule meetings and coordinate activities.

(Source: Courtesy of IBM Corporation.)

Supporting Individual, Group, and Organizational Goals

Every organization relies on the contributions of people, groups, and the entire enterprise to achieve its business objectives. Conversely, the organization also supports people, groups, and the entire enterprise with application software and information systems. One useful way of classifying the many potential uses of information systems is to identify the scope of the problems and opportunities an organization addresses. This scope is called the sphere of

personal sphere of influence
The sphere of influence that serves the needs of an individual user.

personal productivity software
The software that enables users to improve their personal effectiveness, increasing the amount of work they can do and its quality.

workgroup
Two or more people who work together to achieve a common goal.

workgroup sphere of influence
The sphere of influence that serves the needs of a workgroup.

enterprise sphere of influence
The sphere of influence that serves the needs of the firm in its interaction with its environment.

influence. For most companies, the spheres of influence are personal, workgroup, and enterprise. Table 4.1 shows how software can support individuals, workgroups, and enterprises.

Information systems that operate within the **personal sphere of influence** serve the needs of an individual user. These information systems help users improve their personal effectiveness, increasing the amount and quality of work they can do. Such software is often called **personal productivity software**. When two or more people work together to achieve a common goal, they form a **workgroup**. A workgroup might be a large, formal, permanent organizational entity such as a section or department or a temporary group formed to complete a specific project. An information system in the **workgroup sphere of influence** helps a workgroup attain their common goals. Users of such applications must be able to communicate, interact, and collaborate to be successful.

Information systems that operate within the **enterprise sphere of influence** support the firm in its interaction with its environment. The surrounding environment includes customers, suppliers, shareholders, competitors, special interest groups, the financial community, and government agencies. This means the enterprise sphere of influence includes business partners, such as suppliers that provide raw materials, retail companies that store and sell a company's products, and shipping companies that transport raw materials to the plant and finished goods to retail outlets.

Software	Personal	Workgroup	Enterprise
Systems software	Personal computer and workstation operating systems	Network operating systems	Midrange computer and mainframe operating systems
Application software	Word processing, spreadsheet, database, graphics	Electronic mail, group scheduling, shared work	General ledger, order entry, payroll, human resources

Table 4.1

Software Supporting Individuals, Workgroups, and Enterprises

Installing and Removing Software for PCs

Before you can use any type of software, it must be installed on a computer. Installing new software usually involves only a few setup steps (see Figure 4.2). Software for personal computers typically comes on CDs or is downloaded from the Web.

Most operating systems have an add/remove program feature to assist you in removing software. See Figure 4.3. Note that this feature does not work with all software nor does it always remove all elements of the software. Some utility software, such as Norton System-Works and McAfee QuickClean for Microsoft Windows PCs, can help eliminate unwanted elements of software you have removed. Trying to remove program files and icons manually can be a problem with some programs and operating systems, which could make it difficult to use or boot the computer. When possible, it is best to remove software using an add/remove software utility.

Figure 4.2

Installing New Software

Installing software usually involves reading a EULA, selecting a destination folder, beginning the setup, and waiting for the setup program to install the files.

Figure 4.3

Removing Software

Most operating systems have an add/remove program feature to assist you in removing software.

SYSTEMS SOFTWARE

Controlling the operations of computer hardware is one of the most critical functions of systems software. Systems software also supports the application programs' problem-solving capabilities. Types of systems software include operating systems, utility programs, and middleware.

Operating Systems

operating system (OS)
A set of computer programs that controls the computer hardware and acts as an interface with application programs.

An **operating system (OS)** is a set of programs that controls the computer hardware and acts as an interface with applications (see Figure 4.4). Operating systems can control one or more computers, or they can allow multiple users to interact with one computer. The various combinations of OSs, computers, and users include the following:

- **Single computer with a single user.** This system is commonly used in a personal computer or a handheld computer that allows one user at a time.
- **Single computer with multiple users.** This system is typical of larger, mainframe computers that can accommodate hundreds or thousands of people, all using the computer at the same time.
- **Multiple computers.** This system is typical of a network of computers, such as a home network with several computers attached or a large computer network with hundreds of computers attached around the world.
- **Special-purpose computers.** This system is typical of a number of special-purpose computers, such as those that control sophisticated military aircraft, the space shuttle, and some home appliances.

Figure 4.4

The Role of Systems Software

The role of the operating system and other systems software is as an interface or buffer between users and application software and hardware.

The OS, which plays a central role in the functioning of the complete computer system, is usually stored on disk. After you start, or "boot up," a computer system, portions of the OS are transferred to memory as they are needed. You can also boot a computer from a CD, DVD, or even a keychain or thumb drive that plugs into a USB port.[3] A storage device that contains some or all of the OS is often called a "rescue disk" because you can use it to start the computer if you have problems with the primary hard disk.

Some OSs for small computers use an "Instant On" feature that significantly reduces the time needed to boot a computer. The collection of programs that make up the OS performs a variety of activities, including the following:

- Performing common computer hardware functions
- Providing a user interface and input/output management
- Providing a degree of hardware independence
- Managing system memory
- Managing processing tasks
- Providing networking capability
- Controlling access to system resources
- Managing files

kernel
The heart of the operating system, which controls the most critical processes.

The **kernel**, as its name suggests, is the heart of the OS and controls the most critical processes. The kernel ties all of the OS components together and regulates other programs.

Common Hardware Functions

All applications must perform certain tasks, such as the following:

- Get input from the keyboard or another input device
- Retrieve data from disks

- Store data on disks
- Display information on a monitor or printer

Each of these tasks requires a detailed set of instructions. The OS converts a basic request into the set of detailed instructions that the hardware requires. In effect, the OS acts as an intermediary between the application and the hardware. The typical OS performs hundreds of such tasks, translating each into one or more instructions for the hardware. The OS notifies the user if input or output devices need attention, if an error has occurred, and if anything abnormal happens in the system.

User Interface and Input/Output Management

One of the most important functions of any OS is providing a **user interface**. A user interface allows people to access and command the computer system. The first user interfaces for mainframe and personal computer systems were command based. A **command-based user interface** requires you to give text commands to the computer to perform basic activities. For example, the command ERASE 00TAXRTN would cause the computer to erase a file called 00TAXRTN. RENAME and COPY are other examples of commands used to rename files and copy files from one location to another. Many operating systems that use a graphical user interface, discussed next, also have powerful command-based features.[4]

A **graphical user interface (GUI)** uses pictures (called icons) and menus displayed on screen to send commands to the computer system. Many people find that GUIs are easier to use because they intuitively grasp the functions. Today, the most widely used graphical user interface is Microsoft Windows. Alan Kay and others at Xerox PARC (Palo Alto Research Center, located in California) were pioneers in investigating the use of overlapping windows and icons as an interface. As the name suggests, Windows is based on the use of a window, or a portion of the display screen dedicated to a specific application. The screen can display several windows at once. GUIs have contributed greatly to the increased use of computers because users no longer need to know command-line syntax to accomplish tasks. See Figure 4.5.

user interface
The element of the operating system that allows you to access and command the computer system.

command-based user interface
A user interface that requires you to give text commands to the computer to perform basic activities.

graphical user interface (GUI)
An interface that uses icons and menus displayed on screen to send commands to the computer system.

Windows XP

Red Hat Linux

Mac OS X

Figure 4.5

Graphical User Interfaces

Windows XP, Red Hat Linux 8.0, and Mac OS X are examples of graphical user interfaces.

application program interface (API)

An interface that allows applications to make use of the operating system.

Hardware Independence

To run, applications request services from the OS through a defined **application program interface (API)**, as shown in Figure 4.6. Programmers can use APIs to create application software without having to understand the inner workings of the OS.

Suppose that a computer manufacturer designs new hardware that can operate much faster than before. If the same OS for which an application was developed can run on the new hardware, minimal (or no) changes are needed to the application to enable it to run on the new hardware. If APIs did not exist, the application developers might have to completely rewrite the application to take advantage of the new, faster hardware.

Figure 4.6

Application Program Interface Links Application Software to the Operating System

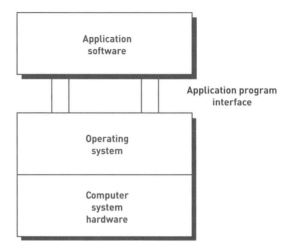

Memory Management

Another OS task is to control how memory is accessed and to maximize available memory and storage. Newer OSs typically manage memory better than older OSs. The memory-management feature of many OSs allows the computer to execute program instructions effectively and speed processing. One way to increase the performance of an old computer is to upgrade to a newer OS and increase the amount of memory.

Most OSs support virtual memory, which allocates space on the hard disk to supplement the immediate, functional memory capacity of RAM. Virtual memory works by swapping programs or parts of programs between memory and one or more disk devices—a concept called paging. This reduces CPU idle time and increases the number of jobs that can run in a given time span.

Processing Tasks

The task-management features of today's OSs manage all processing activities. Task management allocates computer resources to make the best use of each system's assets. Task-management software can permit one user to run several programs or tasks at the same time (multitasking) and allow several users to use the same computer at the same time (time-sharing).

An OS with multitasking capabilities allows a user to run more than one application at the same time. Without having to exit a program, you can work in one application, easily pop into another, and then jump back to the first program, picking up where you left off. Better still, while you're working in the *foreground* in one program, one or more other applications can be churning away, unseen, in the *background*, sorting a database, printing a document, or performing other lengthy operations that otherwise would monopolize your computer and leave you staring at the screen unable to perform other work. Multitasking can save users a considerable amount of time and effort.

Time-sharing allows more than one person to use a computer system at the same time. For example, 15 customer service representatives might be entering sales data into a computer system for a mail-order company at the same time. In another case, thousands of people might be simultaneously using an online computer service to get stock quotes and valuable business news.

The ability of the computer to handle an increasing number of concurrent users smoothly is called *scalability*. This feature is critical for systems expected to handle a large number of users such as a mainframe computer or a Web server. Because personal computer OSs usually are oriented toward single users, they do not need to manage multiple-user tasks often.

Networking Capability

The OS can provide features and capabilities that aid users in connecting to a computer network. For example, Apple computer users have built-in network access, and the Microsoft Windows OSs let users link to other devices and the Internet.

Access to System Resources and Security

Because computers often handle sensitive data that can be accessed over networks, the OS needs to provide a high level of security against unauthorized access to the users' data and programs. Typically, the OS establishes a logon procedure that requires users to enter an identification code and a matching password. If the identification code is invalid or if the password does not match the identification code, the user cannot gain access to the computer. The OS also requires that user passwords change frequently—such as every 20 to 40 days. If the user is successful in logging on to the system, the OS records who is using the system and for how long. The OS also reports any attempted breaches of security.

File Management

The OS manages files to ensure that files in secondary storage are available when needed and that they are protected from access by unauthorized users. Many computers support multiple users who store files on centrally located disks or tape drives. The OS keeps track of where each file is stored and who can access them. The OS must determine what to do if more than one user requests access to the same file at the same time. Even on stand-alone personal computers with only one user, file management is needed to track where files are located, what size they are, when they were created, and who created them.

Current Operating Systems

Early OSs were very basic. Recently, however, more advanced OSs have been developed, incorporating some features previously available only with mainframe OSs.[5] Table 4.2 classifies a few current OSs by sphere of influence.

Microsoft PC Operating Systems

Since a small company called Microsoft developed PC-DOS and MS-DOS to support the IBM personal computer introduced in the 1980s, personal computer OSs have steadily evolved. *PC-DOS* and *MS-DOS* had command-driven interfaces that were difficult to learn and use. Each new version of OS has improved the ease of use, processing capability, reliability, and ability to support new computer hardware devices.

Windows XP (XP reportedly stands for the positive *experience* that you will have with your personal computer) was released in fall 2001. Previous consumer versions of Windows were notably unstable and crashed frequently, requiring frustrating and time-consuming reboots. With XP, Microsoft sought to bring reliability to the consumer. Windows XP includes Microsoft Internet Explorer 6 browser software, which boasts improved security and reliability features, including a one-way firewall that blocks hacker invasions coming in from the Internet. Windows XP N is an alternative version of Windows for the European market.[6] This version of the operating system is available in English, French, German, and Italian. Windows XP Professional X64 takes advantage of computers with newer 64-bit capabilities.[7]

Microsoft Windows XP Media Center Edition incorporates additional multimedia features. Microsoft is also developing software and operating systems to deliver TV programs over cable or the Internet using the *Internet Protocol Television (IPTV)* standard.[8] Windows XP can be run in the *classic style*, which is similar to previous versions of Windows, or the

Personal	Workgroup	Enterprise
Windows Vista, Windows XP, Windows Mobile, and Windows Embedded	Windows NT Server	Windows NT Server
Mac OS	Windows 2003 Server	Windows 2003 Server
Mac OS X	Mac OS Server	Windows Advanced Server, Limited Edition
UNIX	UNIX	UNIX
Solaris	Solaris	Solaris
Linux	Linux	Linux
Red Hat Linux	Red Hat Linux	Red Hat Linux
Palm OS	Netware	
	IBM OS/390	IBM OS/390
	IBM z/OS	IBM z/OS
	HP MPE/iX	HP MPE/iX

Web style, which is similar to how Internet sites work, including letting you start programs or open folders with one mouse click. You can have Windows XP show small images of photos and documents, called *thumbnails*, in folders by clicking the Thumbnails on the folder's View menu (see Figure 4.7). Windows XP also has Windows Error Reporting (WER), which sends error reports to Microsoft when errors occur or the OS crashes. WER can also inform users of fixes or patches to avoid future problems.

Figure 4.7

Windows XP Thumbnails View

Windows XP can show small images of photos and documents, called *thumbnails*, in folders when a user selects Thumbnails view.

Vista is the most recent revision of the Windows OS.[9] Microsoft hopes that Vista is more secure and stable than previous operating systems.[10] The new operating system includes a number of new features. For example, Avalon is a tool to help developers easily construct user interfaces. A new search panel allows people to find important files and documents by author, date, keywords, file types, and text within the file or document. Vista also provides improved graphics and display capabilities.[11] See Figure 4.8. Today, Microsoft has over 90 percent of the PC OS market. Apple, Linux publishers, and other companies account for the rest of the PC OS market.

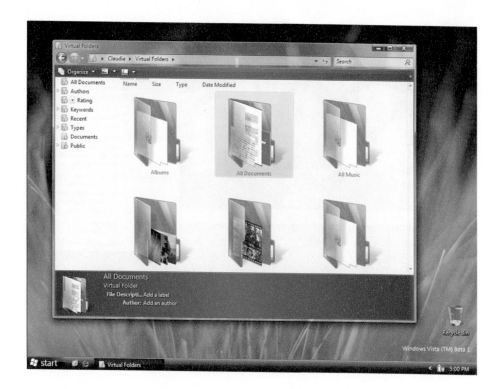

Figure 4.8

Microsoft Vista Operating System

(Source: Courtesy of Microsoft Corporation.)

Apple Computer Operating Systems

Although IBM system platforms traditionally use one of the Windows OSs and Intel microprocessors (often called *Wintel* for this reason), Apple computers in the past have used non-Intel microprocessors designed by Apple, IBM, and Motorola and a proprietary Apple OS—the Mac OS. Newer Apple computers, however, use Intel chips.[12] Although Wintel computers hold the largest share of the business PC market, Apple computers are also popular, especially in the fields of publishing, education, graphic arts, music, movies, and media. Software developed for the Macintosh often provides cutting-edge options for creative people. GarageBand, for example, is Macintosh software that allows you to create your own music the way a professional does, and it can sound like a small orchestra. Pro Tools is another software program used to edit digital music. Although some people believe that this type of software produces "fake" music, others believe software that manipulates digital music will dominate music production and editing in the future.

The Apple OSs have also evolved over a number of years and often provide features not available from Microsoft. Starting in July 2001, the Mac OS X was installed on all new Macs. It includes an entirely new user interface, which provides a new visual appearance for users—including luminous and semitransparent elements, such as buttons, scroll bars, windows, and fluid animation to enhance the user's experience. Since then, OS X has been upgraded with additional releases, named Jaguar (OS X.2) and Panther (OS X.3). Tiger, also called Mac OS X.4, is the most recent version of OS X released in 2005.[13] It has many new features, including support for 64-bit computing.[14] Dashboard is another feature of Tiger that displays tools such as calculators, dictionaries, and calendars so you can access and run them easily. Spotlight for Mac OS X Tiger is Apple's fast search technology that allows you to locate documents, music, images, e-mails, contacts, and other information on your computer by searching the contents of your files.[15] For example, you can use Spotlight to locate all files that include a name, such as Jim Roberts. All e-mails and other documents with Jim Roberts will be displayed. See Figure 4.9. Newer versions of Mac OS X include 3-D effects and better management of windows on a display screen.

Figure 4.9

Mac OS Tiger

(Source: Courtesy of Apple Computer, Inc.)

Linux

Linux is an OS developed by Linus Torvalds in 1991 as a student in Finland. The OS is distributed under the GNU General Public License, and its source code is freely available to everyone. It is, therefore, called an open-source operating system. This doesn't mean however, that Linux and its assorted distributions are free—companies and developers can charge money for a distribution as long as the source code remains available. Linux is actually only the kernel of an OS, the part that controls hardware, manages files, separates processes, and so forth. Several combinations of Linux are available, with various sets of capabilities and applications to form a complete OS. Each of these combinations is called a *distribution* of Linux.

Linux is available on the Internet and from other sources, including Red Hat Linux and Caldera OpenLinux. Many people and organizations use Linux. Panasonic has used Linux to run one of its high-speed television tuners. Grocer Hannaford Brothers uses the Linux operating system in about 140 of its supermarkets.[16] The company expects to save about 30 percent on its software costs by using Linux instead of Windows or another non-open-source operating system. In addition, several large computer vendors, including IBM, Hewlett-Packard, and Intel, support the Linux operating system.[17] For example, IBM has more than 500 programmers working with Linux, primarily because of its security features. Many CIOs are considering switching to Linux and open-source software because of security concerns with Microsoft software.

Linux, however, is not without controversy. One company, for example, is claiming that leading Linux distributors might be infringing on copyright and trade secret rights. Other companies are contributing funds to defend Linux against copyright-infringement lawsuits. In addition, as Linux becomes more popular, people are demanding more support, service, and features.[18]

Workgroup Operating Systems

To keep pace with user demands, the technology of the future must support a world in which network usage, data-storage requirements, and data-processing speeds increase at a dramatic rate. This rapid increase in communications and data-processing capabilities pushes the boundaries of computer science and physics. Powerful and sophisticated OSs are needed to run the servers that meet these business needs for workgroups. Small businesses, for example, often use workgroup OSs to run networks and perform critical business tasks.

Windows Server

Microsoft designed *Windows Server* to perform a host of tasks that are vital for Web sites and corporate Web applications. For example, Microsoft Windows Server can be used to coordinate large data centers. The OS also works with other Microsoft products. It can be used to prevent unauthorized disclosure of information by blocking text and e-mails from being copied, printed, or forwarded to other people.

Microsoft *Windows Advanced Server, Limited Edition* was the first 64-bit version of the Windows Server family. It was designed to run on the 64-bit Itanium processor from Intel (also known as the IA64). This OS enables Microsoft to begin competing with rival Linux vendors (Red Hat, Caldera, SuSE, and TurboLinux), which already have 64-bit Itanium versions of their Linux distributions. In addition, Sun Microsystems and IBM have had 64-bit UNIX OSs for years.

UNIX

UNIX is a powerful OS originally developed by AT&T for minicomputers. UNIX can be used on many computer system types and platforms, from personal computers to mainframe systems. UNIX also makes it much easier to move programs and data among computers or to connect mainframes and personal computers to share resources. There are many variants of UNIX—including HP/UX from Hewlett-Packard, AIX from IBM, UNIX SystemV from UNIX Systems Lab, Solaris from Sun Microsystems, and SCO from Santa Cruz Operations. Sun Microsystems hopes that its open-source Solaris will attract developers to make the software even better.[19]

NetWare

NetWare is a network OS sold by Novell that can support users on Windows, Macintosh, and UNIX platforms. NetWare provides directory software to track computers, programs, and people on a network, helping large companies to manage complex networks. NetWare users can log on from any computer on the network and use their own familiar desktop with all their applications, data, and preferences.

Red Hat Linux

Red Hat Software offers a Linux network OS that taps into the talents of tens of thousands of volunteer programmers who generate a steady stream of improvements for the Linux OS. The *Red Hat Linux* network OS is very efficient at serving Web pages and can manage a cluster of up to eight servers. The film *Lord of the Rings* used Linux and hundreds of servers to deliver many of the special effects shown in the finished film. Linux environments typically have fewer virus and security problems than other OSs. Distributions such as SuSE and Red Hat have proven Linux to be a very stable and efficient OS.

Mac OS X Server

The *Mac OS X Server* is the first modern server OS from Apple Computer. It provides UNIX-style process management. Protected memory puts each service in its own well-guarded chunk of dynamically allocated memory, preventing a single process from going awry and bringing down the system or other services. Under preemptive multitasking, a computer OS uses some criteria to decide how long to allocate to any one task before giving another task a turn to use the OS. Preempting is the act of taking control of the OS from one task and giving it to another. A common criterion for preempting is elapsed time. In more sophisticated OSs, certain applications can be given higher priority than other applications, giving the higher-priority programs longer processing times. Preemptive multitasking ensures that each process gets the right amount of CPU time and the system resources it needs for optimal efficiency and responsiveness.

Enterprise Operating Systems

New mainframe computers provide the computing and storage capacity to meet massive data-processing requirements and offer many users high performance and excellent system availability, strong security, and scalability. In addition, a wide range of application software has been developed to run in the mainframe environment, making it possible to purchase software to address almost any business problem. As a result, mainframe computers remain the computing platform of choice for mission-critical business applications for many companies. z/OS from IBM, MPE/iX from Hewlett-Packard, and Linux are three examples of mainframe OSs.

z/OS

The *z/OS* is IBM's first 64-bit enterprise OS. It supports IBM's z900 and z800 lines of mainframes that can come with up to sixteen 64-bit processors. (The z stands for zero downtime.) The OS provides several new capabilities to make it easier and less expensive for users to run large mainframe computers. The OS has improved workload management and advanced e-commerce security. The IBM zSeries mainframe, like previous generations of IBM mainframes, lets users subdivide a single computer into multiple smaller servers, each of which can run a different application. In recognition of the widespread popularity of a competing OS, z/OS allows partitions to run a version of the Linux OS. This means that a company can upgrade to a mainframe that runs the Linux OS.

MPE/iX, HP-UX, and Linux

Multiprogramming Executive with integrated POSIX (MPE/iX) is the Internet-enabled OS for the Hewlett-Packard e3000 family of computers. MPE/iX is a robust OS designed to handle a variety of business tasks, including online transaction processing and Web applications. It runs on a broad range of HP e3000 servers—from entry-level to workgroup and enterprise servers within the data centers of large organizations. *HP-UX* is a mainframe OS from Hewlett-Packard, and is designed to support Internet, database, and a variety of business applications. It can work with Java programs and Linux applications. The OS comes in four versions: foundation, enterprise, mission critical, and technical. HP-UX supports Hewlett-Packard's computers and those designed to run Intel's Itanium processors. *Red Hat Linux for IBM* mainframe computers is another example of an enterprise operating system.

Operating Systems for Small Computers, Embedded Computers, and Special-Purpose Devices

New OSs and other software are changing the way we interact with personal digital assistants (PDAs), cell phones, digital cameras, TVs, and other appliances. These OSs are also called *embedded operating systems* because they are typically embedded within a device, such as an automobile, TV recorder, or other device. Embedded software is a multibillion dollar industry. Some of these OSs allow you to synchronize handheld devices with PCs using cradles, cables, and wireless connections. Cell phones also use embedded OSs (see Figure 4.10). In addition, some OSs have been developed for special-purpose devices, such as TV set-top boxes, computers on the space shuttle, computers in military weapons, and computers in some home appliances. Following are some of the more popular OSs for such devices.

Figure 4.10

Cell Phones Have Embedded Operating Systems

Many cell phones such as this one from Nokia have an embedded OS and an integrated imaging device. Point, use the color display as a viewfinder, and snap a picture. Images can be stored on the device and sent to a friend.

(Source: Courtesy of Nokia.)

Palm OS

PalmSource makes the Palm operating system that is used in over 30 million handheld computers and smartphones manufactured by Palm, Inc., and other companies. Palm also develops and supports applications, including business, multimedia, games, productivity, reference and education, hobbies and entertainment, travel, sports, utilities, and wireless applications. Today, Palm has about half of the market for PDA or handheld OSs. Microsoft has about 30 percent of the market; Linux and other companies account for the rest of the PDA market.

Windows Embedded

Windows Embedded is a family of Microsoft OSs included with or embedded into small computer devices. Windows Embedded includes Windows CE .Net and *Windows XP Embedded*, an OS used in devices such as handheld computers, TV set-top boxes, and automated industrial machines. *Windows CE .Net* represents a key step in taking Microsoft closer to its vision of anywhere, anytime access to Web-based content and services. It is an embedded OS for use in mobile devices, such as smartphones and PDAs, and other devices, such as digital cameras, thin clients, and automotive computers. PDAs with Windows CE try to bring as much of the functionality of a desktop PC as possible to a handheld device. Such a PDA is a programmable computer that performs most of the tasks of a dedicated device.

Windows Mobile

Windows Mobile is a family of Microsoft OSs for mobile or portable devices. Windows Mobile includes Pocket PC, Pocket PC Phone Edition, and SmartPhone. These OSs offer features such as handwriting recognition, instant messaging technology, support for more secure Internet connections, and the ability to beam information to devices running the

Pocket PC or Palm OS. They also have advanced telecommunications capabilities, discussed in more detail in Chapter 6. Motorola, Samsung, Dell, Hewlett-Packard, and Toshiba have products that run Windows Mobile. Wireless services from Verizon, Cingular, and others are often available with devices that use Windows Mobile. A new Palm Treo uses Microsoft Windows Mobile software, instead of the Palm operating system.[20]

Utility Programs

utility programs

Programs that help to perform maintenance or correct problems with a computer system.

Utility programs help to perform maintenance or correct problems with a computer system. For example, some utility programs merge and sort sets of data, keep track of computer jobs being run, compress files of data before they are stored or transmitted over a network (thus saving space and time), and perform other important tasks. Some utility programs are even used to make computer systems run better and longer without problems. Another type of utility program allows people and organizations to tap into unused computer power over a network. Often called *grid computing*, the approach can be very efficient and less expensive than purchasing additional hardware or computer equipment. In the future, grid computing could become a common feature of OSs and provide inexpensive, on-demand access to computer power and resources.

Utility programs often come installed on computer systems, but you can also purchase utility programs separately. See Figure 4.11. The following sections examine some common types of utilities.

Figure 4.11

WinZip, a File Compression Utility

Hardware Utilities

Some hardware utilities are available from companies such as Symantec, which produces Norton Utilities. Hardware utilities can check the status of all parts of the PC, including hard disks, memory, modems, speakers, and printers. Disk utilities check the hard disk's boot sector, file allocation tables, and directories, and analyze them to ensure that the hard disk is not damaged. Disk utilities can also optimize the placement of files on a crowded disk.

Virus-Detection and Recovery Utilities

Computer viruses from the Internet and other sources can be a nuisance—and sometimes can completely disable a computer. Virus-detection and recovery software can be installed to constantly monitor and protect the computer. If a virus is found, the software can eliminate the virus, or "clean" the computer system. To keep the virus-detection and recovery software current and make sure that the software checks for the latest viruses, it can be easily updated through the Internet. Symantec and McAfee are two companies that make virus-detection and recovery software. Microsoft offers a free malicious software removal tool on its Web site, *www.microsoft.com*.[21] Microsoft has also acquired several antivirus companies and is expected to release antivirus software in the future.[22]

Gas Natural: Improving Customer Service with Mobile Software

An emerging global trend is to open once-regulated markets such as telecommunications and utilities to free competition. The result is typically lower prices and better service for consumers, and greater opportunity and higher level of competition for businesses. In 2002, the European Union began moving to liberate its energy market, and Spain was one of the first members to embrace the concept. Soon after the opening of energy markets in Spain, customers were enjoying better services and savings of 30 percent on utility bills.

Gas Natural is Spain's largest natural gas distributor with 4.5 million domestic customers and $8.5 billion in revenue. While once the only option for natural gas in Spain, Gas Natural suddenly found itself threatened by strong competition. These competitive forces drove Gas Natural to emphasize customer service to retain customers. They focused on managing and fulfilling its customers' service requirements, responding to problems quickly, and getting the job done right the first time.

Gas Natural uses a decentralized approach to field service; that is, it contracts its field service to a broad network of independent franchises to install, repair, and maintain gas service at the customer's location. The system Gas Natural used to communicate with field service workers was archaic, slow, and painstaking. Paper work orders were mailed to the service franchises that dispatched them to field engineers. Upon completion of the service call, the engineer logged the status of the order on paper forms and mailed it back to Gas Natural where data-entry clerks would enter the information into the database system. This lack of real-time communication between service engineers and dispatchers at Gas Natural led to frustration for all, especially customers.

The sales force at Gas Natural was equally hampered. Like the field service workers, they lacked real-time access to corporate information while away from the office and so were unable to leverage the information needed to close deals. Gas Natural required software that would eliminate the need for paper service requests and forms and put their field engineers and sales force in touch with headquarters.

Gas Natural found a solution to their problems with Workforce Management software from the Swedish software company isMobile (*www.ismobile.com*). The isMobile company works with their clients to help them become real-time enterprises with short lead times and high productivity at the lowest possible cost, which is just what Gas Natural needed.

Within three months, isMobile had replaced the clipboards that Gas Natural engineers and sales representatives carried with state-of-the-art PDAs running custom isMobile software connected to back-end systems at Gas Natural. Using the new system, service requests are passed from customer to Gas Natural to service franchise to engineer in moments. The system makes better use of the engineer's time and decreases response time to jobs. Engineers can now finish the job in one visit because they can access useful information through the software running on their mobile PCs.

Gas Natural sales representatives are now armed with real-time information from the company's customer resource management (CRM) systems and mobile sales software tools, empowering them to target new and existing customers with custom-designed service bundles.

By empowering field engineers and sales representatives with the information they need when they need it, Gas Natural has become a customer-centered, responsive business. The new software solution has provided the company with a higher level of market security and allowed it to maintain its monopoly power. Most recently, Gas Natural has been negotiating to take over the electricity firm Endesa. Combined, this company would become one of Europe's biggest companies and the world's third largest energy firm.

Discussion Questions

1. How does a deregulated open market positively affect customers and challenge businesses?
2. What role did software play in empowering Gas Natural to better serve its customers and gain new customers?

Critical Thinking Questions

1. Besides increasing customer satisfaction, what benefits do you think Gas Natural's new system provided for the franchise businesses that provide their service engineers?
2. What other businesses can you think of that have increased customer satisfaction through the use of innovative software to gain a competitive advantage? How?

SOURCES: "Gas Natural Energizes Its Field Force Operations by Enabling Everyplace Access to Realtime Data," IBM Success Story, November 2, 2005, *www-306.ibm.com/software/success/cssdb.nsf/CS/JSTS-6HKRPY? OpenDocument&Site=wsappserv.* "Madrid Backs Gas Natural Takeover," BBC News, February 5, 2006, *http://news.bbc.co.uk/2/hi/business/4677696.stm.* isMobile Web site, accessed March 4, 2006, *www.ismobile.com.*

File-Compression Utilities

File-compression programs can reduce the amount of disk space required to store a file or reduce the time it takes to transfer a file over the Internet. A popular program on Windows PCs is WinZip (*www.winzip.com*), which generates zip files, which are collections of one or more compressed files. A zip file has a .zip extension, and its contents can be easily unzipped to their original size. *MP3 (Motion Pictures Experts Group-Layer 3)* is a popular file-compression format used to store, transfer, and play music and audio files, such as podcasts—audio programs that can be downloaded from the Internet. MP3 can compress files 10 times smaller than the original file with near-CD-quality sound. Software, such as iTunes from Apple, can be used to store, organize, and play MP3 music files.

Spam and Pop-Up Blocker Utilities

Getting unwanted e-mail (spam) and having annoying and unwanted ads pop up on your screen while you are on the Web can be frustrating and a big waste of time. A number of utility programs can be installed to help block unwanted e-mail spam and pop-up ads, including Cloudmark SpamNet, IhateSpam, Spamnix, McAfee SpamKiller, and Ad-aware.

Network and Internet Utilities

A broad range of network- and systems-management utility software is available to monitor hardware and network performance and trigger an alert when a Web server is crashing or a network problem occurs. Although these general management features are helpful, what is needed is a way to pinpoint the cause of the problem. Topaz from Mercury Interactive is an example of software called an *advanced Web-performance monitoring utility*. It is designed to sound an alarm when it detects problems and let network administrators isolate the most likely causes of the problems. Its Auto RCA (root-cause analysis) module uses statistical analysis with built-in rules to measure system and Web performance. Actual performance data is compared with the rules, and the results can help pinpoint where trouble originated—in the application software, database, server, network, or the security features.

Server and Mainframe Utilities

Some utilities enhance the performance of servers and mainframe computers. IBM has created systems-management software that allows a support person to monitor the growing number of desktop computers in a business attached to a server or mainframe computer. With this software, the support people can sit at their personal computers and check or diagnose problems, such as a hard disk failure on a network computer. The support people can even repair individual systems anywhere on the organization's network, often without having to leave their desks. The direct benefit is to the system manager, but the business gains from having a smoothly functioning information system. Utility programs can meet the needs of a single user, workgroup, or enterprise, as listed in Table 4.3. These programs perform useful tasks—from tracking jobs to monitoring system integrity.

Table 4.3

Examples of Utility Programs

Personal	Workgroup	Enterprise
Software to compress data so that it takes less hard disk space	Software to provide detailed reports of workgroup computer activity and status of user accounts	Software to archive contents of a database by copying data from disk to tape
Screen saver	Software that manages an uninterruptible power supply to do a controlled shutdown of the workgroup computer in the event of a loss of power	Software that compares the content of one file with another and identifies any differences
Virus-detection software	Software that reports unsuccessful user logon attempts	Software that reports the status of a particular computer job

Virtualization software can make computers act like or simulate other computers.[23] The result is often called a *virtual machine.* Virtual PC for Mac, for example, allows you to run software written for Windows on an Apple Macintosh computer. Using virtualization software, servers and mainframe computers can run software applications written for different operating systems. For example, you can use a server or mainframe to test and run a number of PC applications, such as spreadsheets, word processors, and databases, at the same time. A virtual machine can also run older programs that cannot run on current operating systems. WMWare Workstation, for example, allows today's Windows computers to run applications written for older operating systems, such as Windows 3.1.[24]

Other Utilities

Utility programs are available for almost every conceivable task or function.[25] For example, you can use Microsoft Windows Rights Management Services with Microsoft Office programs to manage and protect important corporate documents. ValueIT is a utility that can help a company verify the value of investments in information systems and technology. Widgit Software has developed an important software utility that helps people with visual disabilities use the Internet. The software converts icons and symbols into plain text that can be easily seen. Another software utility allows a manager to see every keystroke a worker makes on a computer system. Monitoring software can catalog the Internet sites that employees visit and the time that employees are working at their computer.

In addition, you can use many search tools to find important files and documents.[26] Most of these desktop search tools are free and available from a number of popular Internet sites.[27] Yahoo! Desktop Search, Ask Jeeves Desktop Search, Google Desktop, and MSN Deskbar are examples (see Figure 4.12).

Figure 4.12

Desktop Search Tool

With a desktop search tool, such as Google Desktop, you can search through files on your computer.

Middleware

Middleware is software that allows different systems to communicate and exchange data. Middleware can also be used as an interface between the Internet and older legacy systems. (Legacy software is a previous, major version that continues to be used.) For example, middleware can be used to transfer a request for information from a corporate customer on the corporate Web site to a traditional database on a mainframe computer and return the results to the customer on the Internet.

middleware
Software that allows different systems to communicate and exchange data.

APPLICATION SOFTWARE

As discussed earlier in this chapter, the primary function of application software is to apply the power of the computer to give people, workgroups, and the entire enterprise the ability to solve problems and perform specific tasks. When you need the computer to do something, you use one or more application programs. The application programs interact with systems software, and the systems software then directs the computer hardware to perform the necessary tasks. Applications help you perform common tasks, such as create and format text documents, perform calculations, or manage information, though some applications are more specialized. Pfizer, for example, has developed application software to detect the early signs of Parkinson's disease.[28] The new software detects slight trembling in speech patterns not detectable by the human ear that predicts the disease.

The functions performed by application software are diverse, and range from personal productivity to business analysis. Application software, for example, can be used to help sales managers track sales of a new item in a test market. Another example of innovative applications in business is voice stress software, which is being developed to help detect fraud in the insurance industry. Pilot programs have been very successful in detecting scam insurance companies or people who try to make false claims. In general, applications can complete sales orders, control inventory, pay bills, write paychecks to employees, and provide financial and marketing information to managers and executives. Most of the computerized business jobs and activities discussed in this book involve the use of application software. We begin by investigating the types and functions of application software.

Overview of Application Software

Proprietary software and off-the-shelf software are important types of application software (see Figure 4.13). A company can develop a one-of-a-kind program for a specific application (called **proprietary software**). Proprietary software is not in the public domain. A company can also purchase or acquire an existing software program (sometimes called **off-the-shelf software** because it can literally be purchased or acquired "off the shelf" in a store). The relative advantages and disadvantages of proprietary software and off-the-shelf software are summarized in Table 4.4.

Figure 4.13

Types of Application Software

Some off-the-shelf software can be customized to suit user needs.

Many companies use off-the-shelf software to support business processes. Key questions for selecting off-the-shelf software include the following: (1) Will the software run on the OS and hardware you have selected? (2) Does the software meet the essential business requirements that have been defined? (3) Is the software manufacturer financially solvent and reliable? and (4) Does the total cost of purchasing, installing, and maintaining the software compare favorably to the expected business benefits?

Some off-the-shelf programs can be modified, in effect blending the off-the-shelf and customized approaches. For example, a software developer might write a collection of programs to be used in an auto body shop and include features to generate estimates, order parts, and process insurance. Body shops of all types have these needs. Designed properly—and with provisions for minor tailoring for each user—the same software package can be sold to many users. However, because each body shop has slightly different requirements, some

Proprietary Software		Off-the-Shelf Software	
Advantages	**Disadvantages**	**Advantages**	**Disadvantages**
You can get exactly what you need in terms of features, reports, and so on.	It can take a long time and significant resources to develop required features.	The initial cost is lower because the software firm can spread the development costs over many customers.	An organization might have to pay for features that are not required and never used.
Being involved in the development offers control over the results.	In-house system development staff may become hard pressed to provide the required level of ongoing support and maintenance because of pressure to move on to other new projects.	The software is likely to meet the basic business needs—you can analyze existing features and the performance of the package.	The software might lack important features, thus requiring future modification or customization. This can be very expensive because users must adopt future releases of the software as well.
You can modify features that you might need to counteract an initiative by competitors or to meet new supplier or customer demands. A merger with or acquisition of another firm also requires software changes to meet new business needs.	There is more risk concerning the features and performance of the software that has yet to be developed.	The package is likely to be of high quality because many customer firms have tested the software and helped identify its bugs.	The software might not match current work processes and data standards.

Table 4.4

A Comparison of Proprietary and Off-the-Shelf Software

modifications to the software might be needed. As a result, software vendors often provide a wide range of services, including installing their standard software, modifying the software as the customer requires, training users, and providing other consulting services.

Another approach to obtaining a customized software package is to use an application service provider. An **application service provider** (ASP) is a company that can provide the software, support, and computer hardware on which to run the software from the user's facilities. An ASP can also simplify a complex corporate software package so that it is easier for the users to set up and manage. ASPs provide contract customization of off-the-shelf software, and they speed deployment of new applications while helping IS managers avoid implementation headaches, reducing the need for many skilled IS staff members and decreasing project start-up expenses. Such an approach allows companies to devote more time and resources to more important tasks. Using an ASP makes the most sense for relatively small, fast-growing companies with limited IS resources. It is also a good strategy for companies that want to deploy a single, functionally focused application quickly, such as setting up an e-commerce Web site or supporting expense reporting. Contracting with an ASP might make less sense, however, for larger companies with major systems that have their technical infrastructure already in place.

Using an ASP involves some risks—sensitive information could be compromised in a number of ways, including unauthorized access by employees or computer hackers; the ASP might not be able to keep its computers and network up and running as consistently as necessary; or a disaster could disable the ASP's data center, temporarily putting an organization out of business. These are legitimate concerns that an ASP must address.

application service provider (ASP)
A company that provides software, support, and the computer hardware on which to run the software from the user's facilities.

Personal Application Software

Hundreds of computer applications can help people at school, home, and work. New computer software under development and existing GPS technology, for example, will allow people to see 3-D views of where they are, along with directions and 3-D maps to where they would like to go.[29] The features of personal application software are summarized in Table 4.5. In addition to these general-purpose programs, thousands of other personal computer applications perform specialized tasks: to help you do your taxes, get in shape, lose

Type of Software	Explanation	Example	Vendor
Word processing	Create, edit, and print text documents	Word WordPerfect	Microsoft Corel
Spreadsheet	Provide a wide range of built-in functions for statistical, financial, logical, database, graphics, and date and time calculations	Excel Lotus 1-2-3	Microsoft Lotus/IBM
Database	Store, manipulate, and retrieve data	Access Approach dBASE	Microsoft Lotus/IBM Borland
Online information services	Obtain a broad range of information from commercial services	America Online MSN	America Online Microsoft
Graphics	Develop graphs, illustrations, and drawings	Illustrator FreeHand	Adobe Macromedia
Project management	Plan, schedule, allocate, and control people and resources (money, time, and technology) needed to complete a project according to schedule	Project for Windows On Target Project Schedule Time Line	Microsoft Symantec Scitor Symantec
Financial management	Provide income and expense tracking and reporting to monitor and plan budgets (some programs have investment portfolio management features)	Managing Your Money Quicken	Meca Software Intuit
Desktop publishing (DTP)	Use with personal computers and high-resolution printers to create high-quality printed output, including text and graphics; various styles of pages can be laid out; art and text files from other programs can also be integrated into "published" pages	QuarkXPress Publisher PageMaker Ventura Publisher	Quark Microsoft Adobe Corel
Creativity	Generate innovative and creative ideas and problem solutions. The software does not propose solutions, but provides a framework conducive to creative thought. The software takes users through a routine, first naming a problem, then organizing ideas and "wishes," and offering new information to suggest different ideas or solutions	Organizer Notes	Macromedia Lotus

Table 4.5

Examples of Personal Productivity Software

weight, get medical advice, write wills and other legal documents, repair your computer, fix your car, write music, and edit your pictures and videos. [30] This type of software, often called *user software* or *personal productivity software*, includes the general-purpose tools and programs that support individual needs.

Word Processing

Word processing applications are installed on most PCs today. These applications come with a vast array of features, including those for checking spelling, creating tables, inserting formulas, creating graphics, and much more (see Figure 4.14). This book (and most like it) was entered into a word processing application using a personal computer.

A team of people can use a word processing program to collaborate on a project. The authors and editors who developed this book, for example, used the Track Changes and Reviewing features of Microsoft Word to track and make changes to chapter files. You can

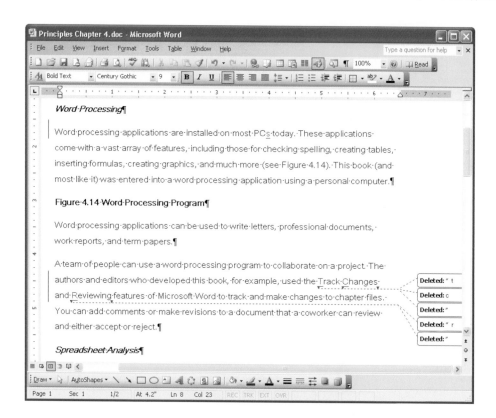

Figure 4.14

Word Processing Program

Word processing applications can be used to write and collaborate on documents such as letters, professional documents, work reports, and term papers.

add comments or make revisions to a document that a coworker can review and either accept or reject.

Spreadsheet Analysis

Spreadsheets are powerful tools for individuals and organizations. Features of spreadsheets include graphics, limited database capabilities, statistical analysis, built-in business functions, and much more (see Figure 4.15). The business functions include calculation of depreciation, present value, internal rate of return, and the monthly payment on a loan, to name a few. Optimization is another powerful feature of many spreadsheet programs. *Optimization* allows the spreadsheet to maximize or minimize a quantity subject to certain constraints. For example, a small furniture manufacturer that produces chairs and tables might want to maximize its profits. The constraints could be a limited supply of lumber, a limited number of workers who can assemble the chairs and tables, or a limited amount of various hardware fasteners that might be required. Using an optimization feature, such as Solver in Microsoft Excel, the spreadsheet can determine what number of chairs and tables to produce with labor and material constraints to maximize profits.

Database Applications

Database applications are ideal for storing, manipulating, and retrieving data. These applications are particularly useful when you need to manipulate a large amount of data and produce reports and documents. Database manipulations include merging, editing, and sorting data. The uses of a database application are varied. You can keep track of a CD collection, the items in your apartment, tax records, and expenses. A student club can use a database to store names, addresses, phone numbers, and dues paid. In business, a database application can help process sales orders, control inventory, order new supplies, send letters to customers, and pay employees. Database management systems can be used to track orders, products, and customers; analyze weather data to make forecasts for the next several days; and summarize medical research results. A database can also be a front end to another application. For example, you can use a database application to enter and store income tax information, then export the stored results to other applications, such as a spreadsheet or tax-preparation application (see Figure 4.16).

Figure 4.15

Spreadsheet Program

Spreadsheet programs should be considered when calculations are required.

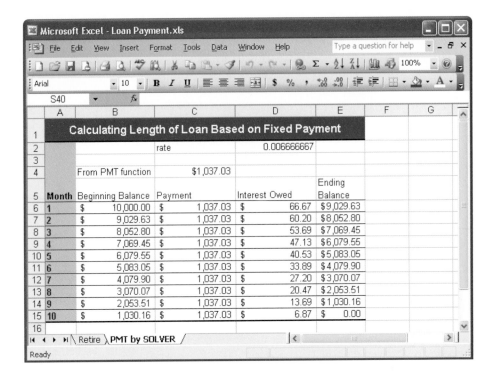

Figure 4.16

Database Program

After being entered into a database application, information can be manipulated and used to produce reports and documents.

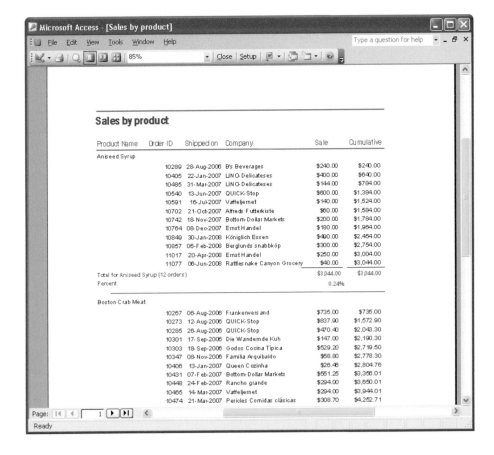

Graphics Program

It is often said that a picture is worth a thousand words. With today's graphics programs, it is easy to develop attractive graphs, illustrations, and drawings. See Figure 4.17. Graphics programs can be used to develop advertising brochures, announcements, and full-color

Figure 4.17

Graphics Program

Graphics programs can help you make a presentation at school or work. They can also be used to develop attractive brochures, illustrations, drawings, and maps, and to organize and edit photographic images.

(Source: Courtesy of Adobe Systems Incorporated.)

presentations, and to organize and edit photographic images. If you need to make a presentation at school or work, you can use a special type of graphics program called a presentation application to develop slides and then display them while you are delivering your presentation. Because of their popularity, many colleges and departments require students to become proficient at using presentation graphics programs.

Many graphics programs, such as PowerPoint by Microsoft, consist of a series of slides. Each slide can be displayed on a computer screen, printed as a handout, or (more commonly) projected onto a large viewing screen for audiences. Powerful built-in features allow you to develop attractive slides and complete presentations. You can select a template for a type of presentation, such as recommending a strategy for managers, communicating bad news to a sales force, giving a training presentation, or facilitating a brainstorming session. The presentation graphics program takes you through the presentation step by step, including applying color and attractive formatting. You can also design a custom presentation. All types of charts, drawings, and formatting are available. Most presentation graphics programs come with many pieces of *clip art*, such as drawings and photos of people meeting, medical equipment, telecommunications equipment, entertainment, and much more.

Personal Information Managers

Personal information managers (PIMs) help people, groups, and organizations store useful information, such as a list of tasks to complete or a list of names and addresses. PIMs usually provide an appointment calendar and a place to take notes. In addition, information in a PIM can be linked. For example, you can link an appointment with a sales manager in the calendar to information on the sales manager in the address book. When you click the appointment in the calendar, a window opens displaying information on the sales manager from the address book. Microsoft Outlook is an example of a PIM software package.

Some PIMs allow you to schedule and coordinate group meetings. If a computer or handheld device is connected to a network, you can upload the PIM data and coordinate it with the calendar and schedule of others using the same PIM software on the network. You can also use some PIMs to coordinate e-mails sent and received over the Internet.

Online Information Services

Online information services allow you to connect a personal computer to the outside world through phone lines, cable, satellite, or power lines in some cases. Using an online service, you can view investment information, make travel plans, and check news from around the world. You can also compare prices and features for most consumer items, research companies, send e-mail to friends and family, learn about degree programs offered by colleges and universities around the world, and search for job openings in your area (see Figure 4.18).

Figure 4.18

Online Information Services

Online services provide instant access to information. Prices of cars and trucks, travel discounts, information about companies, stock market data, and much more is available with a few keystrokes using an online service.

(Source: *www.msn.com*, accessed March 22, 2006.)

software suite

A collection of single application programs packaged in a bundle.

Software Suites and Integrated Software Packages

A **software suite** is a collection of single application programs packaged in a bundle. Software suites can include word processors, spreadsheets, database management systems, graphics programs, communications tools, organizers, and more. Some suites support the development of Web pages, note taking, and speech recognition, where applications in the suite can accept voice commands and record dictation. Software suites offer many advantages. The software programs have been designed to work similarly, so after you learn the basics for one application, the other applications are easy to learn and use. Buying software in a bundled suite is cost-effective; the programs usually sell for a fraction of what they would cost individually.

Microsoft Office, Corel's WordPerfect Office, Lotus SmartSuite, and Sun Microsystems's StarOffice are examples of popular general-purpose software suites for personal computer users (see Figure 4.19). Microsoft Office has the largest market share. The Free Software Foundation offers software similar to Sun Microsystems's StarOffice that includes word processing, spreadsheet, database, presentation graphics, and e-mail applications for the Linux OS. OpenOffice is another Office suite for Linux.[31] Wine can run any Windows application, including those in Microsoft Office, on Linux, although some features might not work as well as a Microsoft OS. Each of these software suites includes a spreadsheet program, word processor, database program, and graphics package with the ability to move documents, data, and diagrams among them (see Table 4.6). Thus, a user can create a spreadsheet and then cut and paste that spreadsheet into a document created using the word processing application.

More than a hundred million people worldwide use the Microsoft Office software suite, with Office 2003 representing the latest version of the productivity software. The next version of Office will be available in 2006.[32] The new version of Office will use a new file format, but it will also support the existing .doc, .xls, .ppt, and other traditional file formats. The updated suite offers many enhancements over earlier versions of Office, including new collaboration features. A SharePoint feature, for example, lets people post files and messages on a central site, making it easier for a group of people to work on common Office documents.

Figure 4.19

Software Suite

A software suite, such as Corel WordPerfect Office, offers a collection of powerful programs, including word processing, spreadsheet, database, graphics, and other programs. The programs in a software suite are designed to be used together. In addition, the commands, icons, and procedures are the same for all programs in the suite.

(Source: Courtesy of Corel Corporation.)

People can work together, viewing documents and making changes in real time. As one person is making changes, other people are locked out from changing the same document—their mouse pointer disappears from their display screen. In addition, information from Office can be integrated with corporate systems and applications. With InfoPath, for example, information can be saved to corporate databases.

Microsoft Office goes beyond its role as a mainstream package of ready-to-run applications with the extensive custom development facilities of Visual Basic for Applications (VBA)—a built-in facility that is part of every Office application. Using VBA, users can enhance off-the-shelf applications to tailor the programs for special tasks.

Personal Productivity Function	Microsoft Office	Lotus SmartSuite Millennium Edition	Corel WordPerfect Office	Sun Microsystems
Word Processing	Word	WordPro	WordPerfect	Writer
Spreadsheet	Excel	Lotus 1-2-3	Quattro Pro	Calc
Presentation Graphics	PowerPoint	Freelance Graphics	Presentations	Impress
Database	Access	Lotus Approach	Paradox	

Table 4.6

Major Components of Leading Software Suites

In addition to suites, some companies produce *integrated application packages* that contain several programs. For example, *Microsoft Works* is one program that contains basic word processing, spreadsheet, database, address book, calendar, and other applications. Although not as powerful as stand-alone software included in software suites, integrated software packages offer a range of capabilities for less money. Some integrated packages cost about $100.

Other Personal Application Software

In addition to the software already discussed, people can use many other interesting and powerful application software tools. In some cases, the features and capabilities of these applications can more than justify the cost of an entire computer system. TurboTax, for example, is a popular tax-preparation program. Other exciting software packages have been developed for training and distance learning. University professors often believe that colleges and universities must invest in distance learning for their students. Using this type of software, some universities offer complete degree programs over the Internet. Engineers, architects, and designers often use computer-aided design (CAD) software to design and develop buildings, electrical systems, plumbing systems, and more. Autosketch, CorelCAD, and AutoCad are examples of CAD software. Other programs perform a wide array of statistical tests. Colleges and universities often have a number of courses in statistics that use this type of application software. Two popular applications in the social sciences are SPSS and SAS.

Workgroup Application Software

**workgroup application
software**
Software that supports teamwork,
whether in one location or around
the world.

Workgroup application software is designed to support teamwork, whether people are in the same location or dispersed around the world. This support can be accomplished with software known as *groupware* that helps groups of people work together effectively. Microsoft Exchange Server, for example, has groupware and e-mail features. Also called *collaborative software*, the approach allows a team of managers to work on the same production problem, letting them share their ideas and work via connected computer systems. The "Three Cs" rule for successful implementation of groupware is summarized in Table 4.7.

Table 4.7

Ernst & Young's "Three Cs" Rule for Groupware

Quality	Description
Convenient	If it's too hard to use, it's not used; it should be as easy to use as the telephone.
Content	It must provide a constant stream of rich, relevant, and personalized content.
Coverage	If it isn't close to everything you need, it might never be used.

Examples of workgroup software include group scheduling software, electronic mail, and other software that enables people to share ideas. Lotus Notes from IBM, for example, lets companies use one software package and one user interface to integrate many business processes. Lotus Notes can allow a global team to work together from a common set of documents, have electronic discussions using threads of discussion, and schedule team meetings. As the program matured, Lotus added services to it and renamed it Domino (Lotus Notes is now the name of the e-mail package), and now an entire third-party market has emerged to build collaborative software based on Domino.

Enterprise Application Software

Software that benefits an entire organization can also be developed or purchased. Some software vendors, such as SAP, specialize in developing software for enterprises.[33] A fast-food chain, for example, might develop a materials ordering and distribution program to make sure that each of its franchises gets the necessary raw materials and supplies during the week. This materials ordering and distribution program can be developed internally using staff and resources in the IS department or purchased from an external software company. Boeing and DaimlerChrysler use enterprise software to design new airplanes and automotive products.[34] The software simulates the effectiveness and safety of designs, allowing the companies to save time and money compared to developing physical prototypes of airplanes and vehicles. Dunkin Brands, owner of Dunkin' Donuts, Baskin-Robbins, and Togo's, uses enterprise software to help it locate new stores.[35] iSite from geoVue (*www.geovue.com*) is site-selection software that lets companies analyze factors to help determine the location of new stores. According to John Dawson, chief development officer for Dunkin, "Years ago, guys like myself did this on gut feelings. Our goal is to get over 1,000 [stores] per year, and we couldn't do it without these tools."

One of the first enterprise applications was a payroll program for Lyons Bakeries in England, developed in 1954 on the Leo 1 computer. Table 4.8 lists some applications that can be addressed with enterprise software. Many organizations are moving to integrated enterprise software that supports supply chain management (movement of raw materials from suppliers through shipment of finished goods to customers), as shown in Figure 4.20.

Organizations can no longer respond to market changes using nonintegrated information systems based on overnight processing of yesterday's business transactions, conflicting data models, and obsolete technology. Wal-Mart and many other companies have sophisticated information systems to speed processing and coordinate communications between stores and their main offices. Many corporations are turning to **enterprise resource planning (ERP)** software, a set of integrated programs that manage a company's vital business operations for an entire multisite, global organization. Thus, an ERP system must be able to support many legal entities, languages, and currencies. Although the scope can vary from vendor to vendor, most ERP systems provide integrated software to support manufacturing and finance. In addition to these core business processes, some ERP systems might support business functions such as human resources, sales, and distribution. The primary benefits of implementing ERP include eliminating inefficient systems, easing adoption of improved work processes, improving access to data for operational decision making, standardizing technology vendors and equipment, and enabling supply chain management. Even small businesses can benefit from enterprise application software. Intuit's QuickBooks and Microsoft's Office Small Business Accounting are accounting and record-keeping programs for small businesses and organizations.[36]

Accounts receivable	Sales ordering
Accounts payable	Order entry
Airline industry operations	Payroll
Automatic teller systems	Human resource management
Cash-flow analysis	Check processing
Credit and charge card administration	Tax planning and preparation
Manufacturing control	Receiving
Distribution control	Restaurant management
General ledger	Retail operations
Stock and bond management	Invoicing
Savings and time deposits	Shipping
Inventory control	Fixed asset accounting

Table 4.8

Examples of Enterprise Application Software

Application Software for Information, Decision Support, and Specialized Purposes

Specialized application software for information, decision support, and other purposes is available in every industry. Genetic researchers, for example, are using software to visualize and analyze the human genome. Music executives use decision support software to help pick the next hit. Sophisticated decision support software is also being used to increase the cure rate for cancer by analyzing about 100 scans of a cancerous tumor to create a 3-D view of the tumor. Software can then consider thousands of angles and doses of radiation to determine the best program of radiation therapy. The software analysis takes only minutes, but the results can save years or decades of life for the patient. As you will see in future chapters, information, decision support, and specialized systems are used in businesses of all sizes and types to increase profits or reduce costs. But how are all these systems actually developed or built? The answer is through the use of programming languages, discussed next.

Figure 4.20

Use of Integrated Supply Chain
Management Software

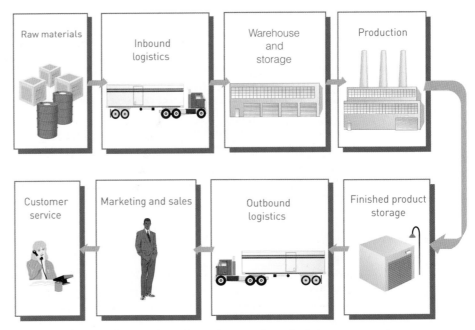

Figure 4.20

Use of Integrated Supply Chain
Management Software

Integrated enterprise software to support supply chain management

PROGRAMMING LANGUAGES

Both OSs and application software are written in coding schemes called *programming languages*. The primary function of a programming language is to provide instructions to the computer system so that it can perform a processing activity. IS professionals work with **programming languages**, which are sets of keywords, symbols, and rules for constructing statements by which people can communicate instructions to be executed by a computer. Programming involves translating what a user wants to accomplish into a code that the computer can understand and execute. *Program code* is the set of instructions that signal the CPU to perform circuit-switching operations. In the simplest coding schemes, a line of code typically contains a single instruction such as, "Retrieve the data in memory address X." As discussed in Chapter 3, the instruction is then decoded during the instruction phase of the machine cycle. Like writing a report or a paper in English, writing a computer program in a programming language requires the programmer to follow a set of rules. Each programming language uses symbols that have special meaning. Each language also has its own set of rules, called the **syntax** of the language. The language syntax dictates how the symbols should be combined into statements capable of conveying meaningful instructions to the CPU. A rule that "Variable names must start with a letter" is an example. A variable is a quantity that can take on different values. Program variable names such as SALES, PAYRATE, and TOTAL follow the rule because they start with a letter, whereas variables such as %INTEREST, $TOTAL, and #POUNDS do not.

programming languages
Sets of keywords, symbols, and a system of rules for constructing statements by which humans can communicate instructions to be executed by a computer.

syntax
A set of rules associated with a programming language.

The Evolution of Programming Languages

The desire to use the power of information processing efficiently in problem solving has pushed the development of newer programming languages. The evolution of programming languages is typically discussed in terms of generations of languages (see Table 4.9).

Visual, Object-Oriented, and Artificial Intelligence Languages

Today, programmers often use visual and object-oriented languages. In the future, they will likely be using artificial intelligence languages to a greater extent. In general, these languages are easier for nonprogrammers to use compared with older generation languages.

Generation	Language	Approximate Development Date	Sample Statement or Action
First	Machine language	1940s	00010101
Second	Assembly language	1950s	MVC
Third	High-level language	1960s	READ SALES
Fourth	Query and database languages	1970s	PRINT EMPLOYEE NUMBER IF GROSS PAY>1000
Beyond Fourth	Natural and intelligent languages	1980s	IF gross pay is greater than 40, THEN pay the employee overtime pay.

Table 4.9

The Evolution of Programming Languages

Visual languages use a graphical or visual interface for program development. Unlike earlier languages that depended on writing detailed programming statements, visual languages allow programmers to "drag and drop" programming elements and icons onto the computer screen. Many of these languages are used to develop Web applications. *Visual Basic* was one of the first visual programming languages. Microsoft Visual Studio .NET is a set of object-oriented programming languages and tools to develop Windows and Web-based applications. You can develop applications that can range from a simple Web-based program to display your résumé to complex business applications that process customer orders, control inventory, and send out bills, using languages such as Visual Basic .NET, Visual C++ .NET, Visual C#, and Visual J#. *C++* is a powerful and flexible programming language used mostly by computer systems professionals to develop applications. *Java* is an object-oriented programming language developed by Sun Microsystems that can run on any OS and on the Internet. Java can be used to develop complete applications or smaller applications, called *Java applets*. Many of these languages are also examples of object-oriented languages, which are discussed next.

The preceding programming languages separate data elements from the procedures or actions that will be performed on them, but another type of programming language ties them together into units called *objects*. An object consists of data and the actions that can be performed on the data. For example, an object could be data about an employee and all the operations (such as payroll calculations) that might be performed on the data. Programming languages that are based on objects are called *object-oriented programming languages*.

Building programs and applications using object-oriented programming languages is like constructing a building using prefabricated modules or parts. The object containing the data, instructions, and procedures is a programming building block. The same objects (modules or parts) can be used repeatedly. One of the primary advantages of an object is that it contains reusable code. In other words, the instruction code within that object can be reused in different programs for a variety of applications, just as the same basic prefabricated door can be used in two different houses. An object can relate to data on a product, an input routine, or an order-processing routine. An object can even direct a computer to execute other programs or to retrieve and manipulate data. So, a sorting routine developed for a payroll application could be used in both a billing program and an inventory control program. By reusing program code, programmers can write programs for specific application problems more quickly (see Figure 4.21). By combining existing program objects with new ones, programmers can easily and efficiently develop new object-oriented programs to accomplish organizational goals.

Programmers often start writing object-oriented programs by developing one or more user interfaces, usually in a Windows or Web environment. You can create programs to run in a Windows or Web environment by using forms to design and develop the type of interface you want. You can select and drag text boxes to add descriptions, buttons that can be clicked and executed, a list box that contains several choices that can be selected, and other input/output features. After creating the Windows interface, you can write programming code to convert tasks a user selects in the interface into actions the computer performs.

Figure 4.21

Reusable Code in Object-Oriented Programming

By combining existing program objects with new ones, programmers can easily and efficiently develop new object-oriented programs to accomplish organizational goals. Note that these objects can be either commercially available or designed internally.

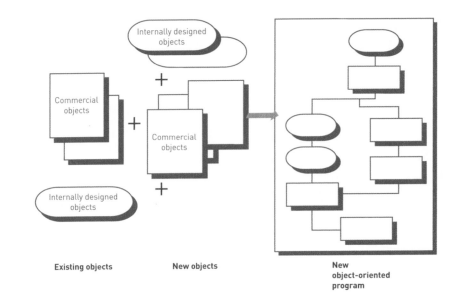

Existing objects   New objects   New object-oriented program

Some of the most popular object-oriented programming languages include Smalltalk, Visual Basic .NET, C++, and Java. See Figure 4.22. Some old languages, such as COBOL, have been modified to support the object-oriented approach. As mentioned earlier, Java is an Internet programming language from Sun Microsystems that can run on a variety of computers and OSs, including UNIX, Windows, and Macintosh OSs.

Figure 4.22

Microsoft Visual Basic

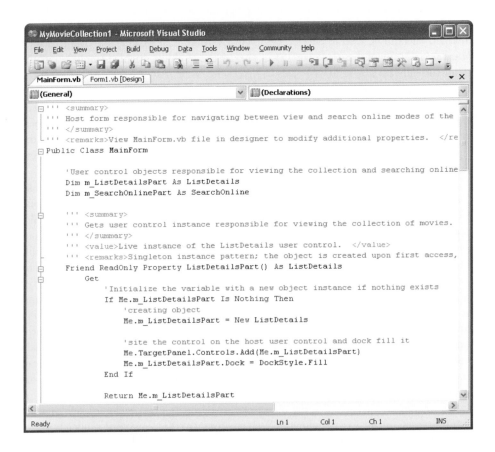

Object-oriented programs often use *methods*, which are instructions to perform a specific task in the program. The following instructions in C++ use a method named ComputeArea to compute the area of a rectangle, given the width and length.

```
// Method to Compute the Area of a Rectangle Given the Width and
// Length
double Rectangle::ComputeArea()
{
    return width * length;
}
// End of the ComputeArea Method
```

After they are developed as part of a C++ program, the instructions or method can be used in other programs to compute the area of a picture frame, a living room, a front lawn, or any other application that requires the area of a rectangle. Following are a few instructions in another C++ program that show how to use the ComputeArea method to compute the area of a picture frame.

```
// Assign Data and Compute Area
frameObject -> SetDimensions (frameWidth, frameLength);
frameArea = frameObject -> ComputeArea();
```

Programming languages used to create artificial intelligence or expert systems applications are often called *fifth-generation languages (5GLs)*. FLEXPERT, for example, is an expert system used to perform plant layout and helps companies determine the best placement for equipment and manufacturing facilities. Fifth-generation languages are sometimes called *natural languages* because they use even more English-like syntax than 4GLs. They allow programmers to communicate with the computer by using normal sentences. For example, computers programmed in fifth-generation languages can understand queries such as, "How many athletic shoes did our company sell last month?"

With third-generation and higher-level programming languages, each statement in the language translates into several instructions in machine language. A special software program called a **compiler** converts the programmer's source code into the machine-language instructions consisting of binary digits, as shown in Figure 4.23. A compiler creates a two-stage process for program execution. First, the compiler translates the program into a machine language; second, the CPU executes that program. Another approach is to use an *interpreter*, which is a language translator that carries out the operations called for by the source code. An interpreter does not produce a complete machine-language program. After the statement executes, the machine-language statement is discarded, the process continues for the next statement, and so on.

compiler

A special software program that converts the programmer's source code into the machine-language instructions consisting of binary digits.

Stage 1:   **Convert program**

Stage 2:   **Execute program**

Program execution

Figure 4.23

How a Compiler Works

A compiler translates a complete program into a complete set of binary instructions (Stage 1). After this is done, the CPU can execute the converted program in its entirety (Stage 2).

SOFTWARE ISSUES AND TRENDS

Because software is such an important part of today's computer systems, issues such as software bugs, licensing, upgrades, and global software support have received increased attention. We highlight several major software issues and trends in this section: software bugs, copyright, software licensing, open-source software, shareware and public domain software, multiorganizational software development, software upgrades, and global software support.

Software Bugs

A software bug is a defect in a computer program that keeps it from performing as it is designed to perform. Some software bugs are obvious and cause the program to terminate unexpectedly. Other bugs are subtler and allow errors to creep into your work. Computer and software vendors say that as long as people design and program hardware and software, bugs are inevitable. In fact, according to the Pentagon and the Software Engineering Institute at Carnegie Mellon University, there are typically 5 to 15 bugs in every 1,000 lines of code—the software instructions that make sense only to computers and programmers. The following list summarizes tips for reducing the impact of software bugs.

- Register all software so that you receive bug alerts, fixes, and patches.
- Check the manual or read-me files for work-arounds.
- Access the support area of the manufacturer's Web site for patches.
- Install the latest software updates.
- Before reporting a bug, make sure that you can re-create the circumstances under which it occurs.
- After you can re-create the bug, call the manufacturer's tech support line.
- Avoid buying the latest release of software for several months or a year until the software bugs have been discovered and removed.

Copyrights and Licenses

Most software products are protected by law using copyright or licensing provisions. Those provisions can vary, however. In some cases, you are given unlimited use of software on one or two computers. This is typical with many applications developed for personal computers. In other cases, you pay for your usage—if you use the software more, you pay more. This approach is becoming popular with software placed on networks or larger computers. Most of these protections prevent you from copying software and giving it to others without restrictions. Some software now requires that you *register* or *activate* it before it can be fully used. Registration and activation sometimes put software on your hard disk that monitors activities and changes your computer system.

Open-Source Software

open-source software
Software that is freely available to anyone in a form that can be easily modified.

Open-source software is freely available to anyone in a form that can be easily modified. The Open Source Initiative (OSI) is a nonprofit corporation dedicated to the development and promotion of open-source software (see the OSI Web site at *www.opensource.org* for more information on the group's efforts). Users can download the source code and build the software themselves, or the software developers can make executable versions available along with the source. Open-source software development is a collaborative process—developers around the world use the Internet to keep in close contact via e-mail and to download and submit new software. Major software changes can occur in days rather than weeks or months. Many open-source software packages are widely used, including the Linux OS; Free BSD, another OS; Apache, a popular Web server; Sendmail, a program that delivers e-mail for most systems on the Internet; and Perl, a programming language used to develop Internet application software. See Table 4.10 for some examples of open-source software.

ETHICAL AND SOCIETAL ISSUES

When Software Goes Wrong

Salesforce.com is the worldwide leader in on-demand customer relationship management (CRM) services. Customer relationship management systems assist companies in all aspects of customer encounters, including marketing, sales, and service. For example, sales representatives can access customer information online no matter their physical location using a Salesforce.com system. On-demand technology is used by several information system companies, such as Salesforce, IBM, and Oracle, and means that technology solutions are available to clients as they are needed, spontaneously, with little to no development time required. Salesforce delivers CRM solutions on demand by hosting their services on their own systems. In so doing, they can help new clients set up and use their system in a matter of weeks or days. Traditional client-server CRM systems take months or years to set up.

CRM and on-demand technologies have become popular in many industries, and competition in this market is high. Companies in the business of providing CRM solutions work hard to improve their systems to win new customers and hang on to existing ones. At the close of 2005, Salesforce instituted a massive upgrade to its software and systems to incorporate new cutting-edge programming techniques. The upgrade required changes throughout their systems. "We had to rebuild our hardware infrastructure to support it and we have rewritten every piece of Salesforce.com software from the application, to integration, caching, database search, and database management," said CEO Marc Benioff. The upgrade included $50 million in hardware.

Rushing to employ cutting-edge technologies can help a business gain an advantage in the market, or prove embarrassing if the new technology fails. Salesforce.com nearly suffered the latter.

Shortly after Salesforce.com set up its system, a software error interrupted their services for about six hours, meaning that clients could not access their customer information during that time. Two weeks later, another outage affected the company's Europe, Middle East, and Africa regions. A week after that, some Salesforce.com servers were down for four hours. Salesforce.com has 399,000 subscribers worldwide, each depending on Salesforce.com to manage their sales, marketing, and customer center and customer service. With such a large business, downtime can be catastrophic.

Salesforce.com was forced to act fast to regain the trust of their clients. By mid-February, the company had posted a new Web site named Trust.Salesforce.com that provides useful and reassuring information to clients concerning the state of the system. Information at Trust.Salesforce.com includes data on the system's performance, throughput, and transaction rates. It also provides information on outages and the cause of any problems.

After the January outage, Marc Benioff was quoted in news reports as saying, "Nothing runs at 100% availability." Salesforce.com claims to be close, however, stating that its availability has never dropped below 99% in its history of operation. Salesforce.com competitors offer service license agreements (SLAs) to their clients. For example, competitor NetSuite guarantees 99.5% uptime, Salesnet guarantees 99.6%, and RightNow guarantees 99.5% availability. So far, Salesforce.com has not provided an SLA, and at least some customers agree that it isn't necessary.

Customer Clifford Bell, CIO of California-based Phoenix Technologies Ltd., said that the outage "was not a significant problem." He was pleased to receive a note from Benioff that detailed Salesforce.com's plans to eliminate the outages. "Funny, but I never have gotten an e-mail from Larry Ellison for any Oracle issues or from Bill Gates when there are Microsoft issues," Bell said, adding that he thinks Salesforce.com is working hard to correct the problems.

Discussion Questions

1. How might Salesforce.com have prevented the problems caused by its upgrade?
2. How did Salesforce.com redeem itself in its customer's eyes?

Critical Thinking Questions

1. Do you feel that software is inherently flawed and people should simply put up with an occasional problem, or should software come with a money-back guarantee?
2. If you were CEO of Salesforce.com, what other actions would you take to qualm the concerns of your customers?

SOURCES: Songini, Marc L., "Salesforce.com's CRM Service Hiccups Again," *Computerworld*, February 20, 2006. Keizer, Gregg, "Salesforce on the Spot—Trust, But Verify," *InformationWeek*, February 27, 2006, p. 19. "Salesforce.com Crashes Again," *ComputerWire*, February 2, 2006, *www.lexis-nexis.com*. Salesforce.com Web site, accessed March 1, 2006, *www.salesforce.com/company*.

Table 4.10

Examples of Open-Source
Software

Software Type	Example
Operating system	Linux
Application software	Open Office
Database software	MySQL
Internet browser	Firefox
Internet messaging	Jabber

Why would an organization run its business using software that's free? Can something that's given away over the Internet be stable or reliable or sufficiently supported to place at the core of a company's day-to-day operations? The answer is surprising—open-source software is often *more* reliable and secure than commercial software.[37] How can this be? First, by making a program's source code readily available, users can fix any problems they discover. A fix is often available within hours of the problem's discovery. Second, with the source code for a program accessible to thousands of people, the chances of a bug being discovered and fixed before it does any damage are much greater than with traditional software packages. Of course, open-source software is usually much less expensive than traditional software that is purchased from a software vendor. The auditor of the state of Massachusetts estimates that the cost savings using open-source software could be as high at $10 million compared to developing software internally, when legal and project delays are included.[38] Some companies are also starting to reveal their source code. IBM recently made about $10 million worth of software patents available to open-source software developers at no charge.[39] Microsoft has offered to reveal some of its source code for its Windows operating system as a result of pressure from some European countries.[40]

There are, however, disadvantages of using open-source software. Although open-source systems can be obtained for next to nothing, the up-front costs are only a small piece of the total cost of ownership that accrues over the years that the system is in place.[41] Some claim that open-source systems contain many hidden costs, particularly for user support or solving problems with the software. Licensed software comes with guarantees and support services that open-source software does not. Still, many businesses appreciate the additional freedom that open-source software provides. The question of software support is the biggest stumbling block to the acceptance of open-source software at the corporate level. Getting support for traditional software packages is easy—you call a company's toll-free support number or access its Web site. But how do you get help if an open-source package doesn't work as expected? Because the open-source community lives on the Internet, you look there for help. Through use of Internet discussion areas, you can communicate with others who use the same software, and you might even reach someone who helped develop it. Users of popular open-source packages can get correct answers to their technical questions within a few hours of asking for help on the appropriate Internet forum. Another approach is to contact one of the many companies emerging to support and service such software—for example, Red Hat for Linux, C2Net for Apache, and Sendmail, Inc., for Sendmail. These companies offer high-quality, for-pay technical assistance.

Shareware, Freeware, and Public Domain Software

Many software users are doing what they can to minimize software costs. Some are turning to **shareware** and **freeware**—software that is very inexpensive or free, usually for use in personal computers, but whose source code cannot be modified. Freeware can be used to perform a variety of tasks.[42] Microsoft Windows Defender, for example, is a program still being tested to find and eliminate spyware. See Figure 4.24. Spybot Search and Destroy is another freeware program to detect and eliminate spyware. Perio Personal Firewall can protect a computer from viruses and intruders. OpenOffice is an office suite that contains a word processor, a spreadsheet, and a presentation program. PhotoPlus 6 is a photo-editing

shareware and freeware
Software that is very inexpensive or free, but whose source code cannot be modified.

program, and Picasa is a photo-editing and managing program. The Web site *www.SourceForge.net* is a resource for programmers to freely exchange programs and program code.[43] It allows programmers to create, participate, and evaluate program code. Over 80,000 programs are at various stages of completion.

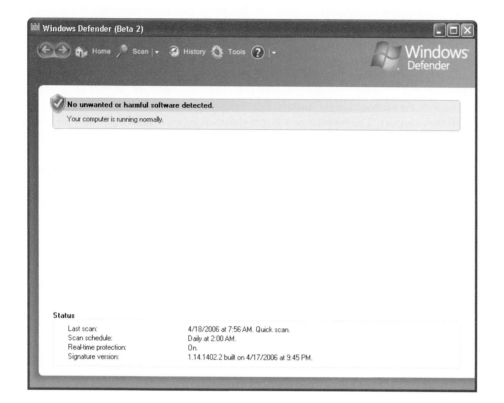

Figure 4.24

Microsoft Windows Defender

Shareware might not be as powerful as commercial software, but provides what some people need at a good price. In some cases, you can try the software before sending a nominal fee to the software developer. Some shareware and freeware is in the public domain, often called *public domain software*. This software is not protected by copyright laws and can be freely copied and used. Although shareware and freeware can be free or inexpensive to acquire, it can be more expensive to use and maintain over time compared with software that is purchased. If the software is hard to use and doesn't perform all the required functions, the cost of wasted time and lost productivity can be far greater than the cost of purchasing better software. Shareware, freeware, and public domain software is often not open source—that is, the source code is not available and cannot be modified.

Software Upgrades

Software companies revise their programs and sell new versions periodically. In some cases, the revised software offers new and valuable enhancements. In other cases, the software uses complex program code that offers little in terms of additional capabilities. In addition, revised software can contain bugs or errors. When software companies stop supporting older software versions or releases, some customers feel forced to upgrade to the newer software. Deciding whether to purchase the newest software can be a problem for corporations and people with a large investment in software. Should the newest version be purchased when it is released? Some users do not always get the most current software upgrades or versions, unless it includes significant improvements or capabilities. Instead, they might upgrade to newer software only when it offers vital new features. Software upgrades usually cost much less than the original purchase price.

Global Software Support

Large global companies have little trouble persuading vendors to sell them software licenses for even the most far-flung outposts of their company. But can those same vendors provide adequate support for their software customers in all locations? Supporting local operations is one of the biggest challenges IS teams face when putting together standardized, company-wide systems. Slower technology growth markets, such as Eastern Europe and Latin America, might not have any official vendor presence. Instead, large vendors such as Sybase, IBM, and Hewlett-Packard typically contract with local providers to support their software.

One approach that has been gaining acceptance in North America is to outsource global support to one or more third-party distributors. The user company can still negotiate its license with the software vendor directly, but it then hands the global support contract to a third-party supplier. The supplier acts as a middleman between software vendor and user, often providing distribution, support, and invoicing. American Home Products Corporation handles global support for both Novell NetWare and Microsoft Office applications this way throughout the 145 countries in which it operates. American Home Products, a pharmaceutical and agricultural products company, negotiated the agreements directly with the vendors for both purchasing and maintenance, but fulfillment of the agreement is handled exclusively by Philadelphia-based Softsmart, an international supplier of software and services.

In today's computer systems, software is an increasingly critical component. Whatever approach people and organizations take to acquire software, everyone must be aware of the current trends in the industry. Informed users are wiser consumers, and they can make better decisions.

SUMMARY

Principle

Systems and application software are critical in helping individuals and organizations achieve their goals.

Software consists of programs that control the workings of the computer hardware. The two main categories of software are systems software and application software. Systems software is a collection of programs that interacts between hardware and application software, and includes operating systems, utility programs, and middleware. Application software can be proprietary or off the shelf, and enables people to solve problems and perform specific tasks.

An operating system (OS) is a set of computer programs that controls the computer hardware to support users' computing needs. An OS converts an instruction from an application into a set of instructions needed by the hardware. This intermediary role allows hardware independence. An OS also manages memory, which involves controlling storage access and use by converting logical requests into physical locations and by placing data in the best storage space, perhaps virtual memory.

An OS manages tasks to allocate computer resources through multitasking and time-sharing. With multitasking, users can run more than one application at a time. Time-sharing allows more than one person to use a computer system at the same time.

The ability of a computer to handle an increasing number of concurrent users smoothly is called *scalability*, a feature critical for systems expected to handle a large number of users.

An OS also provides a user interface, which allows users to access and command the computer. A command-based user interface requires text commands to send instructions; a graphical user interface (GUI), such as Windows, uses icons and menus.

Software applications use the OS by requesting services through a defined application program interface (API). Programmers can use APIs to create application software without having to understand the inner workings of the OS. APIs also provide a degree of hardware independence so that the underlying hardware can change without necessarily requiring a rewrite of the software applications.

Over the years, several popular OSs have been developed. These include several proprietary OSs used primarily on mainframes. MS-DOS is an early OS for IBM-compatibles. Older Windows OSs are GUIs used with DOS. Newer versions, such as Windows Vista and XP, are fully functional OSs that do not need DOS. Apple computers use proprietary OSs such as the Mac OS and Mac OS X. UNIX is a powerful OS that can be used on many computer system types and platforms, from personal computers to mainframe systems. UNIX makes it easy to move programs and data among computers or to

connect mainframes and personal computers to share resources. Linux is the kernel of an OS whose source code is freely available to everyone. Several variations of Linux are available, with sets of capabilities and applications to form a complete OS, for example, Red Hat Linux. z/OS and MPE iX are OSs for mainframe computers. Some OSs have been developed to support consumer appliances such as Palm OS, Windows CE .Net, Windows XP Embedded, Pocket PC, and variations of Linux.

Utility programs can perform many useful tasks and often come installed on computers along with the OS. This software is used to merge and sort sets of data, keep track of computer jobs being run, compress files of data, protect against harmful computer viruses, and monitor hardware and network performance. Middleware is software that allows different systems to communicate and transfer data back and forth.

Principle

Do not develop proprietary application software unless doing so will meet a compelling business need that can provide a competitive advantage.

Application software applies the power of the computer to solve problems and perform specific tasks. One useful way of classifying the many potential uses of information systems is to identify the scope of problems and opportunities addressed by a particular organization or its sphere of influence. For most companies, the spheres of influence are personal, workgroup, and enterprise.

User software, or personal productivity software, includes general-purpose programs that enable users to improve their personal effectiveness, increasing the quality and amount of work that can be done. Software that helps groups work together is often called workgroup application software, and includes group scheduling software, electronic mail, and other software that enables people to share ideas. Enterprise software that benefits the entire organization can also be developed or purchased. Many organizations are turning to enterprise resource planning software, a set of integrated programs that manage a company's vital business operations for an entire multisite, global organization.

Three approaches to developing application software are to build proprietary application software, buy existing programs off the shelf, or use a combination of customized and off-the-shelf application software. Building proprietary software (in-house or on contract) has the following advantages: The organization will get software that more closely matches its needs; by being involved with the development, the organization has further control over the results; and the organization has more flexibility in making changes. The disadvantages include the following: It is likely to take longer and cost more to develop, the in-house staff will be hard pressed to provide ongoing support and maintenance, and there is a greater risk that the software features will not work as expected or that other performance problems will occur.

Purchasing off-the-shelf software has many advantages. The initial cost is lower, there is a lower risk that the software will fail to work as expected, and the software is likely to be of higher quality than proprietary software. Some disadvantages are that the organization might pay for features it does not need, the software might lack important features requiring expensive customization, and the system might require process reengineering.

Some organizations have taken a third approach—customizing software packages. This approach usually involves a mixture of the preceding advantages and disadvantages and must be carefully managed.

An application service provider (ASP) is a company that can provide the software, support, and computer hardware on which to run the software from the user's facilities. ASPs provide contract customization of off-the-shelf software, and they speed deployment of new applications while helping IS managers avoid implementation headaches. Use of ASPs reduces the need for many skilled IS staff members and also lowers a project's start-up expenses.

Although hundreds of computer applications can help people at school, home, and work, the primary applications are word processing, spreadsheet analysis, database, graphics, and online services. A software suite, such as SmartSuite, WordPerfect, StarOffice, or Office, offers a collection of powerful programs.

Principle

Choose a programming language whose functional characteristics are appropriate for the task at hand, considering the skills and experience of the programming staff.

All software programs are written in coding schemes called *programming languages*, which provide instructions to a computer to perform some processing activity. The several classes of programming languages include machine, assembly, high-level, query and database, object-oriented, and visual programming languages.

Programming languages have changed since their initial development in the early 1950s. In the first generation, computers were programmed in machine language, and the second generation of languages used assembly languages. The third generation consists of many high-level programming languages that use English-like statements and commands.

They also must be converted to machine language by special software called a compiler, and include BASIC, COBOL, FORTRAN, and others. Fourth-generation languages include database and query languages such as SQL.

Fifth-generation programming languages combine rules-based code generation, component management, visual programming techniques, reuse management, and other advances. Visual and object-oriented programming languages—such as Smalltalk, C++, and Java—use groups of related data, instructions, and procedures called *objects*, which serve as reusable modules in various programs. These languages can reduce program development and testing time. Java can be used to develop applications on the Internet.

Principle

The software industry continues to undergo constant change; users need to be aware of recent trends and issues to be effective in their business and personal life.

Software bugs, software licensing and copyrighting, open-source software, shareware and freeware, multiorganizational software development, software upgrades, and global software support are all important software issues and trends.

A software bug is a defect in a computer program that keeps it from performing in the manner intended. Software bugs are common, even in key pieces of business software.

Open-source software is software that is freely available to anyone in a form that can be easily modified. Open-source software development and maintenance is a collaborative process with developers around the world using the Internet to keep in close contact via e-mail and to download and submit new software. Shareware and freeware can reduce the cost of software, but sometimes they might not be as powerful as commercial software. Also, their source code usually cannot be modified.

Multiorganizational software development is the process of extending software development beyond a single organization by finding others who share the same business problem and involving them in a common development effort.

Software upgrades are an important source of increased revenue for software manufacturers and can provide useful new functionality and improved quality for software users.

Global software support is an important consideration for large, global companies putting together standardized, company-wide systems. A common solution is outsourcing global support to one or more third-party software distributors.

CHAPTER 4: SELF-ASSESSMENT TEST

Systems and application software are critical in helping individuals and organizations achieve their goals.

1. Which of the following is an example of a command-driven operating system?
   a. XP
   b. Tiger
   c. MS DOS
   d. Vista

2. A command-based user interface uses windows and menus used to control the computer. True or False?

3. _____ is the process of swapping programs or parts of programs between memory and disk.

4. The file manager component of the OS controls how memory is accessed and maximizes available memory and storage. True or False?

Do not develop proprietary application software unless doing so will meet a compelling business need that can provide a competitive advantage.

5. The primary function of application software is to apply the power of the computer to give people, workgroups, and the entire enterprise the ability to solve problems and perform specific tasks. True or False?

6. Software that enables users to improve their personal effectiveness, increasing the amount of work they can do and its quality is called _____.
   a. personal productivity software
   b. operating system software
   c. utility software
   d. graphics software

7. Optimization can be found in which type of application software?
   a. spreadsheets
   b. word processing programs
   c. database programs
   d. presentation graphics programs

8. Software used to solve a unique or specific problem that is usually built in-house, but can also be purchased from an outside company is called _____.

9. A program to detect and eliminate viruses is an example of what type of software?
   a. personal productivity software
   b. operating system software
   c. utility software
   d. applications software

Choose a programming language whose functional characteristics are appropriate for the task at hand, considering the skills and experience of the programming staff.

10. A part of every Microsoft Office application that provides a means of enhancing off-the-shelf applications to allow users to tailor the programs is called _____.
    a. Visual Basic for Applications
    b. SmallTalk
    c. Norton Utilities
    d. Java

11. A class of application software that helps groups work together and collaborate is called _____.

12. Each programming language has its own set of rules, called the _____ of the language.

13. A special software program called an *interpreter* performs the conversion from the programmer's source code into the machine-language instructions consisting of binary digits and results in a machine-language program. True or False?

CHAPTER 4: SELF-ASSESSMENT TEST ANSWERS

(1) c (2) False (3) Paging (4) False (5) True (6) a (7) a (8) proprietary software (9) c (10) a (11) workgroup application software (12) syntax (13) False

REVIEW QUESTIONS

1. What is the difference between systems and application software? Give four examples of personal productivity software.
2. How do software bugs arise?
3. Identify and briefly discuss two types of user interfaces provided by an operating system.

4. What is a software suite? Give several examples.
5. Name four operating systems that support the personal, workgroup, and enterprise spheres of influence.
6. What is middleware?
7. What is multitasking?
8. Define the term *utility* software and give two examples.

9. Identify the two primary sources for acquiring application software.
10. What is an application service provider? What issues arise in considering the use of one?
11. What is open-source software? What is the biggest stumbling block with the use of open-source software?
12. What does the acronym API stand for? What is the role of an API?
13. Briefly discuss the advantages and disadvantages of frequent software upgrades.
14. Describe the term *enterprise resource planning (ERP) system*. What functions does such a system perform?
15. What is freeware? Give two examples.

DISCUSSION QUESTIONS

1. Assume that you must take a computer-programming course next semester. What language do you think would be best for you to study? Why? Do you think that a professional programmer needs to know more than one programming language? Why or why not?
2. You are going to buy a personal computer. What operating system features are important to you? What operating system would you select?
3. Identify the fundamental types of application software. Discuss the advantages and disadvantages of each type.
4. Describe how the OS can manage the computer's memory.
5. You are using a new release of an application software package. You think that you have discovered a bug. Outline the approach that you would take to confirm that it is indeed a bug. What actions would you take if it truly were a bug?
6. How can application software improve the effectiveness of a large enterprise? What are some of the benefits associated with implementation of an enterprise resource planning system? What are some of the issues that could keep the use of enterprise resource planning software from being successful?
7. Define the term *application service provider*. What are some of the advantages and disadvantages of employing an ASP? What precautions might you take to minimize the risk of using one?
8. Describe three personal productivity software packages you are likely to use the most. What personal productivity software packages would you select for your use?
9. Contrast and compare two popular OSs for personal computers.
10. If you were the IT manager for a large manufacturing company, what issues might you have with the use of open-source software? What advantages might there be for use of such software?
11. Identify four types of software licenses frequently used. Which approach does the best job of ensuring a steady, predictable stream of revenue from customers? Which approach is most fair for the small company that makes infrequent use of the software?

PROBLEM-SOLVING EXERCISES

1. Choose an application software package that might be useful for a career that interests you and develop a six-slide presentation of its history, current level of usage, typical applications, ease of use, and so on.
2. Use a spreadsheet package to prepare a simple monthly budget and forecast your cash flow—both income and expenses for the next six months (make up numbers rather than using actual ones). Now use a graphics package to plot the total monthly income and monthly expenses for six months. Cut and paste both the spreadsheet and the graph into a word processing document that summarizes your financial condition.
3. Use a database program to enter five software products you are likely to use at work. List the name, vendor or manufacturer, cost, and features in the columns of a database table. Use a word processor to write a report on the software. Copy the database table into the word processing program.

TEAM ACTIVITIES

1. Form a group of three or four classmates. Find articles from business periodicals, search the Internet, or interview people on the topic of software bugs. How frequently do they occur, and how serious are they? What can software users do to encourage defect-free software? Compile your results for an in-class presentation or a written report.
2. Form a group of three or four classmates. Identify and contact an employee of a local firm. Interview the individual and describe the application software the company uses and the importance of the software to the organization. Write a brief report summarizing your findings.

3. Divide your team into two groups. The first group should prepare a report using a word processing program. Make sure to include a large number of spelling, grammatical, and similar errors in the document. The second group should use the word processing program's features to locate and eliminate the errors. The entire team should write a report on the advantages and limitations of the spelling and grammar checking features of the word processing program you used. What additional features would you like to see in future word processing programs?

WEB EXERCISES

1. Use the Internet to search for three popular freeware utilities that you would find useful. Write a report that describes the features of these three utility programs.
2. Do research on the Web and develop a two-page report summarizing the latest consumer appliance OSs. Which one seems to be gaining the most widespread usage? Why do you think this is the case?

3. Do research on the Web about application software that is used in an industry and is of interest to you. Write a brief report describing how the application software can be used to increase profits or reduce costs.

CAREER EXERCISES

1. What personal computer OS would help you the most in the first job you would like to have after you graduate? Why? What features are the most important to you?
2. Think of your ideal job. Describe five application software packages that could help you advance in your career. If the

software package doesn't exist, describe the kinds of software packages that could help you in your career.

CASE STUDIES

Case One

DreamWorks SKJ Goes Completely Open Source

Steven Spielberg, Jeffrey Katzenberg, and David Geffen launched DreamWorks SKG in October 1994. They have subsequently produced motion picture hits such as *Antz*, *Shrek*, *Madagascar*, *A.I.*, *Galaxy Quest*, *Saving Private Ryan*, and *Wallace and Gromit - The Curse of the Were-Rabbit*.

DreamWorks was originally set up with expensive servers from Sun Microsystems and workstations from SGI running the UNIX operating system, high-end graphics software, and other specialized systems. Around the turn of the millennium, DreamWorks started experimenting with Intel-based servers running Linux, the open-source operating system. Using Intel/Linux servers, DreamWorks produced *Shrek* in 2001 on a system that cost half as much and was four times as powerful as the SGI/UNIX system used to produce *Antz* in 1998. DreamWorks caught the open-source fever.

Since then, DreamWorks has steadily transformed itself into a complete user of open-source software. The biggest challenge has been finding the specialty software required for motion picture animation and production to run on the Linux platform. With little Linux software commercially available, DreamWorks has been writing its own. Working with third-party software partners and HP, DreamWorks has been translating, or porting, its software from its old SGI/UNIX system to Linux. This is no small task considering that its in-house animation software includes millions of lines of code.

After DreamWorks ported its production software to Linux, it focused on its business applications. Recently, DreamWorks replaced a dozen of its core legacy applications with custom-designed software using a service-oriented architecture. A service-oriented architecture, commonly known as SOA, defines the use of software services to support the requirements of software users. In an SOA environment, nodes on a network make resources available to network users as independent services that the participants access in a standardized way. (Chapter 7 discusses this topic in its Web services section.)

The systems that DreamWorks updated using SOA include tasks such as tracking copyright, accessing human resources data, and pulling information from back-end ERP systems. Linux provides the API and tools to make SOA easy to develop. Developers also used the jBoss Enterprise Middleware Suite for software development. JBoss is a global leader in open-source middleware software and provides the industry's leading services and tools to transform businesses to SOA. DreamWorks used these tools to develop a new service that authenticates employee roles and responsibilities against company directories to provide access to applications. It also built a new services-based copyright application that provides authorization and authentication for incoming feature film scripts.

"Having a Linux operating environment and HP Linux servers in racks saves critical data center space," said Abe Wong, DreamWorks' head of IT. "In the animation world, data center space is extremely valuable. We've freed up space for the animation technology group to put in full racks of render-farms space to focus on films as opposed to running servers." Not only has the move to open source freed space, but it's provided a more transparent and modular system architecture that is easy to build on and maintain.

Discussion Questions

1. What benefits has DreamWorks enjoyed since migrating to open-source systems?
2. What price did DreamWorks have to pay for adopting systems that were not standard to the motion picture animation industry?

Critical Thinking Questions

1. Why do you think the SGI special-purpose systems that DreamWorks formerly used were so much more expensive than the Intel/Linux systems it uses now?
2. How has open-source SOA systems provided DreamWorks with a unique edge over its competition?

SOURCES: Havenstein, Heather, "DreamWorks Animation Aims for Open-Source with SOA Project," *Computerworld*, February 28, 2006, *www.computerworld.com*. jBoss Web site, accessed March 4, 2006, *http://jboss.com*. DreamWorks Web site, accessed March 4, 2006, *www.dreamworks.com*.

Case Two

Select Comfort Finds Comfort in ERP

Select Comfort is the bed company that invented the "sleep number" system, which provides a range of mattress firmness settings to accommodate sleeping preferences. Founded in 1987, the Minneapolis, Minnesota-based company delivered net sales of $691 million in 2005. The company has 32 U.S.-issued or pending patents and was ranked by Furniture/Today as the top bedding retailer in the nation for the sixth consecutive year.

Needless to say, a company of this size depends on enterprise-wide software systems to provide access to valuable information throughout the organization. A few years ago, Select Comfort began moving away from its hard-to-maintain legacy systems to integrated enterprise resource planning (ERP) software. The e-Business Suite from Oracle provides ERP services through a convenient Web-based interface. The suite helps Select Comfort coordinate its sophisticated made-to-order manufacturing operations in South Carolina and Utah, and keep mattress orders flowing smoothly from the store to the factory to the customer's home. Select Comfort adopted several e-Business Suite modules to assist in varying parts of its business: an order management module to fulfill the hundreds of mattress orders it receives daily, a customer relationship management (CRM) module for keeping track of customer interaction, and modules that handle typical business needs such as assets management, general ledger, payables, purchasing, and receivables. The ERP system ensures that all these modules and services are synchronized and centralized so they can provide up-to-date information.

Seeking to make use of the latest technologies, Select Comfort adopted business intelligence (BI) software from Siebel Systems, Inc. BI software allows a business to combine its databases and extract useful information to apply to business strategies. The BI software from Siebel caught the interest of Select Comfort because of its power and ease of use. Select Comfort plans to deploy Siebel Business Analytics to 2,500 users company-wide by 2008. The software will deliver alerts and dashboard capabilities to show how the company's 400 stores are performing in real time.

Select Comfort had concerns about using enterprise-wide software from two vendors, Oracle and Siebel. When companies adopt new software, the software must be able to integrate with existing systems. Select Comfort resigned itself to the fact that it would have to work with Siebel on integration issues.

Shortly after Select Comfort purchased the Siebel software, Oracle announced that it was purchasing Siebel. The partnership means that the Siebel BI software will eventually be integrated with Oracle's database and ERP software. David Dobrin, an analyst at B2B Analysts, Inc., in Cambridge, Massachusetts, said Select Comfort will likely have to wait for a strong link between the products. Integration "will take years and years, and probably Oracle will have to do a major revision to data systems," he said.

Discussion Questions

1. What benefits does Select Comfort's ERP system provide that individual software solutions from a variety of vendors could not?
2. What risk did Select Comfort assume when it chose software from a different vendor?

Critical Thinking Questions

1. How might standards assist companies who prefer to use the services of different software vendors?
2. Why do you think most large enterprises outsource their ERP systems to software companies like Oracle, Siebel, and IBM rather than developing their own?

SOURCES: Songini, Marc, "Retailer Hopes for Oracle-Siebel Integration," *Computerworld*, January 16, 2006, *www.computerworld.com*. Oracle Enterprise Management Web site, accessed March 4, 2006, *www.oracle.com/ enterprise_manager/index.html*. Select Comfort Web site, accessed March 4, 2006, *www.selectcomfort.com*.

Questions for Web Case

See the Web site for this book to read about the Whitmann Price Consulting case for this chapter. Following are questions concerning this Web case.

Whitmann Price Consulting: Software Considerations

Discussion Questions

1. What three types of software made the BlackBerry ideal for meeting the needs of Whitmann Price?
2. How did the choice of hardware affect options for software solutions? What difference would it make if Sandra and Josh picked a newly developed handheld device unknown in industry?

Critical Thinking Questions

1. For software other than that which comes on the BlackBerry, should Sandra and Josh look to BlackBerry Alliance Program members or should they just plan to have their own software engineers develop the software? What are the benefits and drawbacks of either option?
2. What process would you use to evaluate software for the Advanced Mobile Communications and Information System?

NOTES

Sources for opening vignette: "El-Al: Optimized Scheduling Produces Significant Savings in Crew Costs," IBM Case Study, March 16, 2005, *http://domino.research.ibm.com/odis/odis.nsf/pages/case.05.html*. El-Al Airlines Web site, accessed February 28, 2006, *www.elal.co.il*.

1   Kandra, Anne, "Software Licenses," *PC World*, July 2005, p. 39.
2   Mendelson, Edward, "The Ultimate USB Key," *PC Magazine*, September 6, 2005, p. 84.
3   Staff, "Make Your Thumb Drive Bootable," *PC Magazine*, September 6, 2005, p. 88.
4   Dunn, Scott, "Windows' Command Line Puts You in Control," *PC World,* September 2005, p. 146.
5   Dahl, Eric, "The Truth About Windows Alternatives," *PC World*, September 2005, p. 75
6   Staff, "PC Makers, Distributors Pass on Windows XP N," *CNN Online*, June 24, 2005.
7   Thurrott, Paul, "64-Bit Windows?" *PC World*, July 2005, p. 24.
8   Greene, Jay, et al., "Microsoft May Be a TV Star Yet," *Business Week*, February 7, 2005, p. 78.
9   Clyman, John, "Microsoft Unleashes Longhorn," *PC Magazine*, August 9, 2005, p. 92.

10   Miller, Michael, "Hands On with the Next Windows," *PC Magazine*, September 6, 2005, p. 104.
11   Spanbauer, Scott, "Longhorn Preview," *PC World*, August 2005, p. 20.
12   Rupley, Sebastian, "Apple Bets on Intel," *PC Magazine*, August 9, 2005, p. 94.
13   Dreier, Troy, "Tiger's Tale," *PC Magazine*, June 28, 2005, p. 32.
14   Mossberg, Walter, "Tiger Leaps Out Front," *The Wall Street Journal*, April 28, 2005, p. B1.
15   Staff, "Spotlight: Find Anything Fast," *www.apple.com/macosx/ features/spotlight*, accessed on September 16, 2005.
16   Hoffman, Thomas, "Grocer Rings Up Savings with Linux Cash Registers," *Computerworld,* January 31, 2005, p. 10.
17   Hamm, Steve, "Linux, Inc.," *Business Week,* January 31, 2005, p. 60.
18   Guth, Robert; Clark, Don, "Linux Feels Growing Pains as Users Demand More Features," *The Wall Street Journal*, August 8, 2005, p. B1.
19   Thibodeau, Patrick, "Sun Begins Its Release of Open-Source Solaris Code," *Computerworld*, January 31, 2005, p. 6.
20   Mossberg, Walter, "A New Palm Treo Uses Microsoft's Software," *The Wall Street Journal*, January 5, 2006, p. A13.
21   Rupley, Sebastian, "Bad Code?" *PC Magazine,* March 8, 2005, p. 19.
22   Guth, Robert, "Aiming to Fix Flaws, Microsoft Buys Another Antivirus Firm," *The Wall Street Journal,* February 9, 2005, p. B1.

23 Babcock, Charles, "Virtual's New Reality," *Information Week,* July 4, 2005, p. 28.

24 Duntemann, Jeff, "Inside the Virtual Machine," *PC Magazine,* September 20, 2005, p. 66.

25 Mossberg, Walter, "Software to Help You Download from iPods," *The Wall Street Journal,* January 12, 2006, p. B1.

26 Dunn, Scott, "Deep File Divers," *PC World*, October 2005, p. 89.

27 Rubenking, Neil, "Yahoo! Enters Desktop Search Fray," *PC Magazine*, February 22, 2005, p. 31.

28 Hensley, Scott, "New Test May Detect Parkinson's Early," *The Wall Street Journal*, January 7, 2005, p. B1.

29 Keeton, Ann, "Software Is Close to Putting Users on the (3-D) Map," *The Wall Street Journal*, January 19, 2006, p. A11.

30 Deibel, Mary, "Filing Free," *Rocky Mountain News*, February 28, 2005, p. 1B.

31 Mendelson, Edward, "Office Software on the Cheap," *PC Magazine*, September 6, 2005, p. 64.

32 Fox, Steve, "Office 12: Easier Data Updates," *PC World,* August 2005, p. 30.

33 Reinhardt, Andy, et al., "SAP: A Sea Change in Software," *Business Week*, July 11, 2005, p. 46.

34 Muller, Joann, "Virtual Real Life," *Forbes*, July 4, 2005, p. 46.

35 Chittum, Ryan, "Location, Location, Location," *The Wall Street Journal*, July 18, 2005, p. R7.

36 Guth, Robert, "Microsoft Hones Software Aimed at Smaller Firms," *The Wall Street Journal*, September 7, 2005, p. B3.

37 Boulanger, Alan, "Open-Source Versus Proprietary Software: Is One More Reliable and Secure Than the Other?" *IBM Systems Journal*, Vol. 44, No. 2, 2005, p. 239.

38 Songini, Marc, "Software Ownership Battle Adds $10M to Cost of Big Dig," *Computerworld*, February 28, 2005, p. 16.

39 Hamm, Steve, "One Way to Hammer at Windows," *Business Week*, January 24, 2005, p. 36.

40 Jacoby, Mary, "Microsoft Bares Code to Avoid Fines," *The Wall Street Journal,* January 26, 2006, p. A3.

41 Joch, Alan, "The Real Cost of Open Source," *FCW.com*, May 3, 2005, accessed at *www.fcw.com*.

42 Steinhart, Michael, "Free Software," *PC Magazine*, October 18, 2005, p. 76.

43 Staff, *www.sourceforce.net*, accessed on January 9, 2006.

CHAPTER
· 5 ·

Organizing Data and Information

PRINCIPLES	LEARNING OBJECTIVES
▪ Data management and modeling are key aspects of organizing data and information.	▪ Define general data management concepts and terms, highlighting the advantages of the database approach to data management. ▪ Describe the relational database model and outline its basic features.
▪ A well-designed and well-managed database is an extremely valuable tool in supporting decision making.	▪ Identify the common functions performed by all database management systems and identify popular user database management systems.
▪ The number and types of database applications will continue to evolve and yield real business benefits.	▪ Identify and briefly discuss current database applications.

▶▶ Information Systems in the Global Economy ▶▶
Valero Energy, United States

Drilling for Valuable Information in the Energy Industry

Valero Energy, based in San Antonio, Texas, is regarded as one of the world's most profitable petrochemical companies. Its global reputation is built on its willingness to buck the status quo and apply information management technology to strengthen its strategic decision making. In 1984, Valero executives predicted that the demand for heavy, sour crude oil would increase during the economic expansion of China and India. At the time, there was little demand for heavy crude because light, premium crude was more abundant and easier to process. Based on its prediction, Valero set up operations to be the first major supplier of heavy crude oil to the global market as demand increased.

Valero's strategy requires keen attention to market data and smart analysis to determine the pricing and production rate for its fuel products. Executives at Valero pore over daily, weekly, and monthly reports generated by business intelligence (BI) tools, which draw valuable information from large databases filled with industry information and indicators. The reports provide the company with competitive intelligence to help them optimize profitability across the hundreds of crude oil, liquid propane, and gasoline products Valero sells in several thousand locations. Valero believes that it invests much more heavily and deeply in its data-mining operation than its competitors.

However, Valero's original database management systems had some problems. Due to rapid growth and corporate acquisitions, Valero was using six different database management systems that occasionally revealed inconsistencies in the data. Although Valero's managers and executives could access detailed data, the information in one manager's database report didn't always match the information in other reports.

Valero launched a project to consolidate its databases and systems and shift its operational reporting onto a business intelligence system from Information Builders Incorporated called WebFocus. Information Builders describes its WebFocus software as a comprehensive and fully integrated enterprise business intelligence platform whose architecture, integration, and simplicity permeate every level of the global organization—executive, analytical, and operational—and make any data available, accessible, and meaningful to every person or application who needs it, when and how they need it.

The consolidation effort eliminated inaccurate data redundancy and restored data integrity. It saved Valero an estimated $191,000 in software licensing fees in 2005 and an expected $478,000 in 2006. Managers can now easily focus on detailed information such as whether a promotional campaign for retail merchandise at service stations helped lift sales. They can also conveniently scroll through daily scheduled reports instead of having to wait for dated monthly information. Vice president of finance for the firm's Canadian marketing operations, Marcel Dupuis, summed it up this way: "Having detailed daily, weekly, and monthly reports that we can generate and refresh ourselves allows us to react faster to the market and offer the best prices to our consumers." By standardizing and consolidating its database management system, Valero can more accurately calculate net profit and loss for gasoline, diesel, and other petroleum products in a tricky market that is constantly fluctuating.

Valero has rapidly grown from a relatively unknown company with only one refinery in 1997 to the biggest oil refiner in the United States with 16 domestic and two international facilities. Valero attributes much of its success to its business sense and investment in databases and the systems that manipulate them. This success was noted by CFO magazine in 2005 when it ranked the company as the most profitable spender among all U.S. petrochemical firms.

As you read this chapter, consider the following:

- What role do databases play in the overall effectiveness of information systems?
- What techniques do businesses use to maximize the value of the information provided from databases?

Why Learn About Database Systems?

A huge amount of data is entered into computer systems every day. Where does all this data go and how is it used? How can it help you on the job? In this chapter, you will learn about database management systems and how they can help you. If you become a marketing manager, you can access a vast store of data on existing and potential customers from surveys, their Web habits, and their past purchases. This information can help you sell products and services. If you become a corporate lawyer, you will have access to past cases and legal opinions from sophisticated legal databases. This information can help you win cases and protect your organization legally. If you become a human resource (HR) manager, you will be able to use databases to analyze the impact of raises, employee insurance benefits, and retirement contributions on long-term costs to your company. Regardless of your major in school, using database management systems will likely be a critical part of your job. In this chapter, you will see how you can use data mining to extract valuable information to help you succeed. This chapter starts by introducing basic concepts of database management systems.

A database is an organized collection of data. Like other components of an information system, a database should help an organization achieve its goals. A database can contribute to organizational success by providing managers and decision makers with timely, accurate, and relevant information based on data. Medicare, for example, is developing a database that can be accessed through the Internet to provide information to doctors, healthcare providers, and patients.[1] The database will give statistics about 17 accepted hospital quality measures on treating heart attacks, pneumonia, and other diseases. According to Mark McClellan, head of the Centers for Medicare and Medicaid, "This is another big step toward supporting and rewarding better quality, rather than just paying more and supporting more services." Databases also help companies generate information to reduce costs, increase profits, track past business activities, and open new market opportunities. In some cases, multiple organizations collaborate in creating and using international databases.[2] Six petroleum organizations, including the Organization of Petroleum Exporting Countries (OPEC), International Energy Agency (IEA), and the United Nations, use a database to monitor the global oil supply. According to Arne Walther, IEA secretary general, over 90 oil-producing and consuming countries have started submitting data.

A database provides an essential foundation for an organization's information and decision support system. Without a well-designed, accurate database, executives and managers do not have the information they need to make good decisions. Wal-Mart and some other retail stores, for example, use their database systems and personnel to react quickly to changing market needs to make sure that the right inventory is at the right stores at the right time. If a snowstorm hits Colorado, Wal-Mart makes sure that its Colorado stores have shovels and snow blowers. During Hurricane Katrina, stores such as Home Depot made sure they had supplies to help people prepare and survive, probably saving lives. Companies like Wal-Mart and Home Depot can consult their databases in their supply chain management systems to locate the inventory or supplies that an individual store needs.

A database is also the foundation of most systems development projects. If the database is not designed properly, the systems development effort can be like a house of cards, collapsing under the weight of inaccurate and inadequate data. Because data is so critical to an organization's success, many firms develop databases to help them access data more efficiently and use it more effectively. This typically requires a well-designed database management system and a knowledgeable database administrator.

A **database management system** (DBMS) consists of a group of programs that manipulate the database and provide an interface between the database and its users and other application programs. Usually purchased from a database company, a DBMS provides a single point of management and control over data resources, which can be critical to maintaining the integrity and security of the data.[3] A database, a DBMS, and the application programs that use the data make up a database environment. A **database administrator** (DBA) is a skilled and trained IS professional who directs all activities related to an organization's database, including providing security from intruders.[4] A now defunct bulk e-mail business, for example, downloaded about 1.6 billion database records from a corporate computer, including names, phone numbers, addresses, and other personal information.[5] According to a corporate representative, "We should have known better." This and similar breaches shows the vulnerability of some databases. DBAs should attempt to protect their organizations from these types of security breaches.[6] Data quality and accuracy also continue to be important issues for DBAs.[7] According to a Gartner study, at least 25 percent of critical data held by Fortune 1,000 companies is inaccurate. In another study, only 34 percent of the respondents reported that they were confident in the quality of their data. To safeguard their data, six large financial institutions, including Bank of America, Citigroup, and J.P. Morgan, have joined together to develop procedures and systems to prevent security breaches.[8] These financial institutions have developed procedures for the telecommunications systems and business partners, in addition to a number of internal controls.

database management system (DBMS)
A group of programs that manipulates the database and provides an interface between the database and the user of the database and other application programs.

database administrator (DBA)
A skilled IS professional who directs all activities related to an organization's database.

DATA MANAGEMENT

Without data and the ability to process it, an organization could not successfully complete most business activities. It could not pay employees, send out bills, order new inventory, or produce information to assist managers in decision making. As you recall, data consists of raw facts, such as employee numbers and sales figures. For data to be transformed into useful information, it must first be organized in a meaningful way.

The Hierarchy of Data

Data is generally organized in a hierarchy that begins with the smallest piece of data used by computers (a bit) and progresses through the hierarchy to a database. A bit (a binary digit) represents a circuit that is either on or off. Bits can be organized into units called *bytes*. A byte is typically eight bits. Each byte represents a **character**, which is the basic building block of information. A character can be an uppercase letter (A, B, C... Z), lowercase letter (a, b, c... z), numeric digit (0, 1, 2... 9), or special symbol (., !, [+], [−], /, ...).

Characters are put together to form a field. A **field** is typically a name, number, or combination of characters that describes an aspect of a business object (such as an employee, a location, or a truck) or activity (such as a sale). In addition to being entered into a database, fields can be computed from other fields. *Computed fields* include the total, average, maximum, and minimum value. A collection of related data fields is a **record**. By combining descriptions of the characteristics of an object or activity, a record can provide a complete description of the object or activity. For instance, an employee record is a collection of fields about one employee. One field includes the employee's name, another field contains the address, and still others her phone number, pay rate, earnings made to date, and so forth. A collection of related records is a **file**—for example, an employee file is a collection of all company employee records. Likewise, an inventory file is a collection of all inventory records for a particular company or organization. Some database software refers to files as tables.

At the highest level of this hierarchy is a *database*, a collection of integrated and related files. Together, bits, characters, fields, records, files, and databases form the **hierarchy of data** (see Figure 5.1). Characters are combined to make a field, fields are combined to make

character
A basic building block of information, consisting of uppercase letters, lowercase letters, numeric digits, or special symbols.

field
Typically a name, number, or combination of characters that describes an aspect of a business object or activity.

record
A collection of related data fields.

file
A collection of related records.

hierarchy of data
Bits, characters, fields, records, files, and databases.

a record, records are combined to make a file, and files are combined to make a database. A database houses not only all these levels of data but the relationships among them.

Hierarchy of data	Example
Database	Personnel file / Department file / Payroll file (Project database)
Files	098 - 40 - 1370 Fiske, Steven 01-05-1985 / 549 - 77 - 1001 Buckley, Bill 02-17-1979 / 005 - 10 - 6321 Johns, Francine 10-07-1997 (Personnel file)
Records	098 - 40 - 1370 Fiske, Steven 01-05-1985 (Record containing SSN, last and first name, hire date)
Fields	Fiske (Last name field)
Characters (bytes)	1000110 (Letter F in ASCII)

Data Entities, Attributes, and Keys

entity
A generalized class of people, places, or things for which data is collected, stored, and maintained.

attribute
A characteristic of an entity.

data item
The specific value of an attribute.

Entities, attributes, and keys are important database concepts. An **entity** is a generalized class of people, places, or things (objects) for which data is collected, stored, and maintained. Examples of entities include employees, inventory, and customers. Most organizations organize and store data as entities.

An **attribute** is a characteristic of an entity. For example, employee number, last name, first name, hire date, and department number are attributes for an employee (see Figure 5.2). The inventory number, description, number of units on hand, and location of the inventory item in the warehouse are attributes for items in inventory. Customer number, name, address, phone number, credit rating, and contact person are attributes for customers. Attributes are usually selected to reflect the relevant characteristics of entities such as employees or customers. The specific value of an attribute, called a **data item**, can be found in the fields of the record describing an entity.

Employee #	Last name	First name	Hire date	Dept. number
005-10-6321	Johns	Francine	10-07-1997	257
549-77-1001	Buckley	Bill	02-17-1979	632
098-40-1370	Fiske	Steven	01-05-1985	598

ENTITIES (records)

KEY FIELD

ATTRIBUTES (fields)

Attributes and data items are used by most organizations. Many governments use attributes and data items to help identify and locate possible terrorists. As part of its Homeland Security effort, the U.S. State Department is sharing its database of 50 million visa applications with the FBI to help identify possible terrorists. Partially as a response to the threat of terrorism, some governments are increasingly using databases to track and prevent unwanted people from entering their country. The U.S. government, for example, is using fingerprint databases to track tens of thousands of suspected terrorists or visitors of "national security concern" as they enter the country. The databases contain visual images of fingerprints. The National Security Entry-Exit Registration System attempts to close the borders to suspected terrorists by comparing the fingerprints of entering visitors against a comprehensive database. Sharing attributes and data items also helps organizations coordinate responses across diverse functional areas.

As discussed, a collection of fields about a specific object is a record. A **key** is a field or set of fields in a record that identifies the record. A **primary key** is a field or set of fields that uniquely identifies the record. No other record can have the same primary key. The primary key is used to distinguish records so that they can be accessed, organized, and manipulated. For an employee record, such as the one shown in Figure 5.2, the employee number is an example of a primary key.

Locating a particular record that meets a specific set of criteria might be easier and faster using a combination of secondary keys. For example, a customer might call a mail-order company to place an order for clothes. If the customer does not know the correct primary key (such as a customer number), a secondary key (such as last name) can be used. In this case, the order clerk enters the last name, such as Adams. If several customers have a last name of Adams, the clerk can check other fields, such as address, first name, and so on, to find the correct customer record. After locating the correct customer record, the order can be completed and the clothing items shipped to the customer.

The Database Approach

At one time, applications, such as payroll and invoicing, used specific files, such as payroll and invoicing files. In other words, each application used files dedicated to that application. This approach to data management, whereby separate data files are created and stored for each application program, is called the **traditional approach to data management**.

Today, most organizations use the **database approach to data management**, where multiple application programs share a pool of related data. A database offers the ability to share data and information resources. Federal databases, for example, often include the results of DNA tests as an attribute for convicted criminals. The Third U.S. Circuit Court of Appeals has ruled that DNA samples taken from criminals on release are legal because of the government's desire to develop a national DNA database.[9] The information can be shared with law enforcement officials around the country.

To use the database approach to data management, additional software—a database management system (DBMS)—is required. As previously discussed, a DBMS consists of a group of programs that can be used as an interface between a database and the user or the database and application programs. Typically, this software acts as a buffer between the application programs and the database itself. Figure 5.3 illustrates the database approach.

Table 5.1 lists some of the primary advantages of the database approach, and Table 5.2 lists some disadvantages.

Many modern databases serve entire enterprises, encompassing much of the data of the organization. Often, distinct yet related databases are linked to provide enterprise-wide databases. For example, Best Buy is a specialty retailer of consumer electronics, personal computers, entertainment software, and appliances. It operates nearly 2,000 retail stores and commercial Web sites under the names Best Buy, Magnolia Hi-Fi, Media Play, On Cue, Sam Goody, and Suncoast. Best Buy uses information about its business and customers to tailor the product mix to its customer base, minimize the time items are held in inventory to reduce costs, and respond quickly to customer needs. At the center of this strategic information is a database, which consolidates information from about 350 sources across the enterprise.

key
A field or set of fields in a record that is used to identify the record.

primary key
A field or set of fields that uniquely identifies the record.

traditional approach to data management
An approach whereby separate data files are created and stored for each application program.

database approach to data management
An approach whereby a pool of related data is shared by multiple application programs.

Figure 5.3

The Database Approach to Data
Management

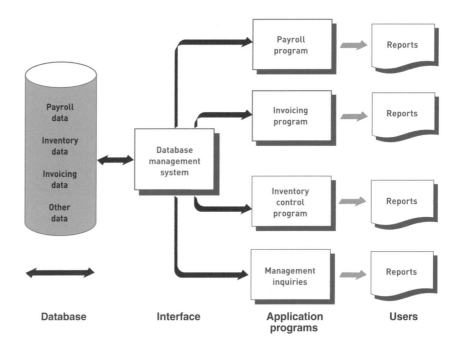

| Database | Interface | Application programs | Users |

Advantages	Explanation
Improved strategic use of corporate data	Accurate, complete, up-to-date data can be made available to decision makers where, when, and in the form they need it. The database approach can also give greater visibility to the organization's data resource.
Reduced data redundancy	Data is organized by the DBMS and stored in only one location. This results in more efficient use of system storage space.
Improved data integrity	With the traditional approach, some changes to data were not reflected in all copies of the data kept in separate files. This is prevented with the database approach because no separate files contain copies of the same piece of data.
Easier modification and updating	The DBMS coordinates updates and data modifications. Programmers and users do not have to know where the data is physically stored. Data is stored and modified once. Modification and updating is also easier because the data is stored in only one location in most cases.
Data and program independence	The DBMS organizes the data independently of the application program, so the application program is not affected by the location or type of data. Introduction of new data types not relevant to a particular application does not require rewriting that application to maintain compatibility with the data file.
Better access to data and information	Most DBMSs have software that makes it easy to access and retrieve data from a database. In most cases, users give simple commands to get important information. Relationships between records can be more easily investigated and exploited, and applications can be more easily combined.
Standardization of data access	A standardized, uniform approach to database access means that all application programs use the same overall procedures to retrieve data and information.
A framework for program development	Standardized database access procedures can mean more standardization of program development. Because programs go through the DBMS to gain access to data in the database, standardized database access can provide a consistent framework for program development. In addition, each application program need address only the DBMS, not the actual data files, reducing application development time.
Better overall protection of the data	Accessing and using centrally located data is easier to monitor and control. Security codes and passwords can ensure that only authorized people have access to particular data and information in the database, thus ensuring privacy.
Shared data and information resources	The cost of hardware, software, and personnel can be spread over many applications and users. This is a primary feature of a DBMS.

Table 5.1

Advantages of the Database
Approach

Disadvantages	Explanation
More complexity	DBMSs can be difficult to set up and operate. Many decisions must be made correctly for the DBMS to work effectively. In addition, users have to learn new procedures to take full advantage of a DBMS.
More difficult to recover from a failure	With the traditional approach to file management, a failure of a file affects only a single program. With a DBMS, a failure can shut down the entire database.
More expensive	DBMSs can be more expensive to purchase and operate. The expense includes the cost of the database and specialized personnel, such as a database administrator, who is needed to design and operate the database. Additional hardware might also be required.

Table 5.2

Disadvantages of the Database Approach

TA MODELING AND THE RELATIONAL DATABASE MODEL

Because today's businesses have so many elements, they must keep data organized so that it can be used effectively. A database should be designed to store all data relevant to the business and provide quick access and easy modification. Moreover, it must reflect the business processes of the organization. When building a database, an organization must carefully consider these questions:

- **Content.** What data should be collected and at what cost?
- **Access.** What data should be provided to which users and when?
- **Logical structure.** How should data be arranged so that it makes sense to a given user?
- **Physical organization.** Where should data be physically located?

Data Modeling

Key considerations in organizing data in a database include determining what data to collect in the database, who will have access to it, and how they might want to use the data. After determining these details, an organization can create a database. Building a database requires two different types of designs: a logical design and a physical design. The *logical design* of a database is an abstract model of how the data should be structured and arranged to meet an organization's information needs. The logical design of a database involves identifying relationships among the data items and grouping them in an orderly fashion. Because databases provide both input and output for information systems throughout a business, users from all functional areas should assist in creating the logical design to ensure that their needs are identified and addressed. *Physical design* starts from the logical database design and fine-tunes it for performance and cost considerations (such as improved response time, reduced storage space, and lower operating cost). For example, Interact, Inc., a telecommunications company, designed a database that uses the computer's memory to speed database operations.[10] The person who fine-tunes the physical design must have an in-depth knowledge of the DBMS. For example, the logical database design might need to be altered so that certain data entities are combined, summary totals are carried in the data records rather than calculated from elemental data, and some data attributes are repeated in more than one data entity. These are examples of **planned data redundancy**, which is done to improve the system performance so that user reports or queries can be created more quickly.

One of the tools database designers use to show the logical relationships among data is a data model. A **data model** is a diagram of entities and their relationships. Data modeling usually involves understanding a specific business problem and analyzing the data and information needed to deliver a solution. When done at the level of the entire organization, this is called enterprise data modeling. **Enterprise data modeling** is an approach that starts by investigating the general data and information needs of the organization at the strategic level, and then examines more specific data and information needs for the various functional areas and departments within the organization. Various models have been developed to help

planned data redundancy
A way of organizing data in which the logical database design is altered so that certain data entities are combined, summary totals are carried in the data records rather than calculated from elemental data, and some data attributes are repeated in more than one data entity to improve database performance.

data model
A diagram of data entities and their relationships.

enterprise data modeling
Data modeling done at the level of the entire enterprise.

managers and database designers analyze data and information needs. An entity-relationship diagram is an example of such a data model.

Entity-relationship (ER) diagrams use basic graphical symbols to show the organization of and relationships between data. In most cases, boxes in ER diagrams indicate data items or entities contained in data tables, and diamonds show relationships between data items and entities. In other words, ER diagrams show data items in tables (entities) and the ways they are related.

ER diagrams help ensure that the relationships among the data entities in a database are correctly structured so that any application programs developed are consistent with business operations and user needs. In addition, ER diagrams can serve as reference documents after a database is in use. If changes are made to the database, ER diagrams help design them. Figure 5.4 shows an ER diagram for an order database. In this database design, one salesperson serves many customers. This is an example of a one-to-many relationship, as indicated by the one-to-many symbol (the "crow's-foot") shown in Figure 5.4. The ER diagram also shows that each customer can place one-to-many orders, each order includes one-to-many line items, and many line items can specify the same product (a many-to-one relationship). This database can also have one-to-one relationships. For example, one order generates one invoice.

entity-relationship (ER) diagrams

Data models that use basic graphical symbols to show the organization of and relationships between data.

Figure 5.4

An Entity-Relationship (ER) Diagram for a Customer Order Database

Development of ER diagrams helps ensure that the logical structure of application programs is consistent with the data relationships in the database.

The Relational Database Model

Although there are a number of different database models, including flat files, hierarchical, and network models, the **relational model** has become the most popular, and use of this model will continue to increase. The relational model describes data using a standard tabular format. In a database structured according to the relational model, all data elements are placed in two-dimensional tables, called *relations*, which are the logical equivalent of files. The tables in relational databases organize data in rows and columns, simplifying data access and manipulation. It is normally easier for managers to understand the relational model (see Figure 5.5) than other database models.

Databases based on the relational model include IBM DB2, Oracle, Sybase, Microsoft SQL Server, Microsoft Access, and MySQL. Oracle is currently the market leader in general-purpose databases, with about 43 percent of the $15 billion database market.[11] Oracle also has developed Database Lite 10g for laptop computers and other mobile or wireless devices.[12] "Companies are constantly looking for new ways of gaining competitive advantage and reducing costs by streamlining their business processes. Making information available directly and in real time to employees on the front line—serving or selling to customers—is an increasingly critical step to gaining such efficiencies," says Oracle's chief technology officer (CTO).

relational model

A database model that describes data in which all data elements are placed in two-dimensional tables, called *relations*, which are the logical equivalent of files.

Data Table 1: Project Table

Project	Description	Dept. number
155	Payroll	257
498	Widgets	632
226	Sales manual	598

Data Table 2: Department Table

Dept.	Dept. name	Manager SSN
257	Accounting	005-10-6321
632	Manufacturing	549-77-1001
598	Marketing	098-40-1370

Data Table 3: Manager Table

SSN	Last name	First name	Hire date	Dept. number
005-10-6321	Johns	Francine	10-07-1997	257
549-77-1001	Buckley	Bill	02-17-1979	632
098-40-1370	Fiske	Steven	01-05-1985	598

Figure 5.5

A Relational Database Model

In the relational model, all data elements are placed in two-dimensional tables, or relations. As long as they share at least one common element, these relations can be linked to output useful information. Note that some organizations might use employee number instead of Social Security number (SSN) in Data Tables 2 and 3.

In the relational model, each row (or record) of a table represents a data entity, with the columns (or fields) of the table representing attributes. Each attribute can accept only certain values. The allowable values for these attributes are called the **domain**. The domain for a particular attribute indicates what values can be placed in each column of the relational table. For instance, the domain for an attribute such as gender would be limited to male or female. A domain for pay rate would not include negative numbers. In this way, defining a domain can increase data accuracy.

domain
The allowable values for data attributes.

Manipulating Data

After entering data into a relational database, users can make inquiries and analyze the data. Basic data manipulations include selecting, projecting, and joining. **Selecting** involves eliminating rows according to certain criteria. Suppose a project table contains the project number, description, and department number for all projects a company is performing. The president of the company might want to find the department number for Project 226, a sales manual project. Using selection, the president can eliminate all rows but the one for Project 226 and see that the department number for the department completing the sales manual project is 598.

Projecting involves eliminating columns in a table. For example, a department table might contain the department number, department name, and Social Security number (SSN) of the manager in charge of the project. A sales manager might want to create a new table with only the department number and the Social Security number of the manager in charge of the sales manual project. The sales manager can use projection to eliminate the department name column and create a new table containing only department number and SSN.

selecting
Manipulating data to eliminate rows according to certain criteria.

projecting
Manipulating data to eliminate columns in a table.

joining

Manipulating data to combine two or more tables.

linking

Manipulating two or more tables that share at least one common data attribute to provide useful information and reports.

Joining involves combining two or more tables. For example, you can combine the project table and the department table to create a new table with the project number, project description, department number, department name, and Social Security number for the manager in charge of the project.

As long as the tables share at least one common data attribute, the tables in a relational database can be **linked** to provide useful information and reports. Being able to link tables to each other through common data attributes is one of the keys to the flexibility and power of relational databases. Suppose the president of a company wants to find out the name of the manager of the sales manual project and the length of time the manager has been with the company. Assume that the company has the manager, department, and project tables shown in Figure 5.5. A simplified ER diagram showing the relationship between these tables is shown in Figure 5.6. Note the crow's-foot by the project table. This indicates that a department can have many projects. The president would make the inquiry to the database, perhaps via a personal computer. The DBMS would start with the project description and search the project table to find out the project's department number. It would then use the department number to search the department table for the manager's Social Security number. The department number is also in the department table and is the common element that links the project table to the department table. The DBMS uses the manager's Social Security number to search the manager table for the manager's hire date. The manager's Social Security number is the common element between the department table and the manager table. The final result is that the manager's name and hire date are presented to the president as a response to the inquiry (see Figure 5.7).

Figure 5.6

A Simplified ER Diagram Showing the Relationship Between the Manager, Department, and Project Tables

One of the primary advantages of a relational database is that it allows tables to be linked, as shown in Figure 5.7. This linkage is especially useful when information is needed from multiple tables. For example, the manager's Social Security number is maintained in the manager table. If the Social Security number is needed, it can be obtained by linking to the manager table.

The relational database model is by far the most widely used. It is easier to control, more flexible, and more intuitive than other approaches because it organizes data in tables. As shown in Figure 5.8, a relational database management system, such as Access, provides tips and tools for building and using database tables. In this figure, the database displays information about data types and indicates that additional help is available. The ability to link relational tables also allows users to relate data in new ways without having to redefine complex relationships. Because of the advantages of the relational model, many companies use it for large corporate databases, such as those for marketing and accounting. The relational model can also be used with personal computers and mainframe systems. A travel reservation company, for example, can develop a fare-pricing system by using relational database technology that can handle millions of daily queries from online travel companies, such as Expedia, Travelocity, and Orbitz.

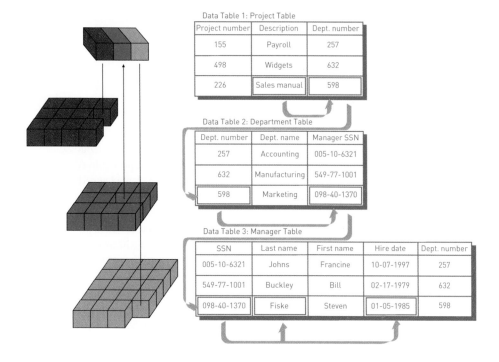

Data Table 1: Project Table

Project number	Description	Dept. number
155	Payroll	257
498	Widgets	632
226	Sales manual	598

Data Table 2: Department Table

Dept. number	Dept. name	Manager SSN
257	Accounting	005-10-6321
632	Manufacturing	549-77-1001
598	Marketing	098-40-1370

Data Table 3: Manager Table

SSN	Last name	First name	Hire date	Dept. number
005-10-6321	Johns	Francine	10-07-1997	257
549-77-1001	Buckley	Bill	02-17-1979	632
098-40-1370	Fiske	Steven	01-05-1985	598

Figure 5.7

Linking Data Tables to Answer
an Inquiry

In finding the name and hire date of
the manager working on the sales
manual project, the president needs
three tables: project, department,
and manager. The project
description (Sales manual) leads to
the department number (598) in the
project table, which leads to the
manager's SSN (098-40-1370) in the
department table, which leads to the
manager's name (Fiske) and hire
date (01-05-1985) in the manager
table. Again, note that some
organizations might use employee
number instead of Social Security
number (SSN).

Figure 5.8

Building and Modifying a
Relational Database

Relational databases provide many
tools, tips, and shortcuts to simplify
the process of creating and
modifying a database.

(Source: Courtesy of Microsoft
Corporation.)

Data Cleanup

As discussed in Chapter 1, valuable data is accurate, complete, economical, flexible, reliable, relevant, simple, timely, verifiable, accessible, and secure. The database must also be properly designed. The purpose of **data cleanup** is to develop data with these characteristics. Consider a database for a fitness center designed to track member dues. The table contains the attribute name, phone number, gender, dues paid, and date paid (see Table 5.3). As the records in Table 5.3 show, Anita Brown and Sim Thomas have paid their dues in September. Sim has paid his dues in two installments. Note that no primary key uniquely identifies each record. As you will see next, this problem must be corrected.

data cleanup
The process of looking for and fixing
inconsistencies to ensure that data
is accurate and complete.

Table 5.3

Fitness Center Dues

Name	Phone	Gender	Dues Paid	Date Paid
Brown, A.	468-3342	Female	$30	September 15
Thomas, S.	468-8788	Male	$15	September 15
Thomas, S.	468-5238	Male	$15	September 25

Because Sim Thomas has paid dues twice in September, the data in the database is now redundant. The name, phone number, and gender for Thomas are repeated in two records. Notice that the data in the database is also inconsistent: Thomas has changed his phone number, but only one of the records reflects this change. Further reducing this database's reliability is the lack of a primary key to uniquely identify Sim Thomas's record. The first Thomas could be Sim Thomas, but the second might be Steve Thomas. These problems and irregularities in data are called *anomalies*. Data anomalies often result in incorrect information, causing database users to be misinformed about actual conditions. Anomalies must be corrected.

To solve these problems in the fitness center's database, we can add a primary key, such as member number, and put the data into two tables: a Fitness Center Members table with gender, phone number, and related information, and a Dues Paid table with dues paid and date paid (see Tables 5.4 and 5.5). Both tables include the member number attribute so that they can be linked.

Table 5.4

Fitness Center Members

Member No.	Name	Phone	Gender
SN123	Brown, A.	468-3342	Female
SN656	Thomas, S.	468-5238	Male

Table 5.5

Dues Paid

Member No.	Dues Paid	Date Paid
SN123	$30	September 15
SN656	$15	September 15
SN656	$15	September 25

The relations in Table 5.4 and Table 5.5 reduce the redundancy and eliminate the potential problem of having two different phone numbers for the same member. Also note that the member number gives each record in the Fitness Center Members table a primary key. Because the Dues Paid table lists two payment entries ($15 each) with the same member number (SN656), one person clearly made the payments, not two different people. Formalized approaches, such as *database normalization*, are often used to clean up problems with data.

DATABASE MANAGEMENT SYSTEMS

Creating and implementing the right database system ensures that the database will support both business activities and goals. But how do we actually create, implement, use, and update a database? The answer is found in the database management system. As discussed earlier, a DBMS is a group of programs used as an interface between a database and application programs or a database and the user. The capabilities and types of database systems, however, vary considerably.

Overview of Database Types

Database management systems can range from small, inexpensive software packages to sophisticated systems costing hundreds of thousands of dollars. The following sections discuss a few popular alternatives. See Figure 5.9 for one example.

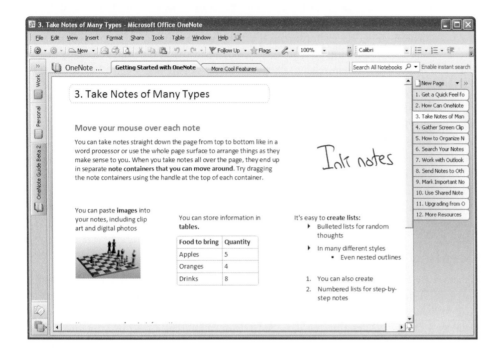

Figure 5.9

Microsoft OneNote

Microsoft OneNote lets you gather any type of information, and then retrieve, copy, and paste the information into other applications, such as word processing and spreadsheet programs.

Flat File

A flat file is a simple database program whose records have no relationship to one another. Flat file databases are often used to store and manipulate a single table or file, and do not use any of the database models discussed previously, such as the relational model. Many spreadsheet and word processing programs have flat file capabilities. These software packages can sort tables and make simple calculations and comparisons. OneNote, developed by Microsoft in 2003 and shown in Figure 5.9, was designed to let people put ideas, thoughts, and notes into a computer file. In OneNote, each note can be placed anywhere on a page or in a box on a page, called a *container*. Pages are organized into sections and subsections that appear as colored tabs. After you enter a note, you can retrieve, copy, and paste it into other applications, such as word processing and spreadsheet programs. Similar to OneNote, EverNote is a free database that can store notes and other pieces of information.[13] Considering the amount of information today's high-capacity hard disks can store, the popularity of databases that can handle unstructured data will continue to grow.

Single User

Databases for personal computers are most often meant for a single user. Only one person can use the database at a time. Access and FileMaker Pro are examples of popular single-user DBMSs, through which users store and manipulate data. Microsoft InfoPath is another example of a single-user database. The database is part of the Microsoft Office suite, and it helps people collect and organize information from a variety of sources. InfoPath has built-in forms that can be used to enter expense information, time-sheet data, and a variety of other information.

Multiple Users

Large mainframe computer systems need multiuser DBMSs. These more powerful, expensive systems allow dozens or hundreds of people to access the same database system at the same time. Popular vendors for multiuser database systems include Oracle, Sybase, and IBM.

All DBMSs share some common functions, such as providing a user view, physically storing and retrieving data in a database, allowing for database modification, manipulating data, and generating reports. These DBMSs can handle the most complex of data-processing tasks. A medical clinic, for example, can use a database like IBM DB2 to develop an all-inclusive database containing patient records, physician notes, demographic data, genetic data, and proteomic (protein-related) data for millions of patients.

Providing a User View

Because the DBMS is responsible for access to a database, one of the first steps in installing and using a large database involves telling the DBMS the logical and physical structure of the data and relationships among the data in the database for each user. This description is called a **schema** (as in schematic diagram). Large database systems, such as Oracle, typically use schemas to define the tables and other database features associated with a person or user. A schema can be part of the database or a separate schema file. The DBMS can reference a schema to find where to access the requested data in relation to another piece of data.

Creating and Modifying the Database

Schemas are entered into the DBMS (usually by database personnel) via a data definition language. A **data definition language (DDL)** is a collection of instructions and commands used to define and describe data and relationships in a specific database. A DDL allows the database's creator to describe the data and relationships that are to be contained in the schema. In general, a DDL describes logical access paths and logical records in the database. Figure 5.10 shows a simplified example of a DDL used to develop a general schema. The *Xs* in Figure 5.10 reveal where specific information concerning the database should be entered. File description, area description, record description, and set description are terms the DDL defines and uses in this example. Other terms and commands can be used, depending on the particular DBMS employed.

schema
A description of the entire database.

data definition language (DDL)
A collection of instructions and commands used to define and describe data and relationships in a specific database.

Figure 5.10

Using a Data Definition Language to Define a Schema

```
SCHEMA DESCRIPTION
SCHEMA NAME IS XXXX
AUTHOR        XXXX
DATE          XXXX
FILE DESCRIPTION
    FILE NAME IS XXXX
        ASSIGN XXXX
    FILE NAME IS XXXX
        ASSIGN XXXX
AREA DESCRIPTION
    AREA NAME IS XXXX
RECORD DESCRIPTION
    RECORD NAME IS XXXX
    RECORD ID IS XXXX
    LOCATION MODE IS XXXX
    WITHIN XXXX AREA FROM XXXX THRU XXXX
SET DESCRIPTION
    SET NAME IS XXXX
    ORDER IS XXXX
    MODE IS XXXX
    MEMBER IS XXXX
    .
    .
    .
```

Another important step in creating a database is to establish a **data dictionary**, a detailed description of all data used in the database. The data dictionary contains the name of the data item, aliases or other names that may be used to describe the item, the range of values that can be used, the type of data (such as alphanumeric or numeric), the amount of storage needed for the item, a notation of the person responsible for updating it and the various users who can access it, and a list of reports that use the data item. A data dictionary can also include a description of data flows, the way records are organized, and the data-processing requirements. Figure 5.11 shows a typical data dictionary entry.

data dictionary
A detailed description of all the data used in the database.

```
                    NORTHWESTERN MANUFACTURING

        PREPARED BY:        D. BORDWELL
        DATE:               04 AUGUST 2007
        APPROVED BY:        J. EDWARDS
        DATE:               13 OCTOBER 2007
        VERSION:            3.1
        PAGE:               1 OF 1

        DATA ELEMENT NAME:  PARTNO
        DESCRIPTION:        INVENTORY PART NUMBER
        OTHER NAMES:        PTNO
        VALUE RANGE:        100 TO 5000
        DATA TYPE:          NUMERIC
        POSITIONS:          4 POSITIONS OR COLUMNS
```

Figure 5.11

A Typical Data Dictionary Entry

For example, the information in a data dictionary for the part number of an inventory item can include the name of the person who made the data dictionary entry (D. Bordwell), the date the entry was made (August 4, 2007), the name of the person who approved the entry (J. Edwards), the approval date (October 13, 2007), the version number (3.1), the number of pages used for the entry (1), the part name (PARTNO), other part names that might be used (PTNO), the range of values (part numbers can range from 100 to 5,000), the type of data (numeric), and the storage required (four positions are required for the part number). The following are some of the typical uses of a data dictionary.

- **Provide a standard definition of terms and data elements.** This standardization can help in programming by providing consistent terms and variables to be used for all programs. Programmers know what data elements are already "captured" in the database and how they relate to other data elements.
- **Assist programmers in designing and writing programs.** Programmers do not need to know which storage devices are used to store needed data. Using the data dictionary, programmers specify the required data elements. The DBMS locates the necessary data. More important, programmers can use the data dictionary to see which programs already use a piece of data and, if appropriate, can copy the relevant section of the program code into their new program, thus eliminating duplicate programming efforts.
- **Simplify database modification.** If for any reason a data element needs to be changed or deleted, the data dictionary would point to specific programs that utilize the data element that may need modification.

A data dictionary helps achieve the advantages of the database approach in these ways:

- **Reduced data redundancy.** By providing standard definitions of all data, it is less likely that the same data item will be stored in different places under different names. For example, a data dictionary reduces the likelihood that the same part number would be stored as two different items, such as PTNO and PARTNO.
- **Increased data reliability.** A data dictionary and the database approach reduce the chance that data will be destroyed or lost. In addition, it is more difficult for unauthorized people to gain access to sensitive data and information.

- **Faster program development.** With a data dictionary, programmers can develop programs faster. They don't need to develop names for data items because the data dictionary does that for them.
- **Easier modification of data and information.** The data dictionary and the database approach make modifications to data easier because users do not need to know where the data is stored. The person making the change indicates the new value of the variable or item, such as part number, that is to be changed. The database system locates the data and makes the necessary change.

Storing and Retrieving Data

As just described, one function of a DBMS is to be an interface between an application program and the database. When an application program needs data, it requests that data through the DBMS. Suppose that to calculate the total price of a new car, an auto dealer pricing program needs price data on the engine option—six cylinders instead of the standard four cylinders. The application program thus requests this data from the DBMS. In doing so, the application program follows a logical access path. Next, the DBMS, working with various system programs, accesses a storage device, such as disk drives, where the data is stored. When the DBMS goes to this storage device to retrieve the data, it follows a path to the physical location (physical access path) where the price of this option is stored. In the pricing example, the DBMS might go to a disk drive to retrieve the price data for six-cylinder engines. This relationship is shown in Figure 5.12.

Figure 5.12

Logical and Physical Access Paths

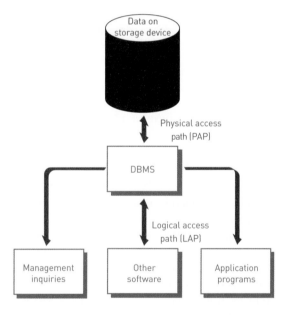

This same process is used if a user wants to get information from the database. First, the user requests the data from the DBMS. For example, a user might give a command, such as LIST ALL OPTIONS FOR WHICH PRICE IS GREATER THAN 200 DOLLARS. This is the logical access path (LAP). Then, the DBMS might go to the options price section of a disk to get the information for the user. This is the physical access path (PAP).

Two or more people or programs attempting to access the same record in the same database at the same time can cause a problem. For example, an inventory control program might attempt to reduce the inventory level for a product by ten units because ten units were just shipped to a customer. At the same time, a purchasing program might attempt to increase the inventory level for the same product by 200 units because more inventory was just received. Without proper database control, one of the inventory updates might not be correctly made, resulting in an inaccurate inventory level for the product. **Concurrency control** can be used to avoid this potential problem. One approach is to lock out all other application programs from access to a record if the record is being updated or used by another program.

concurrency control

A method of dealing with a situation in which two or more people need to access the same record in a database at the same time.

Manipulating Data and Generating Reports

After a DBMS has been installed, employees, managers, and consumers can use it to review reports and obtain important information. The Food Allergen and Consumer Protection Act, effective in 2006, requires that food manufacturing companies generate reports on the ingredients, formulas, and food preparation techniques for the public.[14] The new law requires that food companies clearly label their products for major allergens.

Some databases use *Query-by-Example (QBE)*, which is a visual approach to developing database queries or requests. Like Windows and other GUI operating systems, you can perform queries and other database tasks by opening windows and clicking the data or features you want (see Figure 5.13).

Query by Example

Some databases use Query-by-Example (QBE) to generate reports and information.

In other cases, database commands can be used in a programming language. For example, C++ commands can be used in simple programs that will access or manipulate certain pieces of data in the database. Here's another example of a DBMS query: SELECT * FROM EMPLOYEE WHERE JOB_CLASSIFICATION = "C2". The * tells the program to include all columns from the EMPLOYEE table. In general, the commands that are used to manipulate the database are part of the **data manipulation language (DML)**. This specific language, provided with the DBMS, allows managers and other database users to access, modify, and make queries about data contained in the database to generate reports. Again, the application programs go through schemas and the DBMS before actually getting to the physically stored data on a device such as a disk.

In the 1970s, D. D. Chamberlain and others at the IBM Research Laboratory in San Jose, California, developed a standardized data manipulation language called *Structured Query Language (SQL)*, pronounced like the word *sequel* or spelled out as *SQL*. The EMPLOYEE query shown earlier is written in SQL. In 1986, the American National Standards Institute (ANSI) adopted SQL as the standard query language for relational databases. Since ANSI's acceptance of SQL, interest in making SQL an integral part of relational databases on both mainframe and personal computers has increased. SQL has many built-in functions, such as average (AVG), the largest value (MAX), the smallest value (MIN), and others. Table 5.6 contains examples of SQL commands.

SQL lets programmers learn one powerful query language and use it on systems ranging from PCs to the largest mainframe computers (see Figure 5.14). Programmers and database users also find SQL valuable because SQL statements can be embedded into many

data manipulation language (DML)
The commands that are used to manipulate the data in a database.

Table 5.6

Examples of SQL Commands

SQL Command	Description
SELECT ClientName, Debt FROM Client WHERE Debt > 1000	This query displays all clients (ClientName) and the amount they owe the company (Debt) from a database table called Client for clients who owe the company more than $1,000 (WHERE Debt > 1000).
SELECT ClientName, ClientNum, OrderNum FROM Client, Order WHERE Client.ClientNum=Order.ClientNum	This command is an example of a join command that combines data from two tables: the client table and the order table (FROM Client, Order). The command creates a new table with the client name, client number, and order number (SELECT ClientName, ClientNum, OrderNum). Both tables include the client number, which allows them to be joined. This is indicated in the WHERE clause, which states that the client number in the client table is the same as (equal to) the client number in the order table (WHERE Client.ClientNum=Order.ClientNum).
GRANT INSERT ON Client to Guthrie	This command is an example of a security command. It allows Bob Guthrie to insert new values or rows into the Client table.

programming languages, such as the widely used C++ and COBOL languages. Because SQL uses standardized and simplified procedures for retrieving, storing, and manipulating data in a database system, the popular database query language can be easy to understand and use.

Figure 5.14

Structured Query Language

Structured Query Language (SQL) has become an integral part of most relational databases, as shown by this screen from Microsoft Access 2003.

After a database has been set up and loaded with data, it can produce desired reports, documents, and other outputs (see Figure 5.15). These outputs usually appear in screen displays or hard-copy printouts. The output-control features of a database program allow you to select the records and fields to appear in reports. You can also make calculations specifically for the report by manipulating database fields. Formatting controls and organization options (such as report headings) help you to customize reports and create flexible, convenient, and powerful information-handling tools.

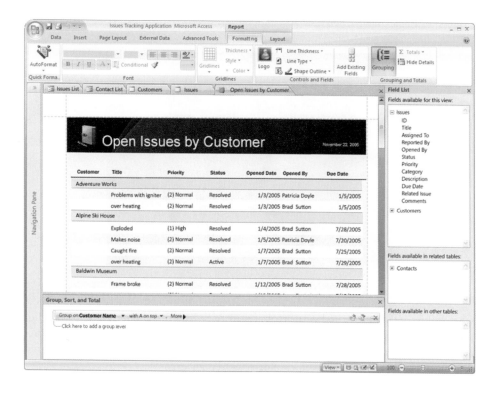

Figure 5.15

Database Output

A database application offers sophisticated formatting and organization options to produce the right information in the right format.

(Source: Courtesy of Microsoft Corporation.)

A DBMS can produce a wide variety of documents, reports, and other outputs that can help organizations achieve their goals. The most common reports select and organize data to present summary information about some aspect of company operations. For example, accounting reports often summarize financial data such as current and past-due accounts. Many companies base their routine operating decisions on regular status reports that show the progress of specific orders toward completion and delivery. Polygon, for example, has developed the ColorNet database to help traders of rare gemstones perform routine processing activities, including the buying and selling of precious gemstones.[15] According to the CEO of Polygon, "If I'm looking for a 2-carat sapphire, [that request] goes out to other users, and I'll also find matches on our own database. It will be that much easier to search for colored stones right on the Internet."

Databases can also provide support to help executives and other people make better decisions. A database by Intellifit, for example, can be used to help shoppers make better decisions and get clothes that fit when shopping online.[16] The database contains true sizes of apparel from various clothing companies that do business on the Web. The process starts when a customer's body is scanned into a database at one of the company's locations, typically in a shopping mall. About 200,000 measurements are taken to construct a 3-D image of the person's body shape. The database then compares the actual body dimensions with sizes given by Web-based clothing stores to get an excellent fit. According to one company executive, "We're 90 percent (accurate) about the sizes and the styles and the brands that will fit you best."

Database Administration

Database systems require a skilled DBA. Lucasfilms, for example, built a 13,500-square foot data center with 18-inch raised floors to accommodate about 600 miles of cabling.[17] The data center includes about 1,500 people, 3,000 processors, and 150 terabytes of storage to deliver state-of-the-art video images to clients around the world. A DBA is expected to have a clear understanding of the fundamental business of the organization, be proficient in the use of selected database management systems, and stay abreast of emerging technologies and new design approaches. The role of the DBA is to plan, design, create, operate, secure, monitor, and maintain databases. Typically, a DBA has a degree in computer science or management information systems and some on-the-job training with a particular database product or more extensive experience with a range of database products. See Figure 5.16.

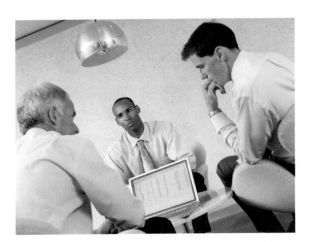

Figure 5.16

Database Administrator

The role of the database administrator (DBA) is to plan, design, create, operate, secure, monitor, and maintain databases.

(Source: BananaStock / Alamy.)

The DBA works with users to decide the content of the database—to determine exactly what entities are of interest and what attributes are to be recorded about those entities. Thus, personnel outside of IS must have some idea of what the DBA does and why this function is important. The DBA can play a crucial role in the development of effective information systems to benefit the organization, employees, and managers.

The DBA also works with programmers as they build applications to ensure that their programs comply with database management system standards and conventions. After the database is built and operating, the DBA monitors operations logs for security violations. Database performance is also monitored to ensure that the system's response time meets users' needs and that it operates efficiently. If there is a problem, the DBA attempts to correct it before it becomes serious.

data administrator
A nontechnical position responsible for defining and implementing consistent principles for a variety of data issues.

Some organizations have also created a position called the *data administrator*, a nontechnical, but important role that ensures that data is managed as an important organizational resource. The **data administrator** is responsible for defining and implementing consistent principles for a variety of data issues, including setting data standards and data definitions that apply across all the databases in an organization. For example, the data administrator would ensure that a term such as "customer" is defined and treated consistently in all corporate databases. This person also works with business managers to identify who should have read or update access to certain databases and to selected attributes within those databases. This information is then communicated to the database administrator for implementation. The data administrator can be a high-level position reporting to top-level managers.

Popular Database Management Systems

Some popular DBMSs for single users include Microsoft Access and FileMaker Pro. The complete database management software market encompasses software used by professional programmers and that runs on midrange, mainframe, and supercomputers. The entire market generates $10 billion per year in revenue, including IBM, Oracle, and Microsoft. Although Microsoft rules in the desktop PC software market, its share of database software on bigger computers is small.

Like other software products, there are a number of open-source database systems, including PostgreSQL and MySQL. Open-source software was introduced in Chapter 4. In addition, many traditional database programs are now available on open-source operating systems. The popular DB2 relational database from IBM, for example, is available on the Linux operating system. The Sybase IQ database and other databases are also available on the Linux operating system.[18]

Special-Purpose Database Systems

In addition to the popular database management systems just discussed, some specialized database packages are used for specific purposes or in specific industries. The Israeli Holocaust Database (*www.yadvashem.org*) is a special-purpose database available through the Internet and contains information on about 3 million people in 14 languages.[19] The Hazmat database is another special-purpose database developed by The National Occupational Health and Safety Commission in Australia and contains information on about 3,500 hazardous materials and various national exposure standards to hazardous materials.[20] Art and Antique Organizer Deluxe is a specialized database for cataloging art works and antiques.[21] Another special-purpose database by Tableau can be used to store and process visual images. According to Christian Chabot, the company's CEO, "Now you can query and walk through a database visually."[22]

Selecting a Database Management System

The database administrator often selects the best database management system for an organization. The process begins by analyzing database needs and characteristics. The information needs of the organization affect the type of data that is collected and the type of database management system that is used. Important characteristics of databases include the following:

- **Database size.** The number of records or files in the database
- **Database cost.** The purchase or lease costs of the database
- **Concurrent users.** The number of people who need to use the database at the same time (the number of concurrent users)
- **Performance.** How fast the database is able to update records
- **Integration.** The ability to be integrated with other applications and databases
- **Vendor.** The reputation and financial stability of the database vendor

A database management system used by online stores must be able to support a large number of concurrent users by quickly checking a customer's credit card and processing his or her order for merchandise.

(Source: *www.potterybarn.com*.)

For many organizations, database size doubles about every year or two.[23] Wal-Mart, for example, adds billions of rows of data to its databases every day. Its database of sales and marketing information is approximately 500 terabytes large. According to Dan Phillips, Wal-Mart's vice president of information systems, "Our database grows because we capture data on every item, for every customer, for every store, every day." Wal-Mart deletes data after two years and doesn't track individual customer purchases. Scientific databases are likely the largest in the world. NASA's Stanford Linear Accelerator Center stores about 1,000 terabytes of data. Britain's forensic DNA database is also huge.[24]

Using Databases with Other Software

Database management systems are often used with other software packages or the Internet. A DBMS can act as a front-end application or a back-end application. A *front-end application* is one that directly interacts with people or users. Marketing researchers often use a database as a front end to a statistical analysis program. The researchers enter the results of market questionnaires or surveys into a database. The data is then transferred to a statistical analysis program to determine the potential for a new product or the effectiveness of an advertising campaign. A *back-end application* interacts with other programs or applications; it only indirectly interacts with people or users. When people request information from a Web site, the Web site can interact with a database (the back end) that supplies the desired information. For example, you can connect to a university Web site to find out whether the university's library has a book you want to read. The Web site then interacts with a database that contains a catalog of library books and articles to determine whether the book you want is available.

DATABASE APPLICATIONS

Today's database applications manipulate the content of a database to produce useful information. David Frayne, for example, uses database analysis tools to find undervalued property in California.[25] Frayne, a former database programmer, has seen a 50 percent return on many of the houses and real estate he has fixed up and sold. His database application, called Scraper, searches through thousands of potential properties to find the best values.

Common manipulations are searching, filtering, synthesizing, and assimilating the data contained in a database using a number of database applications. These applications allow users to link the company databases to the Internet, set up data warehouses and marts, use databases for strategic business intelligence, place data at different locations, use online processing and open connectivity standards for increased productivity, develop databases with the object-oriented approach, and search for and use unstructured data, such as graphics, audio, and video.[26]

Linking Databases to the Internet

Linking databases to the Internet is important for many organizations and people. Yahoo! and several educational partners are scanning books and articles into large Web-based databases.[27] Called the Open-Content Alliance, they plan to offer free access to material that is no longer under copyright agreements. General Electric and Intermountain Health Care are developing a comprehensive Web database on medical treatments and clinical protocols for doctors.[28] The database is expected to cost about $100 million to develop and should help physicians more accurately diagnose patient illnesses. The Health Record Network Foundation, a joint partnership between the medical and business schools of Duke University, is developing a pilot program to make electronic health records (EHR) available over networks and the Internet.[29] The database would allow physicians and other authorized people to access patient records from remote locations. Some banner ads on Web sites are linked directly with sophisticated databases that contain information on products and services.[30] This allows people to get product and service information without leaving their current Web site. LetsGoDigital, a company that sells cameras and other digital products over the Internet based in the Netherlands, uses this approach.

Developing a seamless integration of traditional databases with the Internet is often called a *semantic Web*. A semantic Web allows people to access and manipulate a number of traditional databases at the same time through the Internet. Many software vendors—including IBM, Oracle, Microsoft, Macromedia, Inline Internet Systems, and Netscape Communications—are incorporating the capability of the Internet into their products. Such databases allow companies to create an Internet-accessible catalog, which is nothing more than a database of items, descriptions, and prices.

In addition to the Internet, organizations are gaining access to databases through networks to get good prices and reliable service. Connecting databases to corporate Web sites and networks can lead to potential problems, however. One database expert believes that up to 40 percent of Web sites that connect to corporate databases are susceptible to hackers taking complete control of the database. By typing certain characters in a form on some Web sites, a hacker can issue SQL commands to control the corporate database.

Data Warehouses, Data Marts, and Data Mining

The raw data necessary to make sound business decisions is stored in a variety of locations and formats. This data is initially captured, stored, and managed by transaction processing systems that are designed to support the day-to-day operations of the organization. For decades, organizations have collected operational, sales, and financial data with their online transaction processing (OLTP) systems. The data can be used to support decision making using data warehouses, data marts, and data mining.

Data Warehouses

A **data warehouse** is a database that holds business information from many sources in the enterprise, covering all aspects of the company's processes, products, and customers. The data warehouse provides business users with a multidimensional view of the data they need to analyze business conditions. Data warehouses allow managers to *drill down* to get more detail or *roll up* to take detailed data and generate aggregate or summary reports. A data warehouse is designed specifically to support management decision making, not to meet the needs of transaction processing systems. A data warehouse stores historical data that has been extracted from operational systems and external data sources (see Figure 5.17). This operational and external data is "cleaned up" to remove inconsistencies and integrated to create a new information database that is more suitable for business analysis.

Data warehouses typically start out as very large databases, containing millions and even hundreds of millions of data records. As this data is collected from the various production systems, a historical database is built that business analysts can use. To keep it fresh and accurate, the data warehouse receives regular updates. Old data that is no longer needed is purged from the data warehouse. Updating the data warehouse must be fast, efficient, and automated, or the ultimate value of the data warehouse is sacrificed. It is common for a data warehouse to contain from three to ten years of current and historical data. Data-cleaning tools can merge data from many sources into one database, automate data collection and verification, delete unwanted data, and maintain data in a database management system. Data warehouses can also get data from unique sources. Oracle's Warehouse Management software, for example, can accept information from Radio Frequency Identification (RFID) technology, which is being used to tag products as they are shipped or moved from one location to another.[31] Instead of recalling hundreds of thousands of cars because of a possible defective part, automotive companies could determine exactly which cars had the defective parts and only recall the ten thousand cars with the bad parts using RFID. The savings would be huge.

data warehouse
A database that collects business information from many sources in the enterprise, covering all aspects of the company's processes, products, and customers.

Figure 5.17

Elements of a Data Warehouse

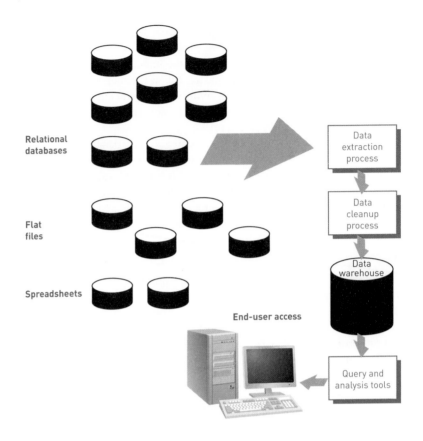

The primary advantage of data warehousing is the ability to relate data in innovative ways. However, a data warehouse can be extremely difficult to establish, with the typical cost exceeding $2 million. Table 5.7 compares online transaction processing (OLTP) and data warehousing.

7-Eleven, a convenience retail chain with more than 25,000 stores worldwide, used Oracle's E-Business Suite to consolidate hundreds of individual business systems into an integrated system that enables the business to track purchasing, assets, costs, and payroll.

(Source: Jack Sullivan / Alamy.)

data mart
A subset of a data warehouse.

Data Marts

A **data mart** is a subset of a data warehouse. Data marts bring the data warehouse concept—online analysis of sales, inventory, and other vital business data that has been gathered from transaction processing systems—to small and medium-sized businesses and to departments within larger companies. Rather than store all enterprise data in one monolithic database, data marts contain a subset of the data for a single aspect of a company's business—for example, finance, inventory, or personnel. In fact, a specific area in the data mart might contain more detailed data than the data warehouse would provide.

Data marts are most useful for smaller groups who want to access detailed data. A warehouse contains summary data that can be used by an entire company. Because data marts typically contain tens of gigabytes of data, as opposed to the hundreds of gigabytes in data warehouses, they can be deployed on less powerful hardware with smaller secondary storage devices, delivering significant savings to an organization. Although any database software can be used to set up a data mart, some vendors deliver specialized software designed and priced

Characteristic	OLTP Database	Data Warehousing
Purpose	Support transaction processing	Support decision making
Source of data	Business transactions	Multiple files, databases—data internal and external to the firm
Data access allowed users	Read and write	Read only
Primary data access mode	Simple database update and query	Simple and complex database queries with increasing use of data mining to recognize patterns in the data
Primary database model employed	Relational	Relational
Level of detail	Detailed transactions	Often summarized data
Availability of historical data	Very limited—typically a few weeks or months	Multiple years
Update process	Online, ongoing process as transactions are captured	Periodic process, once per week or once per month
Ease of process	Routine and easy	Complex, must combine data from many sources; data must go through a data cleanup process
Data integrity issues	Each transaction must be closely edited	Major effort to "clean" and integrate data from multiple sources

specifically for data marts. Already, companies such as Sybase, Software AG, Microsoft, and others have announced products and services that make it easier and cheaper to deploy these scaled-down data warehouses. The selling point: Data marts put targeted business information into the hands of more decision makers.

Table 5.7

Comparison of OLTP and Data Warehousing

Data Mining

Data mining is an information-analysis tool that involves the automated discovery of patterns and relationships in a data warehouse. Like gold mining, data mining sifts through mountains of data to find a few nuggets of valuable information. The Hartford Life Insurance Company, for example, uses data mining to extract detailed information about its customers to increase sales and profits.[32] The data-mining tool helped the company generate record sales, up over 40 percent from the previous year. According to Victoria Severino, chief information officer for Hartford, "We model against these trends and come up with an unexpected risk scenario of the guarantees we offer. We also model risk based on a policyholder's behavior." Data mining has also been used in the airline passenger–profiling system used to block suspected terrorists from flying and in the Total Information Awareness Program, which attempts to detect patterns of terrorist activity. Organizations are also investing in systems for data mining to meet new government regulations.

data mining
An information-analysis tool that involves the automated discovery of patterns and relationships in a data warehouse.

The Hartford Life Insurance Company, for example, uses data mining to extract detailed information about its customers to increase sales and profits.

(Source: © Steven E. Frischling/ Bloomberg News/Landov.)

Google: Protecting User Data from the Government

The Ethical and Societal Issues page in Chapter 3 explained that businesses are instructed by state and federal laws on how to protect customers' private information. Such laws are enacted to protect private information from thieves, to preserve customer privacy, and to deter identity theft. But how should a business react if the government itself wants to examine private records?

In the global war on terror, and with government efforts to find terrorists, many businesses have been placed in that exact position. The United States Patriot Act provides the U.S. government with wide-reaching authority to request information from businesses regarding the activities of suspected terrorists. Privacy advocates are working to limit that reach. In 2006, the Web search company Google found itself stuck between cooperation with U.S. government requests for information and its commitment to preserving its users' privacy. Its refusal to release database records to the federal government landed Google in court.

Google maintains huge databases filled with information on billions of Web sites. When you search the Web using a keyword in the Google Web site, you aren't really searching the Web, but rather Google's database that represents the current state of the Web. Google also maintains databases of search statistics that can be mined to learn about user interests and general search trends. Google and the other search companies also maintain private detailed information that includes the specific searches made by specific IP addresses. IP addresses can be connected to actual users through information provided by Internet service providers. Most search engine companies and Internet service providers keep this information confidential.

The U.S. government requested information from Web search businesses as part of its efforts to obtain data on Internet activity to achieve its law enforcement goals, from national security to the prosecution of online crime. Although Google's three biggest competitors complied with U.S. Department of Justice (DOJ) requests for data, Google refused the subpoena. In addition to wanting records of a week of search queries, which could amount to billions of search terms, the Google subpoena requested a random list of a million Web addresses in its index. Google complained that the request was unnecessary, overly broad, would be onerous to comply with, would jeopardize its trade secrets, and could expose identifying information about its users. The DOJ asked a federal judge to compel Google to turn over records on millions of its users' search queries as part of the government's efforts.

Philip B. Stark, a statistics professor at the University of California, Berkeley, was hired by the DOJ to analyze search engine data in the case, and said in legal documents that search engine data provides crucial insight into information on the Internet. "Google is one of the most popular search engines," he wrote in a court document related to the case. Thus, he said, Google's databases of Web addresses and user searches "are directly relevant."

Google, whose corporate slogan is "Don't Be Evil," complained that providing such information to the government would be detrimental to the trust that users place in the company. "Google's acceding to the request would suggest that it is willing to reveal information about those who use its services," Google said in an October letter to the DOJ. "This is not a perception Google can accept. And one can envision scenarios where queries alone could reveal identifying information about a specific Google user, which is another outcome that Google cannot accept."

Customer trust is a valuable commodity to a company like Google who has recently become more personal with its customers through its Desktop Search software. A new feature in the software allows users to "Search Across Computers" to find information stored on multiple networked computers. This capability is made possible by having Google index personal files in its own database. The digital-rights advocacy organization Electronic Frontier Foundation (EFF) warns users away from the service, stating that it makes it possible for law enforcement officials to examine personal documents from your hard drive without your knowledge. For Google to comply with government requests would mean a serious undermining of customer trust for the company and the possibility of failure of some of its products.

The federal judge in the Google case ruled that Google was required to give the government addresses of 50,000 randomly selected Web sites indexed by its search engine but not any personal information on Google users. The ruling was considered to be a significant victory for Google and privacy rights advocates. "We will always be subject to government subpoenas, but the fact that the judge sent a clear message about privacy is reassuring," Google lawyer Nicole Wong wrote on the company's Web site. "What his ruling means is that neither the government nor anyone else has carte blanche when demanding data from Internet companies."

Discussion Questions

1. How did Google's stance provide the company with an advantage over its competition?
2. What risk did Google take in not initially complying with government requests?

Critical Thinking Questions

1. Do you feel Google was supporting or obstructing justice in its refusal to give up private information to its government? Why?
2. Should a government have the right to force a legitimate business to take an action that might damage its relationship with its customers? What does this imply about the action of the government and support of its citizens?

SOURCES: Hong, Jae C., "Google Resists U.S. Subpoena of Search Data," *New York Times*, January 20, 2006, *www.nytimes.com*. Liedtke, Michael, "Google Avoids Surrendering Search Requests to Government," *The Mercury News*, March 17, 2006, *www.mercurynews.com*. Tweney, Dylan, "Google's Private Lives," *Technology Review*, February 17, 2006, *www.technologyreview.com*.

Data mining's objective is to extract patterns, trends, and rules from data warehouses to evaluate (i.e., predict or score) proposed business strategies, which will improve competitiveness, increase profits, and transform business processes. It is used extensively in marketing to improve customer retention; cross-selling opportunities; campaign management; market, channel, and pricing analysis; and customer segmentation analysis (especially one-to-one marketing). In short, data-mining tools help users find answers to questions they haven't thought to ask.

E-commerce presents another major opportunity for effective use of data mining. Attracting customers to Web sites is tough; keeping them can be next to impossible. For example, when retail Web sites launch deep-discount sales, they cannot easily determine how many first-time customers are likely to come back and buy again. Nor do they have a way of understanding which customers acquired during the sale are price sensitive and more likely to jump on future sales. As a result, companies are gathering data on user traffic through their Web sites and storing the data in databases. This data is then analyzed using data-mining techniques to personalize the Web site and develop sales promotions targeted at specific customers.

Predictive analysis is a form of data mining that combines historical data with assumptions about future conditions to predict outcomes of events such as future product sales or the probability that a customer will default on a loan. Retailers use predictive analysis to upgrade occasional customers into frequent purchasers by predicting what products they will buy if offered an appropriate incentive. Genalytics, Magnify, NCR Teradata, SAS Institute, Sightward, SPSS, and Quadstone have developed predictive analysis tools. Predictive analysis software can be used to analyze a company's customer list and a year's worth of sales data to find new market segments that could be profitable.

predictive analysis
A form of data mining that combines historical data with assumptions about future conditions to predict outcomes of events such as future product sales or the probability that a customer will default on a loan.

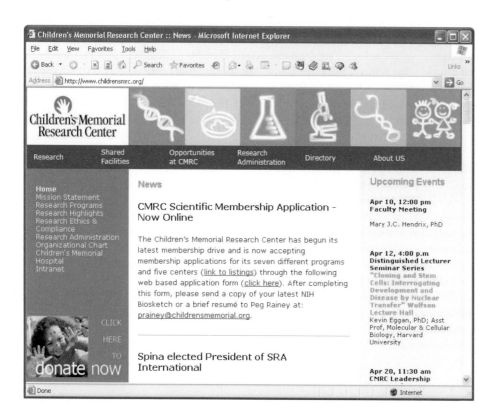

A medical research team at Children's Memorial Research Center uses SPSS's Clementine, a predictive analysis data-mining tool, to help find a cure for pediatric brain tumors.

(Source: *www.childrensmrc.org*.)

Traditional DBMS vendors are well aware of the great potential of data mining. Thus, companies such as Oracle, Sybase, Tandem, and Red Brick Systems are all incorporating data-mining functionality into their products. Table 5.8 summarizes a few of the most frequent applications for data mining.

Application	Description
Branding and positioning of products and services	Enable the strategist to visualize the different positions of competitors in a given market using performance (or other) data on dozens of key features of the product and then to condense all that data into a perceptual map of only two or three dimensions.
Customer churn	Predict current customers who are likely to switch to a competitor.
Direct marketing	Identify prospects most likely to respond to a direct marketing campaign (such as a direct mailing).
Fraud detection	Highlight transactions most likely to be deceptive or illegal.
Market basket analysis	Identify products and services that are most commonly purchased at the same time (e.g., nail polish and lipstick).
Market segmentation	Group customers based on who they are or on what they prefer.
Trend analysis	Analyze how key variables (e.g., sales, spending, promotions) vary over time.

Table 5.8

Common Data-Mining Applications

---

business intelligence
The process of gathering enough of the right information in a timely manner and usable form and analyzing it to have a positive impact on business strategy, tactics, or operations.

Business Intelligence

Closely linked to the concept of data mining is use of databases for business-intelligence purposes. **Business intelligence (BI)** involves gathering enough of the right information in a timely manner and usable form and analyzing it so that it can have a positive effect on business strategy, tactics, or operations.[33] "Right now, we are using our BI tools to generate on-demand statistics and process-control reports," said Steve Snodgrass, CIO of Granite Rock Company.[34] The company uses Business Objects to produce graphic displays of construction supplies, including concrete and asphalt. Business intelligence turns data into useful information that is then distributed throughout an enterprise. Business Objects, a company that sells business intelligence tools, allows managers to connect to databases on Web sites and at other locations to track and analyze data from databases.[35] ARM, a medical transportation company in Denver, uses the new BI tools to effectively schedule ambulances, drivers, and paramedics.

competitive intelligence
One aspect of business intelligence limited to information about competitors and the ways that knowledge affects strategy, tactics, and operations.

Competitive intelligence is one aspect of business intelligence and is limited to information about competitors and the ways that knowledge affects strategy, tactics, and operations. Competitive intelligence is a critical part of a company's ability to see and respond quickly and appropriately to the changing marketplace. Competitive intelligence is not espionage—the use of illegal means to gather information. In fact, almost all the information a competitive-intelligence professional needs can be collected by examining published information sources, conducting interviews, and using other legal, ethical methods. Using a variety of analytical tools, a skilled competitive-intelligence professional can by deduction fill the gaps in information already gathered.

counterintelligence
The steps an organization takes to protect information sought by "hostile" intelligence gatherers.

The term **counterintelligence** describes the steps an organization takes to protect information sought by "hostile" intelligence gatherers. One of the most effective counterintelligence measures is to define "trade secret" information relevant to the company and control its dissemination.

Distributed Databases

distributed database
A database in which the data can be spread across several smaller databases connected via telecommunications devices.

Distributed processing involves placing processing units at different locations and linking them via telecommunications equipment. A **distributed database**—a database in which the data can be spread across several smaller databases connected through telecommunications devices—works on much the same principle. A user in the Milwaukee branch of a clothing manufacturer, for example, might make a request for data that is physically located at corporate headquarters in Milan, Italy. The user does not have to know where the data is physically stored (see Figure 5.18).

Figure 5.18

The Use of a Distributed Database

For a clothing manufacturer, computers might be located at corporate headquarters, in the research and development center, in the warehouse, and in a company-owned retail store. Telecommunications systems link the computers so that users at all locations can access the same distributed database no matter where the data is actually stored.

Distributed databases give corporations and other organizations more flexibility in how databases are organized and used. Local offices can create, manage, and use their own databases, and people at other offices can access and share the data in the local databases. Giving local sites more direct access to frequently used data can improve organizational effectiveness and efficiency significantly. The New York City Police Department, for example, has about 35,000 officers searching for information located in over 70 offices around the city.[36] According to one database programmer, "They had a lot of information available in a lot of different database systems and wanted fingertip access to the information in a very user-friendly front-end." Dimension Data helped the police department by developing an $11 million system to tie their databases together. "Now, we can send them critical details before they even arrive at the scene," said police commissioner Raymond Kelly. The new distributed database is also easier for police officers to use.

Despite its advantages, distributed processing creates additional challenges in integrating different databases (information integration), maintaining data security, accuracy, timeliness, and conformance to standards.[37] Distributed databases allow more users direct access at different sites; thus, controlling who accesses and changes data is sometimes difficult.[38] Also, because distributed databases rely on telecommunications lines to transport data, access to data can be slower.

To reduce telecommunications costs, some organizations build a replicated database. A **replicated database** holds a duplicate set of frequently used data. The company sends a copy of important data to each distributed processing location when needed or at predetermined times. Each site sends the changed data back to update the main database on an update cycle that meets the needs of the organization. This process, often called *data synchronization*, is

replicated database
A database that holds a duplicate set of frequently used data.

used to make sure that replicated databases are accurate, up to date, and consistent with each other. A railroad, for example, can use a replicated database to increase punctuality, safety, and reliability. The primary database can hold data on fares, routings, and other essential information. The data can be continually replicated and downloaded on a read-only basis from the master database to hundreds of remote servers across the country. The remote locations can send back the latest figures on ticket sales and reservations to the main database.

Online Analytical Processing (OLAP)

For nearly two decades, multidimensional databases and their analytical information display systems have provided flashy sales presentations and trade show demonstrations. All you have to do is ask where a certain product is selling well, for example, and a colorful table showing sales performance by region, product type, and time frame appears on the screen. Called **online analytical processing (OLAP)**, these programs are now being used to store and deliver data warehouse information efficiently. The leading OLAP software vendors include Cognos, Comshare, Hyperion Solutions, Oracle, MineShare, WhiteLight, and Microsoft.

The value of data ultimately lies in the decisions it enables. Powerful information-analysis tools in areas such as OLAP and data mining, when incorporated into a data warehousing architecture, bring market conditions into sharper focus and help organizations deliver greater competitive value. OLAP provides top-down, query-driven data analysis; data mining provides bottom-up, discovery-driven analysis. OLAP requires repetitive testing of user-originated theories; data mining requires no assumptions and instead identifies facts and conclusions based on patterns discovered. OLAP, or multidimensional analysis, requires a great deal of human ingenuity and interaction with the database to find information in the database. A user of a data-mining tool does not need to figure out what questions to ask; instead, the approach is, "Here's the data, tell me what interesting patterns emerge." For example, a data-mining tool in a credit card company's customer database can construct a profile of fraudulent activity from historical information. Then, this profile can be applied to all incoming transaction data to identify and stop fraudulent behavior, which might otherwise go undetected. Table 5.9 compares the OLAP and data-mining approaches to data analysis.

online analytical processing (OLAP)
Software that allows users to explore data from a number of perspectives.

Table 5.9

Comparison of OLAP and Data Mining

Characteristic	OLAP	Data Mining
Purpose	Supports data analysis and decision making	Supports data analysis and decision making
Type of analysis supported	Top-down, query-driven data analysis	Bottom-up, discovery-driven data analysis
Skills required of user	Must be very knowledgeable of the data and its business context	Must trust in data mining tools to uncover valid and worthwhile hypotheses

INFORMATION SYSTEMS @ WORK

Boston Celtics and the Business of Basketball

In the business of basketball, ticket sales account for 80 percent of a team's revenue. To maximize profits, basketball team business managers work hard to charge the maximum price possible for tickets while filling as many seats as they can. Not all games are equally popular, and coming up with the right price for tickets is essential but not easy. It requires a real understanding of the fans.

The Boston Celtics have traditionally worked with TicketMaster, their ticketing vendor, to acquire attendance statistics to assist them in ticket pricing. TicketMaster provided the Celtics with synchronized data of both telephone and Internet ticket sales, which the Celtics executives would then import into an Excel spreadsheet to analyze. What the team really needed was a faster, more powerful system for exploring the trends and valuable insight locked in the ticket sales data. Such a system would provide the Celtics with the ability to adjust ticket prices on the fly and offer special deals to fill the arena.

In 2006, the Boston Celtics adopted a new online analytical processing (OLAP) system, referred to as data analytics, in their annual January task of setting prices for the 18,600 seats in TD Banknorth Garden. The StratTix data analytics system, from hosted service provider StratBridge, is designed specifically for sporting and entertainment ticketing, and provides multiple views for event marketing and sales analysis using a graphical presentation. In the ticket office, ticket sellers can see an image of the arena seating chart on a plasma TV screen with color blocks indicating real-time availability and revenue for home games. Sales executives can access this information from their desktops to study buying trends and design new promotions.

At the simplest level, the software gives the Celtics views of seats sold or not sold (green or red, respectively) for each home game. A more advanced, percentage-of-potential-revenue view reveals the extent of discounting based on group sales and promotions. Snapshots over time reveal trends and responses to special promotions such as after-game concerts, ticket discounts, and the popular bobble-head nights. Multiple game views let planners see sales trends and revenues, such as over the last ten games of the season.

The real-time visual insight provided by StratTix enabled the Celtics to promote more effectively and discount more judiciously. "The new tool has helped the organization quickly develop promotions and sales strategies to fill available seats and to analyze revenue based on long-term sales trends," said Daryl Morey, senior vice president of operations and information for the Celtics. For example, multigame ticket packages were easy to set up in the preseason, but it was next to impossible after the season began and the arena was partially sold. With StratTix, special ticket packages could be designed at any time in the season, assigning specific seats based on desirability of the location. Using the analytics tool, for example, planners found that ticket buyers tended to favor aisle seating in certain sections; as a result, the team now focuses on marketing the inner seats.

"Until we had this tool, it was very difficult to create dynamic packages, because our ticket providers didn't have a rapid way to see which seats were open," Morey said. "Now we can actually see in real time every single seat and how much it is sold for."

In its first season of use, StratTix will make a "six-figure" impact on revenue, Morey says, but he adds that the visual history and analysis of this season's ticket sales will lead to even better pricing decisions for next season. For example, besides learning about the value of aisle seats, the team learned that seats near the players' entrance could demand a premium. "Without a visual sense of who was coming into the building and where they wanted to sit, it was like playing chess without a chess board," Morey says. "You could play chess with written notations in Excel, but it makes a big difference if you can actually see the demand and the revenue trends."

Discussion Questions

1. Why were the Celtics executives unable to access the information they required prior to purchasing the data analytics software?
2. What advantages did the data analytics software provide for the Celtics? Why was it important?

Critical Thinking Questions

1. How do you think the new system benefits Boston Celtic fans and makes for better basketball games?
2. What types of useful statistics might be gleaned from the careful evaluation of ticket sales over the course of a season?

SOURCES: Havenstein, Heather, "Celtics Turn to Data Analytics Tool for Help Pricing Tickets," *Computerworld*, January 9, 2006, *www.computerworld.com*. Henschen, Doug, "Visual Analysis Boosts Celtics Ticket Revenue," *Intelligent Enterprise*, February 21, 2006, *www.intelligententerprise.com/showArticle.jhtml?articleID=180205499*.

Object-Oriented and Object-Relational Database Management Systems

object-oriented database
A database that stores both data and its processing instructions.

An **object-oriented database** uses the same overall approach of object-oriented programming that was first discussed in Chapter 4. With this approach, both the data and the processing instructions are stored in the database. For example, an object-oriented database could store monthly expenses and the instructions needed to compute a monthly budget from those expenses. A traditional DBMS might only store the monthly expenses. Starwood Hotels & Resorts, for example, uses an object-oriented database by ObjectStore to process hotel and resort reservations.[39] In an object-oriented database, a *method* is a procedure or action. A sales tax method, for example, could be the procedure to compute the appropriate sales tax for an order or sale—for example, multiplying the total amount of an order by 5 percent, if that is the local sales tax. A *message* is a request to execute or run a method. For example, a sales clerk could issue a message to the object-oriented database to compute sales tax for a new order. Many object-oriented databases have their own query language, called *object query language (OQL)*, which is similar to SQL, discussed previously.

object-oriented database management system (OODBMS)
A group of programs that manipulate an object-oriented database and provide a user interface and connections to other application programs.

An object-oriented database uses an **object-oriented database management system (OODBMS)** to provide a user interface and connections to other programs. A number of computer vendors sell or lease OODBMSs, including eXcelon, Versant, Poet, and Objectivity. Object-oriented databases are used by a number of organizations. Versant's OODBMS, for example, is being used by companies in the telecommunications, financial services, transportation, and defense industries. The *Object Data Standard* is a design standard by the *Object Database Management Group (www.odmg.org)* for developing object-oriented database systems.

object-relational database management system (ORDBMS)
A DBMS capable of manipulating audio, video, and graphical data.

An **object-relational database management system (ORDBMS)** provides a complete set of relational database capabilities plus the ability for third parties to add new data types and operations to the database. These new data types can be audio, images, unstructured text, spatial, or time series data that require new indexing, optimization, and retrieval features. Each of the vendors offering ORDBMS facilities provides a set of application programming interfaces to allow users to attach external data definitions and methods associated with those definitions to the database system. They are essentially offering a standard socket into which users can plug special instructions. DataBlades, Cartridges, and Extenders are the names applied by Oracle and IBM to describe the plug-ins to their respective products. Other plug-ins serve as interfaces to Web servers.

Visual, Audio, and Other Database Systems

In addition to raw data, organizations are increasingly finding a need to store large amounts of visual and audio signals in an organized fashion. Credit card companies, for example, enter pictures of charge slips into an image database using a scanner. The images can be stored in the database and later sorted by customer name, printed, and sent to customers along with their monthly statements. Image databases are also used by physicians to store x-rays and transmit them to clinics away from the main hospital. Financial services, insurance companies, and government branches are using image databases to store vital records and replace paper documents. Drug companies often need to analyze many visual images from laboratories. The PetroView database and analysis tool allows petroleum engineers to analyze geographic information to help them determine where to drill for oil and gas. Recently, a visual-fingerprint database was used to solve a 40-year-old murder case in California. Visual databases can be stored in some object-relational databases or special-purpose database systems. Many relational databases can also store graphic content.

Music companies also need to store and manipulate sound from recording studios. Purdue University has developed an audio database and processing software to give singers a voice makeover. The database software can correct pitch errors and modify voice patterns to introduce vibrato and other voice characteristics.

Combining and analyzing data from different databases is an increasingly important challenge. Global businesses, for example, sometimes need to analyze sales and accounting data stored around the world in different database systems. Companies such as IBM are developing *virtual database systems* to allow different databases to work together as a unified database system. DiscoveryLink, one of IBM's projects, can integrate biomedical data from different sources. The Centers for Disease Control (CDC) also has the problem of integrating more than 100 databases on various diseases.

In addition to visual, audio, and virtual databases, there are a number of other special-purpose database systems. *Spatial data technology* involves using a database to store and access data according to the locations it describes and to permit spatial queries and analysis. MapExtreme is spatial technology software from MapInfo that extends a user's database so that it can store, manage, and manipulate location-based data. Police departments, for example, can use this type of software to bring together crime data and map the data visually so that patterns are easier to analyze. Police officers can select and work with spatial data at a specified location, within a rectangle, a given radius, or a polygon such as a precinct. For example, a police officer can request a list of all liquor stores within a two-mile radius of the precinct. Builders and insurance companies use spatial data to make decisions related to natural hazards. Spatial data can even be used to improve financial risk management with information stored by investment type, currency type, interest rates, and time.

Spatial data technology is used by NASA to store data from satellites and Earth stations. Location-specific information can be accessed and compared.

(Source: Courtesy of NASA.)

SUMMARY

Principle

Data management and modeling are key aspects of organizing data and information.

Data is one of the most valuable resources that a firm possesses. It is organized into a hierarchy that builds from the smallest element to the largest. The smallest element is the bit, a binary digit. A byte (a character such as a letter or numeric digit) is made up of 8 bits. A group of characters, such as a name or number, is called a field (an object). A collection of related fields is a record; a collection of related records is called a file. The database, at the top of the hierarchy, is an integrated collection of records and files.

An entity is a generalized class of objects for which data is collected, stored, and maintained. An attribute is a characteristic of an entity. Specific values of attributes—called data items—can be found in the fields of the record describing an entity. A data key is a field within a record that is used to identify the record. A primary key uniquely identifies a record, while a secondary key is a field in a record that does not uniquely identify the record.

Traditional file-oriented applications are often characterized by program-data dependence, meaning that they have data organized in a manner that cannot be read by other programs. To address problems of traditional file-based data management, the database approach was developed. Benefits of this approach include reduced data redundancy, improved data consistency and integrity, easier modification and updating, data and program independence, standardization of data access, and more-efficient program development.

One of the tools that database designers use to show the relationships among data is a data model. A data model is a map or diagram of entities and their relationships. Enterprise data modeling involves analyzing the data and information needs of an entire organization. Entity-relationship (ER) diagrams can be employed to show the relationships between entities in the organization.

The relational model places data in two-dimensional tables. Tables can be linked by common data elements, which are used to access data when the database is queried. Each row represents a record. Columns of the tables are called attributes, and allowable values for these attributes are called the domain. Basic data manipulations include selecting, projecting, and joining. The relational model is easier to control, more flexible, and more intuitive than the other models because it organizes data in tables.

Principle

A well-designed and well-managed database is an extremely valuable tool in supporting decision making.

A DBMS is a group of programs used as an interface between a database and its users and other application programs. When an application program requests data from the database, it follows a logical access path. The actual retrieval of the data follows a physical access path. Records can be considered in the same way: A logical record is what the record contains; a physical record is where the record is stored on storage devices. Schemas are used to describe the entire database, its record types, and their relationships to the DBMS.

A DBMS provides four basic functions: providing user views, creating and modifying the database, storing and retrieving data, and manipulating data and generating reports. Schemas are entered into the computer via a data definition language, which describes the data and relationships in a specific database. Another tool used in database management is the data dictionary, which contains detailed descriptions of all data in the database.

After a DBMS has been installed, the database can be accessed, modified, and queried via a data manipulation language. A more specialized data manipulation language is the query language, the most common being Structured Query Language (SQL). SQL is used in several popular database packages today and can be installed on PCs and mainframes.

Popular single-user DBMSs include Corel Paradox and Microsoft Access. IBM, Oracle, and Microsoft are the leading DBMS vendors.

Selecting a DBMS begins by analyzing the information needs of the organization. Important characteristics of databases include the size of the database, the number of concurrent users, its performance, the ability of the DBMS to be integrated with other systems, the features of the DBMS, the vendor considerations, and the cost of the database management system.

Principle

The number and types of database applications will continue to evolve and yield real business benefits.

Traditional online transaction processing (OLTP) systems put data into databases very quickly, reliably, and efficiently, but they do not support the types of data analysis needed today. So, organizations are building data warehouses, which are relational database management systems specifically designed to support management decision making. Data marts are subdivisions of data warehouses, which are commonly devoted to specific purposes or functional business areas.

Data mining, which is the automated discovery of patterns and relationships in a data warehouse, is emerging as a practical approach to generating hypotheses about the patterns and anomalies in the data that can be used to predict future behavior.

Predictive analysis is a form of data mining that combines historical data with assumptions about future conditions to forecast outcomes of events such as future product sales or the probability that a customer will default on a loan.

Business intelligence is the process of getting enough of the right information in a timely manner and usable form and analyzing it so that it can have a positive effect on business strategy, tactics, or operations. Competitive intelligence is one aspect of business intelligence limited to information about competitors and the ways that information affects strategy, tactics, and operations. Competitive intelligence is not espionage—the use of illegal means to gather information. Counterintelligence describes the steps an organization takes to protect information sought by "hostile" intelligence gatherers.

With the increased use of telecommunications and networks, distributed databases, which allow multiple users and different sites access to data that may be stored in different physical locations, are gaining in popularity. To reduce telecommunications costs, some organizations build replicated databases, which hold a duplicate set of frequently used data.

Multidimensional databases and online analytical processing (OLAP) programs are being used to store data and allow users to explore the data from a number of different perspectives.

An object-oriented database uses the same overall approach of object-oriented programming, first discussed in Chapter 4. With this approach, both the data and the processing instructions are stored in the database. An object-relational database management system (ORDBMS) provides a complete set of relational database capabilities, plus the ability for third parties to add new data types and operations to the database. These new data types can be audio, video, and graphical data that require new indexing, optimization, and retrieval features.

In addition to raw data, organizations are increasingly finding a need to store large amounts of visual and audio signals in an organized fashion. There are also a number of special-purpose database systems.

CHAPTER 5: SELF-ASSESSMENT TEST

Data management and modeling are key aspects of organizing data and information.

1. A group of programs that manipulate the database and provide an interface between the database and the user of the database and other application programs is called a(n) _____.
   a. GUI
   b. operating system
   c. DBMS
   d. productivity software

2. A(n) _____ has no relationship between its records and is often used to store and manipulate a single table or file.

3. A database administrator is a skilled and trained IS professional who directs all activities related to an organization's database. True or False?

4. The duplication of data in separate files is known as _____.
   a. data redundancy
   b. data integrity
   c. data relationships
   d. data entities

5. _____ is a data-modeling approach that starts by investigating the general data and information needs of the organization at the strategic level and then examining more specific data and information needs for the various functional areas and departments within the organization.

6. What database model places data in two-dimensional tables?
   a. relational
   b. network
   c. normalized
   d. hierarchical

A well-designed and well-managed database is an extremely valuable tool in supporting decision making.

7. _____ involves eliminating columns in a database table.

8. After data has been placed into a relational database, users can make inquiries and analyze data. Basic data manipulations include selecting, projecting, and optimizing. True or False?

9. Because the DBMS is responsible for providing access to a database, one of the first steps in installing and using a database involves telling the DBMS the logical and physical structure of the data and relationships among the data in the database. This description is called a(n) _____.

10. The commands used to access and report information from the database are part of the _____.
    a. data definition language
    b. data manipulation language
    c. data normalization language
    d. schema

11. Access is a popular DBMS for _____.
    a. personal computers
    b. graphics workstations
    c. mainframe computers
    d. supercomputers

12. The ability of a vendor to provide global support for large, multinational companies or companies outside the United States is becoming increasingly important. True or False?

The number and types of database applications will continue to evolve and yield real business benefits.

13. A(n) _____ holds business information from many sources in the enterprise, covering all aspects of the company's processes, products, and customers.

14. An information-analysis tool that involves the automated discovery of patterns and relationships in a data warehouse is called _____.
    a. a data mart
    b. data mining
    c. predictive analysis
    d. business intelligence

15. _____ allows users to explore corporate data from a number of perspectives.

CHAPTER 5: SELF-ASSESSMENT TEST ANSWERS

(1) c (2) flat file (3) True (4) a (5) Enterprise data modeling (6) a (7) Projecting (8) False (9) schema (10) b (11) a (12) True (13) data warehouse (14) b (15) Online analytical processing (OLAP)

REVIEW QUESTIONS

1. What is an attribute? How is it related to an entity?
2. Define the term *database*. How is it different from a database management system?
3. What is a flat file?
4. What is the purpose of data cleanup?
5. What are the advantages of the database approach?
6. What is data modeling? What is its purpose? Briefly describe three commonly used data models.
7. What is a database schema, and what is its purpose?
8. Identify important characteristics in selecting a database management system.
9. What is the difference between a data definition language (DDL) and a data manipulation language (DML)?
10. What is the difference between projecting and joining?
11. What is a distributed database system?
12. What is a data warehouse, and how is it different from a traditional database used to support OLTP?
13. What is data mining? What is OLAP? How are they different?
14. What is an ORDBMS? What kind of data can it handle?
15. What is business intelligence? How is it used?

DISCUSSION QUESTIONS

1. You have been selected to represent the student body on a project to develop a new student database for your school. What actions might you take to fulfill this responsibility to ensure that the project meets the needs of students and is successful?
2. Your company wants to increase revenues from its existing customers. How can data mining be used to accomplish this objective?
3. You are going to design a database for your cooking club to track its recipes. Identify the database characteristics most important to you in choosing a DBMS. Which of the database management systems described in this chapter would you choose? Why? Is it important for you to know what sort of computer the database will run on? Why or why not?
4. Make a list of the databases in which data about you exists. How is the data in each database captured? Who updates each database and how often? Is it possible for you to request a printout of the contents of your data record from each database? What data privacy concerns do you have?

5. What are the advantages of using an object-oriented database instead of a relational database?

6. You are the vice president of information technology for a large, multinational consumer packaged goods company (such as Procter & Gamble, Unilever, or Gillette). You must make a presentation to persuade the board of directors to invest $5 million to establish a competitive-intelligence organization—including people, data-gathering services, and software tools. What key points do you need to make in favor of this investment? What arguments can you anticipate that others might make?

7. Briefly describe how visual and audio databases can be used by companies today.

8. Identity theft, where people steal your personal information, continues to be a threat. Assume that you are the database administrator for a corporation with a large database. What steps would you implement to help you prevent people from stealing personal information from the corporate database?

PROBLEM-SOLVING EXERCISES

1. Develop a simple data model for the music you have on your MP3 player or in your CD collection, where each row is a song. For each row, what attributes should you capture? What will be the unique key for the records in your database? Describe how you might use the database.

2. A video movie rental store is using a relational database to store information on movie rentals to answer customer questions. Each entry in the database contains the following items: Movie ID No. (primary key), Movie Title, Year Made, Movie Type, MPAA Rating, Number of Copies on Hand, and Quantity Owned. Movie types are comedy, family, drama, horror, science fiction, and western. MPAA ratings are G, PG, PG-13, R, X, and NR (not rated). Use a single-user database management system to build a data-entry screen to enter this data. Build a small database with at least ten entries.

3. To improve service to their customers, the salespeople at the video rental store have proposed a list of changes being considered for the database in the previous exercise. From this list, choose two database modifications and modify the data-entry screen to capture and store this new information.

   Proposed changes:
   a. Add the date that the movie was first available to help locate the newest releases.
   b. Add the director's name.
   c. Add the names of three primary actors in the movie.
   d. Add a rating of one, two, three, or four stars.
   e. Add the number of Academy Award nominations.

TEAM ACTIVITIES

1. In a group of three or four classmates, interview a database administrator (DBA) for a company in your area or your school. Describe how this company or your school protects its database from identity theft and privacy concerns.

2. As a team of three or four classmates, interview business managers from three different businesses that use databases to help them in their work. What data entities and data attributes are contained in each database? How do they access the database to perform analysis? Have they received training in any query or reporting tools? What do they like about their database and what could be improved? Do any of them use data-mining or OLAP techniques? Weighing the information obtained, select one of these databases as being most strategic for the firm and briefly present your selection and the rationale for the selection to the class.

3. Imagine that you and your classmates are a research team developing an improved process for evaluating auto loan applicants. The goal of the research is to predict which applicants will become delinquent or forfeit their loan. Those who score well on the application will be accepted; those who score exceptionally well will be considered for lower-rate loans. Prepare a brief report for your instructor addressing these questions:
   a. What data do you need for each loan applicant?
   b. What data might you need that is not typically requested on a loan application form?
   c. Where might you get this data?

d. Take a first cut at designing a database for this application. Using the chapter material on designing a database, show the logical structure of the relational tables for this proposed database. In your design, include the data attributes you believe are necessary for this database, and show the primary keys in your tables. Keep the size of the fields and tables as small as possible to minimize required disk drive storage space. Fill in the database tables with the sample data for demonstration purposes (ten records). After your design is complete, implement it using a relational DBMS.

WEB EXERCISES

1. Use a Web search engine to find information on specific products for one of the following topics: business intelligence, object-oriented databases, or audio databases. Write a brief report describing what you found, including a description of the database products and the companies that developed them.

2. Use a Web search engine to find three companies in an industry that interests you that use a database management system. Describe how databases are used in each company. Could the companies survive without the use of a database management system? Why?

CAREER EXERCISES

1. For a career area of interest to you, describe three databases that could help you on the job.

2. How could you use business intelligence (BI) to do a better job at work? Give some specific examples of how BI can give you a competitive advantage.

CASE STUDIES

Case One

Motorola's Biometrics Solutions Require Large Self-Managing Database

Motorola advertises its Biometric Identification Solution (BIS), named Printrak, as follows: "From fingerprints and palmprints to mugshots, signatures, and case records, Motorola Printrak BIS allows you to enter, search and match, identify and verify." The company's Biometrics Solutions, a division of its Government and Enterprise Mobility Solutions (GEMS) unit, has provided automated fingerprint and palmprint identification systems, inkless live-scan stations, and mobile fingerprint capture solutions used by law enforcement, government, and civil agencies in more than 37 countries.

The BIS is an integrated suite of applications that runs on an Oracle database. Indexing and processing the massive amounts of data that flow through the system requires state-of-the-art database technologies. The system is much more than an automated fingerprint identification system. The database stores graphical descriptive data such as fingerprints, palmprints, facial images, iris images, signatures, and even audio clips. It also holds textual data, including demographic information and criminal investigation information. Some businesses use the BIS to run security checks on job applicants. One valuable aspect of this system is that it can be altered to match a variety of uses.

"Each customer requires a different schema for the demographic and case- or applicant-related information they store, so we use XML for maximum flexibility," says Aris Prassinos of the division's technical staff. XML, the Extensible Markup Language, is a standardized markup language used to describe different types of data. It has become popular for use with database technologies because it provides an ideal method for manipulating database records over the Internet.

Because of the nature of the data it stores, a Printrak BIS database is typically very large. Considering that a 1,000-ppi scan of a palm can comprise several megabytes, it's not uncommon for the database to be anywhere from 1, 2, or 3 terabytes in size. The large BIS databases are often managed by grid computing systems that harness many computers sharing the computational demands. With grid computing power, the system can quickly draw impressive results from the huge database. Three months after deploying Printrak, criminal officials in Palm Beach county, Florida used it to solve 75 crimes, including burglaries.

The BIS division at Motorola appreciates the automated management features of Oracle Database 10g. These features include automatic tuning and management features that allow the system to adjust itself over time for optimum performance. This is important to Printrak customers, who rarely have the staff to properly support the system.

Discussion Questions

1. What unique aspects of Motorola's Biometric Identification Solution (BIS) required special database functionality and support?
2. What unique features of the Motorola BIS and Oracle Database 10g do customers appreciate and why?

Critical Thinking Questions

1. How might a BIS assist in "putting together the pieces" of several related crimes?
2. How might a BIS be useful to businesses? What types of businesses would benefit most from such a system?

SOURCES: Wiseth, Kelli, "Oracle Database 10g Just Got Better," *Oracle Magazine*, September/October 2005, *www.oracle.com/technology/oramag/oracle/ 05-sep/o5510gr2.html*. MOTOROLA PRINTRAK Biometric Identification Solution (BIS) Web site, accessed March 19, 2006, *www.motorola.com/ governmentandenterprise/northamerica/en-us/public/functions/ browsesolution/browsesolution.aspx?navigationpath=id_801i/id_955i / id_1176i*.

Case Two

Consolidating Database Management Systems in Georgia State Courts

As in most U.S. states, Georgia does not mandate how its courts organize their data. Consequently, the design of databases varies from courthouse to courthouse. This posed a considerable challenge to Jorge Basto, the director of technology at the Georgia Administrative Office of the Courts. Basto is working to implement a statewide business intelligence (BI) system for the approximately 1,000 state courts, most of which have different databases, case management applications, and reporting tools.

To consolidate the disparate systems, he first needed to define a unique identifier, a primary key, for each record across all databases. Basto explains, "Our first step is trying to find commonalities, like find a specific, unique identifier for an offender." Basto has the ambitious goal of finishing the new system, to be called Judicial Intelligence, within a year. He selected software from the Business Objects Corporation as the core of the new system. The Business Objects software will give Georgia's judiciary a single reporting, query and analysis tool layered over existing applications.

BI systems include many of the database tools and management systems discussed in this chapter such as online analytical processing (OLAP), data mining, standardized reports, custom report generation, end-user querying, visual analysis tools, and executive dashboards and scorecard functions, which are used for tracking key performance metrics. BI can also include extract, transform, and load tools for moving data into a data warehouse, and integration adapters for connecting directly to an application or database. Generally, though, BI offerings fall into one of two distinct categories: data mining and analysis or business reporting.

Basto expects that the new Judicial Intelligence system will revolutionize the Georgia courts. "There are seven levels of courts, numerous court-related agencies and offices, as well as several executive and legislative agencies that could use this information," he explains. The system will be used by court administrators to dictate case loads. "They could pull statewide case-load statistics, for instance, and analyze them in order to make a case for adding more judges," Basto says.

A single database management system doesn't always satisfy everyone's needs and desires in a large dispersed organization. "In general, standardizing on a single BI platform is a good idea," says Kurt Schlegel, an analyst at Stamford, Connecticut-based Gartner Inc. "However, few organizations have actually done it. Most are hampered by the political realities of replacing a tool from existing projects, all in the name of standardization." According to surveys conducted by Forrester Research, Inc., in Cambridge, Massachusetts, most large organizations have between 5 and 15 different reporting and analysis tools. Consolidating means "taking away those technologies that users feel are most appropriate for their tasks," says Keith Gile, an analyst at Forrester.

However, implementing a consolidated, centralized BI system can be cost-effective. "With a single platform, you can take advantage of caching and clustering [and] blade technology," says Gile. "There really is an economy of scale to managing one BI environment." Consolidation can also provide business users with access to more advanced reporting and querying capabilities and give executives a better view of up-to-date operational data via user-friendly dashboards. "Organizations want a 360-degree view of their customers, employees, and processes—and usually they have to pull that from multiple databases," says Eric Rogge, vice president and research director at Ventana Research, Inc., a business performance management consulting firm in San Mateo, California.

Jorge Basto is unique in his attempt to bring cutting-edge business intelligence technologies to state court information systems. Many in the field consider him brave to be attempting to change the way the court systems operate when so many in the system are content with what they already have. If successful, the Georgia state court system could be the most integrated and organized state court system in the country.

Discussion Questions

1. What benefits would a consolidated BI system provide for the Georgia state courts?
2. What are the primary hurdles that Jorge Basto needs to leap to implement the new system?

Critical Thinking Questions

1. Considering the challenges of updating a state court system, describe the challenges that would face the federal government in implementing a similar system across the federal court system.
2. If you were Jorge Basto, how would you sell the individual state courts on the value of a new system?

SOURCES: Hildreth, Sue, "Central Intelligence: Large Organizations Are Moving to Consolidated BI Suites," *Computerworld*, February 27, 2006, *www.computerworld.com.* "Administrative Office of the Courts of Georgia Standardizes on Business Objects," Business Objects Press Release, January 30, 2006, *www.businessobjects.com/news/press/press2006/20060130_georgia_courts_cust.asp.*

Questions for Web Case

See the Web site for this book to read about the Whitmann Price Consulting case for this chapter. Following are questions concerning this Web case.

Whitmann Price Consulting: Database Considerations

Discussion Questions

1. How will Whitmann Price consultants and the company itself benefit from their ability to call up corporate information in an instant anywhere and at any time?
2. Why will the database itself not require a change to support the new advanced mobile communications and information system?

Critical Thinking Questions

1. The Web has acted as a convenient standard for accessing all types of information from various types of computing platforms. How will this benefit the systems developers of Whitmann Price in developing forms and reports for the new mobile system?
2. What are the suggested limitations of using a BlackBerry device for accessing and interacting with corporate data?

NOTES

Sources for the opening vignette: Hoffman, Thomas, "Drilling Down: Valero Energy Strikes It Rich with BI Tools," *Computerworld*, February 26, 2006, *www.computerworld.com.* Information Builders Incorporated Web site, accessed March 16, 2006, *www.informationbuilders.com.* Whiting, Richard, "Intelligent Spending," *Information Week*, March 6, 2006, Tech Portal. Software Tools, p. 63, *www.lexis-nexis.com.*

1   Rundle, Rhonda, "Government Puts Data Comparing Hospitals onto Public Web Site," *The Wall Street Journal*, April 1, 2005, p. B1.
2   Staff, "Joint Oil-Data Initiative," *Petroleum Economist*, January 4, 2006, p. 41.
3   Brandel, Mary, "Data Scandal: Do You Know How to Respond to the Inevitable Security Breach?" *Computerworld*, October 3, 2005, p. 39.
4   Weber, Harry, "Breach of Data Affects 145,000," *Rocky Mountain News*, February 22, 2005, p. 2B.
5   Carrns, Ann, "Trial Highlights Vulnerability of Databases," *The Wall Street Journal*, August 3, 2005, p. B1.
6   Higgins, Kelly Jackson, "Locked, But Accessible, Data," *Information Week*, October 10, 2005, p. 50.
7   Gilhooly, Kym, "Dirty Data Blights the Bottom Line," *Computerworld*, November 7, 2005, p. 23.
8   Guth, Robert, "Banks Join Forces to Safeguard Data," *The Wall Street Journal*, February 1, 2006, p. A8.
9   Duffy, Shannon, "Taking DNA from Felon Upheld; Court Cites Need for Data," *Delaware Law Weekly*, March 22, 2005, p. D3.
10 Anthes, Gary, "High-Speed Databases," *Computerworld*, January 16, 2006, p. 23.
11 Staff, "Oracle Tops Relational Database Market in 2004," *Computer-Wire*, March 16, 2005.

12 Savvas, Anthony, "Oracle Database Lite Offers Mobile Access," *ComputerWeekly*, January 14, 2005, p. 14.
13 Poor, Alfred, "Three Ways to Get (and Stay) Organized," *PC Magazine*, March 8, 2005, p. 50.
14 Vijayan, Jaikumar, "New Law Prods Food Makers to Focus on Data Management," *Computerworld*, January 24, 2005, p. 16.
15 Novellino, Teresa, "Polygon Launches Colored Gemstone Database," *Business Media*, February 24, 2005.
16 Schuman, Evan, "Company Offers a High Tech Way to Get Clothes to Fit," *CIO Insight*, March 18, 2005.
17 Mitchell, Robert, "Data Center Gets Star Treatment," *Computerworld*, October 3, 2005, p. 25.
18 Thibodeau, Patrick, "Novell, Red Hat Eye Virtualization for Linux," *Computerworld*, January 10, 2005, p. 14.
19 Staff, "Israeli Holocaust Database Online," *Library Journal*, January 15, 2005, p. 38.
20 Staff, "Hazmat Database Released," *Factory Equipment News*, January 2005.
21 Staff, "Art and Antiques Organizer Deluxe," *PC Magazine*, March 22, 2005, p. 12.
22 Zaino, Jennifer, "Analysis That's Easy on the Eyes," *Information-Week*, March 21, 2005, p. 52.
23 Staff, "Data, Data, Everywhere," *Information Week*, January 9, 2006, p. 49.
24 Staff, "Biggest Brother: DNA Evidence," *The Economist*, January 5, 2006.
25 Barron, Kelly, "Hidden Value," *Fortune*, June 27, 2005, p. 184.
26 Gray, Jim; Compton, Mark, "Long Anticipated, the Arrival of Radically Restructured Database Architectures Is Now Finally at Hand," *ACM Queue* vol. 3, no. 3, April 2005.

27 Delaney, Kevin, "Yahoo, Partners Plan Web Database," *The Wall Street Journal*, October 3, 2005, p. B6.

28 Kranhold, Kathryn, "High-Tech Tool Planned for Physicians," *The Wall Street Journal*, February 17, 2005, p. D3.

29 Havenstein, Heather, "Push for Web-based Health Records Launched," *Computerworld*, February 7, 2005, p. 5.

30 Staff, "In-Ad Search," *New Media Age*, January 13, 2006, p. 28.

31 Hall, Mark, "Databases Can't Handle RFID," *Computerworld*, February 7, 2005, p. 8.

32 Staff, "Hartford Life's Condor System Mines Databases for the Gold," *Insurance Networking News*, April 4, 2005, p. 29.

33 Johnson, Avery, "Hotels Take 'Know Your Customers' to a New Level," *The Wall Street Journal*, February 7, 2006, p. D1.

34 McAdams, Jennifer, "Business Intelligence: Power to the People," *Computerworld*, January 2, 2006, p. 24.

35 Havenstein, Heather, "New Tools Aim to Extend Business Intelligence," *Computerworld*, June 6, 2005, p. 14.

36 Murphy, David, "Fighting Crime in Real Time," *PC Magazine*, September 28, 2005, p. 70.

37 Kay, Russell, "Enterprise Information Integration," *Computerworld*, September 19, 2005, p. 64.

38 Babcock, Charles, "Protection Gets Granular," *InformationWeek*, September 23, 2005, p. 58.

39 Lai, Eric, "Starwood Checks in with Object Database for Reservations," *Computerworld*, January 16, 2006, p. 19.

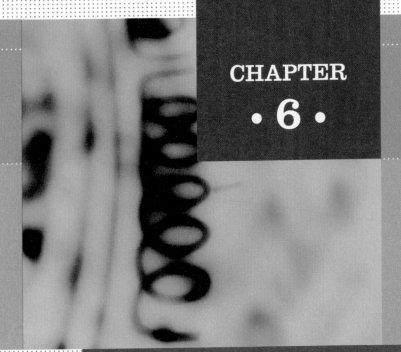

CHAPTER · 6 ·

Telecommunications and Networks

Information Systems in the Global Economy ⟩⟩
Virgin Entertainment Group, United States

Streaming Music to Megastores

Virgin started out as a small mail-order company in the United Kingdom that sold records through a homespun student magazine. Today, Virgin has a presence in four industries and more than 200 companies, such as huge music stores featuring an amazing array of music, movies, books, games, apparel, electronics, and more (Virgin Megastores); a video game company (Virgin Skill Games); a book publishing company (Virgin Books); a cell phone service provider (Virgin Mobile); a winery (Virgin Vines); a travel agency, limousine service, and airline (Virgin Atlantic); and even a spaceship company (Virgin Galactic) and spaceport located in New Mexico. Virgin employs over 25,000 people in dozens of countries and enjoys revenues of over $7.2 billion U.S. annually.

Virgin is a very successful company by anyone's standards. How does Virgin do it? Founder and CEO Richard Branson credits the business' success to innovation and flexibility; Virgin looks for opportunities where it can offer something better, fresher, and more valuable, and seizes them. Recently, Virgin identified opportunities in telecommunications to transform its U.S. Virgin Megastore business.

Virgin Megastores fall under the umbrella of Virgin Entertainment Group, North America. Its recent telecommunications overhaul began as an effort to upgrade its customer listening stations in its music megastores. "Our main goal in 2005 was to maximize the customer experience by improving our listening stations," said Robert Fort, director of information technology for Virgin Entertainment Group, North America. "Up to this point, we only had analog-based stations using CD players behind the scenes. Customers could only access up to four titles per station, and frequently had to wait in line to listen to the most popular titles. Such a limited presentation was not in line with our vision for the stores." Virgin set out to network all of its music listening stations so that they could be controlled by one data center that supplied thousands of song samples, video trailers, and video game demos.

As Virgin explored its options for creating the new system, it uncovered inefficiencies in its entire telecommunications infrastructure. Salespeople were using outdated public telephone network technologies, with limited voice mail, high maintenance and long distance costs, and sometimes unreliable service. In addition to inefficiencies in its voice network, the organization's data network also needed to improve. Network hardware, such as routers and switches, were outdated and slow. The systems in place were not delivering information to Virgin managers in a timely fashion. "We found that we didn't have a good vision of what was happening at any time within our stores, so we were in a very reactive state," commented Ken McNeil, senior manager for administration and support.

Virgin turned to Cisco Systems for a complete telecommunications solution for its Megastores. Using the latest voice-over-IP (VoIP) technology that digitizes voice communications, divides them into packets, and sends them over the Internet as data packets, the company merged its data and voice networks into one efficient, high-speed wide area network (WAN). The system overhaul included new routers and switches, wireless telephony, and security product features that streamline the company's business operations, improve visibility through the supply chain, and enhance the customer experience to meet business-critical objectives.

The new system supports 200 new Virgin Vault Kiosks (listening stations), a series of IBM Anyplace kiosks running Virgin's proprietary kiosk software, which lets shoppers sample more than 250,000 CDs, 11,000 DVDs, and 7,000 games. For example, a shopper interested in the latest *Star Wars* film can instantly preview the sound track CD, watch the video trailer, or view reviews and screenshots from the DVD.

The digital media content is delivered from Virgin headquarters to caches at the stores and served up to the kiosks as needed. Besides music sampling, Virgin's kiosks allow customers to search for additional information on products, such as artist bios, album credits, and reviews.

The kiosk listening stations have been an enormous hit with customers. "Days after the first 100 listening kiosks had been deployed, I walked into our Times Square store," Fort said. "People were lined up with headphones on, bopping along, singing the lyrics, enjoying the experience. It was both humbling and exciting to realize that I was face-to-face with our customers, and the technology had to work, and it was working. It was an incredible experience."

Virgin expects to see an increase in sales revenues due to the promotional impact of the new kiosks. The network upgrade has already allowed the company to achieve a significant cost reduction. The savings that the store achieved by making the transition to VoIP network technology has already paid for the new systems. Conservative estimates indicate that the store might save $700,000 per year by using IP-based voice technology. "All store-to-store calls, and store-to-home office calls, are essentially free, since they are on the data instead of the public telephone network," stated McNeil.

As you read this chapter, consider the following:

- What services do new telecommunications and network technologies offer to assist individuals and businesses in being more effective?
- What role does telecommunications play in connecting businesses and growing the global economy?

Why Learn About Telecommunications and Networks?

Today's decision makers need to access data wherever it resides. They must be able to establish fast, reliable connections to exchange messages, upload and download data and software, route business transactions to processors, connect to databases and network services, and send output to printers. Regardless of your chosen major or future career field, you will need the communications capabilities provided by telecommunications and networks, especially if your work involves the supply chain. Among all business functions, supply chain management might use telecommunications and networks the most because it requires cooperation and communications among workers in inbound logistics, warehouse and storage, production, finished product storage, outbound logistics, and most important, with customers, suppliers, and shippers. All members of the supply chain must work together effectively to increase the value perceived by the customer, so partners must communicate well. Other employees in human resources, finance, research and development, marketing, and sales positions must also use communications technology to communicate with people inside and outside the organization. To be a successful member of any organization, you must be able to take advantage of the capabilities that these technologies offer you. This chapter begins by discussing the importance of effective communications.

In today's high-speed global business world, organizations need always-on, always-connected computing for traveling employees and for network connections to their key business partners and customers. Boeing is one of two major manufacturers of 100-plus seat airplanes for the commercial airline industry. Although most of its operations are in the United States, it is working with over three dozen major partners, key suppliers, and subcontractors around the globe to build the new 787 Dreamliner aircraft that will hold more than 200 passengers and have a cruising range of over 8,000 miles. As the work to complete this monumental project is increasingly distributed, Boeing relies on collaboration technologies and wireless network access.[1] Forward-thinking companies such as Boeing hope to save billions of dollars, reduce time to market, and enable collaboration with their business partners by using telecommunications systems.

AN OVERVIEW OF TELECOMMUNICATIONS

Telecommunications refers to the electronic transmission of signals for communications, by such means as telephone, radio, and television. Telecommunications is creating profound changes in business because it lessens the barriers of time and distance. Telecommunications not only is changing the way businesses operate, but the nature of commerce itself. As networks are connected with one another and transmit information more freely, a competitive marketplace demands excellent quality and service from all organizations.

Figure 6.1 shows a general model of telecommunications. The model starts with a sending unit (1), such as a person, a computer system, a terminal, or another device, that originates the message. The sending unit transmits a signal (2) to a telecommunications device (3). The telecommunications device—a hardware component that facilitates electronic communication—performs many tasks, which can include converting the signal into a different form or from one type to another. The telecommunications device then sends the signal through a medium (4). A **telecommunications medium** is any material substance that carries an electronic signal to support communications between a sending and receiving device. Another telecommunications device (5) connected to the receiving computer (6) receives the signal. The process can be reversed, and the receiving unit (6) can send another message to the original sending unit (1). An important characteristic of telecommunications is the speed at which information is transmitted, which is measured in bits per second (bps). Common speeds are in the range of thousands of bits per second (Kbps) to millions of bits per second (Mbps) and even billions of bits per second (Gbps).

telecommunications medium
Any material substance that carries an electronic signal and serves as an interface between a sending device and a receiving device.

Medium (4)

Signal (2)

Signal (2)

Medium (4)

Telecommunications device (5)

Sending computer system and equipment (1)

Medium (4)

Telecommunications device (3)

Signal (2)

Receiving computer system and equipment (6)

Figure 6.1

Elements of a Telecommunications System

Telecommunications devices relay signals between computer systems and transmission media.

Advances in telecommunications technology allow us to communicate rapidly with clients and coworkers almost anywhere in the world. Telecommunications also reduces the amount of time needed to transmit information that can drive and conclude business actions.

Communications between two people can occur synchronously or asynchronously. With **synchronous communications**, the receiver gets the message instantaneously, when it is sent. Voice and phone communications are examples of synchronous communications. With **asynchronous communications**, the receiver gets the message after some delay—sometimes hours or days after the message is sent. Sending a letter through the post office or e-mail over the Internet are examples of asynchronous communications. Both types of communications are important in business.

Using telecommunications can help businesses solve problems, coordinate activities, and capitalize on opportunities. To use telecommunications effectively, you must carefully analyze telecommunications media and devices.

synchronous communications
A form of communications where the receiver gets the message instantaneously, when it is sent.

asynchronous communications
A form of communications where the receiver gets the message after some delay—sometimes hours or days after the message is sent.

Telecommunications technology enables business people to communicate with coworkers and clients from remote locations.

(Source: © BananaStock / Alamy.)

simplex channel
A communications channel that can transmit data in only one direction.

half-duplex channel
A communications channel that can transmit data in either direction, but not simultaneously.

full-duplex channel
A communications channel that permits data transmission in both directions at the same time, so a full-duplex channel is like two simplex channels.

channel bandwidth
The rate at which data is exchanged over a communications channel, usually measured in bits per second (bps).

broadband communications
A telecommunications system in which a very high rate of data exchange is possible.

narrowband communications
A telecommunications system that supports a much lower rate of data exchange than broadband.

Basic Communications Channel Characteristics

The transmission medium carries messages from the source of the message to its receivers. A transmission medium can be divided into one or more communications channels, each capable of carrying a message. Communications channels can be classified as simplex, half-duplex, or full-duplex.

A **simplex channel** can transmit data in only one direction, and is seldom used for business telecommunications. Doorbells and the radio operate using a simplex channel. A **half-duplex channel** can transmit data in either direction, but not simultaneously. For example, A can begin transmitting to B over a half-duplex line, but B must wait until A is finished to transmit to A. Personal computers are usually connected to a remote computer over a half-duplex channel. A **full-duplex channel** permits data transmission in both directions at the same time, so a full-duplex channel is like two simplex channels. Private leased lines or two standard phone lines are required for full-duplex transmission.

Channel Bandwidth

In addition to the direction of data flow supported by a communications channel, you must consider the speed at which data can be transmitted. Telecommunications **channel bandwidth** refers to the rate at which data is exchanged, usually measured in bits per second (bps)—the broader the bandwidth, the more information can be exchanged at one time. **Broadband communications** can exchange data very quickly, as opposed to **narrowband communications**, which supports a much lower rate of data exchange.

Telecommunications professionals consider the capacity of the channel when they recommend transmission media for a business. In general, today's organizations need more bandwidth for increased transmission speed to carry out their daily functions. To increase bandwidth, first consider the different types of telecommunications media you can use.

Communications Media

You can evaluate each type of communications media according to characteristics such as cost, capacity, and speed. In designing a telecommunications system, the transmission media you select depends on the amount of information to be exchanged, the speed at which data must be exchanged, the level of concern for data privacy, whether the users are stationary or mobile, and many other business requirements. Select the transmission media to support the goals of the information and organizational systems at the lowest cost, but still allow for possible modifications should your business requirements change. Transmission media can be divided into two broad categories: *guided transmission media*, in which communications signals are guided along a solid medium, and *wireless*, in which the communications signal is broadcast over airwaves as a form of electromagnetic radiation.

Guided Transmission Media Types

There are many different guided transmission media types. Table 6.1 summarizes the guided media types by physical media type. Several guided transmission media types are discussed in the sections following the table.

Guided Media Types			
Media Type	**Description**	**Advantages**	**Disadvantages**
Twisted-pair wire	Twisted pairs of copper wire, shielded or unshielded	Used for telephone service; widely available	Transmission speed and distance limitations
Coaxial cable	Inner conductor wire surrounded by insulation	Cleaner and faster data transmission than twisted-pair wire	More expensive than twisted-pair wire
Fiber-optic cable	Many extremely thin strands of glass bound together in a sheathing; uses light beams to transmit signals	Diameter of cable is much smaller than coaxial; less distortion of signal; capable of high transmission rates	Expensive to purchase and install
Broadband over Power Lines	Data is transmitted over standard high-voltage power lines	Can provide Internet service to rural areas where cable and phone service may be non-existent	Can be expensive and may interfere with ham radios and police and fire communications

Table 6.1

Guided Transmission Media Types

Twisted-Pair Wire

Twisted-pair wire contains two or more twisted pairs of wire, usually copper (see Figure 6.2a). Proper twisting of the wire keeps the signal from "bleeding" into the next pair and creating electrical interference. Because the twisted-pair wires are insulated, they can be placed close together and packaged in one group. Hundreds of wire pairs can be grouped into one large wire cable.

(a)

(b)

(c)

Twisted-pair wires are classified by category (Category 1, 2, 3, 4, 5, 5E, and 6). The lower categories are used primarily in homes. Higher categories are sometimes used in smaller networks. Category 1 is traditional telephone cable. Category 3 is the most common type of twisted pair wire found in corporate settings, and it normally contains four pairs of wire. Category 5 is frequently installed in new buildings and is capable of carrying data at speeds faster than 1 Gbps.

10 Gigabit Ethernet (IEEE 802.3an) is an emerging standard for transmitting data at the speed of 10 billion bits per second for limited distances over shielded twisted-pair wires—category 5 or 6. It will be used for the high-speed links that connect groups of computers or to move data stored in large databases on large computers to stand-alone storage devices.[2]

Figure 6.2

Types of Guided Transmission Media

(a) Twisted-pair wire, (b) Coaxial cable, (c) Fiber-optic cable

(Source: a, Fred Bodin; b, Fred Bodin; c, © Greg Pease/Getty Images.)

Coaxial Cable

Figure 6.2b shows a typical coaxial cable, similar to that used in cable television installations. When used for data transmission, coaxial cable falls in the middle of the guided transmission media in terms of cost and performance. The cable itself is more expensive than twisted-pair wire but less so than fiber-optic cable (discussed next). However, the cost of installation and other necessary communications equipment makes it difficult to compare the total costs of each media. Coaxial cable offers cleaner and crisper data transmission (less noise) than twisted-pair wire. It also offers a higher data transmission rate.

Cable companies are aggressively courting customers for telephone service, enticing them away from the phone companies by bundling Internet and phone services along with TV. Comcast, the nation's largest cable TV operator, has over 9 million Internet subscribers and provides digital voice-over-IP (VOIP) service to 19 million homes.[3]

Fiber-Optic Cable

Fiber-optic cable, consisting of many extremely thin strands of glass or plastic bound together in a sheathing (a jacket), transmits signals with light beams (see Figure 6.2c). These high-intensity light beams are generated by lasers and are conducted along the transparent fibers. These fibers have a thin coating, called *cladding*, which effectively works like a mirror, preventing the light from leaking out of the fiber. The much smaller diameter of fiber-optic cable makes it ideal when there is not room for bulky copper wires—for example, in crowded conduits, which can be pipes or spaces carrying both electrical and communications wires. In such tight spaces, the smaller fiber-optic telecommunications cable is very effective. Because fiber-optic cables are immune to electrical interference, they can transmit signals over longer distances with fewer expensive repeaters to amplify or rebroadcast the data. Fiber-optic cable and associated telecommunications devices are more expensive to purchase and install than their twisted-pair wire counterparts, although the cost is decreasing.

Laying thousands of miles of fiber-optic cable across its vast expanses is credited for helping propel India into the high-tech world. With the capability that this infrastructure provided, Indian workers were able to collaborate closely with their Western counterparts even though they were thousands of miles away.[4] As a result, India has emerged as a key business partner to many firms that have outsourced part of their business operations or that use Indian firms for information systems projects.

Broadband over Power Lines

Many utilities, cities, and organizations are experimenting with *broadband over power lines (BPL)* to provide network connections over standard high-voltage power lines. Manassas, Virginia, became the first city in the nation to offer all its citizens this service. To access the Internet, BPL users connect their computer to a special hardware device that plugs into any electrical wall socket. A potential issue with BPL is that transmitting data over unshielded power lines can interfere with both ham radio broadcasts and police and fire radios. However, BPL can provide Internet service in rural areas where broadband access has lagged because electricity is more prevalent in homes than cable or even telephone lines.[5] (As of July 2005, the telephone subscriber penetration rate in the United States was 94.0 percent ranging from a low of 87.4 percent in Arkansas to a high of 97.5 percent in North Dakota.)[6]

Wireless Technologies

Many technologies are used to transmit communications wirelessly. The major technologies include microwave, satellite, radio, and infrared. Their key distinguishing feature is the frequency at which signals are transmitted. These are summarized in Table 6.2 and discussed in the following section.

Microwave Transmission

Microwave is a high-frequency (300 MHz–300 GHz) signal sent through the air (see Figure 6.3). Terrestrial (Earth-bound) microwaves are transmitted by line-of-sight devices, so that the line of sight between the transmitter and receiver must be unobstructed. Typically, microwave stations are placed in a series—one station receives a signal, amplifies it, and

Technology	Description	Advantages	Disadvantages
Microwave—Terrestrial and Satellite	High-frequency radio signal (300 MHz–300 GHz) sent through atmosphere and space (often involves communications satellites)	Avoids cost and effort to lay cable or wires; capable of high-speed transmission	Must have unobstructed line of sight between sender and receiver; signal highly susceptible to interception
Radio	Operates in the 30–300 MHz range	Supports mobile users; costs are dropping	Signal highly susceptible to interception
Infrared	Signals sent through air as light waves	Lets you move, remove, and install devices without expensive wiring	Must have unobstructed line of sight between sender and receiver; transmission effective only for short distances

Table 6.2

Wireless Technologies

retransmits it to the next microwave transmission tower. Such stations can be located roughly 30 miles apart before the curvature of the Earth makes it impossible for the towers to "see one another." Microwave signals can carry thousands of channels at the same time.

Figure 6.3

Microwave Communications

Because they are line-of-sight transmission devices, microwave dishes are frequently placed in relatively high locations, such as atop mountains, towers, or tall buildings.

A communications satellite also operates in the microwave frequency range (see Figure 6.4). The satellite receives the signal from the Earth station, amplifies the relatively weak signal, and then rebroadcasts it at a different frequency. The advantage of satellite communications is that it can receive and broadcast over large geographic regions. Such problems as the curvature of the Earth, mountains, and other structures that block the line-of-sight microwave transmission make satellites an attractive alternative. Geostationary, low earth orbit, and small mobile satellite stations are the most common forms of satellite communications.

A *geostationary satellite* orbits the Earth directly over the equator, approximately 22,000 miles above the Earth so that it appears stationary. Three such satellites, spaced at equal intervals (120 angular degrees apart), can cover the entire world. A geostationary satellite can be accessed using a dish antenna aimed at the spot in the sky where the satellite hovers.

A *low earth orbit (LEO) satellite* system employs many satellites, each in a circular orbit at an altitude of a few hundred miles. The satellites are spaced so that, from any point on the Earth at any time, at least one satellite is on a line of sight. LEO service subscribers can access the satellites using an antenna a little more sophisticated than the old-fashioned television "rabbit ears."

A *very small aperture terminal (VSAT)* is a two-way satellite ground station with a dish antenna smaller than 3 meters in diameter. Many retail chains employ this technology to support point-of-sale transactions, including credit cards. Walgreens, a leader in the retail drugstore industry in the use of information systems, uses VSAT communications to transfer prescriptions among all Walgreens' stores. This allows patients to pick up and drop off their

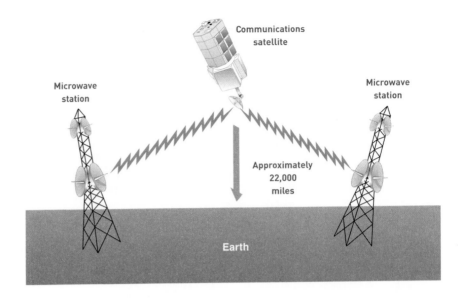

Figure 6.4

Satellite Transmission

Communications satellites are relay stations that receive signals from one Earth station and rebroadcast them to another.

prescriptions at any Walgreens. They use an extensive satellite-based network that connects all Walgreens stores to a centralized customer database. Indeed, Walgreens is the largest private industry user of satellite communications in the world.[7] News organizations employ VSAT dishes that run on battery power to transmit news stories from remote locations. VSAT technology is being used to rebuild the telecommunications infrastructure in Afghanistan and Iraq. Many people are also investing in VSAT technology in their homes to receive TV and send and receive computer communications.

Radio Transmission

Radio transmission operates in the 30 Hz–300 MHz range of the electromagnetic spectrum. At this frequency, radio waves can travel through many obstructions such as walls. Whereas radio transmission usually provides a means to listen to music and talk shows, this form of transmission can also be used to send and receive data. In fact, many of the exciting, new wireless technologies such as RFID chips and Bluetooth and Wi-Fi wireless networks (to be discussed later in this chapter) are based on radio transmission. Satellite radio is a digital radio that receives signals broadcast by a communications satellite. The signal is strong enough that the receiver does not require a satellite dish.

Cellular phones also operate using radio waves to provide two-way communications. With cellular transmission, a local area, such as a city, is divided into cells. As a car or vehicle with a cellular device, such as a mobile phone, moves from one cell to another, the cellular system passes the phone connection from one cell to another (see Figure 6.5). The signals from the cells are transmitted to a receiver and integrated into the regular phone system. Cellular phone users can thus connect to anyone who has access to regular phone service, such as a child at home or a business associate in London. They can also contact other cellular phone users. Because cellular transmission uses radio waves, people with special receivers can listen to cellular phone conversations, so they are not secure.

Ford Motor and its service provider Sprint are pioneers in the conversion of employees from land lines to mobile phone service. Ford converted over 8,000 employees in 2005. The initial deployment covered Ford's product development department, which is a user group that is very collaborative and highly mobile. Although the conversion went well enough that Ford created TV commercials on the subject, there is room for improvement. Ford desires a handset that has a longer battery life. It would also like to see wireless-to-wire line integration, which makes it easier for business users to marry landline PBX features, such as four-digit dialing and a single voice mail box for employees whether they are using VOIP, wireless, or a traditional phone.[8]

Figure 6.5

A Typical Cellular Transmission Scenario

Using a cellular car phone, the caller dials the number (1). The signal is sent from the car's antenna to the low-powered cellular antenna located in that cell (2). The signal is sent to the regional cellular phone switching office, also called the *mobile telephone subscriber office (MTSO)* (3). The signal is switched to the local telephone company switching station located nearest the call destination (4). Now integrated into the regular phone system, the call is switched to the number originally dialed (5), all without the need for operator assistance.

Infrared Transmission

Another mode of transmission, called *infrared transmission*, sends signals through the air via light waves at a frequency of 300 GHz and above. Infrared transmission requires line-of-sight transmission and short distances—under a few hundred yards. Infrared transmission allows handheld computers to transmit data and information to larger computers within the same room and to connect a display screen, a printer, and a mouse to a computer. Some special-purpose phones can also use infrared transmission. You can use this means of transmission to establish a wireless network, with the advantage that devices can be moved, removed, and installed without expensive wiring and network connections.

The Apple Remote is a remote control made for use with Apple products with infrared capabilities. It has just six buttons: Menu, Play/Pause, Volume Up, Volume Down, Previous/Rewind, and Next/Fast-forward. The new Mac Mini features an infrared port designed to work with the Apple Remote and support Front Row, a multimedia application that allows users to access shared iTunes and iPhoto libraries and video throughout their home.[9]

Telecommunications Hardware

Telecommunications hardware devices include modems, multiplexers, and front-end processors.

Modems

At each state of the communications process, telecommunications often uses transmission media of differing types and capacities. If you use an analog telephone line to transfer data, it can only accommodate an **analog signal** (a variable signal continuous in both time and amplitude so that any small fluctuations in the signal are meaningful). Because a computer generates a **digital signal** representing bits, you need a special device to convert the digital signal to an analog signal, and vice versa (see Figure 6.6). Translating data from digital to analog is called *modulation*, and translating data from analog to digital is called *demodulation*. Thus, these devices are modulation/demodulation devices, or **modems**. Penril/Bay Networks, Hayes, Microcom, Motorola, and U.S. Robotics are examples of modem manufacturers.

Modems can dial telephone numbers, originate message sending, and answer incoming calls and messages. Modems can also perform tests and checks on how they are operating. Some modems can vary their transmission rates, commonly measured in bits per second. Cellular modems in laptop personal computers allow people on the go to communicate with other computer systems and devices.

With a cellular modem, you can connect to other computers while in your car, on a boat, or in any area that has cellular transmission service. You can use PC memory card expansion slots for standardized credit card–sized PC modem cards, which work like standard modems. PC modems are becoming increasingly popular with notebook and portable computer users.

Cable companies are promoting the cable modem, which has a low initial cost and transmission speeds up to 10 Mbps. A cable modem can deliver network and Internet access up

analog signal

A variable signal continuous in both time and amplitude so that any small fluctuations in the signal are meaningful.

digital signal

A signal that represents bits.

modem

A telecommunications hardware device that converts (modulates and demodulates) communications signals so they can be transmitted over the communication media.

Figure 6.6

How a Modem Works

Digital signals are modulated into analog signals, which can be carried over existing phone lines. The analog signals are then demodulated back into digital signals by the receiving modem.

Analog transmission

Digital transmission

Digital transmission

Modem

Modem

to 500 times faster than a standard modem and phone line. In addition, a cable modem is always on, so you can be connected to the Internet around the clock. Digital subscriber line is a family of services that provides high-speed digital data communications service over the wires of the local telephone company. Subscribers employ a DSL modem to connect their computers to this service.

A cable modem can deliver network and Internet access up to 500 times faster than a standard modem and phone line.

(Source: Courtesy of D-Link Systems, Inc.)

multiplexer

A device that encodes data from two or more data sources onto a single communications channel, thus reducing the number of communications channels needed and therefore, lowering telecommunications costs.

front-end processor

A special-purpose computer that manages communications to and from a computer system serving hundreds or even thousands of users.

Multiplexers

A **multiplexer** is a device that encodes data from two or more data sources onto a single communications channel, thus reducing the number of communications channels needed and therefore, lowering telecommunications costs (see Figure 6.7).

Front-End Processors

Front-end processors are special-purpose computers that manage communications to and from a computer system serving hundreds or even thousands of users. They poll user devices to see if they have messages to send and facilitate efficient, error-free communications. By performing this work, the front-end processor relieves the primary computer system of much of the overhead processing associated with telecommunications.

Services

Telecommunications carriers organize communications channels, networks, hardware, software, people, and business procedures to provide valuable communications services.

Figure 6.7

Use of a Multiplexer to Consolidate Data Communications onto a Single Communications Link

Telecommunications networks require state-of-the-art computer software technology to continuously monitor the flow of voice, data, and image transmission over billions of circuit miles worldwide.

(Source: © Roger Tully/Getty Images.)

Digital Subscriber Line (DSL)

A **digital subscriber line (DSL)** is a telecommunications service that delivers high-speed Internet access to homes and small businesses over the existing phone lines of the local telephone network (see Figure 6.8). Most home and small business users are connected to an *asymmetric DSL (ADSL)* line designed to provide a connection speed from the Internet to the user (download speed) that is three to four times faster than the connection from the user back to the Internet (upload speed). ADSL does not require an additional phone line and yet provides "always-on" Internet access. A drawback of ADSL is that the farther the subscriber is from the local telephone office, the poorer the signal quality and the slower the transmission speed. ADSL provides a dedicated connection from each user to the phone company's local office, so the performance does not decrease as new users are added. Cable-modem users generally share a network loop that runs through a neighborhood so that adding users means lowering the actual transmission speeds. Verizon provides a 768 Kbps/128 Kbps service for $14.95/month, while its faster 3 Mbps/768 Kbps services runs $29.95/month. *Symmetric DSL (SDSL)* is used mainly by small businesses and does not allow you to use the phone at the same time, but the speed of receiving and sending data is the same.

digital subscriber line (DSL)
A telecommunications service that delivers high-speed Internet access to homes and small businesses over the existing phone lines of the local telephone network.

Figure 6.8

Digital Subscriber Line (DSL)

At the local telephone company's central office, a DSL Access Multiplexer (DSLAM) takes connections from many customers and aggregates them onto a single, high-capacity connection to the Internet. Subscriber phone calls can be routed through a switch at the local telephone central office to the public telephone network.

Wireless Telecommunications

The Internet is revolutionizing how you gather and share information at work, how you collaborate in teams, listen to music or watch video at home, and stay in touch with your families on the road. Add wireless capability and a coffee shop becomes your office, or the bleachers at a ball park become your living room. It is no wonder that all the major long-distance carriers offer wireless telecommunications services that enable you to place phone calls or access the Internet. Indeed, the market growth of landline phones is declining as wireless telephone service plans offer long-distance service at the same per-minute rate as local service. This has resulted in a rapid growth of long-distance telecommunications over wireless networks and a drop in the use of land lines so users can avoid more costly long-distance charges.

A wireless technology application called TechConnect enabled 830 appliance repairers at Maytag Corporation to boost the average number of completed service calls from 5.8 to 6.4 per worker per day. It also eliminated the need for workers to spend nearly 45 minutes at the end of each day updating paper reports. Instead, each night workers dock handheld devices from Internec Technologies Corporation to download information about service calls for the next day. Throughout the day, service people use the device wirelessly to receive updates on service calls and detailed directions and maps to reduce the time to travel to the next service call. TechConnect can also track repair trucks using Global Positioning System technology. Company officials estimate that the wireless application has increased annual revenue by $13 million.[10]

In November 2005, SBC Communications, Inc., acquired AT&T Corporation to form the new AT&T Inc. The combined company has the most wireless customers, provides the most DSL broadband lines, and generates the most revenue from data telecommunications. In addition, AT&T Inc. owns 60 percent of its wireless affiliate, Cingular Wireless.[11]

Adoption of cellular data services is still in its early stages; many business users are taking a trial approach. Wireless data communications will be broadly adopted when providers can offer business users enough bandwidth and connectivity so that they use wireless as their sole connection.

NETWORKS AND DISTRIBUTED PROCESSING

computer network

The communications media, devices, and software needed to connect two or more computer systems and/or devices.

A **computer network** consists of communications media, devices, and software needed to connect two or more computer systems or devices. The computers and devices on the networks are also called *network nodes*. After they are connected, the nodes can share data, information, and processing jobs. Increasingly, businesses are linking computers in networks to streamline work processes and allow employees to collaborate on projects. If a company uses networks effectively, it can grow into an agile, powerful, and creative organization, giving it a long-term competitive advantage. Organizations can use networks to share hardware, programs, and databases. Networks can transmit and receive information to improve organizational effectiveness and efficiency. They enable geographically separated workgroups to share documents and opinions, which fosters teamwork, innovative ideas, and new business strategies.

Network Types

Depending on the physical distance between nodes on a network and the communications and services it provides, networks can be classified as personal area, local area, metropolitan area, or wide area.

Personal Area Networks

A **personal area network (PAN)** is a wireless network that connects information technology devices within a range of 33 feet or so. One device serves as the controller during wireless PAN initialization, and this controller device mediates communication within the PAN. The controller broadcasts a beacon that synchronizes all devices and allocates time slots for the devices. With a PAN, you can connect a laptop, digital camera, and portable printer without physical cables. You could download digital image data from the camera to the laptop and then print it on a high-quality printer—all wirelessly.

Bluetooth is the industry standard for PAN communications. Metro AG is a major German retailer that owns and operates about 2,400 wholesale stores, supermarkets, hypermarkets, department stores, and specialty retailers (such as home improvement and consumer electronics). The firm is testing a new voice-operated, smart tag system equipped with Bluetooth communications to help warehouse personnel quickly and accurately fill merchandise orders. Item pickers wear a headset to receive instructions on which items to pick. They also wear a high-tech glove that gathers product information from the RFID chip on each item that they pick from the shelves. The glove and RFID chip communicate using Bluetooth technology. If workers accidentally pick the wrong product, they receive a message requesting them to repeat the process. The main benefits of the voice-operated RFID picking system are that work sequences aren't interrupted by employees having to manually compare product identification data on the item to the description on a written list.[12]

personal area network (PAN)
A network that supports the interconnection of information technology within a range of 33 feet or so.

Local Area Networks

A network that connects computer systems and devices within a small area, such as an office, home, or several floors in a building is a **local area network (LAN)**. Typically, LANs are wired into office buildings and factories (see Figure 6.9). Although LANs often use unshielded twisted-pair wire, other media—including fiber-optic cable—is also popular. Increasingly, LANs are using some form of wireless communications. You can build LANs to connect personal computers, laptop computers, or powerful mainframe computers. A computer must be equipped with the correct interface for any LAN it is to use. If the necessary interface is not included when the computer is first built, it can always be added later. For example, a salesperson whose notebook computer has a wireless interface card can establish a link to the corporate LAN, and then access the network and download data needed for the next sales call.

local area network (LAN)
A network that connects computer systems and devices within a small area, such as an office, home, or several floors in a building.

An example of a sophisticated LAN is the one that DigitalGlobe uses. DigitalGlobe is the company responsible for the detailed satellite images accessed by millions of Google Earth users. The firm uses a high-speed LAN (10 GB/sec) to connect workers to its huge file storage system (200 TB) so that new images can be quickly captured and added to its rapidly growing repository of Earth photos.[13]

A basic type of LAN is a simple peer-to-peer network that a small business might use to share files and hardware devices such as printers. In a peer-to-peer network, you set up each computer as an independent computer, but let other computers access specific files on its hard drive or share its printer. These types of networks have no server. Instead, each computer is connected to the next machine. Examples of peer-to-peer networks include Windows for Workgroups, Windows NT, Windows 2000, and AppleShare. Performance of the computers on a peer-to-peer network is usually slower because one computer is actually sharing the resources of another computer. However, these networks provide a good foundation from which small businesses can grow. The software cost is minimal, and businesses can use the

Figure 6.9

A Typical LAN

All network users within an office building can connect to each other's devices for rapid communication. For instance, a user in research and development could send a document from her computer to be printed at a printer located in the desktop publishing center.

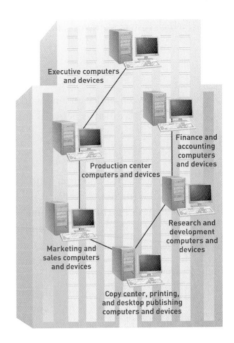

network cards if they decide to enlarge the system. In addition, peer-to-peer networks are becoming cheaper, faster, and easier to use for home-based businesses.

With more people working at home, connecting home computing devices and equipment into a unified network is on the rise. Small businesses are also connecting their systems and equipment. A home or small business can connect network, computers, printers, scanners, and other devices. A person working on one computer, for example, can use data and programs stored on another computer's hard disk. In addition, several computers on the network can share a single printer. To make home and small business networking a reality, many companies are offering standards, devices, and procedures.

Metropolitan Area Networks

metropolitan area network (MAN)

A telecommunications network that connects users and their devices in a geographical area that spans a campus or city.

A **metropolitan area network** (MAN) is a telecommunications network that connects users and their computers in a geographical area that spans a campus or city. Most MANs have a range of roughly 30 to 90 miles. For example, a MAN might redefine the many networks within a city into a single larger network or connect several LANs into a single campus LAN. EasyStreet (an Internet service provider) and OnFiber (a metro network solutions provider) designed a MAN for the city of Portland, Oregon, to provide local businesses fast (more than 1 Gps), low-cost Internet connections.[14]

Sharing Workstations at Sisters of Mercy Health System

Thanks to telecommunications and networking technologies, users no longer need to gather at one location to access their data and information. Using a typical local area network (LAN), an employee can access business records from any workstation on the premises and often from any Internet-connected computer in the world.

Doctors, nurses, and others in the medical profession find this particularly useful as they access patient records from a patient's bedside, nursing station, radiology department, or business office. In a hospital, dozens of clinicians might share a workstation every day. In this setting, users are mostly on their feet and on the move and cannot sit at a computer for any length of time. As doctors and nurses move to a new patient, they stop at the nearest workstation to view the related patient information.

In addition, doctors and staff rarely have time to log on and log off, especially in the emergency room. "Authentication has become a major pain-point for us," says Mick Murphy, chief technology officer at Sisters of Mercy Health System, a five-state, 18-hospital Catholic healthcare system. "Logging onto each of these systems and searching for the records of a specific individual every time a clinician moves to a new patient is completely untenable," says Murphy. "Our users would revolt."

Leaving the workstations logged on using a shared account is not an option. Privacy portions of the Health Insurance Portability and Accountability Act (HIPAA) require strict accountability for every access of an identifiable medical record. Systems use logs and username and password-based security to restrict access to medical records. Even in a secure environment, where only accredited clinicians can enter, everyone accessing the information must log on each time they need to see a patient record and log out when they are finished.

Murphy set out to improve the network in several ways. First, he wanted to automate the process of logging on and off. To avoid leaving sensitive information viewable to unauthorized people, he wanted the screen to black out when the clinician using it moved away a certain distance. Murphy also wanted to organize the account-generation system so that all users were classified as having a particular role such as surgical nurse. Each role would be associated with access restrictions to protect private information from those who did not need it. Murphy also wanted to easily dissolve accounts throughout the systems when clinicians left the organization. In other words, the employee's accounts on all interconnected systems would be deleted when the employee was removed from the payroll.

Mick Murphy found a solution to all of these challenges in Sentillion's Vergence communications software. Vergence integrates role-based provisioning, advanced biometric and proximity-based security, single sign-on, and front-end integration of diverse back-end systems based on a common identifier, such as the patient's name. In addition, the software maintains a fully auditable electronic record of each access to sensitive data showing who accessed it, when, from what terminal, and what tasks that person performed.

The new system uses biometrics—a fingerprint scan—and a password for authentication. Clinicians also wear proximity badges that communicate with the workstation to identify who is approaching and when the clinician leaves. The first time they use a workstation, clinicians log on using a password and fingerprint scan. When a clinician walks away from the terminal, the screen goes blank to preserve privacy, but the background applications continue running. When the clinician returns, if no one else has used that terminal, the system recognizes the clinician from the proximity badge and reactivates the screen. If someone else has used the terminal in the interim, the clinician must use the fingerprint scanner to log on; however, because the applications continue running, access takes only a few seconds rather than several minutes.

After they log on, clinicians use the patient name to access all related records. Patient records are stored on different proprietary systems. By using middleware, clinicians can select records from these disparate systems and merge them together into a single record. As a result, clinicians are unaware of the background systems, reducing the complexity and time needed to access patient information, allowing medical personnel to concentrate on patient care.

Discussion Questions

1. How do network requirements and use differ for hospital clinicians compared to other typical businesses?
2. Why did Mike Murphy want to organize users into particular roles?

Critical Thinking Questions

1. Do you think the solution Mike Murphy chose protects patient privacy as securely as possible? Why or why not?
2. An alternative solution might supply all clinicians with portable wireless Tablet PCs. How might this alternative solution compare with the one used in this case study both negatively and positively?

SOURCES: Latamore, Bert, "When the Doctor Walks Up, the Computer All but Says Hello," *Computerworld*, March 22, 2006, *www.computerworld.com*. Sintillion Web site, accessed April 9, 2006, *www.sentillion.com*. Microsoft Windows Tablet PC Healthcare Web site, accessed April 9, 2006, *www.microsoft.com/windowsxp/tabletpc/evaluation/bymarket/healthcare*.

wide area network (WAN)
A telecommunications network that ties together large geographic regions.

Wide Area Networks

A **wide area network (WAN)** is a telecommunications network that connects large geographic regions. A WAN might be privately owned or rented and includes public (shared users) networks. When you make a long-distance phone call or access the Internet, you are using a WAN. WANs usually consist of computer equipment owned by the user, together with data communications equipment and telecommunications links provided by various carriers and service providers (see Figure 6.10).

Figure 6.10

A Wide Area Network

WANs are the basic long-distance networks used around the world. The actual connections between sites, or nodes (shown by dashed lines), might be any combination of satellites, microwave, or cabling. When you make a long-distance telephone call or access the Internet, you are using a WAN.

North America

One example of a business benefiting from a WAN is Spencer Gifts, a unique retail brand that offers innovative, humorous, and fun merchandise to its primarily 18–25 year old customers. Although it has over 600 mall locations throughout the United States and Canada, the firm had limited communications among their stores. Each store was disconnected from the headquarters except for a once-per-day polling operation to upload store data to headquarters. In the summer of 2005, the firm implemented a WAN by installing a firewall and Internet connection for each store and two network security computers to enable network monitoring from headquarters. Now data flows continuously throughout the day from the stores to headquarters so that managers have up-to-the-minute data for decision making. In addition, credit card transactions can be shortened from 16 seconds to 3 seconds because point-of-sale terminals are always online.[15]

International Networks

international network
A network that links users and systems in more than one country.

Networks that link systems among countries are called **international networks**. However, international telecommunications involves special problems. In addition to requiring sophisticated equipment and software, international networks must meet specific national and international laws regulating the electronic flow of data across international boundaries, often called *transborder data flow*. Some countries have strict laws limiting the use of telecommunications and databases, making normal business transactions such as payroll costly, slow, or even impossible. Other countries have few laws concerning telecommunications and database use. These countries, sometimes called *data havens*, allow other governments and companies to process data within their boundaries. International networks in developing countries can have inadequate equipment and infrastructure that can cause problems and limit the usefulness of the network.

Marks & Spencer PLC is a leading UK retailer of clothing, food, footwear, and home goods. The firm has a substantial international presence with 550 stores worldwide operating in 30 countries. It depends on an international network to link all its stores to capture daily sales and track daily results. It also deals with roughly 2,000 direct suppliers of finished products located in places such as Bangalore, Bangladesh, Delhi, Hong Kong, Istanbul, and Sri Lanka. Reliable international communications are also required for it to work quickly and effectively with its supply base so that it can acquire goods more efficiently while still meeting the firm's strict trading standards. The firm is working with Cable & Wireless PLC to implement an Internet protocol virtual private network to provide VoIP system at its 400 UK stores and a converged voice/data IP international network for all its operations.[16] The stores involved in the first phase of the development have experienced a great improvement in the response times of customer-facing systems, such as payment processing and customer ordering, as well as stock ticketing, e-mail, and personnel management systems.[17]

Mesh Networking

Mesh networking is a way to route communications among network nodes (computers or other device) by allowing for continuous connections and reconfiguration around blocked paths by "hopping" from node to node until a connection can be established. In the *full mesh topology*, each node (workstation or other device) is connected directly to each of the other nodes. In the *partial mesh topology*, some nodes might be connected to all the others, and other nodes are connected only to nodes with which they frequently exchange communications (see Figure 6.11). Mesh networks are very robust: If one node fails, all the other nodes can still communicate with each other, directly or through one or more intermediate nodes.

mesh networking
A way to route communications between network nodes (computers or other device) by allowing for continuous connections and reconfiguration around blocked paths by "hopping" from node to node until a connection can be established.

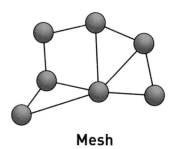

Mesh

Figure 6.11

Partial Mesh Network

Mesh networks are being set up to blanket large areas to provide Internet access, secure connections to corporate networks, and VoIP calls. The city of Philadelphia will provide affordable, universal, wireless, Internet access to its residents through a gigantic wireless mesh network covering some 135 square miles. The mesh will consist of over 3,000 interconnected antennas placed on public buildings, streetlights, and traffic lights. Each node broadcasts a broadband signal, which connects up with other nodes to create a cloud of Internet access for personal computers, laptops, and other wireless devices. EarthLink will lead the construction and operation of the network with several Internet service providers competing to market the service to local residents. To enable as many residents as possible to use the network, the monthly cost will be set at $20/month or less.[18] The Wireless Access Zone Tempe is a mesh network of 550 antennas that serves a 40-square mile area of Tempe, Arizona.[19] Read the Information Systems @ Work to see how GM is using wireless mesh networks to improve its operations.

Wireless Networks at General Motors Connect People, Machines, and Parts

Besides streamlining communications between people, networks can also enable communications among inanimate objects, machines, and people. Using wireless network technologies, sensors, and Radio Frequency ID (RFID) technology, objects on assembly lines and the machines that manipulate them can communicate with the automated systems that control the manufacturing process and the human managers who control the machines. At GM, nearly everything and everybody involved in manufacturing has a voice and is in constant communication over wireless networks.

Through a combination of sensor and wireless networking technologies, GM tracks the health of factory equipment, the movement of products along an assembly line, and the delivery of inventory in real time. At any time, from locations around the world, managers can view the state and location of parts, vehicles, and machinery in various stages of assembly and delivery throughout the value chain.

Today, many companies are using a more distributed manufacturing process and assembling products at dispersed locations. Manufacturing plants are less specialized than they used to be. "Plants are moving away from being a single-purpose facility that produces the same thing all day long for distribution all over the world," says Robert Parker, an analyst with IDC.

This distributed manufacturing environment benefits from managers and specialists monitoring manufacturing operations remotely. In this way, specialists don't need to travel to a variety of plants to offer their expertise. Instead, the necessary data is relayed from the plants to their location. Using sensors, RFID, and wireless network technologies such as the 802.11 family of protocols, a manager can supervise manufacturing in many plants from a central location. Controlling manufacturing remotely through networked technologies has become known in some circles as e-manufacturing.

"The issues that define e-manufacturing are: how real-time data is obtained and what you as a company do with this real-time data," says Pulak Bandyopadhyay, group manager for plant floor systems and control group in GM's Manufacturing Systems Research.

GM has deployed sensor technology to monitor the health of manufacturing equipment such as stamp presses, conveyor belts, and other types of machines. The technology allows supervisors to measure vibration, heat, and other factors that can be used to detect and predict machinery failure. Detecting when a machine in a remote plant might fail from GM's network operations center in Detroit helps the company save maintenance costs.

GM is also adopting mesh technologies to improve operations. In a mesh network, each node (a GM manufacturing machine) acts as a network router. Data passes from node to node rather than from each device to a wireless access point connected to a wired network. The result is a plant where workers can move and position equipment anywhere on the premises without concern for wired connections.

Wireless technologies also assist GM in monitoring inventory through the supply chain. RFID allows managers to track the delivery of components of a car or truck to the factory, and then to track the assembled cars from the plants all the way to the dealership. This provides GM with real-time visibility into its entire supply chain rather than having to wait for monthly reports.

Wireless technologies improve the efficiency and productivity of workers in GM plants who are equipped with Wi-Fi PDAs to monitor equipment from any location in the plant. Systems connected to the sensing devices analyze the collected data and information and deliver useful business information to the plant workers. Pulak Bandyopadhyay estimates that wireless networking technologies save GM from 10 to 20 percent in production costs.

Discussion Questions

1. How do wireless network technologies assist in the remote management of manufacturing plants?
2. How do you think wireless network technologies have changed the role of GM plant workers?

Critical Thinking Questions

1. How does mesh technology extend the usefulness and capabilities of wireless networks? How can mesh capabilities provide solutions for large network environments such as metropolitan area networks?
2. How does real-time monitoring of inventory provide a benefit to businesses over traditional monthly reports?

SOURCES: Hochmuth, Phil, "GM Cuts the Cords to Cut Costs," *Network World*, June 21, 2005, *www.techworld.com*. Thibodeau, Patrick, "GM Drives Dealers Toward Integrated Business Systems," *Computerworld*, January 9, 2006, *www.computerworld.com*, Webster, John, S., "Forecast 2006: RFID," *Computerworld*, January 2, 2006, *www.computerworld.com*.

Distributed Processing

When an organization needs to use two or more computer systems, it can use one of three basic processing alternatives: centralized, decentralized, or distributed. With **centralized processing**, all processing occurs in a single location or facility. This approach offers the highest degree of control because a single centrally managed computer performs all data processing. For example, the LandAmerica Financial Group provides title insurance and real estate transaction services using a global network of more than 12,000 employees and 10,000 active agents who regularly need to access documents and data such as customer title applications and commercial property assessments. LandAmerica has centralized applications and data that are considered business-critical to reduce costs, improve security, and simplify backup and recovery of the more than 140 TB of data.[20]

With **decentralized processing**, processing devices are placed at various remote locations. Each computer system is isolated and does not communicate with another system. Decentralized systems are suitable for companies that have independent operating divisions, such as Valero Energy Corp, which is the biggest oil refiner in the United States with 16 domestic and two international facilities. Maintenance workers at each refinery use handheld devices to test the temperature and vibration levels of equipment. The data is loaded into a refinery-specific database where it is analyzed to determine which pieces of equipment need preventative maintenance to avoid an unexpected breakdown. Using the data and distributed systems means Valero's refineries can run at 92 percent capacity around the clock.[21]

With **distributed processing**, computers are placed at remote locations but connected to each other via telecommunications devices. One benefit of distributed processing is that managers can allocate data to the locations that can process it most efficiently. For example, SimSci-Esscor, the industrial process simulation and control unit of Invensys PLC, uses distributed processing while developing its software products. Project members are chosen from 135 developers in five locations and three time zones so that the best resources can be assigned to the project and the work can be completed efficiently. Each developer has the computing resources required to be productive and can also share code under development with other members of the widespread team. Developers at one site often test code developed at another site so that progress continues without interruption.[22]

The September 11, 2001, terrorist attacks and the current relatively high level of natural disasters such as hurricane Katrina sparked many companies to distribute their workers, operations, and systems much more widely, a reversal of the recent trend toward centralization. The goal is to minimize the consequences of a catastrophic event at one location while ensuring uninterrupted systems availability.

centralized processing
Processing alternative in which all processing occurs at a single location or facility.

decentralized processing
Processing alternative in which processing devices are placed at various remote locations.

distributed processing
Processing alternative in which computers are placed at remote locations but are connected to each other via a network.

Client/Server Systems

Users can share data through file server computing, which allows authorized users to download entire files from certain computers designated as file servers. After downloading data to a local computer, a user can analyze, manipulate, format, and display data from the file (see Figure 6.12).

File downloaded to user

File server computer

Figure 6.12

File Server Connection

The file server sends the user the entire file that contains the data requested. The user can then analyze, manipulate, format, and display the downloaded data with a program that runs on the user's personal computer.

client/server
An architecture in which multiple computer platforms are dedicated to special functions such as database management, printing, communications, and program execution.

In **client/server** architecture, multiple computer platforms are dedicated to special functions such as database management, printing, communications, and program execution. These platforms are called *servers*. Each server is accessible by all computers on the network. Servers can be computers of all sizes; they store both application programs and data files and are equipped with operating system software to manage the activities of the network. The server distributes programs and data to the other computers (clients) on the network as they request them. An application server holds the programs and data files for a particular application, such as an inventory database. The client or the server can do the processing.

A client is any computer (often a user's personal computer) that sends messages requesting services from the servers on the network. A client can converse with many servers concurrently. A user at a personal computer initiates a request to extract data that resides in a database somewhere on the network. A data request server intercepts the request and determines on which database server the data resides. The server then formats the user's request into a message that the database server will understand. When it receives the message, the database server extracts and formats the requested data and sends the results to the client. The database server sends only the data that satisfies a specific query—not the entire file (see Figure 6.13). As with the file server approach, when the downloaded data is on the user's machine, it can then be analyzed, manipulated, formatted, and displayed by a program that runs on the user's personal computer.

Figure 6.13

Client/Server Connection

Multiple computer platforms, called *servers*, are dedicated to special functions. Each server is accessible by all computers on the network. The client requests services from the servers, provides a user interface, and presents results to the user.

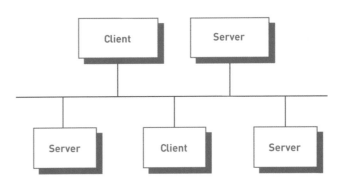

Table 6.3 lists the advantages and disadvantages of client/server architecture.

Table 6.3

Advantages and Disadvantages of Client/ Server Architecture

Advantages	Disadvantages
Moving applications from mainframe computers and terminal-to-host architecture to client/server architecture can yield significant savings in hardware and software support costs.	Moving to client/server architecture is a major two- to five-year conversion process.
Minimizes traffic on the network because only the data needed to satisfy a user query is moved from the database to the client device.	Controlling the client/server environment to prevent unauthorized use, invasion of privacy, and viruses is difficult.
Security mechanisms can be implemented directly on the database server through the use of stored procedures.	Using client/server architecture leads to a multivendor environment with problems that are difficult to identify and isolate to the appropriate vendor.

Communications Software and Protocols

communications protocol
A set of rules that governs the exchange of information over a communications channel.

A **communications protocol** is a set of rules that governs the exchange of information over a communications channel. The goal is to ensure fast, efficient, error-free communications over an imperfect communication channel. Protocols govern several levels of a telecommunications network. For example, some protocols determine data interchange at the hardware device level, and other protocols determine data interchange at the application program level. Many communications protocols govern the international, national, or industry level. Table 6.4 compares common communications protocols.

Protocol	Description
ATM	Asynchronous Transfer Mode is a packet switching technology that multiplexes fixed length packets from each communicating device and then transmits them at different times over the physical transmission medium at rates of 10 Gbps.
Frame relay	Another packet switching protocol designed for efficient transmission of intermittent traffic. Data is placed into variable-sized units called a frame and routed over the network from sender to receiver. Frame relay operates at speeds up to 2 Mbps.
IEEE 802.3 (Ethernet)	A widely used LAN technology that employs coaxial cable or twisted-pair wires as the transmission media. Devices are connected to the medium and compete for access using Carrier Sense Multiple Access with Collision Detection to detect if there was any interference in transmission. Transmission rates of 10 Gbps can be achieved with 10 Gigabit Ethernet.
T-carrier system	The T-carrier system is a digital transmission method that provides communications channels with multiples of the basic 64 Kbps used to transmit a single voice call. T-1 operates at 1.544 Mbps (24 x 64 Kbps), T-2 operates at 6.132 Mbps (96 x), T-3 operates at 44.736 Mbps (672 x), and T-4 operates at 274.176 Mbps (4,032 x).

Table 6.4

Common Telecommunications Protocols

Wireless Communications Protocols

With the spread of wireless network technology to support devices such as PDAs, mobile computers, and cell phones, the telecommunications industry needed new protocols to connect these devices. Wireless communications protocols are still evolving as the industry matures. The Institute for Electrical and Electronics Engineers (IEEE) has been instrumental in defining numerous telecommunications standards. The IEEE 802 series of network standards are summarized in Table 6.5. Figure 6.14 shows a typical wireless network configuration.

Wi-Fi Versus WiMAX

Wi-Fi has evolved and improved over time in terms of its ability to transmit data at higher speeds and over increasing distances. Wi-Fi has proven so popular that "hot spots" are popping up in places such as airports, coffee shops, college campuses, libraries, and restaurants. Dozens of cities and many business organizations are using Wi-Fi technology to connect to people on the go.

WiMAX offers faster data speeds and broader coverage than Wi-Fi. It can support mobile users who are within a range of approximately 30 miles from the WiMAX serving antenna at a transmission rate of up to 70 Mbps. It also offers a higher quality of service that can support applications such as digital life-size videoconferencing where Wi-Fi cannot. Currently, Wi-Fi speeds are less than half those of WiMAX and users must be within a few hundred yards of a hot spot. However, the WiMAX standard for mobile users is still being defined and hardware and software to support it are not yet broadly available at competitive prices. When they are, Wi-Fi adopters will have to decide if the additional benefits of WiMAX can justify the cost of conversion to WiMAX.[24]

Clearwire is a wireless broadband Internet service provider and offers WiMAX services to rural areas and smaller cities such as Visalia, California, and Kileen, Texas. It offers various grades of service, including one with a 765 Kbps downlink for under $35 per month and another with a 1.5 Mbps downlink for under $42 per month, including the cost of the required WiMAX modem. These rates are twice the cost of an equivalent DSL service for stationary Internet users but the Clearwire service supports mobile users.[25]

Table 6.5

Wireless Networks Based on
IEEE 802.*xx* Standards

Protocol	Description
IEEE 802.11a	A standard that provides specifications for wireless ATM systems and supports data transmission at a rate of up to 54 Mbps. It operates in the 5-GHz frequency band, which does not allow the signal to pass through obstacles like walls very easily. As a result, few networks were built using this standard.
IEEE 802.11b	The original standard for wireless networks known as Wi-Fi. Public Wi-Fi "hot spots" exceed 100,000 worldwide[23]—places to access a wireless network with an Internet connection. Current Wi-Fi access points have a maximum range of about 300 feet. 802.11b operates in the 2.4-MHz frequency band and does not have serious signal absorption problems. With a maximum speed of 11 Mbps, it is too slow for many applications.
IEEE 802.11g	A standard that has effectively replaced 802.11b for wireless LANs that supports data transmission at a maximum speed of up to 54 Mbps. It too operates within a frequency band that avoids signal absorption problems. Its much higher speed and greater signal reach have resulted in replacement of many of the older 802.11b networks and equipment.
IEEE 802.11n	A proposed standard for wireless networks that supports data transmission speeds exceeding 100 Mbps over longer distances than even 802.11g.
IEEE 802.15	A standard for personal area networks. The initial version 802.15.1 is called the Bluetooth protocol and describes how cellular phones, computers, faxes, personal digital assistants, printers, and other electronic devices can be interconnected using a short-range (10 to 30 feet) wireless connection.
IEEE 802.16	A group of broadband wireless communications standards for metropolitan area networks often called WiMAX that can transmit at up to 70 Mbps over a distance of 30 miles.
IEEE 802.20	An emerging standard also known as Mobile Broadband Wireless Access. The goal is to provide a packet-based wireless interface for IP-based services.

Figure 6.14

Wi-Fi Network

Wireless network

Existing wired network

Data transmitted and received through airwaves

Wireless access point   Cable modem/router   Internet

Smart Antenna Technology

Conventional wireless communications employs a single antenna at the source and another single antenna at the destination. However, a signal (an electromagnetic field) can suffer from multipath effects when it encounters obstructions (e.g., buildings, hills), which cause the wave fronts to scatter so that they take different paths to reach the destination. The result can be fading, cut-out, and intermittent reception leading to a reduction in the effective data speed and an increase in the number of errors. Using two or more antennas or MIMO (multiple input, multiple output), along with the transmission of multiple signals (one for each antenna) at the source and the destination, eliminates these multipath problems. MIMO technology is being used with digital television (DTV), wireless local area networks, MANs, and mobile communications. MISO (multiple input, single output) and SIMO (single input, multiple output) are other approaches to solving these problems. MIMO enables Wi-Fi devices to use multiple channels, enabling faster speeds much like dual chipsets in a computer.[26]

Ultra Wideband (UWB)

Ultra Wideband (UWB) transmissions consist of a stream of pulses only picoseconds (one trillionth or 10^{-12} of a second) wide. These extremely short impulses result in high frequencies spread over a wide band, which is why this technology is called ultra wideband. UWB could potentially dominate the short range wireless market as it offers the desirable characteristics of very high bandwidth and low power consumption.

3G Wireless Communication

The International Telecommunications Union (ITU) established a single standard for cellular networks in 1999. The goal was to standardize future digital wireless communications and allow global roaming with a single handset. Called IMT-2000, now referred to as 3G, this standard provides for faster transmission speeds in the range of 2–4 Mbps that will enable applications such as VoIP, video telephony, mobile multimedia, and interactive gaming. Originally, 3G was supposed to be a single, unified, worldwide standard, but the 3G standards effort split into several different standards. One standard is the Universal Mobile Telephone System (UMTS) that is the preferred solution for European countries that use Global System for Mobile communications (GSM). GSM is the de facto wireless telephone standard in Europe with more than 120 million users worldwide in 120 countries. Another 3G-based standard is Code-Division Multiple Access (CDMA) that is used in Australia, Canada, China, India, Israel, Mexico, South Korea, the United States, and Venezuela. The wide variety of 3G cellular communications protocols can support many business applications. The challenge is to enable these protocols to intercommunicate and support fast, reliable, global wireless communications.

4G Wireless Communications

4G stands for fourth-generation broadband mobile wireless. 4G is expected to deliver more advanced versions of enhanced multimedia, smooth streaming video, universal access, portability across all types of devices, and hopefully, worldwide roaming capability. 4G will also provide increased data transmission rates in the 20–40 Mbps range. (This is roughly 10 to 20 times faster than the current rates of the ADSL service used in many homes.) Sprint Nextel has already announced plans to start deploying a 4G network beginning in 2008.[27]

Switches, Bridges, Routers, and Gateways

In addition to communications protocols, certain hardware devices switch messages from one network to another at high speeds. A **switch** uses the physical device address in each incoming message on the network to determine to which output port it should forward the message to reach another device on the same network. A **bridge** connects one LAN to another LAN that uses the same telecommunications protocol. A **router** forwards data packets across two or more distinct networks toward their destinations through a process known as routing. Often, an Internet service provider (ISP) installs a router in a subscriber's home that connects the ISP's network to the network within the home. A switch connects various devices in the

switch
A telecommunications device that uses the physical device address in each incoming message on the network to determine to which output port it should forward the message to reach another device on the same network.

bridge
A telecommunications device that connects one LAN to another LAN that uses the same telecommunications protocol.

router
A telecommunications device that forwards data packets across two or more distinct networks toward their destinations, through a process known as routing.

gateway

A telecommunications device that serves as an entrance to another network.

network operating system (NOS)

Systems software that controls the computer systems and devices on a network and allows them to communicate with each other.

network-management software

Software that enables a manager on a networked desktop to monitor the use of individual computers and shared hardware (such as printers), scan for viruses, and ensure compliance with software licenses.

home together to form the home network. Sometimes the switch and the router are combined together in one single package sold as a multiple port router. A **gateway** is a network device that serves as an entrance to another network.

Communications Software

In Chapter 4, you learned that all computers have operating systems that control many functions. When an application program requires data from a disk drive, it goes through the operating system. Now consider a computer attached to a network that connects large disk drives, printers, and other equipment and devices. How does an application program request data from a disk drive on the network? The answer is through the network operating system.

A **network operating system (NOS)** is systems software that controls the computer systems and devices on a network and allows them to communicate with each other. The NOS performs the same types of functions for the network as operating system software does for a computer, such as memory and task management and coordination of hardware. When network equipment (such as printers, plotters, and disk drives) is required, the NOS makes sure that these resources are used correctly. In most cases, companies that produce and sell networks provide the NOS. For example, NetWare is the NOS from Novell, a popular network environment for personal computer systems and equipment. Windows 2000 and Windows 2003 are other common network operating systems.

Software tools and utilities are available for managing networks. With **network-management software**, a manager on a networked personal computer can monitor the use of individual computers and shared hardware (such as printers), scan for viruses, and ensure compliance with software licenses. Network-management software also simplifies the process of updating files and programs on computers on the network—a manager can make changes through a communications server instead of on individual computers. In addition, network-management software protects software from being copied, modified, or downloaded illegally and performs error control to locate telecommunications errors and potential network problems. Some of the many benefits of network-management software include fewer hours spent on routine tasks (such as installing new software), faster response to problems, and greater overall network control.

Network management is one of the most important tasks of IS managers. In fact, poor management of the network can cause a whole company to suffer. Because companies use networks to communicate with customers and business partners, network outages or slow performance can even mean a loss of business. Network management includes a wide range of technologies and processes that monitor the infrastructure and help IS staffs identify and address problems before they affect customers, business partners, or employees.

Fault detection and *performance management* are the two types of network-management products. Both employ the *Simple Network Management Protocol* (SNMP) to obtain key information from individual network components. SNMP allows anything on the network, including switches, routers, firewalls, and even operating systems and server products and utilities, to communicate with management software about its current operations and state of health. SNMP can also control these devices and products, telling them to redirect traffic, change traffic priorities, or even to shut down.

Fault management software alerts IS staff in real time when a device is failing. Equipment vendors place traps (code in a software program for handling unexpected or unallowable conditions) on their hardware to identify problems. In addition, the IS staff can place agents—automated pieces of software—on networks to monitor functions. When a device exceeds a given performance threshold, the agent sends an alarm to the company's IS fault management program. For example, if a CPU registers that it is more than 80 percent busy, the agent might trigger an alarm.

Performance management software sends messages to the various devices (i.e., polls them) to sample their performance and to determine whether they are operating within acceptable levels. The devices reply to the management system with performance data that the system stores in a database. This real-time data is correlated to historical trends and displayed graphically so that the IS staff can identify any unusual variations.

Today, most IS organizations use a combination of fault management and performance management to ensure that their network remains up and running and that every network component and application is performing acceptably. With the two technologies, the IS staff can identify and resolve fault and performance issues before they affect customers and service. The latest network-management technology even incorporates automatic fixes—the network-management system identifies a problem, notifies the IS manager, and automatically corrects the problem before anyone outside the IS department notices it.

Sierra Pacific is a wood products provider that, prior to installing network-management software, learned about network problems in the worst way—from users calling the network operations center to complain. The company has operations in about 50 distributed server locations, including deep in the woods where users are connected through routers to a high-speed network. Sierra Pacific installed Systems Intrusion Analysis and Reporting Environment open-source software on all servers to collect network and performance data around the clock and forward it to a central network server. Now, Sierra Pacific has the data it needs to identify bottlenecks and failed components before users are affected.[28]

TELECOMMUNICATIONS AND NETWORK APPLICATIONS

Telecommunications and networks are a vital part of today's information systems. In fact, it is hard to imagine how organizations could function without them. For example, when a business needs to develop an accurate monthly production forecast, a manager simply downloads sales forecast data gathered directly from customer databases. Telecommunications provides the network link allowing the manager to access the data quickly and generate the production report, which supports the company's objective of better financial planning.

The consumer goods giant Procter & Gamble uses LANs in all its plants to link office and plant workers to common software and shared databases and to provide e-mail services. The result is faster, more cost-effective, higher-quality product manufacturing. Other organizations transfer millions of important and strategic messages from one location to another every day. Telecommunications is a critical component of information systems. In some industries, it is almost a requirement for doing business; most companies could not survive without it. This section looks at some significant business applications of networks.

Linking Personal Computers to Mainframes and Networks

One of the most basic ways that telecommunications connect users to information systems is by connecting personal computers to mainframe computers so that data can be downloaded or uploaded. For example, a user can download a data file or document file from a database to a personal computer. Some communications software programs instruct the computer to connect to another computer on the network, download or send information, and then disconnect from the telecommunications line. These programs are called *unattended systems* because they perform the functions automatically, without user intervention.

Voice Mail

With **voice mail**, users can send, receive, and store verbal messages for and from other people around the world. Some voice mail systems assign a code to a group of people. Suppose the code 100 stands for all 238 sales representatives in a company. If anyone calls the voice mail system, enters the number 100, and leaves a message, all 238 sales representatives receive the same message. Call management systems can be linked to corporate e-mail and instant messaging systems. Calls to employees can be generated from instant messages or converted into e-mail messages to ensure quicker access and response.

voice mail
Technology that enables users to send, receive, and store verbal messages for and from other people around the world.

Home and Small Business Networks

Small businesses and many families own more than one computer and would like to set up a simple network to share printers or an Internet connection; access shared files such as photos, MP3 audio files, spreadsheets, and documents on different machines; play games that support multiple concurrent players; and send the output of network-connected devices such as a security camera or DVD player to a computer.

One simple way is to establish a wireless network that covers your home or small business. To do so, you can buy an 802.11n access point, connect it to your cable modem or DSL modem, and then use it to communicate with all your devices. For less than $100, you can purchase a combined router, firewall, Ethernet hub, and wireless hub in one small device (see Figure 6.15). Computers in your network connect to this box with a wireless card, which is connected by cable or DSL modem to the Internet. This enables each computer in the network to access the Internet. The firewall filters the information coming from the Internet into your network. You can configure it to reject information from offensive Web sites or potential hackers. The router can also encrypt all wireless communications to keep your network secure.

In addition, you can configure your computers to share printers and files. Windows XP includes a Network Setup Wizard that helps with network configuration. Some of the basic configuration steps include assigning each computer to a workgroup and giving it a name, identifying the files you want to share (placing an optional password on some files), and identifying the printers you want to share.

Figure 6.15

A Linksys Wireless-B Broadband Router

(Source: Courtesy of Linksys.)

Electronic Document Distribution

electronic document distribution
A process that enables the sending and receiving of documents in a digital form, without being printed (although printing is possible).

Electronic document distribution lets you send and receive documents in a digital form, without printing them (although printing is possible). It is much faster to distribute electronic documents via networks than to mail printed forms. Viewing documents on screen instead of printing them also saves paper and the document storage space. Accessing and retrieving electronic documents is also much faster. GT Nexus Inc. has launched a new online document distribution system that allows importers and their supply chain partners to create and manage the digital documents required for global trade, including purchase orders, booking requests, commercial invoices, packing lists, cargo manifests, bills of lading, and U.S. Customs entry forms. To access the documents, supply chain partners log on to the system and provide the required information at each step in the supply chain. With this technology, documents are presented, modified, and managed electronically in a timely fashion without the waste of paper and storage space.[29]

Call Centers

A call center is a physical location where an organization handles customer and other telephone calls, usually with some amount of computer automation. Call centers are used by customer service organizations, telemarketing companies, computer product help desks, charitable and political campaign organizations, and any organization that uses the telephone to sell or support products and services. An automatic call distributor (ACD) is a telephone facility that manages incoming calls, handling them based on the number called and an associated database of instructions. Call centers frequently employ an ACD to validate callers, place outgoing calls, forward calls to the right party, allow callers to record messages, gather usage statistics, balance the workload of support personnel, and provide other services.

Complicating the role of the call center is the National Do Not Call Registry set up in 2003 by the U.S. Federal Trade Commission. Telemarketers who call numbers on the list face penalties of up to $11,000 per call, as well as possible consumer lawsuits. Some 60 million consumers have signed up for the list by logging on to the site *www.donotcall.gov*. Although the registry has greatly reduced the number of unwanted calls to consumers, it has created several compliance-related issues for direct marketing companies. For instance, apart from the national registry, most states maintain their own do-not-call lists, which must be combined with the National Registry.

Offshore call centers that provide technical support services are a fact of life for many technology vendors and their customers. However, vendors and users agree that support operations have to balance their desire to reduce labor costs with customer satisfaction considerations. For example, in late 2003, Dell said that it was returning phone-based technical support for its corporate PCs to the United States because of complaints from some users about the quality of service they received from a call center in India. However, other major vendors, including IBM, Hewlett-Packard, Oracle, and Computer Associates International, plan to continue using global support operations.

Following the dramatic and long-lasting impact of Hurricane Katrina and the other hurricanes of 2005, many companies reexamined how they service customers and conduct their business. Property and casualty insurance firms in particular realized they must become better prepared to deal with the surge in customer calls following widespread disasters that affect not only their customers but also their own offices and service centers. Citizens Property Insurance increased its call center capacity by adding geographically dispersed centers in Jacksonville and Tallahassee. It also entered into a partnership to handle fluctuations in call volumes with Lynx Services.[30]

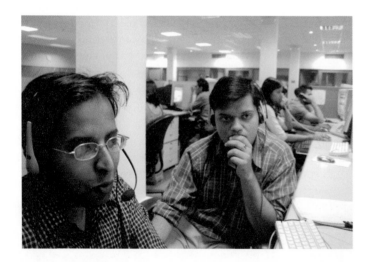

Offshore call centers provide technical support services for many technology vendors and their customers.

(Source: © STR/AFP/Getty Images.)

Telecommuting and Virtual Workers and Workgroups

Employees are increasingly performing work away from the traditional office setting. Many enterprises have adopted policies for **telecommuting** so that employees can work away from the office using computing devices and networks. This means workers can be more effective and companies can save money on office and parking space and office equipment.

Telecommuting is popular among workers for several reasons. Parents find that eliminating the daily commute helps balance family and work responsibilities. Qualified workers who otherwise might be unable to participate in the normal workforce (e.g., those who are physically challenged or who live in rural areas too far from the city office to commute regularly) can use telecommuting to become productive workers. When gas prices soar, telecommuting can help workers reduce significant expenses. Extensive use of telecommuting

telecommuting
A work arrangement whereby employees work away from the office using personal computers and networks to communicate via e-mail with other workers and to pick up and deliver results.

can lead to decreased need for office space, potentially saving a large company millions of dollars. Corporations are also being encouraged by public policy to try telecommuting as a means of reducing traffic congestion and air pollution. Large companies view telecommuting as a means to distribute their workforce and reduce the impact of a disaster at a central facility.

Some types of jobs are better suited for telecommuting than others, including jobs held by salespeople, secretaries, real estate agents, computer programmers, and legal assistants, to name a few. Telecommuting also requires a special personality type to be effective. Telecommuters need to be strongly self-motivated, organized, focused on their tasks with minimal supervision, and have a low need for social interaction. Jobs unsuitable for telecommuting include those that require frequent face-to-face interaction, need much supervision, and have lots of short-term deadlines. Employees who choose to work at home must be able to work independently, manage their time well, and balance work and home life.

One company that uses telecommuting to its advantage is General Electric. Its Energy division in Atlanta has installed over 10,000 steam and heavy-duty gas turbines in more than 120 countries, and is a leader in providing wind and hydro turbines. The Energy division has implemented telecommuting so that workers can access internal corporate systems and data no matter where they are—at home, in an airport, at a hotel, or on a customer's premises. Furthermore, by supplying home computers and a broadband connection for telecommuters, the division has cut its office workstation requirements by 50 systems for an estimated savings of $750,000 per year in real estate and related costs.[31]

Videoconferencing

Videoconferencing enables people to hold a conference by combining voice, video, and audio transmission. Videoconferencing reduces travel expenses and time, and increases managerial effectiveness through faster response to problems, access to more people, and less duplication of effort by geographically dispersed sites. Almost all videoconferencing systems combine video and phone call capabilities with data or document conferencing (see Figure 6.16). You can see the other person's face, view the same documents, and swap notes and drawings. With some systems, callers can change live documents in real time. Many businesses find that the document- and application-sharing feature of the videoconference enhances group productivity and efficiency. It also fosters teamwork and can save corporate travel time and expense.

Figure 6.16

Videoconferencing

Videoconferencing allows participants to conduct long-distance meetings "face to face" while eliminating the need for costly travel.

(Source: © Comstock Images/ Alamy.)

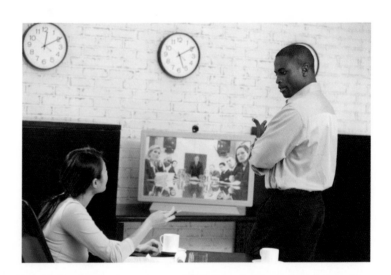

Group videoconferencing is used daily in a variety of businesses as an easy way to connect work teams. Members of a team meet in a specially prepared videoconference room equipped with sound-sensitive cameras that automatically focus on the person speaking, large TV-like monitors for viewing the participants at the remote location, and high-quality speakers and microphones. Videoconferencing costs have declined steadily, while video quality and synchronization of audio to data—once weak points for the technology—have improved.

Medco Health Solutions, Inc., is a large prescription-drug management company whose clients are employers who want to implement an employee prescription-drug plan. One service Medco provides is to ensure that drug-plan members get the right drug, in the right dosage, at the right time, while reducing harmful and wasteful use of prescription drugs. "When a patient is submitting a prescription, we often have a more complete view of [his] medical history [than the pharmacy] because they can see prescription data, medical data, and lab data [that can be used] to apply better judgment in how medication is dispensed," says Ken Klepper, vice president and chief operating officer. Controlling access to all this data is critical to Medco's operation. As a result, it implemented a program to reduce computer viruses and other IT-related problems. As part of the program, IT and business executives meet three times per week from 15 videoconferencing sites nationwide to identify any IT problems and to initiate action to increase system availability.[32]

Electronic Data Interchange

Electronic data interchange (EDI) is a way to communicate data from one company to another and from one application to another in a standard format, permitting the recipient to perform a standard business transaction, such as processing purchase orders. Connecting corporate computers among organizations is the idea behind EDI, which uses network systems and follows standards and procedures that can process output from one system directly as input to other systems, without human intervention. EDI can link the computers of customers, manufacturers, and suppliers (see Figure 6.17). This technology eliminates the need for paper documents and substantially cuts down on costly errors. Customer orders and inquiries are transmitted from the customer's computer to the manufacturer's computer. The manufacturer's computer determines when new supplies are needed and can place orders by connecting with the supplier's computer.

electronic data interchange (EDI)

An intercompany, application-to-application communication of data in a standard format, permitting the recipient to perform a standard business transaction, such as processing purchase orders.

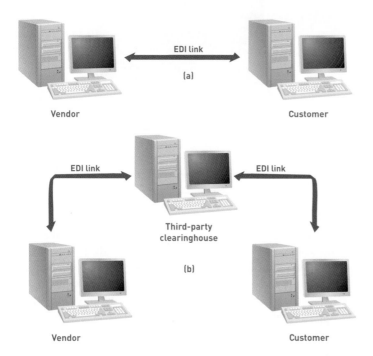

Figure 6.17

Two Approaches to Electronic Data Interchange

Many organizations now insist that their suppliers use EDI systems. Often, the vendor and customer (a) have a direct EDI connection; or (b) the link is provided by a third-party clearinghouse that converts data and performs other services for the participants.

United Parcel Service, Inc. (UPS), worked with the U.S. Customs and Border Protection (CBP) to build an EDI-based system called Automated Commercial Environment (ACE) that automates the flow of information about packages approaching U.S. borders. ACE generates EDI records that not only provide CBP with package information but also data about the vehicle, the driver, who the package is going to, what goods the package contains, who purchased the goods, where the package is being delivered, and who is exporting it out of the country. The EDI records pass through another system that serves as a sort of electronic filter to identify which packages to inspect.[33]

Public Network Services

public network services
Systems that give personal computer users access to vast databases and other services, usually for an initial fee plus usage fees.

Public network services give personal computer users access to vast databases, the Internet, and other services, usually for an initial fee plus usage fees. Public network services allow customers to book airline reservations, check weather forecasts, find information on TV programs, analyze stock prices and investment information, communicate with others on the network, play games, and receive articles and government publications. Fees are based on the services used, and can range from under $15 to over $500 per month (see Figure 6.18). Providers of public network services include Microsoft, America Online, and Prodigy. These companies provide an array of services, including news, electronic mail, and investment information. AOL is the largest provider of public network services.

Figure 6.18

Public Network Services

Public network services provide users with the latest information required to remain competitive. MSN, for example, enables registered users to personalize their home page.

Electronic Funds Transfer

electronic funds transfer (EFT)
A system of transferring money from one bank account directly to another without any paper money changing hands.

Electronic funds transfer (EFT) is a system of transferring money from one bank account directly to another without any paper money changing hands. It is used for both credit transfers, such as payroll payments, and for debit transfers, such as mortgage payments. The benefits of EFT include reduced administrative costs, increased efficiency, simplified book-keeping, and greater security. One of the most widely used EFT programs is Direct Deposit, which deposits employee payroll checks directly into the designated bank accounts. The two primary components of EFT are wire transfer and automated clearing house.

wire transfer
An extremely fast, reliable means to move funds from one account to another.

Wire transfer is an extremely fast, reliable means to move funds from one account to another using one of the three major wire-transfer networks: Fedwire, provided by the U.S. Federal Reserve Bank; Clearing House Inter-Bank Payments System (CHIPS), which is a bank-owned, privately operated system for business-to-business transactions; and the Society for Worldwide Interbank Financial Telecommunication (SWIFT), which is a system that combines messaging and funds transfer.

The **automated clearing house (ACH)** network is a secure private network that connects all U.S. financial institutions to one another by way of the Federal Reserve Board or other ACH operators. It handles and processes electronic payments throughout the 50 states and U.S. territories and Canada. For payments, funds are transferred electronically from one bank account to the billing company's bank, usually less than a day after the scheduled payment date. The National Automated Clearing House Association (NACHA) in Herndon, Virginia, sets the rules and standards for ACH transactions. Many banks and organizations have established payment systems based on ACH. For example, PayPal developed a system that lets people send money to one another via e-mail. Several banks also offer consumer-oriented payment services: Bank One Corporation in Chicago (prior to its merger with JP Morgan in 2004) offered eMoneyMail, Citigroup, Inc., has c2it, and Wells Fargo Bank NA in San Francisco touts BillPoint. Table 6.6 compares the services offered by ACH and wire transfer.

automated clearing house (ACH)
A secure, private network that connects all U.S. financial institutions to one another by way of the Federal Reserve Board or other ACH operators.

Table 6.6

Comparison of ACH Payments and Wire Transfers

	ACH Payments	Wire Transfers
When does payment clear?	Overnight	Immediately
Can payment be canceled?	Yes	No
Is there a guarantee of sufficient funds?	No	Yes
What is the approximate cost per transaction?	$.25	$10–$40

Distance Learning

The 27,000 square mile Navajo Nation spans three states and is made up of more than 210,000 people. Plans are being made to implement a network that will provide them with telephone, video, and data services. One goal is to provide teleconferencing classes offered by schools and universities in Arizona, New Mexico, and Utah to encourage education and economic development.[34] Often called **distance learning** or cyberclasses, such electronic classes are likely to be the wave of the future. With distance learning software and systems, instructors can easily create course home pages on the Internet. Students can access the course syllabus and instructor notes on the Web page. Student e-mail mailing lists allow students and the instructor to e-mail one another for homework assignments, questions, or comments about material presented in the course. It is also possible to form chat groups so that students can work together as a "virtual team," which meets electronically to complete a group project.

distance learning
The use of telecommunications to extend the classroom.

Team Collaboration Systems

Microsoft Office Groove is an example of software designed to enable collaboration and communication within small teams. As a Groove user, you create a workspace and then invite others to join it. You send a copy of the contents of the workspace to those who become members, and they install the workspace on their hard drives. The Groove software keeps all members' copies of the workspace synchronized—when a member changes the workspace, that change is sent to all copies for update. If a member is offline when the change is made, the change is queued and synchronized when the member is online. You can use Groove to set up shared workspaces among employees and others outside your organization. For example, give key suppliers or customers a Groove client for sharing appropriate information and they can stay up to date on rapidly changing developments.[35]

Alliance is a business intelligence solutions consulting firm with six offices in the United States and one in India. It provides a full range of end-to-end business intelligence solutions from data acquisition and warehousing to master data management, analytics, and reporting. Alliance is also involved in applications services, including development, integration, testing, and support delivered through a global delivery model. Alliance consultants at client project sites worldwide use Groove to collaborate on creating documents.[36]

Specialized Systems and Services

With millions of personal computers in businesses across the country, interest in specialized and regional information services is increasing. Specialized services, which can be expensive, include professional legal, patent, and technical information. For example, investment companies can use systems such as Quotron and Shark to get up-to-the-minute information on stocks, bonds, and other investments.

Regional services, also called *metropolitan services*, include local electronic bulletin boards and electronic mail facilities that offer information regarding local club, school, and government activities. An electronic bulletin board is a message center that displays messages in electronic form, much as a bulletin board displays paper messages in schools and offices. Subscribers can access an electronic bulletin board with personal computers, network equipment, and software. In addition to regional bulletin boards, national and international bulletin boards are available for people and groups with special interests or needs. Many groups of people can use these types of bulletin boards, such as users of certain software packages and users with certain hobbies. Many public network services, including Prodigy, America Online, and CompuServe, provide access to hundreds of different bulletin boards on a variety of topics and interest areas.

Global positioning systems (GPSs) provide specialized telecommunications services. The military has long used GPS systems to find locations of troops, equipment, and the enemy—within yards in some cases. Today, companies are using GPSs to survey land and buildings and people use them to locate their positions while camping or exploring. Some auto companies have placed GPSs in their cars to assist travelers in need. Toyota offers its G-Book in-vehicle network information service to Mazda owners in Japan. G-Book provides vehicle owners a wide range of services, including vehicle location and navigation information, maintenance notifications, music and games, emergency roadside assistance, and e-commerce opportunities. The onboard implementation of the G-Book service features a data communications module, a touch screen navigation terminal, and a GPS antenna. Toyota and Mazda believe auto makers and the information industry should work together to promote an industry standard so they can offer customers lower-cost and higher-quality service.[37]

With all these telecommunications systems and services, it is no wonder that managers and workers can perform business tasks from remote locations. Often called *virtual workers*, these employees can conduct business at any time and place. Today, new telecommunications systems and services are being introduced every month. In the future, you can expect even more innovations to dramatically alter how businesses and people stay connected and in touch.

SUMMARY

Principle

Effective communications are essential to organizational success.

Telecommunications refers to the electronic transmission of signals for communications, including telephone, radio, and television. Telecommunications is creating profound changes in business because it removes the barriers of time and distance.

The elements of a telecommunications system include a sending unit, such as a person, a computer system, a terminal, or another device, that originates the message. The sending unit transmits a signal to a telecommunications device, which performs a number of functions such as converting the signal into a different form or from one type to another. A telecommunications device is a hardware component that facilitates electronic communication. The telecommunications device then sends the signal through a medium, which is anything that carries an electronic signal and serves as an interface between a sending device and a receiving device. The signal is received by another telecommunications device that is connected to the receiving computer. The process can then be reversed, and another message can pass from the receiving unit to the original sending unit. With synchronous communications, the receiver gets the message instantaneously, when it is sent. Voice and phone communications are examples. With asynchronous communications, the receiver gets the message hours or days after the message is sent.

A communications channel is the transmission medium that carries a message from the source to its receivers. Communication channels can be classified as simplex, half-duplex, or full-duplex.

Principle

Communications technology lets more people send and receive all forms of information over greater distances at a faster rate than ever before.

The telecommunications media that physically connect data communications devices can be divided into two broad categories: guided transmission media, in which communications signals are guided along a solid medium, and wireless media, in which the communications signal is sent over airwaves. Guided transmission media include twisted-pair wire cable, coaxial cable, fiber-optic cable, and broadband over power lines. Wireless media types include microwave, cellular, and infrared.

A modem is a telecommunications hardware device that converts (modulates and demodulates) communications signals so they can be transmitted over the communication media.

A multiplexer is a device that encodes data from two or more data sources onto a single communications channel, thus reducing the number of communications channels needed and therefore, lowering telecommunications costs.

A front-end processor is a special-purpose computer that manages communications to and from a computer system serving hundreds or even thousands of users.

Telecommunications carriers offer a wide array of phone and dialing services, including digital subscriber line (DSL) and wireless telecommunications.

The effective use of networks can turn a company into an agile, powerful, and creative organization, giving it a long-term competitive advantage. Networks let users share hardware, programs, and databases across the organization. They can transmit and receive information to improve organizational effectiveness and efficiency. They enable geographically separated workgroups to share documents and opinions, which fosters teamwork, innovative ideas, and new business strategies.

The physical distance between nodes on the network and the communications and services provided by the network determines whether it is called a personal area network (PAN), local area network (LAN), metropolitan area network (MAN), or wide area network (WAN). A PAN connects information technology devices within a range of about 33 feet. The major components in a LAN are a network interface card, a file server, and a bridge or gateway. A MAN connects users and their computers in a geographical area larger than a LAN but smaller than a WAN. WANs link large geographic regions, including communications between countries, linking systems from around the world. The electronic flow of data across international and global boundaries is often called transborder data flow.

A mesh network is a way to route communications between network nodes (computers or other device) by allowing for continuous connections and reconfiguration around blocked paths by "hopping" from node to node until a connection can be established.

A client/server system is a network that connects a user's computer (a client) to one or more host computers (servers). A client is often a PC that requests services from the server, shares processing tasks with the server, and displays the results. Many companies have reduced their use of mainframe computers in favor of client/server systems using midrange or personal computers to achieve cost savings, provide more control over the desktop, increase flexibility, and become more responsive to business changes. The start-up costs of these systems can be high, and the systems are more complex than a centralized mainframe computer.

When people on one network want to communicate with people or devices in a different organization on another network, they need a common communications protocol and various network devices to do so. A communications protocol is a set of rules that govern the exchange of information over a communications channel. There are a myriad of communications protocols, including international, national, and industry standards.

ATM (Asynchronous Transfer Mode) is a switching technology that relies on dedicated connections and organizes digital data into 53-byte cell units. These cells are then transmitted over a physical medium using digital signal technology.

Ethernet is the most widely installed LAN technology that typically uses coaxial cable or special grades of twisted-pair wires. The most common Ethernet systems are called 10Base-T, which provide transmission speeds up to 10 Mbps. Devices are connected to the cable and compete for access using a Carrier Sense Multiple Access with Collision Detection (CSMA/CD) protocol.

Frame relay is a packet switching protocol for efficient data transmission of intermittent traffic between LANs and end points in a WAN. Data is placed into variable-sized units called a frame and routed over the network from sender to receiver. Frame relay operates at speeds up to 2 Mbps.

The IEEE 802.*xx* is a family of technical specifications for wireless local area networks developed by a working group of the Institute of Electrical and Electronics Engineers (IEEE). All use the Ethernet protocol and CSMA/CA (Carrier Sense Multiple Access with Collision Avoidance) for path sharing.

Using two or more antennas or MIMO (multiple input, multiple output) and transmitting multiple signals (one for each antenna) at the source and the destination eliminates multipath problems that occur when electromagnetic radiation encounters barriers.

UWB transmissions consist of a stream of pulses only picoseconds (one trillionth or 10^{-12} of a second) wide. These extremely short impulses result in high frequencies spread over a wide band.

3G refers to the various evolutions of mobile wireless networks used to support users and their various devices and communications needs with transmission speeds of 2 to 4 Mbps.

4G stands for fourth-generation broadband mobile wireless, which will support telecommunications at 20 to 40 Mbps.

In addition to communications protocols, telecommunications uses various devices. A switch uses the physical device address in each incoming message on the network to determine which output port to forward the message to in order to reach another device on the same network. A bridge is a device that connects one LAN to another LAN that uses the same telecommunications protocol. A router forwards data packets across two or more distinct networks toward their destinations, through a process known as routing. A gateway is a network device that serves as an entrance to another network.

When an organization needs to use two or more computer systems, it can follow one of three basic data-processing strategies: centralized, decentralized, or distributed. With centralized processing, all processing occurs in a single location or facility. This approach offers the highest degree of control. With decentralized processing, processing devices are placed at various remote locations. The individual computer systems are isolated and do not communicate with each other. With distributed processing, computers are placed at remote locations but are connected to each other via telecommunications devices. This approach helps minimize the consequences of a catastrophic event at one location while ensuring uninterrupted systems availability.

Communications software performs important functions, such as error checking and message formatting. A network operating system controls the computer systems and devices on a network, allowing them to communicate with one another. Network-management software enables a manager to monitor the use of individual computers and shared hardware, scan for viruses, and ensure compliance with software licenses.

Some of the many applications of telecommunications include the following: linking personal computer to mainframes, voice mail, home and small business networks, electronic document distribution, call centers, telecommuting, videoconferencing, electronic data interchange, public network services, electronic funds transfer, distance learning, and specialized systems and services. Personal-computer-to-mainframe links enable people to upload and download data. Voice mail users can send, receive, and store messages for other people around the world. Home and small business networks enable people to share printers or an Internet connection; access shared files such as photos, MP3 audio files, spreadsheets, and documents on different machines; play games that support multiple concurrent players; and send the output of network-connected devices such as a security camera or DVD player to a computer. Electronic document distribution allows organizations to transmit documents without the use of paper, thus cutting costs and saving time. Call centers efficiently handle many inbound and outbound telephone calls. Telecommuting uses information technology to enable employees to work away from the office. Videoconferencing brings groups together in voice, video, and audio calls. Electronic data interchange (EDI), another rapidly growing area, enables customers, suppliers, and manufacturers to exchange data electronically. EDI reduces the need for manual paper systems, while speeding up the rate at which business can be transacted. Electronic funds transfer is a system of transferring money from one bank account directly to another without any paper money changing hands. Public network services give users access to vast databases and services, usually for an initial fee plus usage fees. Distance learning is a way to support education of students who are unable to meet frequently with their instructor. Specialized services, which are more expensive, include legal, patent, and technical information. Regional services include local electronic bulletin boards that offer e-mail facilities and information regarding local activities.

CHAPTER 6: SELF-ASSESSMENT TEST

Effective communications are essential to organizational success.

1. Voice and phone communications are examples of asynchronous communications. True or False?

2. Gbps stands for billions of bytes per second. True or False?

3. A full-duplex simplex channel can transmit data in both directions at the same time. True or False?

4. Two broad categories of transmission media are _____.
   a. guided and wireless
   b. shielded and unshielded
   c. twisted and untwisted
   d. infrared and microwave

Communications technology lets more people send and receive all forms of information over greater distances at a faster rate than ever before.

5. Some utilities, cities, and organizations are experimenting with the use of _____ to provide network connections over standard high-voltage power lines.

6. Walgreens is the largest private industry user of satellite communications in the world. True or False?

7. Which of the following is a telecommunications service that delivers high-speed Internet access to homes and small businesses over existing phone lines?
   a. BPL
   b. DSL
   c. Wi-Fi
   d. Ethernet

8. A device that encodes data from two or more devices onto a single communications channel is called a(n) _____.

9. A(n) _____ is a network that can connect technology devices within a range of 33 feet or so.

10. Which of the following statements about call centers is true?
    a. They are used to handle incoming calls only, not outgoing calls.
    b. They frequently employ an ACD to forward calls to the correct party and balance the workload of support personnel.
    c. It is illegal for charitable organizations to employ them.
    d. Do Not Call legislation has had a minimal effect on their operation.

11. Telecommuting enables employees to save money because less office space, parking space, and office equipment is required. True or False?

12. _____ is a system of transferring money from one bank account directly to another without any paper money changing hands.

CHAPTER 6: SELF-ASSESSMENT TEST ANSWERS

(1) False (2) False (3) True (4) a (5) broadband over power lines (6) True (7) d (8) multiplexer (9) personal area network (10) b (11) True (12) Electronic funds transfer

REVIEW QUESTIONS

1. What is the difference between synchronous and asynchronous communications?
2. Describe the elements and steps involved in the telecommunications process.
3. What characteristic of a channel determines its information-carrying capacity?
4. Identify four types of guided telecommunications media.
5. Define the term *computer network*.
6. What advantages and disadvantages are associated with the use of client/server computing?
7. What is a DSL line? What capabilities does it provide?
8. What is the difference between DSL and ADSL service?
9. What is mesh networking? What are some of its advantages?
10. What role do the bridge, router, gateway, and switch play in a network?
11. Describe a local area network and its various components.
12. What is a metropolitan area network?
13. What is EDI? Why are companies using it?
14. What is telecommuting? Why are many companies using it?

DISCUSSION QUESTIONS

1. How might you use a local area network in your home? What devices might eventually connect to such a network?

2. Why is an organization that employs centralized processing likely to have a different management decision-making philosophy than an organization that employs distributed processing?

3. What are the pros and cons of distributed processing versus centralized processing for a large retail chain?

4. Briefly discuss the pros and cons of e-mail versus voice mail. Under what circumstances would you use one and not the other?

5. If it were available and you could afford it, which would you rather have in your home—T1, ADSL, or DSL? Why?

6. You need to send cash to a friend across county—quickly! What are your options? Which one would you choose and why? Describe the fundamental network that supports this service.

7. Do you think that this course is a good candidate for a distance learning course? Why or why not?

8. Identify at least three wireless telecommunications protocols. Why do you think that there are so many protocols? Will the number of protocols increase or shrink over time?

PROBLEM-SOLVING EXERCISES

1. You have been hired as a telecommunications consultant to help an organization assess the benefits and potential cost savings associated with installing videoconferencing systems at each of the firm's six work locations. You have determined that the cost to establish each videoconference facility is $60,000 and that each facility will have an annual operating and support cost of $20,000. In the first year, you estimate that videoconferencing will save the entire firm approximately $400,000 in travel expenses. The savings are expected to increase at a rate of 5 percent per year. Develop a spreadsheet to analyze the costs and savings over a three-year period. Write a recommendation to management based on your findings and any other factors that might support the installation of a videoconferencing system.

2. As the CIO of a hospital, you are convinced that installing a wireless network and portable computers is a necessary step to reduce costs and improve patient care. Use PowerPoint or similar software to make a convincing presentation to management for adopting such a program. Your presentation must address points such as defining the benefits and potential issues that make such a program a success.

TEAM ACTIVITIES

1. With a group of your classmates, develop a proposal to install videoconferencing equipment in one of your school's classrooms so that students can view lectures at a distant videoconferencing facility. What sort of equipment is required, who provides this equipment, and what does it cost to install and operate?

2. Form a team to identify the public locations (such as an airport, public library, or café) in your area where wireless LAN connections are available. Visit at least two locations and write a brief paragraph discussing your experience at each location trying to connect to the Internet.

WEB EXERCISES

1. Do research on the Web on wireless communications standards for MANs often called WiMAX. What is its current status of the various standards and what are the current major issues? Write a short report on what you found.

2. Go online to investigate how the banking industry is currently using wire transfer and ACH services. Find two different organizations that provide these services and compare their systems.

CAREER EXERCISES

1. Consider an industry with which you are familiar through work experience, coursework, or a study of industry performance. How could this industry use high-speed, wireless network communications? What limitations would such networks have in this industry?
2. One of the many online job-search companies includes Monster.com. Investigate one or more of these companies and research the positions available in the telecommunications industry, including the Internet. You might be asked to summarize your findings for your class in a written or verbal report.

CASE STUDIES

Case One

U.S. Census-Takers Go Wireless

The next census to be taken in 2010 is expected to cost the United States between 11 and 12 billion dollars. In an effort to reduce this cost and the amount of errors in census data, census takers will be switching from pencil and paper to wireless networks.

Every ten years, the U.S. Census Bureau mails packages of forms to U.S. citizens to complete and return. Roughly 65 percent of people actually follow through and return the forms. Those that don't can expect a visit from a Census Bureau employee to collect the required information. In past censuses, census takers write down the subjects' answers on paper and then enter them into a computer at day's end. In 2010 census takers will be recording and submitting census data with specially designed wireless handheld computers. "We are revolutionizing the census," Census Bureau director Louis Kincannon said in a statement announcing the new system. "The handheld computers are an integral part of a re-engineered 2010 Census."

The pocket-size handhelds, which are being designed by Harris Corporation and manufactured by High Tech Computer Corporation in Taiwan, will run Windows Mobile 5.0 operating system. Windows Mobile 5.0 was chosen primarily because it can run customized off-the-shelf applications that are being developed by Harris. This includes software for security and wireless data synchronization. The handheld computers have a rugged design to withstand rough fieldwork, and include a 10-hour battery to support a full day on the job.

The devices have no keyboard or numeric keypad. Census workers will enter data using a stylus on a touch screen. The displays are designed to be extra bright and easy to see in sunlight. The devices use a cellular data radio to send collected data to a Census office over a Sprint Nextel private network. Although the devices use a regular cell phone service, they will not include voice communication capability. Still, census takers will be able to receive data, such as advisories that say a resident has already filed a paper census survey and can be skipped.

The data collected at each home will be encrypted and stored on the device and transmitted over the private network immediately or at set intervals. If a census taker fails to send the data, the handheld device will do so on its own. To further protect the data and the privacy of citizens, no removable storage device is provided nor is it possible to upload the data to any location other than the Census Bureau. A traditional phone line connection port will be custom-installed for backup connections to use in areas where cell service is not available.

The purchase of 500,000 mobile devices from Harris Corporation is considered by experts to be the largest deployment of wireless handheld devices in a system to date.

Discussion Questions

1. What benefits does the handheld wireless system provide the Census Bureau over its previous methods?
2. How will the new system affect the hiring of census takers and the level of satisfaction of employees?

Critical Thinking Questions

1. What types of changes will the new system require on the back end, where the data is received and recorded in the database?
2. Are you more or less confident in the reliability of census data in 2010? Why?

SOURCES: Hamblen, Matt, "Census Bureau to Deploy a Half-Million Wireless Handhelds," *Computerworld*, April 5, 2006, *www.computerworld.com*. Associated Press Staff, "Census Takers to Trade Pens for Handheld PCs," MSNBC Online, April 4, 2006, *www.msndb.msn.com*. U.S. Census Bureau Web site, accessed April 9, 2006, *www.census.gov*.

Case Two

Delivering People More Efficiently Through Telecommunications

Scott Boyes has worked in the Canadian transportation industry for years. The company he worked for previously shuttled railway crew members to train locations where they are needed. Scott found that this company was slow to adopt new telecommunications technologies. The van drivers who shuttled the railway workers could not communicate with the dispatcher while in route. At each stop, the drivers would have to call the dispatcher to find out where to go next. This problem frustrated all involved especially the railways that paid for the shuttle service. "With an eye to controlling their margins and cutting costs, railways want more accurate ETAs from their crew transportation providers and the flexibility to accept new trip orders and reroute quickly," Boyes said. He added, "This is something the current providers can't give them."

Boyes recognized that the business was in a quagmire and not about to change. He identified an opportunity for a new tech-savvy business to enter the market and provide the level of service that the railways wanted. Boyes felt that he was the person that could deliver that service. In May 2005, he founded Toronto-based RailCrewXpress and hired some of his former coworkers to help build the company. From day one, Scott Boyes and his associates saw telecommunications technology and information systems as the key to set the company apart from the competition.

"The whole premise of our pitch to our investors and the railways was the technology platform on which we intended to serve the railways," said Boyes. He worked with a mobile computing hardware company, Mississauga, and a mobile wireless software provider, MobileDataforce, to create a unique wireless telecommunications system that connects drivers with dispatchers and the company information systems. The system tracks drivers from home base and provides dispatchers with the power to manage routing dynamically while vehicles are in route.

Not only does the system allow for dynamic routing changes, but it also collects information and develops a history that includes routes and driving times. It can then analyze this data to optimize the efficiency of dispatches and routing. The system also automates the tasks of billing customers and paying drivers.

The biggest challenge in building the system was using the most economical networking technologies based on the location of the van. The system is designed to automatically switch between satellite and General Packet Radio Service (GPRS) data transmission, depending on the van's location. The less costly GPRS is used in populated areas where coverage is available, and expensive satellite service is used outside of GPRS coverage areas. The system can intelligently decide how much information to send and when so that it does not lose information when switching between technologies.

RailCrewXpress expects to gain $75 million U.S. in revenue in its first full year in business. The technology team isn't resting on its laurels, however. Their next project involves using magnetic-strip readers or RFID scanners in vans to scan passenger ID badges, and track their locations much the way UPS and other shipping companies track the movement of packages.

Discussion Questions

1. What advantages does RailCrewXpress have over its competition, Scott's former employer?
2. How does the concept of real-time monitoring relate to RailCrewXpress? In general, what is the benefit of real-time monitoring?

Critical Thinking Questions

1. What mistakes did Scott Boyes' previous employer make? How could they have prevented Scott leaving and creating formidable competition?
2. How do you think Scott Boyes' previous employer will react to the new competition? What is RailCrewXpress doing to ensure its dominance in the market?

SOURCES: Jedras, Jeff, "A Tech Team Effort Keeps the Trains Running on Time," *Computer World Canada*, February 17, 2006, *www.itworldcanada.com*. Press release: "RailCrewXpress Selects MobileDataForce as Part of Team Delivering Mobile Computing Solution for Transport Fleet," January 18, 2006, *www.prweb.com/releases/2006/1/prweb334473.htm*.

Questions for Web Case

See the Web site for this book to read about the Whitmann Price Consulting case for this chapter. Following are questions concerning this Web case.

Whitmann Price Consulting: Telecommunications and Networks

Discussion Questions

1. What role does bandwidth play in the successful delivery of the Advanced Mobile Communications and Information System?
2. When does functionality transform the standard BlackBerry device into an Advanced Mobile Communications and Information System?

Critical Thinking Questions

1. Describe three telecommunications and network technologies used to connect the BlackBerry with other devices.
2. At this stage in the process, what actions might Sandra and Josh take to reduce the overall costs of the Advanced Mobile Communications and Information System?

NOTES

Sources for the opening vignette: Staff, "Virgin Megastores Transforms the Customer Experience Through Powerful Converged Networking," 2005, Cisco Systems Customer Case Study, *www.cisco.com*. Virgin Web site, accessed April 9, 2006, *www.virgin.com*. "Virgin Megastore Times Square Features 150 New IBM Anyplace Kiosks That Help Create an 'Emotionally Exciting' Shopping Experience," IBM News Release, September 20, 2005, *www.ibm.com*.

1 Mitchell, Robert L., "Sidebar: Waiting for Wireless at Boeing," *Computerworld*, February 13, 2006.

2 Webster, John S., "Ethernet Over Copper," *Computerworld*, March 27, 2006.

3 Yao, Deborah, "Comcast Posts Huge Q1 Profit," *E-Commerce Times*, April 28, 2006.

4 Di Stefano, Theodore F., "India's Hi-Tech Dominance: How Did IT Happen?" *E-Commerce Times*, April 14, 2006.

5 Orzech, Dan, "Surfing Through the Power Grid," *Wired News*, October 20, 2005.

6 Staff, "FCC Releases New Telephone Subscribership Report," NEWS, Federal Communications Commission, November 7, 2005, *www.fcc.gov*.

7 Taylor, Trent, "Walgreens: Trent Taylor on IT Careers – Our Advanced Technology," *www.walgreens.com*.

8 Pappalardo, Denise, "Ford Not Quite in Cruise Control," *Network World*, February 27, 2006, *www.networkworld.com*.

9 Honan, Mathew, "Apple Unveils Intel-Powered Mac Minis," *Computerworld*, February 28, 2006.

10 Hamblem, Matt, "Mobile & Wireless World: Maytag, Safelite Cite Project ROI," *Computerworld*, June 15, 2005.

11 Staff, Chairman's Letter, 2005 AT&T Annual Report.

12 Blau, John, "Metro Shows Voice-Operated RFID Device at CeBIT," *Computerworld*, March 8, 2006.

13 Mearian, Lucas, "Google Earth's Photographer Builds Out Infrastructure," *Computerworld*, March 1, 2006.

14 Staff, "EasyStreet and OnFiber Bring State-of-the-Art Network to Portland," January 20, 2005, Press Releases, OnFiber Web site, *www.onfiber.com*.

15 Bosavage, Jennifer, "Case Study: Spencer's Gives Itself the Gift of Security," *InformationWeek*, January 3, 2006.

16 Betts, Mitch, "Global Dispatches: Hitachi Replacing PCs with Thin Clients to Boost Security," *Computerworld*, May 30, 2005.

17 About Us – Strategic Relationships - Marks & Spencer - Cable & Wireless Web site, *www.cw.com/US/about_us/strategic_relationships/cisco/cisco_customers_ms.html*.

18 McChensey, Robert; Podesta, John, "Let There Be Wi-Fi," *Washington Monthly*, January 2006.

19 Cox, John, "Choosing Sides: Wi-Fi or EV-DO?" *Computerworld*, March 3, 2006.

20 Horwitt, Elizabeth, "LandAmerica Moves Beyond Consolidated Storage to Centralized Backup," *Computerworld*, June 22, 2005.

21 Hoffman, Thomas, "Drilling Down: Valero Energy Strikes It Rich with BI Tools," *Computerworld*, February 27, 2006.

22 Robb, Drew, "Global Workgroups," *Computerworld*, August 15, 2005.

23 Hamblen, Matt, "Global Wi-Fi Hot Spots Top 100,000," *Computerworld*, January 23, 2006.

24 Malykhina, Elena, "Imax Versus Wi-Fi, Which One Will Be the King Kong of Wireless?" *InformationWeek*, December 18, 2005.

25 Taylor, Steve, "Imax Could Provide Broadband Competition in Rural Areas," *Network World*, January 23, 2006.

26 Britt, Phil, "10-Minute Guide to Wi-Fi Standards," *InformationWeek*, December 12, 2005.

27 Lawson, Stephen, "CTIS-Sprint Nears Decision on 4G," *NetworkWorld*, April 7, 2006, *www.networkworld.com*.

28 Dubie, Denise, "Sierra Pacific Taps Open Source Management Tools," *Network World*, December 12, 2005.

29 Rosencrance, Linda, "GT Nexus Helps Importers Go Paperless," *Computerworld*, July 1, 2005.

30 Amato-McCoy, Deena M., "Ready for the Next Big One," *InformationWeek*, April 17, 2006.

31 Hamblen, Matt; Thibodeau, Patrick, "Gas Price Increases Spur Interest in Telecommuting," *Computerworld*, August 22, 2005.

32 Staff, "No More Spin Control," *Computerworld*, January 23, 2006.

33 Rosencrance, Linda, "UPS Deploys New Customs Security Technology," *Computerworld*, August 31, 2005.

34 Gardner, David W., "Navajo Nation Gets Networked," *InformationWeek*, April 18, 2006.

35 Gaskin, James, E., "Groove for Small Businesses," *Network World*, May 19, 2005.

36 Fontana, John, "Microsoft Moves to Groove," *Network World*, March 6, 2006.

37 Rosencrance, Linda, "Toyota to Provide Mazda Owners in Japan In-Vehicle Info Service," *Computerworld*, February 10, 2004, *www.computerworld.com*.

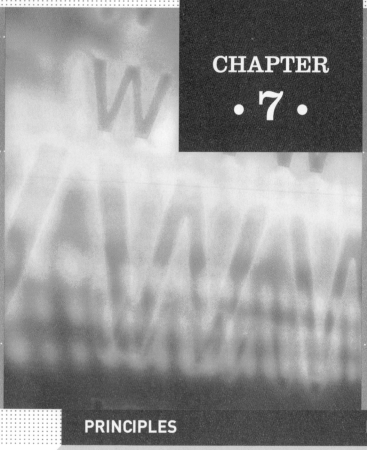

CHAPTER · 7 ·

The Internet, Intranets, and Extranets

PRINCIPLES

- The Internet is like many other technologies—it provides a wide range of services, some of which are effective and practical for use today, others are still evolving, and still others fade away from lack of use.

- Originally developed as a document-management system, the World Wide Web is a menu-based system that is easy to use for personal and business applications.

- Because the Internet and the World Wide Web are becoming more universally used and accepted for business use, management, service and speed, privacy, and security issues must continually be addressed and resolved.

LEARNING OBJECTIVES

- Briefly describe how the Internet works, including alternatives for connecting to it and the role of Internet service providers.

- Describe the World Wide Web and the way it works.

- Explain the use of Web browsers, search engines, and other Web tools.

- Identify and briefly describe the applications associated with the Internet and the Web.

- Identify who is using the Web to conduct business and discuss some of the pros and cons of Web shopping.

- Outline a process for creating Web content.

- Describe Java and discuss its potential impact on the software world.

- Define the terms *intranet* and *extranet* and discuss how organizations are using them.

- Identify several issues associated with the use of networks.

Information Systems in the Global Economy ⟫
Firwood Paints, United Kingdom

Firwood Paints Looks to Web for Expanded Customer Service and Order Processing

Firwood Paints in Bolton, England, manufactures specialty paints and surface coatings for industrial applications. Established in 1925, the company operates from three UK sites servicing over 2,000 global customers. Firwood employs 50 people who research, develop, test, and manufacture high-performance products to precise British Standards.

Until recently, customers relied on telephone and fax communications to place orders with Firwood. As the company grew, the cost of processing orders became a concern to Firwood management. Most disturbing was that the cost of processing small orders was as high as for large orders. Firwood wanted to implement a system that would increase its customer base, while decreasing the amount it was spending on transaction processing. Martin Wallen, managing director at Firwood, explains: "The market was getting increasingly competitive as volumes and margins shrank, and we needed a strategy that would reverse this trend, help us reach new markets, improve customer service to retain existing accounts, and cut the costs of doing business on smaller accounts."

Firwood engaged Stratagem, an information systems consulting firm, which recommended a Web solution. An interactive Web site would allow Firwood to offer better service and information to all of its customers and provide self-service sales for smaller orders, freeing its sales force to focus on big accounts.

Firwood had recently installed new ERP systems to improve the company's efficiency. The challenge was to develop a Web-based system for its customers that would seamlessly connect to its current information infrastructure. Because its ERP systems and servers were designed by IBM, Firwood looked to IBM solutions for its new Web-based system.

IBM recommended adding two servers: an iSeries 800 server to support the manufacturing control systems and an xSeries 225 server running Linux and IBM WebSphere software to manage the product catalog for the Web site. "The new iSeries server has given us the capacity to handle our production and ERP systems, while serving live data to WebSphere on the x225," says Martin Wallen. "When customers place orders online, the integrated solution generates all the necessary internal documentation, avoiding human error, saving time, and giving us a clearer audit trail and analysis tool than we have ever had before."

The introduction of Web-based ordering and account management is expected to lead to a dramatic reduction in administrative workload for Firwood, saving time and money. Small customers can order single and lower-value items at low transaction costs to the company thanks to the automated sales data-entry system. The Web site also provides enhanced customer service with online account information served directly from the internal ERP system.

Internet and Web technologies have provided valuable communications links between people, businesses, and organizations. Data, voice, and video are streaming over the Internet backbone as people and automated systems communicate, access information, and conduct business. Like Firwood, most businesses are doing away with older, less efficient, and more expensive methods of communication and turning to new Internet technologies.

As you read this chapter, consider the following:

- What unique features of the Internet and Web make them popular choices for many business communications applications?
- In what ways are the Internet and Web used by individuals to improve our quality of life and by businesses to improve the bottom line?

Why Learn About the Internet?

To say that the Internet has had a big impact on organizations of all types and sizes would be a huge understatement. Since the early 1990s when the Internet was first used for commercial purposes, it has affected all aspects of business. Businesses use the Internet to sell and advertise their products and services, reaching out to new and existing customers. If you are undecided about a career, you can use the Internet to help you investigate career opportunities and salaries using sites such as *www.monster.com* and HotJobs at *www.yahoo.com*. Most companies have Internet sites that list job opportunities, descriptions, qualifications, salaries, and benefits. If you have a job, you probably use the Internet daily to communicate with coworkers and your boss. People working in every field and at every level use the Internet in their jobs. Purchasing agents use the Internet to save millions of dollars in supplies every year. Travel and events-management agents use the Internet to find the best deals on travel and accommodations. Automotive engineers use the Internet to work with other engineers around the world developing designs and specifications for new automobiles and trucks. Property managers use the Internet to find the best prices and opportunities for commercial and residential real estate. Whatever your career, you will probably use the Internet daily. This chapter starts by exploring how the Internet works and then investigates the many exciting opportunities for using the Internet to help you achieve your goals.

Internet

A collection of interconnected networks, all freely exchanging information.

The Internet is the world's largest computer network. Actually, the **Internet** is a collection of interconnected networks, all freely exchanging information (see Figure 7.1). Research firms, colleges, and universities have long been part of the Internet, and businesses, high schools, elementary schools, and other organizations have joined it as well. Nobody knows exactly how big the Internet is because it is a collection of separately run, smaller computer networks. There is no single place where all the connections are registered. Figure 7.2 shows the staggering growth of the Internet, as measured by the number of Internet host sites or domain names. Domain names are discussed later in the chapter.

Figure 7.1

Routing Messages over the Internet

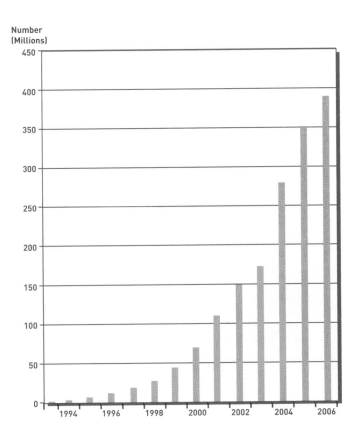

Figure 7.2

Internet Growth: Number of Internet Domain Names

(Source: Data from "The Internet Domain Survey," *www.isc.org*.)

USE AND FUNCTIONING OF THE INTERNET

The Internet is truly international in scope, with users on every continent—including Antarctica. However, the United States has the most usage by far. More than 100 million people in the United States (roughly one-third the population) have used the Internet. Although the United States still claims more Web activity than other countries, the Internet is expanding around the globe but at differing rates for different countries. China has spent almost $140 billion in its telecommunications infrastructure in the last five years.[1] In 2005, almost 100 million Chinese were connected to the Internet. China, however, restricts the use of the Internet.[2] In 2005, for example, China implemented new Internet rules. According to the Xinhua News Agency of China, "[Only] healthy and civilized news and information that is beneficial to the improvement of the quality of the nation, beneficial to its economic development and conductive to social progress will be allowed. The sites are prohibited from spreading news and information that goes against state security and public interest." In Russia, using the Internet's e-mail capabilities provides a timely mail service; it might take weeks for a Russian airmail letter to reach the United States. Most Internet usage in South Korea is through high-speed broadband connections. International use of the Internet is expected to continue its growth.

The ancestor of the Internet was the **ARPANET**, a project started by the U.S. Department of Defense (DoD) in 1969. The ARPANET was both an experiment in reliable networking and a means to link DoD and military research contractors, including many universities doing military-funded research. (*ARPA* stands for the Advanced Research Projects Agency, the branch of the DoD in charge of awarding grant money. The agency is now known as DARPA—the added *D* is for *Defense*.) The ARPANET was highly successful, and every university in the country wanted to use it. This wildfire growth made it difficult to manage the ARPANET, particularly its large and rapidly growing number of university sites. So, the ARPANET was broken into two networks: MILNET, which included all military sites, and a new, smaller ARPANET, which included all the nonmilitary sites. The two

ARPANET

A project started by the U.S. Department of Defense (DoD) in 1969 as both an experiment in reliable networking and a means to link DoD and military research contractors, including many universities doing military-funded research.

More than 600 million people use the Internet around the world, and the number of users could surge to more than 1 billion in a few years.

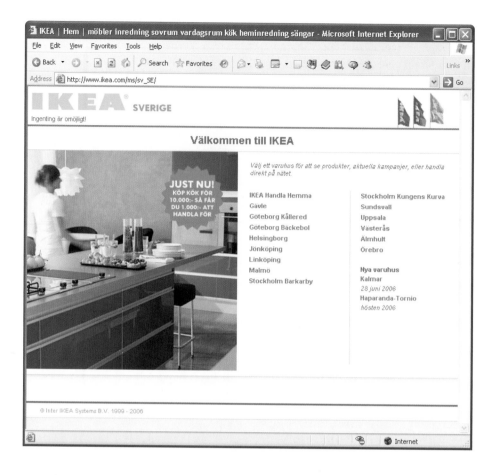

Internet Protocol (IP)

A communication standard that enables traffic to be routed from one network to another as needed.

networks remained connected, however, through use of the **Internet Protocol (IP)**, which enables traffic to be routed from one network to another as needed. All the networks connected to the Internet speak IP, so they all can exchange messages. Table 7.1 outlines a brief history of the Internet. Katie Hafner's book, *Where Wizards Stay Up Late: The Origins of the Internet*, gives a more detailed description of the history of the Internet.[3]

Table 7.1

A Brief History of the Internet

Event	Date
ARPANET is created	1969
TCP/IP becomes the protocol for ARPANET	1982
Domain Name System (DNS) is created	1984
Tim Berners-Lee creates the World Wide Web	1991
Commercial Internet Exchange (CIX) Association is established to allow businesses to connect to the Internet	1991

Today, people, universities, and companies are attempting to make the Internet faster and easier to use. Robert Kahn, who managed the early development of the ARPANET, wants to take the Internet to the next level. He is president of the nonprofit organization National Research Initiatives, which provides guidance and funding for the development of a national information infrastructure. The organization is looking into using "digital objects," which allow all types of computer systems to use and share programs and data. To speed Internet access, a group of corporations and universities called the University Corporation for Advanced Internet Development (UCAID) is working on a faster, new Internet. Called Internet2 (I2), Next Generation Internet (NGI), and Abilene, depending on the universities or corporations involved, the new Internet offers the potential of faster Internet speeds, up to 2 Gbps per second or more.[4] Some I2 connections can transmit data at 100 Mbps per second, which is about 10 times faster than many cable modems and 200 times faster than dial-up connections. The speed allows you to transfer the contents of a DVD in less than a minute.

How the Internet Works

The Internet transmits data from one computer (called a *host*) to another (see Figure 7.1). If the receiving computer is on a network to which the first computer is directly connected, it can send the message directly. If the receiving and sending computers are not directly connected to the same network, the sending computer relays the message to another computer that can forward it. The message might be sent through a router (see Chapter 6) to reach the forwarding computer. The forwarding host, which presumably is attached to at least one other network, delivers the message directly if it can or passes it to another forwarding host. A message can pass through a dozen or more forwarders on its way from one part of the Internet to another.

The various networks that are linked to form the Internet work much the same way—they pass data around in chunks called *packets*, each of which carries the addresses of its sender and its receiver. The set of conventions used to pass packets from one host to another is known as the Internet Protocol (IP). Many other protocols are used in connection with IP. The best known is the **Transmission Control Protocol (TCP)**, which operates at the Transport layer. Many people use TCP/IP as an abbreviation for the combination of TCP and IP used by most Internet applications. After a network following these standards links to a **backbone**—one of the Internet's high-speed, long-distance communications links—it becomes part of the worldwide Internet community.

Each computer on the Internet has an assigned address called its **Uniform Resource Locator**, or **URL**, to identify it to other hosts. The URL gives those who provide information over the Internet a standard way to designate where Internet elements such as servers, documents, and newsgroups can be found. Consider the URL for Course Technology, *http://www.course.com*.

The "http" specifies the access method and tells your software to access a file using the Hypertext Transport Protocol. This is the primary method for interacting with the Internet. In many cases, you don't need to include http:// in a URL because it is the default protocol. Thus, *http://www.course.com* can be abbreviated as *www.course.com*.

The "www" part of the address signifies that the address is associated with the World Wide Web service, discussed later. The "course.com" part of the address is the domain name that identifies the Internet host site. Domain names must adhere to strict rules. They always have at least two parts, with each part separated by a dot (period). For some Internet addresses, the far right part of the domain name is the country code (such as au for Australia, ca for Canada, dk for Denmark, fr for France, and jp for Japan). Many Internet addresses have a code denoting affiliation categories. (Table 7.2 contains a few popular categories.) The far left part of the domain name identifies the host network or host provider, which might be the name of a university or business.

Transmission Control Protocol (TCP)
The widely used Transport-layer protocol that most Internet applications use with IP.

backbone
One of the Internet's high-speed, long-distance communications links.

Uniform Resource Locator (URL)
An assigned address on the Internet for each computer.

Affiliation ID	Affiliation
com	Business organizations
edu	Educational sites
gov	Government sites
net	Networking organizations
org	Organizations

Table 7.2

U.S. Top-Level Domain Affiliations

Originally, Herndon, Virginia-based Network Solutions, Inc. (NSI), was the sole company in the world with the direct power to register addresses using .com, .net, or .org domain names. But its government contract ended in October 1998, as part of the U.S. government's move to turn management of the Web's address system over to the private sector. Today, other companies, called *registrars*, can register domain names, and additional companies are seeking accreditation to register domain names from the Internet Corporation for Assigned Names and Numbers (ICANN). Some registrars are concentrating on large corporations, where the profit margins might be higher, compared with small businesses or individuals. Note that some other countries outside the United States use different top-level domain affiliations than the ones described in Table 7.2.

There are millions of registered domain names. Some people, called *cyber-squatters*, have registered domain names in the hope of selling the names to corporations or people at a later date. The domain name Business.com, for example, sold for $7.5 million. In one case, a federal judge ordered the former owner of one Web site to pay the person who originally registered the domain name $40 million in compensatory damages and an additional $25 million in punitive damages. But some companies are fighting back, suing people who register domain names only to sell them to companies. Today, the ICANN has the authority to resolve domain-name disputes. Under new rules, if an address is found to be "confusingly similar" to a registered trademark, the owner of the domain name has no legitimate interest in the name. The rule was designed in part to prevent cyber-squatters.

Accessing the Internet

Although you can connect to the Internet in numerous ways (see Figure 7.3), Internet access is not distributed evenly throughout the world. Which access method you choose is determined by the size and capability of your organization or system.

Figure 7.3

Several Ways to Access the Internet

Users can access the Internet in several ways, including using a LAN server, dialing into the Internet using SLIP or PPP, or using an online service with Internet access.

Connect via LAN Server

This approach requires you to install a network adapter card and Open Datalink Interface (ODI) or Network Driver Interface Specification (NDIS) packet drivers on your computer. These drivers allow multiple transport protocols to run on one network card simultaneously. LAN servers are typically connected to the Internet at 56 Kbps or faster. In addition, you can share the higher cost of this service among several dozen LAN users to allow a reasonable cost per user.

Connect via SLIP/PPP

This approach requires a modem and the TCP/IP protocol software plus **Serial Line Internet Protocol (SLIP)** or **Point-to-Point Protocol (PPP)** software. SLIP and PPP are two communications protocols that transmit packets over telephone lines, allowing dial-up access to the Internet. If you are running Windows, you also need Winsock. You must also have an Internet service provider that lets you dial into a SLIP/PPP server. SLIP/PPP accounts can be purchased for $30 per month or less from regional providers. With all this in place, you use a modem to call into the SLIP/PPP server. After the connection is made, you are on the Internet and can access any of its resources. The expenses include the cost of the modem and software, plus the service provider's charges for access to the SLIP/PPP server. The speed of this Internet connection is limited to the slowest modem in the connection—your computer's modem or the modem of the SLIP/PPP server to which you connect.

Connect via an Online Service

This approach requires nothing more than what is required to connect to any of the online information services, such as a modem, standard communications software, and an online information service account. Online services usually offer digital subscriber line (DSL), satellite, or cable connection to the Internet, which provide faster speeds. These technologies were discussed in Chapter 6. You are normally charged a fixed monthly cost for basic services, including e-mail. Additional fees usually apply for DSL, satellite, or cable access, although these costs are falling. The online information services provide a wide range of services, including e-mail and the World Wide Web. America Online and Microsoft Network are examples of such services.

Other Ways to Connect

In addition to computers, many other devices can be connected to the Internet, including cell phones, PDAs, and home appliances. These devices also require specific protocols and approaches to connect. For example, *wireless application protocol (WAP)* is used to connect cell phones and other devices to the Internet.

Internet Service Providers

An **Internet service provider** (ISP) is any company that provides people and organizations with access to the Internet. ISPs do not offer the extended information services offered by commercial online services such as America Online or EarthLink. Thousands of organizations serve as Internet service providers, ranging from universities making unused communications line capacity available to students and faculty to major communications giants such as AT&T and MCI. To use this type of connection, you must have an account with the service provider and software that allows a direct link via TCP/IP.

In most cases, ISPs charge a monthly fee that can range from $15 to $30 for unlimited Internet connection through a standard modem. The fee normally includes e-mail. Some ISPs are experimenting with low-fee or no-fee Internet access, though strings are attached to the no-fee offers in most cases. Some free ISPs require that customers provide detailed demographic and personal information. In other cases, customers must put up with extra advertising when using the Web. For example, a *pop-up ad* is a window that is displayed when someone visits a Web site. It pops up and advertises a product or service. Some e-commerce retailers have posted ads that resemble computer-warning messages and have been sued for deceptive advertising. A *banner ad* appears as a banner or advertising window that you can ignore or click to go to the advertiser's Web site. Table 7.3 identifies several corporate Internet service providers.

Many ISPs and online services offer broadband Internet access through DSLs, cable, or satellite transmission. Most broadband users pay $50 or less per month for unlimited service. Broadband use has spread globally. Some businesses and universities use the very fast T-1 or T-3 lines to connect to the Internet. Table 7.4 compares the speed of

Serial Line Internet Protocol (SLIP)
A communications protocol that transmits packets over telephone lines.

Point-to-Point Protocol (PPP)
A communications protocol that transmits packets over telephone lines.

Internet service provider (ISP)
Any company that provides people or organizations with access to the Internet.

modem, DSL, cable, and T-1 Internet connections to perform basic tasks. These technologies were discussed in Chapter 6.

Table 7.3

A Representative List of Internet Service Providers

Internet Service Provider	Web Address
AT&T WorldNet Service	www.att.net
BellSouth	www.bellsouth.com
EarthLink	www.earthlink.net
Sprint	www.sprint.com

Table 7.4

Approximate Times to Perform Basic Tasks with Various Internet Connections

Task	Modem	DSL or Cable	T-1 or T-3
Send 20-page term paper	30 seconds	3 seconds	Almost instantaneous
Send a four-minute song as an MP3 file	30 minutes	2 minutes	Almost instantaneous
Send a full-length motion picture as a compressed file	About 2 weeks	About 2 days	20 seconds

To use the full capabilities of an online service, such as AOL, you must have an account with the service provider and software that allows a direct link via TCP/IP.

Liz Claiborne Moves to IP Technologies

Recently, fashion company Liz Claiborne found itself in a risky situation. The bulk of its sales, design, and manufacturing personnel are located in New York and New Jersey offices and depend on an aging circuit-switched PBX voice communications network and a managed data network service. After the terrorist attacks of 2001, the company was concerned that its data and voice communications channels could be completely taken out by one well-placed attack. A second concern was that the data network was being strained by the demands of its increasing business. Liz Claiborne IT specialists looked to Internet technologies to upgrade and consolidate their legacy PBX voice communications and data network.

"We needed more bandwidth just for the data," John Kovac, vice president of IT explains. While looking at new IP technologies to upgrade the data network, Kovac decided that it made good sense to piggyback voice onto the data network and replace the PBX infrastructure with a more redundant IP infrastructure.

Liz Claiborne now runs a four-server Cisco CallManager IP PBX cluster in the New York and New Jersey region. "I could lose three CallManagers and still deliver phone service to everyone," claims Kovac. The new servers can be restored in "a couple of hours" rather than the seven days required to restore the old system.

The new IP-based system also supports useful voice and data convergence applications. It hosts 3,600 IP handsets, connecting seven buildings in the New York/New Jersey area, plus offices in Atlanta and California. Through a feature called extension mobility, employees can use personalized settings at any phone on the system. An employee attending a meeting at another Liz Claiborne office can "log onto" a Cisco 7960 IP phone in a conference room and use their personalized buttons to retrieve messages from the Cisco Unity voice-mail system as if they were in their own offices.

Because the phones include sizable displays, they can handle text communications and even video communications. "If we want

to deliver an announcement to an employee or conference room without a PC, we can send messages to the phone screen," explains Anthony Ladisernia, director of IT. "We also have an XML application tied to our global directory, so users can bring up a directory, click on a name or function and call anyone in the company." Liz Claiborne is also testing Cisco VT Advantage, an application that layers videoconferencing onto Cisco CallManager and IP phones to help reduce travel time and expenses.

By converging data and voice networks onto a routed IP network, Liz Claiborne has reduced its long-distance calling charges and the number of dedicated T-1 lines. Overall, it says it has reduced operations, telecommunications, and management expenses by 16 percent.

Discussion Questions

1. How is Liz Claiborne's new packet-switched IP network more secure and reliable than the previous circuit-switched voice network?
2. What benefits are Liz Claiborne employees enjoying by using a network that combines data and voice communications?

Critical Thinking Questions

1. How will an IP network benefit Liz Claiborne as the company grows and expands operations into new locations?
2. How is this new IP network limited? For example, can it be extended to mobile cell phones and used away from the office?

SOURCES: Wexler, Joanie, "Fashion Designer Gets Hip to IP," *Computerworld*, January 17, 2005, *www.computerworld.com*. Cisco CallManager Web site, accessed April 22, 2006, *www.cisco.com/en/US/products/sw/voicesw/ps556/index.html*. Liz Claiborne Web site, accessed April 22, 2006, *www.lizclaiborne.com*.

THE WORLD WIDE WEB

World Wide Web (WWW or W3)
A collection of tens of thousands of independently owned computers that work together as one in an Internet service.

The World Wide Web was developed by Tim Berners-Lee at CERN, the European Organization for Nuclear Research in Geneva. He originally conceived of it as an internal document-management system. This server can be located at *http://public.web.cern.ch/public*. From this modest beginning, the **World Wide Web** (the Web, WWW, or W3) has grown to a collection of tens of thousands of independently owned computers that work together as one in an Internet service. These computers, called *Web servers*, are scattered all over the world and contain every imaginable type of data. Thanks to the high-speed Internet circuits connecting them and some clever cross-indexing software, users can jump from one Web computer to another effortlessly—creating the illusion of using one big computer. Because of its ability to handle multimedia objects, including linking multimedia objects distributed on Web servers around the world, the Web has become the most popular means of information access on the Internet today.

The Web is a menu-based system that uses the client/server model. It organizes Internet resources throughout the world into a series of menu pages, or screens, that appear on your computer. Each Web server maintains pointers, or links, to data on the Internet and can retrieve that data. However, you need the right hardware and telecommunications connections, or the Web can be painfully slow. Traditionally, graphics and photos have taken a long time to materialize on the screen, and an ordinary phone line connection might not always provide sufficient speed to use the Web effectively. Serious Web users need to connect via the LAN server, DSL, cable, or other approaches discussed earlier. Web *plug-ins* can help provide additional features to standard Web sites. Macromedia's Flash and Real Player are examples of Web plug-ins.

home page
A cover page for a Web site that has graphics, titles, and text.

Data can exist on the Web as ASCII characters, word processing files, audio files, graphic and video images, or any other sort of data that can be stored in a computer file. A Web site is like a magazine, with a cover page called a **home page** that has color graphics, titles, and text. All the highlighted type (sometimes underlined) is hypertext, which links the on-screen page to other documents, or Web sites. **Hypermedia** connects the data on pages, allowing users to access topics in whatever order they want. As opposed to a regular document, which you read linearly, hypermedia documents are more flexible, letting you explore related documents at your own pace and navigate in any direction. For example, if a document mentions the Egyptian pharaohs, you might be able to choose to see a picture of the pyramids, jump into a description of the building of the pyramids, and then jump back to the original document. Hypertext links are maintained using URLs. Table 7.5 lists some interesting Web sites. Many PC and business magazines also publish interesting and useful Web sites, and Web sites are often evaluated and reviewed in print media and online.

hypermedia
Tools that connect the data on Web pages, allowing users to access topics in whatever order they want.

A *Web portal* is an entry point or doorway to the Internet. Web portals include AOL, MSN, Yahoo!, and others. For example, some people use Yahoo.com as their Web portal, which means they have set Yahoo! as their starting point. When they enter the Internet, the Yahoo! Web site appears. You can use Yahoo! to search the Internet, send e-mail, get directions for a trip, buy products and services, find the address and phone number of friends or relatives, and more. A *corporate Web portal* refers to the company's Internet site that is a gateway or entry point to corporate data and resources.

Hypertext Markup Language (HTML)
The standard page description language for Web pages.

Hypertext Markup Language (HTML) is the standard page description language for Web pages. One way to think about HTML is as a set of highlighter pens that you use to mark up plain text to make it a Web page—red for the headings, yellow for bold, and so on. The **HTML tags** let the browser know how to format the text: as a heading, as a list, or as body text, for example. HTML also tells whether images, sound, and other elements should be inserted. Users mark up a page by placing HTML tags before and after a word or words. For example, to turn a sentence into a heading, you place the <h1> tag at the start of the sentence. At the end of the sentence, you place the closing tag </h1>. When you view this page in your browser, the sentence will be displayed as a heading. So, a Web page is made up of two things: text and tags. The text is your message, and the tags are codes that mark

HTML tags
Codes that let the Web browser know how to format text—as a heading, as a list, or as body text—and whether images, sound, and other elements should be inserted.

Site	Description	URL
Monster	A job-hunting site where you can search for a job by type or company, list your résumé, and perform basic company research. One feature, Talent Market, allows people to put their skills up for bid.	www.monster.com
Centers for Disease Control (CDC)	A government site that provides a wealth of information on a wide variety of health topics.	www.cdc.gov
NASA Human SpaceFlight	A site from NASA that gives information about past and present missions into space.	www.spaceflight.nasa.gov
Yahoo! Maps	A service that offers street addresses and driving directions.	www.yahoo.com
eBay	A popular auction site.	www.ebay.com
Amazon.com	A popular site that sells books, videos, music, furniture, and much more.	www.amazon.com
Travelocity	A large site that offers travel information, reservations, and bargains.	www.travelocity.com
WebMD	A site that provides medical information and advice.	www.webmd.com

Table 7.5

Several Interesting Web Sites

the way words will be displayed. All HTML tags are enclosed in a set of angle brackets (< and >), such as <h2>. The closing tag has a forward slash in it, such as </b> for closing bold. Consider the following text and tags:

```
<h1 align="center">Principles of Information Systems</h1>
```

This HTML code centers Principles of Information Systems as a major, or level 1, heading. The "h1" in the HTML code indicates a first-level heading. On some browsers, the heading might be 14-point type size with a Times Roman font. On other browsers, it might be a larger 18-point size in a different font. Figure 7.4 shows a simple document and its corresponding HTML tags. Notice the <html> tag at the top indicating the beginning of the HTML code. The <title> indicates the beginning of the title: "Course Technology – Leading the Way in IT Publishing." The </title> tag indicates the end of the title.

Some newer Web standards are gaining in popularity, including Extensible Markup Language (XML), Extensible Hypertext Markup Language (XHTML), Cascading Style Sheets (CSS), Dynamic HTML (DHTML), and Wireless Markup Language (WML), which can display Web pages on small screens, such as smartphones and PDAs. XHTML is a combination of XML and HTML that has been approved by the World Wide Web Consortium (W3C).

Extensible Markup Language (XML) is a markup language for Web documents containing structured information, including words and pictures. XML does not have a predefined tag set. With HTML, for example, the <h1> tag always means a first-level heading. The content and formatting are contained in the same HTML document. XML Web documents contain the content of a Web page. The formatting of the content is contained in a style sheet. A few typical instructions in XML follow:

<chapter>Hardware
<topic>Input Devices
<topic>Processing and Storage Devices
<topic>Output Devices

How the preceding content is formatted and displayed on a Web page is contained in a style sheet, such as the following cascading style sheet (CSS). Note that the chapter title Hardware is displayed on the Web page in a large font (18 points). Hardware will appear in bold blue text. The Input Devices title will appear in a smaller font (12 points). Input devices will appear in italic red text.

Extensible Markup Language (XML)

The markup language for Web documents containing structured information, including words, pictures, and other elements.

Figure 7.4

Sample Hypertext Markup Language

Shown at the left on the screen is a document, and at the right are the corresponding HTML tags.

chapter: (font-size: 18pt; color: blue; font-weight: bold; display: block; font-family: Arial; margin-top: 10pt; margin-left: 5pt)

topic: (font-size: 12pt; color: red; font-style: italic; display: block; font-family: Arial; margin-left: 12pt)

XML includes the capabilities to define and share document information over the Web. A company can use XML to exchange ordering and invoicing information with its customers. CSS improves Web page presentation, and DHTML provides dynamic presentation of Web content. These standards move more of the processing for animation and dynamic content to the Web browser, discussed next, and provide quicker access and displays.

Web Browsers

Web browser
Software that creates a unique, hypermedia-based menu on a computer screen, providing a graphical interface to the Web.

A **Web browser** creates a unique, hypermedia-based menu on your computer screen that provides a graphical interface to the Web. The menu consists of graphics, titles, and text with hypertext links. The hypermedia menu links you to Internet resources, including text documents, graphics, sound files, and newsgroup servers. As you choose an item or resource, or move from one document to another, you might be accessing various computers on the Internet without knowing it, while the Web handles all the connections. The beauty of Web browsers and the Web is that they make surfing the Internet fun. Clicking with a mouse on a highlighted word or graphic whisks you effortlessly to computers halfway around the world. Most browsers offer basic features such as support for backgrounds and tables, displaying a Web page's HTML source code, and a way to create hot lists of your favorite sites. Web browsers enable net surfers to view more complex graphics and 3-D models, as well as audio and video material, and to run small programs embedded in Web pages called **applets**. (Applets are discussed in more detail in a later section.) A Web browser *plug-in* is an external program that is executed by a Web browser when it is needed. For example, if you are working with a Web page and encounter an Adobe .pdf file, the Web browser will typically run the external Adobe .pdf reader program or plug-in to allow you to open the Adobe .pdf file.

applet
A small program embedded in Web pages.

Microsoft Internet Explorer and Netscape Navigator are examples of Web browsers for PCs. Safari is a popular Web browser from Apple Computer for their Macintosh computer, and Mozilla Firefox is a Web browser available in numerous languages that can be used on PCs, computers with the Linux operating system, and Apple Mac computers. See Figure 7.5.

Figure 7.5

Mozilla Firefox

Search Engines and Web Research

Looking for information on the Web is like browsing in a library—without the alphabetic listing of books in the card catalog, it is difficult to find information. Web search tools—called **search engines**—take the place of the card catalog. Most search engines, such as Yahoo.com and Google.com, are free. They make money by charging advertisers to put ad banners on their search engines. Google has almost 60 percent of search volume, and Yahoo! has almost 30 percent.[5] Other search engines make up the rest of the search-engine volume. Companies often pay a search engine for a *sponsored link*, which is usually displayed at the top of the list of links for an Internet search.

> Search engines that use keyword indexes produce an index of all the text on the sites they examine. Typically, the engine reads at least the first few hundred words on a page, including the title, the HTML "alt text" coded into Web-page images, and any keywords or descriptions that the author has built into the page structure. The engine throws out words such as "and," "the," "by," and "for." The engine assumes remaining words are valid page content; it then alphabetizes these words (with their associated sites) and places them in an index where they can be searched and retrieved. Some companies include a *meta tag* in the HTML header for search engine robots from sites such as Yahoo! and Google to find and use. Meta tags are not shown on the Web page when it is displayed; they only help search engines discover and display a Web site. When searching the Web, you can try using more than one search engine to expand the total number of potential Web sites of interest. In addition, searches can use words, such as AND and OR, to refine the search. Searches can also use filters, such as displaying Web sites in English only. Filters limit searches to a language, certain file formats, a range of dates, and more. Some search engines also have a subject directory that allows

search engine
A Web search tool.

Google is a popular search engine on the Web.

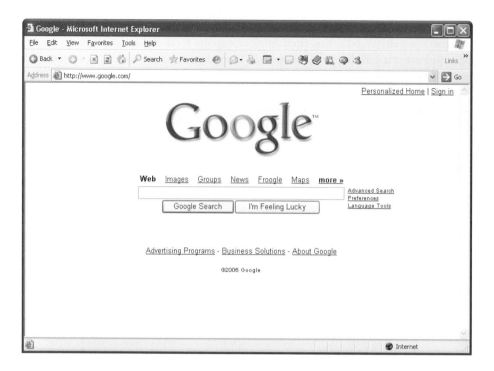

people to get information on various industries and organizations. Table 7.6 lists a few popular Web search tools.

Table 7.6

Popular Search Engines

Search Engine	Web Address
AltaVista	www.altavista.com
Ask Jeeves	www.ask.com
Google	www.google.com
HotBot	www.hotbot.lycos.com
Northern Light	www.northernlight.com
Yahoo!	www.yahoo.com

Today's search engines do more than look for words, phrases, or sentences on the Web. For example, you can use Google to search for images and video.[6] You can even search for geographic locations to get a view from the skies using satellites.[7] Google, for example, offers Maps and Earth to provide aerial views. After downloading and installing Google Earth, you can type an address and Google will show you the neighborhood or even a house in some cases. Microsoft Virtual Earth and Local Search also give aerial views and close-ups of some locations, including retail stores in some cases.[8]

meta-search engine

A tool that submits keywords to several search engines and returns the results from all search engines queried.

Another option is to use a meta-search engine. A **meta-search engine** submits keywords to several search engines and returns the results from all it queried. Ixquick (*www.ixquick.com*), ProFusion (*www.profusion.com*), and Dogpile (*www.dogpile.com*) are examples of meta-search engines. Dogpile, for example, searches Ask Jeeves, Google, Fast, and other search engines. Meta-search engines do not query all search engines.

In addition to search engines, you can use other Internet sites to research information. For example, *www.findarticles.com* contains millions of articles on a variety of topics, including business and finance, health and fitness, sports, and reference and education. You can use many news organizations, including CNN (*www.cnn.com*) and FoxNews (*www.foxnews.com*), to access current information on a variety of topics. Some Web sites maintain versions in different languages, especially for research purposes. Others offer programs such as Flexnet that combine online learning with face-to-face instruction. Wikipedia, an encyclopedia that has been created by thousands of people, is another example of a Web site that can be used to research information.[9] In Hawaiian, Wiki means quick. The Web site is both open source and open editing, which means that people can add or edit entries in the encyclopedia at any time. Because thousands of people are monitoring Wikipedia, the Web-based encyclopedia is self-regulating. Incorrect, outdated, or offensive material is usually removed, although people with an axe to grind have distorted information on Wikipedia intentionally. Jimmy Wales, the founder of Wikipedia, would like to expand the concept into books, manuals, and quotations. Some believe that the approach of Wikipedia can be used to allow people to collaborate on important projects. Squidoo (*www.squidoo.com*) is a Web site you can use to find information about a person's view of a particular topic, often called a "lens." Lenses exist on a wide variety of topics, including the arts, computers and technology, education, health, movies, music, news, and much more.

Web Programming Languages

There are a number of important Web programming languages. **Java**, for example, is an object-oriented programming language from Sun Microsystems based on the C++ programming language, which allows small programs—the applets mentioned earlier—to be embedded within an HTML document. When the user clicks the appropriate part of an HTML page to retrieve an applet from a Web server, the applet is downloaded onto the client workstation, where it begins executing. Unlike other programs, Java software can run on any type of computer. Programmers use Java to make Web pages come alive, adding splashy graphics, animation, and real-time updates.

The relationship among Java applets, a Java-enabled browser, and the Web is shown in Figure 7.6. To develop a Java applet, the author writes the code for the client computer and installs that on the Web server. The user accesses the Web page on a personal computer, which serves as a client. The Web page contains an additional HTML tag called APP, which refers to the Java applet. A rectangle on the page is occupied by the Java applet. If the user clicks the rectangle to execute the Java applet, the client computer checks to see whether a copy of the applet is already stored locally on its hard drive. If it is not, the computer accesses the Web server and requests that the applet be downloaded. The applet can be located anywhere on the Web. If the user has a Java-enabled browser (such as Sun HotJava or Netscape Navigator), the applet is downloaded to the user's computer and is executed in the browser.

Java
An object-oriented programming language from Sun Microsystems based on C++ that allows small programs (applets) to be embedded within an HTML document.

Figure 7.6

Downloading an Applet from a Web Server

The user accesses the Web page from a Web server. If the user clicks the APP rectangle to execute the Java applications, the client's computer checks for a copy on its local hard drive. If it does not find the applet, the client requests the Web server to download the applet.

The Web server that delivers the Java applet to the Web client cannot determine what kind of hardware or software environment the client is running, and the developer who creates the Java applet does not have to worry about whether it will work correctly on Windows, UNIX, and Mac OS. Java is thus often described as a "cross-platform" programming language.

In addition to Java, companies use a variety of other programming languages and tools to develop Web sites. JavaScript, VBScript, and ActiveX (used with Internet Explorer) are Internet languages used to develop Web pages and perform important functions, such as accepting user input. *Hypertext Preprocessor*, or *PHP*, is an open-source programming language. PHP code or instructions can be embedded directly into HTML code. Unlike some other Internet languages, PHP can run on a Web server, with the results being transferred to a client computer. PHP can be used on a variety of operating systems, including Microsoft Windows, Macintosh OS X, HP-UX, and others. It can also be used with a variety of database management systems, such as DB2, Oracle, Informix, MySQL, and many others. These characteristics—running on different operating systems and database management systems, and being an open-source language—make PHP popular with many Web developers.

Developing Web Content

The art of Web design involves working within the technical limitations of the Web and using a set of tools to make appealing designs. A study at Glamorgan University Business School in Wales, for example, concluded that women prefer Web pages with more color in the background and informal pictures and images.[10] Men prefer darker colors and liked 3-D, moving images. The following are suggestions for creating a Web page.

1. Your computer must be linked to a Web server, which can deliver Web pages to other browsers.
2. You need a Web browser program to view the HTML pages you create.
3. The actual design can take one of the following approaches: (a) Write your copy with a word processor, and then use an HTML converter to convert the page into HTML format complete with tags so the browser knows how it should format the page. (b) Use an HTML editor to write text and add HTML tags at the same time. (c) Edit an existing HTML template (with all the tags ready to use) to meet your needs. (d) Use an ordinary text editor and type the start and end tags for each item.
4. Open the HTML page with the browser and see the result. You can correct mistakes by correcting the tags.
5. Add links to your home page to allow your readers to click a word and open a related home page. The new page can be either a part of your Web site or a home page on a different Web site.
6. To add pictures, you must first store them as files on your hard drive. You can do this in one of several ways: Draw them yourself using a graphics software package, copy pictures from other Web pages (with permission from the artist if the pictures are not in the public domain), buy a disk of clip art, scan photos, or use a digital camera.
7. You can add sound by using a microphone connected to your computer to record a sound file; add links to the page so that those who access your Web page can hear the audio.
8. Upload the HTML file to your Web site using e-mail or File Transfer Protocol (FTP), discussed later in this chapter.
9. Review the Web page to make sure that all links are correctly established to other Web sites.
10. Advertise your Web page to others and encourage them to stop, take a look, and send feedback by e-mail.

After you develop Web content, your next step is to place, or publish, the content on a Web site or home page. Popular publishing options include using ISPs, free sites, and Web hosting. Web hosting services provide space on their Web site for people and businesses that don't have the financial resources, time, or skills to host their own Web site. A Web host can charge $15 or more per month, depending on services. Some Web hosting sites include domain name registration, Web authoring software, and activity reporting and monitoring of the Web site. Some ISPs also provide limited Web space, typically 1 to 6 MB, as part of their monthly fee. If more disk space is needed, there are additional charges. Free sites offer limited space for a Web site. In return, free sites often require the user to view advertising or agree to other terms and conditions.

Some Web developers are creating programs and procedures to combine two or more Web sites into one Web site, called a *mash-up*.[11] A mash-up is named for the process of mixing two or more hip-hop songs into one song. A Web site containing crime information, for example, can be mashed-up with a mapping Web site to produce a Web site with crime information placed on top of a map of a metropolitan area. People are becoming very creative in how they mash-up several Web sites into new ones. According to Jeff Bezos, chief executive of Amazon, "They're taking little bits and pieces from a number of companies and stitching them together in some clever ways." Mashing-up Web sites is becoming popular, but not everyone is happy with the practice. Some companies are trying to block the mash-up of the content on their Web site without permission.

After a Web site has been constructed, a *content management system (CMS)* can keep the Web site running smoothly. CMS consists of both software and support. Companies that provide CMS can charge from $15,000 to more than $500,000 annually, depending on the complexity of the Web site being maintained and the services being performed. Leading CMS vendors include BroadVision, Focumentum, EBT, FileNet, Open Market, and Vignette.

Many products make it easy to develop Web content and interconnect Web services, discussed in the next section. Microsoft, for example, has introduced a development and Web services platform called .NET. The .NET platform allows developers to use different programming languages to create and run programs, including those for the Web. The .NET platform also includes a rich library of programming code to help build XML Web applications. Bubbler is another example of a Web design and building tool.[12] Bubbler helps you obtain a domain name, develop attractive Web pages by dragging and dropping features and options, host your Web site, and maintain it. Other Web publishing packages include Homestead QuickSites and JobSpot. See Figure 7.7. Many of these products are popular with a newer approach to developing and maintaining Web content called Web services, discussed next.

Figure 7.7

Developing a Web Page Using Microsoft Visual Studio

The window on the left shows the Web page being developed using Microsoft Visual Studio, part of the .NET Web development platform. The window on the right shows the Web page in a browser.

Web Services

Web services consist of standards and tools that streamline and simplify communication among Web sites, promising to revolutionize the way we develop and use the Web for business and personal purposes. Internet companies, including Amazon, eBay, and Google, are now using Web services. Amazon, for example, has developed Amazon Web Services (AWS) to make the contents of its huge online catalog available by other Web sites or software applications. Mitsubishi Motors of North America uses Web services to link about 700 automotive dealers on the Internet.

The key to Web services is XML. Just as HTML was developed as a standard for formatting Web content into Web pages, XML is used within a Web page to describe and

Web services

Standards and tools that streamline and simplify communication among Web sites for business and personal purposes.

transfer data between Web service applications. XML is easy to read and has wide industry support. Besides XML, three other components are used in Web service applications:

1. SOAP (Simple Object Access Protocol) is a specification that defines the XML format for messages. SOAP allows businesses, their suppliers, and their customers to communicate with each other. It provides a set of rules that makes it easier to move information and data over the Internet.
2. WSDL (Web Services Description Language) provides a way for a Web service application to describe its interfaces in enough detail to allow a user to build a client application to talk to it. In other words, it allows one software component to connect to and work with another software component on the Internet.
3. UDDI (Universal Discovery Description and Integration) is used to register Web service applications with an Internet directory so that potential users can easily find them and carry out transactions over the Web.

INTERNET AND WEB APPLICATIONS

The types of Internet and Web applications available are vast and ever expanding. Here are a few examples.

Amazon.com lets you buy books, music, videos, software, electronics, and other products online.

Business Uses of the Web

In 1991, the Commercial Internet Exchange (CIX) Association was established to allow businesses to connect to the Internet. Since then, firms have been using the Internet for a number of applications, many of them discussed next in the section on Internet and Web Applications.

E-Mail, Instant Messaging, and Push Technology

E-mail is no longer limited to simple text messages. Depending on your hardware and software and the hardware and software of your recipient, you can embed sound and images in your message and attach files that contain text documents, spreadsheets, graphs, or executable programs. The authors of this book, for example, attached chapter files to e-mail messages that were sent to editors and reviewers for feedback. E-mail travels through the systems and networks that make up the Internet. Gateways can receive e-mail messages from the Internet and deliver them to users on other networks. Thus, you can send e-mail messages to anyone in the world if you know that person's e-mail address and you have access to the Internet or another system that can send e-mail. In addition, some Internet companies offer various services when switching accounts.[13] Some help you send out e-mails to your contact list telling them you have a new e-mail address; others offer free e-mail forwarding to another address.

For large organizations whose operations span a country or the world, e-mail allows people to work around the time zone changes. Some users of e-mail estimate that they eliminate two hours of verbal communications for every hour of e-mail use. But the person at the other end still must check the mailbox to receive messages.

To personalize messages, e-mail authors can include typographical expressions or abbreviations. Table 7.7 lists some abbreviations commonly used in personal e-mail messages. These abbreviations are normally not appropriate for business correspondence.

Expressions	Abbreviations
;-] Smile with a wink	AAMOF—As a matter of fact
;-(Frown with a wink	AFAIK—As far as I know
:-# My lips are sealed	BTW—By the way
:-D Laughing	CUL8R—See you later
:-0 Shocked	F2F—Face to face
:-] Blockhead	LOL—Laughing out loud
:-@ Screaming	OIC—Oh, I see
:-& Tongue-tied	TIA—Thanks in advance
%-) Brain-dead	TTFN—Ta-Ta for now

Table 7.7

Some Common Abbreviations Used in Personal E-Mail

Some companies use bulk e-mail to send legitimate and important information to sales representatives, customers, and suppliers around the world. With its popularity and ease of use, however, some people feel they are drowning in too much e-mail. Over a trillion e-mail messages are sent from businesses in North America each year. This staggering number is up from 40 billion e-mail messages in 1995. Many e-mails are copies sent to a large list of corporate users. Users are taking a number of steps to cope with and reduce the mountain of e-mail. Some companies have banned the use of copying others on e-mails unless it is critical. Some e-mail services scan for possible junk or bulk mail, called *spam*, and delete it or place it in a separate file. As much as half of all e-mail can be considered spam. Some business executives receive 300 or more spam e-mails in their corporate mailboxes every morning. Software products can help users sort and answer large amounts of e-mail. This software can recognize key words and phrases and respond to them.

Instant messaging is online, real-time communication between two or more people who are connected to the Internet. With instant messaging, two or more windows or panes open, with each one displaying what a person is typing. Because the typing is displayed in real-time, instant messaging is like talking to someone using the keyboard. See Figure 7.8.

instant messaging
A method that allows two or more people to communicate online using the Internet.

Instant Messaging

Instant messaging lets you converse with another Internet user by exchanging messages instantaneously.

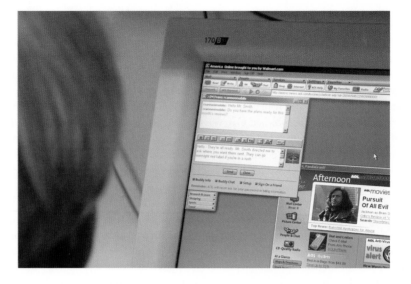

Many companies offer instant messaging, including America Online, Yahoo!, and Microsoft. America Online is one of the leaders in instant messaging, with about 40 million users of its Instant Messenger and about 50 million people using its client program ICQ. In addition to being able to type messages on a keyboard and have the information instantly displayed on the other person's screen, some instant messaging programs are allowing voice communication or connection to cell phones. A wireless service provider announced that it has developed a technology that can detect when a person's cell phone is turned on. With this technology, someone on the Internet can use instant messaging to communicate with someone on a cell phone anywhere in the world. Today, instant messaging can be delivered over the Internet, through the cell phone services, and other telecommunications services first discussed in Chapter 6.

Instant messaging services often use a *buddy list* that alerts people when their friends are also online. This feature makes instant messaging even more useful. Instant messaging is so popular that it helps Internet service providers and online services draw new customers and keep old ones. Buddy lists, however, can lead to inadvertent and unwanted sales pitches. **Push technology** is used to send information automatically over the Internet rather than make users search for it with their browsers.

push technology

The automatic transmission of information over the Internet rather than making users search for it with their browsers.

Internet Cell Phones and Handheld Computers

Increasingly, cell phones, handheld computers, and other devices are being connected to the Internet. Some cell phones, for example, can be connected to the Internet to allow people to search for information, buy products, and chat with business associates and friends. A sales manager for a computer company can use a cell phone to check the company's Internet site to see whether there are enough desktop computers in inventory to fill a large order for an important customer. Using Short Message Service (SMS), people can send brief text messages of up to 160 characters between two or more cell phone users. The service is often called *texting*. Some cell phones also come equipped with digital cameras, FM radios, video games, and small color screens to watch TV. Using enhanced message service (EMS) and multimedia messaging service (MMS), people can send pictures, video, and audio over cell phones to other cell phones or Internet sites.

Text messaging has exploded in popularity in the United States and around the world. The number of text messages sent has jumped from about 250 million messages in 2001 to more than 4 billion messages in 2005.[14] Shirley Jackson, for example, uses text messaging to conduct prayer groups using her cell phone. According to Jackson, "I can't carry my computer everywhere, but I can carry my cell phone." During the London bombings in 2005, when phone lines were tied up and cell phone service was overloaded, many people communicated with text messages.[15] Text messages are transmitted over the part of the cell phone network that only transmits data, so text messages don't compete with cell phones for space on cell phone networks.

In addition to cell phones, handheld computers and other devices can be connected to the Internet using phone lines or wireless connections, such as WiFi, discussed in Chapter 6. Once connected, these devices have full access to the Internet and all its applications discussed in this chapter and throughout the book. Managers use handheld computers, such as the BlackBerry or Treo handheld computer, and the Internet to check business e-mail when they are out of the office; sales representatives use them to demonstrate products to customers, check product availability and pricing, and upload customer orders.

Career Information and Job Searching

The Internet is an excellent source of job-related information. People looking for their first job or seeking information about new job opportunities can find a wealth of information. Search engines can be a good starting point for searching for specific companies or industries. You can use a directory on Yahoo's home page, for example, to explore industries and careers. Most medium and large companies have Internet sites that list open positions, salaries, benefits, and people to contact for further information. The IBM Web site, *www.ibm.com*, has a link to "Jobs at IBM." When you click this link, you can find information on jobs with IBM around the world. Some Internet sites specialize in certain careers or industries. The site *www.directmarketingcareers.com* lists direct marketing jobs and careers, for example. Some sites can help you develop a good résumé and find a good job. They can also help you develop an effective cover letter for a résumé, prepare for a job interview, negotiate a good employment contract, and more. In addition, several Internet sites specialize in helping you find job information and even apply for jobs online, including *www.monster.com*, *www.hotjobs.com*, and *www.careerbuilder.com*. You must be careful when applying for jobs online, however. Some bogus companies or Web sites will steal your identity by asking for personal information. People eager to get a job often give their Social Security number, birth date, and other personal information. The result can be no job, large bills on your credit card, and ruined credit.

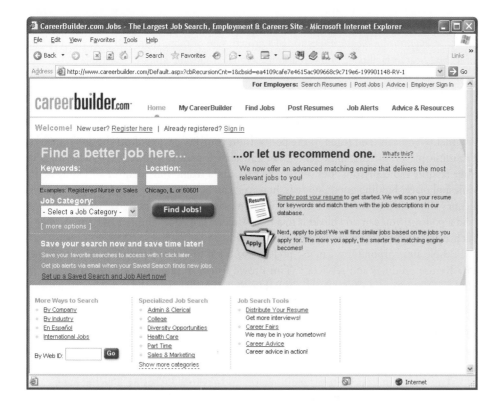

Several Internet sites specialize in helping people get job information and even apply for jobs online.

Telnet and FTP

Telnet is a terminal emulation protocol that enables you to log on to other computers on the Internet to gain access to their publicly available files. Telnet is particularly useful for perusing library holdings and large databases. It is also called *remote logon.*

File Transfer Protocol (FTP) is a protocol that describes a file transfer process between a host and a remote computer. Using FTP, users can copy files from one computer to another. Companies, for example, use it to transfer vast amounts of business transactional data to the computers of its customers and suppliers. You can also use FTP to gain access to a wealth of free software on the Internet. FTP can be used to upload or download content to a Web site. The authors and editors of this book used an FTP site provided by the publisher, Course Technology, to share and transfer important files during the publication process. Chapter files and artwork, for example, were uploaded to a Course Technology Web site and downloaded by authors and editors to review.

Web Log (Blog), Video Log (Vlog), and Podcasting

A **Web log**, also called a **blog**, is a Web site that people can create and use to write about their observations, experiences, and feelings on a wide range of topics. The community of blogs and bloggers is often called the *blogosphere.*[16] A *blogger* is a person who creates a blog, while *blogging* refers to the process of placing entries on a blog site. A blog is like a journal. When people post information to a blog, it is placed at the top of the blog.[17] Blogs can contain links to original or source material. Other links can send a blogger to comments made by others. Video content can also be placed on the Internet using the same overall approach as a blog. This is often called a *video log* or *vlog.* Blogs are easy to post, but they can cause problems when people tell or share too much.[18] People have been fired for blogging about work, and the daughter of a politician embarrassed her father when she made personal confessions on a blog.

Blog sites, such as *www.blogger.com* and *www.globeofblogs.com* can include information and tools to help people create and use Web logs. To set up a blog, you can go to a blog service provider, such as *www.livejournal.com*, create a username and password, select a theme, choose a URL, follow any other instructions, and start making your first entry. Blog search engines include Technorati, Feedster, and Blogdigger.[19] Technorati (*www.technorati.com/blogs/*) hopes to see a growth rate of more than 400 percent in the use of its blog search facility.[20] You can also use Google to locate a blog.

A *podcast* is an audio broadcast over the Internet. The name podcast comes from the word iPod, Apple's portable music player, and the word broadcasting.[21] A podcast is an audio blog, like a personal radio station on the Internet, and extends blogging by adding audio messages.[22] Using PCs and microphones, you can record audio messages and place them on the Internet.[23] You can then listen to the podcasts on your PC or download the audio material to a music player, such as Apple's iPod. You can also use podcasting to listen to TV programs, your favorite radio personalities, music, and messages from your friends and family at any time and place. Finding good podcasts, however, can be challenging. Apple's new version of iTunes allows you to download free software to search for podcasts by keyword. After you find a podcast, you can download it to a PC (Windows or Mac) and to a MP3 music player such as the iPod for future listening.

People and corporations can use podcasts to listen to audio material, increase revenues, or advertise products and services.[24] Tim Brown listens to podcasts of radio programs, including some programs from National Public Radio (NPR), while he is driving. "I basically don't listen to radio anymore when I drive around." Clear Channel Communications, a radio broadcasting corporation, sells memberships to podcasts of popular radio shows and personalities, including Rush Limbaugh and others. ABC News uses podcasts to allow people to listen to some TV programs, such as Nightline. Colleges and universities often use blogs and podcasts to deliver course material to students.

Many blogs and podcasts offer automatic updates to a PC using a technology, called Really Simple Syndication (RSS). RSS is a collection of Web formats to help provide Web content or summaries of Web content. With RSS, you can get a blog update without actually

Telnet
A terminal emulation protocol that enables users to log on to other computers on the Internet to gain access to public files.

File Transfer Protocol (FTP)
A protocol that describes a file transfer process between a host and a remote computer and allows users to copy files from one computer to another.

Web log (blog)
A Web site that people can create and use to write about their observations, experiences, and feelings on a wide range of topics.

visiting the blog Web site. RSS can also be used to get other updates on the Internet from news Web sites and podcasts.

Usenet and Newsgroups

Usenet is an older technology that uses e-mail to provide a centralized news service. It is actually a protocol that describes how groups of messages can be stored on and sent between computers. Following the Usenet protocol, e-mail messages are sent to a host computer that acts as a Usenet server. This server gathers information about a single topic into a central place for messages. A user sends e-mail to the server, which stores the messages. The user can then log on to the server to read these messages or have software on the computer log on and automatically download the latest messages to be read at leisure. Thus, Usenet forms a virtual forum for the electronic community, and this forum is divided into newsgroups.

Newsgroups make up Usenet, a worldwide discussion system classified by subject. Articles or messages are posted to newsgroups using newsreader software and are then broadcast to other interconnected computer systems via a wide variety of networks. A newsgroup is essentially an online discussion group that focuses on a particular topic. Newsgroups are organized into various hierarchies by general topic, and each topic can contain many subtopics. On the Internet, tens of thousands of newsgroups cover topics from astrology to zoology (see Table 7.8). Discussions take place via e-mail, which is sent to the newsgroup's address. A newsgroup can be moderated or unmoderated. If a newsgroup is moderated, e-mail is automatically routed to the moderator, a person who screens all incoming e-mail to make sure that it is appropriate before posting it to the newsgroup. People posting messages on newsgroups should be careful. Some Internet sites display millions of newsgroup messages that go back to 1980s online for anyone to see.

Usenet
A system closely allied with the Internet that uses e-mail to provide a centralized news service; a protocol that describes how groups of messages can be stored on and sent between computers.

newsgroups
Online discussion groups that focus on specific topics.

Chat Rooms

A **chat room** is a facility that enables two or more people to engage in interactive "conversations" over the Internet. When you participate in a chat room, dozens of people might be participating from around the world. Multiperson chats are usually organized around specific topics, and participants often adopt nicknames to maintain anonymity. One form of chat

chat room
A facility that enables two or more people to engage in interactive "conversations" over the Internet.

Table 7.8

Selected Usenet Newsgroups

Newsgroup Address	Description
alt.airline	Current schedules of various airlines
alt.books	Index of book reviews
alt.current-events.net-abuse.spam	Reports of e-mail and newsgroup spam abuse
alt.fan	Fans of various performers, artists, and others
alt.politics	Index of political discussions
alt.sports.baseball	Major league baseball
biz.ecommerce	Internet retailers
misc.legal	Miscellaneous discussions of legal matters
news.software	Usenet software
rec.backcountry	Activities in the great outdoors
rec.food.restaurants	Discussion of dining out

room, Internet Relay Chat (IRC), requires participants to type their conversation rather than speak. Voice chat is also an option, but you must have a microphone, sound card and speakers, a fast modem or broadband, and voice-chat software compatible with the other participants'.

Internet Phone and Videoconferencing Services

Internet phone service enables you to communicate with others around the world. This service is relatively inexpensive and can make sense for international calls. With some services, you can use the Internet to call someone who is using a standard phone. You can also keep your phone number when you move to another location. According to one Internet phone user who moved from Madison, Wisconsin, to California, "I was so happy about that. Nothing changed for my customers. For all they knew I was still in Madison." Cost is often a big factor for those using Internet phones—a call can be as low as 1 cent per minute for calls within the United States. Low rates are also available for calling outside the United States. In addition, voice mail and fax capabilities are available. Some cable TV companies, for example, are offering cable TV, phone service, and caller ID for under $40 a month.

Using *voice-over-IP (VoIP)* technology, as described in Chapter 6, network managers can route phone calls and fax transmissions over the same network they use for data—which means no more separate phone bills. See Figure 7.9. Gateways installed at both ends of the communications link convert voice to IP packets and back. With the advent of widespread, low-cost Internet telephone services, traditional long-distance providers are being pushed to either respond in kind or trim their own long-distance rates. VoIP (pronounced *voyp*) is growing rapidly.

Although the technology for VoIP has existed for decades, the widespread use of VoIP is just beginning. In 2005, VoIP customers totaled about 3 million people, three times the number of a few years ago.[25] Vonage Holding, which started in 2002 with about 50 employees, now has more than 1,500 employees with about 800,000 subscribers.[26] In 2005, VoIP services had about 4 million subscribers. This number is expected to surge to almost 20 million in three or four years. Even so, there are obstacles to using VOIP. Some service providers, for example, might have trouble connecting their customers to emergency 911 service.[27] In 2005, the FCC issued an order that Web-phone services must notify their customers of the potential problems with making emergency 911 calls. Today, many companies offer Web phone service using VOIP, including AT&T, Comcast, Vonage, Verizon, AOL, Packet8, Callserve, Net2Phone, and WebPhone.[28]

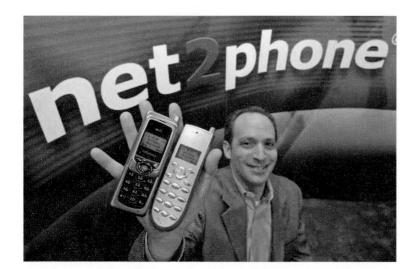

Figure 7.9

Voice-over-IP (VoIP) Technology

With VoIP, network managers can route phone calls and fax transmissions over the same network they use for data.

Internet videoconferencing, which supports both voice and visual communications, is another important Internet application. Microsoft's NetMeeting, a utility within Windows, is an inexpensive and easy way for people to meet and communicate on the Web. The Internet can also be used to broadcast group meetings, such as sales seminars, using presentation software and videoconferencing equipment. These Internet presentations are often called Webcasts or Webinars. The ideal video product will support multipoint conferencing, in which users appear simultaneously on multiple screens.

Content Streaming

Content streaming is a method for transferring multimedia files, radio broadcasts, and other content over the Internet so that the data stream of voice and pictures plays more or less continuously, without a break, or with very few breaks. It also enables users to browse large files in real time. For example, rather than wait for an entire 5-MB video clip to download before they can play it, users can begin viewing a streamed video as it is being received. Content streaming works best when the transmission of a file can keep up with the playback of the file.

content streaming
A method for transferring multimedia files over the Internet so that the data stream of voice and pictures plays more or less continuously without a break, or very few breaks; enables users to browse large files in real time.

Shopping on the Web

Shopping on the Web for books, clothes, cars, medications, and even medical advice can be convenient, easy, and cost effective. Amazon.com, for example, sells short stories by popular authors for 49 cents per story.[29] The service, called Amazon Shorts, has stories that vary in length from 2,000 to 10,000 words by authors such as Danielle Steel, Terry Brooks, and others. The company also sells traditional books and other consumer products. To add to their other conveniences, many Web sites offer free shipping and pickup for returned items that don't fit or otherwise meet a customer's needs.

Increasingly, people are using bots to help them search for information or shop on the Internet. A **bot** is a software tool that searches the Web for information, products, or prices. A bot, short for *robot*, can find the best prices or features from multiple Web sites. Hotbot.com is an example of an Internet bot.

bot
A software tool that searches the Web for information, products, or prices.

Web Auctions

A **Web auction** is a way to connect buyers and sellers. Web auction sites are a place where businesses are growing their markets or reaching customers for a low cost per transaction. Web auctions are transforming the customer-supplier relationship.

One of the most popular auction sites is eBay, which often has millions of auctions occurring at the same time. The eBay site is easy to use, and includes thousands of products and services in many categories. eBay remains a good way to get rid of things you don't need or find bargains on things you do need.[30] eBay drop-off stores allow people who are

Web auction
An Internet site that matches buyers and sellers.

inexperienced with Internet auctions or too busy to develop their own listings to sell items on the popular Web site.[31] One Denver, Colorado, resident found a 20-year-old Gregory backpack in his garage. He gave it to an eBay drop-off store hoping to get about $20 for it. He ended up getting almost $3,000 from a Japanese hiking collector. In addition to eBay, you can find a number of other auction sites on the Web. Traditional companies are even starting their own auction sites.

Although auction Web sites are excellent for matching buyers and sellers, they can present problems. Auction sites cannot always determine whether the people and companies listing products and services are legitimate. In addition, some Web sites have had illegal or questionable items offered. Many Web sites have an aggressive fraud investigation system to prevent and help prosecute fraudulent use of their sites. Even with these potential problems, the use of Web auction sites is expected to continue to grow rapidly.

eBay is a popular auction Web site.

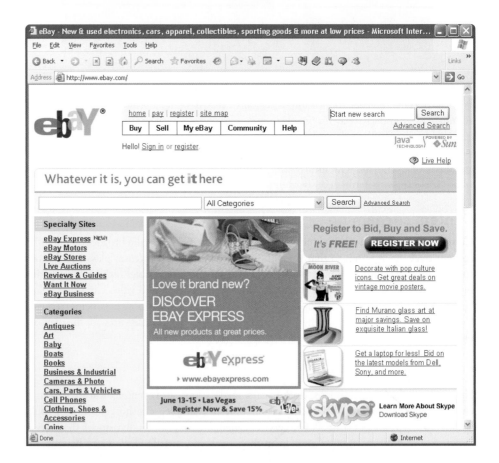

Music, Radio, Video, and TV on the Internet

Music, radio, and video are hot growth areas on the Internet. Audio and video programs can be played on the Internet, or files can be downloaded for later use. Using music players and music formats such as MP3, discussed in Chapter 3, you can download music from the Internet and listen to it anywhere using small, portable music players. In 2005, Yahoo! launched its Music Unlimited service for an introductory price of less than $100 per year.[32] Similar services can charge more than $100 per year. A subscriber to these music services can download an unlimited number of songs from the site, as long as they pay the annual subscription service fee. The Internet has also helped many musicians launch their careers without a lucrative music contract.[33] According to Geoff Byrd, who started his music career on the Internet, "In order to get a contract, you have to have been shot 10 times or have a thug reputation." Musicians like Byrd are increasingly using the Internet to get exposure. Byrd has four hits on Garage-Band.com, a listener-rated Internet site.

The Recording Industry Association of America (RIAA) won a legal battle against companies such as Napster and Grokster that once offered a way for people to freely share copies of music files over the Internet. In 2005, the U.S. Supreme Court ruled that Internet sites could be held liable if they helped users violate copyright laws.[34]

Radio broadcasts are now available on the Internet. Entire audio books can also be downloaded for later listening, using devices such as the Audible Mobile Player. This technology is similar to the popular books-on-tape or CD medium, except you don't need a cassette tape, CD, or player.

The Internet makes it possible to listen to radio broadcasts. Users can select a music genre and hear streaming audio.

Video and TV are also becoming available on the Internet. One way to put TV programming on the Internet is to use the Internet Protocol television (IPTV) protocol.[35] Akimbo, a start-up company, offers a subscription that delivers TV programming from a broadband Internet connection to your TV using IPTV. With the potential of offering an almost unlimited number of programs, IPTV can serve a vast array of programming on specialty areas, such as yoga, vegetarian food, unusual sporting events, and news from a city or region of a state. Google now has a service that allows people to download selected television shows, movies, and other video.[36] Some TV episodes will cost users $1.99 to download. According to Jennifer Feikin, director of Google Video, "It's the biggest marketplace of content that was previously off-line and is now brought online."[37] Devices such as Apple's video-enabled iPod can be used to view the video content.

A number of new, innovative devices let you record TV programs and view them at any time and place.[38] A California company called Sling Media (*www.slingmedia.com*), for example, offers a device that can broadcast any TV program coming into your home to a broadband Internet-connected PC. Once on the Internet, you can watch the TV program at any time and place that has broadband Internet service. The device, called a Slingbox, costs about $250 and doesn't require monthly service fees. See Figure 7.10.

Many content providers offer their programs over the Internet. Nickelodeon, for example, gives free access to some of its popular TV programs over the Internet in a program called TurboNick. The programs include *SpongeBob Square Pants* and *Jimmy Neutron*.[39]

Office on the Web

Having an Internet office with access to office files and information can be critical for people who travel frequently or work at home. An Internet office is a Web site that contains files, phone numbers, e-mail addresses, an appointment calendar, and more. Using a standard Web browser, you can access important files and information. An Internet office allows your desktop computer, phone books, appointment schedulers, and other important information to be with you wherever you are.

Many services and software products give you remote access to your files and programs over the Internet, including Avvenu, EasyReach, and BeInSync.[40] If you are traveling thousands of miles from your home or office, have access to a PC, and find that you need an important document or file; you can still get the documents or run the programs you want from your home or office computer.[41] Remote access requires a service fee, typically about $10 per month, and installed software on your home or office PC. Companies that offer this type of remote access include MyWebEx PC Pro, Citrix Online, Laplink, and others. Office on the Web works best with broadband Internet access. With slower dial-up access, it can take a very long time to download or upload large files.

Internet Sites in Three Dimensions

Some Web sites offer three-dimensional views of places and products. For example, a 3-D Internet auto showroom allows people to get different views of a car, simulating the experience of walking around in a real auto showroom. When looking at a 3-D real estate site on the Web, people can tour the property, go into different rooms, look at the kitchen appliances, and even take a virtual walk in the garden. Without a doubt, 3-D Internet sites will become common in the future.

Other Internet Services and Applications

Other Internet services are constantly emerging. A vast amount of information is available over the Internet from libraries. Many articles that served as the basis of the sidebars, cases, and examples used throughout this book were obtained from university libraries online. Movies can be ordered and even delivered over the Internet. The Internet can provide critical information during times of disaster or terrorism. During a medical emergency, critical medical information can be transmitted over the Internet. People wanting to consolidate their credit card debt or to obtain lower payments on their existing home mortgages have turned to sites such as Quicken Loan, E-Loan, and LendingTree for help.

The Internet can also be used to translate words, sentences, or complete documents from one language into another. For example, Babel Fish Translation (*www.world.altavista.com*) and Free Translation (*www.freetranslation.com*) can translate a block of text from one language into another. Some search engines have translation capabilities and allow you to search for Web sites published in certain languages or countries. Clicking Language Tools from the home page of Google, for example, provides these capabilities.

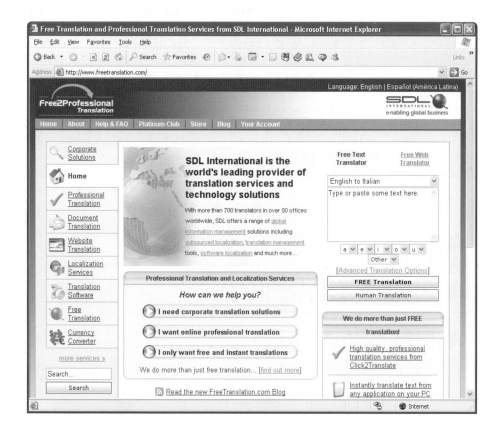

Web sites such as Free Translation (*www.freetranslation.com*) can translate a block of text from one language into another.

The Internet also facilitates distance learning, which has dramatically increased in the last several years. Many colleges and universities now allow students to take courses without visiting the campus. In fact, you might be taking this course online. Businesses are also taking advantage of distance learning through the Internet. Video cameras can be attached to computers and connected to the Internet. Internet cameras can be used to conduct job interviews, hold group meetings with people around the world, monitor young children at daycare centers, check rental properties and second homes from a distance, and more. People can use the Internet to connect with friends or others with similar interests. Internet sites, such as *www.zerodegrees.com, www.tribe.net,* and *www.spoke.net* are examples of social networking Internet sites. Today, manufacturers of sound systems are putting Internet and network capabilities into their devices.

INTRANETS AND EXTRANETS

An **intranet** is an internal corporate network built using Internet and World Wide Web standards and products. Employees of an organization use it to gain access to corporate information. After getting their feet wet with public Web sites that promote company products and services, corporations are seizing the Web as a swift way to streamline—even transform—their organizations. These private networks use the infrastructure and standards of the Internet and the World Wide Web. A big advantage of using an intranet is that many people are already familiar with Internet technology, so they need little training to make effective use of their corporate intranet.

An intranet is an inexpensive yet powerful alternative to other forms of internal communication, including conventional computer setups. One of an intranet's most obvious

intranet
An internal corporate network built using Internet and World Wide Web standards and products; used by employees to gain access to corporate information.

virtues is its ability to reduce the need for paper. Because Web browsers run on any type of computer, the same electronic information can be viewed by any employee. That means that all sorts of documents (such as internal phone books, procedure manuals, training manuals, and requisition forms) can be inexpensively converted to electronic form on the Web and be constantly updated. An intranet provides employees with an easy and intuitive approach to accessing information that was previously difficult to obtain. For example, it is an ideal solution to providing information to a mobile sales force that needs access to rapidly changing information.

An intranet is an internal corporate network used by employees to gain access to company information.

(Source: Trillium sample intranet, *www.trilliumgrp.com.*)

A rapidly growing number of companies offer limited network access to selected customers and suppliers. Such networks are referred to as extranets, which connect people who are external to the company. An **extranet** is a network that links selected resources of the intranet of a company with its customers, suppliers, or other business partners. Again, an extranet is built around Web technologies.

Security and performance concerns are different for an extranet than for a Web site or network-based intranet. User authentication and privacy are critical on an extranet so that information is protected. Obviously, performance must be good to provide quick response to customers and suppliers. Table 7.9 summarizes the differences between users of the Internet, intranets, and extranets.

extranet

A network based on Web technologies that links selected resources of a company's intranet with its customers, suppliers, or other business partners.

Table 7.9

Summary of Internet, Intranet, and Extranet Users

Type	Users	Need User ID and Password?
Internet	Anyone	No
Intranet	Employees and managers	Yes
Extranet	Business partners	Yes

Secure intranet and extranet access applications usually require the use of a virtual private network (VPN). A **virtual private network (VPN)** is a secure connection between two points on the Internet. VPNs transfer information by encapsulating traffic in IP packets and sending the packets over the Internet, a practice called **tunneling**. Most VPNs are built and run by ISPs. Companies that use a VPN from an ISP have essentially outsourced their networks to save money on wide area network equipment and personnel.

virtual private network (VPN)
A secure connection between two points on the Internet.

tunneling
The process by which VPNs transfer information by encapsulating traffic in IP packets over the Internet.

NET ISSUES

The topics raised in this chapter apply not only to the Internet and intranets but also to LANs, private WANS, and every type of network. Control, access, hardware, and security issues affect all networks, so you should be familiar with the following issues.

- **Management issues.** Although the Internet is a huge, global network, it is managed at the local level; no centralized governing body controls the Internet. Preventing attacks is always an important management issue. Increasingly, states are proposing legislation to help collect sales tax from Internet sales. An economist at the University of Tennessee estimates that states and local governments lose about $18 billion in online sales tax annually.[42]

- **Service and speed issues.** The growth in Internet traffic continues to be phenomenal. Traffic volume on company intranets is growing even faster than the Internet. Companies setting up an Internet or intranet Web site often underestimate the amount of computing power and communications capacity they need to serve all the "hits" (requests for pages) they get from Web cruisers.

- **Privacy, fraud, security, and unauthorized Internet sites.** As the use of the Internet grows, privacy, fraud, and security issues become even more important. People and companies are reluctant to embrace the Internet unless these issues are successfully addressed. Unauthorized and unwanted Internet sites are also problems some companies face. A competitor or an unhappy employee can create an Internet site with an address similar to a company's. When someone searches for information about the company, he might find an unauthorized site instead. While the business use of the Web has soared, online scams have put the brakes on some Internet commerce. According to a study of 5,000 Internet users, 42 percent of online shoppers and 28 percent of people who bank online have cut back on their Internet shopping and banking because of potential Internet scams and concerns about privacy and identity theft.[43] According to a Gartner analyst, "These attacks and disclosures are taking a steep toll on consumer confidence. The only place [consumers] can show their concern is in their online behavior." In a business setting, the Web can also be a distraction to doing productive work. Some believe that about 50 percent of American workers spend up to five hours per day on the Internet for personal, not business purposes.[44] Although many businesses block certain Web sites at work, others monitor Internet usage. Workers have been fired for inappropriate or personal use of the Internet while on the job.

Surfing the Web in Anonymity

The Web provides detailed information about businesses to consumers and other interested parties. Many businesses use the Web to keep track of their competition. For example, U.S. Airways might use competitor Southwest Airlines' Web site to learn about its rates and routing strategies. Some businesses have automated this form of competitive intelligence gathering with software that continuously monitors competitors' Web sites for the latest pricing information.

Just as the Web is used for competitive intelligence, so too can Internet technologies be used for counterintelligence. All devices connected to the Internet are assigned a unique IP address. This means that visitors to Web sites can be tracked through their IP address. Any particular IP address that is behaving suspiciously can be blocked from a Web site. For example, if Southwest Airlines noticed 1,000 queries coming from the same IP address over a short period, it might suspect automated intelligence gathering from its competitor and block that IP address.

An IP address can give away a significant amount of information about the person to whom it is assigned. For home users, who might be assigned varying IP addresses by an Internet service provider (ISP), the IP address can be traced back to the ISP, who will typically protect the identity of the user unless otherwise required by law. Even though a Web site owner cannot identify a visitor by IP address, it can build a general profile based on how the IP address interacts with the Web site and through the use of cookies.

For a business, which typically owns its permanent IP addresses, it is possible for a Web site owner to trace the IP address directly to the business that owns it to discover the identity of the visitor. This provides a certain defensive capability for businesses that want to find out who is visiting their Web site and protect their data from competitors—unless the competitor can assume a false identity.

Some software tools disguise the IP address of any user who visits Web sites. The Anonymizer company provides a service called Enterprise Chameleon that distributes dummy IP addresses to users within an enterprise to use when they want to visit a Web site anonymously. The service works by establishing an encrypted connection to Anonymizer's own secure virtual private network (VPN). Web activities are then routed through the Anonymizer network before being routed to the actual destination address. The Anonymizer network assigns temporary IP addresses that cannot be traced to the company that originated the request.

Tools such as those provided by Anonymizer have valuable uses outside of competitive intelligence. Law enforcement agencies and law firms also need covert intelligence gathering. For example, intellectual property attorney Adam Phillip uses products from Anonymizer in his daily work. "As a patent attorney who deals with software and electronics quite a bit, I have always been very aware that we all leave tracks wherever we go on the Internet," he said. Reverse lookups to discover who is accessing a Web site are easy to do, and "as an attorney, if I worked for a company engaged in a lawsuit, the first thing I would do is make sure opposing counsel could not get any information from our Web site."

Anonymizer has political uses as well. "We helped people in Kosovo who were reporting on the activities of the Serbs protect their anonymity on the Web," said Lance Cottrell, founder and president of Anonymizer. "We have helped a member of the British House of Lords who is investigating human rights abuses in the PRC." Anonymizer is being used by Chinese Internet users to access sites blocked by the Chinese government's censorship campaign. The United States government has become a large Anonymizer customer and is expected to exceed the consumer market in total sales in 2006.

People, businesses, organizations, and governments have many reasons for seeking anonymity when accessing information on the Web. As Web technologies become increasingly sophisticated, new technologies will emerge for invading and protecting privacy.

Discussion Questions

1. How is the Web used for competitive intelligence and counterintelligence?
2. How are IP addresses used to identify visitors, both individual users and businesses, to Web sites?

Critical Thinking Questions

1. For what Internet and Web activities might people, businesses, and governments want to remain anonymous?
2. How might a user's right to privacy be invaded by IP address tracking?

SOURCES: Latamore, Bert, "Attorney Finds Personal (and Corporate) Privacy in a Very Public World," *Computerworld*, accessed April 5, 2006, *www.computerworld.com*. Anonymizer Web site, accessed April 22, 2006, *www.anonymizer.com*. Enterprise Chameleon Web site, accessed April 22, 2006, *www.anonymizer.com/enterprise/solutions/enterprise_chameleon*.

SUMMARY

Principle

The Internet is like many other technologies—it provides a wide range of services, some of which are effective and practical for use today, others are still evolving, and still others will fade away from lack of use.

The Internet started with ARPANET, a project sponsored by the U.S. Department of Defense (DoD). Today, the Internet is the world's largest computer network. Actually, it is a collection of interconnected networks, all freely exchanging information. The Internet transmits data from one computer (called a host) to another. The set of conventions used to pass packets from one host to another is known as the Internet Protocol (IP). Many other protocols are used with IP. The best known is the Transmission Control Protocol (TCP). TCP is so widely used that many people refer to the Internet protocol as TCP/IP, the combination of TCP and IP used by most Internet applications. Each computer on the Internet has an assigned address to identify it from other hosts, called its Uniform Resource Locator (URL). There are several ways to connect to the Internet: via a LAN whose server is an Internet host, via SLIP or PPP, and via an online service that provides Internet access.

An Internet service provider is any company that provides access to the Internet. To use this type of connection, you must have an account with the service provider and software that allows a direct link via TCP/IP. Among the value-added services ISPs provide are electronic commerce, intranets, and extranets, Web-site hosting, Web transaction processing, network security and administration, and integration services.

Principle

Originally developed as a document-management system, the World Wide Web is a menu-based system that is easy to use for personal and business applications.

The Web is a collection of independently owned computers that work together as one. High-speed Internet circuits connect these computers, and cross-indexing software is employed to enable users to jump from one Web computer to another effortlessly. Because of its ability to handle multimedia objects and hypertext links between distributed objects, the Web is emerging as the most popular means of information access on the Internet today.

A Web site is like a magazine, with a cover page called a *home page* that has graphics, titles, and black and highlighted text. Web pages are loosely analogous to chapters in a book. Hypertext links are maintained using URLs, a standard way of coding the locations of the HTML (Hypertext Markup Language) documents. In addition to HTML, a number of newer Web standards are gaining in popularity, including Extensible Markup Language (XML), Extensible Hypertext Markup Language (XHTML), Cascading Style Sheets (CSS), and Dynamic HTML (DHTML).

A Web browser reads HTML and creates a unique, hypermedia-based menu on your computer screen that provides a graphical interface to the Web. The browser uses data about links to accomplish this; the data is stored on the Web server. The hypermedia menu links you to other Internet resources, not just text documents, graphics, and sound files. Internet Explorer and Netscape are examples of Web browsers. A search engine helps find information on the Internet. Popular search engines include Yahoo! and Google. A meta-search engine submits keywords to several individual search engines and returns the results. Push technology is used to send information automatically over the Internet rather than making users search for it with their browsers.

Principle

Because the Internet and the World Wide Web are becoming more universally used and accepted for business use, management, service and speed, privacy, and security issues must continually be addressed and resolved.

A rapidly growing number of companies are doing business on the Web and enabling shoppers to search for and buy products online. For many people, it is easier to shop on the Web than search through catalogs or trek to the shopping mall.

The steps to creating a Web page include getting space on a Web server; getting a Web browser program; writing your copy with a word processor, using an HTML editor, editing an existing HTML document, or using an ordinary text editor to create your page; opening the page using a browser, viewing the result, and correcting any tags; adding links to your home page to take viewers to another home page; adding pictures and sound; uploading the HTML file to your Web site; reviewing the Web page to make sure that all links are working correctly; and advertising your Web page. A number of products make developing and maintaining Web content easier, such as Microsoft .NET Framework. Google's free Web site analytics allows Web developers to monitor the number of Web page visits per day, product merchandising, and conversion—the number of Web site visits that are converted to buyers. After a Web site has been constructed, a content management system (CMS) can be used to keep the Web site running smoothly. Web services are also used to develop Web content. Web services consist of a collection of standards and tools that streamline and simplify communication among Web sites, which could revolutionize the way people develop and use the Web for business and personal purposes. In addition to XML, Web services use the Simple Object Access Protocol (SOAP), Web Services Description Language (WSDL), and Universal Discovery Description and Integration (UDDI).

Java is an object-oriented programming language from Sun Microsystems based on C++ that allows small programs—applets—to be embedded within an HTML document. When the user clicks on the appropriate part of the HTML page to retrieve it from a Web server, the applet is downloaded onto the client workstation, where it begins executing. The development of Java has had a major impact on the software industry. In addition to Java, there are a number of other programming languages and tools used to develop Web sites, including Hypertext Preprocessor (PHP).

Internet and Web applications include e-mail; career information and job searching; Telnet; FTP; Web logs (blogs); podcasts; Usenet and newsgroups; chat rooms; Internet phone; Internet video; content streaming; instant messaging; shopping on the Web; Web auctions; music, radio, and video; office on the Web; 3-D Internet sites; free software; and other applications. E-mail is used to send messages. The Internet offers a vast amount of career and job search information. Telnet enables you to log on to remote computers. FTP is used to transfer a file from another computer to your computer. Web logs (blogs) are Internet sites that people and organizations can create and use to write about their observations, experiences, and feelings on a wide range of topics. A podcast is an audio broadcast over the Internet. Usenet supports newsgroups, which are online discussion groups focused on a particular topic. Chat rooms let you talk to dozens of people at one time, who can be located all over the world. Internet phone service enables you to communicate with others around the world. Internet video enables people to conduct virtual meetings. Content streaming is a method of transferring multimedia files over the Internet so that the data stream of voice and pictures plays continuously. Instant messaging allows people to communicate in real time using the Internet. Shopping on the Web is popular for a host of items and

services. Web auctions are a way to match people looking for products and services with people selling these products and services. The Web can also be used to download and play music, listen to radio, and view video programs. With office on the Web, it is possible to store important files and information on the Internet. When telecommuting or traveling, these files and this information can be downloaded or sent to other people. Some Internet sites are three-dimensional, allowing people to manipulate the site to see different views of products and images on the Internet. A wealth of free software and services are available through the Internet. Some of the free information, however, might be misleading or even false. Some of the other Internet services include information about space exploration, fast information transfer, obtaining a home loan, and distance learning.

An intranet is an internal corporate network built using Internet and World Wide Web standards and products. It is used by the employees of an organization to gain access to corporate information. Computers using Web server software store and manage documents built on the Web's HTML format. With a Web browser on your PC, you can call up any Web document—no matter what kind of computer it is on. Because Web browsers run on any type of computer, the same electronic information can be viewed by any employee. That means that all sorts of documents can be converted to electronic form on the Web and constantly be updated.

An extranet is a network that links selected resources of the intranet of a company with its customers, suppliers, or other business partners. It is also built around Web technologies. Security and performance concerns are different for an extranet than for a Web site or network-based intranet. User authentication and privacy are critical on an extranet. Obviously, performance must be good to provide quick response to customers and suppliers.

Management issues and service and speed affect all networks. No centralized governing body controls the Internet. Also, because the amount of Internet traffic is so large, service bottlenecks often occur. Privacy, fraud, and security issues must continually be addressed and resolved.

CHAPTER 7: SELF-ASSESSMENT TEST

The Internet is like many other new technologies—it provides a wide range of services, some of which are effective and practical for use today, others are still evolving, and still others will fade away from lack of use.

1. The _____ was the ancestor of the Internet, and was developed by the U.S. Department of Defense.

2. On the Internet, what enables traffic to flow from one network to another?

   a. Internet Protocol
   b. ARPANET
   c. Uniform Resource Locator
   d. LAN server

3. Each computer on the Internet has an address called the *Transmission Control Protocol*. True or False?

4. Which of the following is *not* a way to gain access to the Internet?
   a. LAN server
   b. Usenet
   c. Point-to-Point Protocol, or PPP
   d. online service

5. _____ is a company that provides people and organizations with access to the Internet.

Originally developed as a document-management system, the World Wide Web is a menu-based system that is easy to use for personal and business applications.

6. A podcast is an online Web site that people can create and use to write about their observations, experiences, and feelings on a wide range of topics. True or False?

7. Netscape and Internet Explorer are examples of _____.
   a. Web browsers
   b. chat rooms
   c. search engines
   d. Web programming languages

8. _____ can be used to route phone calls over networks and the Internet.

9. What is the standard page description language for Web pages?
   a. Home Page Language
   b. Hypermedia Language
   c. Java
   d. Hypertext Markup Language (HTML)

Because the Internet and the World Wide Web are becoming more universally used and accepted for business use, management, service and speed, privacy, and security issues must continually be addressed and resolved.

10. A(n) _____ is a network based on Web technology that links customers, suppliers, and others to the company.

11. An intranet is an internal corporate network built using Internet and World Wide Web standards and products. True or False?

CHAPTER 7: SELF-ASSESSMENT TEST ANSWERS

(1) ARPANET (2) a (3) False (4) b (5) Internet service provider (ISP) (6) False (7) a (8) Voice over IP (VoIP) (9) d (10) extranet (11) True

REVIEW QUESTIONS

1. What is the Internet? Who uses it and why?
2. What is ARPANET?
3. What is TCP/IP? How does it work?
4. Explain the naming conventions used to identify Internet host computers.
5. What is a domain name?
6. Briefly describe three different ways to connect to the Internet. What are the advantages and disadvantages of each approach?
7. What is an Internet service provider? What services do they provide?
8. What are the advantages and disadvantages of e-mail?
9. What is a podcast?
10. For what are Telnet and FTP used?
11. What is an Internet chat room?
12. What is content streaming?
13. What is instant messaging?
14. What is the Web? Is it another network like the Internet or a service that runs on the Internet?
15. What is a URL and how is it used?
16. What is HTML and how is it used?
17. What is a Web browser? How is it different from a Web search engine?
18. What is an intranet? Provide three examples of the use of an intranet.
19. What is an extranet? How is it different from an intranet?
20. Describe at least three important Internet issues.

DISCUSSION QUESTIONS

1. Instant messaging is being widely used today. Describe how this technology could be used in a business setting. Are there any drawbacks or limitations to using instant messaging in a business setting?

2. Your company is about to develop a new Web site. Describe how you could use Web services for your site.
3. How can you protect yourself from unwanted e-mail?
4. Describe how a company could use a blog and podcasting.

5. Briefly describe how the Internet phone service operates. Discuss the potential impact that this service could have on traditional telephone services and carriers.
6. Identify three companies with which you are familiar that are using the Web to conduct business. Describe their use of the Web.
7. What is voice over IP (VoIP), and how could it be used in a business setting?
8. Outline a process to create a Web page. Briefly describe some of the tools that can be used to build a Web page.
9. One of the key issues associated with the development of a Web site is getting people to visit it. If you were developing a Web site, how would you inform others about it and make it interesting enough that they would return and also tell others about it?
10. Getting music, radio, and video programs from the Internet is easier than in the past, but some companies are still worried that people will illegally obtain copies of this programming without paying the artists and producers royalties. If you were an artist or producer, what would you do?
11. How could you use the Internet if you were a traveling salesperson?
12. Briefly summarize the differences in how the Internet, a company intranet, and an extranet are accessed and used.

PROBLEM-SOLVING EXERCISES

1. Do research on the Web to find several popular Web auction sites. After researching these sites, use a word processor to write a report on the advantages and potential problems of using a Web auction site to purchase a product or service. Also discuss the advantages and potential problems of selling a product or service on a Web auction site. How could you prevent scams on an auction Web site?
2. Develop a brief proposal for creating a business Web site. How could you use Web services to make creating and maintaining the Web site easier and less expensive? Develop a simple spreadsheet to analyze the income you need to cover your Web site and other business expenses.
3. You have been hired to research the use of a blog for a company. Develop a brief report on the advantages and disadvantages of using a blog to advertise corporate products and services. Using a graphics program, prepare a slide show to help you make a verbal presentation.

TEAM ACTIVITIES

1. With your teammates, identify a company that is making effective use of a company extranet. Find out all you can about its extranet. Try to speak with one or more of the customers or suppliers who use the extranet and ask what benefits it provides from their perspective.
2. Have each team member use a different search engine to find information about podcasting. Meet as a team and decide which search engine was the best for this task. Write a brief report to your instructor summarizing your findings.

WEB EXERCISES

1. This chapter covers a number of powerful Internet tools, including Internet phones, search engines, browsers, e-mail, newsgroups, Java, intranets, and much more. Pick one of these topics and get more information from the Internet. You might be asked to develop a report or send an e-mail message to your instructor about what you found.
2. The Internet can be a powerful source of information about various industries and organizations. Locate several industry or organization Web sites. Which Web site is the best designed? Which one provides the most amount of information?
3. Research some of the potential disadvantages of using the Internet, such as privacy, fraud, or unauthorized Web sites. Write a brief report on what you found.

CAREER EXERCISES

1. Use the Internet to explore possible starting salaries, benefits, and job descriptions for a career in developing or managing a Web site.

2. Describe how the Internet can be used on the job for two careers that interest you.

CASE STUDIES

Case One

Web Portal for State of Michigan Provides Services and Savings

Governments and their agencies often are pressured to improve services to the public with little funding for those improvements. State and federal agencies have found the Web to be an ideal tool to provide public services for a relatively small investment. Over time, services and information provided on government Web sites end up saving money by streamlining operations and automating processes that were traditionally handled by government workers.

The state of Michigan provides an excellent example of a successful implementation of a government Web site. It recently launched an initiative to save costs by offering more services over the Internet. Michigan's government agencies had more than 100 different Web sites operating independently and lacking a standard design. The state needed to unify its Web resources across all of its agencies, providing a consistent, user-friendly resource that would enable constituents to access government services online.

The plan for a new state of Michigan Web site included several criteria: It needed to be a Web-based portal that was user friendly, had a citizen-centric design, and would make all information, services, and transactions available within three mouse clicks. The portal needed to be developed using open standards, have uncompromising security, deliver availability around the clock, and be scalable for future expected growth. Lastly, the Web site portal needed to be up and running in 90 days.

The project had two primary challenges: First was the development and design of the portal, and the second was the on-going hosting of the portal. The State of Michigan decided that they preferred not to host the Web portal on government-owned servers. They decided to outsource the project and Web hosting to IBM Global Services.

The IBM team worked closely with the State of Michigan to refine the portal specifications and requirements and to develop portal application components that would allow citizens to interact with government information systems.

The resulting Web portal system included numerous components:

- Sun Web servers and application servers running Vignette Content Management Suite software
- IBM pSeries application servers running an Oracle Database Server cluster
- IBM xSeries servers to support the state's travel site
- IBM xSeries Web server running IBM WebSphere Application Server to support e-Michigan's e-ticket application
- Sun application servers running an IBM WebSphere Application Server cluster supporting IBM WebSphere Commerce and IBM WebSphere Payment Manager software
- Nokia Checkpoint Firewalls, Altheon load-balancing appliances, Cisco switches, and intrusion detection devices to support the security and high availability requirements of the portal infrastructure

In addition to working with the state to design and set up the state of Michigan Web portal system, IBM provides server hosting, middleware support, database support, security services, backup and restore services, network support services, firewall support services, server monitoring services, help desk support, host and network intrusion detection services, account management, and other services.

The result is a user-friendly Web portal developed for the state of Michigan within the 90-day deadline. The portal offers citizens and businesses a single point of access to Michigan's government resources. The solution meets the state's requirements for ease of use and intuitive navigation. The site has achieved near 100-percent availability, and the flexibility of the portal application enables the state to launch new sites without developing new code. Since the introduction of the new portal, the average daily volume of traffic to *www.michigan.gov* has increased more than tenfold. More citizens and businesses are finding information via the portal instead of by mail or phone calls. The state is saving money through higher employee productivity and reduced mailings and Michigan residents can find information and services that they need more easily.

Discussion Questions

1. How can a government Web site improve the government's service to citizens for less money than traditional methods?
2. Why do you think the State of Michigan decided to outsource the Web portal project rather than doing it themselves?

Critical Thinking Questions

1. Why do you think the State of Michigan found itself with 100+ different Web sites operating independently of one another and lacking a standard design?
2. How do the pressures on governments to improve service differ from those of a profit-driven business?

SOURCES: IBM Success Stories, "State of Michigan Enhances E-government with Help from IBM Global Services," February 14, 2006, *www.ibm.com*. State of Michigan Web site, accessed April 22, 2006, *www.michigan.gov*.

Case Two
FUJI Market Switches to Web-Based Ordering System

FUJI Market is the largest supermarket chain by sales and size in Shikoku, one of Japan's four main islands. FUJI has nearly 90 stores and 10,000 employees and is growing rapidly. It has set a goal of $3.4 billion USD in sales by 2008, a 31 percent increase from its 2004 revenue.

Many companies are making use of networking technologies, databases, and information systems to centralize business operations. Although centralization provides a unified view of business activities, some business functions are best left decentralized. As businesses grow, it takes more time to route all transactions through a central authority. When FUJI made this discovery, it decided to allow individual stores to make their own operating decisions and to replace old ways of doing business with more efficient processes.

"After we grew beyond 80 stores, it became almost impossible for us to continue making operational and marketing decisions centrally from our headquarters," says Toshihiko Yamanaka, manager of M2 Systems Co., Ltd., a subsidiary of FUJI Company Ltd. that provides IT services to the supermarket chain. For example, FUJI's order placement system for perishable goods was inefficient and wasteful. FUJI headquarters ordered perishable goods for all 80 stores. "We had to call or fax written requests to our suppliers by noon. Employees estimated how much of each item they thought their store would need, even though it was still too early to accurately predict what the inventory level would be later in the day," explains Yamanaka.

FUJI's inflexible mainframe-based system made it difficult for employees at the individual stores to order their own stock. To do so would require installing 1,000 terminals, each requiring proprietary client software and dedicated lines for data communications in the stores and at the vendor locations. FUJI searched for a more practical alternative.

FUJI restructured its store operations with new systems that use Internet and Java technology-based Web applications. Web-based inventory and merchandise applications were installed based on IBM WebSphere software. New application and database servers were installed to support the system, named the SEISEN ordering system. The SEISEN ordering system allows employees at each store to order produce, deli items, dairy, and meat via the Web. It provides employees with an electronic order book to confirm sales results to determine how much merchandise to order. Employees can review estimated customer visits, budgets, and losses resulting from discarded items to assist them in placing orders.

"Having a Web-based application helps employees interact quickly and easily with our suppliers and place orders later in the day when they can better estimate the stock they need," explains Yamanaka. The system has provided multiple benefits. By avoiding the installation of specialized terminals and software at each store, the company has saved $600,000 USD, and avoids the associated ongoing maintenance costs. With one vendor for all its hardware and software components, FUJI can keep system management simple, speed application development, and ensure that it can easily integrate new components as its infrastructure grows. Resource costs are expected to decline because the new Web-based applications are easier to use and do not require additional specialized training or knowledge.

With its new SEISEN ordering system in place, FUJI is now looking to restructure other store operations such as its supply chain management (SCM) system and, eventually, customer relationship management (CRM) system. It is planning to use Internet and Web technologies for all of its information systems to minimize operational waste, enable decentralized decision making at each store, and lower IT costs.

Discussion Questions

1. Why was the Internet and Web a better alternative to a private network for FUJI's order processing system?
2. Why do you think FUJI decided to use a system based on Java technology and Web applications?

Critical Thinking Questions

1. What concerns might FUJI executives have about sending private business transactions over a public network? What precautions can they apply to address those concerns?
2. How do you think the vendors that supply FUJI's perishable goods have been affected by this new system?

SOURCES: IBM Success Stories, "FUJI Company Rings Up Operational Savings with IBM Web-based Solutions," February 28, 2006, *www.ibm.com*. "Fuji Co., Ltd.: Company Snapshot," accessed May 2, 2006, *www.corporateinformation.com/snapshot.asp?Cusip=C39236560*.

Questions for Web Case

See the Web site for this book to read about the Whitmann Price Consulting case for this chapter. Following are questions concerning this Web case.

Whitmann Price Consulting: Telecommunications and Network Considerations

Discussion Questions

1. Why do you think it is easiest and most economical to develop custom designed applications using Web standards?
2. What additional security concerns arise when providing access to private information over a wireless public network?

Critical Thinking Questions

1. In what situations might Whitmann Price consider developing an extranet?
2. After the Advanced Mobile Communications and Information System is up and running, what would happen if the organization decided to switch to Palm Treo devices to replace the Blackberries? The Palm Treo has many of the same features as a Blackberry, including a Web browser. Do you think Whitmann Price custom designed applications would still work on the new devices? Why or why not?

NOTES

SOURCES for the opening vignette: Staff, "Firwood Paints Brightens Up Sales with Web Innovation," IBM Success Story, January 19, 2005, *www.ibm.com*. Firwood Paints Web site, accessed April 21, 2006, *www.firwood.co.uk*.

1   Miller, Matthew, "China's Internet Boom," *Rocky Mountain News*, August 15, 2005, p. 2B.
2   Staff, "China Tightens Web-Content Rules," *The Wall Street Journal*, September 26, 2005, B3.
3   Hafner, Katie, *Where Wizards Stay Up Late: The Origins of the Internet*, Touchstone, Rockefeller Center, New York, New York, 1996.
4   Internet2 Web site, *www.internet2.edu*, accessed November 29, 2005.
5   Miller, Michael, "Web Portals Make a Comeback," *PC Magazine*, October 4, 2005, p. 7.
6   Pike, Sarah, "The Expert's Guide to Google, Yahoo!, MSN, and AOL," *PC Magazine*, October 4, 2005, p 112.
7   Guth, Robert, et al., "Sky-High Search Wars," *The Wall Street Journal*, May 24, 2005, p. B1.
8   Staff, "Microsoft Looks for a Place Among Competitors with MSN Local Search," *Rocky Mountain News*, June 21, 2005, p. 6B.
9   Wikipedia Web site, *www.wikipedia.org*, accessed on January 12, 2006.
10 Staff, "Study: Web Site's Appearance Matters," *CNN Online*, August 11, 2005.
11 Hof, Robert, "Mix, Match, and Mutate," *Business Week*, July 25, 2005, p. 72.
12 Broida, Rick, "New Ways to Web," *PC Magazine*, September 20, 2005, p. 28.
13 Kinnersley, Hannah Kate, "Switching Your Email Account," *The Wall Street Journal*, September 22, 2005, p. D5.
14 Yuan, Li, "Text Messages Sent by Cell Phone Finally Catch on in U.S.," *The Wall Street Journal*, August 11, 2005, p. B1.
15 Searcey, Dionne, et al., "Amid Crisis, Phones Jammed, but Text Messages Worked," *The Wall Street Journal*, July 8, 2005, p. B1.
16 McCormack, Karyn; Stone, Amey, "Blogging for Dollars," *Business Week*, July 11, 2005, p. 83.
17 Dvorak, John, "Dvorak's Blogging Primer," *PC Magazine*, October 18, 2005, p. 60.

18 Staff, "Bloggers Learn the Price of Telling Too Much," *CNN Online*, July 11, 2005.
19 Technorati Web site, *www.technorati.com*, accessed January 12, 2006.
20 Baker, Stephen, "Looking for a Blog in a Haystack," *Business Week*, July 25, 2005, p. 38.
21 Tynan, Dan, "Singing the Blog Electric," *PC World*, August 2005, p. 120.
22 Mossberg, Walter, "Podcasting Is Still Not Quite Ready for the Masses," *The Wall Street Journal*, July 6, 2005, p. D5.
23 Lewis, Peter, "Invasion of the Podcast People," *Fortune*, July 25, 2005, p. 204.
24 McBride, Sarah; Wingfield, Nick, "As Podcasts Boom, Big Media Rushes to Stake a Claim," *The Wall Street Journal*, October 10, 2005, p. A1.
25 Armstrong, David, "VoIP and Vanish," *Forbes*, September 19, 2005, p. 58.
26 Young, Shawn, "Market for Internet Calling," *The Wall Street Journal*, August 26, 2005, p. A1.
27 Staff, "Web-Phone Firms Face Deadline Over 911 Calling," *The Wall Street Journal*, August 25, 2005, p. D2.
28 Bertolucci, Jeff, "Net Phones Grow Up," *PC World*, September 2005, p. 103.
29 Mangalindan, Mylene, "New on Amazon: Short Stories for 49 Cents," *The Wall Street Journal*, August 22, 2005, p. B1.
30 Kandra, Anne, "Use Auctions to Save Money on Tech Gear," *PC World*, September 2005, p. 37.
31 Brand, Rachel, "eBay Drop-Off Stores Popping Up," *Rocky Mountain News*, June 20, 2005, p. 1B.
32 Burrows, Peter, et al., "Online Music: Rewriting the Score," *Business Week*, May 30, 2005, p. 34.
33 Bourdreau, John, "Bands Break Out of Garage," *The News Herald*, September 30, 2005, p. 10.
34 Staff, "Grokster to Stop Distributing Service," *CNN Online*, November 7, 2005.
35 Wildstrom, Stephen, "From the Internet to Your TV," *Business Week*, June 13, 2005, p. 24.
36 Delaney, Kevin, "Google Moves Beyond the Web," *The Wall Street Journal*, January 8, 2006, p. A3.

37 Wong, May, "Google Lays Down Gauntlet," *Rocky Mountain News,* January 7, 2006, p. 1C.

38 Mossberg, Walter, "Device Lets You Watch Shows on a Home TV, TiVo from Elsewhere," *The Wall Street Journal,* June 30, 2005, A1.

39 Hansell, Saul, "With a Click of the Mouse, Watch the Latest of Must-See (PC) TV," *Rocky Mountain News,* August 8, 2005, p. 2B.

40 Staff, "New Ways to Access Your Files Anywhere," *PC World,* February 2006, p. 26.

41 Waring, Becky, "PC in a Browser," *PC World,* August 2005, p. 105.

42 Matthews, Robert Guy, "Some States Push to Collect Sales Tax from Internet Stores," *The Wall Street Journal,* September 30, 2005, p. B1

43 Richmond, Riva, "Internet Scams, Breaches Drive Buyers Off the Web, Survey Finds," *The Wall Street Journal,* June 23, 2005, p. B3.

44 Shropshire, Corilyn, "Net Effects," *Rocky Mountain News,* June 13, 2005, p. 2B.

Australian Film Studio Benefits from Broadband Networks...and Location

Sigi Goode
The Australian National University

It's been a hectic 24 hours at Fire Is a Liquid Films, a small Australian film-production studio. The firm's 18 staff members have just finished rendering scenes from three feature films, pitched ideas for two TV commercials, and delivered two milestone (progress) updates to production clients. Thanks to the power of broadband networking, the employees have accomplished most of this work without leaving their small studio. Of her company's tight deadlines and advanced technology, Toni Brasting, managing director, explains, "We play to our strengths in terms of location and technology.

"Our work ranges from cleaning up footage to improve color or contrast, up to full-blown CGI modeling and rendering. We work with TV commercials right up to feature film houses, and from one or two frames up to entire scenes. We can capitalize on our geographic location. With most feature films being shot in the U.S. or Europe, [clients] can send us their digital footage before they go to bed, and we can work on the scene during our daytime. By the time they wake up, the processed footage is ready and waiting for them. We rely heavily on our broadband network infrastructure to move digital data around. Depending on the quality and duration of the footage, the upload process can take anywhere from a few minutes for a brief scene to a few hours for an entire feature at DVD quality. Most of our clients prefer to deal with individual scenes in this way because it allows them to shuffle their work flow priorities around.

"Most films and commercials used to be shot on 35mm analogue film, which can take time to process. The trend is moving more and more towards digital video because it can be easily edited and manipulated. With 35mm, if you wanted to bid on work with a film studio, it helped if you were fairly close by so [that] you could have access to the film reels for digitizing purposes. With the advent of digital video and broadband networking, we can bid on projects anywhere in the world, from the *Matrix* to *28 Days Later*. The actors can get a better feel for each scene, and the director can place each scene in greater context: Their vision comes to life quicker.

"For editing, we mostly use Final Cut Pro on our Macintosh systems—the Macs are easy to use, and they look good, too (which impresses clients when they tour the studio). They run Mac OS X, which was originally based on BSD UNIX. Occasionally, we need to develop a new technique for a project, so it's useful to have access to software source code. Mac OS X gives us good support for our own open-source Linux tools.

"Another technique we use is laptop imaging. Clients like the feeling that we're taking care of them and that they are our focus. When one of our teams goes to meet a client for a pitch or a milestone update, they can pull a laptop off the shelf and build a customized hard-drive image right from their desktop with Norton Ghost or one of our home-brewed imaging applications. The image software 'bakes' a hard-drive image on the fly, which includes the OS, editing software, and relevant film footage. If necessary, we can also include correspondence and storyboards—we can have that client's entire relationship history right there in the meeting. We're also about to start phasing in tablet PCs with scribe pens so that we can work on potential storyboards during the meeting itself.

"The other important part of our processing platform is our render farm. It's basically a room full of identical Linux-based PCs, which work on a problem in parallel. We use commodity PC hardware and high-speed networking to allow each machine to work on the problem at once. It means we can get supercomputer performance at a fraction of the cost. We use our desktop machines to edit in draft resolution (which is of inferior quality but much faster to render or preview). When we're happy with the cut, we 'rush it' to the render farm to process the final piece of footage. We also offer render farm leasing services, so other film companies can send us their raw model files, use the render farm over the network, and then download their completed footage. We then invoice them for processing time—they never have to leave their offices."

Discussion Questions

1. How might open-source software benefit a small business? What problems might a small business encounter when using open-source software?
2. The studio profiled in this case processes tasks during a client firm's downtime. What other tasks could be accomplished more inexpensively or more efficiently by using providers in other geographic locations?
3. What other types of firms could also benefit from a render farm? What firms do you know of that actively make use of this technology?

Critical Thinking Questions

1. Are modern firms too reliant on electronic networking? If so, what can be done about this drawback? What paper-based systems could the firm use if its electronic systems fail?
2. How might wireless networking improve the operation of systems such as this firm's render farm?
   Note: All names have been changed at the request of the interviewee.

Virtual Learning Environment Provides Instruction Flexibility

Vida Bayley and Kathy Courtney
Coventry University, United Kingdom

One of the major components of Coventry University's teaching and learning strategy has been the introduction and use of a Virtual Learning Environment (VLE) using the WebCT (Web Course Tools) product platform. The Joint Information Systems Committee (JISC) in the UK has set out a definition of a Virtual Learning Environment and states that it refers to the components that support online interactions of various kinds taking place between learners and tutors.

A central aim of the university was to enhance face-to-face teaching and to offer greater flexibility to both teaching staff and students. It created a central support unit—CHED (Centre for Higher Education Development)—to introduce and develop new initiatives in teaching and learning across the university. In addition, a task force of academic staff was established to lead and facilitate these initiatives in collaboration with CHED. Through these developments and with the support of the university's Computing Services unit, WebCT was implemented as a core instrument in achieving the university's educational vision. The WebCT application allows course material and resources to be placed on the Web and provides staff and students with a set of tools that can be customized to course and module requirements. In brief, it also contains the following functions, which JISC considers to be essential features of a VLE:

- Controlled access to curriculum with mapping to elements that can be individually assessed and recorded
- Tracking mechanisms for student activities and performance
- Support of online learning through content development
- Communication and feedback mechanisms between various participants in the learning process
- Incorporation of links to other internal/external systems

In line with the aims and goals of the VLE strategy, Coventry Business School has been active for a number of years in the development of curricula and modules to create innovative learning communities and to engage students in the learning process. One of the most successful programs the Business School developed was the B.A. in Business Enterprise (B.A.B.E) degree. The Business Enterprise course was a learning program designed to develop graduates who would be able to proactively manage business-related tasks, solve business problems, and influence the business environment in which they work. A central element of the program was that it would enable students to lay the foundation for lifelong learning, in which students take responsibility for continual learning and personal development through a variety of media. The incorporation of WebCT into the B.A.B.E program marked a significant departure from established methods of teaching, learning, and assessment in the Business School.

Although the school had previously used a variety of computer-based activities within the learning environment, their use was limited to individual modules and in many cases even restricted. In accounting and information technology courses, software products such as EQL tutorials (EQL International Ltd. is a developer of e-learning and computer-based assessment solutions) were used as supplementary teaching material and for online assessment or self-test exercises. The WinEcon package (PC-based introductory economics software) had been developed with Bristol University through a consortium of eight UK economics departments and was used by the Economics subject group in course teaching. In other subject areas, the teaching staff provided students with Web URLs related to particular reading or assessment topics, which could be easily accessed via the Internet. In contrast, the Business Enterprise program was designed for delivery via the World Wide Web, so module delivery, program management, and staff/student communication were all built around WebCT.

This program was the first to be offered by the university via the World Wide Web. And it was significant because it was not designed solely for distance learning but rather to make use of the range of tools offered by the WebCT platform for teaching and learning both within a classroom setting and off campus via the Net. WebCT tools were used to structure and to provide Web-based access to course and module materials, and communications facilities were adapted to dovetail with existing classroom-based teaching and learning. Of particular importance was the consideration by the teaching team of how the Web software features could add value to the traditional teaching/learning modes and methods that they would be supporting or replacing. This evaluation in turn spurred individual lecturers to examine their teaching practice, stimulated ideas about the applicability of Web-based teaching, and encouraged staff to develop their knowledge of and skills related to the university's intranet and the Internet. After these opportunities were identified and placed within the context of the module and course objectives, lecturers then customized the WebCT features and integrated them within the overall teaching and learning framework. The benefit to students is flexibility: Course and module content is accessible from outside the university and outside of teaching hours. Course materials either can be placed within WebCT as HTML documents (so they can be edited directly online) or can be uploaded as individual applications, such as PowerPoint slides, Word documents, and Excel spreadsheets.

Effective communication and feedback are important aspects of a successful learning environment. One way of facilitating peer group communication and communication between students and staff was to use the discussion forum/bulletin board facilities provided by the software. The B.A.B.E program used this feature as the primary mode of delivering information to the student community. All course- and module-related announcements, such as class rescheduling, changes to assessment deadlines, and meeting arrangements, for example, were made via the main bulletin board. The flexibility that the software provides for creating discussion areas allowed lecturers to construct arenas for students to work individually on specific projects yet be involved as group members in discussions centered on common themes. Being asynchronous, these discussion groups allowed students to become involved in an ongoing exchange of ideas, reflect on their own and others' contributions, and thereby facilitate their own learning. In addition to the discussion forum, WebCT incorporates a chat facility that can enable a small group of students to participate in online, real-time conversations. An additional advantage of this type of synchronous communication is that both students and lecturers could develop group management and e-moderation skills. In addition, the Business School linked the Web e-mail facilities available in WebCT to the e-mail provided to all students by Coventry University, allowing messages to be transmitted within modules. This meant that messages were module specific and did not get lost in the larger system—especially in modules with large numbers of students.

The student presentation area of WebCT was used to allow students to create their own Web pages, which could be displayed and shared, thus providing them with an opportunity to gain new skills and an understanding of an important part of the processes involved in managing a business enterprise.

The Business School also uses STile, a portal that was developed to provide students with a range of resources that are available on the university's intranet. The portal contains an easy-to-use link to WebCT.

In the course of the development of the B.A.B.E program, it became apparent that for the success of new IT innovations, such as Web-based teaching, the project staff must identify and involve key stakeholders, particularly when a radical restructuring of teaching and learning methods is needed. These stakeholders include administrative personnel, teaching professionals, and also those who are at the very center of learning activity—the student body. The university benefited greatly by actively involving students and staff in the design, use, and evaluation of their educational environment. The experience of the B.A.B.E venture highlighted the fact that in order to transform the students' learning experience, all the stakeholders needed to understand how to construct a fit between the needs and expectations of the learner, the skills and pedagogical tools of the practitioners, and the tools available within the VLE. The rewards have been great. The establishment of a VLE, with WebCT as its center, has generated a new excitement in the approach to university teaching, stimulating reflection, reinvention, and transformation and creating a "journey of learning" for both students and teaching staff.

Although the B.A.B.E degree program has now been modified under the current course structure, many of its innovative features have been transferred and embedded into teaching and learning processes across the whole of the undergraduate business program.

In addition, WebCT has enabled the Business School to introduce and develop with business partners new, forward-looking projects, such as the development and delivery of the Post Graduate Certificate, Post Graduate Diploma, and MA in Communications Management via the Cable and Wireless Virtual Academy. It has also played a significant role in the establishment of learning communities and development of work-based learning projects with public-sector organizations such as the National Health Service, local authorities, and private-sector enterprises.

Discussion Questions

1. Why do you think that a discussion forum would motivate you to use the Internet to access learning resources?
2. What do you think the use of e-mail facilities would contribute to your learning?

Critical Thinking Questions

1. Discuss the advantages that a VLE with Internet access could offer students in developing their learning. Are there any disadvantages?
2. Identify and discuss the different types of contributions that various stakeholders could make to a student's learning process.

PART
· 3 ·

Business Information Systems

Chapter 8 Electronic and Mobile Commerce

Chapter 9 Enterprise Systems

Chapter 10 Information and Decision Support Systems

Chapter 11 Knowledge Management and Specialized Information Systems

CHAPTER · 8 ·

Electronic and Mobile Commerce

PRINCIPLES	LEARNING OBJECTIVES
▪ Electronic commerce and mobile commerce are evolving, providing new ways of conducting business that present both opportunities for improvement and potential problems.	▪ Describe the current status of various forms of e-commerce, including B2B, B2C, and C2C. ▪ Outline a multistage purchasing model that describes how e-commerce works. ▪ Define m-commerce and identify some of its unique challenges.
▪ E-commerce and m-commerce can be used in many innovative ways to improve the operations of an organization.	▪ Identify several e-commerce and m-commerce applications. ▪ Identify several advantages associated with the use of e-commerce and m-commerce.
▪ Although e-commerce and m-commerce offer many advantages, users must be aware of and protect themselves from many threats associated with this technology.	▪ Identify the major issues that represent significant threats to the continued growth of e-commerce and m-commerce.
▪ Organizations must define and execute a strategy to be successful in e-commerce.	▪ Outline the key components of a successful e-commerce strategy.
▪ E-commerce and m-commerce require the careful planning and integration of a number of technology infrastructure components.	▪ Identify the key components of technology infrastructure that must be in place for e-commerce and m-commerce to work. ▪ Discuss the key features of the electronic payment systems needed to support e-commerce.

Information Systems in the Global Economy
Nike, United States

Nike Empowers Customers with Online Customization

Shopping and selling online provide sellers and buyers with an intimate environment in which to do business. Through the use of dynamic Web sites that are designed to uniquely appeal to each visitor, electronic retailers, or e-tailers, can provide customers with exactly what they desire. Unlike brick-and-mortar retail stores with promotions designed to appeal to the lowest common denominator, e-tailers can provide visitors with an environment customized to their tastes and interests. For example, frequent Amazon.com shoppers are not surprised when they are greeted at Amazon's Web site by name and shown only items that are likely to appeal to them.

Customization is a powerful tool for online businesses selling to individual consumers (B2C) and to other businesses (B2B). Customization eliminates the frustration consumers feel when they must navigate through massive amounts of unrelated information to find items of interest. Consider, for example, searching the Sunday newspaper for advertisements and coupons of interest, or the percentage of commercials on network TV that are for products that actually appeal to you.

Realizing the power of customization and providing shoppers with exactly what they are looking for has inspired Nike and other companies to take advantage of the concept. Going beyond customizing the online shopping experience, Nike is empowering shoppers to custom-design the products themselves. Visitors to the recently upgraded Nikeid.com Web site can design their own footwear, sports apparel, sports bags, balls, and wristwatches. For example, you can select colors from an extensive palette for the nine elements of a Nike athletic shoe, including a base color and colors for the tip and heel, lining, tongue, and even the famous Nike swoosh. Furthermore, you can emboss the shoe with your name or slogan, and add a national symbol or flag to the heel tab.

Market analysts believe that the appeal of product customization will increase over time, especially to young shoppers. "It is really a democratic desire," said Sharon Lee, cofounder of a Los Angeles based consumer research and trend consulting company called Look-Look, Inc. "Every person wants to say this is much more me and I'm not part of this kind of mass culture." Customization also connects shoppers more closely with a brand and helps companies attract fickle but lucrative young shoppers by giving them the power to put their personal stamp on what they purchase.

The level of customization provided by Nike is only made possible through e-commerce technologies and supply chain management. E-commerce systems stream the order information from Nike's Web site to Nike's order processing system and from there to Nike's suppliers and manufacturers, all within seconds of the moment the order is placed. Nike's manufacturing process has been adjusted to accommodate individual products in addition to mass production of its stock items. The result is a wider and happier customer base that is getting exactly what it wants.

As you read this chapter, consider the following:

- What advantages does e-commerce and m-commerce offer buyers and sellers over traditional shopping venues?
- What are the limitations of e-commerce and m-commerce? What doesn't sell well online, and why are some shoppers uncomfortable shopping online?

Why Learn About Electronic and Mobile Commerce?

Electronic and mobile commerce have transformed many areas of our lives and careers. One fundamental change has been the manner in which companies interact with their suppliers, customers, government agencies, and other business partners. As a result, most organizations today have or are considering setting up business on the Internet. To be successful, all members of the organization need to participate in that effort. As a sales or marketing manager, you will be expected to help define your firm's e-commerce business model. Customer service employees can expect to participate in the development and operation of their firm's Web site. As a human resource or public relations manager, you will likely be asked to provide content for a Web site directed to potential employees and investors. Analysts in finance need to know how to measure the business impact of their firm's Web operations and how to compare that to competitors' efforts. Clearly, as an employee in today's organization, you must understand what the potential role of e-commerce is, how to capitalize on its many opportunities, and how to avoid its pitfalls. The emergence of m-commerce adds an exciting new dimension to these opportunities and challenges. This chapter begins by providing a brief overview of the dynamic world of e-commerce and defines its various components.

AN INTRODUCTION TO ELECTRONIC COMMERCE

electronic commerce
Conducting business activities (e.g., distribution, buying, selling, marketing, and servicing of products or services) electronically over computer networks such as the Internet, extranets, and corporate networks.

Electronic commerce is the conducting of business activities (e.g., distribution, buying, selling, marketing, and servicing of products or services) electronically over computer networks such as the Internet, extranets, and corporate networks. Business activities that are strong candidates for conversion to e-commerce are paper based, time-consuming, and inconvenient for customers. Thus, some of the first business processes that companies converted to an e-commerce model were those related to buying and selling. For example, after Cisco Systems, the maker of Internet routers and other telecommunications equipment, put its procurement operation online in 1998, the company reported that it halved cycle times and saved an additional $170 million in material and labor costs. Similarly, Charles Schwab & Co. slashed transaction costs by more than half by shifting brokerage transactions from traditional channels such as retail and phone centers to the Internet.

Business-to-Business (B2B) E-Commerce

business-to-business (B2B) e-commerce
A subset of e-commerce where all the participants are organizations.

Business-to-business (B2B) e-commerce is a subset of e-commerce where all the participants are organizations. B2B e-commerce is a useful tool for connecting business partners in a virtual supply chain to cut resupply times and reduce costs. Although the business-to-consumer market grabs more of the news headlines, the B2B market is considerably larger and is growing more rapidly. Over 80 percent of U.S. companies have experimented with some form of B2B online procurement.

Fairchild Semiconductor is a global supplier of power products used in the computing, communications, and automotive industries. Arrow Electronics is a global distributor of electronic components and computer products for Fairchild Semiconductor and more than 600 other suppliers plus 150,000 original equipment manufacturers, contract manufacturers, and value-added resellers. The two companies reduced the time involved with their order management and quoting process by over 50 percent by implementing a B2B e-commerce system that automates the exchange of information between the two firms. As a result, both companies benefited from a more efficient supply chain, improved productivity, and a faster response time for mutual electronic components customers. The system also frees up valuable time to spend with customers. The firms relied on communication and transaction standards developed by RosettaNet to implement this system. RosettaNet is a nonprofit, globally supported standards organization with over 400 members that promotes B2B commerce by developing standards for the global supply chain. Billions of dollars are transacted each year following RosettaNet standards.[1]

Business-to-Consumer (B2C) E-Commerce

Early **business-to-consumer (B2C) e-commerce** pioneers competed with the traditional "brick-and-mortar" retailers in an industry selling their products directly to consumers. For example, in 1995, upstart Amazon.com challenged well-established booksellers Waldenbooks and Barnes and Noble. Although Amazon did not become profitable until 2003, the firm has grown from selling only books from a U.S.-based Web site to selling a wide variety of products (including apparel, CDs, DVDs, home and garden supplies, and consumer electronic devices) from international Web sites in Canada, China, France, Germany, Japan, and the United Kingdom.[2] According to Forrester Research, consumers are expected to spend $329 billion per year online by 2010 with the percentage of U.S. households shopping online expected to grow from 39 percent in 2006 to 48 percent in 2010.[3] Although it is estimated that B2C e-commerce represents around 2.2 percent of all U.S. retail sales, the rate of growth of online purchases is three times faster than the growth in total retail sales.[4] One reason for the rapid growth is that shoppers find that many goods and services are cheaper when purchased via the Web—for example, stocks, books, newspapers, airline tickets, and hotel rooms. They can also find information about automobiles, cruises, loans, insurance, and home prices to cut better deals.

More than just a tool for placing orders, the Internet is an extremely useful way to compare prices, features, and value. Internet shoppers can, for example, unleash shopping bots or access sites such as eBay's Shopping.com, Google's Froogle, Shopzilla, PriceGrabber, Yahoo! Shopping, or Excite to browse the Internet and obtain lists of items, prices, and merchants. Yahoo! is adding what it calls "social commerce" to its Web site by creating a new section of Yahoo! where users can go to see only those products that have been reviewed and listed by other shoppers.[5] As mentioned in Chapter 7, bots are software programs that can follow a user's instructions; they can also be used for search and identification.

By using B2C e-commerce to sell directly to consumers, producers or providers of consumer products can eliminate the middlemen, or intermediaries, between them and the consumer. In many cases, this squeezes costs and inefficiencies out of the supply chain and can lead to higher profits and lower prices for consumers.[6] The elimination of intermediate organizations between the producer and the consumer is called *disintermediation*.

Dell is an example of a manufacturer that has successfully embraced this model to achieve a strong competitive advantage. People can specify their own unique computer online and Dell assembles the components and ships the computer directly to the consumer within five days. Dell does not inventory computers and does not sell through intermediate resellers or distributors. The savings are used to increase Dell profits and reduce consumer prices.

Consumer-to-Consumer (C2C) E-Commerce

Consumer-to-consumer (C2C) e-commerce is a subset of e-commerce that involves consumers selling directly to other consumers. eBay is an example of a C2C e-commerce site; customers buy and sell items directly to each other through the site. Founded in 1995, eBay has become one of the most popular Web sites in the world where 181 million users buy and sell items valued at more than $44 billion.[7]

Other popular online auction Web sites include Craigslist, uBid, Yahoo! Auctions, Onsale, WeBidz, and many others. The growth of C2C is responsible for reducing the use of the classified pages of a newspaper to advertise and sell personal items.

eGovernment

eGovernment is the use of information and communications technology to simplify the sharing of information, speed formerly paper-based processes, and improve the relationship between citizen and government. Government-to-consumer (G2C), government-to-business (G2B), and government-to-government (G2G) are all forms of eGovernment, each with different applications. For example, citizens can use G2C applications to submit their

business-to-consumer (B2C) e-commerce
A form of e-commerce in which customers deal directly with an organization and avoid intermediaries.

consumer-to-consumer (C2C) e-commerce
A subset of e-commerce that involves consumers selling directly to other consumers.

eGovernment
The use of information and communications technology to simplify the sharing of information, speed formerly paper-based processes, and improve the relationship between citizen and government.

state and federal tax returns online, renew auto licenses, apply for student loans, and make campaign contributions. G2B applications support the purchase of materials and services from private industry by government procurement offices, enable firms to bid on government contracts, and help businesses receive current government regulations related to their operations. G2G applications are designed to improve communications among the various levels of government. For example, the E-Vital initiative establishes common electronic processes for federal and state agencies to collect, process, analyze, verify, and share death record information. Geospatial One-Stop's Web portal, GeoData.gov, makes it easier, faster, and less expensive to find, share, and access geospatial information for all levels of government. Disaster Management run by the Department of Homeland Security provides federal, state, and local emergency responders online access to disaster management information, planning, and response tools.

The next section describes a basic model that supports products for purchase via e-commerce methods.

Multistage Model for E-Commerce

A successful e-commerce system must address the many stages that consumers experience in the sales life cycle. At the heart of any e-commerce system is the user's ability to search for and identify items for sale; select those items and negotiate prices, terms of payment, and delivery date; send an order to the vendor to purchase the items; pay for the product or service; obtain product delivery; and receive after-sales support. Figure 8.1 shows how e-commerce can support each of these stages. Product delivery can involve tangible goods delivered in a traditional form (e.g., clothing delivered via a package service) or goods and services delivered electronically (e.g., software downloaded over the Internet).

Figure 8.1

Multistage Model for
E-Commerce (B2B and B2C)

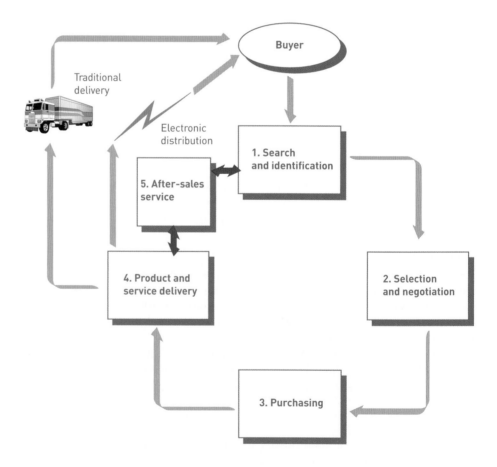

Search and Identification

An employee ordering parts for a storeroom at a manufacturing plant would follow the steps shown in Figure 8.1. Such a storeroom stocks a wide range of office supplies, spare parts, and maintenance supplies. The employee prepares a list of needed items—for example, fasteners, piping, and plastic tubing. Typically, for each item carried in the storeroom, a corporate buyer has already identified a preferred supplier based on the vendor's price competitiveness, level of service, quality of products, and speed of delivery. The employee then logs on to the Internet and goes to the Web site of the preferred supplier.

From the supplier's home page, the employee can access a product catalog and browse until finding the items that meet the storeroom's specifications. The employee fills out a request-for-quotation form by entering the item codes and quantities needed. When the employee completes the quotation form, the supplier's Web application prices the order with the most current prices and shows the additional cost for various forms of delivery— overnight, within two working days, or the next week. The employee might elect to visit other suppliers' Web home pages and repeat this process to search for additional items or obtain competing prices for the same items.

Selection and Negotiation

After the price quotations have been received from each supplier, the employee examines them and indicates, by clicking the request-for-quotation form, which items to order from a given supplier. The employee also specifies the desired delivery date. This data is used as input into the supplier's order processing TPS. In addition to price, an item's quality and the supplier's service and speed of delivery can be important in the selection and negotiation process.

B2B e-commerce systems need to support negotiation between buyer and the selected seller over the final price, delivery date, delivery costs, and any extra charges. However, this is not a fundamental requirement of most B2C systems, which offer their products for sale on a "take-it-or-leave-it basis."

Purchasing Products and Services Electronically

The employee completes the purchase order specifying the final agreed-to terms and prices by sending a completed electronic form to the supplier. Complications can arise in paying for the products. Typically, a corporate buyer who makes several purchases from the supplier each year has established credit with the supplier in advance, and all purchases are billed to a corporate account. But when individual consumers make their first, and perhaps only, purchase from the supplier, additional safeguards and measures are required. Part of the purchase transaction can involve the customer providing a credit card number. Another approach to paying for goods and services purchased over the Internet is using electronic money, which can be exchanged for hard cash, as discussed later in the chapter.

Product and Service Delivery

Electronic distribution can be used to download software, music, pictures, video, and written material through the Internet faster and for less expense than shipping the items via a package delivery service. Most products cannot be delivered over the Internet, so they are delivered in a variety of other ways: overnight carrier, regular mail service, truck, or rail. In some cases, the customer might elect to drive to the supplier and pick up the product.

Many manufacturers and retailers have outsourced the physical logistics of delivering merchandise to cybershoppers—the storing, packing, shipping, and tracking of products. To provide this service, DHL, Federal Express, United Parcel Service, and other delivery firms have developed software tools and interfaces that directly link customer ordering, manufacturing, and inventory systems with their own system of highly automated warehouses, call centers, and worldwide shipping networks. The goal is to make the transfer of all information and inventory—from the manufacturer to the delivery firm to the consumer—fast and simple.

For example, when a customer orders a printer at the Hewlett-Packard Web site, that order actually goes to FedEx, which stocks all the products that HP sells online at a dedicated e-distribution facility in Memphis, a major FedEx shipping hub. FedEx ships the order, which triggers an e-mail notification to the customer that the printer is on its way and an inventory notice to HP that the FedEx warehouse now has one less printer in stock (see Figure 8.2). For product returns, HP enters return information into its own system, which is linked to FedEx. This signals a FedEx courier to pick up the unwanted item at the customer's house or business. Customers don't need to fill out shipping labels or package the item. Instead, the FedEx courier uses information transmitted over the Internet to a computer in his truck to print a label from a portable printer attached to his belt. FedEx has control of the return, and HP can monitor its progress from start to finish.

Figure 8.2

Product and Information Flow for HP Printers Ordered over the Web

After-Sales Service

In addition to capturing the information to complete the order, comprehensive customer information is captured from the order and stored in the supplier's customer database. This information can include customer name, address, telephone numbers, contact person, credit history, and some order details. For example, if the customer later contacts the supplier to complain that not all items were received, that some arrived damaged, or even that product use instructions are not clear, all customer service representatives can retrieve the order information from the database via a computing/communications device. Companies are adding the capability to answer many after-sales questions to their Web sites, such as how to maintain a piece of equipment, how to effectively use the product, and how to receive repairs under warranty.

E-Commerce Challenges

A company must overcome many challenges to convert its business processes from the traditional form to e-commerce processes, especially for B2C e-commerce. This section summarizes a few key challenges.

The first major challenge is for the company to define an effective e-commerce model and strategy. Although companies can select from a number of different approaches, the most successful e-commerce models include three basic components: community, content, and commerce, as shown in Figure 8.3. Message boards and chat rooms are used to build a loyal *community* of people who are interested in and enthusiastic about the company and its products and services. Providing useful, accurate, and timely *content*—such as industry and economic news and stock quotes—is a sound approach to get people to return to your Web site time and again. *Commerce* involves consumers and businesses paying to purchase physical goods, information, or services that are posted or advertised online.

A major challenge for companies moving to business-to-consumer e-commerce is the need to change distribution systems and work processes to manage shipments of individual units directly to consumers. Traditional distribution systems send complete cases of a product to a store. The store opens the cases, takes the units out, and stacks them on a shelf. Then, consumers walk through the aisles and pick up what they need. In business-to-consumer

Content
Industry news
Economic news
Stock prices

Commerce
Consumers and
businesses buying
and selling

Community
Message boards
Chat rooms

Figure 8.3

Three Basic Components of a
Successful E-Commerce Model

e-commerce, companies need a distribution system that can manage **split-case distribution**, in which cases of goods are split open on the receiving dock and the individual items are stored on shelves or in bins in the warehouse. The distribution system must also be able to ship and track individual items. The demands of business-to-consumer e-commerce fulfillment are so great that many online vendors outsource the function to companies such as FedEx and UPS.

Another tough challenge for e-commerce is the integration of new Web-based order processing systems with traditional mainframe computer–based inventory control and production planning systems. It's one thing to allow sales reps to place orders over the Web, but another to let them find out what products are in (or soon to be in) inventory and available for sale. That means that front-end Web-enabled applications such as order taking need to be tightly integrated to traditional back-end applications such as inventory control and production planning, as shown in Figure 8.4.

split-case distribution
A distribution system that requires
cases of goods to be opened on the
receiving dock and the individual
items from the cases to be stored in
the manufacturer's warehouse.

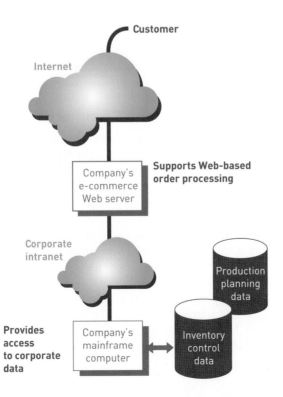

Customer

Internet

**Company's
e-commerce
Web server**

**Supports Web-based
order processing**

**Corporate
intranet**

**Production
planning
data**

**Provides
access
to corporate
data**

**Company's
mainframe
computer**

**Inventory
control
data**

Figure 8.4

Web-Based Order Processing
Must Be Linked to Traditional
Back-End Systems

AN INTRODUCTION TO MOBILE COMMERCE

As discussed briefly in Chapter 1, mobile commerce (m-commerce) relies on the use of wireless devices, such as personal digital assistants, cell phones, and smartphones, to place orders and conduct business. Handset manufacturers such as Ericsson, Motorola, Nokia, and Qualcomm are working with communications carriers such as AT&T, Cingular, Sprint/Nextel, and Verizon to develop such wireless devices, related technology, and services. In addition, content providers and mobile service providers are working together more closely than ever. Content providers recognize that customers want access to their content whenever and wherever they go, and mobile service providers seek out new forms of content to send over their networks.

Mobile Commerce in Perspective

According to the GSM Association, there are 1.8 billion mobile phone users in the world but only 12 to 14 percent have ever used the Web from their phones.[8] (GSM stands for global system for mobile communication, which is a European digital standard for mobile telephony.) The Internet Corporation for Assigned Names and Numbers (ICANN) created a .mobi domain in late 2005 to help attract mobile users to the Web. mTlD Top Level Domain Ltd of Dublin, Ireland, is responsible for administration of this domain and helping to ensure that the .mobi destinations work fast, efficiently, and effectively with user handsets.[9]

The market for m-commerce in North America is maturing much later than in Western Europe and Japan for several reasons. In North America, responsibility for network infrastructure is fragmented among many providers, consumer payments are usually done by credit card, and many Americans are unfamiliar with mobile data services. In most Western European countries, communicating via wireless devices is common, and consumers are much more willing to use m-commerce. Japanese consumers are generally enthusiastic about new technology and are much more likely to use mobile technologies for making purchases.

The market research firm In-Stat MDR predicts greatly expanding revenues for mobile gaming as users seek a diversion for those times they would otherwise be inactive, such as waiting for an appointment or while commuting. Capitalizing on this need, Electronic Arts is making some of its titles (e.g., Madden Football, Tiger Woods PGA, The Sims) available via direct download to Sprint and Verizon wireless network customers. The Yankee Group estimates that mobile phone gaming in the United States will increase from about $380 million in 2006 to more than $1 billion by 2009. Mobile gaming is growing even faster overseas where high-speed wireless networks are in broader use.[10] Read the Ethical and Societal Issues special interest box to learn more about m-commerce applications and some of the associated issues.

Technology Needed for Mobile Commerce

For m-commerce to work effectively, the interface between the wireless device and its user must improve to the point that it is nearly as easy to purchase an item on a wireless device as it is to purchase it on a PC. In addition, network speed must improve so that users do not become frustrated. Security is also a major concern, particularly in two areas: the security of the transmission itself and the trust that the transaction is being made with the intended party. Encryption can provide secure transmission. Digital certificates, discussed later in this chapter, can ensure that transactions are made between the intended parties.

The handheld devices used for m-commerce have several limitations that complicate their use. Their screens are small, perhaps no more than a few square inches, and might be able to display only a few lines of text. Their input capabilities are limited to a few buttons, so entering data can be tedious and error prone. They also have less processing power and less bandwidth than desktop computers, which are usually hardwired to a high-speed LAN. They also operate on limited-life batteries. For these reasons, it is currently impossible to directly access many Web sites with a handheld device. Web developers must rewrite Web applications so that users with handheld devices can access them.

The Power of the Cell Phone

The cell phone is increasingly being considered a remote control to the digital world. Banks, marketing agencies, those in the entertainment industry, and a host of other businesses are focused on creating useful m-commerce applications that will increase revenues. The ability to carry out transactions from any location any time using a cell phone is revolutionizing the manner in which we conduct our day-to-day activities. Here are four examples of how m-commerce is affecting societies worldwide.

First National Bank (FNB) of South Africa offers banking services via cell phone. Bank members can purchase prepaid cell phone airtime, view purchase records and account balances, transfer funds, and make payments using their cell phone menu system. The mobile banking platform is built on SMS (Short Messaging Service) and USSD (Unstructured Supplementary Service Data) combined with custom designed open-source software. The service supported 1.8 million transactions valued at $33 million in its first year. "We are proud to say that FNB Cellphone Banking is now profitable on a month-to-month basis and is outperforming all initial expectations," said Len Pienaar, CEO of FNB Mobile and Transact Solutions.

Meanwhile, on the opposite side of the globe in Japan, Aeon Co., which operates shopping malls, supermarkets, and general merchandise retailers, is installing terminals at cash registers that will allow customers to make payments using a cell phone. Japan's cell phone giant DoCoMo is introducing cell phones equipped with Near Field Communications (NFC) technology. Using this technology, DoCoMo customers can make credit card payments by bringing a compatible cell phone handset close to a terminal installed in Aeon Group shops. Aeon will support this new method of payment in 1,350 of its convenience stores across Japan by the end of 2006.

In the United States, online payment specialist PayPal, a unit of Internet auction giant eBay, has introduced PayPal Mobile. The wireless version of its service enables users to buy goods and exchange money using their phones. As with the online service offered by PayPal, PayPal Mobile allows cell phone users to transfer funds from bank accounts or credit cards to merchants or other user's bank accounts. Transactions are conducted by secure text message. "PayPal Mobile is an important indicator of the broader changes now occurring in the mobile content/payments space," says Ed Kountz, senior financial services analyst with Jupiter Research.

U.S. cell phone users can expect television-style advertising to arrive on their cell phone displays in the near future. The New York Times claims that "It is part of a broader push by marketers to create a new generation of 'up-close-and-personal' ads by delivering video, audio, banner displays and text clips over a device carried by most American adults." Marketers are especially keen on increasing numbers of cell phones equipped with a GPS. Using this technology, they will be able to send targeted advertisements to cell phone users based on location. For example, as you pull into a Wal-Mart parking lot, your cell phone might deliver a text message about items on sale at Wal-Mart.

Discussion Questions

1. Of the cell phone services discussed in this article, which would you find useful? Why?
2. What social concerns do you have when you consider these cell phone services?

Critical Thinking Questions

1. What activities that you perform on a computer could be transferred to a cell phone and which could not? Why?
2. What security concerns surround the use of cell phones for business transactions? Consider a cell phone that provides all the services outlined in this article. What might happen if that phone were lost? How might a user's identity and money be protected from theft?

SOURCES: Richtel, Matt, "Marketers Interested in Small Screen," *New York Times*, January 16, 2006, *www.nytimes.com*. Staff, "Cell Phone Banking Pays Off for South African Bank," *Computing South Africa*, February 24, 2006, *www.computerworld.com*. "Will That Be Cash, Credit, or a Cell Phone?" Reuters, April 15, 2006, *www.reuters.com*. Williams, Martyn, "Big Retailer Backs NTT DoCoMo's Phone Payments," *InfoWorld*, March 27, 2006, *www.infoworld.com*.

To address the limitations of wireless devices, the industry has undertaken a standardization effort for their Internet communications. The Wireless Application Protocol (WAP) is a standard set of specifications for Internet applications that run on handheld, wireless devices. It effectively serves as a Web browser for such devices. WAP is a key underlying technology of m-commerce that is supported by an entire industry association of over 200 vendors of wireless devices, services, and tools. In the future, devices and service systems based on WAP and its derivatives (including WAP 2.0 and Wireless Internet Protocol) will be able to interoperate. Japan's largest wireless network provider, DoCoMo, developed a competing standard called the i-mode system.

For equipment and service providers, the existence of competing standards makes it much more difficult to meet the needs of their customers. In many cases, the providers must develop their services or products based on one standard and forfeit the market for customers who elect to adopt the competing standard. Of course, multiple standards also create problems for customers who must make a decision on which set of services and equipment to adopt. Early adopters may find to their dismay they have chosen a standard that falls out of favor.

WAP uses the Wireless Markup Language (WML), which is designed for effectively displaying information on small devices. A user with a WAP-complaint device uses the built-in microbrowser to make a WML request. The request is forwarded to a special WAP gateway to fetch the information from the appropriate Internet server. If the information is already in WML format, it can be passed from the Internet server through the gateway directly to the user's device. If the information is in HTML format, the gateway translates the HTML content into WML so it can be displayed on the user's device.

ELECTRONIC AND MOBILE COMMERCE APPLICATIONS

E-commerce and m-commerce are being used in innovative and exciting ways. This section examines a few of the many B2B, B2C, C2C, and m-commerce applications in the retail and wholesale, manufacturing, marketing, investment and finance, and auction arenas.

Retail and Wholesale

electronic retailing (e-tailing)
The direct sale from business to consumer through electronic storefronts, typically designed around an electronic catalog and shopping cart model.

cybermall
A single Web site that offers many products and services at one Internet location.

E-commerce is being used extensively in retailing and wholesaling. **Electronic retailing**, sometimes called *e-tailing*, is the direct sale of products or services by businesses to consumers through electronic storefronts, which are typically designed around the familiar electronic catalog and shopping cart model. Companies such as Office Depot, Wal-Mart, and many others have used the same model to sell wholesale goods to employees of corporations. Tens of thousands of electronic retail Web sites sell everything from soup to nuts. In addition, cybermalls are another means to support retail shopping. A **cybermall** is a single Web site that offers many products and services at one Internet location—similar to a regular shopping mall. An Internet cybermall pulls multiple buyers and sellers into one virtual place, easily reachable through a Web browser. For example, Cybermall New Zealand (*www.cybermall.co.nz*) is a virtual shopping mall that offers retail shopping, travel, and infotainment products and services.

A key sector of wholesale e-commerce is spending on manufacturing, repair, and operations (MRO) goods and services—from simple office supplies to mission-critical equipment, such as the motors, pumps, compressors, and instruments that keep manufacturing facilities running smoothly. MRO purchases often approach 40 percent of a manufacturing company's total revenues, but the purchasing system can be haphazard, without automated controls. In addition to these external purchase costs, companies face significant internal costs resulting from outdated and cumbersome MRO management processes. For example, studies show that a high percentage of manufacturing downtime is often caused by not having the right part at the right time in the right place. The result is lost productivity and capacity. E-commerce software for plant operations provides powerful comparative searching

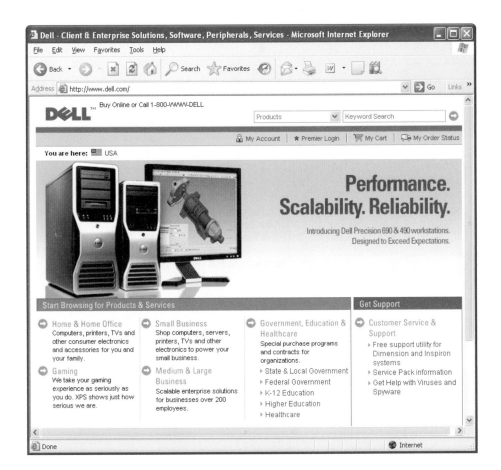

capabilities to enable managers to identify functionally equivalent items, helping them spot opportunities to combine purchases for cost savings. Comparing various suppliers, coupled with consolidating more spending with fewer suppliers, leads to decreased costs. In addition, automated workflows are typically based on industry best practices, which can streamline processes.

Manufacturing

One approach taken by many manufacturers to raise profitability and improve customer service is to move their supply chain operations onto the Internet. Here they can form an **electronic exchange** to join with competitors and suppliers alike, using computers and Web sites to buy and sell goods, trade market information, and run back-office operations, such as inventory control, as shown in Figure 8.5. With such an exchange, the business center is not a physical building but a network-based location where business interactions occur. This approach has greatly speeded up the movement of raw materials and finished products among all members of the business community, thus reducing the amount of inventory that must be maintained. It has also led to a much more competitive marketplace and lower prices. Private exchanges are owned and operated by a single company. The owner uses the exchange to trade exclusively with established business partners. Public exchanges are owned and operated by industry groups. They provide services and a common technology platform to their members and are open, usually for a fee, to any company that wants to use them.

National Foods (a wholly owned subsidiary of San Miguel Corporation) is one of Australia's largest food companies and sells milk, juice, and fresh dairy products. It employs 3,500 people at 20 processing plants around Australia and New Zealand.[11] National Foods recognized an opportunity to expand its sales and streamline its operations by joining Quadrem, an electronic exchange used heavily by firms in Australia's booming mining industry. The Quadrem exchange connects over 24,500 suppliers and 422 buyers transacting nearly $7 billion in orders per year. National Foods has seen administrative processing errors

electronic exchange
An electronic forum where manufacturers, suppliers, and competitors buy and sell goods, trade market information, and run back-office operations.

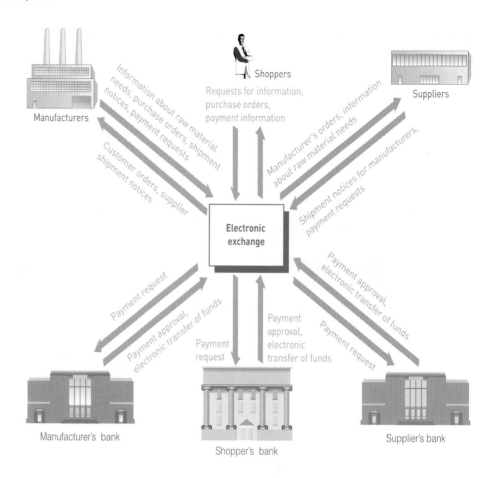

Figure 8.5

Model of an Electronic Exchange

plummet by 90 percent while order processing efficiency has increased by over 10 percent. The efficiency gain has come by reducing the amount of phoning and faxing and manual entering of order data. Instead, customers can place orders simply and directly by logging on to the Quadrem electronic exchange.[12]

Several strategic and competitive issues are associated with the use of exchanges. Many companies distrust their corporate rivals and fear they might lose trade secrets through participation in such exchanges. Suppliers worry that the online marketplaces and their auctions will drive down the prices of goods and favor buyers. Suppliers also can spend a great deal of money in the setup to participate in multiple exchanges. For example, more than a dozen new exchanges have appeared in the oil industry, and the printing industry is up to more than 20 online marketplaces. Until a clear winner emerges in particular industries, suppliers are more or less forced to sign on to several or all of them. Yet another issue is potential government scrutiny of exchange participants—when competitors get together to share information, it raises questions of collusion or antitrust behavior.

Many companies that already use the Internet for their private exchanges have no desire to share their expertise with competitors. At Wal-Mart, the world's number-one retail chain, executives turned down several invitations to join exchanges in the retail and consumer goods industries. Wal-Mart is pleased with its in-house exchange, Retail Link, which connects the company to 7,000 worldwide suppliers that sell everything from toothpaste to furniture.

Marketing

The nature of the Web allows firms to gather much more information about customer behavior and preferences than they could using other marketing approaches. Marketing organizations can measure many online activities as customers and potential customers gather information and make their purchase decisions. Analysis of this data is complicated because of the Web's interactivity and because each visitor voluntarily provides or refuses to provide personal data such as name, address, e-mail address, telephone number, and demographic data. Internet advertisers use the data they gather to identify specific portions of their markets and target them with tailored advertising messages. This practice, called **market segmentation**, divides the pool of potential customers into subgroups, which are usually defined in terms of demographic characteristics, such as age, gender, marital status, income level, and geographic location.

Technology-enabled relationship management is a new twist on establishing direct customer relationships made possible when firms promote and sell on the Web. **Technology-enabled relationship management** occurs when a firm obtains detailed information about a customer's behavior, preferences, needs, and buying patterns and uses that information to set prices, negotiate terms, tailor promotions, add product features, and otherwise customize its entire relationship with that customer.

DoubleClick is a leading global Internet advertising company that leverages technology and media expertise to help advertisers use the power of the Web to build relationships with customers. The DoubleClick Network is its flagship product, a collection of high-traffic and well-recognized sites on the Web, including MSN, Sports Illustrated, Continental Airlines, the Washington Post, CBS, and more than 1,500 others. This network of sites is coupled with DoubleClick's proprietary DART targeting technology, which allows advertisers to target their best prospects based on the most precise profiling criteria available. DoubleClick then places a company's ad in front of those best prospects. DART powers over 60 billion ads per month and is trusted by top advertising agencies. Comprehensive online reporting lets advertisers know how their campaign is performing and what type of users are seeing and clicking on their ads. This high-level targeting and real-time reporting provide speed and efficiency not available in any other medium. The system is also designed to track advertising transactions, such as impressions and clicks, to summarize these transactions in the form of reports and to compute DoubleClick Network member compensation.

Investment and Finance

The Internet has revolutionized the world of investment and finance. Perhaps the changes have been so great because this industry had so many built-in inefficiencies and so much opportunity for improvement.

market segmentation
The identification of specific markets to target them with advertising messages.

comScore Networks is a global information provider to large companies seeking information on consumer behavior to boost their marketing, sales, and trading strategies.

technology-enabled relationship management
Occurs when a firm obtains detailed information about a customer's behavior, preferences, needs, and buying patterns and uses that information to set prices, negotiate terms, tailor promotions, add product features, and otherwise customize its entire relationship with that customer.

The brokerage business adapted to the Internet faster than any other arm of finance. The allure of online trading that enables investors to do quick, thorough research and then buy shares in any company in a few seconds and at a fraction of the cost of a full-commission firm has brought many investors to the Web. Recently, online brokerage firms have consolidated, with Ameritrade acquiring TD Waterhouse and E-Trade acquiring Harrisdirect and the online brokerage services of JP Morgan. In spite of the wealth of information available online, the average consumer buys stocks based on a tip or a recommendation rather than as the result of research and analysis. It is the more sophisticated investor that really takes advantage of the data and tools available on the Internet.[13]

E-Trade is an online brokerage site that offers information, tools, and account-management services for investors.

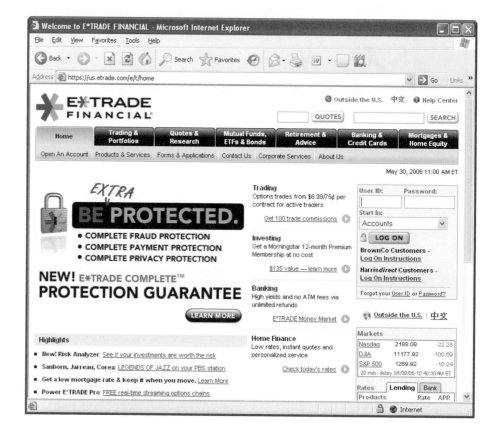

Online banking customers can check balances of their savings, checking, and loan accounts; transfer money among accounts; and pay their bills. These customers enjoy the convenience of not writing checks by hand, tracking their current balances, and reducing expenditures on envelopes and stamps. The Information Systems @ Work feature highlights consumer preferences in researching and negotiating mortgages online.

All of the nation's major banks and many of the smaller banks enable their customers to pay bills online; many support bill payment via a cell phone or other wireless device. Banks are eager to gain more customers who pay bills online because such customers tend to stay with the bank longer, have higher cash balances, and use more of the bank's products. To encourage the use of this service, many banks have eliminated all fees associated with online bill payment. Bank of America went so far as to allow customers to earn interest on their money until the payment is delivered to the recipient rather than having funds deducted from their accounts immediately. It also rolled out person-to-person online payment.[14]

The next advance in online bill paying is **electronic bill presentment**, which eliminates all paper, right down to the bill itself. With this process, the vendor posts an image of your statement on the Internet and alerts you by e-mail that your bill has arrived. You then direct your bank to pay it. Consolidated Edison, the Northeast energy conglomerate, achieved less than 1 percent of its customers converting to online bill payment and more than 15 percent converting with electronic bill presentment.[15]

electronic bill presentment
A method of billing whereby a vendor posts an image of your statement on the Internet and alerts you by e-mail that your bill has arrived.

INFORMATION SYSTEMS @ WORK

Acquiring a Mortgage Off and Online

This chapter reveals the benefits of using e-commerce to sell products and services online. However, a recent study by Forrester Research and Compete, Inc., indicates that for some types of purchases, consumers prefer closing the deal offline.

As increasing amounts of consumers shop online, and as e-commerce matures in its ability to meet consumers' needs, models for successful e-commerce services are coming into focus. Forrester-Compete's research on online mortgage services shed some light on how consumers are using e-commerce to secure mortgage loans. Researchers discovered that although consumers are flocking to online mortgage services to do research, most find interaction with a live professional reassuring when it comes to sealing the deal. Consumers researching mortgages online have increased 17 percent, according to the Forrester-Compete report, but of those using the Web for mortgage research, 61 percent end up applying for loans either by phone or by visiting a brick-and-mortar location.

Fifty-two percent of online researchers who chose an offline channel to apply for a mortgage told the Forrester-Compete researchers that they wanted human assistance to validate their decision. "That human touch, that wanting to validate that decision is the number one reason why consumers don't apply for these products online," stated Forrester Senior Analyst Brad Strothkamp. Complexity is also an issue. Frank Destra, senior vice president for national sales for ditech.com in Santa Ana, California, stated, "The mortgage process, in and of itself, is complex, and consumers want to talk to a live person to help guide them through that process."

Consumers find Web sites such as LendingTree.com, LowerMyBills.com, and NexTag.com useful for educating themselves on mortgage rates, policies, and procedures. More than 75 percent of the survey's 283 respondents said they were going online to look for rates and fees. These three Web sites are called aggregators; they don't lend money but provide a connection between the consumer and lenders. You might have heard LendingTree's winning slogan: "When banks compete, you win." "Aggregators do the best job of converting prospects into leads," Forrester's Strothkamp stated. "They are the most popular, with LendingTree leading the way. It has done a ton of advertising and really built up a brand awareness."

Consumers also make heavy use of online mortgage calculators. "There are at least a dozen types of tools and calculators for mortgages, but the highest usage comes from just a handful," the report said. A payment calculator, an affordability calculator, and a loan selection tool are must-haves for a mortgage site. Online mortgage aggregators and companies place links to these calculators on high-level pages to drive prospects deeper into the site.

Armed with this valuable research information, those in the mortgage industry can improve the design of their online resources to increase the level of comfort for consumers interested in "signing the bottom line" online. Those in the industry believe that increasing amounts of people will be doing just that. Frank Destra, from ditech.com, predicts that "...as lenders enhance their online functionality that need to contact a live person will diminish." Only time will tell.

Discussion Questions

1. Outline the concerns that consumers have about securing a mortgage online. How might these concerns vary from consumer to consumer? How does consumer experience play into the equation?
2. Why do you think most customers are attracted to mortgage aggregator Web sites rather than individual lender's Web sites?

Critical Thinking Questions

1. Do you think that information security affects consumer willingness to submit a loan application online? How?
2. If you were responsible for the design of LendingTree.com, what changes might you make to increase the number of prospects who apply online? If none, what do you think LendingTree has already done to accomplish this goal?

SOURCES: Mello, John P. Jr., "Study: Online Mortgage Seekers Seal Deals Offline," *E-Commerce Times*, May 16, 2006, *www.ecommercetimes.com*. Forrester Research Web site, accessed May 16, 2006, *www.forrester.com*. LendingTree.com Web site, accessed May 16, 2006, *www.lendingtree.com*.

Auctions

eBay has become synonymous with online auctions for both private sellers and small companies. However, hundreds of online auction sites cater to newcomers to online auctions and to unhappy eBay customers. The most frequent complaints are increases in fees and problems with unscrupulous buyers. As a result, eBay is constantly trying to expand and improve its services. eBay spent $2.6 billion to acquire Skype, a pioneer in Voice over IP (VoIP) services with the goal of improving communications between sellers and potential buyers for "high-involvement" items such as automobiles, business equipment, and high-end collectibles. eBay might also provide a pay-for-call service to provide a lead generation service for sellers based on the Skype technology.[16] eBay purchased the payment gateway system of security company VeriSign to provide a payment solution to tens of thousands of new small and midsized businesses. Under the deal, eBay will also receive 2 million VeriSign security tokens, physical devices like keychain-sized USB plug-ins that are used to create two-factor security where users must provide both a security password and the physical token.[17]

Priceline.com is a patented Internet bidding system that enables consumers to save money by naming their own price for goods and services.

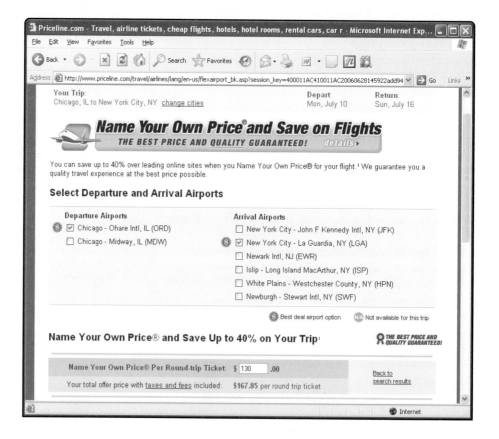

Anywhere, Anytime Applications of Mobile Commerce

Because m-commerce devices usually have a single user, they are ideal for accessing personal information and receiving targeted messages for a particular consumer. Through m-commerce, companies can reach individual consumers to establish one-to-one marketing relationships and communicate whenever it is convenient—in short, anytime and anywhere. Following are just a few examples of potential m-commerce applications:

- Banking customers can use their wireless handheld devices to access their accounts and pay their bills.
- Clients of brokerage firms can view stock prices and company research as well as conduct trades to fit their schedules.
- Information services such as financial news, sports information, and traffic updates can be delivered to people whenever they want.
- On-the-move retail consumers can place and pay for orders instantaneously.

- Telecommunications service users can view service changes, pay bills, and customize their services.
- Retailers and service providers can send potential customers advertising, promotions, or coupons to entice them to try their services as they move past their place of business.

The most successful m-commerce applications suit local conditions and people's habits and preferences. Most people do their research online and then buy offline at a local retailer. As a result, a growing market for local search engines is designed to answer the question, where do I buy product x at a brick-and-mortar retailer near me? Consumers provide their zip code and begin by asking a basic question—"What local stores carry a particular category of items" (e.g., flat-panel televisions). Consumers typically don't start searching knowing that they want a specific model Panasonic flat-panel TV. The local search engine then provides a list of local stores, including those with a Web site and those without, which sell this item. Yokel is a local-commerce search engine that plans to expand its in-depth local retailer coverage from Boston to additional cities.[18]

As with any new technology, m-commerce will only succeed if it provides users with real benefits. Companies involved in m-commerce must think through their strategies carefully and ensure that they provide services that truly meet customers' needs.

Advantages of Electronic and Mobile Commerce

According to the Council of Supply Chain Management Professionals, "Supply Chain Management encompasses the planning and management of all activities involved in sourcing and procurement, conversion, and all logistics management activities. Importantly, it also includes coordination and collaboration with channel partners, which can be suppliers, intermediaries, third-party service providers, and customers."[19] Conversion to an e-commerce driven supply chain provides businesses with an opportunity to achieve operational excellence by enabling consumers and companies to gain a global reach to worldwide markets, reduce the cost of doing business, speed the flow of goods and information, increase the accuracy of order processing and order fulfillment, and improve the level of customer service.

Global Reach

As discussed in Chapter 1, e-commerce offers enormous opportunities. It allows manufacturers to buy at a low cost worldwide, and it offers enterprises the chance to sell to a global market right from the very start-up of their business. Moreover, e-commerce offers great promise for developing countries, helping them to enter the prosperous global marketplace, and hence helping reduce the gap between rich and poor countries. (Recall that Chapter 1 introduced globalization and its challenges.)

Reduce Costs

By eliminating or reducing time-consuming and labor-intensive steps throughout the order and delivery process, more sales can be completed in the same period and with increased accuracy. With increased speed and accuracy of customer order information, companies can reduce the need for inventory—from raw materials, to safety stocks, to finished goods—at all the intermediate manufacturing, storage, and transportation points.

Speed the Flow of Goods and Information

When organizations are connected via e-commerce, the flow of information is accelerated because of the already established electronic connections and communications processes. As a result, information can flow easily, directly, and rapidly from buyer to seller.

Increased Accuracy

By enabling buyers to enter their own product specifications and order information directly, human data-entry error on the part of the supplier is eliminated.

Improve Customer Service

Increased and more detailed information about delivery dates and current status can increase customer loyalty. In addition, the ability to consistently meet customers' desired delivery dates with high-quality goods and services eliminates any incentive for customers to seek other sources of supply.

THREATS TO ELECTRONIC AND MOBILE COMMERCE

Businesses must deal with a host of issues to ensure that e-commerce and m-commerce transactions are safe and consumers are protected. The following sections summarize a number of threats to the continued growth and success of e-commerce and m-commerce and present practical ideas on how to minimize their impact.

Security

Many organizations that accept credit cards to pay for items purchased via e-commerce have adopted the Payment Card Industry security standard. This standard spells out measures and security procedures to safeguard the card issuer, the cardholder, and the merchant. Some of the measures include installing and maintaining a firewall configuration to control access to computers and data; never using software/hardware vendor-supplier defaults for system passwords; and requiring merchants to protect stored data, encrypt transmission of cardholder information across public networks, use and regularly update antivirus software, and restrict access to sensitive data on a need-to-know basis.[20]

Various measures are being implemented to increase the security associated with the use of credit cards at the time of purchase. Address Verification System is a check built into the payment authorization request that compares the address on file with the card issuer to the billing address provided by the cardholder. The Card Verification Number technique is a check of the additional digits printed on the back of the card. Visa has Advanced Authorization, a Visa-patented process that provides an instantaneous rating of that transaction's potential for fraud to the financial institution that issued the card. The card issuer can then send an immediate response to the merchant whether to accept or decline the transaction. The technology is now being applied to every Visa credit and check card purchase today. Visa estimates that this technique will prevent about $164 million in fraudulent credit card charges over the next five years.[21]

The Federal Financial Institutions Examination Council has developed a new set of guidelines called "Authentication in an Internet Banking Environment," recommending two-factor authorization. This approach adds another identity check along with the password system. A number of multifactor authentication schemes can be used such as hardware tokens that plug into a USB port on the computer and generate a password that matches the ones used by a bank's security system, biometrics, or one-time passwords.[22] Currently, the use of biometric technology to secure online transactions is rare for both cost and privacy reasons. It can be expensive to outfit every merchant with a biometric scanner, and it is difficult to convince consumers to supply something as personal and distinguishing as a fingerprint. In spite of this, a growing number of financial service firms from large (e.g., Citibank) to small (e.g., Perdue Employees Federal Credit Union) are considering biometric systems.

Theft of Intellectual Property

intellectual property
Includes works of the mind such as books, films, music, processes, and software, which are distinct somehow and are owned and/or created by a single entity.

Intellectual property includes works of the mind such as books, films, music, processes, and software, which are distinct somehow and are owned or created by a single entity. The owner of the intellectual property is entitled to certain rights in relation to the subject matter of the intellectual property. Thus, copyright law protects authored works such as books, film, images, music, and software from unauthorized copying. Patents can also protect software as well as business processes, formulae, compounds, and inventions. Information that is of great importance to a firm and for which strong measures are taken to protect it are trade secrets.

They too are protected under various laws. Although concerns about intellectual property and Digital Rights Management (discussed next) apply to creative works distributed traditionally through brick-and-mortar retailers and libraries, these issues are more urgent for e-commerce because computers and the Internet make it easy to access, copy, and distribute digital content.

Digital Rights Management (DRM) refers to the use of any of several technologies to enforce policies for controlling access to digital media such as movies, music, and software. Many digital content publishers state that DRM technologies are needed to prevent revenue loss due to illegal duplication of their copyrighted works. For example, the Motion Picture Association of America (MPAA) estimates that the film industry lost approximately $3.5 billion in movie piracy in 2004.[23] On the other hand, many digital content users argue that DRM and associated technologies lead to a loss of user rights. For example, users can purchase a music track online for under a dollar through Apple's iTunes Music Store. They can then burn that song to a CD and transfer it to an iPod. However, the purchased music files are encoded in the AAC format supported by iPods and protected by FairPlay, a DRM technology developed by Apple. To the consternation of music lovers, many music devices are not compatible with the AAC format and cannot play iTunes' protected files.

Fraud

The first wave of Internet crime consisted mostly of online versions of offline hoaxes, the usual get-rich-quick schemes. For example, many people received pleas from desperate Nigerians trying to enlist their help in transferring funds out of their country. More recently, however, fraud artists have begun to exploit the Internet to execute more sophisticated ploys, using fake Web sites and spam.

Phishing entails sending bogus messages purportedly from a legitimate institution to pry personal information from customers by convincing them to go to a "spoofed" Web site. The spoofed Web site appears to be a legitimate site but actually collects personal information from unsuspecting victims. Phishers have simulated the messages and Web sites of banks such as Citibank, Lloyds TSB Bank, NatWest, Visa, Halifax Bank, and Westpac Bank. As with all scams, phishing requires a degree of gullibility by the victims. Criminals hacked into a Chinese bank's server and used it to host phishing sites in an attempt to steal personal data from customers of eBay and Chase Bank. One of the phishing sites offered Chase Bank customers $20.00 if they would complete a survey asking for the user's ID and password so the money could be deposited to their account.[24] The U.S. Internal Revenue Service set up an e-mail address (*phishing@irs.gov*) that taxpayers can use to send copies of suspicious e-mails that claim to be from the IRS. The IRS expects such a large volume of these e-mails that it has informed taxpayers that it will be unable to acknowledge receipt of the e-mails or reply to senders of such e-mails.[25]

Click fraud can arise in a pay-per-click online advertising environment when additional clicks are generated beyond those that come from actual, legitimate users. In pay-per-click advertising, the advertiser pays when a user clicks its ad to visit its Web site. The additional clicks may be generated by an illegitimate user, automated script, or some other means. These bogus clicks generate revenue for the advertising network such as Google or Yahoo!. Although no one knows how much click fraud affects Web advertisers, some have estimated the impact to be over $1 billion per year.[26] Google set aside $90 million in the form of account credits for future advertising to repay advertisers who can prove they fell victim to click fraud.[27]

Online auction fraud represents a major source of complaints both in the United States and abroad. In 2005, the FTC received 80,450 complaints related to Internet auctions, or about 12 percent of the total number of complaints.[28] The majority of that fraud comes from so-called person-to-person auctions, which account for roughly half the auction sites. On these sites, it is up to the buyer and seller to resolve details of payment and delivery; the auction sites offer no guarantees. Sticking with auction sites like eBay (*www.ebay.com*) that ensure the delivery and quality of all the items up for bidding can help buyers avoid trouble. Another problem is fake goods that find their way to virtually all of the online auctions. Tiffany & Company has sued eBay for providing a venue for the sale of counterfeit Tiffany jewelry. eBay argues that it is not liable and that it is the trademark holder who must monitor

Digital Rights Management (DRM)
Refers to the use of any of several technologies to enforce policies for controlling access to digital media such as movies, music, and software.

phishing
A practice that entails sending bogus messages purportedly from a legitimate institution to pry personal information from customers by convincing them to go to a "spoofed" Web site.

click fraud
A problem arising in a pay-per-click online advertising environment where additional clicks are generated beyond those that come from actual, legitimate users.

listings for infringement. eBay states it is not possible for them to distinguish between a legitimate designer item and a fake. If eBay wins the dispute, the responsibility will fall on brand owners to police Web sites for counterfeit offerings. If eBay loses, hundreds of online auctions will be under pressure to monitor the millions of auctions they host for fake goods and take preventative action.[29]

Invasion of Consumer Privacy

clickstream data
The data gathered based on the Web sites you visit and the items you click.

Online consumers are more at risk today than ever before. One of the primary factors causing higher risk is *online profiling*—the practice of Web advertisers recording online behavior for the purpose of producing targeted advertising. **Clickstream data** is the data gathered based on the Web sites you visit and the items you click. From the marketers' perspective, the use of online profiling allows businesses to market to customers electronically one to one. The benefit to customers is better, more-effective service; the benefit to providers is the increased business that comes from building relationships and encouraging customers to return for subsequent purchases. From the consumers' perspective, online profiling squeezes out anonymity, which remains crucial to protecting privacy on the Internet. Consumers also fear that the personal information gathered will be shared with others without their knowledge.

The Apple iTunes service called MiniStore makes recommendations on songs currently in the software's active playlist. Some consumer advocates became upset when they learned the feature was included by default in iTunes and that it transmitted data associated with the iTune customer back to Apple. iTunes has expanded what it offers to include books and episodes of TV shows. Much could be deduced about an individual's tastes and shopping habits by accessing this information. To avoid upsetting customers, Apple modified the service so that it asks customers for permission to activate this feature.[30]

Lack of Internet Access

The *digital divide* is a term that describes the difference between people who do and people who don't have the access or the capability to use high-quality, modern information and communications technology such as computers, the Internet, telephone, and television to improve their standard of living. For example, it is estimated that of the roughly 1 billion Internet users worldwide, only 20 million (2 percent) are in the less-developed nations. The lack of universal Internet access makes it impossible to conduct e-commerce with many of the world's people. The digital divide exists not only between more and less developed countries but within countries between economic classes, the educated and uneducated, and those who live in cities and those who live in rural areas. Obviously, those who lack Internet access form a barrier to further e-commerce expansion.

Return on Investment

Often the investment required for a large firm to establish and operate a B2B or B2C Web site can be in the millions of dollars. For example, Starwood Hotels and Resorts Worldwide plans to move its 700 hotels to a new Web-based reservation system at an estimated cost of between $10 and $60 million with annual cost savings of $15 million.[31] Using the low cost estimate, the payback period is $10 million/$15 million or .67 years while the high cost estimate yields a payback period of four years, not nearly as economically attractive. The example illustrates a common problem with determining return on investment—it is difficult to forecast project costs and benefits.

Legal Jurisdiction

Companies engaging in e-commerce must be careful that their sales do not violate the rules of various county, state, or country legal jurisdictions. For example, New York and six other states forbid the possession of stun guns and similar devices. The New York state attorney general received a tip that such guns were being sold to New Yorkers though eBay. A subsequent investigation led to the arrest of 16 sellers allegedly responsible for the sale of more than 1,100 stun guns and Tasers.[32] Other examples of illegal sales are sales to those who would not be able to obtain cigarettes or wine because of their age.[33]

Taxation

United States businesses and consumers must be aware of taxation issues when conducting e-commerce. Based on U.S. Supreme Court rulings (Quill Corp. vs. North Dakota), Internet-based merchants must apply sales tax only when buyers live in a state where the company has physical facilities, or "nexus." Most businesses want to avoid the complexity of dealing with the nonstandard rules of the more than 7,500 taxing districts nationwide. To avoid this complexity of paying sales taxes, businesses set up their Internet sales operations as legally separate companies with no physical presence outside of where its computers and warehouses are located. This leaves it up to the consumers to voluntarily remit the sales taxes; but because it is almost impossible to enforce this practice, few people pay them. Thus, despite having a legal basis to do so, the states find it very difficult to collect sales taxes on Internet purchases. The amount of lost sales tax was estimated to be over $18 billion in 2005.[34]

STRATEGIES FOR SUCCESSFUL E-COMMERCE

With all the constraints to e-commerce just covered, a company must develop an effective Web site—one that is easy to use and accomplishes the goals of the company, yet is safe, secure, and affordable to set up and maintain. The next sections examine several issues for a successful e-commerce site.

Defining the Web Site Functions

When building a Web site, you should first decide which tasks the site must accomplish. Most people agree that an effective Web site is one that creates an attractive presence and that meets the needs of its visitors, including the following:

- Obtaining general information about the organization
- Obtaining financial information for making an investment decision
- Learning the organization's position on social issues
- Learning about the products or services that the organization sells
- Buying the products or services that the company offers
- Checking the status of an order
- Getting advice or help on effective use of the products
- Registering a complaint about the organization's products
- Registering a complaint concerning the organization's position on social issues
- Providing a product testimonial or idea for a product improvement or new product
- Obtaining information about warranties or service and repair policies for products
- Obtaining contact information for a person or department in the organization

After a company determines which objectives its site should accomplish, it can proceed to the details of actually developing a site.

As the number of e-commerce shoppers increases and they become more comfortable—and more selective—making online purchases, you might need to redefine your site's basic business model to capture new business opportunities. For example, consider the major travel sites such as Expedia, Travelocity, Cheap Tickets, Orbitz, and Priceline. These sites used to specialize in one area of travel—inexpensive airline tickets. Now they offer a full range of travel products, including airline tickets, auto rentals, hotel rooms, tours, and last-minute trip packages. Expedia provides in-depth hotel details to help comparison shoppers and even offers 360-degree visual tours and expanded photo displays. It also entices flexible travelers to search for rates, compare airfares, and configure hotel and air prices at the same time. Expedia also has developed numerous hotel partnerships to reduce costs and help secure great values for consumers. Meanwhile, Orbitz has launched a special full-service program for corporate business travelers.

Establishing a Web Site

Companies large and small can establish Web sites. Previously, companies had to develop their sites in house or find and hire contractors to develop their sites. But no longer must you learn the intricacies of HTML or Java, master Web design software, or hire someone to build your site. Web site hosting services and the use of storefront brokers are two options to designing, building, operating, and maintaining a Web site. Both options offer the advantage of getting the Web site running faster and cheaper than doing it yourself, especially for a firm with few or no experienced Web developers.

Web site hosting companies such as Bigstep.com and freemerchant.com make it possible to set up a Web page and conduct e-commerce within a matter of days and with little upfront cost. These companies can also provide free hosting for your store, but to allow visitors to pay for merchandise with credit cards, you need a merchant account with a bank. If your company doesn't already have one, it must establish one.

storefront broker
Companies that act as intermediaries between your Web site and online merchants who have the products and retail expertise.

Another model for setting up a Web site is the use of a **storefront broker**, which serves as an intermediary between your Web site and online merchants who have the actual products and retail expertise. At sites such as Bigstep (*www.bigstep.com*), you select what to sell to match the themes of your site and the products you think might interest your visitors. The storefront broker deals with the details of the transactions, including who gets paid for what, and is responsible for bringing together merchants and reseller sites. The storefront broker is similar to a distributor in standard retail operations, but in this case no product moves— only electronic data flows back and forth. Products are ordered by a customer at your site, orders are processed through a user interface provided by the storefront broker, and the product is shipped by the merchant.

Building Traffic to Your Web Site

The Internet includes hundreds of thousands of e-commerce Web sites. With all those potential competitors, a company must take strong measures to ensure that the customers it wants to attract can find its Web site. The first step is to obtain and register a domain name, and your domain name should say something about your business. For instance, stuff4u might seem to be a good catchall, but it doesn't describe the nature of the business—it could be anything. If you want to sell soccer uniforms and equipment, then you'd try to get a domain name like *www.soccerstuff4u.com*, *www.soccerequipment.com*, or *www.stuff4soccercoaches.com*. The more specific the Web address, the better.

meta tag
A special HTML tag, not visible on the displayed Web page, that contains keywords representing your site's content, which search engines to use to build indexes pointing to your Web site.

The next step to attracting customers is to make your site search-engine-friendly by including a meta tag in your store's home page. A **meta tag** is a special HTML tag, not visible on the displayed Web page, that contains keywords representing your site's content, which search engines to use to build indexes pointing to your Web site. Again, selecting keywords is critical to attracting customers, so they should be chosen carefully.

You can also use Web site traffic data analysis software to turn the data captured in the Web log file into useful information. This data can tell you the URLs from which your site is being accessed, the search engines and keywords that find your site, and other useful information. Using this data can help you identify search engines to which you need to market your Web site, allowing you to submit your Web pages to them for inclusion in the search engine's index.

The use of the Internet is growing rapidly in markets throughout Europe, Asia, and Latin America. Obviously, companies that want to succeed on the Web cannot ignore this global shift. Companies must be aware that consumers outside the United States will access sites with different devices and modify their site design accordingly. In Europe, for example, closed-system iDTVs (interactive digital televisions) are becoming popular for accessing online content, with some 80 million European households now using them. Because such devices have better resolution and more screen space than the PC monitors that U.S. consumers use to access the Internet, iDTV users expect more ambitious graphics. Successful global firms operate with a portfolio of sites designed for each market, with shared sourcing and infrastructure to support the network of stores, and with local marketing and business development teams to take advantage of local opportunities. Service providers continue to emerge to solve the cross-border logistics, payments, and customer service needs of these global retailers.

Maintaining and Improving Your Web Site

Web site operators must constantly monitor the traffic to their site and the response times experienced by visitors. Internet shoppers expect service to be better than or equal to their in-store experience, says AMR Research (a Boston-based, independent research analysis firm).

Nothing will drive potential customers away faster than if they experience unbearable delays while trying to view or order your products or services. To keep pace with technology and increasing traffic, it might be necessary over time to modify the software, databases, or hardware on which the Web site runs to ensure good response times.

Web site operators must also continually be alert to new trends and developments in the area of e-commerce and be prepared to take advantage of new opportunities. For example, recent studies show that customers more frequently visit Web sites they can customize. **Personalization** is the process of tailoring Web pages to specifically target individual consumers. The goal is to meet the customer's needs more effectively, make interactions faster and easier, and, consequently, increase customer satisfaction and the likelihood of repeat visits. Building a better understanding of customer preferences also can aid in cross-selling related products and more expensive products. The most basic form of personalization involves using the consumer's name in an e-mail campaign or in a greeting on the Web page. Amazon uses a more advanced form of personalization, in which each repeat customer is greeted by name, and a list of new products is recommended based on the customer's previous purchases.

personalization
The process of tailoring Web pages to specifically target individual consumers.

Businesses use two types of personalization techniques to capture data and build customer profiles. *Implicit personalization* techniques capture data from actual customer Web sessions—primarily based on which pages were viewed and which weren't. *Explicit personalization* techniques capture user-provided information, such as information from warranties, surveys, user registrations, and contest entry forms completed online. Data can also be gathered through access to other data sources such as the Bureau of Motor Vehicles, Bureau of Vital Statistics, and marketing affiliates (firms that share marketing data). Marketing firms aggregate this information to build databases containing a huge amount of consumer behavioral data. During each customer interaction, powerful algorithms analyze both types of data in real time to predict the consumer's needs and interests. This analysis makes it possible to deliver new, targeted information before the customer leaves the site. Because personalization depends on gathering and using personal user information, privacy issues are a major concern.

These tips and suggestions are only a few ideas that can help a company set up and maintain an effective e-commerce site. With technology and competition changing constantly, managers should read articles in print and on the Web to keep up to date on ever-evolving issues.

Now that we've examined how to establish e-commerce effectively, let's look at some of the technical issues related to e-commerce systems and technology that make it possible.

TECHNOLOGY INFRASTRUCTURE REQUIRED TO SUPPORT E-COMMERCE AND M-COMMERCE

Successful implementation of e-business requires significant changes to existing business processes and substantial investment in IS technology. These technology components must be chosen carefully and integrated to support a large volume of transactions with customers, suppliers, and other business partners worldwide. Online consumers complain that poor Web site performance (e.g., slow response time, poor customer support, and lost orders) drives them to abandon some e-commerce sites in favor of those with better, more reliable performance. This section provides a brief overview of the key technology infrastructure components (see Figure 8.6).

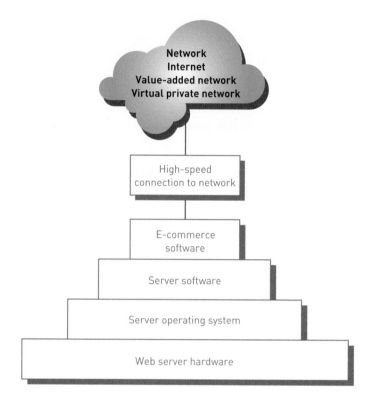

Hardware

A Web server hardware platform complete with the appropriate software is a key e-commerce infrastructure ingredient. The amount of storage capacity and computing power required of the Web server depends primarily on two things: the software that must run on the server and the volume of e-commerce transactions that must be processed. Although IS staff can sometimes define the software to be used, they can only estimate how much traffic the site will generate. As a result, the most successful e-commerce solutions are designed to be highly scalable so that they can be upgraded to meet unexpected user traffic.

A key decision facing new e-commerce companies is whether to host their own Web site or to let someone else do it. Many companies decide that using a third-party Web service provider is the best way to meet initial e-commerce needs. The third-party company rents space on its computer system and provides a high-speed connection to the Internet, which minimizes the initial out-of-pocket costs for e-commerce start-up. The third party can also provide personnel trained to operate, troubleshoot, and manage the Web server. Of course, many companies decide to take full responsibility for acquiring, operating, and supporting the Web server hardware and software themselves, but this approach requires considerable up-front capital and a set of skilled and trained workers. No matter which approach a company takes, it must have adequate hardware backup to avoid a major business disruption in case of a failure of the primary Web server.

Web Server Software

In addition to the Web server operating system, each e-commerce Web site must have Web server software to perform fundamental services, including security and identification, retrieval and sending of Web pages, Web site tracking, Web site development, and Web-page development. The two most popular Web-server software packages are Apache HTTP Server and Microsoft Internet Information Server.

Security and Identification

Security and identification services are essential for intranet Web servers to identify and verify employees accessing the system from the Internet. Access controls provide or deny access to files based on the username or URL. Web servers support encryption processes for transmitting private information securely over the public Internet.

Retrieving and Sending Web Pages

The fundamental purpose of a Web server is to process and respond to client requests that are sent using HTTP. In response to such a request, the Web server program locates and fetches the appropriate Web page, creates an HTTP header, and appends the HTML document to it. For dynamic pages, the server involves other programs, retrieves the results from the back-end process, formats the response, and sends the pages and other objects to the requesting client program.

Web-Site Tracking

Web servers capture visitors' information, including who is visiting the Web site (the visitor's IP address), what search engines and keywords they used to find the site, how long their Web browser viewed the site, the date and time of each visit, and which pages were displayed. This data is placed into a Web log file for future analysis.

Web Site Development

Web site development tools include features such as an HTML/visual Web page editor (e.g., Microsoft's FrontPage, Macromedia's Dreamweaver, NetStudio's NetStudio, and SoftQuad's HoTMetaL Pro), software development kits that include sample code and code development instructions for languages such as Java or Visual Basic, and Web page upload support to move Web pages from a development PC to the Web site. The tools bundled with the Web server software depends on which Web server software you select.

Web site development tools
Tools used to develop a Web site, including HTML or visual Web page editor, software development kits, and Web page upload support.

Web Page Construction

Web page construction software uses Web editors and extensions to produce Web pages—either static or dynamic. **Static Web pages** always contain the same information—for example, a page that provides text about the history of the company or a photo of corporate headquarters. **Dynamic Web pages** contain variable information and are built to respond to a specific Web site visitor's request. For example, if a Web site visitor inquires about the availability of a certain product by entering a product identification number, the Web server searches the product inventory database and generates a dynamic Web page based on the current product information it found, thus fulfilling the visitor's request. This same request by another visitor later in the day might yield different results due to ongoing changes in product inventory. A server that handles dynamic content must be able to access information from a variety of databases. The use of open database connectivity enables the Web server to assemble information from different database management systems, such as SQL Server, Oracle, and Informix.

Web page construction software
Software that uses Web editors and extensions to produce both static and dynamic Web pages.

static Web pages
Web pages that always contain the same information.

dynamic Web pages
Web pages containing variable information that are built to respond to a specific Web visitor's request.

E-Commerce Software

After you have located or built a host server, including the hardware, operating system, and Web server software, you can begin to investigate and install e-commerce software. E-commerce software must support five core tasks: catalog management, product configuration, shopping cart facilities, e-commerce transaction processing, and Web traffic data analysis.

The specific e-commerce software you choose to purchase or install depends on whether you are setting up for B2B or B2C transactions. For example, B2B transactions do not include sales tax calculations if they involve items purchased for resale, and software to support B2B must incorporate electronic data transfers between business partners, such as purchase orders, shipping notices, and invoices. B2C software, on the other hand, must handle the complication of accounting for sales tax based on the current state laws and rules. However, it does not need to support negotiation between buyer and seller.

Catalog Management

Any company that offers a wide range of products requires a real-time interactive catalog to deliver customized content to a user's screen. *Catalog management software* combines different product data formats into a standard format for uniform viewing, aggregating, and integrating catalog data. It also provides a central repository for easy access, retrieval, and updating of pricing and availability changes. The data required to support large catalogs is almost always stored in a database on a computer that is separate from, but accessible to, the e-commerce server machine.

Product Configuration

Customers need help when an item they are purchasing has many components and options. *Product configuration software* tools were originally developed in the 1980s to assist B2B salespeople to match their company's products to customer needs. Buyers use the new Web-based product configuration software to build the product they need online with little or no help from salespeople. For example, Dell customers use product configuration software to build the computer that meets their needs. Such software is also used in the service arena to help people decide what sort of consumer loan or insurance is best for them.

Shopping Cart

Today many e-commerce sites use an *electronic shopping cart* to track the items selected for purchase, allowing shoppers to view what is in their cart, add new items to it, or remove items from it, as shown in Figure 8.7. To order an item, shoppers simply click an item. All the details about it—including its price, product number, and other identifying information—are stored automatically. If shoppers later decide to remove one or more items from the cart, they can view the cart's contents and remove any unwanted items. When shoppers are ready to pay for the items, they click a button (usually labeled "proceed to checkout") and begin a purchase transaction. Clicking the "Checkout" button opens another window that usually asks shoppers to fill out billing, shipping, and payment method information and to confirm the order.

Figure 8.7

Electronic Shopping Cart

An electronic shopping cart (or bag) allows online shoppers to view their selections and add or remove items.

Web Services

Web services are software modules supporting specific business processes that users can interact with over a network (such as the Internet) as necessary. Web services can combine software and services from different companies to provide an integrated way to communicate. For example, an organization could use a supplier provided Web service to streamline the payment of vendor invoices. The Web service could be developed so that when the user moves the mouse over a purchase order number in an e-mail from the supplier, the amount of funds remaining in the purchase order are displayed and then the user can approve payment by clicking a button or link.

Software manufacturers are scrambling to meet customer demands by offering software applications for use over the Web as services supported by advertising or subscription fees. SAP, for example, offers more than 500 components that run as Web services to support business functions such as finance, human resources, logistics, manufacturing, procurement, and product development.[35]

Kenexa, a software provider specializing in supporting HR-related business functions, offers a set of Web services that automates the recruiting process, conducts and analyzes workforce surveys, and assists in employee performance management. Although the Web services can be used to streamline and support the HR business processes, Kenexa offers its customers another interesting service based on the concept of sharing data. Its customers can "opt in" and *anonymously* share the results of their frequently asked questions on the hiring process to create a large database of responses. One of its customers, GlaxoSmithKline, found that by using the predictive results of community intelligence, it could hire sales reps who performed better and stayed with the company longer.[36]

Electronic Payment Systems

Electronic payment systems are a key component of the e-commerce infrastructure. Current e-commerce technology relies on user identification and encryption to safeguard business transactions. Actual payments are made in a variety of ways, including electronic cash, electronic wallets, and smart, credit, charge, and debit cards. Web sites that accept multiple payment types convert more visitors to purchasing customers than merchants who offer only a single payment method.[37]

Authentication technologies are used by many organizations to confirm the identity of a user requesting access to information or assets. A **digital certificate** is an attachment to an e-mail message or data embedded in a Web site that verifies the identity of a sender or Web site. A **certificate authority (CA)** is a trusted third-party organization or company that issues digital certificates. The CA is responsible for guaranteeing that the people or organizations granted these unique certificates are, in fact, who they claim to be. Digital certificates thus create a trust chain throughout the transaction, verifying both purchaser and supplier identities.

Secure Sockets Layer

All online shoppers fear the theft of credit card numbers and banking information. To help prevent this type of identity theft, the **Secure Sockets Layer (SSL)** communications protocol is used to secure sensitive data. The SSL communications protocol includes a handshake stage, which authenticates the server (and the client, if needed), determines the encryption and hashing algorithms to be used, and exchanges encryption keys. Following the handshake stage, data might be transferred. The data is always encrypted, ensuring that your transactions are not subject to interception or "sniffing" by a third party. Although SSL handles the encryption part of a secure e-commerce transaction, a digital certificate is necessary to provide server identification.

Electronic Cash

Electronic cash is an amount of money that is computerized, stored, and used as cash for e-commerce transactions. Typically, consumers must open an account with an electronic cash service provider by providing identification information. When the consumers want to withdraw electronic cash to make a purchase, they access the service provider via the Internet

Web services
Software modules supporting specific business processes that users can interact with over a network (such as the Internet) on an as-needed basis.

digital certificate
An attachment to an e-mail message or data embedded in a Web site that verifies the identity of a sender or Web site.

certificate authority (CA)
A trusted third-party organization or company that issues digital certificates.

Secure Sockets Layer (SSL)
A communications protocol is used to secure sensitive data during e-commerce.

electronic cash
An amount of money that is computerized, stored, and used as cash for e-commerce transactions.

The SSL (Secure Sockets Layer encryption) communications protocol assures customers that information they provide to retailers, such as credit card numbers, cannot be viewed by anyone else on the Web.

and present proof of identity—a digital certificate issued by a certification authority or a username and password. After verifying a consumer's identity, the system debits the consumer's account and credits the seller's account with the amount of the purchase. PayPal, BillMeLater, and MoneyZap are three popular forms of electronic cash.

The PayPal service of eBay enables any person or business with an e-mail address to securely, easily, and quickly send and receive payments online. To send money, you enter the recipient's e-mail address and the amount you want to send. You can pay with a credit card, debit card, or funds from a checking account. The recipient gets an e-mail that says, "You've Got Cash!" Recipients can then collect their money by clicking a link in the e-mail that takes them to *www.paypal.com*. To receive the money, the user also must have a credit card or checking account to accept fund transfers. To request money for an auction, invoice a customer, or send a personal bill, you enter the recipient's e-mail address and the amount you are requesting. The recipient gets an e-mail and instructions on how to pay you using PayPal.[38] PayPal has more than 8.6 million registered users and offers 100 percent protection against unauthorized payments sent from a user's account.[39]

Bill Me Later by I4 Commerce is for customers who do not have a credit card or prefer not to use a credit card online. To make a purchase, an existing account owner provides basic information, such as the last four digits of a Social Security number and date of birth. Within seconds, Bill Me Later qualifies the customer, completes the purchase, and sends a bill. The customer can pay the cost in full or finance the purchase over time. Walmart.com, Overstock.com, and SmartBargains accept Bill Me Later payments.[40]

MoneyZap is a service offered by Western Union that enables consumers and businesses to pay retailers from an existing checking account. They must complete a one-time registration and provide their name, address, checking account information, Social Security number, e-mail address, and home phone number. After successfully registering, they can authorize payments using their username and password. An electronic funds transfer is initiated to debit money from their account and transfer money to the merchant.[41]

Credit, Charge, Debit, and Smart Cards

Many online shoppers use credit and charge cards for most of their Internet purchases. A credit card, such as Visa or MasterCard, has a preset spending limit based on the user's credit history, and each month the user can pay part or all of the amount owed. Interest is charged on the unpaid amount. A charge card, such as American Express, carries no preset spending limit, and the entire amount charged to the card is due at the end of the billing period. Charge cards do not involve lines of credit and do not accumulate interest charges. American Express became the first company to offer disposable credit card numbers in 2000. Other banks, such as Citibank, protect the consumer by providing a unique number for each transaction.[42] Debit cards look like credit cards or automated teller machine (ATM) cards, but they operate like cash or a personal check. Credit, charge, and debit cards currently store limited information about you on a magnetic strip. This information is read each time the card is swiped to make a purchase. All credit card customers are protected by law from paying more than $50 for fraudulent transactions.

The **smart card** is a credit card–sized device with an embedded microchip to provide electronic memory and processing capability. Smart cards can be used for a variety of purposes, including storing a user's financial facts, health insurance data, credit card numbers, and network identification codes and passwords. They can also store monetary values for spending.

Smart cards are better protected from misuse than conventional credit, charge, and debit cards because the smart-card information is encrypted. Conventional credit, charge, and debit cards clearly show your account number on the face of the card. The card number, along with a forged signature, is all that a thief needs to purchase items and charge them against your card. A smart card makes credit theft practically impossible because a key to unlock the encrypted information is required, and there is no external number that a thief can identify and no physical signature a thief can forge.

Smart cards have been around for over a decade and are widely used in Europe, Australia, and Japan, but they have not caught on in the United States. Use has been limited because there are few smart-card readers to record payments, and U.S. banking regulations have slowed smart-card marketing and acceptance as well. You can use a smart-card reader that attaches to your PC monitor to make online purchases with your American Express Blue smart card. You must visit the American Express Web site to get an electronic wallet to store your credit card information and shipping address. When you want to buy something online, you open the checkout window of a Web merchant, swipe your Blue card through the reader, and then enter a password. The digital wallet tells the vendor your credit card number, its expiration date, and your shipping information.

smart card
A credit card–sized device with an embedded microchip to provide electronic memory and processing capability.

SUMMARY

Principle

Electronic commerce and mobile commerce are evolving, providing new ways of conducting business that present both opportunities for improvement and potential problems.

E-commerce is the conducting of business activities electronically over networks. Business-to-business (B2B) e-commerce allows manufacturers to buy at a low cost worldwide, and it offers enterprises the chance to sell to a global market. B2B e-commerce is currently the largest type of e-commerce. Business-to-consumer (B2C) e-commerce enables organizations to sell directly to consumers, eliminating intermediaries. In many cases, this squeezes costs and inefficiencies out of the supply chain and can lead to higher profits and lower prices for consumers. Consumer-to-consumer (C2C) e-commerce involves consumers selling directly to other consumers. Online auctions are the chief method by which C2C e-commerce is currently conducted.

A successful e-commerce system must address the many stages consumers experience in the sales life cycle. At the heart of any e-commerce system is the ability of the user to search for and identify items for sale; select those items; negotiate prices, terms of payment, and delivery date; send an order to the vendor to purchase the items; pay for the product or service; obtain product delivery; and receive after-sales support.

A firm must overcome three key challenges to convert its business processes from the traditional form to e-commerce processes: 1) It must define an effective e-commerce model and strategy, 2) it needs to change its distribution systems and work processes to be able to manage shipments of individual units directly to consumers and deal with split-case distribution, and 3) it must integrate new Web-based order processing systems with traditional mainframe computer-based inventory control and production planning systems.

Mobile commerce is the use of wireless devices such as PDAs, cell phones, and smartphones to facilitate the sale of goods or services—anytime, anywhere. The market for m-commerce in North America is expected to mature much later than in Western Europe and Japan. Although some industry experts predict great growth in this arena, several hurdles must be overcome, including improving the ease of use of wireless devices, addressing the security of wireless transactions, and improving network speed. The Wireless Application Protocol (WAP) is a standard set of specifications to enable development of m-commerce software for wireless devices. WAP uses the Wireless Markup Language, which is designed for effectively displaying information on small devices. The development of the Wireless Application Protocol (WAP) and its derivatives addresses many m-commerce issues.

Principle

E-commerce and m-commerce can be used in many innovative ways to improve the operations of an organization.

Electronic retailing (e-tailing) is the direct sale from a business to consumers through electronic storefronts designed around an electronic catalog and shopping cart model.

A cybermall is a single Web site that offers many products and services at one Internet location.

Manufacturers are joining electronic exchanges, where they can work with competitors and suppliers to use computers and Web sites to buy and sell goods, trade market information, and run back-office operations such as inventory control. They are also using e-commerce to improve the efficiency of the selling process by moving customer queries about product availability and prices online.

The Web allows firms to gather much more information about customer behavior and preferences than they could using other marketing approaches. This new technology has greatly enhanced the practice of market segmentation and enabled companies to establish closer relationships with their customers. Detailed information about a customer's behavior, preferences, needs, and buying patterns allow companies to set prices, negotiate terms, tailor promotions, add product features, and otherwise customize a relationship with a customer.

The Internet has also revolutionized the world of investment and finance, especially online stock trading and online banking. The Internet has also created many options for electronic auctions, where geographically dispersed buyers and sellers can come together.

M-commerce transactions can be used in all these application arenas. M-commerce provides a unique opportunity to establish one-on-one marketing relationships and support communications anytime and anywhere.

Principle

Although e-commerce and m-commerce offer many advantages, users must be aware of and protect themselves from many threats associated with this technology.

Businesses and people use e-commerce to reduce transaction costs, speed the flow of goods and information, improve the level of customer service, and enable the close coordination of actions among manufacturers, suppliers, and customers. E-commerce also enables consumers and companies to gain access to worldwide markets. E-commerce offers great promise for developing countries, helping them to enter the prosperous global marketplace, and hence helping to reduce the gap between rich and poor countries.

Revolutionary change always raises new issues, and e-commerce is no exception. Among the issues that must be addressed are security, theft of intellectual property, fraud, invasion of consumer privacy, lack of Internet access, return on investment, legal jurisdiction, and taxation.

Principle

Organizations must define and execute a strategy to be successful in e-commerce.

Most people agree that an effective Web site is one that creates an attractive presence and meets the needs of its visitors. E-commerce start-ups must decide whether they will build and operate the Web site themselves or outsource this function. Web site hosting services and storefront brokers provide alternatives to building your own Web site.

To build traffic to your Web site, you should register a domain name that is relevant to your business, make your site search-engine-friendly by including a meta tag in your home page, use Web site traffic data analysis software to attract additional customers, and modify your Web site so that it supports global commerce. Web site operators must constantly monitor the traffic and response times associated with their site and adjust software, databases, and hardware to ensure that visitors have a good experience when they visit the site.

Principle

E-commerce and m-commerce require the careful planning and integration of a number of technology infrastructure components.

A number of infrastructure components must be chosen and integrated to support a large volume of transactions with customers, suppliers, and other business partners worldwide. These components include hardware, Web server software, and e-commerce software.

Electronic payment systems are a key component of the e-commerce infrastructure. A digital certificate is an attachment to an e-mail message or data embedded in a Web page that verifies the identity of a sender or a Web site. To help prevent the theft of credit card numbers and banking information, the Secure Sockets Layer (SSL) communications protocol is used to secure all sensitive data. There are several electronic cash alternatives that require the purchaser to open an account with an electronic cash service provider and to present proof of identity whenever payments are to be made. Payments can also be made by credit, charge, debit, and smart cards.

CHAPTER 8: SELF-ASSESSMENT TEST

Electronic commerce and mobile commerce are evolving, providing new ways of conducting business that present both opportunities for improvement and potential problems.

1. Successful implementation of e-business requires _____ and _____.
   a. changes to existing business processes; substantial investment in IS technology
   b. conversion to XML software standards; Java programming scripts
   c. implementation of tight security standards; Web site personalization
   d. market segmentation; Web site globalization

2. eBay is an example of which of the following forms of e-commerce?
   a. A2B
   b. B2B
   c. B2C
   d. C2C

3. Which form of e-commerce generates the greatest dollar volume of sales?

4. What is the elimination of intermediate organizations between the producer and the consumer called?

5. A multistage model for purchasing over the Internet includes search and identification, _____ and _____, purchasing, product or service delivery, and after-sales support.

6. The market for m-commerce in North America is far advanced relative to Western Europe and Japan. True or false?

E-commerce and m-commerce can be used in many innovative ways to improve the operations of an organization.

7. A(n) _____ is a single Web site that offers many products and services at one Internet location.
   a. e-tailer
   b. Web service
   c. cybermall
   d. none of the above

8. Buyers and sellers alike can use an electronic exchange to _____.
   a. buy and sell goods
   b. trade market information
   c. run back-office operations
   d. all of the above

9. The practice of _____ divides the pool of potential customers into subgroups, which are usually defined in terms of demographic characteristics.

10. An advancement in online bill payment that uses e-mail for the biller to post an image of your statement on the Internet so you can direct your bank to pay it is called _____.

Although e-commerce and m-commerce offer many advantages, users must be aware of and protect themselves from many threats associated with this technology.

11. Which of the following is a frequent advantage of converting to an e-commerce supply chain?
   a. a decrease in transportation costs
   b. an increase in available product inventory
   c. acquisition of expensive information systems technology
   d. an improved level of customer service

12. The use of any of several technologies to enforce policies for controlling access to digital media is called _____.

Organizations must define and execute a strategy to be successful in e-commerce.

13. After your Web site is established and successful, there is no need to redefine your site's basic business model. True or false?

14. A(n) ____ is a special HTML tag, not visible on the display Web page, that contains keywords representing your site's content, which search engines use to build indexes pointing to your Web site.

E-commerce and m-commerce require the careful planning and integration of a number of technology infrastructure components.

15. Poor Web site performance can drive consumers to abandon your Web site in favor of those with better, more reliable performance. True or false?

16. An attachment to an e-mail message or data embedded in a Web site that verifies the identity of a sender or Web site is called a(n) _____.

CHAPTER 8: SELF-ASSESSMENT TEST ANSWERS

(1) a (2) d (3) B2B (4) disintermediation (5) selection and negotiation (6) false (7) c (8) d (9) market segmentation (10) electronic bill presentment (11) d (12) digital rights management (13) false (14) meta tag (15) true (16) digital certificate

REVIEW QUESTIONS

1. Define the term *m-commerce*. What forms of e-commerce can it support?
2. Identify and briefly describe three limitations that complicate the use of handheld devices used for m-commerce.
3. What is an electronic exchange? What business issues are associated with its use?
4. Identify the six stages consumers experience in the sales life cycle that must be supported by a successful e-commerce system.
5. Define the term *supply chain management*. Which business processes does it encompass?
6. What are Web services? Provide a brief example of how you might use a Web service in conducting e-commerce.
7. What benefits can a firm achieve by converting to an e-commerce supply chain system?
8. What is the Wireless Application Protocol? Is it universally accepted? Why or why not?

9. Why is it necessary to continue to maintain and improve an existing Web site?
10. What role do digital certificates and certificate authorities play in e-commerce?
11. What is the Secure Sockets Layer and how does it support e-commerce?
12. Briefly explain the differences among smart, credit, charge, and debit cards.
13. What is a cybermall? How is it different from an electronic exchange?
14. What is technology-enabled relationship management?
15. Identify the key elements of the technology infrastructure required to successfully implement e-commerce within an organization.

DISCUSSION QUESTIONS

1. Describe the process of electronic bill presentment. Outline some potential problems in using this form of billing customers.

2. Why are many manufacturers and retailers outsourcing the physical logistics of delivering merchandise to shoppers? What advantages does such a strategy offer? Are there any potential issues or disadvantages?

3. What does it mean to define the functions of a Web site? What are some of the possible functions?

4. What do you think are the biggest barriers to wide-scale adoption of m-commerce by consumers? Who do you think is working on solutions to these problems and what might the solutions entail?

5. Wal-Mart, the world's largest retail chain, has turned down several invitations to join exchanges in the retail and consumer goods industries. Is this good or bad for the overall U.S. economy? Why?

6. Identify and briefly describe three m-commerce applications you or a friend has used.

7. Discuss the use of e-commerce to improve spending on manufacturing, repair, and operations (MRO) of goods and services.

8. Discuss the pros and cons of e-commerce companies capturing data about you as you visit their Web sites.

9. Outline the key steps in developing a corporate global e-commerce strategy.

10. Discuss the difference between electronic bill payment and electronic bill presentment. What are the benefits and some issues associated with each?

11. Discuss how you might gather Web traffic data for analysis of your firm's Web site. What decisions might this data be useful in making?

PROBLEM-SOLVING EXERCISES

1. Develop a set of criteria you would use to evaluate various business-to-consumer Web sites based on factors such as ease of use, protection of consumer data, and security of payment process. Develop a simple spreadsheet containing these criteria. Evaluate five popular Web sites using the criteria you developed. What changes would you recommend to the Web developer of the site that scored lowest?

2. Research the current state of the two-factor authorization process recommended by the Federal Financial Examination Council. Find data that defines the rate of adoption of

this recommendation and plot the growth over time. Write a paragraph discussing the barriers and issues affecting the rate of adoption.

3. Do research to learn more about the use of WAP and other specifications being developed to support m-commerce. Briefly describe the specifications you uncover. Who is behind the development of these standards? Which standards seem to be gaining the broadest acceptance? Prepare a one- to two-page report for your instructor.

TEAM ACTIVITIES

1. Imagine that your team has been hired as consultants to provide recommendations to boost the traffic to a Web site that sells health food and supplements. Identify as many ideas as possible for how you can increase traffic to this Web site. Next, rank your ideas from best to worst.

2. As a team, develop a set of criteria that you would use in selecting a Web service provider. Identify and evaluate three

different providers of Web services to support business processes associated with the Human Resources business function. Prepare a table outlining the pros and cons of each service provider.

WEB EXERCISES

1. Visit e-Bay or another online auction Web site and choose an item on which to bid. Before entering the bid process, research the site for information about any rules associated with bidding and how to bid effectively. Follow the suggested processes and record your results. Write a brief memo to your instructor summarizing your experience.

2. Do research and document the current status of the Motion Picture Association of America (MPAA) efforts to prevent the illegal copying and/or sharing of movies. Write a brief report for your instructor. Include a discussion of your opinion on the legality of sharing movies.

CAREER EXERCISES

1. Do research to identify those industries where m-commerce is making the biggest impact and the least impact. Prepare a brief report about several interesting m-commerce applications.

2. For your chosen career field, describe how you might use or be involved with e-commerce. If you have not chosen a career yet, answer this question for someone in marketing, finance, or human resources.

CASE STUDIES

Case One

Wells Fargo's Web-Enabled ATMs

For e-commerce to be effective, the systems in the supply chain must be networked and well integrated. When a transaction occurs, all of the related systems should be affected without human involvement. When consumers purchase a product, the system should automatically decrement a central inventory database, alert the supplier when a reorder is necessary, and trigger related activities in billing and other systems—all initiated by the shopper's click of a button on a Web site or swipe of a debit card.

To make this possible, the systems need to effectively communicate using standardized languages and technologies. As businesses work to streamline their information systems and business procedures, many old proprietary systems are being retired in favor of new systems designed with the latest standards. Wells Fargo Bank recently completed a five-year project doing just that.

Wells Fargo wanted to Web-enable its 6,200 ATMs in 23 states so that they could seamlessly communicate with their back-end information systems over the Internet. "We want to make sure our ATMs are integrated with every other channel so when I do a deposit in a [branch] I want to be able to go to [an] ATM immediately and see that deposit," stated Jonathan Velline, head of Wells Fargo's ATM Banking.

Velline also wanted to add services to ATMs and upgrade ATM software from a remote location. Most ATMs require direct on-site software upgrades. Upgrading the software on all ATMs with one command from an Internet-connected computer would save Wells Fargo personnel 6,200 trips.

Velline and his technical staff decided to build new ATM systems based on Web services technologies. Web services use XML, SOAP, and other standardized technologies to create communication between systems. (See Chapter 7 for more on Web services.) Unfortunately, the ATM processors were outdated and could not support the demands of the new software. The company needed to upgrade all of the ATMs to new, more powerful machines based on the Microsoft Windows platform.

Jonathan Velline said the biggest challenge in the project was the amount of internal software development needed to migrate the bank's ATM back-end systems operating system from OS/2 to Windows. Another hurdle was tightly integrating the ATM back-end systems with other business units, such as branch and online banking.

Five years later, with the project completed, Wells Fargo can now update and add services such as new languages and envelope-free deposits to its entire WebATM network remotely. Wells Fargo's new WebATM machines feature six language screen options, provide access to 22 different financial accounts, and claims to offer the highest level of security the bank has had to date.

An obvious concern to outsiders is security. Avivah Litan, an analyst at Gartner, Inc., in Stamford, Connecticut, said ATM fraud will likely increase as most banks move to Web-enabled systems, "because of the combination of stealing ATM numbers online and creating counterfeit ATM cards to use offline." Litan also said using Windows-based systems is "not great news for the security of the system. I'm sure there's a lot of holes that will be created because of this." According to research by TowerGroup, 30 percent of the world's ATMs will be running on Windows by 2006.

Wells Fargo addresses these concerns by citing the use of the Triple Data Encryption Standard, an upgrade required by credit card companies and related funds-transfer networks, such as MasterCard International, Inc., and Visa U.S.A., Inc. Triple Data Encryption is much more powerful than the current Double Encryption Standard, which has become vulnerable to hacking attacks as a result of increases in computing power.

Wells Fargo has also recently installed more than 3,000 online banking stations in nearly all of its 6,046 branch locations. Visiting customers can carry out transactions from these workstations rather than waiting for a teller. The WebATM machines and online stations are part of the company's strategy to integrate all channels—stores, phone, ATM, and Internet. The bank clearly believes in the power of online banking. Velline says that 51 percent of the bank's checking accounts are accessed online, and its customers are conducting as many transactions at ATMs as they do with branch tellers. Wells Fargo claims it is the first bank to completely Web-enable its ATM infrastructure.

Discussion Questions

1. What benefits does the WebATM platform provide to Wells Fargo and its customers?
2. What benefits does live interaction with a human teller provide that ATMs and banking stations might not?

Critical Thinking Questions

1. A new ATM machine can cost as much as $50,000. Wells Fargo bought 6,200 ATM machines. How do you think those who proposed this project justified the cost?
2. Why are WebATMs less secure than those not connected to the Internet? What are Avivah Litan's concerns?

SOURCES: Mearian, Lucas, "Wells Fargo Web-Enables 6,200 ATMs," *Computerworld*, March 3, 2005, *www.computerworld.com*. "Wells Fargo: All ATMs Now Web-Enabled, All Banking Stores with Online Stations," Wells Fargo News Release, March 1, 2005, *www.wellsfargo.com/wf/press/20050301_ATMS?year=2005*. "Wells Fargo Adds Korean, Vietnamese Language Screens to WebATM Machines," ATM Marketplace.com, March 9, 2005, *www.atmmarketplace.com/news_story_22466.htm*.

Case Two

Japanese Mobile Music Service Provides Peek into M-Commerce Future

Today's most popular m-commerce products are digital media goods such as ring tones, music, games, cell phone wallpaper graphics, and video clips. Many analysts foresee

that digital music will be the product that skyrockets in popularity over the next few years. "Certain things just naturally go together and satellite radio, and music in general, and cell phones together are a perfect combination for certain customer groups," observes telecom analyst Jeff Kagan.

A number of cell phone delivery models for music are being explored by various carriers with no clear indication as to which will find the most success. The following three methods are being explored:

- **Streaming music.** These services allow you to listen to a song without storing it on your handset; rather like on-demand radio.
- **Direct to handset downloads.** This method allows you to download a music file direct to your handset, where the file becomes your property to listen to repeatedly.
- **Transfer from PC.** Personal computer-based music services such as iTunes allow you to transfer music files from your computer collection to your handset.

Part of the challenge of translating music e-commerce models to m-commerce are the limitations of the interface. It is tedious and frustrating to search for music from huge online catalogs through the tiny display of the cell phone. A new music service launched by Japan's second-largest mobile carrier, KDDI, provides an intriguing solution that might provide a model for m-commerce applications of all sorts.

If you are in Japan and subscribe to KDDI's "Listen and Search" service, acquiring music that you hear and like anywhere anytime involves a few button pushes on your handset. For example, suppose you're in a club and the DJ plays a song you've never heard before but want to add to your private collection. Using your cell phone, you select "Listen and Search" on your menu, hold your phone in the air while the song is playing, and then press the Select button on your phone. The KDDI service analyzes the sound bite you provided, identifies the song, and provides a list of vendors from which you can purchase the song. Selecting the vendor with the best deal, you download the song and import it into your private collection to listen to whenever you like.

Three technologies are combined to create KDDI's instant-access music service. California-based Gracenote's Mobile MusicID technology uses a database of six million popular song "fingerprints" and successfully matches song clips as short as three seconds to a song in the database to provide song title and artist. The Mobile MusicID technology then delivers the name of the song and artist to the handset. Another Gracenote product, known as Link, executes commerce applications that enable the transaction. Software designed by California-based Media Socket is used to transfer the music file to the handset.

Since its introduction in mid 2005, the "Listen and Search" service has been tremendously successful in Japan. "We've found that offering instant access to music and purchases is very compelling," Makoto Takahashi, vice president and general manager of the Content Division at KDDI, said. "For the first time, our customers will be able to identify songs they're hearing on the television, on the radio or other sources, and buy related content, such as ring songs, full songs or albums, at exactly the moment they're listening to that song." After releasing the new service, KDDI sold over 2 million handsets that double as MP3 players, exceeding Apple iPod sales in Japan for the month. KDDI reported 1.8 million full-song downloads in the first month.

New cutting-edge cell phone technologies are often launched in countries such as Japan and Korea, where cell phone networks are further evolved and support high-speed network services. It is expected that "Listen and Search" or a similar service will arrive in the United States over the next two years as high-speed 3G networks become increasingly popular.

Discussion Questions

1. Would you subscribe to a music service like the one offered by KDDI? Why or why not?
2. How do you think KDDI profits from its music service?

Critical Thinking Questions

1. KDDI's "Listen and Search" service most likely lets customers find music in ways other than holding the handset up to a music source. What alternative methods of finding music would you build into this system?
2. KDDI's music service involves several steps: 1) Identify the desired product, 2) provide a list of vendors for the prod-

uct, 3) enable commerce between consumer and vendor, and 4) deliver the product. Apply this process to another m-commerce product besides music. Describe what takes place in each step.

SOURCES: Regan, Keith, "Japanese Mobile Song-ID Feature Hints at M-Commerce Future," *E-Commerce Times*, June 14, 2005, *www.ecommercetimes.com*. "Gracenote Acquires Cutting Edge Audio Fingerprinting Technology from Philips Electronics and Announces Long-Term Research Agreement with Philips Research," PRNewswire, August 30, 2005, *www.prnewswire.com*.

Whitmann Price Consulting: E-Commerce Considerations

1. How will the AMCI system contribute to the e-commerce components of Whitmann Price?
2. How can the new system save Whitmann Price employees time and improve the reliability of billing data?

Critical Thinking Questions

1. How might the new system be extended to partners who contract with Whitmann Price to serve its clients?
2. What types of security and privacy concerns might arise over the ability to send billing information directly over the wireless network?

NOTES

Sources for the opening vignette: Kahn, Michael, "Nike Says Just Do It Yourself," Reuters, May 30, 2005, *www.reuters.com*. Hallett, Vicky, "Satisfied Customizers," *U.S. News & World Report*, November 21, 2005, DIVERSIONS; Vol. 139, No. 19; Pg. D2, D4, or *www.lexis-nexis.com*. Nike Customization Web site, accessed May, 04, 2006, *http://nikeid.nike.com/nikeid/*.

1   Member press releases, "Fairchild Semiconductor and Arrow Electronics Significantly Streamline Quoting Process," RosettaNet Web site at *www.rosettanet.org*, May 13, 2005.
2   Perez, Juan Carlos, "Amazon Turns 10, Helped by Strong Tech, Service," *Computerworld*, July 15, 2005.
3   Patton, Susannah, "The ABCs of B2C," *CIO*, February 1, 2006.
4   Chabrow, Eric, "Taste of Online Sales Gets Sweeter," *InformationWeek*, August 22, 2005.
5   Perez, Juan Carlos, "Yahoo Shopping Gets 'Social Commerce' Features," *Computerworld*, November 15, 2005.
6   Javed, Naseem, "Move Over B2B, B2C – It's M2E Time," *E-Commerce Times*, August 17, 2005.
7   Investor Relations, eBay Web site at *http://investor.ebay.com/fundamentals.cfm*, accessed May 4, 2006.
8   Mello, John P. Jr., "New .mobi Domain Approved but Challenges Remain," *TechNewsWorld*, May 11, 2006.
9   Mello, John P. Jr., "New .mobi Domain Approved but Challenges Remain," *TechNewsWorld*, May 11, 2006.

10 Regan, Keith, "Sprint, Verizon Sign on With EA for Mobile Gaming," *E-Commerce News*, May 11, 2006.
11 Company Profile, National Foods Web site, *www.natfoods.com.au* accessed May 26, 2006.
12 Trading Solutions – Case Studies – National Foods, Quadrem Web site, *www.quadrem.com*, May 26, 2006.
13 Rosencrance, Linda, "Survey: User Satisfaction with E-Commerce Sites Rises Slightly," *Computerworld*, February 21, 2006.
14 Mohl, Bruce, "Bank Sweetens Paying Bills Online," *E-Commerce Times*, February 20, 2006.
15 Shermach, Kelly, "Can E-Billing Solve the Phishing Problem?" *E-Commerce Times*, March 21, 2005.
16 Mello, Jr, John P., "Skype-eBay Deal Draws Mixed Reaction," *E-Commerce Times*, September 12, 2005.
17 Regan, Keith, "Eyeing Expansion of PayPal, eBay Buys VeriSign Payment Gateway," *E-Commerce Times*, October 11, 2005.
18 Regan, Keith, "Yokel Aims for Local Shopping Search Niche," *E-Commerce Times*, May 6, 2006.
19 Council of Supply Chain Management Professionals Web site at *www.cscmp.org/Website/AboutCSCMP/Definitions/Definitions.asp*, accessed May 4, 2006.
20 "LeClaire, Jennifer, "Security Key When Choosing Credit Cards for Online Shopping," *E-Commerce Times*, December 14, 2005.
21 LeClaire, Jennifer, "Credit Card Security: Where Are We Now?", *E-Commerce Times*, December 7, 2005.

22 Korzeniowski, Paul, "U.S. Government Pushes Banks to Tighten Up Online Security," *E-Commerce Times*, January 4, 2006.

23 LeClaire, Jennifer, "Bit Torrent Strikes Anti-Piracy Deal with MPAA," *E-Commerce Times*, November 23, 2005.

24 Kirk, Jeremy, "Hacked Bank Server Hosts Phishing Sites," *Computerworld*, March 13, 2006.

25 Rosencrance, Linda, "IRS Targets Bogus Phishing E-Mails," *Computerworld*, March 28, 2006.

26 Helm, Burt, "Click Fraud Gets Smarter, Faces Crackdown," *Business Week Online*, February 28, 2006.

27 Regan, Keith, "Google Seeks to Settle Click Fraud Suit for $90 Million," *E-Commerce Times*, March 9, 2006.

28 Regan, Keith, "eBay Disputes Report on Rampant Fraud," *E-Commerce Times*, February 27, 2006.

29 Evans, Meryl K., "Online Auction Venues Sullied by Counterfeit Goods," *E-Commerce Times*, May 10, 2006.

30 Morphy, Erika, "Apple Backs Down in Privacy Flap," *E-Commerce Times*," January 19, 2006.

31 Havenstein, Heather, "Web Services Mature, See More B2B Transaction Use," *Computerworld*, July 25, 2005.

32 Jones, K.C., "eBay Cooperates with Authorities in New York Stun-Gun Sales Crackdown," *InformationWeek*, October 11, 2005.

33 Regan, Keith, "Phillip Morris to Curb Online Cigarette Sales," *E-Commerce Times*, January 27, 2006.

34 Gardner, David W., "Some Voluntary Collection of Internet Sales Taxes Started Saturday," *InformationWeek*, September 30, 2005.

35 Whiting, Rick, "SAP Debuts ERP Apps, Web Services," *Computerworld*, May 22, 2006.

36 Case Studies – Glaxo-Smith-Klein on the Kenexa Web site, *www.kenexa.com*, accessed May 27, 2006.

37 LeClaire, Jennifer, "Online Merchants Choosing Alternative Payment Options," *E-Commerce Times*, December 21, 2005.

38 "Send Money and Request Money" page, PayPal Web site, *www.paypal.com*, accessed May 11, 2006.

39 LeClaire, Jennifer, "Online Merchants Choosing Alternative Payment Options," *E-Commerce Times*, December 21, 2005.

40 Evans, Meryl K., "Online Auction Venues Sullied by Counterfeit Goods," *E-Commerce Times*, May 10, 2006.

41 "How Does the Western Union MoneyZap Service Work?" page, Western Union MoneyZap Web site, *www.moneyzap.com/Forward?cat=help&pn=customercare.jsp#1*, accessed June 7, 2006.

42 LeClaire, Jennifer, "Online Merchants Choosing Alternative Payment Options," *E-Commerce Times*, December 21, 2005.

CHAPTER · 9 ·

Enterprise Systems

Information Systems in the Global Economy ▶
Barilla Wasa, Germany

Barilla Tracks Transactions and Customer Interaction

Transaction processing systems (TPSs), particularly order processing systems, are the backbone of an organization. TPSs assist businesses in paying bills from suppliers, paying employees, and receiving payments from customers. Using these systems, businesses collect valuable information about which products are being sold and gain insight into the desires of their customers. The efficiency of a TPS can dramatically affect customer satisfaction. A robust TPS can become a powerful tool for a sales force. The Wasa company, a member of the Barilla Group, is well aware of these benefits.

The Barilla Group exports to more than 100 countries around the world. Although many consumers are familiar with the Barilla logo on pasta boxes and Italian sauce jars, few are aware of Barilla's reach into other types of food. For example, in 1999 Barilla acquired Wasa, a Swedish company known for its whole grain Crispbreads, and moved the headquarters of Wasa to Cologne, Germany. Wasa as a member of the Barilla Group unites the production of Mediterranean pasta and Nordic baked goods under one roof. The company employs around 350 people at its Cologne and Celle German facilities.

Wasa has a reputation of maintaining excellent customer orientation and focusing intensely on customer needs. It also takes pride in the service provided by its external sales force. The Barilla Group sales force can meet its customer needs by conducting transactions in an efficient manner, anywhere, anytime, using the SAP Mobile Sales TPS. Salespeople also gain an advantage by having ready access to past transaction information accessed through a customer relationship management (CRM) system.

The recently adopted SAP TPS provides a high level of automation for easy entry of point-of-sale data. Transaction data from the mobile TPS is collected into the Wasa transactions database where it serves as valuable input into the CRM system and other systems that are key to the Barilla Group's successful business strategies.

The Barilla Group uses its transaction data in a variety of ways. For example, it relies on the CRM system to plan organization-wide promotions. Wasa takes advantage of a trade promotion management system built in to the CRM system to evaluate individual promotions using measurable criteria. Wasa managers can then systematically optimize the promotion portfolio. The TPS and CRM systems also enable Wasa to carry out its annual sales and demand planning with a high level of accuracy.

The TPS and CRM systems assist Barilla Group sales representatives and regional account managers in providing better, smarter service to customers. Using these systems, the sales force can access previous orders and evaluate a customer's needs to extend sales into new markets. The integrated sales call and route planning tools allow the external sales reps and regional managers to adjust visit and route schedules. They can also automatically transmit standardized market data and daily reports on sales calls made, which increases productive sales time. Order changes and customer-specific assortment lists can be updated centrally and are available to those needing them immediately. Because customer analyses are transparent and accessible, the sales force can efficiently recognize additional sales potential.

Employees throughout the organization have access to up-to-the-minute statistics regarding the details of what is being sold. The Barilla Group's TPS and CRM systems link to its organization-wide enterprise resource planning (ERP) system. The ERP ties together an array of Barilla's information systems through an integrated central user interface.

Using its ERP system, Barilla managers can view current critical information from all of Barilla's business units.

As you read this chapter, consider the following:

- How can an effective enterprise system affect the overall well-being of a business?
- What types of information systems are critical to the success of a business and how are the systems related to one another?

Why Learn About Enterprise Systems?

Organizations today are moving from a collection of nonintegrated transaction processing systems to highly integrated enterprise systems to perform routine business processes and maintain records about them. These systems support a wide range of business activities associated with supply chain management and customer relationship management as mentioned in Chapter 1. Although they were initially thought to be cost-effective only for very large companies, even small and midsized companies are now implementing these systems to reduce costs and improve service.

In our increasingly service-oriented economy, outstanding customer service has become a goal of virtually all companies. Employees who work directly with customers—whether in sales, customer service, or marketing—require high-quality transaction processing systems and their associated information to provide good customer service. Such workers might use an enterprise system to check the inventory status of ordered items, view the production planning schedule to tell the customer when the item will be in stock, or enter data to schedule delivery to the customer.

No matter what your role, it is very likely that you will provide input to or use the output from your organization's enterprise systems. Your effective use of these systems will be essential to raise the productivity of your firm, improve customer service, and enable better decision making. Thus, it is important that you understand how these systems work and what their capabilities and limitations are.

enterprise system

A system central to the organization that ensures information can be shared across all business functions and all levels of management to support the running and managing of a business.

An **enterprise system** is central to an organization and ensures information can be shared across all business functions and all levels of management to support the running and managing of a business. Enterprise systems employ a database of key operational and planning data that can be shared by all. This eliminates the problems of lack of information and inconsistent information caused by multiple transaction processing systems that support only one business function or one department in an organization. Examples of enterprise systems include enterprise resource planning systems that support supply-chain processes, such as order processing, inventory management, and purchasing and customer relationship management systems that support sales, marketing, and customer service–related processes.

As demonstrated in the opening vignette, businesses rely on such systems to perform many of their daily activities in areas such as product supply, distribution, sales, marketing, human resources, manufacturing, accounting, and taxation so that work is performed quickly, while avoiding waste and mistakes. Without such systems, recording and processing business transactions would consume huge amounts of an organization's resources. This collection of processed transactions also forms a storehouse of data invaluable to decision making. The ultimate goal is to satisfy customers and provide a competitive advantage by reducing costs and improving service.

This chapter begins by discussing the various TPSs of a typical organization and also covers the use of enterprise systems to provide a set of integrated and coordinated systems to meet the needs of the firm.

AN OVERVIEW OF ENTERPRISE SYSTEMS: TRANSACTION PROCESSING SYSTEMS AND ENTERPRISE RESOURCE PLANNING

Every organization has many *transaction processing systems*, which capture and process the detailed data necessary to update records about the fundamental business operations of the organization. These systems include order entry, inventory control, payroll, accounts payable, accounts receivable, and the general ledger, to name just a few. The input to these systems includes basic business transactions, such as customer orders, purchase orders, receipts, time cards, invoices, and customer payments. The processing activities include data collection, data editing, data correction, data manipulation, data storage, and document production. The result of processing business transactions is that the organization's records are updated to reflect the status of the operation at the time of the last processed transaction.

A TPS also provides employees involved in other business processes—via management information system/decision support system (MIS/DSS) and the special-purpose information systems—with data to help them achieve their goals. (MIS/DSS systems are discussed in Chapter 10.) A transaction processing system serves as the foundation for these other systems (see Figure 9.1).

Figure 9.1

TPS, MIS/DSS, and Special Information Systems in Perspective

Transaction processing systems support routine operations associated with customer ordering and billing, employee payroll, purchasing, and accounts payable. The amount of support for decision making that a TPS directly provides managers and workers is low.

TPSs work with a large amount of input and output data and use this data to update the official records of the company about such things as orders, sales, customers, and so on. As systems move from transaction processing to management information/decision support and special-purpose information systems, they involve less routine, more decision support, less input and output, and more sophisticated and complex processing and analysis. These higher-level systems require the basic business transaction data captured by the TPS.

Because TPSs often perform activities related to customer contacts—such as order processing and invoicing—these information systems play a critical role in providing value to the customer. For example, by capturing and tracking the movement of each package, shippers such as Federal Express and United Parcel Service (UPS) can provide timely and accurate data on the exact location of a package. Shippers and receivers can access an online database and, by providing the airbill number of a package, find the package's current location. If the package has been delivered, they can see who signed for it (a service that is especially useful in large companies where packages can become "lost" in internal distribution systems and mailrooms). Such a system provides the basis for added value through improved customer service.

FedEx adds value to its service by providing timely and accurate data online on the exact location of a package.

Traditional Transaction Processing Methods and Objectives

batch processing system
A form of data processing where business transactions are accumulated over a period of time and prepared for processing as a single unit or batch.

With **batch processing systems**, business transactions are accumulated over a period of time and prepared for processing as a single unit or batch (see Figure 9.2a). Transactions are accumulated for the length of time needed to meet the needs of the users of that system. For example, it might be important to process invoices and customer payments for the accounts receivable system daily. On the other hand, the payroll system might receive time cards and process them biweekly to create checks, update employee earnings records, and distribute labor costs. The essential characteristic of a batch processing system is that there is some delay between an event and the eventual processing of the related transaction to update the organization's records.

online transaction processing (OLTP)
A form of data processing where each transaction is processed immediately, without the delay of accumulating transactions into a batch.

With **online transaction processing (OLTP)**, each transaction is processed immediately, without the delay of accumulating transactions into a batch (see Figure 9.2b). Consequently, at any time, the data in an online system reflects the current status. This type of processing is essential for businesses that require access to current data such as airlines, ticket agencies, and stock investment firms. Many companies find that OLTP helps them provide faster, more efficient service—one way to add value to their activities in the eyes of the customer. Increasingly, companies are using the Internet to capture and process transaction data such as customer orders and shipping information from e-commerce applications.

Although the technology is advanced enough, TPS applications do not always run using online processing. For many applications, batch processing is more appropriate and cost effective. Payroll transactions and billing are typically done via batch processing. Specific goals of the organization define the method of transaction processing best suited for the various applications of the company.

Figure 9.3 shows the flow of key pieces of information from one TPS to another for a typical manufacturing organization. TPSs can be designed so that the flow of information from one system to another is automatic and requires no manual intervention or reentering of data. Such a set of systems is called an *integrated information system*. Many organizations have limited or no integration among its TPSs. In this case, data input to one TPS must be printed out and manually reentered into other systems. Of course, this increases the amount of effort required and introduces the likelihood of processing delays and errors.

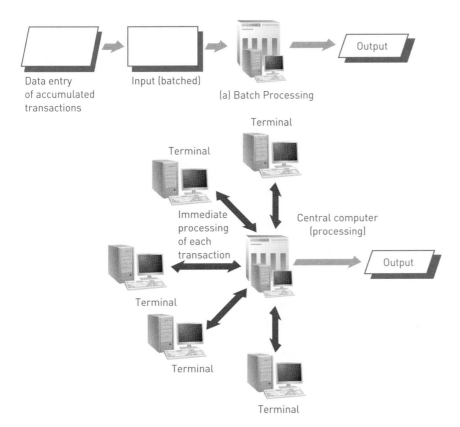

Figure 9.2

Batch Versus Online
Transaction Processing

(a) Batch processing inputs and
processes data in groups. (b) In
online processing, transactions are
completed as they occur.

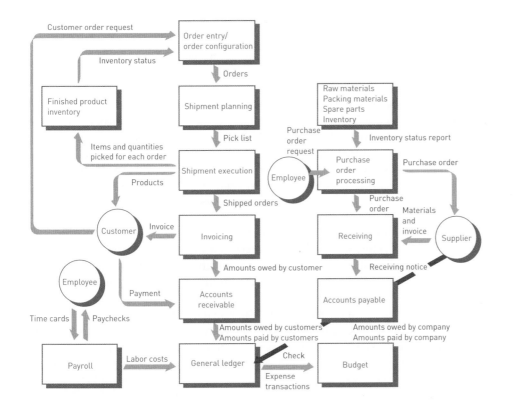

Figure 9.3

Integration of a Firm's TPSs

Because of the importance of transaction processing, organizations expect their TPSs to accomplish a number of specific objectives, including the following:

- **Process data generated by and about transactions.** The primary objective of any TPS is to capture, process, and update databases of business data required to support routine business activities. Utilities, telecommunications companies, and financial-services organizations especially are under pressure to process ever-larger volumes of online transactions. The giant telecommunications firm, Sprint Nextel Corporation, runs a mission-critical TPS to support its text messaging service that processes up to 800 transactions per second and up to 15 million transactions per day.[1]

- **Maintain a high degree of accuracy and integrity.** Ensuring that the data is processed accurately and completely is critical because reports generated by the TPS are used to execute key operational activities such as filling customer orders and scheduling shipments to various customer locations.

- **Avoid processing fraudulent transactions.** Related to data integrity is the need to avoid processing fraudulent transactions. Standard Chartered, a London-based international bank, recently implemented anti-money-laundering software to monitor banking transactions against terrorist watch lists, and flags transactions between individuals, organizations, or countries deemed high-risk.[2]

- **Produce timely user responses and reports.** The ability to conduct business transactions quickly can be essential for an organization's bottom line. For instance, if bills (invoices) are sent to customers a few days later than planned, payment is delayed, possibly forcing the firm to seek costly short-term borrowing to avoid cash flow problems. As a result, firms employ monitoring systems to measure and ensure system performance. Liberty Mutual Group (LMG), the sixth largest property and casualty insurer in the United States, relies on its claims-processing TPS to provide essential customer services. When the system experienced periodic slowdowns, the firm installed Mercury Interactive Corporation's Business Availability Center, a set of monitoring applications for optimizing production operations. The IT operations staff now uses this software to measure the user response time, and then search performance data to quickly identify any computer, network, or application problem that can affect business availability. Using this tool, Liberty Mutual can identify the cause of problems and make more computer memory and application processing resources available to claims processing during its peak processing periods.[3]

- **Increase labor efficiency.** Before businesses used computers, manual processes often required rooms full of clerks and office equipment to process the necessary business transactions. Today, TPSs substantially reduce clerical and other labor requirements. For many small businesses, a personal computer networked to the firm's cash registers has replaced a room full of clerks, typewriters, and filing cabinets.

- **Help improve customer service.** One objective of a TPS is to assist an organization in providing fast, efficient service. Some companies' EDI systems (see Chapter 6) allow customers to place orders electronically, thus bypassing slower and more error-prone methods of written or oral communication.

- **Help build and maintain customer loyalty.** A firm's TPSs are often the means for customers to communicate. Customer interaction with these systems must, therefore, keep customers satisfied and returning. A recent Web study by Allulent, Inc., found that 55 percent of consumers surveyed said a frustrating online shopping experience diminishes their overall opinion of that retailer. Surprisingly, nearly 33 percent said they might stop shopping at the retailer's brick-and-mortar store as well.[4]

- **Achieve competitive advantage.** A goal common to almost all organizations is to gain and maintain a competitive advantage. As discussed in Chapter 2, a competitive advantage provides a significant and long-term benefit for the organization. When a TPS is developed or modified, the personnel involved should carefully consider the significant and long-term benefits the new or modified system might provide. Table 9.1 summarizes some of the ways that companies can use transaction processing systems to achieve competitive advantage.

Competitive Advantage	Example
Customer loyalty increased	Customer interaction system to monitor and track each customer interaction with the company
Superior service provided to customers	Tracking systems that customers can access to determine shipping status
Better relationship with suppliers	Internet marketplace to allow the company to purchase products from suppliers at discounted prices
Superior information gathering	Order configuration system to ensure that products ordered will meet customer's objectives
Costs dramatically reduced	Warehouse management system employing RFID technology to reduce labor hours and improve inventory accuracy
Inventory levels reduced	Collaborative planning, forecasting, and replenishment to ensure the right amount of inventory is in stores

Table 9.1

Examples of Transaction Processing Systems for Competitive Advantage

Depending on the specific nature and goals of the organization, any of these objectives might be more important than others. By meeting these objectives, TPSs can support corporate goals such as reducing costs; increasing productivity, quality, and customer satisfaction; and running more efficient and effective operations. For example, overnight delivery companies such as FedEx expect their TPSs to increase customer service. These systems can locate a client's package at any time—from initial pickup to final delivery. This improved customer information allows companies to produce timely information and be more responsive to customer needs and queries.

TRANSACTION PROCESSING ACTIVITIES

Along with having common characteristics, all TPSs perform a common set of basic data-processing activities. TPSs capture and process data that describes fundamental business transactions. This data is used to update databases and to produce a variety of reports people both within and outside the enterprise use (see Figure 9.4). The business data goes through

Figure 9.4

A Simplified Overview of a Transaction Processing System

Data entry and input

Processing

Documents and reports

Internally Generated Transactions

Shipped orders
Purchase orders
Employee time cards

Externally Generated Transactions

Customer orders
Vendor invoices
Customer payments

Databases

Database Update

Customer orders
Inventory
Purchase orders
Customers
Suppliers

Documents

Pick list
Checks to vendors
Receiving notices
Paychecks

Operational Reports

Finished product inventory status
Inventory status of raw materials, packing materials, and spare parts

transaction processing cycle
The process of data collection, data editing, data correction, data manipulation, data storage, and document production.

a **transaction processing cycle** that includes data collection, data editing, data correction, data manipulation, data storage, and document production (see Figure 9.5).

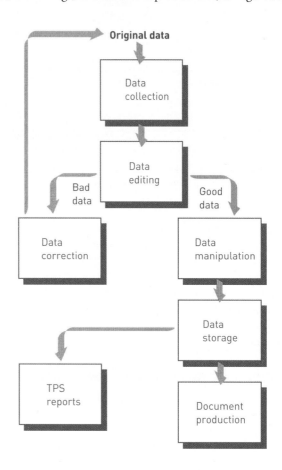

Figure 9.5

Data-Processing Activities Common to Transaction Processing Systems

data collection
Capturing and gathering all data necessary to complete the processing of transactions.

Data Collection

Capturing and gathering all data necessary to complete the processing of transactions is called **data collection**. In some cases, it can be done manually, such as by collecting handwritten sales orders or changes to inventory. In other cases, data collection is automated via special input devices such as scanners, point-of-sale devices, and terminals.

Data collection begins with a transaction (e.g., taking a customer order) and results in data that serves as input to the TPS. Data should be captured at its source and recorded accurately in a timely fashion, with minimal manual effort, and in an electronic or digital form that can be directly entered into the computer. This approach is called *source data automation*. An example of source data automation is an automated device at a retail store that speeds the checkout process—either UPC codes read by a scanner or RFID signals picked up when the items approach the checkout stand. Using both UPC bar codes and RFID tags is quicker and more accurate than having a clerk enter codes manually at the cash register. The product ID for each item is determined automatically, and its price retrieved from the item database. The point-of-sale TPS uses the price data to determine the customer's bill. The store's inventory and purchase databases record the number of units of an item purchased, the date, the time, and the price. The inventory database generates a management report notifying the store manager to reorder items that have fallen below the reorder quantity. The detailed purchases database can be used by the store or sold to marketing research firms or manufacturers for detailed sales analysis (see Figure 9.6).

The purchase of items at the checkout stand updates a store's inventory database and its database of purchases.

Many grocery stores combine point-of-sale scanners and coupon printers. The systems are programmed so that each time a specific product—for example, a box of cereal—crosses

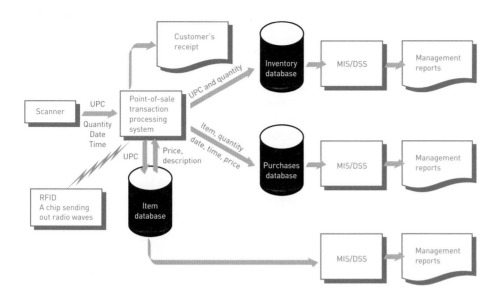

Figure 9.6

Point-of-Sale Transaction Processing System

The purchase of items at the checkout stand updates a store's inventory database and its database of purchases.

a checkout scanner, an appropriate coupon—perhaps a milk coupon—is printed. Companies can pay to be promoted through the system, which is then reprogrammed to print those companies' coupons if the customer buys a competitive brand. These TPSs help grocery stores increase profits by improving their repeat sales and bringing in revenue from other businesses.

With the high price of oil and an increasing demand for their products, oil producing firms can improve their profitability by millions of dollars by efficiently capturing data. Valero Energy is a major petroleum company that processes some 3 million barrels of oil per day. Using their TPS systems, Valero's managers can quickly access reliable data on wholesale and retail pricing and the availability of crude oil, inventory, and refinery operations. The company's transportation group then gathers real-time data on the movement of ocean-going supertankers from a Web service and uses the data to schedule dock workers and equipment. The firm estimates that it can reduce expenses by $7.5 million per year by cutting dock time by 30 minutes per vessel.[5] Shell uses sensors on the pumps that bring oil to the surface of wells in the Gulf of Mexico to capture real-time data on pump pressure, flow, and temperature. Software analyzes the data, and if it detects measurements outside the ordinary (such as a flow rate slower than expected), an alert is sent to engineers' handheld devices and remote PCs to warn them about problems.[6] Read the Information Systems @ Work special interest feature to learn more about current data collection processes used by retailers.

Data Editing

An important step in processing transaction data is to perform **data editing** for validity and completeness to detect any problems. For example, quantity and cost data must be numeric and names must be alphabetic; otherwise, the data is not valid. Often, the codes associated with an individual transaction are edited against a database containing valid codes. If any code entered (or scanned) is not present in the database, the transaction is rejected.

Information Resources Inc. (IRI) is a leading provider of enterprise market information solutions and services that help its business clients grow profitably in a complex marketplace. IRI collects over 600 million point-of-sale records per week from 40,000 grocery, drug, and convenience stores. This data is supplemented with data from 100,000 consumer households plus information about the promotional activities of consumer packaged goods manufacturers and retailers. All this data is sold to consumer packaged goods manufacturers for their analysis in determining market share, setting prices, and planning marketing campaigns. To maintain its business and reputation, IRI must ensure the accuracy and completeness of its data. To do so, IRI applies its own special data-editing algorithms to detect any data anomalies such as unusual changes in day-to-day sales volume or shelf price.[7]

data editing
The process of checking data for validity and completeness.

Retailers Turn to Smart Carts

The retail industry is going through an extraordinary metamorphosis as transactions are increasingly supported by a wide variety of digital technologies. The previous chapter provided many examples of businesses expanding to the Web to reach more customers. New forms of transaction data collection are also evident in brick-and-mortar stores. For example, consider the rapidly expanding number of self-service checkout systems in popular grocery stores, department stores, super discount stores, home warehouse stores, and even fast food restaurants.

Fujitsu calls it the Pervasive Retailing Environment: the use of digital technologies to integrate wired and wireless network devices to facilitate transactions in retail stores. Self-serve checkouts are only the tip of the iceberg. Soon customers will have access to product information from any location in the store through devices like Fujitsu's U-Scan Shopper. Mounted on a shopping cart, the U-Scan Shopper is a rugged wireless computer with an integral bar code scanner. The device provides services to shoppers as well as retailers.

The device reduces checkout time by allowing customers to scan and bag items themselves as they pick them off the shelves. Shoppers can view the running total to see exactly how much is being spent as they shop. No more surprises at the checkout counter. If an item is missing a price, the device can be used as a price-checker. Consumers can also use the U-Scan Shopper to place orders with departments in the store for pickup. For example, you can place a deli or prescription order when you arrive at the store and pick it up at the deli counter or pharmacy. The U-Scan Shopper also provides a store directory so you can easily find the department or goods you want.

U-Scan devices are integrated into the store network and Internet. This means customers can upload a shopping list to the store's Web site before leaving home, and then download the list to the shopping cart upon arriving at the store. When shopping is completed, the U-Scan device uploads information to the self-serve checkout and the shopper is out the door after a quick swipe of a debit or credit card.

For retailers, the U-Scan device offers what Fujitsu calls "true 1:1 marketing" that enables personalized in-store advertisement campaigns that are relevant both to shoppers' preferences and to their location in the store. Location is determined by shelf-mounted, battery-powered infrared transmitters that track the movement of U-Scan devices through the store. As a shopper passes the condiments aisle, for example, the shopping cart display might post a message stating, "It has been over a month since you purchased mustard. If you want to pick some up today, turn down this aisle." A retailer can offer special deals to each consumer. For example, as a shopper passes the condiments aisle, a message on the U-Scan device might state, "You have just won an electronic coupon for $0.89 off mustard. Turn now to take advantage of this special deal!" The 89 cents would be deducted as the item is scanned on the U-Scan device.

Discussion Questions

1. What transaction processing services does the U-Scan Shopper provide for consumers?
2. How does U-Scan technology provide retailers with a competitive advantage? Why might you choose a U-Scan store over one without U-Scan devices?

Critical Thinking Questions

1. What security concerns might be raised by retailers who adopt U-Scan technology? How do they compare to other TPSs? How might they be addressed?
2. What other types of services might be provided for customers and retailers through U-Scan devices?

SOURCES: Wallace, Brice, "U-Scan Could Be Your New Shopping Pal," *Deseret Morning News*, September 18, 2005, *http://deseretnews.com/dn/view/ 0,1249,605154892,00.html*. Staff, "Wireless Shopping Cart Runs Windows CE," Windows for Devices.com, February 18, 2005, *www.windowsfordevices.com*. Fujitsu Transaction Solutions Inc. Web site, accessed May 4, 2006, *www.fujitsu.com/us/services/retailing*.

Data Correction

It is not enough simply to reject invalid data. The system should also provide error messages that alert those responsible for editing the data. Error messages must specify the problem so proper corrections can be made. A **data correction** involves reentering data that was not typed or scanned properly. For example, a scanned UPC code must match a code in a master table of valid UPCs. If the code is misread or does not exist in the table, the checkout clerk is given an instruction to rescan the item or type the information manually.

Avaya is a telecommunications company that specializes in VoIP, mobile voice and data services, and the operation of contact centers. It stores over 100 terabytes of customer, vendor, service, financial, and pricing data. The company established a Data Quality Center of Excellence to correct low-quality data that was driving up expenses and reducing revenue. For example, in some cases, Avaya serviced customers' telecom equipment but did not bill them for the gear because it was erroneously left out of service agreements. As another example, some customers who paid only for standard service were getting premium service because of data errors. Avaya's Data Quality Center of Excellence provides the tools for improving and maintaining data quality and is a heavy user of the Business Objects IQ Insight profiling tool for identifying bad data.[8] This software can assess data from disparate sources, measure and track data quality over time, and provide information about data defects to help focus data correction efforts.[9]

data correction
The process of reentering data that was not typed or scanned properly.

Data Manipulation

Another major activity of a TPS is **data manipulation**, the process of performing calculations and other data transformations related to business transactions. Data manipulation can include classifying data, sorting data into categories, performing calculations, summarizing results, and storing data in the organization's database for further processing. In a payroll TPS, for example, data manipulation includes multiplying an employee's hours worked by the hourly pay rate. Overtime pay, federal and state tax withholdings, and deductions are also calculated.

data manipulation
The process of performing calculations and other data transformations related to business transactions.

Data Storage

Data storage involves updating one or more databases with new transactions. After being updated, this data can be further processed and manipulated by other systems so that it is available for management reporting and decision making. Thus, although transaction databases can be considered a by-product of transaction processing, they have a pronounced effect on nearly all other information systems and decision-making processes in an organization.

data storage
The process of updating one or more databases with new transactions.

Document Production and Reports

Document production involves generating output records, documents, and reports. These can be hard-copy paper reports or displays on computer screens (sometimes referred to as *soft copy*). Printed paychecks, for example, are hard-copy documents produced by a payroll TPS, whereas an outstanding balance report for invoices might be a soft-copy report displayed by an accounts receivable TPS. Often, results from one TPS flow downstream to become input to other systems (as shown in Figure 9.6), which might use the results of updating the inventory database to create the stock exception report (a type of management report) of items whose inventory level is below the reorder point.

In addition to major documents such as checks and invoices, most TPSs provide other useful management information and decision support, such as printed or on-screen reports that help managers and employees perform various activities. A report showing current inventory is one example; another might be a document listing items ordered from a supplier to help a receiving clerk check the order for completeness when it arrives. A TPS can also produce reports required by local, state, and federal agencies, such as statements of tax withholding and quarterly income statements. American Electric Power Co. (AEP) uses software from Oversight Systems Inc. to monitor transactions in its accounts payable group. If a

document production
The process of generating output records and reports.

manager authorizes a purchase above a specified spending limit, the system recognizes it and creates an exception report. The capabilities of this data-monitoring software helped AEP comply with Section 404 requirements of the Sarbanes-Oxley Act.[10]

Throughout this chapter we will look at some ways companies have employed TPSs to help them meet organizational goals.

CONTROL AND MANAGEMENT ISSUES

Transaction processing systems process the fundamental business transactions that are the lifeblood of the firm's operation. They capture facts about basic business operations of the organization—facts without which orders cannot be shipped, customers cannot be invoiced, and employees and suppliers cannot be paid. In addition, the data captured by TPSs flows downstream to other systems in the organization where they are used to support analysis and decision making. TPSs are so critical to the operation of most firms that many business activities would come to a halt if the supporting TPSs failed. Because firms must ensure the reliable operation of their TPSs, they must also engage in disaster recovery planning and TPS audits.

Disaster Recovery Plan

disaster recovery plan (DRP)
A formal plan describing the actions that must be taken to restore computer operations and services in the event of a disaster.

Unfortunately, recent history reminds us of the need to be prepared in the event of a natural or man-made accident or disaster. The **disaster recovery plan** (**DRP**) is a firm's plan to recover data, technology, and tools that support critical information systems and necessary information systems components such as the network, databases, hardware, software, and operating systems.

Those TPSs that directly affect the cash flow of the firm (such as order processing, accounts receivable, accounts payable, and payroll) are typically identified as critical business information systems. A lengthy disruption in the operation of any of those systems could create a serious cash flow problem for the firm and potentially put it out of business. Companies vary widely in the thoroughness and effectiveness of their disaster recovery planning, and, as a result, some have a harder time resuming business than others. According to AT&T's fifth-annual Business Continuity Survey of U.S. companies, although 30 percent of the firms studied suffered from some form of disaster, a full 28 percent admitted they did not have adequate plans in place to cope with any sort of disaster.[11]

SunGuard and Hewlett-Packard are two of many firms that provide disaster recovery services, including those for infrastructure and staff. SunGuard has established dozens of recovery centers around the United States.[12] Hewlett-Packard invested more than $100 million to create 60-some recovery centers worldwide.[13]

In the aftermath of the 9/11 attacks, Arch Chemicals, a global specialty chemicals company with annual sales exceeding $1 billion, was one of many firms determined to improve their disaster preparedness. Because roughly 60 percent of the Arch Chemicals business runs through its enterprise system, any downtime to the system affects the entire organization. A formal risk assessment determined that Arch Chemicals needed to restore the system within 48 hours of any interruption. Arch Chemicals business managers worked with SunGuard disaster recovery consultants to devise a solution to meet this tight recovery deadline. A key component of the solution was to install special software on the computers of both Arch Chemicals and SunGuard that automatically relays any changes in the user's configuration (such as the operating systems and applications) to ensure a quicker, more reliable recovery process.[14]

Disaster recovery service provider AppRiver has partnered with the Florida Chamber of Commerce to offer a service meant to protect and preserve e-mail traffic for hurricane-prone Florida-based firms. AppRiver monitors the e-mail server activity of member organizations. If service from the receiving e-mail server is interrupted, AppRiver saves the firm's incoming e-mail until the server is running again or redirects the e-mail elsewhere.[15]

Read the Ethical and Societal Issues special interest feature to learn more about the need for disaster recovery planning. This important topic is discussed further in Chapter 13.

Transaction Processing System Audit

The Sarbanes-Oxley Act, enacted as a result of several major accounting scandals, requires public companies to implement procedures to ensure their audit committees can document financial data, validate earnings reports, and verify the accuracy of information. The Financial Services Modernization Act (Gramm-Leach-Bliley) requires systems security for financial service providers, including specific standards to protect customer privacy. The Health Insurance Portability and Accountability Act (HIPAA) defines regulations covering healthcare providers to ensure that their patient data is adequately protected. Many organizations conduct ongoing **transaction processing system audits** to prevent the kind of accounting irregularities or loss of data privacy that can put their firm in violation of these acts and erase investor confidence. The audit can be performed by the firm's own internal audit group or an outside auditor might be hired to provide a higher degree of objectivity. A transaction processing system audit attempts to answer four basic questions:

transaction processing system audit
A check of a firm's TPS systems to prevent accounting irregularities and/or loss of data privacy.

- Does the system meet the business need for which it was implemented?
- What procedures and controls have been established?
- Are these procedures and controls being used properly?
- Are the information systems and procedures producing accurate and honest reports?

A typical audit also examines the distribution of output documents and reports, determines if only appropriate people can execute key system functions (e.g., approve the payment of an invoice), assesses the training and education associated with existing and new systems, and determines the effort required to perform various tasks and to resolve problems in the system. General areas of improvement are also identified and reported during the audit.

Battening Down the Systems for the Next Hurricane Season

The U.S. Gulf Coast took a beating during the 2005 hurricane season. Included in the massive amounts of damage to property was damage to the infrastructure of hundreds of private and public information systems. Before life on the Gulf Coast could return to normal, information systems that supported transactions and stored customer records needed to be brought back online.

The engineering firm URS Corporation's New Orleans offices were inaccessible for about a month after Hurricane Katrina struck on August 29, 2005, and caused devastating flooding. Because systems were out of reach, URS had to buy new hardware, especially PCs for its users, said David Scripter, regional IT manager of URS Corporation's New Orleans operations. The company set up remote offices in Baton Rouge, Louisiana, and in Houston and Austin, Texas. It took URS two weeks after the storm to restore its IT operations.

The Federal Credit Union in Biloxi, Mississippi, lost its roof and most of its windows to Katrina, but the credit union's data center was spared because it was located between two cement floors in the middle of the building. Still the credit union was unable to offer service to its customers because of the widespread failure of voice and data networks in the Gulf Coast region.

The day after Katrina struck, Neal Hennegan, director of technology at insurer Gilsbar Inc., had to drive 60 miles from his office in Covington, Louisiana, to find a working phone so he could declare a disaster and authorize a switch to Gilsbar's Chicago data center.

Blue Cross and Blue Shield of Louisiana had invested heavily in business continuity capabilities. The health insurer used diverse routing techniques on its networks but wasn't aware that at one point all networking traffic passed through a common physical site in New Orleans, which was knocked out during the storm bringing the entire network down.

A big challenge in bringing back information systems after a disaster is in bringing the staff back to work. At the University of New Orleans, IT workers could not begin returning to the school's campus until nearly three months after the storm. Jim Burgard, assistant vice chancellor for university computing and communications, was still trying to replace eight IT workers, one-fifth of his staff, a year after the storm.

The home of URS Corporation's David Scripter was located near a breached canal in the city. "I lost my house and pretty much everything I own," Scripter said, adding that at least he, his wife, and their three young children were all safe. IT professionals can't repair business systems when they are busy repairing their own lives.

As the 2006 hurricane season approached, IT specialists worked hard to analyze how they could improve their disaster recovery and business continuity planning. "I have a great concern for this coming hurricane season," said James Mehaffey, manager of the business continuity program at Blue Cross and Blue Shield of Louisiana. "In the Gulf Coast, our infrastructure is severely weakened."

There was a particularly strong push to increase data backup capabilities, especially via online replication to other sites, and to set up wireless and satellite communications systems. URS Corporation installed a storage area network (SAN) in New Orleans and is investigating how to mirror the data on the SAN to an off-site location in another state. URS is also installing a system that will replicate separate Oracle databases used in 3-D engineering applications to remote locations.

Keesler Federal Credit Union is installing satellite communications links as network backups at its 12 locations in the United States and three offices in the United Kingdom. Gilsbar Inc. plans to send IT employees to Chicago in advance of any future storms "and empower them to declare a disaster without talking to us," says Neal Hennegan. The University of New Orleans is installing an uninterruptible power supply and a new generator powered by natural gas.

Ochsner Clinic Foundation, which operates a 600-bed hospital on its main campus in New Orleans and 35 clinics throughout Louisiana, is installing a redundant Freon and chilled-water cooling system in the company's primary data center and setting up a remote data center at least 100 miles away from New Orleans.

Jan Rideout, CIO at Northrop Grumman Corporation's Ship Systems unit, points out that "disaster recovery plans have to be flexible and must take an organization's ability to respond into account." For instance, many plans assume that employees will be available after a disaster occurs, she said. But when the air conditioners in Northrop Grumman's data center stopped working after Katrina, no employees remained who knew how to gracefully shut down the servers.

Discussion Questions

1. What were the biggest IT and system problems for businesses affected by Hurricane Katrina?
2. Outline the ways in which Gulf Coast businesses are working to improve their business continuity planning.

Critical Thinking Questions

1. What suggestions might you provide to a business to ensure their staff will return to work after a disaster?
2. Jan Rideout, CIO at Northrop Grumman's Ship Systems, believes that "There is not ever going to be the perfect disaster recovery plan." Do you agree? Why or why not?

SOURCES: Thibodeau, Patrick; Havenstein, Heather, "Mardi Gras Is Back—What About IT?" *Computerworld*, March 3, 2006, *www.computerworld.com*. Staff, "James Gaskin, More Disaster Planning Justification," ITWorld.com, September 09, 2005, *www.itworld.com*. Gross, Grant, "FCC Head: Hurricane Shows Need for Redundant Telecom," ITWorld.com, Sept 9, 2005, *www.itworld.com*.

TRADITIONAL TRANSACTION PROCESSING APPLICATIONS

This section presents an overview of several common transaction processing systems that support the order processing, purchasing, and accounting business functions (see Table 9.2).

Order Processing	Purchasing	Accounting
• Order processing • Sales configuration • Shipment planning • Shipment execution • Inventory control (finished product) • Accounts receivable	• Inventory control (raw materials, packing materials, spare parts, and supplies) • Purchase order processing • Receiving • Accounts payable	• Budget • Accounts receivable • Payroll • Asset management • General ledger

Table 9.2

Systems That Support Order Processing, Purchasing, and Accounting Functions

Order Processing Systems

The traditional TPSs for order processing include order entry, sales configuration, shipment planning, shipment execution, inventory control, and accounts receivable. Running these systems efficiently and reliably is so critical to an enterprise that the order processing systems are sometimes referred to as the "lifeblood of the organization." Figure 9.7 is a system-level flowchart that shows the various systems and the information that flows among them. A rectangle represents a system, a line represents the flow of information from one system to another, and a circle represents any entity outside the system—in this case, the customer.

Figure 9.7

Traditional TPS Systems That Support the Order Processing Business Function

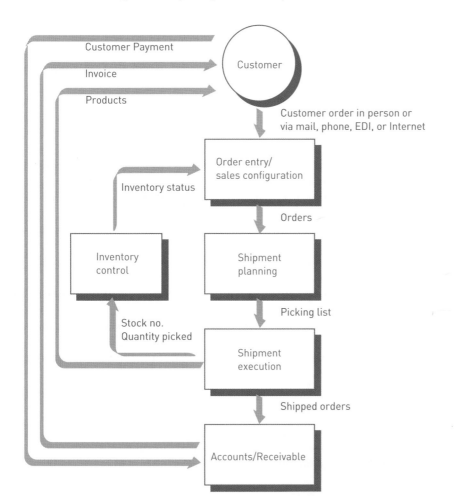

Table 9.3 summarizes the input, processing, and output (IPO) of the essential systems that include the traditional order processing systems.

System	Input	Processing	Output
Order entry	Customer order information via a variety of means: data entry by sales rep, customer input, mail, phone, e-commerce, or computer to computer via EDI or XML formats	Order is checked for completeness and accuracy. On-hand Inventory is checked to ensure each item can be shipped in the quantity ordered or a substitute item is suggested.	An open order record
Sales configuration	Customer order information including model and options desired	Review customer order information and ensure the configuration will meet the customer's needs; suggest additional options and features when appropriate.	Revised customer order
Shipment planning	Open orders, i.e., orders received but not yet shipped	Determine which open orders will be filled when and from which location each order will be shipped to minimize delivery costs and meet customer desired delivery dates.	Pick list for each order to be filled from each shipping location showing the items and quantities needed to fill the order
Shipment execution	Pick list and data entered by warehouse operations personnel as they fill the order	Data entered by warehouse operations personnel captured and used to update record of what was shipped to the customer.	A shipped order record specifying exactly what was shipped to the customer—this can be different than what was ordered
Inventory control (finished product)	Record of each item picked to fill a customer order	Inventory records are updated to reflect current quantity of each item.	Updated inventory database and various management reports
Accounts receivable	Shipped order records received from shipment execution that show precisely what was shipped on each order; payments from customers	Determine amount owed by each customer for each order placed	Invoice statement containing details of each order and its associated costs; customers' accounts receivable data is updated

Table 9.3

IPO of the Traditional TPS Systems That Support Order Processing

Beaulieu Group LLC is the third-largest carpet manufacturer in the world. Its major customers include home improvement chains The Home Depot and Lowe's Companies. Its most popular brands are Beaulieu, Coronet, Hollytex, and Laura Ashley Home. In an effort to streamline its traditional order processing process, the firm equipped 250 of its commercial accounts sales staff with an order entry application that runs on a Pocket PC. With the new system, salespeople enter customer orders, access the company's pricing databases, and make changes to orders over a wireless network. If a wireless connection cannot be made at the customer's site, the salesperson can enter orders on the Pocket PC and then transmit the data later when communications can be established. The new process has improved the way salespeople interact with customers and reduced the time they spend filling out paperwork. Previously, orders had to be written out at a customer's site and then sent to the company's central office, where clerical workers keyed them into an order processing system. As a result, the salespeople spent too much time on administrative work entering and correcting orders and not enough time selling.[16]

Stars Sales and Service is an automotive parts distributor with four wholesale stores with some 30,000 parts in inventory valued at around $2 million. Star deployed an integrated suite of supply chain and e-commerce applications that enabled customers to order parts online, cutting the number of manual orders flowing into the order desk. The new system also has improved inventory management by identifying nearly $500,000 of surplus parts in inventory.[17]

Purchasing Systems

The traditional TPS systems that support the purchasing business function include inventory control, purchase order processing, receiving, and accounts payable (see Figure 9.8). Table 9.4 shows the input, processing, and output associated with this collection of systems.

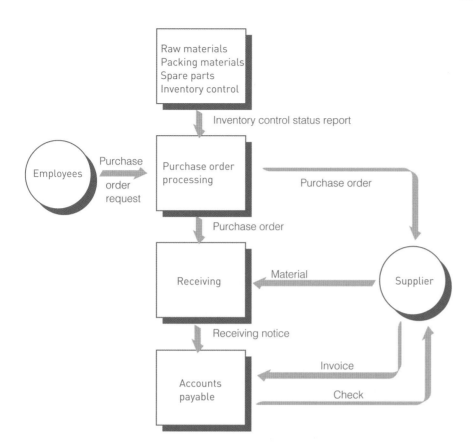

Figure 9.8

Traditional TPS Systems That Support the Purchasing Business Function

System	Input	Processing	Output
Inventory control	Records reflecting any increase or decrease in the inventory of specific items of raw materials, packing materials, or spare parts	Withdrawals are subtracted from inventory counts of specific items; additions are added to the inventory count	The inventory record of each item is updated to reflect its current count
Purchase order processing	Inventory records, employee-prepared purchase order requests, information on preferred suppliers	Items that need to be ordered are identified, quantities to be ordered are determined, qualified supplier with whom to place the order is identified	Purchase orders are placed with preferred suppliers for items
Receiving	Information on the quantity and quality of items received	Receipt is matched to purchase order, input data is edited for accuracy and completeness	Receiving report is created, inventory records are updated to reflect new receipts
Accounts payable	Purchase orders placed, information on receipts, supplier invoices	Supplier invoice matched to original purchase order and receiving report	Payment generated to supplier

Table 9.4

IPO for the Traditional TPS Systems that Support Purchasing

Accounting Systems

The primary **accounting systems** include the budget, accounts receivable, payroll, asset management, and general ledger (see Figure 9.9). Table 9.5 shows the input, processing, and output associated with these systems.

accounting systems
Systems that include budget, accounts receivable, payroll, asset management, and general ledger.

Figure 9.9

Traditional TPS Systems That Support the Accounting and Finance Business Function

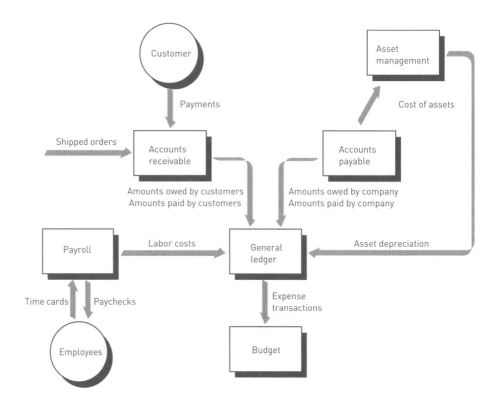

Table 9.5

IPO for the Traditional TPS Systems That Support Accounting

System	Input	Processing	Output
Budget	Amounts budgeted for various categories of expense	Accumulates amount spent in each budget category	Budget status report showing amount under/over budget
Accounts receivable	Shipment records specifying exactly what was shipped to a customer	Determines amount to be paid by customer including delivery costs and taxes	Customer bills and monthly statements, management reports summarizing customer payments
Payroll	Number of hours worked by each employee, employee pay rate, employee tax and withholding information	Calculates employee gross pay and net pay and amount to be withheld for various taxing agencies and employee benefit programs	Payroll check and stub, payroll register (a summary report of all payroll transactions), W-2 forms
Asset management	Data regarding the purchase of capital assets	Calculates depreciation and net value of all corporate assets	Listing of all assets showing purchase price and current value after depreciation
General ledger	All transactions affecting the financial standing of the firm	Posts financial transactions to appropriate accounts specified in the firms chart of accounts (see Table 9.6)	Financial reports such as the profit and loss statement, balance sheet

A key to the proper recording and reporting of financial transactions is the corporation's chart of accounts (see Table 9.6). This chart provides codes for each type of expense or revenue. By entering transactions consistent with the chart of accounts, financial data can be reported in a simple and consistent fashion across all organizations of the enterprise, even if it is a multinational corporation.

Major Account Name	Type of Expense	Subaccount Code Used to Identify Transaction
Wages and Benefits	Management salaries and benefits	MSALSB
	Nonmanagement salaries and benefits	NMALSB
	Overtime	OVT
Travel and Training	Travel-related expenses	TRAVEL
	Tuition for training classes	TUITION
Professional Services	Fees paid to consultants, contractors, trainers, and other professionals	PROFSV
Maintenance Expense	Maintenance labor	MAINTL
	Maintenance parts	MAINTP
	Maintenance supplies	MAINTS

Table 9.6

Sample Partial Chart of Accounts

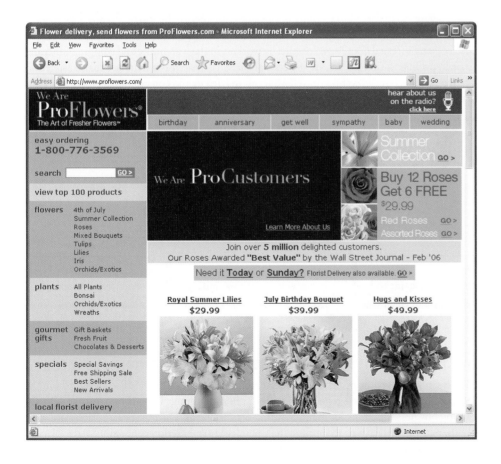

ProFlowers uses a sophisticated order processing system that relays a consumer's online order for fresh flowers directly to the grower. Because their flowers bypass extended stays in a warehouse, truck, or retail florist's cooler, ProFlowers can guarantee its customers the freshest flowers available. Throughout the process, the ProFlowers system sends e-mail messages to the customer regarding order status, shipping, and delivery.

Many companies use RFID tags to shorten order processing time and improve inventory accuracy.

(Source: Courtesy of Intermec Technologies.)

ENTERPRISE RESOURCE PLANNING, SUPPLY CHAIN MANAGEMENT, AND CUSTOMER RELATIONSHIP MANAGEMENT

As defined in Chapter 4, enterprise resource planning (ERP) is a set of integrated programs that manage a company's vital business operations for an entire multisite, global organization. Recall that a business process is a set of coordinated and related activities that takes one or more kinds of input and creates an output of value to the customer of that process. The customer might be a traditional external business customer who buys goods or services from the firm. An example of such a process is capturing a sales order, which takes customer input and generates an order. The customer of a business process might also be an internal customer such as a worker in another department of the firm. For example, the shipment process creates the necessary internal documents needed by workers in the warehouse and shipping functions to pick, pack, and ship orders. At the core of the ERP system is a database that is shared by all users so that all business functions have access to current and consistent data for operational decision making and planning, as shown in Figure 9.10.

Figure 9.10

Enterprise Resource Planning System

An ERP integrates business processes and the ERP database.

An Overview of Enterprise Resource Planning

ERP systems evolved from materials requirement planning systems (MRP) developed in the 1970s. These systems tied together the production planning, inventory control, and purchasing business functions for manufacturing organizations. During the late 1980s and early 1990s, many organizations recognized that their legacy transaction processing systems lacked the integration needed to coordinate activities and share valuable information across all the business functions of the firm. As a result, costs were higher and customer service poorer than

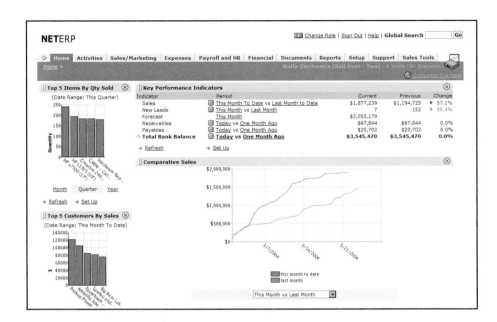

desired. The impending year 2000 (Y2K) problem that people expected to cause date-related processing to operate incorrectly after January 1, 2000, provided further impetus for organizations all over the world to review, modify, and upgrade their computer systems. Many firms used the Y2K issue to justify scrapping large parts of their existing information systems and converting to new ERP systems. Large organizations, members of the *Fortune* 1000, were the first to take on the challenge of implementing ERP. As they did, they uncovered many advantages as well as some disadvantages summarized in the following sections.

Advantages of ERP

Increased global competition, new needs of executives for control over the total cost and product flow through their enterprises, and ever-more-numerous customer interactions drive the demand for enterprise-wide access to real-time information. ERP offers integrated software from a single vendor to help meet those needs. The primary benefits of implementing ERP include improved access to data for operational decision making, elimination of inefficient or outdated systems, improvement of work processes, and technology standardization. ERP vendors have also developed specialized systems for specific applications and market segments.

Improved Access to Data for Operational Decision Making

ERP systems operate via an integrated database, using one set of data to support all business functions. The systems can support decisions on optimal sourcing or cost accounting, for instance, for the entire enterprise or business units from the start, rather than gathering data from multiple business functions and then trying to coordinate that information manually or reconciling data with another application. The result is an organization that looks seamless, not only to the outside world but also to the decision makers who are deploying resources within the organization. The data is integrated to facilitate operational decision making and allows companies to provide greater customer service and support, strengthen customer and supplier relationships, and generate new business opportunities.

For example, success in the retail industry is determined by a retailer's ability to have the right products on store shelves and priced correctly when customers come to shop. As a result, retailers need accurate, current, point-of-sale data and inventory data to match the merchandise assortment in their stores to the needs of their local markets. Watch and leather goods maker Fossil, Inc., is positioning itself for long-term growth and concentrating on cost reductions in its supply chain. As a key part of its plans, Fossil consolidated inventory for all sales channels (wholesale, retail, e-commerce) into a single 500,000 square foot warehouse and distribution center in Dallas. It also implemented the SAP ERP Apparel and Footwear software to support its wholesale business that ships to department store retailers such as

Bloomingdale's, Macy's, and Nordstrom. The software will link point-of-sale data and inventory information across the 140 Fossil boutique stores and outlets worldwide that sell jeans, shirts, belts, and other accessories, with its separate wholesale and e-commerce business. The goal is to reduce the amount of inventory required to meet customer demand and increase the speed at which product flows through the warehouse and into the stores.[18]

Elimination of Costly, Inflexible Legacy Systems

Adoption of an ERP system enables an organization to eliminate dozens or even hundreds of separate systems and replace them with a single, integrated set of applications for the entire enterprise. In many cases, these systems are decades old, the original developers are long gone, and the systems are poorly documented. As a result, the systems are extremely difficult to fix when they break, and adapting them to meet new business needs takes too long. They become an anchor around the organization that keeps it from moving ahead and remaining competitive. An ERP system helps match the capabilities of an organization's information systems to its business needs—even as these needs evolve.

Improvement of Work Processes

best practices
The most efficient and effective ways to complete a business process.

Competition requires companies to structure their business processes to be as effective and customer oriented as possible. ERP vendors do considerable research to define the best business processes. They gather requirements of leading companies within the same industry and combine them with research findings from research institutions and consultants. The individual application modules included in the ERP system are then designed to support these **best practices**, the most efficient and effective ways to complete a business process. Thus, implementation of an ERP system ensures good work processes based on best practices. For example, for managing customer payments, the ERP system's finance module can be configured to reflect the most efficient practices of leading companies in an industry. This increased efficiency ensures that everyday business operations follow the optimal chain of activities, with all users supplied the information and tools they need to complete each step.

Upgrade of Technology Infrastructure

In an ERP system, an organization can upgrade and simplify the information technology it employs. When implementing ERP, a company must determine which hardware, operating systems, and databases it wants to use. While centralizing and formalizing these decisions, the organization can eliminate the hodgepodge of multiple hardware platforms, operating systems, and databases it is currently using—most likely from a variety of vendors. Standardizing on fewer technologies and vendors reduces ongoing maintenance and support costs as well as the training load for those who must support the infrastructure.

Disadvantages of ERP Systems

Unfortunately, implementing ERP systems can be difficult and error-prone. Some of the major disadvantages of ERP systems are the expense and time required for implementation, the difficulty in implementing the many business process changes that accompany the ERP system, the problems with integrating the ERP system with other systems, the risks associated with making a major commitment to a single vendor, and the risk of implementation failure.

Expense and Time in Implementation

Getting the full benefits of ERP takes time and money. Although ERP offers many strategic advantages by streamlining a company's TPSs, large firms typically need three to five years and spend tens of millions of dollars to implement a successful ERP system.

Difficulty Implementing Change

In some cases, a company has to radically change how it operates to conform to the ERP's work processes—its best practices. These changes can be so drastic to long-time employees that they retire or quit rather than go through the change. This exodus can leave a firm short of experienced workers. Sometimes, the best practices simply are not appropriate for the firm and cause great work disruptions.

Difficulty Integrating with Other Systems

Most companies have other systems that must be integrated with the ERP system, such as financial analysis programs, e-commerce operations, and other applications. Many companies have experienced difficulties making these other systems operate with their ERP system. Other companies need additional software to create these links.

Risks in Using One Vendor

The high cost to switch to another vendor's ERP system makes it extremely unlikely that a firm will do so. After a company has adopted an ERP system, the vendor has less incentive to listen and respond to customer concerns. The high cost to switch also increases risk—in the event the ERP vendor allows its product to become outdated or goes out of business. Selecting an ERP system involves not only choosing the best software product but also the right long-term business partner. It was unsettling for many companies that had implemented PeopleSoft, J.D. Edwards, or Siebel Systems enterprise software when these firms were acquired by Oracle.

Risk of Implementation Failure

Implementing an ERP system is extremely challenging and requires tremendous amounts of resources, the best IS and businesspeople, and plenty of management support. Unfortunately, ERP installations occasionally fail, and problems with an ERP implementation can require expensive solutions.

- The 11-campus North Dakota University System spent $49 million ($14 million more than budgeted) trying to implement PeopleSoft ERP software to handle grants and contracts management as well as payroll and human resources functions. The implementation remains incomplete and is more than two years late.[19]
- King County in Seattle, Washington, spent $39 million on an ERP system that failed to meet management's goals for the project. The system included PeopleSoft human resources software and SAP AG R/3 financial application. They plan to make a second implementation attempt at an estimated cost of $50 million.[20]
- The Government Accountability Office claims the U.S. Navy wasted $1 billion on four flawed ERP pilot projects intended to help modernize the Navy's acquisition and financial management, supply chain, and other operations. The pilot projects were based on SAP software.[21]

The following list provides tips for avoiding many common causes for failed ERP implementations:

- Assign a full-time executive to manage the project.
- Appoint an experienced, independent resource to provide project oversight and to verify and validate system performance.
- Allow sufficient time for transition from the old way of doing things to the new system and new processes.
- Plan to spend a lot of time and money training people; many project managers recommend that $10,000–$20,000 per employee be budgeted for training of personnel.
- Define metrics to assess project progress and to identify project-related risks.
- Keep the scope of the project well defined and contained to essential business processes.
- Be wary of modifying the ERP software to conform to your firm's business practices.

Although ERP implementation can be difficult, many efforts have been successful. Table 9.7 summarizes some of the largest and most successful ERP projects.

It is not only large *Fortune* 1000 companies that are successful in implementing ERP. Even small companies can achieve real business benefits from their ERP efforts. Take the case of Bedford Industries in Adelaide, Australia. This nonprofit coordinates employment for 800 people with disabilities who work in one of three divisions: Bedford Furniture, which manufactures ready-to-assemble furniture, Bedford Packaging Services, which provides packaging services for other firms, and Adelaide Property and Gardens, which provides a range of horticultural services from lawn mowing to litter collection. The company's legacy systems could not provide consolidated information for planning and scheduling, production, purchasing, financials, and customer relationship management. The desire to

Table 9.7

Large, Successful ERP
Implementations

Source: "10 Top ERP
Implementations," *Information
Week*, July 11, 2005.

Company	ERP Provider	Number of Users	Initial ERP Implementation	
			Year Begun	Year Ended
Alcoa	Oracle	> 10,000	2001	N/A
BASF Aktiengesellschaft	SAP	35,000	Late 1980s	Early 1990s
Capital One Financial	Oracle	18,000	2000	2002
Colgate-Palmolive	SAP	15,000	1996	Ongoing
DaimlerChrysler	SAP	50,000	1996	Ongoing
Del Monte Foods	SSA Global Technologies	1,200	2003	2004
Dow Corning	SAP	1,500	1996	1998
FedEx	Oracle	>20,000	1997	N/A
Home Depot	SAP	N/A	N/A	N/A
Whirlpool	SAP	10,000	1999	On going

improve these core activities led Bedford to evaluate various solutions to meet their needs. They appointed a full-time project manager from the business to lead the effort to select an ERP system. Epicor Vantage was chosen because of its reputation for keeping information system costs at an absolute minimum and because the software is designed for rapid installation, low training costs, and simple operation. Bedford decided to use experienced resources from Vantage distributor Cogita to help implement the system. Another key decision was to follow the business processes as they were implemented in the ERP system, which were based on industry "best practices," rather than customize the software to conform to existing Bedford business practices. Only one month after implementation, the company's financial managers were able to produce accurate, end-of-financial-year reports. The ERP database provides a valuable source of data to enable more timely decision making and significant improvement in the efficiency and flexibility of the company's operations. Bedford expects to recover the cost of the system within five years with most savings coming from the company's leaner and more accurate operations.[22]

The following sections will outline how an ERP system can support the various major business processes.

Production and Supply Chain Management

ERP systems follow a systematic process for developing a production plan that draws on the information available in the ERP system database.

The process starts with *sales forecasting* to develop an estimate of future customer demand. This initial forecast is at a fairly high level with estimates made by product group rather than by each individual product item. The sales forecast extends for months into the future. The sales forecast might be developed using an ERP software module or it might be produced by other means using specialized software and techniques. Many organizations are moving to a collaborative process with major customers to plan future inventory levels and production rather than relying on an internally generated sales forecast. For example, Whirlpool installed the i2 Tradematrix Collaborative Planning, Forecasting, and Replenishment system, a Web-based collaboration tool for sharing and combining the sales forecasts of Whirlpool and its major trade partners—Sears, Lowe's, and Best Buy. The change has yielded outstanding results, reducing forecasting errors by half, boosting product availability from 83 to 97 percent, cutting finished product inventory by 20 percent, and trimming 5 percent from freight and warehouse costs.[23]

The *sales and operations plan* takes demand and current inventory levels into account and determines the specific product items that need to be produced and when to meet the forecast future demand. Production capacity and any seasonal variability in demand must also be considered. The result is a high-level production plan that balances market demand to production capacity. Panasonic and other companies have outsourced the development of a sales and operation plan to i2 Technologies in India. Best Buy, a major Panasonic customer, collects information on sales of Panasonic items at its stores' checkout stations and sends the data to i2. i2 processes the data and sends manufacturing recommendations to Panasonic, which become the basis for factory schedules.[24]

Demand management refines the production plan by determining the amount of weekly or daily production needed to meet the demand for individual products. The output of the demand management process is the master production schedule, which is a production plan for all finished goods.

Detailed scheduling uses the production plan defined by the demand management process to develop a detailed production schedule specifying production scheduling details, such as which item to produce first and when production should be switched from one item to another. A key decision is how long to make the production runs for each product. Longer production runs reduce the number of machine setups required, thus reducing production costs. Shorter production runs generate less finished product inventory and reduce inventory holding costs.

Materials requirement planning determines the amount and timing for placing raw material orders with suppliers. The types and amounts of raw materials required to support the planned production schedule are determined based on the existing raw material inventory and the bill of materials or BOM, a sort of "recipe" of ingredients needed to make each product item. The quantity of raw materials to order also depends on the lead time and lot sizing. Lead time is the time it takes from the time a purchase order is placed until the raw materials arrive at the production facility. Lot size has to do with discrete quantities that the supplier will ship and the amount that is economical for the producer to receive and/or store. For example, a supplier might ship a certain raw material in units of 80,000 pound rail cars. The producer might need 95,000 pounds of the raw material. A decision must be made to order one or two rail cars of the raw material.

Purchasing uses the information from materials requirement planning to place purchase orders for raw materials and transmit them to qualified suppliers. Typically, the release of these purchase orders is timed so that raw materials arrive just in time to be used in production and minimize warehouse and storage costs. Often, producers will allow suppliers to tap into data via an extranet that enables them to determine what raw materials the supplier needs, thus minimizing the effort and lead time to place and fill purchase orders.

Production uses the detailed schedule to plan the details of running and staffing the production operation.

ERP systems do not interface directly with production machines so some means must be provided to capture information about what was produced. This data must be passed onto the ERP accounting modules to keep an accurate count of finished product inventory. Many companies have personal computers on the production floor that count the number of cases of each product item produced by scanning a UPC code on the packing material. Other approaches for capturing production quantities include the use of RFID chips and manually entering the data via a PDA.

Separately, production-quality data can be added based on the results of quality tests run on a sample of the product for each batch of product produced. Typically, this data includes the batch identification number, which identifies this production run versus any other production run and the results of various product quality tests.

Customer Relationship Management and Sales Ordering

Customer Relationship Management

As discussed in Chapter 2, a **customer relationship management (CRM) system** helps a company manage all aspects of customer encounters, including marketing and advertising, sales, customer service after the sale, and programs to keep and retain loyal customers (see Figure 9.11). The goal of CRM is to understand and anticipate the needs of current and potential customers to increase customer retention and loyalty while optimizing the way that products and services are sold. Businesses implementing CRM systems report business benefits such as improved customer satisfaction (73 percent), increased customer retention (56 percent), reduced operating costs (51 percent), and the ability to meet customer demand

customer relationship management (CRM) system
A system that helps a company manage all aspects of customer encounters, including marketing and advertising, sales, customer service after the sale, and programs to retain loyal customers.

Figure 9.11

Customer Relationship
Management System

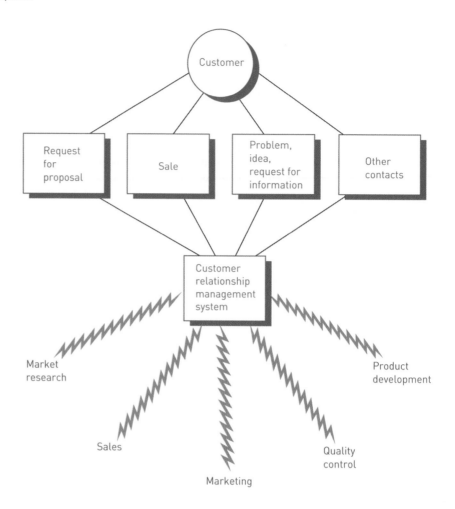

(51 percent), according to a survey of 200 business and technology professionals by Optimize Research.[25]

CRM software automates and integrates the functions of sales, marketing, and service in an organization. The objective is to capture data about every contact a company has with a customer through every channel and store it in the CRM system so the company can truly understand customer actions. CRM software helps an organization build a database about its customers that describes relationships in sufficient detail so that management, salespeople, customer service providers—and even customers—can access information to match customer needs with product plans and offerings, remind them of service requirements, and know what other products they have purchased. Figure 9.12 shows contact manager software from SAP that fills this CRM role.

ISM, Inc., is a CRM strategic advisor that rigorously tests the available CRM packages each year. Table 9.8 lists ISM's top-rated packages in alphabetical order for both large enterprises and small and medium-sized businesses (SMB). These packages scored the highest according to 217 selection criteria: 110 business functions, 49 technical features, 36 implementation capabilities, 5 real-time criteria, and 17 user-support features.

The focus of CRM involves much more than installing new software. Moving from a culture of simply selling products to placing the customer first is essential to a successful CRM deployment. Before any software is loaded onto a computer, a company must retrain employees. Who handles customer issues and when must be clearly defined, and computer systems need to be integrated so that all pertinent information is available immediately, whether a customer calls a sales representative or customer service representative. In addition to using stationary computers, most CRM systems can now be accessed via wireless devices.

The Salvation Army raises around $1.5 billion annually with donations collected in each of its four regions distributed locally. Each regional office runs its own accounting and fundraising software and manages its own database of donors. The current donation tracking

Figure 9.12

SAP Contact Manager

(Source: Copyright (c) by SAP AG.)

ISM Top CRM for Enterprise	ISM Top CRM for SMB
Amdocs CRM v. 6, Amdocs Limited	Ardexus MODE v. 5.0, Ardexus, Inc.
Pivotal CRM Suite v. 5.7, CDC Software	Powertrak v. 7.1, Axonom, Inc.
C2 CRM v. 7.08.08, Clear Technologies, Inc.	C2 CRM v. 7.08.08, Clear Technologies, Inc.
Firstwave CRM 2005, Firstwave Technologies, Inc.	Goldmine v. 7.0 & HEAT v. 8.37, FrontRange Solutions, Inc.
CMS 9.0/ONCV 5.1.2, Oncontact Software Corporation	Salesplace CRM for Lotus Notes & BlackBerry 2005.3.2, Interchange Solutions
Onyx Customer Management, Onyx Software Corporation	Maximizer Enterprise 9.0, Maximizer Software Inc.
ExSellence 4.9, Optima Technologies, Inc.	NetCRM v. 10.6 & NetSuite v. 10.6, NetSuite, Inc.
PeopleSoft CRM, Oracle/PeopleSoft, Inc.	CMS 9.0/ONCV 5.1.2, Oncontact Software Corporation
mySAP CRM 4.0, SAP AG	Relavis v. 6.5.1, Relavis Corporation
Saratoga CRM 6.5, Saratoga Systems, Inc.	RightNow CRM v. 7.5, RightNow Technologies, Inc.
Siebel 7.8, Siebel Systems, Inc.	Sage CRM 100/200 v. 5.8, Sage Software
growBusiness Solutions v. 3.0, Software Innovation ASA	Sage CRM SalesLogix v. 6.2.3, Sage Software
SSA CRM, SSA Global	Salesforce.com, Salesforce.com
Tibco Process RM v. 9.0, Tibco	SalesPage v. 4.7, SalesPage Technologies, LLC
marketing.manager 7.0, update software AG	Siebel CRM OnDemand v. 9.0, Siebel Systems, Inc.
Stayin Front, Inc.	StayinFront CRM v. 9.3, StayinFront, Inc.

Table 9.8

Top-Rated CRM Systems

Source: "ISM Announces Winners of Top 15 CRM Software Awards for Enterprise and Small & Medium Business Solutions" ISM Web site at *www.ismguide.com/www*, accessed July 3, 2006.

system used in the 13-state Western Region was developed by a software firm no longer in business and is obsolete. In addition, the donation data itself is difficult to access and integrate; separate databases track donations made through direct mail and the Internet and when goods are collected by a Salvation Army truck. Thus, it is not easy to tell if a donor has given in multiple ways or to determine the total amount of his contributions. Officials complain that donors who are not recognized think they are not appreciated. The Western Territory will spend about $2 million to install a new CRM software package to centralize its donor data and make it easier to access and use.[26]

Sales Ordering

Sales ordering is the set of activities that must be performed to capture a customer sales order. A few of the essential steps include recording the items to be purchased, setting the sales price, recording the order quantity, determining the total cost of the order including delivery costs, and confirming the customer's available credit. The determination of the sales prices can become quite complicated and include quantity discounts, promotions, and incentives. After the total cost of the order is determined, it is necessary to check the customer's available credit to see if this order puts the customer over his credit limit. Figure 9.13 shows a sales order entry window in SAP business software.

Figure 9.13

Sales Order Entry Window

(Source: Copyright (c) by SAP AG.)

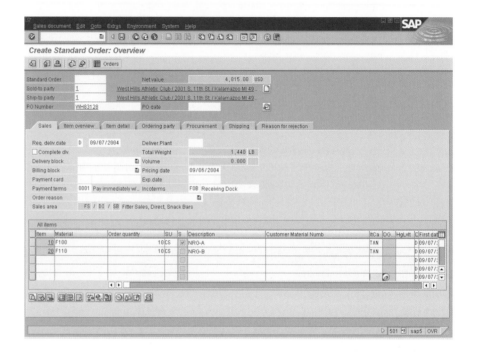

Many small-to-midsize businesses are turning to ERP software to make it easier for their large customers to place orders with them. Vetco International Inc. is a small supplier of safety equipment to major oil firms such as ExxonMobile and British Petroleum. The firm uses SAP's Business One suite, which has modules that automate purchasing, sales, and distribution; sales management; and other business functions. It cost Vetco about $150,000 to implement the software because it is compatible with the SAP software used by many of its customers. The software enables Vetco to connect its product catalogs via easy Web access to the purchasing systems of its much larger customers. The goal is to capture more business by ensuring that its offerings are just a click away from the oil companies' purchasing departments.[27]

Financial and Managerial Accounting

The general ledger is the main accounting record of a business. It is often divided into different categories, including assets, liabilities, revenue, expenses, and equity. These categories, in turn, are subdivided into subledgers to capture details such as cash, accounts payable, accounts receivable, and so on. In an ERP system, input to the general ledger occurs simultaneously with the input of a business transaction to a specific module. Here are several examples of how this occurs:

- An order clerk records a sale and the ERP system automatically creates an accounts receivable entry indicating that a customer owes money for goods received.
- A buyer enters a purchase order and the ERP system automatically creates an accounts payable entry in the general ledger registering that the company has an obligation to pay for goods that will be received at some time in the future.

- A dock worker enters a receipt of purchased materials from a supplier and the ERP system automatically creates a general ledger entry to increase the value of inventory on hand.
- A production worker withdraws raw materials from inventory to support production and the ERP system generates a record to reduce the value of inventory on hand.

Thus, the ERP system captures transactions entered by workers in all functional areas of the business. The ERP system then creates the associated general ledger record to track the financial impact of the transaction. This set of records is an extremely valuable resource that companies can use to support financial accounting and managerial accounting.

Financial accounting consists of capturing and recording all the transactions that affect a company's financial state and then using these documented transactions to prepare financial statements to external decision makers, such as stockholders, suppliers, banks, and government agencies. These financial statements include the profit and loss statement, balance sheet, and cash flow statement. They must be prepared in strict accordance to rules and guidelines of agencies such as the Securities and Exchange Commission, the Internal Revenue Service, and the Financial Accounting Standards Board. Data gathered for financial accounting can also form the basis for tax accounting because this involves external reporting of a firm's activities to the local, state, and federal tax agencies.

Managerial accounting involves the use of "both historical and estimated data in providing information that management uses in conducting daily operations, in planning future operations, and in developing overall business strategies."[28] Managerial accounting provides data to enable the firm's managers to assess the profitability of a given product line or specific product, identify underperforming sales regions, establish budgets, make profit forecasts, and measure the effectiveness of marketing campaigns.

All transactions that affect the financial state of the firm are captured and recorded in the database of the ERP system. This data is used in the financial accounting module of the ERP system to prepare the statements required by various constituencies. The data can also be used in the managerial accounting module of the ERP system along with various assumptions and forecasts to perform various analyses such as generating a forecasted profit and loss statement to assess the firm's future profitability.

The U.S. Department of Justice will spend $150 million to convert from a collection of nonintegrated and fragmented systems to a single, integrated set of financial management software. It will install CGI-ASM's Momentum software, including software modules for Core Accounting/General Ledger, Financial Management Reporting, Payment Management, Receivables Management, Funds Management, Cost Management, and Procurement. The goal is to standardize processes, improve efficiency, promote information sharing, allow enterprise-wide views of income and expenses, enhance security, and increase accountability.[29]

Hosted Software Model for Enterprise Software

Business application software vendors are experimenting with the hosted software model to see if the approach meets customer needs and is likely to generate significant revenue. The target market is primarily small-to-medium businesses employing five to 75 people with revenue up to $20 million. NetSuite, Salesforce.Com Inc., Everest Software, and SAP are among the software vendors who offer hosted versions of their ERP and CRM software at a cost of $75–$150 per month per user.[30] This pay-as-you-go approach is appealing to small businesses because they can then experiment with powerful software capabilities without making a major financial investment. Also, using the hosted software model means the small business firm does not need to employ a full-time IT person to maintain key business applications. The small business firm can expect additional savings from reduced hardware costs and costs associated with maintaining an appropriate computer environment (such as air conditioning, power, and an uninterruptible power supply).

Not only is the hosted software model attractive to small and medium-sized firms, even some large companies are experimenting with it. DuPont, the large, multinational chemical company with recent annual sales of $30 billion, is one of the early adopters of the hosted software model. The firm is retooling its sales force by leveraging best practices and focusing its e-business and marketing capabilities into 16 high-powered global centers. As part of the change, DuPont plans to use the hosted SAP Sales on Demand software across the enterprise to provide a common systems platform and a common set of business processes for DuPont's entire sales force. It hopes to integrate the hosted system with its SAP ERP software and retire some of its legacy CRM applications. The business goal is to make sure that the firm presents itself as one DuPont to customers who buy from different DuPont businesses. Its largest customers are served as "corporate accounts" with a point of contact who can manage all their interactions with DuPont to ensure the maximum benefit to the customer.[31]

INTERNATIONAL ISSUES ASSOCIATED WITH ENTERPRISE SYSTEMS

Enterprise systems must support businesses that interoperate with customers, suppliers, business partners, shareholders, and government agencies in multiple countries. Different languages and cultures, disparities in IS infrastructure, varying laws and customs rules, and multiple currencies are among the challenges that must be met by an enterprise system of a multinational company. The following sections highlight these issues.

Different Languages and Cultures

Teams composed of people from several countries speaking different languages and familiar with different cultures might not agree on a single work process. In some cultures, people do not routinely work in teams in a networked environment. Despite these complications, many multinational companies can establish close connections with their business partners and roll out standard IS applications for all to use. However, those standard applications often don't account for all the differences among business partners and employees operating in other parts of the world. So, sometimes they require extensive and costly customization. For example, even though English has become a standard business language among executives and senior managers, many people within organizations do not speak English. As a result, software might need to be designed with local language interfaces to ensure the successful implementation of a new system. Customization might also be needed for date fields: The U.S. date format is month/day/year, the European format is day/month/year, and Japan uses year/month/day. Sometimes, users might also have to implement manual processes to override established formatting to enable systems to function correctly.

Disparities in Information System Infrastructure

The lack of a robust or a common information infrastructure can also create problems. The U.S. telecommunications industry is highly competitive, with many options for high-quality service at relatively low rates. Many other countries' telecommunications services are controlled by a central government or operated as a monopoly, with no incentives to provide fast and inexpensive customer service. For example, much of Latin America lags the rest of the world in Internet usage, and online marketplaces are almost nonexistent there. This gap makes it difficult for multinational companies to get online with their Latin American business partners. Even something as mundane as the power plug on a piece of equipment built in one country might not fit into the power socket of another country.

Varying Laws and Customs Rules

Numerous laws can affect the collection and dissemination of data. For example, labor laws in some countries prohibit the recording of worker performance data. Also, some countries have passed laws limiting the transborder flow of data linked to individuals. Specifically, European Community Directive 95/96/EC of 1998 requires that any company doing business within the borders of the 25 European Union member nations protect the privacy of customers and employees. It bars the export of data to countries that do not have data-protection standards comparable to the European Union's.

Trade custom rules between nations are international laws that set practices for two or more nations' commercial transactions. They cover imports and exports and the systems and procedures dealing with quotas, visas, entry documents, commercial invoices, foreign trade zones, payment of duty and taxes, and many other related issues. For example, the North American Free Trade Agreement (NAFTA) of 1994 created trade custom rules to address the flow of goods throughout the North American continent. Most of these custom rules and their changes over time create nightmares for people who must keep enterprise systems consistent with the rules.

Multiple Currencies

The enterprise system of multinational companies must conduct transactions in multiple currencies. To do so, a set of exchange rates is defined, and the information systems apply these rates to translate from one currency to another. The systems must be current with foreign currency exchange rates, handle reporting and other transactions such as cash receipts, issue vendor payments and customer statements, record retail store payments, and generate financial reports in the currency of choice.

ERP software vendors are working to help meet these challenges. For example, the Global Trade Services software from SAP helps customers who do business globally to cut through the morass of international trade laws and procedures. The software can automatically create a letter of credit, a document commonly used in trade in Europe and Asia that is issued by a financial institution and acts as an irrevocable guarantee of payment to a supplier. About 70 percent of all transactions based on manually prepared letters of credit are rejected because of errors. Proper standardization of the processes will boost both efficiency and cut costs. The software also automates customs warehousing processes that allow companies to defer or eliminate duties payments. The software also supports multiple languages, including Chinese.[32]

Comparison of Several ERP Systems

ERP systems are commonly used in manufacturing companies, colleges and universities, professional service organizations, retailers, and healthcare organizations. Small and midsized companies, which generally include those ranging from $50 million to $500 million in annual revenue, represent the greatest growth for ERP companies. Table 9.9 summarizes a few of the leading vendors of ERP systems for such midsized firms.

Table 9.9

Some ERP Software Vendors

Software Vendor	Product	Industries Where Used	Strengths	Weaknesses
IFS	IFS Applications™	Aviation and defense, automotive, medical device, process industries	Offers many software components so users can avoid large-scale upgrades	Lacks global recognition, firm not profitable
Microsoft (Axapta)	Dynamics AX 4.0	Too soon to tell who will adopt this software	Has easy linkages to other Microsoft tools	Too new to have full functionality needed by many firms
Oracle	Oracle E-business Suite	Aerospace and defense, automotive, consumer packaged goods, industrial manufacturing, oil and gas	Strong manufacturing functionality	Likely to be replaced by completely new Oracle product by 2013
Oracle	JDE Edwards Enterprise One	Chemical industries, consumer packaged goods, life sciences, oil and gas	Product widely used in target industries	Likely to be replaced by completely new Oracle product by 2013
QAD	QAD Global Enterprise Edition	Automotive supply, food and beverage, and medial instrumentation	Noted for global reach, ease of implementation at individual plant level, and industry experience	Product not well aligned with Microsoft or IBM

Interestingly, although Microsoft and SAP market competing enterprise software, they also partnered with one another in the development of Duet, a set of technologies that enables SAP users to access and interact with their back-end ERP system via a familiar Microsoft Outlook interface. The goal is to boost worker productivity without additional training. Easy access to key back-end data helps SAP/Microsoft users to make faster and yet more informed business decisions.[33]

SUMMARY

Principle

An organization must have information systems that support the routine, day-to-day activities that occur in the normal course of business and help a company add value to its products and services.

Transaction processing systems (TPSs) are at the heart of most information systems in businesses today. A TPS is an organized collection of people, procedures, software, databases, and devices used to capture fundamental data about events that affect the organization (transactions) and use that data to update the official records of the organization. All TPSs perform the following basic activities: data collection, which involves the capture of source data to complete a set of transactions; data editing, which checks for data validity and completeness; data correction, which involves providing feedback of a potential problem and enabling users to change the data; data manipulation, which is the performance of calculations, sorting, categorizing, summarizing, and storing data for further processing; data storage, which involves placing transaction data into one or more databases; and document production, which involves outputting records and reports.

The methods of transaction processing systems include batch and online. Batch processing involves the collection of transactions into batches, which are entered into the system at regular intervals as a group. Online transaction processing (OLTP) allows transactions to be entered as they occur.

Organizations expect TPSs to accomplish a number of specific objectives, including processing data generated by and about transactions, maintaining a high degree of accuracy and information integrity, compiling accurate and timely reports and documents, increasing labor efficiency, helping provide increased and enhanced service, and building and maintaining customer loyalty. In some situations, an effective TPS can help an organization gain a competitive advantage.

Because of the importance of TPSs to ongoing operations, organizations must develop a disaster recovery plan that focuses on the actions that must be taken to restore computer operations and services in the event of a disaster. Although companies have known about the importance of disaster planning and recovery for decades, many do not adequately prepare.

Principle

Traditional TPSs support the various business functions of organizations that have not yet implemented enterprise systems.

Many organizations conduct ongoing TPS audits to prevent accounting irregularities or loss of data privacy that can violate federal acts and diminish investor confidence. The audit can be performed by the firm's internal audit group or by an outside auditor for greater objectivity. The TPS audit attempts to answer four basic questions: (1) Does the system meet the business need for which it was implemented? (2) What procedures and controls have been established? (3) Are these procedures and controls being used properly? and (4) Are the information systems and procedures producing accurate and honest reports?

The traditional TPS systems that support the order processing business functions include order entry, sales configuration, shipment planning, shipment execution, inventory control, and accounts receivable.

The traditional TPSs that support the purchasing function include inventory control, purchase order processing, accounts payable, and receiving.

The traditional TPSs that support the accounting business function include the budget, accounts receivable, payroll, asset management, and general ledger.

Principle

A company that implements an enterprise resource planning system is creating a highly integrated set of systems, which can lead to many business benefits.

Enterprise resource planning (ERP) is software that supports the efficient operation of business processes by integrating activities throughout a business, including sales, marketing, manufacturing, logistics, accounting, and staffing. Implementation of an ERP system can provide many advantages, including providing access to data for operational decision making; elimination of costly, inflexible legacy systems; providing improved work processes; and creating the opportunity to upgrade technology infrastructure. Some of the disadvantages associated with an ERP system are that they are time consuming, difficult, and expensive to implement.

Although the scope of ERP implementation can vary from firm to firm, most firms use ERP systems to support production and supply chain management, customer relationship management and sales ordering, and financial and managerial accounting.

The production and supply chain management process starts with sales forecasting to develop an estimate of future customer demand. This initial forecast is at a fairly high level with estimates made by product group rather than by each individual product item. The sales and operations plan takes demand and current inventory levels into account and determines the specific product items that need to be produced and when to meet the forecast future demand. Demand management refines the production plan by determining the amount of weekly or daily production needed to meet the demand for individual products. Detailed scheduling uses the production plan defined by the demand management process to develop a detailed production schedule specifying production scheduling details such as which item to produce first and when production should be switched from one item to another. Materials requirement planning determines the amount and timing for placing raw material orders with suppliers. Purchasing uses the information from materials requirement planning to place purchase orders for raw materials and transmit them to qualified suppliers. Production uses the detailed schedule to plan the details of running and staffing the production operation.

The individual application modules included in the ERP system are designed to support best practices, the most efficient and effective ways to complete a business process.

Business application software vendors are experimenting with the hosted software model to see if the approach meets customer needs and is likely to generate significant revenue.

Numerous complications arise that multinational corporations must address in planning, building, and operating their TPSs. These challenges include dealing with different languages and cultures, disparities in IS infrastructure, varying laws and customs rules, and multiple currencies.

CHAPTER 9: SELF-ASSESSMENT TEST

An organization must have information systems that support the routine, day-to-day activities that occur in the normal course of business and help a company add value to its products and services.

1. Identify the missing TPS basic activity: data collection, data editing, data _____, data manipulation, data storage, and document production.

2. The primary objective of any TPS is to capture, process, and store transactions and to produce a variety of documents related to routine business activities. True or False?

3. Which of the following is *not* one of the basic components of a TPS?
   a. databases
   b. networks
   c. procedures
   d. analytical models

4. A form of TPS where business transactions are entered as soon as they occur is called _____.

5. Data should be captured at its source and recorded accurately in a timely fashion, with minimal manual effort, and in an electronic or digital form that can be directly entered into the computer are the principles behind _____.

6. Which of the following statements is *true*?
   a. Disaster recovery focuses on the actions that must be taken to restore computer operations and services in the event of a disaster.
   b. Companies have known about the importance of disaster planning and recovery for decades.
   c. Companies vary widely in the thoroughness and effectiveness of their business continuity planning.
   d. all of the above

Traditional TPSs support the various business functions of organizations that have not yet implemented enterprise systems.

7. Many organizations conduct ongoing transaction processing system _____ to prevent accounting irregularities or loss of data privacy that might violate federal acts.

8. The _____ system determines which open orders will be filled and from which location they will be shipped.

9. Inventory control, purchase order processing, receiving, and accounts payable systems make up a set of systems that support the _____ business function.

10. The _____ transaction processing system manages the cash flow of the company by keeping track of the money owed the company.

A company that implements an enterprise resource planning system is creating a highly integrated set of systems, which can lead to many business benefits.

11. Many multinational companies roll out standard IS applications for all to use. However, those standard applications often don't account for all the differences among business partners and employees operating in other parts of the world. Which of the following is a frequent modification that is needed to standard software?
   a. Software might need to be designed with local language interfaces to ensure the successful implementation of a new IS.
   b. Customization might be needed to handle date fields correctly.
   c. Users might also have to implement manual processes and overrides to enable systems to function correctly.
   d. all of the above

12. Which of the following is a primary benefit of implementing an ERP system?
   a. elimination of inefficient systems
   b. easing adoption of improved work processes

c. improving access to data for operational decision making
d. all of the above

13. The individual application modules included in an ERP system are designed to support the _____ _____, the most efficient and effective ways to complete a business process.

14. Because it is so critical to the operation of an organization, most companies are able to implement an ERP system without major difficulty. True or False?

15. Only large, multinational companies can justify the implementation of ERP systems. True or False?

CHAPTER 9: SELF-ASSESSMENT TEST ANSWERS

(1) correction (2) True (3) d (4) online processing (5) source data automation (6) d (7) audit (8) shipment planning (9) purchasing (10) accounts receivable (11) d (12) d (13) best practices (14) False (15) False

REVIEW QUESTIONS

1. List several characteristics that distinguish a TPS from an MIS.
2. What basic transaction processing activities are performed by all transaction processing systems?
3. Distinguish between a batch processing system and an online processing system.
4. An ERP system follows a systematic process for developing a production plan that draws on the information available in the ERP system database. Outline this process and identify the software modules that are used to support it.
5. What is source data automation? Give an example.
6. Identify four complications that multinational corporations must address in planning, building, and operating their ERP systems.
7. How does materials requirement planning support the purchasing process? What are some of the issues and complications that arise in materials requirement planning?
8. A disaster recovery plan focuses on what two key issues?
9. What is the role of a CRM system? What sort of business benefits can such a system produce?
10. What systems are included in the traditional TPS systems that support the accounting business function?
11. What is the purpose of the shipment planning system?
12. Why is the general ledger application key to the generation of accounting information and reports?
13. What is the difference between managerial and financial accounting?
14. What is the chart of accounts? How is it used?
15. List and briefly describe the set of activities that must be performed by the sales ordering module of an ERP system to capture a customer sales order.

DISCUSSION QUESTIONS

1. Do you think that it is possible for a TPS to provide a firm with a competitive advantage? Why or why not?
2. Assume that you are the owner of a large landscaping firm serving hundreds of customers in your tristate area. Identify the kinds of customer information you would like to have captured by your firm's CRM system. How might this information be used to provide better service or increase revenue?
3. Imagine that you are the new IS manager for a *Fortune* 1000 company. You have uncovered the complete lack of a disaster recovery plan for many of the firm's critical systems. Prepare a brief outline of a talk you will make to senior company managers to convince them that substantial resources must be assigned to developing a disaster recovery plan.
4. Identify five or six key elements or pieces of information that should be included in a firm's disaster recovery plan. What steps would you take to define your firm's disaster recovery plan? Who should be involved?
5. What are the advantages of implementing ERP as an integrated solution to link multiple business processes versus a series of nonintegrated TPS systems? Can you identify any disadvantages?
6. What are some of the challenges and potential problems of implementing an ERP system for a large, multinational corporation?
7. You are the key user of the firm's accounts receivable system and have been asked to lead an internal audit of this system. Outline the steps you would take to complete the audit. Identify specific problems you would look for.

8. Identify a line of business that would benefit from the use of a sales configuration system. Describe how such a system might work for this company.

9. Your friend has been appointed the project manager of your firm's ERP implementation plan. What advice would you offer to help ensure the success of the project?

10. What sort of benefits should the suppliers and customers of a firm that has successfully implemented an ERP system see? What sort of issues might arise for suppliers and customers during an ERP implementation?

11. Many organizations are moving to a collaborative process with their major customers to get their input on planning future inventory levels and production rather than relying on an internally generated demand forecast. Explain how such a process might work. What issues and concerns might a customer have in entering into an agreement to do this?

PROBLEM-SOLVING EXERCISES

1. Assume that you are starting an online store for athletic clothing (such as shoes, jerseys, wind breakers, sweatshirts, hats, and t-shirts) that will allow users to "browse" your aisles electronically via the Web and make their purchase selections. After an order is complete, the items are pulled from a warehouse, packed, and shipped overnight. Using a graphics program, draw a diagram that shows the different ways you will interact with your customers. Use a word processing program to develop a list of key facts you would like to capture about each customer and about each contact with a customer.

2. The order processing application in your online athletic clothing store has three main databases: items, orders, and customers. The item database contains information about each item in stock. The order database identifies basic information about each order placed, including customer, date, and items ordered. The customer database contains each customer's ID number, address, and contact information.

Use a database management system to build a simple transaction processing system to support the online athletic clothing store operation. Enter the complete database definitions into your database management software and create a data-entry screen to efficiently capture customer selections, build an order, and ship and bill for the order.

   a. Enter several items, including description and price, into the item database.
   b. Enter three customers into the customer database.
   c. Enter the data necessary to process two different orders to two different customers.

TEAM ACTIVITIES

1. Assume that your team has formed a consulting firm to perform an external audit of a firm's accounting information systems. Develop a list of at least ten questions you would ask as part of your audit. What specific inputs and outputs from various systems would you want to see? Visit a company and perform the audit based on these questions or role play the scenario if a live visit is impossible.

2. Your team members should interview several business managers at a firm that has implemented an ERP system. Try to define the scope and schedule for the overall project. Make a list of what they see as the primary benefits of the implementation. What were the biggest hurdles they had to overcome?

WEB EXERCISES

1. Do research on the Web and identify the most popular ERP system solution for small and medium-sized businesses. Why is this solution the most popular? Develop a one-page report or send an e-mail message to your instructor about what you found.

2. Using the Web, identify several companies that have implemented an ERP system in the last two years. Classify the implementations as success, partial success, or failure. What is your basis for making this classification? Do you see any common reasons for success? For failure?

CAREER EXERCISES

1. Initially thought to be cost-effective for only very large companies, ERP systems are now being implemented in small and midsized companies to reduce costs and improve service. A firm's operations and accounting personnel play a dual role in the implementation of such a system: (1) They must ensure a good payback on the investment in ERP, and (2) they must also ensure that the system meets the needs of the operations and accounting organizations. Identify three or four tasks that the operations and accounting personnel need to perform to ensure that both these goals are met.

2. CRM software vendors need business systems analysts that understand both information systems and business processes. Make a list of six or more specific qualifications needed to be a strong business systems analyst supporting a CRM implementation within a medium-sized, but global organization.

CASE STUDIES

Case One

Pico Electronics Develops TPS to Support Rapid-Fire Transactions

Malaysian electronic components distributor Pico Electronics had a problem that many midsized businesses would like to have: Orders were coming in too quickly; they had to process an order every few minutes. Pico didn't mind its order volume, but it was having difficulty keeping an accurate picture of current inventory. The company wanted to avoid a scenario in which three salespeople are accepting orders for the same item with only enough inventory to support two of the orders. In 2006, Pico Electronics went shopping for a new order processing system that would provide real-time visibility to avoid such problems.

Pico's procurement division needed to see what was being ordered as soon as it was ordered. The division could then purchase the required materials as necessary to fulfill customer's orders on time. The company wanted to be more nimble in its ability to stay ahead of the competition.

The project evaluation team, headed by finance manager Sharyne Tee, set several primary goals for the new system. The new system should:

- Go beyond order processing to include all relevant business process–related data.
- Generate reports to give the management an accurate snapshot of the company's performance in real time.
- Accurately track inventory and costs in real time so that they could anticipate demand and have reasonable inventory reserves.
- Allow for fast access to both overseas customer and vendor histories.
- Provide a stable and fast connection to link their overseas offices in China, Hong Kong, Malaysia, and Thailand.

To support these needs, Pico purchased IBM BladeCentre HS20 servers for their Singapore and China offices. Pico decided on SAP Business One software as a platform for the new TPS and ERP system. Since the implementation, SAP Business One has allowed Pico management to view current inventory in the warehouse. The software also lets them analyze the performance of the separate divisions in the company. Because this software automates accounting-related data collection, Pico can better handle its accounts receivable, which improves cash flow.

Pico purchased a Jobshop add-on module designed by Sabre that allows the company to add more functionality as it grows. The module manages the entire production environment, from bill of material to shop floor control. With the Jobshop module, Pico can monitor all planning and production control activities and integrate sales, purchasing, inventory, and financial activities.

This system development project at Pico Electronics illustrates how improvements to one system often involve changes to many systems. It would be unwise to examine changes to one system, such as a TPS or CRM, without evaluating their effect on related systems, such as ERP.

Discussion Questions

1. Explain the problem that Pico wanted to address, and why Pico considered it a problem.
2. What is meant by real-time visibility? How does it relate to the pace at which products are sold?

Critical Thinking Questions

1. Why do changes to one type of system, such as a TPS, affect other systems in an enterprise?
2. What additional challenges do you think Pico faced in extending real-time visibility to its overseas offices?

SOURCES: Liew, Kenneth, "Pico Electronics Gains Real-Time Visibility," *Computerworld Malaysia*, March 23, 2006, *www.computerworld.com.my*.

Case Two

Catholic Healthcare West Implements ERP

As mentioned in this chapter, the implementation of an ERP system can be difficult and time consuming. No one recognizes this fact better than Rick Canning, vice president of administrative systems at San Francisco-based healthcare system, Catholic Healthcare West (CHW). CHW is nearing the end of a rocky ten-year effort to standardize packaged ERP applications in its 40 hospitals and medical centers. After ten years, the healthcare system is finally seeing some return on its significant investment in terms of cost reductions and streamlined business processes.

The project had a rough start in 1996 when it rolled out SAP ERP, finance, and supply chain systems in seven of its hospitals. After five years and $120 million, CHW determined that the new system was a failure. Users found SAP supply chain and financial applications "extremely complex" and not intuitive, Canning said.

In 2001, a new CEO oversaw a corporate reorganization to stem operational losses in 1999 and 2000, and that led to a decision to outsource CHW's IT operations to Perot Systems Corporation. Perot alleged that the SAP system was geared more toward manufacturing industries and suggested ERP systems from Lawson, which generates 40 percent of its revenue from healthcare organizations. An SAP spokeswoman blamed the failure of the project on executive changes at CHW.

Moving forward with their new plan, CHW took six months to find the right staff to support the Lawson applications and then began the implementation. CHW formed a project management office and established four committees to develop data standards for the Lawson human resources, finance, supply chain, and payroll systems. In 2002, CHW deployed a pilot project using the Lawson systems, which was declared a huge success. Over the following four years, CHW worked hard to install the systems in all of their hospitals. By the end of 2006, all CHW hospitals in California, Arizona, and Nevada were on board and using the new systems.

The new Lawson systems replace legacy systems that include 200 internally developed and packaged systems, financial applications from six vendors, and materials management software from four vendors. The ERP overhaul has already cut supply costs by $1.5 million and annual IT support and paper costs by $1 million apiece. CHW officials are projecting a 144 percent return on investment over the next eight years, or $94 million in savings.

The new software has helped CHW reduce the time it takes to close its books from 14 days at some hospitals to one day. A recent project to consolidate the management of employee retirement plans from multiple vendors to one took four months; the same task would have taken a year using the old systems.

The new systems empower users. Linda Pike, the Lawson materials management system administrator at St. Rose Dominican Hospital in Henderson, Nevada, said the new software lets her department generate reports to monitor pur-chasing compliance under established contracts. Previously, her unit had to rely on administrators from a regional office in Phoenix to run the reports. "We can retrieve that data from our system relatively easily using Lawson canned reports and tools we use with Excel," Pike said.

Discussion Questions

1. What was the primary reason the ERP systems took so long to implement at Catholic Healthcare West? How might that problem have been avoided?
2. What did Catholic Healthcare West do right in implementing the Lawson ERP system?

Critical Thinking Questions

1. What types of difficulties, as exemplified in this article, would motivate a business to endure the painstaking effort of ERP system implementation?
2. Do you expect that Catholic Healthcare West will need to overhaul their systems again? Why or why not?

SOURCES: Havenstein, Heather, "Health Care Provider Nears End of 10-Year ERP Journey," *Computerworld*, December 19, 2005, *www.computerworld.com/action/article.do?command=viewArticleBasic&articleId=107180*. Catholic Healthcare West Success Story, Novell Cool Solutions: Feature, May 17, 2006, *www.novell.com/coolsolutions/feature/17201.html*. Lawson Web site, accessed May 23, 2006, *www.lawson.com*. "About Pico – Chairman's Message," Pico Electronics Web site, *www.picodenshi.com/abt_pmsg.htm*, accessed February 16, 2006.

Questions for Web Case

See the Web site for this book to read about the Whitmann Price Consulting case for this chapter. Following are questions concerning this Web case.

Whitmann Price Consulting Inc.: Enterprise Systems

Discussion Questions

1. What would be the danger of Josh and Sandra developing the Advanced Mobile Communications and Information System without considering other systems within Whitmann Price?
2. What are the advantages of ERP systems that provide an integrated one-vendor approach over multiple systems from multiple vendors?

Critical Thinking Questions

1. What are the pros and cons of buying predesigned software from a vendor such as SAP instead of a company developing software itself to exactly meet its own needs?
2. How might a CRM system assist Whitmann Price consultants in the field?

NOTES

Sources for the opening vignette: Staff, "Barilla Wasa: Identifying Measurable Value With MySAP CRM", SAP Case Study, 2005, *www.sap.com/usa/solutions/business-suite/crm/customersuccess/index.epx.* Barilla Wasa Web site, accessed May 21, 2006, *www.wasa.com.* Barilla Group Web site, accessed May 21, 2006, *www.barillaus.com/The_Barilla_Group.aspx.* SAP Mobile Sales Web site, accessed May 21, 2006, *www.sap.com/solutions/xapps/mobilebusiness/mobilesales.epx.*

1 Lai, Eric, "ANTS Updates Database, Touts Increased Transaction Speeds," *Computerworld*, November 14, 2005.
2 Marlin, Steven, "Bank Deploys Anti-Money Laundering System," *Information Week*, October 11, 2005.
3 Babcock, Charles, "Liberty Mutual Gains Insight into Performance Slowdowns," *Information Week*, October 14, 2005.
4 Sullivan, Laurie, "Bad Online Shopping Experiences Are Bad for Business," *Information Week*, January 24, 2005.
5 Chabrow, Eric, "Oil Companies Turn to IT to Shave Costs, Improve Efficiency," *Information Week*, June 5, 2006.
6 Chabrow, Eric, "Oil Companies Turn to IT to Shave Costs, Improve Efficiency," *Information Week*, June 5, 2006.
7 Whiting, Rick, "Acxiom, IRI Tackle Data Accuracy," *Information Week*, May 8, 2006.
8 Whiting, Rick, "Avaya Group Safeguards Internal Information Quality," *Information Week*, May 8, 2006.
9 IQ Insight® Functionality, Business Objects Web site at *www.firstlogic.com/dataquality*, accessed July 12, 2006.
10 Hoffman, Thomas, "Double Dipping on SOX," *Computerworld*, November 7, 2005.
11 Sullivan, Laurie, "Small Companies Unprepared for Disasters: AT&T Survey," *Information Week*, June 29, 2006.
12 DeFelice, Alexandra, "How to Keep Your Data from Blowing in the Wind," *E-Commerce Times*, June 6, 2006.
13 Campbell, Scott, "New HP Tool Assesses Disaster Preparedness," *Information Week*, June 9, 2006.
14 Staff, "Customer Success Story: Arch Chemicals," SunGuard Web site, *www.SunGuard*, accessed July 8, 2006.
15 DeFelice, Alexandra, "How to Keep Your Data from Blowing in the Wind," *E-Commerce Times*, June 6, 2006.
16 Malykhina, Elena, "Beyond Contact Management," *Information Week*, October 3, 2005.
17 Hulme, George V., "Supply-Chain Software Cuts Costs for Auto-Parts Firm," *Information Week*, May 6, 2005.
18 Sullivan, Laurie, "ERP Vendors Battle for Retailers," *Computerworld*, June 13, 2005.
19 Songini, Marc, "PeopleSoft Apps Vex N.D, Colleges," *Computerworld*, June 23, 2006.
20 Songini, Marc, "King County Ready to Restart ERP Effort," *Computerworld*, September 5, 2005.
21 Songini, Marc, "GAO Says Navy Sank $1B into Four Failed ERP Pilot Projects," *Computerworld*, November 14, 2005.
22 Braue, David, "ERP Boosts Reporting, ROI for Manufacturer," *Computerworld*, April 12, 2006.
23 Anthes, Gary, "Supply Chain Whirl," *Computerworld*, June 6, 2005.
24 Anthes, Gary, "Sidebar: It's All Global Now," *Computerworld*, February 20, 2006.
25 Violino, Bob, "Focus on Customer Experience," *Information Week*, September 5, 2005.
26 Lai, Eric, "Salvation Army West Updates Fund-Raising Software," *Computerworld*, January 17, 2006.
27 McDougall, Paul, "Closing the Last Supply Gap," *Information Week*, November 8, 2005.
28 Glossary of terms at *www.crfonline.org/orc/glossary/m.html*, accessed June 25, 2006.
29 Jones, K.C., "IBM Aims to Streamline DOJ's Financial Management Systems," *Information Week*, January 19, 2006.
30 Martens, China, "Everest Enters On-Demand Market, Takes on Net Suite," *Computerworld*, May 10, 2006.
31 Songini, Marc, "SAP Launches First Piece of Hosted CRM Service," *Computerworld*, February 16, 2006.
32 Songini, Marc, "SAP Seeks to Cut Global Trading Risks with Updated App," *Computerworld*, June 14, 2005.
33 Songini, Marc, "SAP, Microsoft Set to Ship Integration Technology," *Computerworld*, May 2, 2006.

CHAPTER
· 10 ·

Information and Decision Support Systems

PRINCIPLES

LEARNING OBJECTIVES

PRINCIPLES	LEARNING OBJECTIVES
▪ Good decision-making and problem-solving skills are the key to developing effective information and decision support systems.	▪ Define the stages of decision making. ▪ Discuss the importance of implementation and monitoring in problem solving.
▪ The management information system (MIS) must provide the right information to the right person in the right format at the right time.	▪ Explain the uses of MISs and describe their inputs and outputs. ▪ Discuss information systems in the functional areas of business organizations.
▪ Decision support systems (DSSs) are used when the problems are unstructured.	▪ List and discuss important characteristics of DSSs that give them the potential to be effective management support tools. ▪ Identify and describe the basic components of a DSS.
▪ Specialized support systems, such as group support systems (GSSs) and executive support systems (ESSs), use the overall approach of a DSS in situations such as group and executive decision making.	▪ State the goals of a GSS and identify the characteristics that distinguish it from a DSS. ▪ Identify the fundamental uses of an ESS and list the characteristics of such a system.

Information Systems in the Global Economy ⟫
Beauté Prestige International, France

French Perfume Company Relies on Management Information Systems to Unify Employees

Beauté Prestige International (BPI) is a French company best known for its three brands of perfume: Issey Miyake, Jean Paul Gaultier, and Narciso Rodriguez. BPI is a small to midsized enterprise (SME) with an international presence through ten subsidiaries located around the world. BPI employs 1,300 people who sell its perfumes in 112 countries.

Recently, BPI found itself in need of a centralized system from which managers could produce meaningful business reports to guide decision making. "Everyone was producing spreadsheets in their own little corner of the company, which of course brought IT maintenance problems with it, especially in terms of consistency for company figures," BPI's CIO Christophe Davy explained. The company needed to shepherd its managers away from individual spreadsheets toward more centralized decision making based on reports generated by a management information system (MIS). Christophe Davy set off to develop a new system that would include standardized decision support tools, a user-friendly interface, easy implementation, and access to all key indicators—facts most important to decision making, from a single portal.

Christophe found his solution in the ReportNet system from the Cognos Corporation. Cognos worked with the IS staff at BPI to import corporate data into a data warehouse, which could then be manipulated by the ReportNet MIS to develop useful reports. One year later, the company placed its first batch of financial indicators online for use by the sales force and management control. Six months later, they rolled out merchandising data reports that were used to conduct analysis on the position of its products compared with the contracts negotiated with its customers, and to compare the companies positioning in relation to its competitors.

Importing the corporate data from the many individual spreadsheets was the largest hurdle for developers—but well worth the effort. "The work to clean up and amalgamate data was exactly what was needed to implement our strategy of having a single unified portal where employees could find reliable information that was shared throughout the entire company," explained Christophe Davy.

The resulting system has vastly improved business processes and decision making in the organization. Processes that used to take hours in Microsoft Excel are now done virtually in real time. Also BPI no longer requires external companies to provide annual reports on product positioning. Such reports are generated anytime they are needed by the new system. The biggest benefit of the new MIS is the ability to access reliable important corporate information anytime it is needed for any use within the company. Additional benefits include enhanced business monitoring, more effective queries, lower costs, shorter lead times, and enhanced interactivity. BPI is working to extend the power of the portal beyond finance and merchandising to other business areas.

Management information systems are keys to a smooth-running and unified organization. A well-implemented MIS puts everyone in the organization on the same page working toward common goals.

As you read this chapter, consider the following:

- How is an MIS used in the various functional areas of businesses?
- How do an MIS and DSS affect a company's business practices and its ability to compete in the market?

Why Learn About Information and Decision Support Systems?

You have seen throughout this book how information systems can make you more efficient and effective through database systems, the Internet, e-commerce, transaction processing systems, and many other technologies. The true potential of information systems, however, is in helping you and your coworkers make more informed decisions. This chapter shows you how to slash costs, increase profits, and uncover new opportunities for your company using management information and decision support systems. Transportation coordinators can use management information reports to find the least expensive way to ship products to market and to solve bottlenecks. A loan committee at a bank or credit union can use a group support system to help them determine who should receive loans. Store managers can use decision support systems to help them decide what and how much inventory to order to meet customer needs and increase profits. An entrepreneur who owns and operates a temporary storage company can use vacancy reports to help determine what price to charge for new storage units. Everyone wants to be a better problem solver and decision maker. This chapter shows you how information systems can help. It begins with an overview of decision making and problem solving.

As shown in the opening vignette, information and decision support are the lifeblood of today's organizations. Thanks to information and decision support systems, managers and employees can obtain useful information in real time. As discussed in Chapter 9, TPS and ERP systems capture a wealth of data. When this data is filtered and manipulated, it can provide powerful support for managers and employees. The ultimate goal of management information and decision support systems is to help managers and executives at all levels make better decisions and solve important problems. The result can be increased revenues, reduced costs, and the realization of corporate goals. Many of today's information and decision support systems are built into the organization's TPS or ERP systems. In other cases, they are developed separately. No matter what type of information and decision support system you use, a primary goal is to help you and others become better decision makers and problem solvers.

DECISION MAKING AND PROBLEM SOLVING

Every organization needs effective decision making. In most cases, strategic planning and the overall goals of the organization set the course for decision making, helping employees and business units achieve their objectives and goals. Often, information systems also assist with strategic planning and problem solving, helping people make better decisions and save lives. For example, an information system at Hackensack University Medical Center in New Jersey analyzes possible drug interactions.[1] In one case, an AIDS patient taking drugs for depression avoided an AIDS drug that could have dangerously interacted with the depression drug. According to one doctor, "There's no way I would have picked that up. It was totally unexpected." The hospital has invested millions of dollars into its information system.

Decision Making as a Component of Problem Solving

In business, one of the highest compliments you can receive is to be recognized by your colleagues and peers as a "real problem solver." Problem solving is a critical activity for any business organization. After identifying a problem, you begin the problem-solving process with decision making. A well-known model developed by Herbert Simon divides the **decision-making phase** of the problem-solving process into three stages: intelligence, design, and choice. This model was later incorporated by George Huber into an expanded model of the entire problem-solving process (see Figure 10.1). Increasingly, people are using computers to analyze decision-making styles and approaches.[2] Using computer-based magnetic resonance imaging (MRI) techniques, you can look inside the brain to understand the decision-making process and why some people make logical decisions and others make emotional decisions not based on logic. Dr. Laibson of Harvard University, for example, is using this technique to explore why some people make logical long-term decisions but emotional short-term decisions.

decision-making phase
The first part of problem solving, including three stages: intelligence, design, and choice.

How Decision Making Relates to Problem Solving

The three stages of decision making—intelligence, design, and choice—are augmented by implementation and monitoring to result in problem solving.

The first stage in the problem-solving process is the **intelligence stage**. During this stage, you identify and define potential problems or opportunities. For example, you might learn about the need for an intervention or change in an unsatisfactory situation. During the intelligence stage, you also investigate resource and environmental constraints. For example, if you were a Hawaiian farmer, during the intelligence stage you would explore the possibilities of shipping tropical fruit from your farm in Hawaii to stores in Michigan. The perishability of the fruit and the maximum price that consumers in Michigan are willing to pay for the fruit are problem constraints. Aspects of the problem environment that you must consider include federal and state regulations regarding the shipment of food products.

intelligence stage
The first stage of decision making, in which potential problems or opportunities are identified and defined.

In the **design stage**, you develop alternative solutions to the problem. In addition, you evaluate the feasibility of these alternatives. In the tropical fruit example, you would consider the alternative methods of shipment, including the transportation times and costs associated with each. During this stage, you might determine that shipment by freighter to California and then by truck to Michigan is not feasible because the fruit would spoil.

design stage
The second stage of decision making, in which alternative solutions to the problem are developed.

The last stage of the decision-making phase, the **choice stage**, requires selecting a course of action. In the tropical fruit example, you might select the method of shipping fruit by air from your Hawaiian farm to Michigan as the solution. The choice stage would then conclude with selection of an air carrier. As you will see later, various factors influence choice; the act of choosing is not as simple as it might first appear.

choice stage
The third stage of decision making, which requires selecting a course of action.

problem solving
A process that goes beyond decision making to include the implementation and monitoring stages.

implementation stage
A stage of problem solving in which a solution is put into effect.

monitoring stage
The final stage of the problem-solving process, in which decision makers evaluate the implementation.

programmed decision
A decision made using a rule, procedure, or quantitative method.

Problem solving includes and goes beyond decision making. It also includes the **implementation stage**, when the solution is put into effect. For example, if your decision is to ship tropical fruit to Michigan as air freight using a specific air freight company, implementation involves informing your farming staff of the new activity, getting the fruit to the airport, and actually shipping the product to Michigan.

The final stage of the problem-solving process is the **monitoring stage**. In this stage, decision makers evaluate the implementation to determine whether the anticipated results were achieved and to modify the process in light of new information. Monitoring can involve feedback and adjustment. For example, after the first shipment of fruit from Hawaii to Michigan, you might learn that the flight of your chosen air freight firm routinely stops in Phoenix, Arizona, where the plane sits on the runway for a number of hours while loading additional cargo. If this unforeseen fluctuation in temperature and humidity adversely affects the fruit, you might have to readjust your solution to include a new air freight firm that does not make such a stop, or perhaps you would consider a change in fruit packaging.

Programmed versus Nonprogrammed Decisions

In the choice stage, various factors influence the decision maker's selection of a solution. One such factor is whether the decision can be programmed. **Programmed decisions** are made using a rule, procedure, or quantitative method. For example, to say that inventory should be ordered when inventory levels drop to 100 units is a programmed decision because it adheres to a rule. Programmed decisions are easy to computerize using traditional information systems. For example, you can easily program a computer to order more inventory when inventory levels for a certain item reach 100 units or less. Most of the processes automated through enterprise resource planning or transaction processing systems share this characteristic: The relationships between system elements are fixed by rules, procedures, or numerical relationships. Management information systems can also reach programmed decisions by providing reports on problems that are routine and in which the relationships are well defined. (In other words, they are structured problems.)

Ordering more inventory when inventory levels drop to specified levels is an example of a programmed decision.

(Source: Courtesy of Symbol Technologies.)

nonprogrammed decision
A decision that deals with unusual or exceptional situations that can be difficult to quantify.

Nonprogrammed decisions, however, deal with unusual or exceptional situations. In many cases, these decisions are difficult to quantify. Determining the appropriate training program for a new employee, deciding whether to start a new type of product line, and weighing the benefits and drawbacks of installing a new pollution control system are examples. Each of these decisions contains unique characteristics, and standard rules or procedures might not apply to them. Today, decision support systems help solve many nonprogrammed decisions, in which the problem is not routine and rules and relationships are not well defined (unstructured or ill-structured problems). These problems can include deciding the best location for a new manufacturing plant or whether to rebuild a hospital that was severely damaged from a hurricane or tornado.

Optimization, Satisficing, and Heuristic Approaches

In general, computerized decision support systems can either optimize or satisfice. An **optimization model** finds the best solution, usually the one that will best help the organization meet its goals. For example, an optimization model can find the appropriate number of products that an organization should produce to meet a profit goal, given certain conditions and assumptions. Optimization models use problem constraints. A limit on the number of available work hours in a manufacturing facility is an example of a problem constraint. Some spreadsheet programs, such as Excel, have optimizing features (see Figure 10.2). A business such as an appliance manufacturer can use an optimization program to reduce the time and cost of manufacturing appliances and increase profits by millions of dollars. For example, the Scheduling Appointments at Trade Events (SATE) software package is an optimization program that schedules appointments between buyers and sellers at trade shows and meetings. Optimization software also allows decision makers to explore various alternatives.

Consider a few examples of how you can use optimization to achieve huge savings. Bombardier Flexjet, a company that sells fractional ownership of jets, used an optimization program to save almost $30 million annually to better schedule its aircraft and crews.[3] Hutchinson Port Holdings, the world's largest container terminal, saved even more—over $50 million annually.[4] The company processes a staggering 10,000 trucks and 15 ships every day, and used optimization to maximize the use of its trucks. Deere & Company, a manufacturer of commercial vehicles and equipment, increased shareholder value by over $100 million annually by using optimization to minimize inventory levels and by enhancing customer satisfaction.[5]

optimization model
A process to find the best solution, usually the one that will best help the organization meet its goals.

Figure 10.2

Optimization Software

Some spreadsheet programs, such as Microsoft Excel, have optimizing routines. This figure shows Solver, which can find an optimal solution given certain constraints.

satisficing model
A model that will find a good—but not necessarily the best—problem solution.

A **satisficing model** is one that finds a good—but not necessarily the best—problem solution. Satisficing is usually used because modeling the problem properly to get an optimal decision would be too difficult, complex, or costly. Satisficing normally does not look at all possible solutions but only at those likely to give good results. Consider a decision to select a location for a new manufacturing plant. To find the optimal (best) location, you must consider all cities in the United States or the world. A satisficing approach is to consider only five or ten cities that might satisfy the company's requirements. Limiting the options might not result in the best decision, but it will likely result in a good decision, without spending the time and effort to investigate all cities. Satisficing is a good alternative modeling method because it is sometimes too expensive to analyze every alternative to find the best solution.

heuristics
Commonly accepted guidelines or procedures that usually find a good solution.

Heuristics, often referred to as "rules of thumb"—commonly accepted guidelines or procedures that usually find a good solution—are often used in decision making. A heuristic that baseball team managers use is to place batters most likely to get on base at the top of the lineup, followed by the power hitters who can drive them in to score. An example of a heuristic used in business is to order four months' supply of inventory for a particular item when the inventory level drops to 20 units or less; although this heuristic might not minimize total inventory costs, it can serve as a good rule of thumb to avoid stockouts without maintaining excess inventory. Trend Micro, a provider of antivirus software, has developed an antispam product that is based on heuristics. The software examines e-mails to find those most likely to be spam. It doesn't examine all e-mails.

Sense and Respond

Sense and Respond (SaR) involves determining problems or opportunities (sense) and developing systems to solve the problems or take advantage of the opportunities (respond).[6] SaR often requires nimble organizations that replace traditional lines of authority with those that are flexible and dynamic. IBM, for example, used SaR with its Microelectronics Division to help with inventory control. The Microelectronics Division, located in Burlington, Vermont, used mathematical models and optimization routines to control inventory levels. The models sensed when a shortage of inventory for customers was likely and responded by backlogging and storing extra inventory to avoid the shortages. In this application, SaR identified potential problems and solved them before they became a reality. SaR can also identify opportunities, such as new products or marketing approaches, and then respond by building the new products or starting new marketing campaigns. One way to implement the SaR approach is through management information and decision support systems, discussed next.

AN OVERVIEW OF MANAGEMENT INFORMATION SYSTEMS

A management information system (MIS) is an integrated collection of people, procedures, databases, and devices that provides managers and decision makers with information to help achieve organizational goals. MISs can often give companies and other organizations a competitive advantage by providing the right information to the right people in the right format and at the right time. For example, a shipping department could develop a spreadsheet to generate a report on possible delays to increase the number of on-time deliveries for the day. A music store might use a database system to develop a report that summarizes profits and losses for the month to make sure that the store is on track to make a 10 percent profit for the year.

Management Information Systems in Perspective

The primary purpose of an MIS is to help an organization achieve its goals by providing managers with insight into the regular operations of the organization so that they can control, organize, and plan more effectively. One important role of the MIS is to provide the right information to the right person in the right format at the right time. In short, an MIS provides managers with information, typically in reports, that supports effective decision making and provides feedback on daily operations.[7] Three California utilities, Pacific Gas & Electric, Southern California Edison, and San Diego Gas & Electric, are investigating a new MIS that uses high-tech electric meters to generate utility reports, giving accurate and up-to-date information on electricity usage for homes and small businesses.[8] If implemented, the new MIS would cost about $3.6 billion and allow the utilities to monitor utility usage and charge customers more during peak usage and less during other times of the day. The proposed MIS, which is controversial to some, would also reduce the number of "meter-reader" jobs.

Figure 10.3 shows the role of MISs within the flow of an organization's information. Note that business transactions can enter the organization through traditional methods or via the Internet or an extranet connecting customers and suppliers to the firm's ERP or transaction processing systems. The use of MISs spans all levels of management. That is, they provide support to and are used by employees throughout the organization.

Figure 10.3

Sources of Managerial Information

The MIS is just one of many sources of managerial information. Decision support systems, executive support systems, and expert systems also assist in decision making.

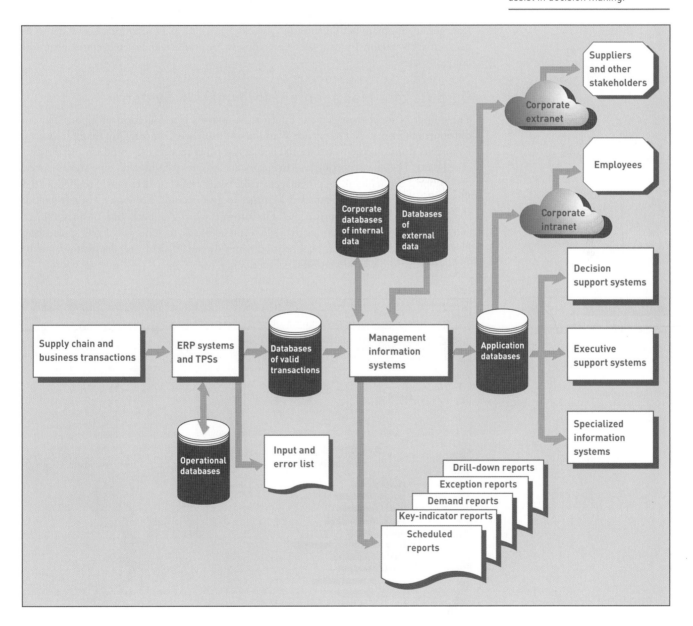

Inputs to a Management Information System

As shown in Figure 10.3, data that enters an MIS originates from both internal and external sources, including the company's supply chain, first discussed in Chapter 2. The most significant internal data sources for an MIS are the organization's various TPS and ERP systems and related databases. As discussed in Chapter 5, companies also use data warehouses and data marts to store valuable business information. Business intelligence, also discussed in Chapter 5, can be used to turn a database into useful information throughout the organization. Other internal data comes from specific functional areas throughout the firm.

External sources of data can include customers, suppliers, competitors, and stockholders, whose data is not already captured by the TPS, as well as other sources, such as the Internet. In addition, many companies have implemented extranets to link with selected suppliers and other business partners to exchange data and information.

The MIS uses the data obtained from these sources and processes it into information more usable by managers, primarily in the form of predetermined reports. For example, rather than simply obtaining a chronological list of sales activity over the past week, a national sales manager might obtain her organization's weekly sales data in a format that allows her to see sales activity by region, by local sales representative, by product, and even in comparison with last year's sales.

Outputs of a Management Information System

The output of most management information systems is a collection of reports that are distributed to managers. Providence Washington Insurance Company, for example, is using ReportNet from Cognos to reduce the number of paper reports they produce and the associated costs. The new reporting system creates an "executive dashboard" that shows current data, graphs, and tables to help managers make better real-time decisions. See Figure 10.4. Microsoft makes Business Scorecard Manager to give decision makers timely information about sales and customer information.[9] The software, which competes with Business Objects and Cognos, can integrate with other Microsoft software, including its Excel spreadsheet. Hewlett-Packard's OpenView Dashboard can quickly and efficiently render pictures, graphs, and tables that show how a business is functioning.[10]

Figure 10.4

An Executive Dashboard

This MIS reporting system puts many kinds of real-time information at managers' fingertips to aid in decision making.

Management reports can come from various company databases, data warehouses, and other sources. These reports include scheduled reports, key-indicator reports, demand reports, exception reports, and drill-down reports (see Figure 10.5).

(a) Scheduled Report

Daily Sales Detail Report

Prepared: 08/10/08

Order #	Customer ID	Salesperson ID	Planned Ship Date	Quantity	Item #	Amount
P12453	C89321	CAR	08/12/08	144	P1234	$3,214
P12453	C89321	CAR	08/12/08	288	P3214	$5,660
P12454	C03214	GWA	08/13/08	12	P4902	$1,224
P12455	C52313	SAK	08/12/08	24	P4012	$2,448
P12456	C34123	JMW	08/13/08	144	P3214	$720
......	......	......	......	......	......	......

(b) Key-Indicator Report

Daily Sales Key-Indicator Report

	This Month	Last Month	Last Year
Total Orders Month to Date	$1,808	$1,694	$1,914
Forecasted Sales for the Month	$2,406	$2,224	$2,608

(c) Demand Report

Daily Sales by Salesperson Summary Report

Prepared: 08/10/08

Salesperson ID	Amount
CAR	$42,345
GWA	$38,950
SAK	$22,100
JWN	$12,350
......	......

(d) Exception Report

Daily Sales Exception Report—Orders Over $10,000

Prepared: 08/10/08

Order #	Customer ID	Salesperson ID	Planned Ship Date	Quantity	Item #	Amount
P12345	C89321	GWA	08/12/08	576	P1234	$12,856
P22153	C00453	CAR	08/12/08	288	P2314	$28,800
P23023	C32832	JMN	08/11/08	144	P2323	$14,400
......	......	......	......	......	......	......

(e) First-Level Drill-Down Report

Earnings by Quarter (Millions)

		Actual	Forecast	Variance
2nd Qtr.	2008	$12.6	$11.8	6.8%
1st Qtr.	2008	$10.8	$10.7	0.9%
4th Qtr.	2008	$14.3	$14.5	-1.4%
3rd Qtr.	2008	$12.8	$13.3	-3.8%

(f) Second-Level Drill-Down Report

Sales and Expenses (Millions)

Qtr: 2nd Qtr. 2008	Actual	Forecast	Variance
Gross Sales	$110.9	$108.3	2.4%
Expenses	$ 98.3	$ 96.5	1.9%
Profit	$ 12.6	$ 11.8	6.8%

(g) Third-Level Drill-Down Report

Sales by Division (Millions)

Qtr: 2nd Qtr. 2008	Actual	Forecast	Variance
Beauty Care	$ 34.5	$ 33.9	1.8%
Health Care	$ 30.0	$ 28.0	7.1%
Soap	$ 22.8	$ 23.0	-0.9%
Snacks	$ 12.1	$ 12.5	-3.2%
Electronics	$ 11.5	$ 10.9	5.5%
Total	$110.9	$108.3	2.4%

(h) Fourth-Level Drill-Down Report

Sales by Product Category (Millions)

Qtr: 2nd Qtr. 2008 Division: Health Care	Actual	Forecast	Variance
Toothpaste	$12.4	$10.5	18.1%
Mouthwash	$ 8.6	$ 8.8	-2.3%
Over-the-Counter Drugs	$ 5.8	$ 5.3	9.4%
Skin Care Products	$ 3.2	$ 3.4	-5.9%
Total	$30.0	$28.0	7.1%

Scheduled Reports

Scheduled reports are produced periodically, or on a schedule, such as daily, weekly, or monthly. For example, a production manager could use a weekly summary report that lists total payroll costs to monitor and control labor and job costs. A manufacturing report generated once per day to monitor the production of a new item is another example of a scheduled report. Other scheduled reports can help managers control customer credit, performance of sales representatives, inventory levels, and more.

A **key-indicator report** summarizes the previous day's critical activities and is typically available at the beginning of each workday. These reports can summarize inventory levels, production activity, sales volume, and the like. Key-indicator reports are used by managers and executives to take quick, corrective action on significant aspects of the business.

Demand Reports

Demand reports are developed to give certain information upon request. In other words, these reports are produced on demand. Like other reports discussed in this section, they often come from an organization's database system. For example, an executive might want to know the production status of a particular item—a demand report can be generated to provide the requested information by querying the company's database. Suppliers and customers can also use demand reports. FedEx, for example, provides demand reports on its Web site to allow its customers to track packages from their source to their final destination. The Laurel Pub

---

Figure 10.5

Reports Generated by an MIS

The types of reports are (a) scheduled, (b) key indicator, (c) demand, (d) exception, and (e–h) drill down.

(Source: George W. Reynolds, *Information Systems for Managers*, Third Edition. St. Paul, MN: West Publishing Co., 1995.)

scheduled report
A report produced periodically, or on a schedule, such as daily, weekly, or monthly.

key-indicator report
A summary of the previous day's critical activities; typically available at the beginning of each workday.

demand report
A report developed to give certain information at someone's request.

Company, a bar and pub chain in England with over 630 outlets, uses demand reports to generate important sales data when requested. Other examples of demand reports include reports requested by executives to show the hours worked by a particular employee, total sales to date for a product, and so on. The Atlanta Veterans Administration is putting some medical records on the Internet to make them available on demand.[11] Dr. David Bower, chief of staff of the hospital, used the new system to look at a patient's x-ray, read the medical files online, make a diagnosis, and prescribe treatment from his home, saving valuable time.

exception report
A report automatically produced when a situation is unusual or requires management action.

Exception Reports

Exception reports are reports that are automatically produced when a situation is unusual or requires management action. For example, a manager might set a parameter that generates a report of all inventory items with fewer than the equivalent of five days of sales on hand. This unusual situation requires prompt action to avoid running out of stock on the item. The exception report generated by this parameter would contain only items with fewer than five days of sales in inventory. Detroit-based financial services provider Comerica uses exception reports to improve its customer service. The reports list customer inquiries that have been open a period of time without some progress or closure. Exception reports are also used to help fight terrorism. The Matchmaker System scans airline passenger lists and displays an exception report of passengers who could be a threat, so authorities can remove the suspected passengers before the plane takes off. The system was developed in England as part of its Defence Evaluation Research Agency.

As with key-indicator reports, exception reports are most often used to monitor aspects important to an organization's success. In general, when an exception report is produced, a manager or executive takes action. Parameters, or *trigger points*, for an exception report should be set carefully. Trigger points that are set too low might result in too many exception reports; trigger points that are too high could mean that problems requiring action are overlooked. For example, if a manager wants a report that contains all projects over budget by $100 or more, the system might retrieve almost every company project. The $100 trigger point is probably too low. A trigger point of $10,000 might be more appropriate.

drill-down report
A report providing increasingly detailed data about a situation.

Drill-Down Reports

Drill-down reports provide increasingly detailed data about a situation. Through the use of drill-down reports, analysts can see data at a high level first (such as sales for the entire company), then at a more detailed level (such as the sales for one department of the company), and then a very detailed level (such as sales for one sales representative). Boehringer Ingelheim, a large German drug company with over $7 billion in revenues and thousands of employees in 60 countries, uses a variety of drill-down reports so it can respond rapidly to changing market conditions. Managers can drill down into more levels of detail to individual transactions if they want.

Developing Effective Reports

Management information system reports can help managers develop better plans, make better decisions, and obtain greater control over the operations of the firm, but in practice, the types of reports can overlap. For example, a manager can demand an exception report or set trigger points for items contained in a key-indicator report. In addition, some software packages can be used to produce, gather, and distribute reports from different computer systems. Certain guidelines should be followed in designing and developing reports to yield the best results. Table 10.1 explains these guidelines.

Characteristics of a Management Information System

Scheduled, key-indicator, demand, exception, and drill-down reports have all helped managers and executives make better, more timely decisions. When the guidelines for developing effective reports are followed, organizations can increase revenues and lower costs. In general, MISs perform the following functions:

Guidelines	Reason
Tailor each report to user needs.	The unique needs of the manager or executive should be considered, requiring user involvement and input.
Spend time and effort producing only reports that are useful.	After being instituted, many reports continue to be generated even if no one uses them anymore.
Pay attention to report content and layout.	Prominently display the information that is most desired. Do not clutter the report with unnecessary data. Use commonly accepted words and phrases. Managers can work more efficiently if they can easily find desired information.
Use management-by-exception reporting.	Some reports should be produced only when a problem needs to be solved or an action should be taken.
Set parameters carefully.	Low parameters might result in too many reports; high parameters mean valuable information could be overlooked.
Produce all reports in a timely fashion.	Outdated reports are of little or no value.
Periodically review reports.	Review reports at least once per year to make sure they are still needed. Review report content and layout. Determine whether additional reports are needed.

Table 10.1

Guidelines for Developing MIS Reports

- **Provide reports with fixed and standard formats.** For example, scheduled reports for inventory control can contain the same types of information placed in the same locations on the reports. Different managers can use the same report for different purposes.
- **Produce hard-copy and soft-copy reports.** Some MIS reports are printed on paper and are considered hard-copy reports. Most output soft copy, using visual displays on computer screens. Soft-copy output is typically formatted in a report format. In other words, a manager might display an MIS report directly on the computer screen, but the report would still appear in the standard hard-copy format.
- **Use internal data stored in the computer system.** MIS reports use primarily internal sources of data that are contained in computerized databases. Some MISs use external sources of data about competitors, the marketplace, and so on. The Internet is a frequently used source for external data.
- **Allow users to develop their own custom reports.** Although analysts and programmers might be involved in developing and implementing complex MIS reports that require data from many sources, users are increasingly developing their own simple programs to query a database and produce basic reports. This capability, however, can result in several users developing the same or similar reports, which can increase the total time expended and require more storage, compared with having an analyst develop one report for all users.
- **Require user requests for reports developed by systems personnel.** When IS personnel develop and implement MIS reports, they typically require others to submit a formal request to the IS department. If a manager, for example, wants a production report to be used by several people in his department, a formal request for the report is often required. User-developed reports require much less formality.

Management Information Systems and National Security

Management information systems are valuable to professionals in the fields of security and law enforcement. Assembling the pieces of crime investigations relies on the ability of MISs to scan large amounts of information and identify relationships among them, relationships that combine to create incriminating evidence.

On the federal level, the United States National Security Agency (NSA) requires information systems that cross state boundaries to provide information about U.S. citizens and visitors. Travelers encounter a national security MIS when they fly. Traveler's names are entered into a system that checks them against records in two databases of terrorist suspects: one known as the No-fly list, the other known as the Selectee list. Combined, these are referred to as the Transportation Security Administration's (TSA) watch lists. If your name is found on either list, you might be refused transport or subject to a thorough search. People who are mistakenly included on the watch lists can request an investigation and clearance, but often are inconvenienced until the error is corrected.

While the TSA watch lists are designed to strengthen security, the new TSA Registered Traveler system is designed to offer convenience. This MIS is being piloted in several high-congestion airports to speed frequent travelers through security gates. Frequent flyers are invited to provide identifying information so the TSA can conduct a background check and security assessment to determine eligibility for participation in the program. If approved, the volunteer becomes a "trusted traveler" and is eligible for an expedited screening process at TSA security checkpoints. Participants provide both fingerprint and iris biometrics, allowing either biometric to be used for positive identity verification at the airport.

The NSA made headlines in 2006 with an MIS that collected phone call records of tens of millions of Americans and analyzed them for ties to terrorist activities. *USA Today* reported that the NSA had been building a database using records provided by three major phone companies—AT&T Inc., Verizon Communications Inc., and BellSouth Corp.—but that the system "does not involve the NSA listening to or recording conversations." The MIS evaluates the mountains of phone call data for associations to the phone numbers of suspected terrorists and provides reports used in investigations.

Future generations of national MISs might provide an inventory of everyone living in the United States, including name, birth date, sex, ID number, digital photograph, and address. This inventory was made nearly inevitable recently when the Real ID Act was signed into law. The law provides standards and specifications for state driver's licenses that link the driver data in one large distributed database. After collecting the data, it can be mined by federal and state government information systems to reveal useful information. The federally approved ID card will be required to travel on an airplane, open a bank account, collect Social Security payments, or take advantage of nearly any government service. The information provided by the system will allow people in the United States to be monitored to an extent that has never before been possible.

Privacy advocates take issue with all MISs mentioned in this article. They suspect that the MISs allow government to invade our private lives, and are concerned with the accuracy and security of such systems. They fear that others outside the government might access the information to build revealing databases that track spending, travel, and other private aspects of our personal lives.

Discussion Questions

1. How have management information systems provided safety for U.S. citizens?
2. How have management information systems inconvenienced U.S. citizens?

Critical Thinking Questions

1. How much information about yourself are you comfortable sharing with the government for a national ID system?
2. Would you be willing to undergo an extensive background check to obtain shorter security checks at airports? Do you think this is a fair system? Why or why not?

SOURCES: McCullagh, Declan, "FAQ: How Real ID Will Affect You," *CNet News*, May 6, 2006, *http://news.com.com*. Staff, "Report: NSA Has Database of Domestic U.S. Phone Calls," *Reuters*, May 11, 2006, *www.reuters.com*. Singel, Ryan, "Stuck on the No-Fly List," *Wired News*, September 26, 2005, *www.wired.com*. Registered Traveler Pilot Program Web site, accessed June 12, 2006, *www.tsa.gov/public/interapp/tsa_policy/tsa_policy_0043.xml*.

FUNCTIONAL ASPECTS OF THE MIS

Most organizations are structured along functional lines or areas. This functional structure is usually apparent from an organization chart, which typically shows vice presidents under the president. Some traditional functional areas include finance, manufacturing, marketing, human resources, and other specialized information systems. The MIS can also be divided along those functional lines to produce reports tailored to individual functions (see Figure 10.6).

Figure 10.6

An Organization's MIS

The MIS is an integrated collection of functional information systems, each supporting particular functional areas.

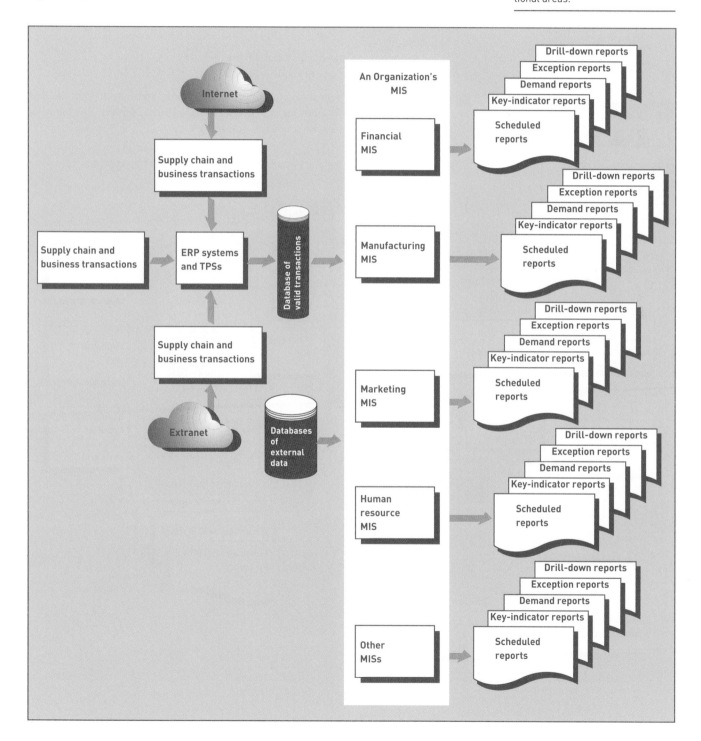

Financial Management Information Systems

financial MIS
An information system that provides financial information not only for executives but also for a broader set of people who need to make better decisions on a daily basis.

A **financial MIS** provides financial information not only for executives but also for a broader set of people who need to make better decisions on a daily basis. Financial MISs are used to streamline reports of transactions. Most financial MISs perform the following functions:

- Integrate financial and operational information from multiple sources, including the Internet, into a single system.
- Provide easy access to data for both financial and nonfinancial users, often through the use of a corporate intranet to access corporate Web pages of financial data and information.
- Make financial data immediately available to shorten analysis turnaround time.
- Enable analysis of financial data along multiple dimensions—time, geography, product, plant, customer.
- Analyze historical and current financial activity.
- Monitor and control the use of funds over time.

Figure 10.7 shows typical inputs, function-specific subsystems, and outputs of a financial MIS, including profit and loss, auditing, and uses and management of funds. The following are some of the financial MIS subsystems and outputs.

Figure 10.7

Overview of a Financial MIS

- **Profit/loss and cost systems.** Many departments within an organization are **profit centers**, which means that they focus on generating profits. An investment division of a large insurance or credit card company is an example of a profit center. Other departments can be **revenue centers**, which are divisions within the company that focus primarily on sales or revenues, such as a marketing or sales department. Still other departments can be **cost centers**, which are divisions within a company that do not directly generate revenue, such as manufacturing or research and development. In most cases, information systems are used to compute revenues, costs, and profits.
- **Auditing.** Auditing involves analyzing the financial condition of an organization and determining whether financial statements and reports produced by the financial MIS are accurate. **Internal auditing** is performed by individuals within the organization. For example, the finance department of a corporation might use a team of employees to perform an audit. **External auditing** is performed by an outside group, such as an accounting or consulting firm such as PricewaterhouseCoopers, Deloitte & Touche, or one of the other major, international accounting firms. Computer systems are used in all aspects of internal and external auditing.
- **Uses and management of funds.** Internal uses of funds include purchasing additional inventory, updating plants and equipment, hiring new employees, acquiring other companies, buying new computer systems, increasing marketing and advertising, purchasing raw materials or land, investing in new products, and increasing research and development. External uses of funds are typically investment related. Companies often invest excess funds in such external revenue generators as bank accounts, stocks, bonds, bills, notes, futures, options, and foreign currency using financial MISs.

profit center
A department within an organization that focuses on generating profits.

revenue center
A division within a company that generates sales or revenues.

cost center
A division within a company that does not directly generate revenue.

auditing
Analyzing the financial condition of an organization and determining whether financial statements and reports produced by the financial MIS are accurate.

internal auditing
Auditing performed by individuals within the organization.

external auditing
Auditing performed by an outside group.

Manufacturing Management Information Systems

More than any other functional area, advances in information systems have revolutionized manufacturing. As a result, many manufacturing operations have been dramatically improved over the last decade. Also, with the emphasis on greater quality and productivity, having an effective manufacturing process is becoming even more critical. The use of computerized systems is emphasized at all levels of manufacturing—from the shop floor to the executive suite. People and small businesses, for example, can benefit from manufacturing MISs that once were only available to large corporations. Personal fabrication systems, for example, can make circuit boards, precision parts, radio tags, and more.[12] Personal fabrication systems include precise machine tools, such as milling machines and cutting tools and sophisticated software. The total system can cost $20,000. For example, in a remote area of Norway, Maakon Karlson uses a personal fabrication system that makes radio tags to track sheep and other animals. The use of the Internet has also streamlined all aspects of manufacturing. Figure 10.8 gives an overview of some of the manufacturing MIS inputs, subsystems, and outputs.

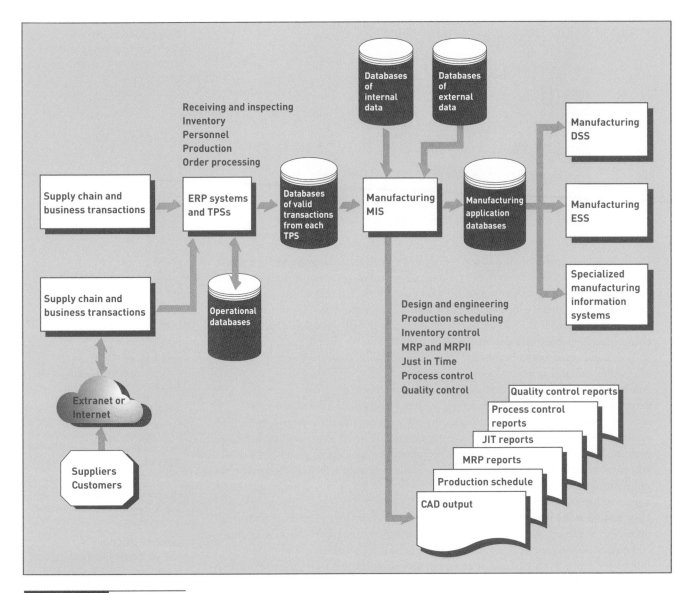

Receiving and inspecting
Inventory
Personnel
Production
Order processing

Databases of internal data

Databases of external data

Manufacturing DSS

Supply chain and business transactions

ERP systems and TPSs

Databases of valid transactions from each TPS

Manufacturing MIS

Manufacturing application databases

Manufacturing ESS

Specialized manufacturing information systems

Supply chain and business transactions

Operational databases

Design and engineering
Production scheduling
Inventory control
MRP and MRPII
Just in Time
Process control
Quality control

Extranet or Internet

Suppliers
Customers

Quality control reports
Process control reports
JIT reports
MRP reports
Production schedule
CAD output

Figure 10.8

Overview of a Manufacturing MIS

The manufacturing MIS subsystems and outputs monitor and control the flow of materials, products, and services through the organization. As raw materials are converted to finished goods, the manufacturing MIS monitors the process at almost every stage. New technology could make this process easier. Using specialized computer chips and tiny radio transmitters, companies can monitor materials and products through the entire manufacturing process. Procter & Gamble, Gillette, Wal-Mart, and Target have funded research into this manufacturing MIS. Car manufacturers, which convert raw steel, plastic, and other materials into a finished automobile, also monitor their manufacturing processes. Auto manufacturers add thousands of dollars of value to the raw materials they use in assembling a car. If the manufacturing MIS also lets them provide additional service, such as customized paint colors, on any of their models, it has added further value for customers. In doing so, the MIS helps provide the company the edge that can differentiate it from competitors. The success of an organization can depend on the manufacturing function. Some common information subsystems and outputs used in manufacturing are discussed next.

- **Design and engineering.** Manufacturing companies often use computer-aided design (CAD) with new or existing products. For example, Boeing uses a CAD system to develop a complete digital blueprint of an aircraft before it ever begins its manufacturing process. As mock-ups are built and tested, the digital blueprint is constantly revised to reflect the most current design. Using such technology helps Boeing reduce its manufacturing costs and the time to design a new aircraft.

Computer-aided design (CAD) is used in the development and design of complex products or structures.

(Source: © Christophe Bosset/ Bloomberg News /Landov.)

- **Master production scheduling and inventory control.** Scheduling production and controlling inventory are critical for any manufacturing company. The overall objective of master production scheduling is to provide detailed plans for both short-term and long-range scheduling of manufacturing facilities. Many techniques are used to minimize inventory costs. Most determine how much and when to order inventory. One method of determining how much inventory to order is called the **economic order quantity (EOQ)**. This quantity is calculated to minimize the total inventory costs. The "When to order?" question is based on inventory usage over time. Typically, the question is answered in terms of a **reorder point (ROP)**, which is a critical inventory quantity level. When the inventory level for a particular item falls to the reorder point, or critical level, the system generates a report so that an order is immediately placed for the EOQ of the product. Another inventory technique used when the demand for one item depends on the demand for another is called **material requirements planning (MRP)**. The basic goal of MRP is to determine when finished products, such as automobiles or airplanes, are needed and then to work backward to determine deadlines and resources needed, such as engines and tires, to complete the final product on schedule. **Just-in-time (JIT) inventory** and manufacturing is an approach that maintains inventory at the lowest levels without sacrificing the availability of finished products. With this approach, inventory and materials are delivered just before they are used in a product. A JIT inventory system would arrange for a car windshield to be delivered to the assembly line only a few moments before it is secured to the automobile, rather than storing it in the manufacturing facility while the car's other components are being assembled. JIT, however, can result in some organizations running out of inventory when demand exceeds expectations.[13]

- **Process control.** Managers can use a number of technologies to control and streamline the manufacturing process. For example, computers can directly control manufacturing equipment, using systems called **computer-assisted manufacturing (CAM)**. CAM systems can control drilling machines, assembly lines, and more. **Computer-integrated manufacturing (CIM)** uses computers to link the components of the production process into an effective system. CIM's goal is to tie together all aspects of production, including order processing, product design, manufacturing, inspection and quality control, and shipping. A **flexible manufacturing system (FMS)** is an approach that allows manufacturing facilities to rapidly and efficiently change from making one product to another. In the middle of a production run, for example, the production process can be changed to make a different product or to switch manufacturing materials. By using an FMS, the time and cost to change manufacturing jobs can be substantially reduced, and companies can react quickly to market needs and competition.

economic order quantity (EOQ)
The quantity that should be reordered to minimize total inventory costs.

reorder point (ROP)
A critical inventory quantity level.

material requirements planning (MRP)
A set of inventory-control techniques that help coordinate thousands of inventory items when the demand of one item is dependent on the demand for another.

just-in-time (JIT) inventory
A philosophy of inventory management in which inventory and materials are delivered just before they are used in manufacturing a product.

computer-assisted manufacturing (CAM)
A system that directly controls manufacturing equipment.

computer-integrated manufacturing (CIM)
Using computers to link the components of the production process into an effective system.

flexible manufacturing system (FMS)
An approach that allows manufacturing facilities to rapidly and efficiently change from making one product to making another.

Computer-assisted manufacturing systems control complex processes on the assembly line and provide users with instant access to information.

(Source: © PHOTOTAKE Inc. / Alamy.)

- **Quality control and testing.** With increased pressure from consumers and a general concern for productivity and high quality, today's manufacturing organizations are placing more emphasis on **quality control**, a process that ensures that the finished product meets the customers' needs. Information systems are used to monitor quality and take corrective steps to eliminate possible quality problems.

quality control
A process that ensures that the finished product meets the customers' needs.

Marketing Management Information Systems

A **marketing MIS** supports managerial activities in product development, distribution, pricing decisions, promotional effectiveness, and sales forecasting. Marketing functions are increasingly being performed on the Internet. Many companies are developing Internet marketplaces to advertise and sell products. The amount spent on online advertising is worth billions of dollars annually. Software can measure how many customers see the advertising. Some companies use a software product called SmartLoyalty to analyze customer loyalty. Some marketing departments are actively using blogs, first introduced in Chapter 7, to keep customers happy.[14] Greencine, a DVD rental business, uses blogs to encourage its customers to review movies and motivate them to rent more. Greencine's Web site realized a 20-fold increase in customer visits as a result. General Motors, Anheuser-Busch, and others use video blogs to advertise products and services.[15]

marketing MIS
An information system that supports managerial activities in product development, distribution, pricing decisions, and promotional effectiveness.

Customer relationship management (CRM) programs, available from some ERP vendors, help a company manage all aspects of customer encounters. CRM software can help a company collect customer data, contact customers, educate customers on new products, and sell products to customers through a Web site. An airline, for example, can use a CRM system to notify customers about flight changes. New Zealand's Jade Stadium, for example, uses CRM software from GlobalTech Solutions to give a single entry point to its marketing efforts and customer databases, instead of using about 20 spreadsheets.[16] The CRM software will help Jade Stadium develop effective marketing campaigns, record and track client contacts, and maintain an accurate database of clients. Yet, not all CRM systems and marketing sites on the Internet are successful. Customization and ongoing maintenance of a CRM system can be expensive. Figure 10.9 shows the inputs, subsystems, and outputs of a typical marketing MIS.

Subsystems for the marketing MIS include marketing research, product development, promotion and advertising, and product pricing. These subsystems and their outputs help marketing managers and executives increase sales, reduce marketing expenses, and develop plans for future products and services to meet the changing needs of customers.

- **Marketing research.** The purpose of marketing research is to conduct a formal study of the market and customer preferences. Computer systems are used to help conduct and analyze the results of surveys, questionnaires, pilot studies, and interviews. Harrah's Entertainment uses programs from SAS and Cognos to get sales and marketing information.[17] The marketing MIS allows Harrah's to integrate its Total Rewards customer loyalty program with marketing and revenue management programs.

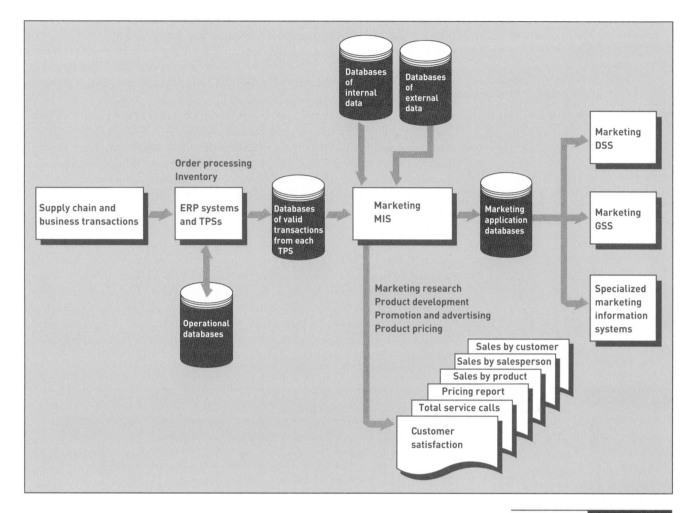

Figure 10.9

Overview of a Marketing MIS

Marketing research data yields valuable information for the development and marketing of new products.

(Source: © Michael Newman/Photo Edit.)

- **Product development.** Product development involves the conversion of raw materials into finished goods and services and focuses primarily on the physical attributes of the product. Many factors, including plant capacity, labor skills, engineering factors, and materials are important in product development decisions. In many cases, a computer program analyzes these various factors and selects the appropriate mix of labor, materials, plant and equipment, and engineering designs. Make-or-buy decisions can also be made with the assistance of computer programs.

- **Promotion and advertising.** One of the most important functions of any marketing effort is promotion and advertising. Product success is a direct function of the types of advertising and sales promotion done. Increasingly, organizations are using the Internet to advertise and sell products and services. Trying to figure out how to pay for college, Alex Tew decided he would try to sell pixels (the dots of light or color that make up text and images) on the Internet.[18] The 21-year-old British student decided to sell a pixel for a dollar. Within a few weeks, he sold $40,000 worth of pixels to companies and people promoting and advertising their products and services on his Web site.

- **Product pricing.** Product pricing is another important and complex marketing function. Retail price, wholesale price, and price discounts must be set. Most companies try to develop pricing policies that will maximize total sales revenues. Computers are often used to analyze the relationship between prices and total revenues.

- **Sales analysis.** Computerized sales analysis is important to identify products, sales personnel, and customers that contribute to profits and those that do not. Several reports can be generated to help marketing managers make good sales decisions (see Figure 10.10). The sales-by-product report lists all major products and their sales for a period of time, such as a month. This report shows which products are doing well and which need improvement or should be discarded altogether. The sales-by-salesperson report lists total sales for each salesperson for each week or month. This report can also be subdivided by product to show which products are being sold by each salesperson. The sales-by-customer report is a tool that can be used to identify high- and low-volume customers.

Figure 10.10

Reports Generated to Help Marketing Managers Make Good Decisions

(a) This sales-by-product report lists all major products and their sales for the period from August to December. (b) This sales-by-salesperson report lists total sales for each salesperson for the same time period. (c) This sales-by-customer report lists sales for each customer for the period. Like all MIS reports, totals are provided automatically by the system to show managers at a glance the information they need to make good decisions.

(a) Sales by Product

Product	August	September	October	November	December	Total
Product 1	34	32	32	21	33	152
Product 2	156	162	177	163	122	780
Product 3	202	145	122	98	66	633
Product 4	345	365	352	341	288	1,691

(b) Sales by Salesperson

Salesperson	August	September	October	November	December	Total
Jones	24	42	42	11	43	162
Kline	166	155	156	122	133	732
Lane	166	155	104	99	106	630
Miller	245	225	305	291	301	1,367

(c) Sales by Customer

Customer	August	September	October	November	December	Total
Ang	234	334	432	411	301	1,712
Braswell	56	62	77	61	21	277
Celec	1,202	1,445	1,322	998	667	5,634
Jung	45	65	55	34	88	287

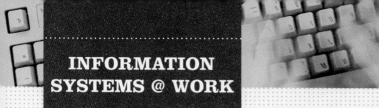
Fully Integrated Decision Support

Decision support tools are increasingly being integrated with management information systems and day-to-day business activities. Embedded DSSs are popular for helping to speed decision-making processes. Companies such as Briggs & Stratton, Pharmascience Inc., and Alaska Airlines often use DSS software tools and utilities to help them make better decisions and get more from their data.

Briggs & Stratton Corporation, based in Wauwatosa, Wisconsin, manufactures lawn mower and garden tiller engines, and recently began using operational BI systems from SAS. The new system will provide Briggs & Stratton employees with BI information embedded in accounting, production, and sales processes. "It is such a hot button for us right now," said Grant Felsing, decision support manager at Briggs & Stratton. "Show me those things that are within my area that are not within norms or ... are heading for a collision course."

Felsing explained that the new system can alert accountants that correct accounting procedures are not in place to handle orders as a new engine is set to be shipped. Before the BI was embedded in its processes, the company would have to take orders out of the system, reenter the correct accounting information, and then reenter orders the next day to ensure that the products would ship correctly.

"Businesses want to get more value out of all of the data, not just the data warehouse. Many of the real-time decisions that need to be made must be made while the process is happening, like while the customer is on the phone or when the patient is being treated," said Keith Gile, an analyst at Forrester Research Inc.

Montreal-based Pharmascience Inc. is hoping that the new operational BI integration features will help the pharmaceutical company better manage inventory. Without the embedded BI system, inventory information can be delayed by as much as a week, explained Jonathan Despres, manager of information access. The new system with embedded BI will allow inventory information to be included in product warehouse businesses process. "If [users] get information delayed by a week, it's almost impossible to reduce the inventory level," Despres said.

Alaska Airlines Inc. has recently begun deploying business analytics tools from Siebel Systems Inc. in its marketing organization. The tools will be integrated with Alaska Air's customer management system and will incorporate data from Sabre Holdings Corporation's Sabre reservations system, according to James Archuleta, director of CRM at the Seattle-based airline. Call center representatives will have quick access to the best options for customers to assist in travel decisions.

Decision support systems are evolving from stand-alone software tools to utilities embedded within commonly used information systems. In business and in our personal lives, computers will be advising us on many types of decisions that we make, big or small. As these systems play a larger role in our day-to-day lives, the accuracy and intelligence built into embedded decision support systems becomes increasingly important.

Discussion Questions

1. Why are the decision support systems described in this article considered institutional?
2. How does the information provided by an operational BI system differ from a standard MIS report?

Critical Thinking Questions

1. How does using operational BI affect the day-to-day lives of business decision makers?
2. How might operational BI concepts be applied to popular personal information management software such as a calendar application? Provide an explanation and example.

SOURCES: Havenstein, Heather, "Users Turn to Operational Business Intelligence Tools," *Computerworld*, March 21, 2005, *www.computerworld.com*. Webfocus7 Operational BI Web site, accessed June 16, 2006, *www.webfocus7.com*. SAS9 Web site, accessed June 16, 2006, *www.sas.com/software/sas9*.

Human Resource Management Information Systems

human resource MIS
An information system that is concerned with activities related to employees and potential employees of an organization, also called a personnel MIS.

A **human resource MIS (HRMIS)**, also called the *personnel MIS*, is concerned with activities related to previous, current, and potential employees of the organization. Because the personnel function relates to all other functional areas in the business, the human resource (HR) MIS plays a valuable role in ensuring organizational success. Some of the activities performed by this important MIS include workforce analysis and planning, hiring, training, job and task assignment, and many other personnel-related issues. An effective human resource MIS allows a company to keep personnel costs at a minimum, while serving the required business processes needed to achieve corporate goals. Although human resource information systems focus on cost reduction, many of today's HR systems concentrate on hiring and managing existing employees to get the total potential of the human talent in the organization. According to the HR manager at Oregon Health & Science University, "More and more, HR is being called upon to be a strategic partner. We have to come to the table laden with data, with plans of where we're heading and where the organization should be heading."[19] According to the High Performance Workforce Study conducted by Accenture, the most important HR initiatives include improving worker productivity, improving adaptability to new opportunities, and facilitating organizational change. Figure 10.11 shows some of the inputs, subsystems, and outputs of the human resource MIS.

Figure 10.11

Overview of a Human Resource MIS

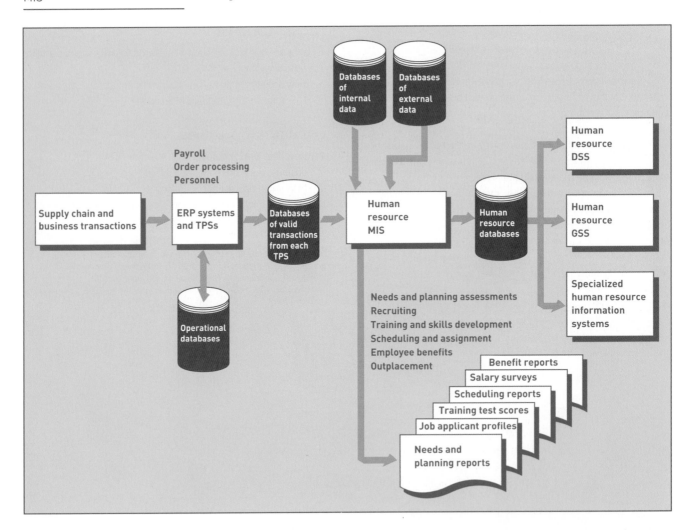

Human resource subsystems and outputs range from the determination of human resource needs and hiring through retirement and outplacement. Most medium and large organizations have computer systems to assist with human resource planning, hiring, training and skills inventorying, and wage and salary administration. Outputs of the human resource

MIS include reports, such as human resource planning reports, job application review profiles, skills inventory reports, and salary surveys discussed next.

- **Human resource planning.** One of the first aspects of any human resource MIS is determining personnel and human needs. The overall purpose of this MIS subsystem is to put the right number and kinds of employees in the right jobs when they are needed. Effective human resource planning often requires computer programs, such as SPSS and SAS, to forecast the future number of employees needed and anticipating the future supply of people for these jobs. IBM is using an HR pilot program, called Professional Marketplace, to plan for workforce needs, including the supplies and tools the workforce needs to work efficiently.[20] The program and related ones have saved IBM about $6 billion per year in costs. Professional Marketplace helps IBM to catalog employees into a glossary of skills and abilities. Like many other companies, HR and workforce costs are IBM's biggest expense.
- **Personnel selection and recruiting.** If the human resource plan reveals that additional personnel are required, the next logical step is recruiting and selecting personnel. Companies seeking new employees often use computers to schedule recruiting efforts and trips and to test potential employees' skills. Many companies now use the Internet to screen for job applicants. Applicants use a template to load their résumé onto the Internet site. HR managers can then access these résumés and identify applicants they are interested in interviewing.
- **Training and skills inventory.** Some jobs, such as programming, equipment repair, and tax preparation, require very specific training for new employees. Other jobs may require general training about the organizational culture, orientation, dress standards, and expectations of the organization. When training is complete, employees often take computer-scored tests to evaluate their mastery of skills and new material.
- **Scheduling and job placement.** Employee schedules are developed for each employee, showing his job assignments over the next week or month. Job placements are often determined based on skills inventory reports, which show which employee might be best suited to a particular job. Sophisticated scheduling programs are often used in the airline industry, the military, and many other areas to get the right people assigned to the right jobs at the right time.
- **Wage and salary administration.** Another human resource MIS subsystem involves determining wages, salaries, and benefits, including medical payments, savings plans, and retirement accounts. Wage data, such as industry averages for positions, can be taken from the corporate database and manipulated by the human resource MIS to provide wage information and reports to higher levels of management.
- **Outplacement.** Employees leave a company for a number of reasons. Outplacement services are offered by many companies to help employees make the transition. *Outplacement* can include job counseling and training, job and executive search, retirement and financial planning, and a variety of severance packages and options. Many employees use the Internet to plan their future retirement or to find new jobs, using job sites such as *www.monster.com*.

Other Management Information Systems

In addition to finance, manufacturing, marketing, and human resource MISs, some companies have other functional management information systems. For example, most successful companies have well-developed accounting functions and a supporting accounting MIS. Also, many companies use geographic information systems for presenting data in a useful form.

Accounting MISs

In some cases, accounting works closely with financial management. An **accounting MIS** performs a number of important activities, providing aggregate information on accounts payable, accounts receivable, payroll, and many other applications. The organization's ERP system and TPS capture accounting data, which is also used by most other functional information systems.

accounting MIS
An information system that provides aggregate information on accounts payable, accounts receivable, payroll, and many other applications.

Organizations are constantly auditing their books and computer records for potential fraud.[21] In one case, an employee was caught padding his expenses and was expected of deeper corporate fraud. According to the director of a forensic advisory and commercial damages auditing organization, "We returned in the dark of the night and started searching through the guy's computer... By morning, we suspected that he was making payments to overseas law firms that didn't exist, then cashing the checks himself." The results of the audit were turned over to the U.S. Department of Justice for possible prosecution.

Some smaller companies hire outside accounting firms to assist them with their accounting functions. These outside companies produce reports for the firm using raw accounting data. In addition, many excellent integrated accounting programs are available for personal computers in small companies. Depending on the needs of the small organization and its personnel's computer experience, using these computerized accounting systems can be a very cost-effective approach to managing information.

Geographic Information Systems

geographic information system (GIS)

A computer system capable of assembling, storing, manipulating, and displaying geographic information, that is, data identified according to its location.

Increasingly, managers want to see data presented in graphical form. A **geographic information system (GIS)** is a computer system capable of assembling, storing, manipulating, and displaying geographically referenced information, that is, data identified according to its location. A GIS enables users to pair maps or map outlines with tabular data to describe aspects of a particular geographic region. For example, sales managers might want to plot total sales for each county in the states they serve. Using a GIS, they can specify that each county be shaded to indicate the relative amount of sales—no shading or light shading represents no or little sales, and deeper shading represents more sales. Staples Inc., the large office supply store chain, used a geographic information system to select about 100 new store locations, after considering about 5,000 possible sites.[22] Finding the best location is critical. It can cost up to $1 million for a failed store because of a poor location. Staples uses a GIS tool from Tactician Corporation along with software from SAS. Although many software products have seen declining revenues, the use of GIS software is increasing.

We saw earlier in this chapter that management information systems (MISs) provide useful summary reports to help solve structured and semistructured business problems. Decision support systems (DSSs) offer the potential to assist in solving both semistructured and unstructured problems. These systems are discussed next.

AN OVERVIEW OF DECISION SUPPORT SYSTEMS

A DSS is an organized collection of people, procedures, software, databases, and devices used to help make decisions that solve problems. The focus of a DSS is on decision-making effectiveness when faced with unstructured or semistructured business problems. As with a TPS and an MIS, a DSS should be designed, developed, and used to help an organization achieve its goals and objectives. Decision support systems offer the potential to generate higher profits, lower costs, and better products and services. For example, healthcare organizations use DSSs to improve patient care and reduce costs. Decision support systems can also monitor and improve patient care. One diabetes Web site developed a DSS to provide customized treatment plans and reports. The Web site uses powerful DSS software and grid computing, where the idle capacity of a network of computers is harnessed.

Decision support systems, although skewed somewhat toward the top levels of management, are used at all levels. To some extent, today's managers at all levels are faced with less structured, nonroutine problems, but the quantity and magnitude of these decisions increase as a manager rises higher in an organization. Many organizations contain a tangled web of complex rules, procedures, and decisions. DSSs are used to bring more structure to these problems to aid the decision-making process. In addition, because of the inherent flexibility of decision support systems, managers at all levels are able to use DSSs to assist in some relatively routine, programmable decisions in lieu of more formalized management information systems. DSSs are also used in nonprofit organizations.

Decision support systems are also used in government and by nonprofit organizations. The Los Angeles Police Department (LAPD) uses a $35 million DSS to track complaints related to its officers.[23] Supervisors use the system to spot "bad" cops. The information system analyzes complaints, lawsuits, the use of force, and related police department records. See Figure 10.12.

Figure 10.12

Decision support systems are also used by nonprofit organizations and in government, such as in police departments.

(Source: © Spencer C. Grant/Photo Edit.)

Characteristics of a Decision Support System

Decision support systems have many characteristics that allow them to be effective management support tools. Of course, not all DSSs work the same. The following are goals of some DSSs:

- **Provide rapid access to information.** Some DSSs provide fast and continuous access to information; for example, the gauges on the dashboard of a car or truck are used to see how the vehicle is running. Craig Erickson, VP of Erickson Retirement Communities, for example, uses a DSS *dashboard* to instantly see food costs, staffing costs, and other performance measures.[24] By clicking icons on his PC, Erickson can display the information he wants. His DSS dashboard also displays numbers in green if everything is fine, yellow to signal possible problems, and red if immediate action is needed. Erickson uses a DSS dashboard from Hyperion Solutions.
- **Handle large amounts of data from different sources.** For instance, advanced database management systems and data warehouses have allowed decision makers to search for information with a DSS, even when some data resides in different databases on different computer systems or networks. Other sources of data can be accessed via the Internet or over a corporate intranet. Using the Internet, an oil giant can use a decision support system to save hundreds of millions of dollars annually by coordinating a large amount of drilling and exploration data from around the globe.
- **Provide report and presentation flexibility.** Managers can get the information they want, presented in a format that suits their needs. Furthermore, output can be displayed on computer screens or printed, depending on the needs and desires of the problem solvers.
- **Offer both textual and graphical orientation.** Today's DSSs can produce text, tables, line drawings, pie charts, trend lines, and more. By using their preferred orientation, managers can use a DSS to get a better understanding of a situation and to convey this understanding to others.
- **Support drill-down analysis.** A manager can get more levels of detail when needed by drilling down through data. For example, a manager can get more detailed information for a project—viewing the overall project cost or drilling down and seeing the cost for each phase, activity, and task.
- **Perform complex, sophisticated analysis and comparisons using advanced software packages.** Marketing research surveys, for example, can be analyzed in a variety of ways using programs that are part of a DSS. Many of the analytical programs associated with a DSS are actually stand-alone programs, and the DSS brings them together.

- **Support optimization, satisficing, and heuristic approaches.** By supporting all types of decision-making approaches, a DSS gives the decision maker a great deal of flexibility in computer support for decision making. For example, **what-if analysis**, the process of making hypothetical changes to problem data and observing the impact on the results, can be used to control inventory. Given the demand for products, such as automobiles, the computer can determine the necessary parts and components, including engines, transmissions, windows, and so on. With what-if analysis, a manager can make changes to problem data (the number of automobiles needed for the next month) and immediately see the impact on the parts requirements.

Goal-seeking analysis is the process of determining the problem data required for a given result. For example, a financial manager might be considering an investment with a certain monthly net income, and the manager might have a goal to earn a return of 9 percent on the investment. Goal seeking allows the manager to determine what monthly net income (problem data) is needed to yield a return of 9 percent (problem result). Some spreadsheets can be used to perform goal-seeking analysis (see Figure 10.13).

what-if analysis
The process of making hypothetical changes to problem data and observing the impact on the results.

goal-seeking analysis
The process of determining the problem data required for a given result.

Figure 10.13

With a spreadsheet program, a manager can enter a goal, and the spreadsheet will determine the input needed to achieve the goal.

simulation
The ability of the DSS to duplicate the features of a real system.

Simulation is the ability of the DSS to duplicate the features of a real system. In most cases, probability or uncertainty is involved. For example, the number of repairs and the time to repair key components of a manufacturing line can be calculated to determine the impact on the number of products that can be produced each day. Engineers can use this data to determine which components need to be reengineered to increase the mean time between failures and which components need to have an ample supply of spare parts to reduce the mean time to repair. Drug companies are using simulated trials to reduce the need for human participants and reduce the time and costs of bringing a new drug to market. Drug companies are hoping that this use of simulation will help them identify successful drugs earlier in development. Corporate executives and military commanders often use computer simulations to allow them to try different strategies in different situations. Corporate executives, for example, can try different marketing decisions under various market conditions. Military commanders often use computer war games to fine-tune their military strategies in different warfare conditions. The Turkish Army, for example, uses simulation to help coordinate its fuel-supply system.[25]

Capabilities of a Decision Support System

Developers of decision support systems strive to make them more flexible than management information systems and to give them the potential to assist decision makers in a variety of situations. Table 10.2 lists a few DSS applications. DSSs can assist with all or most problem-solving phases, decision frequencies, and different degrees of problem structure. DSS approaches can also help at all levels of the decision-making process. A single DSS might provide only a few of these capabilities, depending on its uses and scope.

Company or Application	Description
ING Direct	The financial services company uses a DSS to summarize the bank's financial performance. The bank needed a measurement and tracking mechanism to determine how successful it was and to make modifications to plans in real time.
Cinergy Corporation	The electric utility developed a DSS to reduce lead time and effort required to make decisions in purchasing coal.
U.S. Army	It developed a DSS to help recruit, train, and educate enlisted forces. The DSS uses a simulation that incorporates what-if features.
National Audubon Society	It developed a DSS called Energy Plan (EPLAN) to analyze the impact of U.S. energy policy on the environment.
Hewlett-Packard	The computer company developed a DSS called Quality Decision Management to help improve the quality of its products and services.
Virginia	The state of Virginia developed the Transportation Evacuation Decision Support System (TEDSS) to determine the best way to evacuate people in case of a nuclear disaster at its nuclear power plants.

Table 10.2

Selected DSS Applications

Support for Problem-Solving Phases

The objective of most decision support systems is to assist decision makers with the phases of problem solving. As previously discussed, these phases include intelligence, design, choice, implementation, and monitoring. A specific DSS might support only one or a few phases. By supporting all types of decision-making approaches, a DSS gives the decision maker a great deal of flexibility in getting computer support for decision-making activities.

Support for Different Decision Frequencies

Decisions can range on a continuum from one-of-a-kind to repetitive decisions. One-of-a-kind decisions are typically handled by an **ad hoc DSS**. An ad hoc DSS is concerned with situations or decisions that come up only a few times during the life of the organization; in small businesses, they might happen only once. For example, a company might need to decide whether to build a new manufacturing facility in another area of the country. Repetitive decisions are addressed by an institutional DSS. An **institutional DSS** handles situations or decisions that occur more than once, usually several times per year or more. An institutional DSS is used repeatedly and refined over the years. Examples of institutional DSSs include systems that support portfolio and investment decisions and production scheduling. These decisions might require decision support numerous times during the year. For example, DSSs are used to help solve computer-related problems that can occur multiple times throughout the day. With this approach, the DSS monitors computer systems second by second for problems and takes action to prevent problems, such as slowdowns and crashes, and to recover from them when they occur. One IBM engineer believes that this approach, called *autonomic computing*, is the key to the future of computing. Between these two extremes are decisions managers make several times, but not regularly or routinely.

ad hoc DSS

A DSS concerned with situations or decisions that come up only a few times during the life of the organization.

institutional DSS

A DSS that handles situations or decisions that occur more than once, usually several times per year or more. An institutional DSS is used repeatedly and refined over the years.

highly structured problems
Problems that are straightforward and require known facts and relationships.

semistructured or unstructured problems
More complex problems in which the relationships among the pieces of data are not always clear, the data might be in a variety of formats, and the data is often difficult to manipulate or obtain.

Support for Different Problem Structures

As discussed previously, decisions can range from highly structured and programmed to unstructured and nonprogrammed. **Highly structured problems** are straightforward, requiring known facts and relationships. **Semistructured** or **unstructured problems**, on the other hand, are more complex. The relationships among the pieces of data are not always clear, the data might be in a variety of formats, and it is often difficult to manipulate or obtain. In addition, the decision maker might not know the information requirements of the decision in advance. For example, a DSS has been used to support sophisticated and unstructured investment analysis and make substantial profits for traders and investors. Some DSS trading software is programmed to place buy and sell orders automatically without a trader manually entering a trade, based on parameters set by the trader.[26]

Support for Various Decision-Making Levels

Decision support systems can provide help for managers at different levels within the organization. Operational managers can get assistance with daily and routine decision making. Tactical decision makers can use analysis tools to ensure proper planning and control. At the strategic level, DSSs can help managers by providing analysis for long-term decisions requiring both internal and external information (see Figure 10.14).

Figure 10.14

Decision-Making Level

Strategic managers are involved with long-term decisions, which are often made infrequently. Operational managers are involved with decisions that are made more frequently.

A Comparison of DSS and MIS

A DSS differs from an MIS in numerous ways, including the type of problems solved, the support given to users, the decision emphasis and approach, and the type, speed, output, and development of the system used. Table 10.3 lists brief descriptions of these differences.

COMPONENTS OF A DECISION SUPPORT SYSTEM

dialogue manager
A user interface that allows decision makers to easily access and manipulate the DSS and to use common business terms and phrases.

At the core of a DSS are a database and a model base. In addition, a typical DSS contains a user interface, also called **dialogue manager**, that allows decision makers to easily access and manipulate the DSS and to use common business terms and phrases. Finally, access to the Internet, networks, and other computer-based systems permits the DSS to tie into other powerful systems, including the TPS or function-specific subsystems. Internet software agents, for example, can be used in creating powerful decision support systems. Figure 10.15 shows a conceptual model of a DSS. Specific DSSs might not have all the components shown in Figure 10.15.

The Database

The database management system allows managers and decision makers to perform *qualitative analysis* on the company's vast stores of data in databases, data warehouses, and data marts, discussed in Chapter 5. A *data-driven DSS* primarily performs qualitative analysis based on the company's databases. Data-driven DSSs tap into vast stores of information contained in the corporate database, retrieving information on inventory, sales, personnel, production, finance, accounting, and other areas.[27] Data mining and business intelligence, introduced in Chapter 5, are often used in a data-driven DSS. Airline companies, for example, use a data-driven DSS to help it identify customers for round-trip flights between major cities. The data-driven DSS can be used to search a data warehouse to

Factor	DSS	MIS
Problem Type	A DSS can handle unstructured problems that cannot be easily programmed.	An MIS is normally used only with structured problems.
Users	A DSS supports individuals, small groups, and the entire organization. In the short run, users typically have more control over a DSS.	An MIS supports primarily the organization. In the short run, users have less control over an MIS.
Support	A DSS supports all aspects and phases of decision making; it does not replace the decision maker—people still make the decisions.	This is not true of all MIS systems—some make automatic decisions and replace the decision maker.
Emphasis	A DSS emphasizes actual decisions and decision-making styles.	An MIS usually emphasizes information only.
Approach	A DSS is a direct support system that provides interactive reports on computer screens.	An MIS is typically an indirect support system that uses regularly produced reports.
System	The computer equipment that provides decision support is usually online (directly connected to the computer system) and related to real time (providing immediate results). Computer terminals and display screens are examples—these devices can provide immediate information and answers to questions.	An MIS, using printed reports that might be delivered to managers once per week, cannot provide immediate results.
Speed	Because a DSS is flexible and can be implemented by users, it usually takes less time to develop and is better able to respond to user requests.	An MIS's response time is usually longer.
Output	DSS reports are usually screen oriented, with the ability to generate reports on a printer.	An MIS, however, typically is oriented toward printed reports and documents.
Development	DSS users are usually more directly involved in its development. User involvement usually means better systems that provide superior support. For all systems, user involvement is the most important factor for the development of a successful system.	An MIS is frequently several years old and often was developed for people who are no longer performing the work supported by the MIS.

Table 10.3

Comparison of DSSs and MISs

Figure 10.15

Conceptual Model of a DSS

DSS components include a model base; database; external database access; access to the Internet and corporate intranet, networks, and other computer systems; and a user interface or dialogue manager.

contact thousands of customers who might be interested in an inexpensive flight. A casino can use a data-driven DSS to search large databases to get detailed information on patrons. It can tell how much each patron spends per day on gambling, and more. Opportunity International uses a data-driven DSS to help it make loans and provide services to tsunami victims and others in need around the world.[28] According to the information services manager of Opportunity International, "We need to pull all the data … to one central database that we can analyze, and we need a way to get that information back out to people in the field." Data-driven DSSs can also be used in emergency medical situations to make split-second, life-or-death treatment decisions.[29] Data-driven medical DSSs allow doctors to have access to a complete medical record of a patient. Some of these medical record systems also allow patients to enter their own health information into the database, such as medicines, allergies, and family health histories. WebMD, iHealthRecord, Walgreens, and PersonalHealthKey allow people to put their medical records online for rapid access.[30]

A database management system can also connect to external databases to give managers and decision makers even more information and decision support. External databases can include the Internet, libraries, government databases, and more. The combination of internal and external database access can give key decision makers a better understanding of the company and its environment.

The Model Base

model base
Part of a DSS that provides decision makers access to a variety of models and assists them in decision making.

The **model base** allows managers and decision makers to perform *quantitative analysis* on both internal and external data.[31] A *model-driven DSS* primarily performs mathematical or quantitative analysis. The model base gives decision makers access to a variety of models so that they can explore different scenarios and see their effects. Ultimately, it assists them in the decision-making process. Procter & Gamble, maker of Pringles potato chips, Pampers diapers, and hundreds of other consumer products, uses a model-driven DSS to streamline how raw materials and products flow from its suppliers to its customers.[32] The model-driven DSS has saved the company hundreds of millions of dollars in supply chain related costs. Scientists and mathematicians also use model-driven DSSs.[33] Model-driven DSSs are excellent at predicting customer behaviors.[34] LoanPerformance, for example, uses models to help it forecast which customers will be late with payments or might default on their loans. Other financial services and insurance firms, such as health insurer HighMark, also use model-driven DSSs to predict fraud.

The models and algorithms used in a model-driven DSS are often reviewed and revised over time.[35] As a result of Hurricane Katrina, for example, software developers, such as EQECAT and Air Worldwide, and insurance companies, such as State Farm and Allstate, plan to revise their models about storm damage and insurance requirements.[36] Modeling storm damage intensified after Hurricane Andrew, which devastated Florida in 1992.

model management software
Software that coordinates the use of models in a DSS.

Model management software (MMS) is often used to coordinate the use of models in a DSS, including financial, statistical analysis, graphical, and project-management models. Depending on the needs of the decision maker, one or more of these models can be used (see Table 10.4).

Model Type	Description	Software
Financial	Provides cash flow, internal rate of return, and other investment analysis	Spreadsheet, such as Microsoft Excel
Statistical	Provides summary statistics, trend projections, hypothesis testing, and more	Statistical program, such as SPSS or SAS
Graphical	Assists decision makers in designing, developing, and using graphic displays of data and information	Graphics programs, such as Microsoft PowerPoint
Project Management	Handles and coordinates large projects; also used to identify critical activities and tasks that could delay or jeopardize an entire project if they are not completed in a timely and cost-effective fashion	Project management software, such as Microsoft Project

Table 10.4

Model Management Software

DSSs often use financial, statistical, graphical, and project-management models.

The User Interface or Dialogue Manager

The user interface or dialogue manager allows users to interact with the DSS to obtain information. It assists with all aspects of communications between the user and the hardware and software that constitute the DSS. In a practical sense, to most DSS users, the user interface is the DSS. Upper-level decision makers are often less interested in where the information came from or how it was gathered than that the information is both understandable and accessible.

GROUP SUPPORT SYSTEMS

The DSS approach has resulted in better decision making for all levels of individual users. However, many DSS approaches and techniques are not suitable for a group decision-making environment. Although not all workers and managers are involved in committee meetings and group decision-making sessions, some tactical and strategic-level managers can spend more than half their decision-making time in a group setting. Such managers need assistance with group decision making. A **group support system (GSS)**, also called a *group decision support system* and a *computerized collaborative work system*, consists of most of the elements in a DSS, plus software to provide effective support in group decision-making settings (see Figure 10.16).[37]

group support system (GSS)
Software application that consists of most elements in a DSS, plus software to provide effective support in group decision making; also called *group support system* or *computerized collaborative work system*.

Figure 10.16

Configuration of a GSS

A GSS contains most of the elements found in a DSS, plus software to facilitate group member communications.

Group support systems are used in most industries. Architects are increasingly using GSSs to help them collaborate with other architects and builders to develop the best plans and to compete for contracts. Manufacturing companies use GSSs to link raw material suppliers to their own company systems. Engineers can use Mathcad Enterprise, another GSS. The software allows engineers to create, share, and reuse calculations.

Characteristics of a GSS That Enhance Decision Making

It is often said that two heads are better than one. When it comes to decision making, a GSS's unique characteristics have the potential to result in better decisions. Developers of these systems try to build on the advantages of individual support systems while adding new approaches unique to group decision making. For example, some GSSs can allow the exchange of information and expertise among people without direct face-to-face interaction. The following sections describe some characteristics that can improve and enhance decision making.

Special Design

The GSS approach acknowledges that special procedures, devices, and approaches are needed in group decision-making settings. These procedures must foster creative thinking, effective communications, and good group decision-making techniques.

Ease of Use

Like an individual DSS, a GSS must be easy to learn and use. Systems that are complex and hard to operate will seldom be used. Many groups have less tolerance than do individual decision makers for poorly developed systems.

Flexibility

Two or more decision makers working on the same problem might have different decision-making styles and preferences. Each manager makes decisions in a unique way, in part because of different experiences and cognitive styles. An effective GSS not only has to support the different approaches that managers use to make decisions, but also must find a means to integrate their different perspectives into a common view of the task at hand.

Decision-Making Support

A GSS can support different decision-making approaches, including the **delphi approach**, in which group decision makers are geographically dispersed throughout the country or the world. This approach encourages diversity among group members and fosters creativity and original thinking in decision making. Another approach, called **brainstorming**, in which members offer ideas "off the top of their heads," fosters creativity and free thinking. The **group consensus approach** forces members in the group to reach a unanimous decision. The Shuttle Project Engineering Office at the Kennedy Space Center has used the consensus-ranking organizational-support system (CROSS) to evaluate space projects in a group setting. The group consensus approach analyzes the benefits of various projects and their probabilities of success. CROSS is used to evaluate and prioritize advanced space projects. With the **nominal group technique**, each decision maker can participate; this technique encourages feedback from individual group members, and the final decision is made by voting, similar to a system for electing public officials.

Anonymous Input

Many GSSs allow anonymous input, where the person giving the input is not known to other group members. For example, some organizations use a GSS to help rank the performance of managers. Anonymous input allows the group decision makers to concentrate on the merits of the input without considering who gave it. In other words, input given by a top-level manager is given the same consideration as input from employees or other members of the group. Some studies have shown that groups using anonymous input can make better decisions and have superior results compared with groups that do not use anonymous input. Anonymous input, however, can result in flaming, where an unknown team member posts insults or even obscenities on the GSS.

Reduction of Negative Group Behavior

One key characteristic of any GSS is the ability to suppress or eliminate group behavior that is counterproductive or harmful to effective decision making. In some group settings, dominant individuals can take over the discussion, which can prevent other members of the group from presenting creative alternatives. In other cases, one or two group members can sidetrack or subvert the group into areas that are nonproductive and do not help solve the problem at hand. Other times, members of a group might assume they have made the right decision without examining alternatives—a phenomenon called *groupthink*. If group sessions are

delphi approach

A decision-making approach in which group decision makers are geographically dispersed; this approach encourages diversity among group members and fosters creativity and original thinking in decision making.

brainstorming

A decision-making approach that often consists of members offering ideas "off the top of their heads."

group consensus approach

A decision-making approach that forces members in the group to reach a unanimous decision.

nominal group technique

A decision-making approach that encourages feedback from individual group members, and the final decision is made by voting, similar to the way public officials are elected.

poorly planned and executed, the result can be a tremendous waste of time. Today, many GSS designers are developing software and hardware systems to reduce these types of problems. Procedures for effectively planning and managing group meetings can be incorporated into the GSS approach. A trained meeting facilitator is often employed to help lead the group decision-making process and to avoid groupthink. See Figure 10.17.

Figure 10.17

Using the GSS Approach

A trained meeting facilitator can help lead the group decision-making process and avoid groupthink.

(Source: © Bill Bachmann/Getty Images.)

Parallel Communication

With traditional group meetings, people must take turns addressing various issues. One person normally talks at a time. With a GSS, every group member can address issues or make comments at the same time by entering them into a PC or workstation. These comments and issues are displayed on every group member's PC or workstation immediately. Parallel communication can speed meeting times and result in better decisions.

Automated Recordkeeping

Most GSSs can keep detailed records of a meeting automatically. Each comment that is entered into a group member's PC or workstation can be anonymously recorded. In some cases, literally hundreds of comments can be stored for future review and analysis. In addition, most GSS packages have automatic voting and ranking features. After group members vote, the GSS records each vote and makes the appropriate rankings.

GSS Software

GSS software, often called *groupware* or *workgroup software*, helps with joint work group scheduling, communication, and management. Software from Autodesk, for example, is helping design the 1,776-foot tall Freedom Tower to replace the twin towers of the World Trade Center in New York City.[38] Autodesk and other software products have GSS capabilities that allow groups to work together on the design. The designers, for example, are using Autodesk's Buzzsaw Professional Online Collaboration Service, which works with AutoCAD, a design and engineering software product from Autodesk. The U.S. Navy uses Virtual Office from Groove Networks to help it manage critical information in delivering humanitarian relief in disaster areas.[39] The software is used for collaboration and communications in transmitting critical information between field offices. Virtual Office also has encryption capabilities to keep sensitive information safe and secure.

One popular package, Lotus Notes, can capture, store, manipulate, and distribute memos and communications that are developed during group projects. It can also incorporate knowledge management, discussed in Chapter 5, into the Lotus Notes Package. Some companies standardize on messaging and collaboration software, such as Lotus Notes. Microsoft has invested billions of dollars in GSS software to incorporate collaborative features into its Office suite and related products.[40] Office Communicator, for example, is a Microsoft product to allow better and faster collaboration. Other companies are also heavily investing in GSS software. In addition to Lotus Notes, IBM has developed Workplace to allow workers to collaborate more efficiently in doing their jobs. Microsoft's NetMeeting product supports application sharing in multiparty calls. Microsoft Exchange is another example of groupware. This software allows users to set up electronic bulletin boards, schedule group meetings, and use e-mail in a group setting. NetDocuments Enterprise can be used for Web collaboration. The groupware is intended for legal, accounting, and real-estate businesses. A Breakout Session feature allows two people to take a copy of a document to a shared folder for joint revision and work. The software also permits digital signatures and the ability to download and work on shared documents on handheld computers. Other GSS software packages include Collabnet, Collabra Share, OpenMind, and TeamWare. All of these tools can aid in group decision making.

WebEx, Genesys Meeting Center, and GoToMeeting Corporate are examples of groupware products available on the Web.[41] In 2005, GoToMeeting Corporate was a *PC Magazine*'s Editor's Choice for workgroups and small businesses. Using groupware gives every employee rapid access to a vast source of information. Wyndam International, for example, uses collaboration and e-meeting software from Centra Software to save time and money.[42] The company estimates that it has saved more than $1 million so far in travel and communications costs. According to a company manager, "We use it for weekly and monthly conferences, and because it's voice over IP, we save about $10,000 to $15,000 a month on our telephone bill."

In addition to stand-alone products, GSS software is increasingly being incorporated into existing software packages. Today, some transaction processing and enterprise resource planning packages include collaboration software. Some ERP producers (see Chapter 9), for example, have developed groupware to facilitate collaboration and to allow users to integrate applications from other vendors into the ERP system of programs. Today, groupware can interact with wireless devices. Research In Motion, the maker of BlackBerry software, offers mobile communications, access to group information, meeting schedules, and other services that can be directly tied to groupware software and servers. In addition to groupware, GSSs use a number of tools discussed previously, including the following:

- E-mail and instant messaging (IM)
- Videoconferencing
- Group scheduling
- Project management
- Document sharing

GSS software allows work teams to collaborate and reach better decisions—even if they work across town, in another region, or on the other side of the globe.

(Source: © Fisher/Thatcher/Getty Images.)

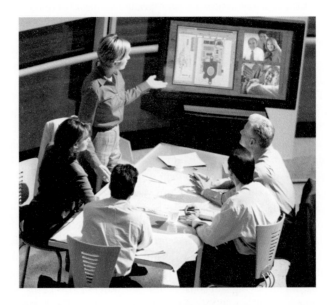

GSS Alternatives

Group support systems can take on a number of network configurations, depending on the needs of the group, the decision to be supported, and the geographic location of group members. GSS alternatives include a combination of decision rooms, local area networks, teleconferencing, and wide area networks.

- The **decision room** is ideal for situations in which decision makers are located in the same building or geographic area and the decision makers are occasional users of the GSS approach. In some cases, the decision room might have a few computers and a projector for presentations. In other cases, the decision room can be fully equipped with a network of computers and sophisticated GSS software. A typical decision room is shown in Figure 10.18.

decision room
A room that supports decision making, with the decision makers in the same building, combining face-to-face verbal interaction with technology to make the meeting more effective and efficient.

The GSS Decision Room

For group members who are in the same location, the decision room is an optimal GSS alternative. This approach can use both face-to-face and computer-mediated communication. By using networked computers and computer devices, such as project screens and printers, the meeting leader can pose questions to the group, instantly collect their feedback, and, with the help of the governing software loaded on the control station, process this feedback into meaningful information to aid in the decision-making process.

- The *local area decision network* can be used when group members are located in the same building or geographic area and under conditions in which group decision making is frequent. In these cases, the technology and equipment for the GSS approach is placed directly into the offices of the group members.
- *Teleconferencing* is used when the decision frequency is low and the location of group members is distant. These distant and occasional group meetings can tie together multiple GSS decision-making rooms across the country or around the world.
- The *wide area decision network* is used when the decision frequency is high and the location of group members is distant. In this case, the decision makers require frequent or constant use of the GSS approach. This GSS alternative allows people to work in **virtual workgroups**, where teams of people located around the world can work on common problems.

virtual workgroups
Teams of people located around the world working on common problems.

EXECUTIVE SUPPORT SYSTEMS

Because top-level executives often require specialized support when making strategic decisions, many companies have developed systems to assist executive decision making. This type of system, called an **executive support system** (ESS), is a specialized DSS that includes all hardware, software, data, procedures, and people used to assist senior-level executives within the organization. In some cases, an ESS, also called an *executive information system (EIS)*, supports

executive support system (ESS)
Specialized DSS that includes all hardware, software, data, procedures, and people used to assist senior-level executives within the organization.

decision making of members of the board of directors, who are responsible to stockholders. These top-level decision-making strata are shown in Figure 10.19.

Figure 10.19

The Layers of Executive Decision Making

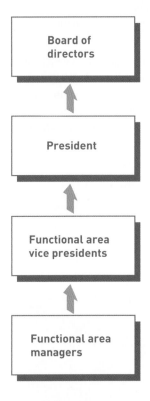

An ESS can also be used by individuals at middle levels in the organizational structure. Once targeted at the top-level executive decision makers, ESSs are now marketed to—and used by—employees at other levels in the organization. In the traditional view, ESSs give top executives a means of tracking critical success factors. Today, all levels of the organization share information from the same databases. However, for our discussion, assume ESSs remain in the upper-management levels, where they highlight important corporate issues, indicate new directions the company might take, and help executives monitor the company's progress.

Executive Support Systems in Perspective

An ESS is a special type of DSS, and, like a DSS, an ESS is designed to support higher-level decision making in the organization. The two systems are, however, different in important ways. DSSs provide a variety of modeling and analysis tools to enable users to thoroughly analyze problems—that is, they allow users to *answer* questions. ESSs present structured information about aspects of the organization that executives consider important. In other words, they allow executives to *ask* the right questions.

The following are general characteristics of ESSs:

- **Are tailored to individual executives.** ESSs are typically tailored to individual executives; DSSs are not tailored to particular users. An ESS is an interactive, hands-on tool that allows an executive to focus, filter, and organize data and information.

- **Are easy to use.** A top-level executive's most critical resource can be his time. Thus, an ESS must be easy to learn and use and not overly complex.

- **Have drill-down abilities.** An ESS allows executives to drill down into the company to determine how certain data was produced. Drilling down allows an executive to get more detailed information if needed.

- **Support the need for external data.** The data needed to make effective top-level decisions is often external—information from competitors, the federal government, trade associations and journals, consultants, and so on. An effective ESS can extract data useful to the decision maker from a wide variety of sources, including the Internet and other electronic publishing sources such as legal and public business information from Lexis/Nexis.

- **Can help with situations that have a high degree of uncertainty.** Most executive decisions involve a high degree of uncertainty. Handling these unknown situations using modeling and other ESS procedures helps top-level managers measure the amount of risk in a decision.
- **Have a future orientation.** Executive decisions are future oriented, meaning that decisions will have a broad impact for years or decades. The information sources to support future-oriented decision making are usually informal—from organizing golf partners to tying together members of social clubs or civic organizations.
- **Are linked with value-added business processes.** Like other information systems, executive support systems are linked with executive decision making about value-added business processes. For instance, executive support systems can be used by car-rental companies to analyze trends.

Capabilities of Executive Support Systems

The responsibility given to top-level executives and decision makers brings unique problems and pressures to their jobs. Following is a discussion of some of the characteristics of executive decision making that are supported through the ESS approach. ESSs take full advantage of data mining, the Internet, blogs, podcasts, executive dashboards, and many other technological innovations. As you will note, most of these decisions are related to an organization's overall profitability and direction. An effective ESS should have the capability to support executive decisions with components such as strategic planning and organizing, crisis management, and more.

Support for Defining an Overall Vision

One of the key roles of senior executives is to provide a broad vision for the entire organization. This vision includes the organization's major product lines and services, the types of businesses it supports today and in the future, and its overriding goals.

Support for Strategic Planning

ESSs also support strategic planning. **Strategic planning** involves determining long-term objectives by analyzing the strengths and weaknesses of the organization, predicting future trends, and projecting the development of new product lines. It also involves planning the acquisition of new equipment, analyzing merger possibilities, and making difficult decisions concerning downsizing and the sale of assets if required by unfavorable economic conditions.

strategic planning
Determining long-term objectives by analyzing the strengths and weaknesses of the organization, predicting future trends, and projecting the development of new product lines.

Support for Strategic Organizing and Staffing

Top-level executives are concerned with organizational structure. For example, decisions concerning the creation of new departments or downsizing the labor force are made by top-level managers. Overall direction for staffing decisions and effective communication with labor unions are also major decision areas for top-level executives. ESSs can be employed to help analyze the impact of staffing decisions, potential pay raises, changes in employee benefits, and new work rules.

Support for Strategic Control

Another type of executive decision relates to strategic control, which involves monitoring and managing the overall operation of the organization. Goal seeking can be done for each major area to determine what performance these areas need to achieve to reach corporate expectations. Effective ESS approaches can help top-level managers make the most of their existing resources and control all aspects of the organization.

Support for Crisis Management

Even with careful strategic planning, a crisis can occur. Major disasters, including hurricanes, tornadoes, floods, earthquakes, fires, and terrorist activities, can totally shut down major parts of the organization. Handling these emergencies is another responsibility for top-level executives. In many cases, strategic emergency plans can be put into place with the help of an ESS. These contingency plans help organizations recover quickly if an emergency or crisis occurs.

Decision making is a vital part of managing businesses strategically. IS systems such as information and decision support, group support, and executive support systems help employees by tapping existing databases and providing them with current, accurate information. The increasing integration of all business information systems—from TPSs to MISs to DSSs—can help organizations monitor their competitive environment and make better-informed decisions. Organizations can also use specialized business information systems, discussed in the next chapter, to achieve their goals.

SUMMARY

Principle

Good decision-making and problem-solving skills are the key to developing effective information and decision support systems.

Every organization needs effective decision making and problem solving to reach its objectives and goals. Problem solving begins with decision making. A well-known model developed by Herbert Simon divides the decision-making phase of the problem-solving process into three stages: intelligence, design, and choice. During the intelligence stage, potential problems or opportunities are identified and defined. Information is gathered that relates to the cause and scope of the problem. Constraints on the possible solution and the problem environment are investigated. In the design stage, alternative solutions to the problem are developed and explored. In addition, the feasibility and implications of these alternatives are evaluated. Finally, the choice stage involves selecting the best course of action. In this stage, the decision makers evaluate the implementation of the solution to determine whether the anticipated results were achieved and to modify the process in light of new information learned during the implementation stage.

Decision making is a component of problem solving. In addition to the intelligence, design, and choice steps of decision making, problem solving also includes implementation and monitoring. Implementation places the solution into effect. After a decision has been implemented, it is monitored and modified if needed.

Decisions can be programmed or nonprogrammed. Programmed decisions are made using a rule, procedure, or quantitative method. Ordering more inventory when the level drops to 100 units or fewer is an example of a programmed decision. A nonprogrammed decision deals with unusual or exceptional situations. Determining the best training program for a new employee is an example of a nonprogrammed decision.

Decisions can use optimization, satisficing, or heuristic approaches. Optimization finds the best solution. Optimization problems often have an objective such as maximizing profits given production and material constraints. When a problem is too complex for optimization, satisficing is often used. Satisficing finds a good, but not necessarily the best, decision. Finally, a heuristic is a "rule of thumb" or commonly used guideline or procedure used to find a good decision.

Principle

The management information system (MIS) must provide the right information to the right person in the right format at the right time.

A management information system is an integrated collection of people, procedures, databases, and devices that provides managers and decision makers with information to help achieve organizational goals. An MIS can help an organization achieve its goals by providing managers with insight into the regular operations of the organization so that they can control, organize, and plan more effectively and efficiently. The primary difference between the reports generated by the TPS and those generated by the MIS is that MIS reports support managerial decision making at the higher levels of management.

Data that enters the MIS originates from both internal and external sources. The most significant internal sources of data for the MIS are the organization's various TPSs and ERP systems. Data warehouses and data marts also provide important input data for the MIS. External sources of data for the MIS include extranets, customers, suppliers, competitors, and stockholders.

The output of most MISs is a collection of reports that are distributed to managers. These reports include scheduled reports, key-indicator reports, demand reports, exception reports, and drill-down reports. Scheduled reports are produced periodically, or on a schedule, such as daily, weekly, or monthly. A key-indicator report is a special type of scheduled report. Demand reports are developed to provide certain information at a manager's request. Exception reports are automatically produced when a situation is unusual or requires management action. Drill-down reports provide increasingly detailed data about situations.

Management information systems have a number of common characteristics, including producing scheduled, demand, exception, and drill-down reports; producing reports with fixed and standard formats; producing hard-copy and soft-copy reports; using internal data stored in organizational computerized databases; and having reports developed and implemented by IS personnel or end users.

Most MISs are organized along the functional lines of an organization. Typical functional management information systems include financial, manufacturing, marketing, human resources, and other specialized systems. Each system is composed of inputs, processing subsystems, and outputs. The primary sources of input to functional MISs include the corporate strategic plan, data from the ERP system and TPS, information from supply chain and business transactions, and external sources including the Internet and extranets. The primary output of these functional MISs are summary reports that assist in managerial decision making.

A financial management information system provides financial information to all financial managers within an organization, including the chief financial officer (CFO). Subsystems are profit/loss and cost systems, auditing, and use and management of funds.

A manufacturing MIS accepts inputs from the strategic plan, the ERP system and TPS, and external sources, such as supply chain and business transactions. The systems involved support the business processes associated with the receiving and inspecting of raw material and supplies; inventory tracking of raw materials, work in process, and finished goods; labor and personnel management; management of assembly lines, equipment, and machinery; inspection and maintenance; and order processing. The subsystems involved are design and engineering, master production scheduling and inventory control, process control, and quality control and testing.

A marketing MIS supports managerial activities in the areas of product development, distribution, pricing decisions, promotional effectiveness, and sales forecasting. Subsystems include marketing research, product development, promotion and advertising, and product pricing.

A human resource MIS is concerned with activities related to employees of the organization. Subsystems include human resource planning, personnel selection and recruiting, training and skills inventories, scheduling and job placement, wage and salary administration, and outplacement.

An accounting MIS performs a number of important activities, providing aggregate information on accounts payable, accounts receivable, payroll, and many other applications. The organization's ERP system or TPS captures accounting data, which is also used by most other functional information systems. Geographic information systems provide regional data in graphical form.

Principle

Decision support systems (DSSs) are used when the problems are unstructured.

A decision support system (DSS) is an organized collection of people, procedures, software, databases, and devices working to support managerial decision making. DSS characteristics include the ability to handle large amounts of data; obtain and process data from different sources; provide report and presentation flexibility; support drill-down analysis; perform complex statistical analysis; offer textual and graphical orientations; support optimization, satisficing, and heuristic approaches; and perform what-if, simulation, and goal-seeking analysis.

DSSs provide support assistance through all phases of the problem-solving process. Different decision frequencies also require DSS support. An ad hoc DSS addresses unique, infrequent decision situations; an institutional DSS handles routine decisions. Highly structured problems, semistructured problems, and unstructured problems can be supported by a DSS. A DSS can also support different managerial levels, including strategic, tactical, and operational managers. A common database is often the link that ties together a company's TPS, MIS, and DSS.

The components of a DSS are the database, model base, user interface or dialogue manager, and a link to external databases, the Internet, the corporate intranet, extranets, networks, and other systems. The database can use data

warehouses and data marts. A data-driven DSS primarily performs qualitative analysis based on the company's databases. Data-driven DSSs tap into vast stores of information contained in the corporate database, retrieving information on inventory, sales, personnel, production, finance, accounting, and other areas. Data mining is often used in a data-driven DSS. The model base contains the models used by the decision maker, such as financial, statistical, graphical, and project-management models. A model-driven DSS primarily performs mathematical or quantitative analysis. Model management software (MMS) is often used to coordinate the use of models in a DSS. The dialogue manager provides a dialogue management facility to assist in communications between the system and the user. Access to other computer-based systems permits the DSS to tie into other powerful systems, including the TPS or function-specific subsystems.

Principle

Specialized support systems, such as group support systems (GSSs) and executive support systems (ESSs), use the overall approach of a DSS in situations such as group and executive decision making.

A group support system (GSS), also called a *computerized collaborative work system*, consists of most of the elements in a DSS, plus software to provide effective support in group decision-making settings. GSSs are typically easy to learn and use and can offer specific or general decision-making support. GSS software, also called *groupware*, is specially designed to help generate lists of decision alternatives and perform data analysis. These packages let people work on joint documents and files over a network.

The frequency of GSS use and the location of the decision makers will influence the GSS alternative chosen. The decision room alternative supports users in a single location who meet infrequently. Local area networks can be used when group members are located in the same geographic area and users meet regularly. Teleconferencing is used when decision frequency is low and the location of group members is distant. A wide area network is used when the decision frequency is high and the location of group members is distant.

Executive support systems (ESSs) are specialized decision support systems designed to meet the needs of senior management. They serve to indicate issues of importance to the organization, indicate new directions the company might take, and help executives monitor the company's progress. ESSs are typically easy to use, offer a wide range of computer resources, and handle a variety of internal and external data. In addition, the ESS performs sophisticated data analysis, offers a high degree of specialization, and provides flexibility and comprehensive communications abilities. An ESS also supports individual decision-making styles. Some of the major decision-making areas that can be supported through an ESS are providing an overall vision, strategic planning and organizing, strategic control, and crisis management.

CHAPTER 10: SELF-ASSESSMENT TEST

Good decision-making and problem-solving skills are the key to developing effective information and decision support systems.

1. The last stage of the decision making process is the _____.
   a. initiation stage
   b. intelligence stage
   c. design stage
   d. choice stage

2. Problem solving is one of the stages of decision making. True or False?

3. The final stage of problem solving is _____.

4. A decision that inventory should be ordered when inventory levels drop to 500 units is an example of a(n) _____.
   a. synchronous decision
   b. asynchronous decision
   c. nonprogrammed decision
   d. programmed decision

5. A(n) _____ model will find the best solution, usually the one that will best help the organization meet its goals.

6. A satisficing model is one that will find a good problem solution, but not necessarily the best problem solution. True or False?

The management information system (MIS) must provide the right information to the right person in the right format at the right time.

7. What summarizes the previous day's critical activities and is typically available at the beginning of each workday?
   a. key-indicator report
   b. demand report
   c. exception report
   d. database report

8. MRP and JIT are a subsystem of the _____.
   a. marketing MIS
   b. financial MIS
   c. manufacturing MIS
   d. auditing MIS

9. Another name for the _____ MIS is the personnel MIS because it is concerned with activities related to employees and potential employees of the organization.

Decision support systems (DSSs) are used when the problems are unstructured.

10. The focus of a decision support system is on decision-making effectiveness when faced with unstructured or semistructured business problems. True or False?

11. The process of determining the problem data required for a given result is called _____ analysis.

12. What component of a decision support system allows decision makers to easily access and manipulate the DSS and to use common business terms and phrases?
   a. the knowledge base
   b. the model base
   c. the user interface or dialogue manager
   d. the expert system

Specialized support systems, such as group support systems (GSSs) and executive support systems (ESSs), use the overall approach of a DSS in situations such as group and executive decision making.

13. What allows a person to give his input without his identity being known to other group members?
   a. groupthink
   b. anonymous input
   c. nominal group technique
   d. delphi

14. A type of software that helps with joint work group scheduling, communication, and management is called _____.

15. The local area decision network is the ideal GSS alternative for situations in which decision makers are located in the same building or geographic area and the decision makers are occasional users of the GSS approach. True or False?

16. A(n) _____ supports the actions of members of the board of directors, who are responsible to stockholders.

CHAPTER 10: SELF-ASSESSMENT TEST ANSWERS

(1) d (2) False (3) monitoring (4) d (5) optimization (6) True (7) a (8) c (9) human resource (10) True (11) goal seeking (12) c (13) b (14) groupware or workgroup software (15) False (16) executive information system (EIS)

REVIEW QUESTIONS

1. What is a satisficing model? Describe a situation when it should be used.
2. What is the difference between intelligence and design in decision making?
3. What is the difference between a programmed decision and a nonprogrammed decision? Give several examples of each.
4. What are the basic kinds of reports produced by an MIS?
5. What guidelines should be followed in developing reports for management information systems?
6. What are the functions performed by a financial MIS?
7. Describe the functions of a marketing MIS.
8. What is a human resource MIS? What are its outputs?
9. List and describe some other types of MISs.
10. What are the stages of problem solving?
11. What is the difference between decision making and problem solving?
12. What is a geographic information system?
13. Describe the difference between a structured and an unstructured problem and give an example of each.
14. Define *decision support system*. What are its characteristics?
15. Describe the difference between a data-driven and a model-driven DSS.
16. What is the difference between what-if analysis and goal-seeking analysis?
17. What are the components of a decision support system?
18. State the objective of a group support system (GSS) and identify three characteristics that distinguish it from a DSS.
19. Identify three group decision-making approaches often supported by a GSS.
20. What is an executive support system? Identify three fundamental uses for such a system.

DISCUSSION QUESTIONS

1. Select an important problem you had to solve during the last two years. Describe how you used the decision-making and problem-solving steps discussed in this chapter to solve the problem.
2. Describe how a MIS can be used at your school or university.
3. How can management information systems be used to support the objectives of the business organization?
4. Describe a financial MIS for a *Fortune* 1000 manufacturer of food products. What are the primary inputs and outputs? What are the subsystems?
5. How can a strong financial MIS provide strategic benefits to a firm?
6. Why is auditing so important in a financial MIS? Give an example of an audit that failed to disclose the true nature of the financial position of a firm. What was the result?
7. Describe two industries where a marketing MIS is critical to sales and success.
8. You have been hired to develop a management information system and a decision support system for a manufacturing company. Describe what information you would include in printed reports and what information you would provide using a screen-based decision support system.
9. Pick a company and research its human resource management information system. Describe how its system works.

What improvements could be made to its human resource MIS?
10. You have been hired to develop a DSS for a car company such as Ford or GM. Describe how you would use both data-driven and model-driven DSSs.
11. Imagine that you are the CFO for a service organization. You are concerned with the integrity of the firm's financial data. What steps might you take to ascertain the extent of problems?
12. What functions do decision support systems support in business organizations? How does a DSS differ from a TPS and an MIS?
13. How is decision making in a group environment different from individual decision making, and why are information systems that assist in the group environment different? What are the advantages and disadvantages of making decisions as a group?
14. You have been hired to develop group support software. Describe the features you would include in your new GSS software.
15. The use of ESSs should not be limited to the executives of the company. Do you agree or disagree? Why?
16. Imagine that you are the vice president of manufacturing for a *Fortune* 1000 manufacturing company. Describe the features and capabilities of your ideal ESS.

PROBLEM-SOLVING EXERCISES

1. You have been asked to develop an effective executive support system (ESS) for the vice president of marketing for a large retail chain. Use a word processor to describe the features of the ESS. Develop a set of slides using a graphics program to deliver a presentation on your recommendations.

2. Review the summarized consolidated statement of income for the manufacturing company whose data is shown here. Use graphics software to prepare a set of bar charts that shows the data for this year compared with the data for last year.

   a. This year, operating revenues increased by 3.5 percent, while operating expenses increased 2.5 percent.
   b. Other income and expenses decreased to $13,000.
   c. Interest and other charges increased to $265,000.

Operating Results (in millions)

Operating Revenues	$2,924,177
Operating Expenses (including taxes)	2,483,687
Operating Income	440,490
Other Income and Expenses	13,497
Income before Interest and Other Charges	453,987
Interest and Other Charges	262,845
Net Income	191,142
Average Common Shares Outstanding	147,426
Earnings per Share	1.30

If you were a financial analyst tracking this company, what detailed data might you need to perform a more complete analysis? Write a brief memo summarizing your data needs.

3. As the head buyer for a major supermarket chain, you are constantly being asked by manufacturers and distributors to stock their new products. Over 50 new items are introduced each week. Many times, these products are launched with national advertising campaigns and special promotional allowances to retailers. To add new products, the amount of shelf space allocated to existing products must be reduced or items must be eliminated altogether.

   Develop a marketing MIS that you can use to estimate the change in profits from adding or deleting an item from inventory. Your analysis should include input such as estimated weekly sales in units, shelf space allocated to stock an item (measured in units), total cost per unit, and sales price per unit. Your analysis should calculate total annual profit by item and then sort the rows in descending order based on total annual profit.

TEAM ACTIVITIES

1. Using only the Internet to do research and communicate, have your team work together to select five companies that are most likely to have the best stock price increase over the next year. Your team might be asked to prepare a report on your selection and how difficult it was to use only the Internet for communication.

2. Have your team make a group decision about how to solve the most frustrating aspect of college or university life. Appoint one or two members of the team to disrupt the meeting with negative group behavior. After the meeting,

have your team describe how to prevent this negative group behavior. What GSS software features would you suggest to prevent the negative group behavior your team observed?

3. Imagine that you and your team have decided to develop an ESS software product to support senior executives in the music recording industry. What are some of the key decisions these executives must make? Make a list of the capabilities that such a system must provide to be useful. Identify at least six sources of external information that will be useful to its users.

WEB EXERCISES

1. Use a search engine, such as Yahoo! or Google, to explore two or more companies that produce and sell MIS or DSS software. Describe what you found and any problems you had in using search engines on the Internet to find information. You might be asked to develop a report or send an e-mail message to your instructor about what you found.

2. Use the Internet to explore two or more software packages that can be used to make group decisions easier. Summarize your findings in a report.

3. Software, such as Microsoft Excel, is often used to find an optimal solution to maximize profits or minimize costs. Search the Internet using Yahoo!, Google, or another search engine to find other software packages that offer optimization features. Write a report describing one or two of the optimization software packages. What are some of the features of the package?

CAREER EXERCISES

1. What decisions are critical for success in a career that interests you? What specific types of reports could help you make better decisions on the job? Give three specific examples.

2. For a career area of your choice, describe how top-level executives can use an information system to help them make better decisions. Give two specific examples of actual executives in your career area who have benefited from the use of information systems.

CASE STUDIES

Case One

American National Insurance Speeds Information to Customers with MIS and BPM

The goal of an effective management information system (MIS) is to improve a company's business processes. A business process is a set of coordinated tasks and activities that people and equipment perform to accomplish a specific organizational goal. Working to make business processes more effective and more capable of adapting to an ever-changing environment is commonly referred to as business process management (BPM). MISs are at the core of BPM. A poorly designed MIS leads to problems with business processes.

The American National Insurance Company had a problem with its MISs and business processes in its customer service department. The company uses many legacy management information systems. When a customer phones the company asking for information regarding a policy, a customer service representative needs 10 or 11 minutes to find the required information stored in several systems. The rep needs to drill down into one system for part of the answer, log out, drill down into the next system, and so on until the full answer is assembled. In the mean time, the rep needs to keep the customer informed during the wait.

American National hired Pegasystems to develop an MIS to link all of its legacy systems. Pegasystems business process specialists studied the types of queries the customer service reps made and developed process models that covered the most common customer requests. Then they designed a properly modeled and easy-to-use system to automate the collection of data from the legacy systems.

Today, when American National reps field calls from customers, they enter the customer ID into the system, select the information requested from menus and wizards provided by a custom interface, and typically have the desired information within 15 seconds. "Depending on what the customer asks for, different business rules are enacted that take the customer service rep down different paths," says Gary Kirkham, vice president and director of planning and support at American National.

The new system has satisfied the company's primary goal: to put the customer service rep in a position to help the customer as quickly as possible. Now the company is working on its secondary goal: to optimize and automate business processes. For example, when a customer dies, it used to require an effort from three American National Insurance employees to handle the paperwork and business processes to manage the life insurance claim. These processes have been optimized and automated. "Once Pegasystems knows an insured has expired, it processes automatically all the things three people used to do," Kirkham says.

Kirkham stated that the new system has increased sales in its annuity division from $750 million to $2.2 billion two years in a row. He credited the increase to the ability to keep customers happy and to use the system to identify callers who will produce the most sales and providing those clients with the priority they deserve.

Discussion Questions

1. How were poorly designed management information systems responsible for problems in American National Insurance Company's customer service business process?
2. How did American National Insurance Company use an MIS to turn a bad situation into an advantage?

Critical Thinking Questions

1. How do you think American National found itself with information systems that were so difficult to navigate?
2. Based on what you have learned throughout this book, what would you suggest as an alternative but much more costly solution to American National's dilemma (rather than the one provided by Pegasystems)?

SOURCES: Schwartz, Ephraim, "Building a Workflow for Insurance Reps," *Infoworld*, February 20, 2006, *www.infoworld.com*. Pegasystems Web site, accessed June 8, 2006, *www.pega.com/AboutPega/AboutPega.asp*. "A New Foundation for Document Creation," Group1 Software Case Study, accessed June 8, 2006, *www.g1.com/PDF/CaseStudy/ANICOCaseStudy.pdf*.

Case Two

Transit New Zealand Reroutes Information to Avoid Traffic Jams

New Zealand's state highway network takes drivers through 11,000 kilometers of some of the most beautiful scenery in the world. Transit New Zealand (TNZ) manages those highways, which carry half the country's traffic, to make sure that drivers have smooth and efficient routes to their destinations. Unfortunately, until recently, the flow of information within TNZ was anything but smooth and efficient. TNZ was in dire need of an upgrade to its management information systems.

The public face of TNZ provides drivers with information about the condition and safety of its roads. Its main mission, however, is to work with external contractors and government bodies that decide on highway building and maintenance projects. TNZ needed to freely exchange information with all these constituencies but lacked the integrated information systems to do so.

"Our content was stored in several siloed systems," says Geoff Yeats, TNZ's chief information officer. "This made it dif-

ficult for our employees to find the information they needed to do their jobs. For example, to initiate a road repair project using the same contractor that had worked on the road previously, we had to consult a number of disparate systems to find contractor information, spec sheets for the old job and the new repairs, and the government road construction standards that were current at the time. It took a lot of time."

A siloed system is one that stands alone, disconnected from surrounding systems. Such systems typically stand in the way of establishing one true integrated enterprise information system. TNZ realized that what they needed was an enterprise-wide information management solution so they could centrally manage information from all departments and save it economically for easy and timely retrieval. Because the information would be viewed by different groups of users for different purposes, the system would need to present information in multiple ways.

If TNZ accessed organization-wide data from one central interface, the walls dividing its siloed systems would have to come down. TNZ worked with an information system provider to develop a document, records, and content management system within one repository. The new system provided a consistent and customizable framework that let users access information through scheduled, demand, and drill-down reports. The new MIS could also provide an interface to the corporate applications and services that TNZ developed over time as needed.

The first phase of the system provides access to documents of all types as well as contact data from its customer relationship management and information held in its Road Asset Maintenance Management database. This database is one of TNZ's largest repositories of information, containing millions of items of data relating to everything from the technical condition of various stretches of roads and maintenance records to details on traffic lights and signage. "You can go into this database and find out what sort of gravel has been used to build a road and even what the road surface characteristics are like," says Yeats.

The new system also provides many other benefits. TNZ is better able to control costs and manage the growth of its information while ensuring the currency and accuracy of that information. Searches made across multiple systems are completed with subsecond response times. Because the new system uses commonly recognized standards, TNZ road information can be published through third parties simply by giving the vendor access to the information system.

The new system provides solid business value by integrating, analyzing, and optimizing heterogeneous types and sources of information throughout its life cycle to manage risk and create new business insights. It is designed to get the right information to the right people or process at the right time to take advantage of opportunities. "By responding to opportunities and threats with information on demand, we can lower our costs, optimize our infrastructure, gain control of our master data and manage information complexity," says Yeats.

Future initiatives include electronic collaboration, browser-based content creation, and a link with TNZ's geographical information system. Because the system is designed to comply with Java Specification Request (JSR) 170, an emerging standard for accessing content repositories, it will be easy for TNZ to continue to realize benefits from the system.

Discussion Questions

1. How were siloed systems affecting the flow of information throughout TNZ?
2. What business value does the new system provide for TNZ?

Critical Thinking Questions

1. Provide two reasons that TNZ wanted to use commonly recognized standards such as JSP 170 in the design of its new system.
2. Why does TNZ find it beneficial to allow third parties to publish information stored in its systems?

SOURCES: Staff, "Transit New Zealand Drives Business Transformation with IBM Enterprise Content Management Solution," IBM Case Study, January 13, 2006, *www-306.ibm.com/software/success/cssdb.nsf/CS/HSAZ-6J728L? OpenDocument&Site=cmportal.* "Transit NZ Picks IBM for Enterprise Content Management," *iStart,* September 2005, *www.istart.co.nz.* Transit New Zealand Web site, accessed June 10, 2006, *www.transit.govt.nz.*

Questions for Web Case

See the Web site for this book to read about the Whitmann Price Consulting case for this chapter. Following are questions concerning this Web case.

Whitmann Price Consulting: MIS and DSS Considerations

Discussion Questions

1. What different types of needs can MISs and DSSs fulfill in Whitmann Price's new system?
2. Why did Whitmann Price decide to design its own Calendar and Contacts MIS rather than using standard BlackBerry software?

Critical Thinking Questions

1. How does the source of input differ between the MISs being designed for all consulting areas and the DSSs uniquely designed for each consulting area?
2. How might a GSS be useful for consultants at Whitmann Price?

NOTES

Sources for the opening vignette: Staff, "Companies Worldwide Choose Cognos 8 Business Intelligence," Press Release, CNW Group, February 14, 2006, *www.newswire.ca/en/releases/archive/ February2006/14/c2182.html.* Beauté Prestige International Case Study, Cognos Web site, accessed June 6, 2006, *www.cognos.com/products/ cognos8businessintelligence/success-stories.html.* Cognos ReportNet Web site, accessed June 6, 2006, *www.cognos.com/products/ cognos8businessintelligence/reporting.html.*

1 Mullaney, Timothy; Weintraub, Arlene, "The Digital Hospital," *Business Week,* March 28, 2005, p. 77.
2 Coy, Peter, "Why Logic Often Takes a Back Seat," *Business Week,* March 28, 2005, p. 94.
3 Lacroix, Yvan, et al., "Bombardier Flexjet Significantly Improves Its Fractional Aircraft Ownership Operations," *Interfaces,* January-February, 2005, p. 49.
4 Murty, Katta, et al., "Hongkong International Terminals Gains Elastic Capacity," *Interfaces,* January-February, 2005, p. 61.
5 Troyer, Loren, et al., "Improving Asset Management and Order Fulfillment at Deere," *Interfaces,* January-February, 2005, p. 76.
6 Kapoor, S., et al., "A Technical Framework for Sense-and-Respond Business Management," IBM Systems Journal, Vol. 44, 2005, p. 5.
7 Havenstein, Heather, "Business Objects Adds Crystal Reports Tools," *Computerworld,* January 17, 2005, p. 4.
8 Smith, Rebecca, "California Taps High-Tech Meters," *The Wall Street Journal,* May 11, 2005, p. A1.
9 Clark, Don, "Microsoft Unveils Tool to Analyze Business Data," *The Wall Street Journal,* October 24, 2005, p. B4.

10 Staff, "HP Enables Greater Insight into IT with HP OpenView Management Software," *www.hp.com/hpinfo/newsroom/press/ 2005/051205xa.html,* accessed January 22, 2006.
11 Cropper, Carol, "Between You, the Doctor, and the PC," *Business Week,* January 31, 2005, p. 90.
12 Port, Otis, "Desktop Factories," *Business Week,* May 2, 2005, p. 22.
13 Wysocki, Bernard, et al., "Just-In-Time Inventories Make U.S. Vulnerable in a Pandemic," *The Wall Street Journal,* January 12, 2006, p. A1.
14 Richmond, Rita, "Blogs Keep Internet Customers Coming Back," *The Wall Street Journal,* March 1, 2005, p. B8.
15 Vranica, Suzanne, "Marketers Aim New Ads at Video iPod Users," *The Wall Street Journal,* January 31, 2006, B1.
16 Peart, Mark, "Service Excellence & CRM," *New Zealand Management,* May 2005, p. 68.
17 Havenstein, Heather, "Harrah's Set to Replace Caesars BI Analytics Tools," *Computerworld,* October 10, 2005, p. 12.
18 Bounds, Gwendolyn, "How Selling Pixels May Yield a Million Bucks," *The Wall Street Journal,* November 22, 2005, p. B1.
19 Brandel, Mary, "HR Gets Strategic," *Computerworld,* January 24, 2005, p. 34.
20 Forelle, Charles, "IBM Tool Deploys Employees Efficiently," *The Wall Street Journal,* July 14, 2005, p. B3.
21 Bennett, Julie, "The Search is On for Forensic Accountants," *The Wall Street Journal,* October 11, 2005, p. B11.
22 Anthes, Gary, "Beyond Zip Codes," *Computerworld,* September 19, 2005, p. 56.

23 Marquez, Jeremiah, "Computers Policing L.A. Cops," *The Denver Post*, July 24, 2005, p. 10A.

24 Hymowitz, Carol, "Dashboard Technology," *The Wall Street Journal*, September 26, 2005, p. B1.

25 Sabuncuoglu, Ihsan, et al., "The Turkish Army Uses Simulation to Model and Optimize Its Fuel-Supply System," *Interfaces*, November-December, 2005, p. 474.

26 Hadi, Mohammed, "Algorithm Trading Software Takes Leap into Options," *The Wall Street Journal*, January 25, 2006, p. B11A.

27 Havenstein, Heather, "Celtics Turn to Data Analytics Tools for Help Pricing Tickets," *Computerworld*, January 9, 2006, p. 43.

28 Havenstein, Heather, "Business Intelligence Tools Help Nonprofit Group Make Loans to Tsunami Victims," *Computerworld*, March 14, 2005, p. 19.

29 Rubenstein, Sarah, "Next Step Toward Digitized Health Records," *The Wall Street Journal*, May 9, 2005, p. B1.

30 Rubenstein, Sarah, "Putting Your Health History Online," *The Wall Street Journal*, June 21, 2005, p. D7.

31 Bhattacharya, K., et al., "A Model-Driven Approach to Industrializing Discovery Processes in Pharmaceutical Research," *IBM Systems Journal*, Vol. 44, No. 1, 2005, p. 145.

32 Anthes, Gary, "Modeling Magic," *Computerworld*, February 7, 2005, p. 26.

33 Port, Otis, "Simple Solutions," *Business Week*, October 3, 2005, p. 24.

34 Mitchell, Robert, "Anticipation Game," *Computerworld*, June 13, 2005, p. 23.

35 Aston, Adam, "The Worst Isn't Over," *Business Week*, January 16, 2006, p. 29.

36 Babcock, Charles, "A New Model for Disasters," *Information Week*, October 10, 2005, p. 47.

37 Majchrak, Ann, et al., "Perceived Individual Collaboration Know-How Development Through Information Technology-Enabled Contextualization," *Information Systems Research*, March 2005, p. 9.

38 Hamblen, Matt, "Project at the World Trade Center Site Puts Advanced Design Tools to the Test," *Computerworld*, March 7, 2005, p. 7.

39 Rosencrance, Linda, "Collaborating Angels," *Computerworld*, January 24, 2005, p. 25.

40 Greene, Jay, "Combat Over Collaboration," *Business Week*, April 18, 2005, p. 64.

41 Lipschutz, Robert, "Instant Meeting: Easy Web Conferencing," *PC Magazine*, March 22, 2005, p. 120.

42 Rosencrance, Linda, "Meet Me in Cyber Space," *Computerworld*, February 21, 2005, p. 23.

CHAPTER · 11 ·

Knowledge Management and Specialized Information Systems

PRINCIPLES	LEARNING OBJECTIVES
▪ Knowledge management allows organizations to share knowledge and experience among their managers and employees.	▪ Discuss the differences among data, information, and knowledge. ▪ Describe the role of the chief knowledge officer (CKO). ▪ List some of the tools and techniques used in knowledge management.
▪ Artificial intelligence systems form a broad and diverse set of systems that can replicate human decision making for certain types of well-defined problems.	▪ Define the term *artificial intelligence* and state the objective of developing artificial intelligence systems. ▪ List the characteristics of intelligent behavior and compare the performance of natural and artificial intelligence systems for each of these characteristics. ▪ Identify the major components of the artificial intelligence field and provide one example of each type of system.
▪ Expert systems can enable a novice to perform at the level of an expert but must be developed and maintained very carefully.	▪ List the characteristics and basic components of expert systems. ▪ Identify at least three factors to consider in evaluating the development of an expert system. ▪ Outline and briefly explain the steps for developing an expert system. ▪ Identify the benefits associated with the use of expert systems.
▪ Virtual reality systems can reshape the interface between people and information technology by offering new ways to communicate information, visualize processes, and express ideas creatively.	▪ Define the term *virtual reality* and provide three examples of virtual reality applications.
▪ Specialized systems can help organizations and individuals achieve their goals.	▪ Discuss examples of specialized systems for organizational and individual use.

Information Systems in the Global Economy »
Lafarge, France
Global Building Supply Leader Gets the Recipe Right Every Time with Artificial Intelligence

Lafarge is a world leader in building-supply products: It is the third largest global supplier of gypsum, the second largest global supplier of aggregates and concrete, and the largest global supplier of cement and roofing. Lafarge has a presence in 76 countries and a workforce of 80,000. If any company knows the most efficient way of turning rocks into building materials, it is Lafarge. The company has built its dominant market position in part by implementing cutting-edge automated systems controlled by intelligent software systems.

Creating high-quality cement is similar to baking a cake. Ingredients are mixed in a specific and precise manner in the plant's raw feed mill, and then heated at exact temperatures in preheaters and a rotary kiln. After heating, baking gypsum and other ingredients are added to the mix to produce the finished dry concrete solution. If ingredients are mixed too little or too long, or if the temperature in the heaters or kiln is too high or too low, imperfections are introduced to the batch, which can cause engineering and construction problems later.

Lafarge engineers know the value of producing a consistent, high-quality product. To achieve that consistency, Lafarge depends on intelligent computer systems to monitor and control every step of cement production. The automated system produces the perfect recipe for production each time by reducing the variability in the process and optimizing production.

The intelligent system named LUCIE automates processes for controlling kilns, grate coolers, and mills at over 40 Lafarge cement plants worldwide. LUCIE uses several forms of artificial intelligence (AI) discussed in this chapter. Expert knowledge of the cement-making process is built in to the system using rules. This makes LUCIE a form of AI called an expert system. Kiln automation is achieved by blending fuzzy logic, process modeling, and statistical auto-adaptation of key parameters. These techniques allow LUCIE to react to various situations in the production plant, even when the variables are unexpected. A neural network—computer software that functions like the human brain—and rules-based reasoning are used to judge the process conditions in the plant's raw feed mill, preheaters, rotary kiln, and cement mill. If conditions are not optimum, LUCIE applies short-term and long-term actions to make sure that the cement mixture is always produced in perfect fashion.

Lafarge engineers continue to add valuable functionality to LUCIE. Recently, LUCIE was upgraded to include more statistical and graphical ways to diagnose and optimize key parameters. A built-in fuel manager now improves stability and performance by managing the secondary fuels used in kilns.

Artificial intelligence and other specialized business information systems support many of today's top global companies by providing automation that can control processes in a more consistent and reliable manner than people can. These systems extend human capabilities by providing machines, which process quickly and tirelessly with humanlike intelligence.

As you read this chapter, consider the following:

- How does computer intelligence compare to human intelligence?
- How can people and businesses make the best use of artificial intelligence and other specialized systems?

Why Learn About Specialized Information Systems?

Knowledge management and specialized information systems are used in almost every industry. If you are a manager, you might use a knowledge management system to support decisive action to help you correct a problem. If you are a production manager at an automotive company, you might oversee robots that attach windshields to cars or paint body panels. As a young stock trader, you might use a special system called a *neural network* to uncover patterns and make millions of dollars trading stocks and stock options. As a marketing manager for a PC manufacturer, you might use virtual reality on a Web site to show customers your latest laptop and desktop computers. If you are in the military, you might use computer simulation as a training tool to prepare you for combat. In a petroleum company, you might use an expert system to determine where to drill for oil and gas. You will see many additional examples of using these specialized information systems throughout this chapter. Learning about these systems will help you discover new ways to use information systems in your day-to-day work.

Like other aspects of an information system, the overall goal of knowledge management and the specialized systems discussed in this chapter is to help people and organizations achieve their goals. In some cases, knowledge management and these specialized systems can help an organization achieve a long-term, strategic advantage. In this chapter, we explore knowledge management, artificial intelligence, and many other specialized information systems, including expert systems, robotics, vision systems, natural language processing, learning systems, neural networks, genetic algorithms, intelligent agents, and virtual reality.

KNOWLEDGE MANAGEMENT SYSTEMS

Chapter 1 defines and discusses data, information, and knowledge. Recall that *data* consists of raw facts, such as an employee number, number of hours worked in a week, inventory part numbers, or sales orders. A list of the quantity available for all items in inventory is an example of data. When these facts are organized or arranged in a meaningful manner, they become information. *Information* is a collection of facts organized so that they have additional value beyond the value of the facts themselves. An exception report of inventory items that might be out of stock in a week because of high demand is an example of information. *Knowledge* is the awareness and understanding of a set of information and the ways that information can be made useful to support a specific task or reach a decision. Knowing the procedures for ordering more inventory to avoid running out is an example of knowledge. In a sense, information tells you what has to be done (low inventory levels for some items), while knowledge tells you how to do it (make two important phone calls to the right people to get the needed inventory shipped overnight). See Figure 11.1. A *knowledge management system (KMS)* is an organized collection of people, procedures, software, databases, and devices used to create, store, share, and use the organization's knowledge and experience.[1] KMSs cover a

Figure 11.1

The Differences Among Data, Information, and Knowledge

Data	There are 20 PCs in stock at the retail store.
Information	The store will run out of inventory in a week unless more is ordered today.
Knowledge	Call 800-555-2222 to order more inventory.

wide range of systems from software that contains some KMS components to dedicated systems designed specifically to capture, store, and use knowledge.

Overview of Knowledge Management Systems

Like the other systems discussed throughout the book, including information and decision support systems, knowledge management systems attempt to help organizations achieve their goals. For businesses, this usually means increasing profits or reducing costs. For nonprofit organizations, it can mean providing better customer service or providing special needs to people and groups. Many types of firms use KMSs to increase profits or reduce costs. Ryder Systems, for example, worked with the Accenture consulting company to streamline its transportation and logistics business.[2] According to Gene Tyndall, Ryder's executive vice president for global solutions and e-commerce, "People—the knowledge they have and the new knowledge they create— are the corporate assets that impact Ryder's performance more than any other form of capital." According to a survey of CEOs, firms that use KMSs are more likely to innovate and perform better.[3]

A KMS can involve different types of knowledge. *Explicit knowledge* is objective and can be measured and documented in reports, papers, and rules. For example, knowing the best road to take to minimize drive time from home to the office when a major highway is closed due to an accident is explicit knowledge. It can be documented in a report or a rule, as in "If I-70 is closed, take Highway 6 to town and the office." *Tacit knowledge*, on the other hand, is hard to measure and document and typically is not objective or formalized. Knowing the best way to negotiate with a foreign government about nuclear disarmament or a volatile hostage situation often requires a lifetime of experience and a high level of skill. These are examples of tacit knowledge. It is difficult to write a detailed report or a set of rules that would always work in every hostage situation. Many organizations actively attempt to convert tacit knowledge to explicit knowledge to make the knowledge easier to measure, document, and share with others.

Data and Knowledge Management Workers and Communities of Practice

The personnel involved in a KMS include data workers and knowledge workers. Secretaries, administrative assistants, bookkeepers, and similar data-entry personnel are often called *data workers*. As mentioned in Chapter 1, *knowledge workers* are people who create, use, and disseminate knowledge. They are usually professionals in science, engineering, or business, and work in offices and belong to professional organizations. Other examples of knowledge workers include writers, researchers, educators, and corporate designers. See Figure 11.2.

Knowledge Workers

Knowledge workers are people who create, use, and disseminate knowledge, including professionals in science, engineering, business, and other areas.

(Source: Photo by Scott Bauer/ USDA.)

chief knowledge officer (CKO)
A top-level executive who helps the organization use a KMS to create, store, and use knowledge to achieve organizational goals.

The **chief knowledge officer (CKO)** is a top-level executive who helps the organization work with a KMS to create, store, and use knowledge to achieve organizational goals. The CKO is responsible for the organization's KMS, and typically works with other executives and vice presidents, including the chief executive officer (CEO), chief financial officer (CFO), and others.

Some organizations and professions use *communities of practice (COP)* to create, store, and share knowledge. A COP is a group of people dedicated to a common discipline or practice, such as open-source software, auditing, medicine, or engineering. A group of oceanographers investigating climate change or a team of medical researchers looking for new ways to treat lung cancer are examples of COPs. COPs excel at obtaining, storing, sharing, and using knowledge.

Obtaining, Storing, Sharing, and Using Knowledge

Obtaining, storing, sharing, and using knowledge is the key to any KMS.[4] Using a KMS often leads to additional knowledge creation, storage, sharing, and usage. See Figure 11.3. A meteorologist, for example, might develop sophisticated mathematical models to predict the path and intensity of hurricanes. Business professors often conduct research in marketing strategies, management practices, corporate and individual investments and finance, effective accounting and auditing practices, and much more. Drug companies and medical researchers invest billions of dollars in creating knowledge on cures for diseases. Although knowledge workers can act alone, they often work in teams to create or obtain knowledge.

Figure 11.3

Knowledge Management System

Obtaining, storing, sharing, and using knowledge is the key to any KMS.

After knowledge is created, it is often stored in a *knowledge repository* that includes documents, reports, files, and databases. The knowledge repository can be located both inside the organization and outside. Some types of software can store and share knowledge contained in documents and reports. Adobe Acrobat PDF files, for example, allow you to store corporate reports, tax returns, and other documents and send them to others over the Internet. This publisher and the authors of this book used PDF files to store, share, and collaborate on each chapter. You can use hardware devices and software to store and share audio and video material.[5] Traditional databases, data warehouses, and data marts discussed in Chapter 5 often store the organization's knowledge. Specialized knowledge bases in expert systems, discussed later in the chapter, can also be used.

Because knowledge workers often work in groups or teams, they can use collaborative work software and group support systems discussed in Chapter 10 to share knowledge, such as groupware, meeting software, and collaboration tools. Intranets and password-protected Internet sites also provide ways to share knowledge. The social services department of the Surrey County Council in the United Kingdom, for example, used an intranet to help it create and manipulate knowledge.[6] Because knowledge can be critical in maintaining a competitive advantage, businesses should be careful in how they share knowledge. Although they want important decision makers inside and outside the organization to have complete and easy access to knowledge, they also need to protect knowledge from competitors, hackers, and others who shouldn't obtain the organization's knowledge. As a result, many businesses use patents, copyrights, trade secrets, Internet firewalls, and other measures to keep prying eyes from seeing important knowledge that is often expensive and hard to create.

In addition to using information systems and collaborative software tools to share knowledge, some organizations use nontechnical approaches. These include corporate retreats and gatherings, sporting events, informal knowledge worker lounges or meeting places, kitchen facilities, daycare centers, and comfortable workout centers.

Using a knowledge management system begins with locating the organization's knowledge. This is often done using a *knowledge map* or directory that points the knowledge worker to the needed knowledge. Drug companies have sophisticated knowledge maps that include database and file systems to allow scientists and drug researchers to locate previous medical studies. Lawyers can use powerful online knowledge maps, such as the legal section of Lexis-Nexis to research legal opinions and the outcomes of previous cases. Medical researchers, university professors, and even textbook authors use Lexis-Nexis to locate important knowledge. Corporations often use the Internet or corporate Web portals to help their knowledge workers find knowledge stored in documents and reports. The following are examples of profit and nonprofit organizations that use knowledge and knowledge management systems.

- CACI International, in Arlington, Virginia, received a multimillion dollar contract from the U.S. federal government to develop a knowledge management system to support intelligence and national security efforts.[7]
- China Netcom Corporation uses KM software from Autonomy Corporation to search the records of up to 100 million telecommunications customers and create knowledge about its customers and marketing operations.[8]
- Feilden, Clegg, Bradley, and Aedas, an architectural firm, uses KM to share best practices among its architects.[9] According to one designer, "Knowledge management was one of those ideas that sprang up in the 1990s, along with fads such as total quality management and the concept of the learning organization. But knowledge management (KM) appears to have had staying power, and it is still firmly on the business agenda."
- Munich Re Group, a German insurance organization, uses KM to share best practices and knowledge.[10] "It was always important to us that knowledge management isn't just an IT platform," said Karen Edwards, knowledge management consultant in Munich Re's Knowledge Management Center of Competence in Munich, Germany. "The Munich Re people, they really were the assets. They're the things you try to bring together."

Technology to Support Knowledge Management

KMSs use a number of tools discussed throughout the book. In Chapter 2, for example, we explored the importance of *organizational learning* and *organizational change*. An effective KMS is based on learning new knowledge and changing procedures and approaches as a result.[11] A manufacturing company, for example, might learn new ways to program robots on the factory floor to improve accuracy and reduce defective parts. The new knowledge will likely cause the manufacturing company to change how it programs and uses its robots. In Chapter 5 on database systems, we investigated the use of *data mining* and *business intelligence*. These powerful tools can be important in capturing and using knowledge. Enterprise resource planning tools, such as SAP, include knowledge management features.[12] In Chapter 10, we showed how *groupware* could improve group decision making and collaboration. Groupware can also be used to help capture, store, and use knowledge. Of course, hardware, software, databases, telecommunications, and the Internet discussed in Part 2 are important technologies used to support most knowledge management systems.

Hundreds of companies provide specific KM products and services.[13] In addition, researchers at colleges and universities have developed tools and technologies to support knowledge management.[14] See Figure 11.4. Companies such as IBM have many knowledge management tools in a variety of products, including Lotus Notes and Domino.[15] Lotus Notes is a collection of software products that help people work together to create, share, and store important knowledge and business documents. Its knowledge management features include domain search, content mapping, and Lotus Sametime. Domain search allows people to perform sophisticated searches for knowledge in Domino databases using a single simple query. Content mapping organizes knowledge by categories, like a table of contents for a book. Lotus Sametime helps people communicate, collaborate, and share ideas in real time. Lotus Domino Document Manager, formerly called Lotus Domino, helps people and organizations store, organize, and retrieve documents.[16] The software can be used to write, review, archive, and publish documents throughout the organization. Morphy Richards, a leading supplier of small home appliances in Britain, uses Domino for e-mail, collaboration, and document management.[17] According to one executive, "Rather than relying on groups of employees emailing each other, we are putting in place a business application through which documents will formally flow—to improve the efficiency of the supply chain and create more transparent working practices."

Figure 11.4

Knowledge Management Technology

Lotus Sametime helps people communicate, collaborate, and share ideas in real time.

(Source: Courtesy of IBM Corporation)

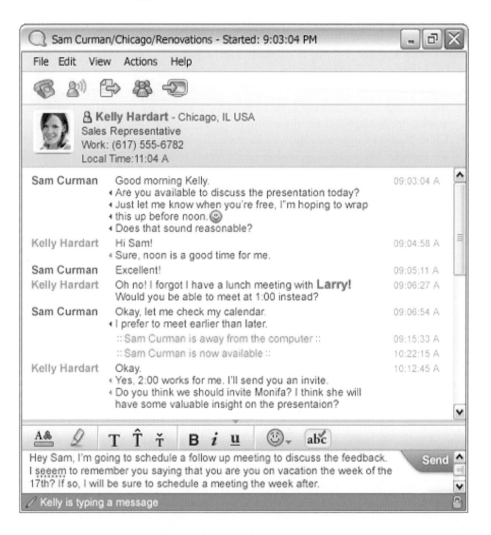

Microsoft offers a number of knowledge management tools, including Digital Dashboard, which is based on the Microsoft Office suite.[18] Digital Dashboard integrates information from different sources, including personal, group, enterprise, and external information and documents. "Microsoft has revolutionized the way that people use technology to create and share information. The company is the clear winner in the knowledge management business," according to Rory Chase, managing director of Teleos, an independent knowledge management research company based in the United Kingdom. Other tools from Microsoft include Web Store Technology, which uses wireless technology to deliver knowledge to any location at any time; Access Workflow Designer, which helps database developers create effective systems to process transactions and keep work flowing through the organization; and related products. Some additional knowledge management organizations and resources are summarized in Table 11.1. In addition to these tools, several artificial intelligence and special-purpose technologies and tools can be used in a KMS, discussed next.

Company	Description	Web Site
CortexPro	Knowledge management collaboration tools	www.cortexpro.com
Delphi Group	A knowledge management consulting company	www.delphigroup.com
Knowledge Associates	A knowledge management consulting company	www.knowledgeassociates.com
Knowledge management resource center	Knowledge management sites, products and services, magazines, and case studies	www.kmresource.com
Knowledge Management Solutions, Inc.	Tools to create, capture, classify, share, and manage knowledge	www.kmsi.us
Knowledge Management Web Directory	A directory of knowledge management Web sites	www.knowledge-manage.com
KnowledgeBase	Content creation and management	www.knowledgebase.net
Law Clip Knowledge Manager	A service that collects and organizes text, Web links, and more from law-related Web sites	www.lawclip.com
Meta KM	Knowledge management articles, resources, and opinions	www.metakm.com

Table 11.1

Additional Knowledge Management Organizations and Resources

AN OVERVIEW OF ARTIFICIAL INTELLIGENCE

At a Dartmouth College conference in 1956, John McCarthy proposed the use of the term **artificial intelligence (AI)** to describe computers with the ability to mimic or duplicate the functions of the human brain. For example, advances in AI have led to systems that work like the human brain to recognize complex patterns.[19]

Many AI pioneers attended this first conference; a few predicted that computers would be as "smart" as people by the 1960s. The prediction has not yet been realized, but the benefits of artificial intelligence in business and research can be seen today, and research continues.

Science fiction novels and popular movies have featured scenarios of computer systems and intelligent machines taking over the world. Steven Hawking, who is the Lucasian professor of mathematics at Cambridge University (a position once held by Isaac Newton) and author of *A Brief History of Time*, said, "In contrast with our intellect, computers double their performance every 18 months. So the danger is real that they could develop intelligence and take over the world." Computer systems such as Hal in the classic movie *2001: A Space Odyssey* and those in the movie *A.I.* are futuristic glimpses of what might be. These accounts are fictional, but they show the real application of many computer systems that use the notion of AI. These systems help to make medical diagnoses, explore for natural resources, determine what is wrong with mechanical devices, and assist in designing and developing other computer systems.

Artificial Intelligence in Perspective

Artificial intelligence systems include the people, procedures, hardware, software, data, and knowledge needed to develop computer systems and machines that demonstrate characteristics of intelligence. Researchers, scientists, and experts on how human beings think are often involved in developing these systems.

artificial intelligence (AI)
The ability of computers to mimic or duplicate the functions of the human brain.

artificial intelligence systems
People, procedures, hardware, software, data, and knowledge needed to develop computer systems and machines that demonstrate the characteristics of intelligence.

intelligent behavior
The ability to learn from experiences and apply knowledge acquired from experience, handle complex situations, solve problems when important information is missing, determine what is important, react quickly and correctly to a new situation, understand visual images, process and manipulate symbols, be creative and imaginative, and use heuristics.

The Nature of Intelligence

From the early AI pioneering stage, the research emphasis has been on developing machines with **intelligent behavior**. In a book called *The Singularity Is Near*, Ray Kurzweil predicts computers will have humanlike intelligence in 20 years.[20] The author also foresees that by 2045 human and machine intelligence might merge. Machine intelligence, however, is hard to achieve.

The *Turing Test* attempts to determine whether the responses from a computer with intelligent behavior are indistinguishable from responses from a human being. No computer has passed the Turing Test, developed by Alan Turing, a British mathematician (see *www.turing.org.uk*). Some of the specific characteristics of intelligent behavior include the ability to do the following:

- **Learn from experience and apply the knowledge acquired from experience.** Learning from past situations and events is a key component of intelligent behavior and is a natural ability of humans, who learn by trial and error. This ability, however, must be carefully programmed into a computer system. Today, researchers are developing systems that can learn from experience. For instance, computerized AI chess software can learn to improve while playing human competitors. In one match, Garry Kasparov competed against a personal computer with AI software developed in Israel, called Deep Junior. This match was a 3-3 tie, but Kasparov picked up something the machine would have no interest in—$700,000. The 20 questions (20q) Web site, *www.20q.net*, is another example of a system that learns.[21] The Web site is an artificial intelligence game that learns as people play.

- **Handle complex situations.** People are often involved in complex situations. World leaders face difficult political decisions regarding terrorism, conflict, global economic conditions, hunger, and poverty. In a business setting, top-level managers and executives must handle a complex market, challenging competitors, intricate government regulations, and a demanding workforce. Even human experts make mistakes in dealing with these situations. Developing computer systems that can handle perplexing situations requires careful planning and elaborate computer programming.

- **Solve problems when important information is missing.** The essence of decision making is dealing with uncertainty. Often, decisions must be made with little information or inaccurate information because obtaining complete information is too costly or impossible. Today, AI systems can make important calculations, comparisons, and decisions even when information is missing.

- **Determine what is important.** Knowing what is truly important is the mark of a good decision maker. Developing programs and approaches to allow computer systems and machines to identify important information is not a simple task.

- **React quickly and correctly to a new situation.** A small child, for example, can look over a ledge or a drop-off and know not to venture too close. The child reacts quickly and correctly to a new situation. Computers, on the other hand, do not have this ability without complex programming.

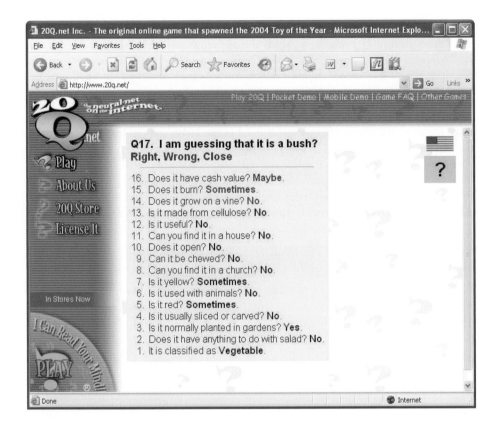

20Q is an online game where users play the popular game, *Twenty Questions*, against an artificial intelligence foe.

(Source: *www.20q.net.*)

- **Understand visual images.** Interpreting visual images can be extremely difficult, even for sophisticated computers. Moving through a room of chairs, tables, and other objects can be trivial for people but extremely complex for machines, robots, and computers. Such machines require an extension of understanding visual images, called a **perceptive system**. Having a perceptive system allows a machine to approximate the way a person sees, hears, and feels objects. Military robots, for example, use cameras and perceptive systems to conduct reconnaissance missions to detect enemy weapons and soldiers. Detecting and destroying them can save lives.

perceptive system
A system that approximates the way a person sees, hears, and feels objects.

- **Process and manipulate symbols.** People see, manipulate, and process symbols every day. Visual images provide a constant stream of information to our brains. By contrast, computers have difficulty handling symbolic processing and reasoning. Although computers excel at numerical calculations, they aren't as good at dealing with symbols and three-dimensional objects. Recent developments in machine-vision hardware and software, however, allow some computers to process and manipulate symbols on a limited basis.

- **Be creative and imaginative.** Throughout history, some people have turned difficult situations into advantages by being creative and imaginative. For instance, when shipped defective mints with holes in the middle, an enterprising entrepreneur decided to market these new mints as LifeSavers instead of returning them to the manufacturer. Ice cream cones were invented at the St. Louis World's Fair when an imaginative store owner decided to wrap ice cream with a waffle from his grill for portability. Developing new and exciting products and services from an existing (perhaps negative) situation is a human characteristic. Few computers can be imaginative or creative in this way, although software has been developed to enable a computer to write short stories.

- **Use heuristics.** For some decisions, people use heuristics (rules of thumb arising from experience) or even guesses. In searching for a job, you might rank the companies you are considering according to profits per employee. Today, some computer systems, given the right programs, obtain good solutions that use approximations instead of trying to search for an optimal solution, which would be technically difficult or too time consuming.

This list of traits only partially defines intelligence. Unlike the terminology used in virtually every other field of IS research, in which the objectives can be clearly defined, the term *intelligence* is a formidable stumbling block. One of the problems in AI is arriving at a working definition of real intelligence against which to compare the performance of an AI system.

The Difference Between Natural and Artificial Intelligence

Since the term *artificial intelligence* was defined in the 1950s, experts have disagreed about the difference between natural and artificial intelligence. Can computers be programmed to have common sense? Profound differences separate natural from artificial intelligence, but they are declining in number (see Table 11.2). One of the driving forces behind AI research is an attempt to understand how people actually reason and think. Creating machines that can reason is possible only when we truly understand our own processes for doing so.

Table 11.2

A Comparison of Natural and Artificial Intelligence

Ability to	Natural Intelligence (Human)		Artificial Intelligence (Machine)	
	Low	**High**	**Low**	**High**
Use sensors (eyes, ears, touch, smell)		√	√	
Be creative and imaginative		√	√	
Learn from experience		√	√	
Adapt to new situations		√	√	
Afford the cost of acquiring intelligence		√	√	
Acquire a large amount of external information		√		√
Use a variety of information sources		√		√
Make complex calculations	√			√
Transfer information	√			√
Make a series of calculations rapidly and accurately	√			√

The Major Branches of Artificial Intelligence

AI is a broad field that includes several specialty areas, such as expert systems, robotics, vision systems, natural language processing, learning systems, and neural networks (see Figure 11.5). Many of these areas are related; advances in one can occur simultaneously with or result in advances in others.

Expert Systems

expert system
Hardware and software that stores knowledge and makes inferences, similar to a human expert.

An **expert system** consists of hardware and software that stores knowledge and makes inferences, similar to those of a human expert. Because of their many business applications, expert systems are discussed in more detail in the next several sections of the chapter.

Robotics

robotics
Mechanical or computer devices that perform tasks requiring a high degree of precision or that are tedious or hazardous for humans.

Robotics involves developing mechanical or computer devices that can paint cars, make precision welds, and perform other tasks that require a high degree of precision or are tedious

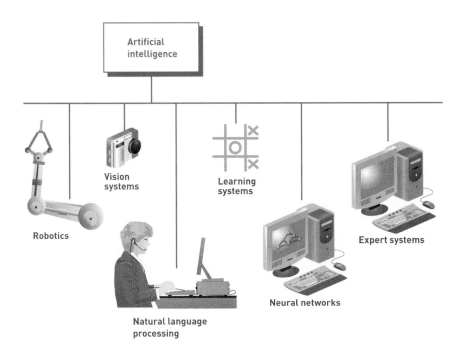

Figure 11.5

A Conceptual Model of Artificial Intelligence

or hazardous for human beings. Some robots are mechanical devices that don't use the AI features discussed in this chapter. Others are sophisticated systems that use one or more AI features or characteristics, such as vision systems, learning systems, or neural networks discussed later in the chapter. For many businesses, robots are used to do the three Ds—dull, dirty, and dangerous jobs.[22] Manufacturers use robots to assemble and paint products. The NASA shuttle crash of the early 2000s, for example, has led some people to recommend using robots instead of people to explore space and perform scientific research. Some robots, such as the ER-1, can be used for entertainment. Placing a portable computer on a wheeled platform, attaching a camera, and installing the necessary software allows the ER-1 to maneuver around objects. With an optional gripper arm, the robot can pick up small objects. Contemporary robotics combine both high-precision machine capabilities and sophisticated controlling software. The controlling software in robots is what is most important in terms of AI.

The field of robotics has many applications, and research into these unique devices continues. The following are a few examples:

- The Robot Learning Laboratory, part of the computer science department and the Robotics Institute at Carnegie Mellon University, conducts research into the development and use of robotics.[23]
- IRobot is a company that builds a number of robots, including the Roomba Floorvac for cleaning floors and the PackBot, an unmanned vehicle used to assist and protect soldiers.[24] Manufacturers use robots to assemble and paint products.
- The Porter Adventist Hospital in Denver, Colorado, uses a $1.2 million Da Vinci Surgical System to perform surgery on prostate cancer patients.[25] The robot has multiple arms that hold surgical tools. According to one doctor at Porter, "The biggest advantage is it improves recovery time. Instead of having an eight-inch incision, the patient has a 'band-aid' incision. It's much quicker."
- DARPA (The Defense Advanced Research Project Agency) sponsors the DARPA Grand Challenge, a 132-mile race over rugged terrain for computer-controlled cars.[26] For the first time, a computer-controlled car finished the race.[27] First place went to Stanford University.
- Because of an age limit on camel jockeys, the state of Qatar decided to use robots in its camel races.[28] Developed in Switzerland, the robots have a human shape and only weigh 59 pounds. The robots use Global Positioning Systems, a microphone to deliver voice commands to the camel, and cameras. A camel trainer uses a joystick to control the robot's

movements on the camel. Camel racing is as popular in Qatar as the Kentucky Derby in the United States.

- In military applications, robots are moving beyond movie plots to become real weapons. The Air Force is developing a smart robotic jet fighter. Often called *unmanned combat air vehicles (UCAVs)*, these robotic war machines, such as the X-45A, will be able to identify and destroy targets without human pilots. UCAVs send pictures and information to a central command center and can be directed to strike military targets. These new machines extend the current Predator and Global Hawk technologies the military used in Afghanistan after the September 11th terrorist attacks.

Although robots are essential components of today's automated manufacturing and military systems, future robots will find wider applications in banks, restaurants, homes, doctors' offices, and hazardous working environments such as nuclear stations. The Repliee Q1 and Q2 robots from Japan are ultra-humanlike robots or androids that can blink, gesture, speak, and even appear to breathe.[29] See Figure 11.6. Microrobotics, also called *micro-electro-mechanical systems (MEMS)*, are the size of a grain of salt and are also being developed. MEMS can be used in a person's blood to monitor the body, and for other purposes in air bags, cell phones, refrigerators, and more. A robot must not only execute tasks programmed by the user, but must also interact with its environment.

Robots can be used in situations that are hazardous or inaccessible to humans. The Rover was a remote-controlled robot used by NASA to explore the surface of Mars.

(Source: Courtesy of NASA.)

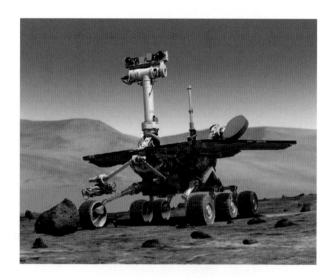

Figure 11.6

The Repliee Q2 Robot from Japan

(Source: AP Images)

Vision Systems

Another area of AI involves vision systems. **Vision systems** include hardware and software that permit computers to capture, store, and manipulate visual images. The U.S. Justice Department uses vision systems to perform fingerprint analysis with almost the same level of precision as human experts. The speed with which the system can search a huge database of fingerprints has brought quick resolution to many long-standing mysteries. Vision systems are also effective at identifying people based on facial features. In another application, a California wine bottle manufacturer uses a computerized vision system to inspect wine bottles for flaws. The company produces about 2 million wine bottles per day, and the vision system saves the bottle producer both time and money.

Vision systems can be used with robots to give these machines "sight." Factory robots typically perform mechanical tasks with little or no visual stimuli. Robotic vision extends the capability of these systems, allowing the robot to make decisions based on visual input. Generally, robots with vision systems can recognize black and white and some gray shades but do not have good color or three-dimensional vision. Other systems concentrate on only a few key features in an image, ignoring the rest. It might take years before a robot or other computer system can "see" in full color and draw conclusions from what it perceives the way that people do. Even with recent breakthroughs in vision systems, computers cannot see and understand visual images the way people can.

Natural Language Processing and Voice Recognition

As discussed in Chapter 4, **natural language processing** allows a computer to understand and react to statements and commands made in a "natural" language, such as English. Restoration Hardware, for example, has developed a Web site that uses natural language processing to allow its customers to quickly find what they want. The natural language processing system corrects spelling mistakes, converts abbreviations into words and commands, and allows people to ask questions in English. In addition to making it easier for customers, Restoration Hardware has seen an increase in revenues as a result of the use of natural language processing.

> **vision systems**
> The hardware and software that permit computers to capture, store, and manipulate visual images.

> **natural language processing**
> Processing that allows the computer to understand and react to statements and commands made in a "natural" language, such as English.

> Dragon Systems' Naturally Speaking 9 Essentials uses continuous voice recognition, or natural speech, allowing the user to speak to the computer at a normal pace without pausing between words. The spoken words are transcribed immediately onto the computer screen.
>
> (Source: Courtesy of Nuance Communications, Inc.)

In some cases, voice recognition is used with natural language processing. *Voice recognition* involves converting sound waves into words. After converting sounds into words, natural language processing systems can be used to react to the words or commands by performing a variety of tasks. Brokerage services are a perfect fit for voice-recognition and natural language processing technology to replace the existing "press 1 to buy or sell a stock" touchpad telephone menu system. People buying and selling stock use a vocabulary too varied for easy access through menus and touchpads, but still small enough for software to process in real time. Several brokerages—including Charles Schwab & Company, Fidelity Investments, DLJdirect, and TD Waterhouse Group—offer these services. These systems use voice recognition and natural language processing to let customers access retirement accounts, check balances, and find stock quotes. Eventually, the technology will allow people to make transactions using voice commands over the phone and to use search engines to have their questions answered through the brokerage firm's call center. One of the big advantages is that the number of calls routed to the customer service department drops considerably after new voice features are added. That is desirable to brokerages because it helps them staff their call centers correctly—even in volatile markets. Whereas a typical person uses a vocabulary of about 20,000 words or fewer, voice-recognition software can have a built-in vocabulary of 85,000 words. Some companies claim that voice-recognition and natural language processing software is so good that customers forget they are talking to a computer and start discussing the weather or sports scores.

Learning Systems

learning systems
A combination of software and hardware that allows the computer to change how it functions or reacts to situations based on feedback it receives.

Another part of AI deals with **learning systems**, a combination of software and hardware that allows a computer to change how it functions or reacts to situations based on feedback it receives. For example, some computerized games have learning abilities. If the computer does not win a game, it remembers not to make the same moves under the same conditions again. Tom Mitchell, director of the Center for Automated Learning and Discovery at Carnegie Mellon University, is experimenting with two learning software packages that help each other learn.[30] He believes that two learning software packages that cooperate are better than separate learning packages. Mitchell's learning software helps Internet search engines do a better job in finding information.

Learning systems software requires feedback on the results of actions or decisions. At a minimum, the feedback needs to indicate whether the results are desirable (winning a game) or undesirable (losing a game). The feedback is then used to alter what the system will do in the future.

Neural Networks

neural network
A computer system that can simulate the functioning of a human brain.

An increasingly important aspect of AI involves neural networks, also called neural nets. A **neural network** is a computer system that can act like or simulate the functioning of a human brain. The systems use massively parallel processors in an architecture that is based on the human brain's own meshlike structure. In addition, neural network software simulates a neural network using standard computers. Neural networks can process many pieces of data at the same time and learn to recognize patterns. A chemical company, for example, can use neural network software to analyze a large amount of data to control chemical reactors. Some of the specific abilities of neural networks include the following:

- Retrieving information even if some of the neural nodes fail
- Quickly modifying stored data as a result of new information
- Discovering relationships and trends in large databases
- Solving complex problems for which all the information is not present

A particular skill of neural nets is analyzing detailed trends. Large amusement parks and banks use neural networks to determine staffing needs based on customer traffic—a task that requires precise analysis, down to the half-hour. Increasingly, businesses are firing up neural nets to help them navigate ever-thicker forests of data and make sense of a myriad of customer traits and buying habits. Computer Associates has developed Neugents, neural intelligence agents that "learn" patterns and behaviors and predict what will happen next. For example, Neugents can track the habits of insurance customers and predict which ones will not renew an automobile policy, for example. They can then suggest to an insurance agent what changes to make in the policy to persuade the consumer to renew it. The technology also can track individual users at e-commerce sites and their online preferences so that they don't have to enter the same information each time they log on—their purchasing history and other data is recalled each time they access a Web site.

AI Trilogy is a neural network software program that can run on a standard PC. The software can make predictions with NeuroShell Predictor and classify information with NeuroShell Classifier. See Figure 11.7. The software package also contains GeneHunter, which uses a special type of algorithm called a genetic algorithm to get the best result from the neural network system. (Genetic algorithms are discussed later in this chapter.) Some pattern-recognition software uses neural networks to analyze hundreds of millions of bank, brokerage, and insurance accounts involving a trillion dollars to uncover money laundering and other suspicious money transfers.

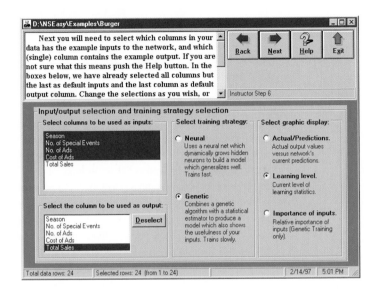

Figure 11.7

Neural Network Software

NeuroShell Predictor uses recognized forecasting methods to look for future trends in data.

Other Artificial Intelligence Applications

A few other artificial intelligence applications exist in addition to those just discussed. A **genetic algorithm**, also called a *genetic program*, is an approach to solving large, complex problems in which many repeated operations or models change and evolve until the best one emerges. The approach is based on the theory of evolution that requires (1) variation and (2) natural selection. The first step is to change or vary competing solutions to the problem. This can be done by changing the parts of a program or by combining different program segments into a new program, mimicking the evolution of species, in which the genetic makeup of a plant or animal mutates or changes over time. The second step is to select only the best models or algorithms, which continue to evolve. Programs or program segments that are not as good as others are discarded, similar to natural selection or "survival of the fittest," in which only the best species survive and continue to evolve. This process of variation and natural selection continues until the genetic algorithm yields the best possible solution to the original problem. For example, some investment firms use genetic algorithms to help select the best stocks or bonds. Genetic algorithms are also used in computer science and mathematics. Genetic algorithms can help companies determine which orders to accept for maximum profit. This approach helps companies select the orders that will increase profits and take full advantage of the company's production facilities. Genetic algorithms are also being used to make better decisions in developing inputs to neural networks.

An **intelligent agent** (also called an *intelligent robot* or *bot*) consists of programs and a knowledge base used to perform a specific task for a person, a process, or another program. Like a sports agent who searches for the best endorsement deals for a top athlete, an intelligent agent often searches to find the best price, schedule, or solution to a problem. The programs used by an intelligent agent can search large amounts of data as the knowledge base refines the search or accommodates user preferences. Often used to search the vast resources of the Internet, intelligent agents can help people find information on an important topic or the best price for a new digital camera. Intelligent agents can also be used to make travel arrangements, monitor incoming e-mail for viruses or junk mail, and coordinate meetings and schedules of busy executives. In the human resources field, intelligent agents help with online training. The software can look ahead in training materials and know what to start next. Intelligent agents have been used by the U.S. Army to route security clearance information for soldiers to the correct departments and people. What used to take days when done manually now takes hours.

genetic algorithm
An approach to solving large, complex problems in which a number of related operations or models change and evolve until the best one emerges.

intelligent agent
Programs and a knowledge base used to perform a specific task for a person, a process, or another program; also called *intelligent robot* or *bot*.

Online Matchmaker eHarmony

Chances are you haven't heard of Dr. Neil Clark Warren, but you'd probably recognize him if you saw him. He is one of America's best-known experts on singles issues, mate selection, and developing healthy relationships. He has written nine books, published hundreds of articles, and appeared on more than 4,000 radio and television programs. Even with all these credentials you probably still don't know who he is by his name alone. But if you watch television, you know him as the smiling guy on the eHarmony commercials who can't wait to help you meet your perfect mate through his scientific patented Compatibility Matching System.

The eHarmony Compatibility Matching System is an expert system that has been programmed with Dr. Warren's knowledge about what makes a good relationship. Its goal is to match partners for successful relationships using 29 key dimensions that help predict compatibility and the potential for relationship success. The 29 key dimensions are organized into two general categories: Core Traits that include emotional temperament, social style, cognitive mode, and physicality, and Vital Attributes that include relationship skills, values and beliefs, and key experiences.

An eHarmony applicant fills out a 436-item Relationship Questionnaire that allows the expert system to categorize that person's personality and build a compatibility profile. Heuristics are used to search the eHarmony database for compatible partners whose personality attributes make a good match according to Dr. Warren's studies. An ordered list of potential partners is produced with the best candidates listed first. The service then provides methods to get in touch with prospective mates.

Is the system successful? eHarmony has more than 11 million registered users and, according to an independent poll, more than 90 eHarmony members on average marry every day as a result of being matched through the service.

eHarmony uses artificial intelligence throughout its organization. Technology called predictive analytics, from SPSS, Inc., is used in various areas of the eHarmony business, including scientific research, brand development, product research, compatibility models, customer satisfaction and retention, and projective analysis. Predictive analytics is a form of data mining that employs artificial intelligence techniques to make assumptions about the future based on historical data and to predict outcomes of events. You can imagine how such technology can be applied to matchmaking as well as traditional business activities.

eHarmony's senior director of research and product development is Steve Carter, a strong believer in AI for business. He believes that predictive analytics is the lens through which eHarmony views all of its data. The numerous analytic and data management tools provided by the SPSS systems enables eHarmony to understand important information in more novel, forward-thinking ways. eHarmony is a prime example of how artificial intelligence and expert system tools are taking a lot of the guesswork out of life.

Discussion Questions

1. What role does artificial intelligence play in the matchmaking process at eHarmony?
2. How does eHarmony use predictive analytics in all of its business units?

Critical Thinking Questions

1. Do you think the scientific methods provided by eHarmony are superior or inferior to traditional random chance encounters for finding a mate? Why?
2. What dangers to privacy and safety, if any, are involved in using a service like eHarmony?

SOURCES: DM Review Editorial Staff, "eHarmony Expands SPSS Deployment Company-Wide for Research and Development," DM Direct Newsletter, June 9, 2006, *www.dmreview.com/article_sub.cfm?articleID=1055723*. eHarmony Web site, accessed June 30, 2006, *www.eharmony.com*.

A new hearing aid with artificial intelligence (shown next to a fingernail) contains a tiny microprocessor that works the way the brain does in detecting and distinguishing sounds while filtering out distractions.

(Source: AP Images.)

AN OVERVIEW OF EXPERT SYSTEMS

As mentioned earlier, an expert system behaves similarly to a human expert in a particular field. Computerized expert systems have been developed to diagnose problems, predict future events, and solve energy problems. They have also been used to design new products and systems, develop innovative insurance products, determine the best use of lumber, and increase the quality of healthcare. Like human experts, computerized expert systems use heuristics, or rules of thumb, to arrive at conclusions or make suggestions. Expert systems have also been used to determine credit limits for credit cards. An agricultural company can use expert systems to find the best fertilizer mix to use on certain soils to improve crops while minimizing costs. The research conducted in AI during the past two decades is resulting in expert systems that explore new business possibilities, increase overall profitability, reduce costs, and provide superior service to customers and clients.

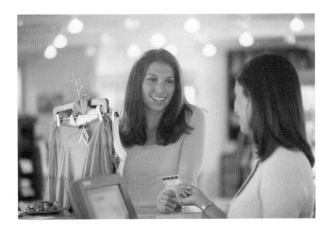

Credit card companies often use expert systems to determine credit limits for credit cards.

(Source: Ariel Skelley/Getty Images.)

When to Use Expert Systems

Sophisticated expert systems can be difficult, expensive, and time consuming to develop. This is especially true for large expert systems implemented on mainframes. The following is a list of factors that normally make expert systems worth the expenditure of time and money. Develop an expert system if it can do any of the following:

- Provide a high potential payoff or significantly reduce downside risk.
- Capture and preserve irreplaceable human expertise.
- Solve a problem that is not easily solved using traditional programming techniques.
- Develop a system more consistent than human experts.

- Provide expertise needed at a number of locations at the same time or in a hostile environment that is dangerous to human health.
- Provide expertise that is expensive or rare.
- Develop a solution faster than human experts can.
- Provide expertise needed for training and development to share the wisdom and experience of human experts with many people.

Components of Expert Systems

An expert system consists of a collection of integrated and related components, including a knowledge base, an inference engine, an explanation facility, a knowledge base acquisition facility, and a user interface. A diagram of a typical expert system is shown in Figure 11.8. In this figure, the user interacts with the interface, which interacts with the inference engine. The inference engine interacts with the other expert system components. These components must work together to provide expertise. This figure shows the inference engine coordinating the flow of knowledge to other components of the expert system. Note that there can be different knowledge flows, depending on what the expert system is doing and the specific expert system involved.

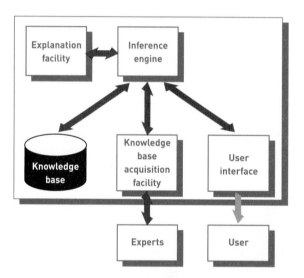

The Knowledge Base

The **knowledge base** stores all relevant information, data, rules, cases, and relationships that the expert system uses. As shown in Figure 11.9, a knowledge base is a natural extension of a database (presented in Chapter 5) and an information and decision support system (presented in Chapter 10). A knowledge base must be developed for each unique application. For example, a medical expert system contains facts about diseases and symptoms. The following are some tools and techniques that can be used to create a knowledge base.

knowledge base
A component of an expert system that stores all relevant information, data, rules, cases, and relationships used by the expert system.

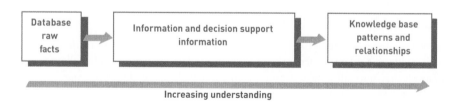

- **Assembling human experts.** One challenge in developing a knowledge base is to assemble the knowledge of multiple human experts. Typically, the objective in building a knowledge base is to integrate the knowledge of people with similar expertise (for example, many doctors might contribute to a medical diagnostics knowledge base).

- **Using fuzzy logic.** Another challenge for expert system designers and developers is capturing knowledge and relationships that are not precise or exact. Instead of the black-and-white, yes/no, or true/false conditions of typical computer decisions, fuzzy logic allows shades of gray, or what is known as "fuzzy sets." Fuzzy logic rules help computers evaluate the imperfect or imprecise conditions they encounter and make educated guesses based on the probability of correctness of the decision.

- **Using rules.** A **rule** is a conditional statement that links conditions to actions or outcomes. In many instances, these rules are stored as **IF-THEN statements**, such as "If a certain set of network conditions exists, then a certain network problem diagnosis is appropriate." In an expert system for a weather forecasting operation, for example, the rules could state that if certain temperature patterns exist with a given barometric pressure and certain previous weather patterns over the last 24 hours, then a specific forecast will be made, including temperatures, cloud coverage, and wind-chill factor. Figure 11.10 shows how to use expert system rules in determining whether a person should receive a mortgage loan from a bank. These rules can be placed in almost any standard program language discussed in Chapter 4 using "IF-THEN" statements or into special expert systems shells and products, discussed later in the chapter. In general, as the number of rules that an expert system knows increases, the precision of the expert system also increases.

IF-THEN statements
Rules that suggest certain conclusions.

Figure 11.10

Rules for a Credit Application

> **Mortgage Application for Loans from $100,000 to $200,000**
>
> > If there are no previous credit problems and
> >
> > If monthly net income is greater than 4 times monthly loan payment and
> >
> > If down payment is 15% of the total value of the property and
> >
> > If net assets of borrower are greater than $25,000 and
> >
> > If employment is greater than three years at the same company
>
> > Then accept loan application
>
> > Else check other credit rules

- **Using cases.** An expert system can use cases in developing a solution to a current problem or situation. This process involves (1) finding cases stored in the knowledge base that are similar to the problem or situation at hand and (2) modifying the solutions to the cases to fit or accommodate the current problem or situation. For example, a company might use an expert system to determine the best location for a new service facility in the state of New Mexico. The expert system might identify two previous cases involving the location of a service facility where labor and transportation costs were also important—one in the state of Colorado and the other in the state of Nevada. The expert system can modify the solution to these two cases to determine the best location for a new facility in New Mexico.

The Inference Engine

inference engine
Part of the expert system that seeks information and relationships from the knowledge base and provides answers, predictions, and suggestions the way a human expert would.

The overall purpose of an **inference engine** is to seek information and relationships from the knowledge base and to provide answers, predictions, and suggestions the way a human expert would. In other words, the inference engine is the component that delivers the expert advice. To provide answers and give advice, expert systems can use backward and forward chaining. **Backward chaining** is the process of starting with conclusions and working backward to the supporting facts. If the facts do not support the conclusion, another conclusion is selected and tested. This process is continued until the correct conclusion is identified. **Forward chaining** starts with the facts and works forward to the conclusions. Consider the expert system that forecasts future sales for a product. Forward chaining starts with a fact such as "The demand for the product last month was 20,000 units." With the forward-chaining approach, the expert system searches for rules that contain a reference to product demand. For example, "IF product demand is over 15,000 units, THEN check the demand for competing products." As a result of this process, the expert system might use information on the demand for competitive products. Next, after searching additional rules, the expert system might use information on personal income or national inflation rates. This process continues until the expert system can reach a conclusion using the data supplied by the user and the rules that apply in the knowledge base.

backward chaining
The process of starting with conclusions and working backward to the supporting facts.

forward chaining
The process of starting with the facts and working forward to the conclusions.

The Explanation Facility

explanation facility
Component of an expert system that allows a user or decision maker to understand how the expert system arrived at certain conclusions or results.

An important part of an expert system is the **explanation facility**, which allows a user or decision maker to understand how the expert system arrived at certain conclusions or results. A medical expert system, for example, might reach the conclusion that a patient has a defective heart valve given certain symptoms and the results of tests on the patient. The explanation facility allows a doctor to find out the logic or rationale of the diagnosis made by the expert system. The expert system, using the explanation facility, can indicate all the facts and rules that were used in reaching the conclusion. This facility allows doctors to determine whether the expert system is processing the data and information correctly and logically.

The Knowledge Acquisition Facility

A difficult task in developing an expert system is the process of creating and updating the knowledge base. In the past, when more traditional programming languages were used, developing a knowledge base was tedious and time consuming. Each fact, relationship, and rule had to be programmed into the knowledge base. In most cases, an experienced programmer had to create and update the knowledge base.

knowledge acquisition facility
Part of the expert system that provides convenient and efficient means of capturing and storing all the components of the knowledge base.

Today, specialized software allows users and decision makers to create and modify their own knowledge bases through the knowledge acquisition facility (see Figure 11.11). The purpose of the **knowledge acquisition facility** is to provide a convenient and efficient means for capturing and storing all components of the knowledge base. Knowledge acquisition software can present users and decision makers with easy-to-use menus. After entering the appropriate attributes, the knowledge acquisition facility correctly stores information and relationships in the knowledge base, making the knowledge base easier and less expensive to set up and maintain. Knowledge acquisition can be a manual process or a mixture of manual and automated procedures. Regardless of how the knowledge is acquired, the knowledge base must be validated and updated frequently to make sure that it is still accurate.

The User Interface

Specialized user interface software is employed for designing, creating, updating, and using expert systems. The main purpose of the user interface is to make an expert system easier for

Figure 11.11

Knowledge Acquisition Facility

The knowledge acquisition facility acts as an interface between experts and the knowledge base.

users and decision makers to develop and use. At one time, skilled computer personnel created and operated most expert systems; today, the user interface permits decision makers to develop and use their own expert systems. Because expert systems place more emphasis on directing user activities than do other types of systems, text-oriented user interfaces (using menus, forms, and scripts) might be more common in expert systems than the graphical interfaces often used with DSSs. A typical user interface between a user and the expert system for the credit application shown in Figure 11.12 follows.

EXPERT SYSTEM: Are there any previous credit problems?

USER: No.

EXPERT SYSTEM: Is the applicant's net income greater than four times the monthly loan payment?

USER: Yes.

EXPERT SYSTEM: Is the down payment greater than 15% of the property value?

USER: Yes.

EXPERT SYSTEM: Are the net assets of the borrower greater than $25,000?

USER: Yes.

EXPERT SYSTEM: Has the borrower worked more than three years at the same company?

USER: Yes.

EXPERT SYSTEM: **Accept the loan application.**

Expert Systems Development

Like other computer systems, expert systems require a systematic development approach for best results (see Figure 11.12). This approach includes determining the requirements for the expert system, identifying one or more experts in the area or discipline under investigation, constructing the components of the expert system, implementing the results, and maintaining and reviewing the complete system.

Participants in Developing and Using Expert Systems

Typically, several people are involved in developing and using an expert system (see Figure 11.13).

The Domain Expert

Because of the time and effort involved in the task, an expert system is developed to address only a specific area of knowledge. This area of knowledge is called the **domain**. The **domain expert** is the person or group with the expertise or knowledge the expert system is trying to capture. In most cases, the domain expert is a group of human experts. The domain expert (individual or group) usually can do the following:

- Recognize the real problem.
- Develop a general framework for problem solving.
- Formulate theories about the situation.
- Develop and use general rules to solve a problem.

domain
The area of knowledge addressed by the expert system.

domain expert
The individual or group who has the expertise or knowledge one is trying to capture in the expert system.

- Know when to break the rules or general principles.
- Solve problems quickly and efficiently.
- Learn from experience.
- Know what is and is not important in solving a problem.
- Explain the situation and solutions of problems to others.

Figure 11.12

Steps in the Expert System Development Process

Determining requirements

↓

Identifying experts

↓

Constructing expert system components

↓

Implementing results

↓

Maintaining and reviewing system

Figure 11.13

Participants in Expert Systems Development and Use

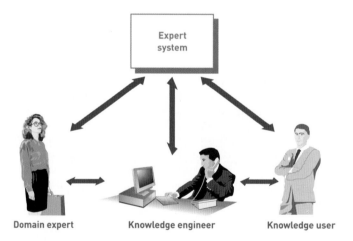

Domain expert   Knowledge engineer   Knowledge user

knowledge engineer
A person who has training or experience in the design, development, implementation, and maintenance of an expert system.

knowledge user
The person or group who uses and benefits from the expert system.

The Knowledge Engineer and Knowledge Users

A **knowledge engineer** is a person who has training or experience in the design, development, implementation, and maintenance of an expert system, including training or experience with expert system shells. The **knowledge user** is the person or group who uses and benefits from the expert system. Knowledge users do not need any previous training in computers or expert systems.

Expert Systems Development Tools and Techniques

Theoretically, expert systems can be developed from any programming language. Since the introduction of computer systems, programming languages have become easier to use, more powerful, and increasingly able to handle specialized requirements. In the early days of expert systems development, traditional high-level languages, including Pascal, FORTRAN, and COBOL, were used (see Figure 11.14). LISP was one of the first special languages developed and used for artificial intelligence applications. PROLOG was also developed for AI applications. Since the 1990s, however, other expert system products (such as shells) have become available that remove the burden of programming, allowing nonprogrammers to develop and benefit from the use of expert systems.

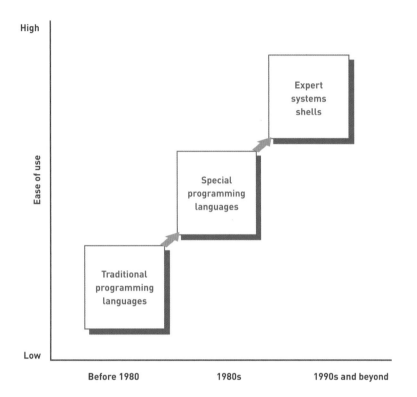

Figure 11.14

Expert Systems Development

Software for expert systems development has evolved greatly since 1980, from traditional programming languages to expert system shells.

Expert System Shells and Products

As discussed, an expert system shell is a collection of software packages and tools used to design, develop, implement, and maintain expert systems. Expert system shells are available for both personal computers and mainframe systems. Some shells are inexpensive, costing less than $500. In addition, off-the-shelf expert system shells are complete and ready to run. The user enters the appropriate data or parameters, and the expert system provides output to the problem or situation.

Some expert system products can analyze LAN networks, monitor air quality in commercial buildings, and evaluate oil and drilling operations. Table 11.3 lists a few expert system products.

Applications of Expert Systems and Artificial Intelligence

Expert systems and artificial intelligence have wide applications in business and government. For example, expert systems have been used to help power plants reduce pollutants while maintaining profits. They have also been used to determine the best way to distribute weight in a ferryboat to reduce the risk of capsizing or sinking. Other applications of these systems are summarized next.

- **Credit granting and loan analysis.** Many banks employ expert systems to review a customer's credit application and credit history data from credit bureaus to make a decision on whether to grant a loan or approve a transaction. KPMG Peat Marwick uses an expert system called Loan Probe to review its reserves to determine whether sufficient funds have been set aside to cover the risk of some uncollectible loans.
- **Stock picking.** Some expert systems help investment professionals pick stocks and other investments.

Table 11.3

Popular Expert System Products

Name of Product	Application and Capabilities
Financial Advisor	Analyzes financial investments in new equipment, facilities, and the like; requests the appropriate data and performs a complete financial analysis.
G2	Assists in oil and gas operations. Transco, a British company, uses it to help in the transport of gas to more than 20 million commercial and domestic customers.
HazMat Loader	Analyzes hazardous materials in truck shipments.
LSI Indicator	Helps determine property values; developed by one of the largest residential title and closing companies.
MindWizard	Enables development of compact expert systems ranging from simple models that incorporate business decision rules to highly sophisticated models; PC-based and inexpensive.
RAMPART	Analyzes risk. The U.S. General Services Administration uses it to analyze risk to the approximately 8,000 federal buildings it manages.

- **Catching cheats and terrorists.** Some gambling casinos use expert system software to catch gambling cheats. The CIA is testing the software to see whether it can detect possible terrorists when they make hotel or airline reservations.
- **Budgeting.** Automotive companies can use expert systems to help budget, plan, and coordinate prototype testing programs to save hundreds of millions of dollars.
- **Games.** Some expert systems are used for entertainment. For example, Proverb is an expert system designed to solve standard American crossword puzzles given the grid and clues.
- **Information management and retrieval.** The explosive growth of information available to decision makers has created a demand for devices to help manage the information. Expert systems can aid this process through the use of bots. Businesses might use a bot to retrieve information from large distributed databases or a vast network like the Internet. Expert system agents help managers find the right data and information while filtering out irrelevant facts that might impede timely decision making.
- **AI and expert systems embedded in products.** The antilock braking system on today's automobiles is an example of a rudimentary expert system. A processor senses when the tires are beginning to skid and releases the brakes for a fraction of a second to prevent the skid. AI researchers are also finding ways to use neural networks and robotics in everyday devices, such as toasters, alarm clocks, and televisions.
- **Plant layout and manufacturing.** FLEXPERT is an expert system that uses fuzzy logic to perform plant layout. The software helps companies determine the best placement for equipment and manufacturing facilities. Expert systems can also spot defective welds during the manufacturing process. The expert system analyzes radiographic images and suggests which welds could be flawed.
- **Hospitals and medical facilities.** Some hospitals use expert systems to determine a patient's likelihood of contracting cancer or other diseases. Hospitals, pharmacies, and other healthcare providers can use CaseAlert by MEDecision to determine possible high-risk or high-cost patients. MYCIN is an expert system developed at Stanford University to analyze blood infections. UpToDate is another expert system used to diagnose patients. A medical expert system used by the Harvard Community Health Plan allows members of the HMO to get medical diagnoses via home personal computers. For minor problems, the system gives uncomplicated treatments; for more serious conditions, the system schedules appointments. The system is highly accurate, diagnosing 97 percent of the patients correctly (compared with the doctors' 78 percent accuracy rating). To help doctors in the diagnosis of thoracic pain, MatheMEDics has developed THORASK, a straightforward, easy-to-use program, requiring only the input of carefully obtained clinical information. The program helps the less experienced to distinguish the three principal categories of chest pain from each other. It does what a true medical expert

system should do without the need for complicated user input. The user answers basic questions about the patient's history and directed physical findings, and the program immediately displays a list of diagnoses. The diagnoses are presented in decreasing order of likelihood, together with their estimated probabilities. The program also provides concise descriptions of relevant clinical conditions and their presentations, as well as brief suggestions for diagnostic approaches. For purposes of recordkeeping, documentation, and data analysis, there are options for saving and printing cases.

Seagate Technology implemented an expert system to monitor its disk drive components-manufacturing processes and improve yields.

(Source: REUTERS/Kin Cheung / Landov.)

- **Help desks and assistance.** Customer service help desks use expert systems to provide timely and accurate assistance. Kaiser Permanente, a large HMO, uses an expert system and voice response to automate its help desk function. The automated help desk frees up staff to handle more complex needs while still providing more timely assistance for routine calls.
- **Employee performance evaluation.** An expert system developed by Austin-Hayne, called Employee Appraiser, provides managers with expert advice for use in employee performance reviews and career development.
- **Virus detection.** IBM is using neural network technology to help create more advanced software for eradicating computer viruses, a major problem in American businesses. IBM's neural network software deals with "boot sector" viruses, the most prevalent type, using a form of artificial intelligence that mimics the human brain and generalizes by looking at examples. It requires a vast number of training samples, which in the case of antivirus software are 3-byte virus fragments.
- **Repair and maintenance.** ACE is an expert system used by AT&T to analyze the maintenance of telephone networks. IET-Intelligent Electronics uses an expert system to diagnose maintenance problems related to aerospace equipment. General Electric Aircraft Engine Group uses an expert system to enhance maintenance performance levels at all sites and improve diagnostic accuracy.
- **Shipping.** CARGEX cargo expert system is used by Lufthansa, a German airline, to help determine the best shipping routes.
- **Marketing.** CoverStory is an expert system that extracts marketing information from a database and automatically writes marketing reports.
- **Warehouse optimization.** United Distillers uses an expert system to determine the best combinations of liquor stocks to produce its blends of Scotch whiskey. This information is then supplemented with information about the location of the casks for each blend. The system optimizes the selection of required casks, keeping to a minimum the number of "doors" (warehouse sections) from which the casks must be taken and the number of casks that need to be moved to clear the way. Other constraints must be satisfied, such as the current working capacity of each warehouse and the maintenance and restocking work that may be in progress.

Call Centers Use Artificial Intelligence to Improve Service to Customers

Call centers, sometimes referred to as contact centers, are emerging as an important tactical weapon for corporations in the battle for the customer dollar. A call center is a unit within an organization or an outside firm that handles remote customer communications. Most 800 numbers you dial for placing orders and after-sales support are handled by a call center. Telephone solicitation is also handled by call centers.

Internet-based technologies, such as e-mail, discussion boards, and chat, are increasing the duties and boundaries of call centers. Research shows that about 70 percent of real-time customer interactions are handled by call centers. Business executives are picking up on these statistics and investing heavily in new technology to empower call center operators and provide convenience to customers. Artificial intelligence technology plays a key role.

"Contact centers are entering a new phase and becoming more intelligent," analyst Catriona Wallace says. "New technologies will allow them to take an even more central role in organizations." Wallace, whose research company Callcentres.net tracks growth in the sector, says increasing use of the Internet by consumers, and technologies such as VoIP and speech recognition are driving change.

Increasingly, the software used by call centers is being integrated with powerful CRM systems. This integration provides a solid foundation on which a range of new applications and agent techniques can be built. With ready access to everything from a customer's purchasing history to personal details and demographics, selling can take on a new dimension.

For example, the outbound calling function of call centers has been hampered by consumer resistance to unsolicited calls and the rise of do-not-call lists. Companies need to be smarter about when and why they contact the public. New tools enable companies to optimize the lists they are using, and customize them for specific campaigns. Previous histories and details can be automatically checked to increase the chance of finding a receptive customer, rather than randomly dialing people.

Intelligent software tools can examine demographic profiles to decide the best time to reach customers and whether to call their work, home, or mobile number. Users of the software claim a 40 to 50 percent improvement in speaking to the right person at the right time and the right place.

Intelligent software is also being used to accurately match the number of outgoing calls being made to the number of agents in a call center. New versions of automatic dialers can predict how often agents will become available and have a caller waiting on the line as soon as they are free, increasing agent utilization.

New intelligent software also helps companies respond smarter when customers or prospective customers call them.

James Brooks, senior vice president at call center specialist Genesys, says leading companies are examining a technique called psychographic routing. This involves routing incoming calls to the agent that the system deems is the best match for the caller. Based on caller line ID, a psychographic system can instantly assess known details such as the age, gender, and previous history of a caller, and automatically route them to the appropriate agent. For example, a 55-year-old shopping for life insurance would be routed to an operator of a similar age who is more likely to understand the concerns of the caller.

Such systems can also provide important customer information relevant to targeted sales. For example, the software can alert an agent that a customer is qualified for an increase in their credit limit, which can then be offered during the call. The success of such targeted sales pitches is high.

More complicated information could be calculated and provided. For example, bank call center agents can be provided with real-time credit scoring for customers seeking loans. "You can also undertake things like predictive claims," says Tim Macdermid, Australia manager for analytical software specialist SPSS. "An insurance company call center agent can collect information from a customer and the system will predict whether that claim is fraudulent."

From the customer's perspective, this call center evolution will lead to big changes in the way they interact with large organizations. Increasing amounts of transactions and business communications will be conducted through call centers. Call center agents will be more empowered as their tools become more intelligent and customers will have better experiences when they dial 800 numbers.

Discussion Questions

1. How is intelligent software helping call center agents be more effective?
2. Why are call centers becoming the primary conduit for communication between a business and its customers?

Critical Thinking Questions

1. What is psychographic routing and how might some interpret it as an infringement on their privacy, whereas others consider it a valuable service?
2. List communications technologies that a customer might use to communicate with a call center agent along with the pros and cons of each method. How might intelligent systems assist with the cons?

SOURCES: Grayson, Ian, "Digging Deeper into Data," *Australian IT*, April 25, 2006, *http://australianit.news.com.au*. Genesys Corporation Web site, accessed June 30, 2006, *www.genesyslab.com*. SPSS Web site, accessed June 30, 2006, *www.spss.com*.

VIRTUAL REALITY

The term *virtual reality* was initially coined by Jaron Lanier, founder of VPL Research, in 1989. Originally, the term referred to *immersive virtual reality* in which the user becomes fully immersed in an artificial, three-dimensional world that is completely generated by a computer. Immersive virtual reality can represent any three-dimensional setting, real or abstract, such as a building, an archaeological excavation site, the human anatomy, a sculpture, or a crime scene reconstruction. Through immersion, the user can gain a deeper understanding of the virtual world's behavior and functionality.

A **virtual reality system** enables one or more users to move and react in a computer-simulated environment. Virtual reality simulations require special interface devices that transmit the sights, sounds, and sensations of the simulated world to the user. These devices can also record and send the speech and movements of the participants to the simulation program, enabling users to sense and manipulate virtual objects much as they would real objects. This natural style of interaction gives the participants the feeling that they are immersed in the simulated world. For example, an auto manufacturer can use virtual reality to help it simulate and design factories.

virtual reality system
A system that enables one or more users to move and react in a computer-simulated environment.

Interface Devices

To see in a virtual world, often the user wears a head-mounted display (HMD) with screens directed at each eye. The HMD also contains a position tracker to monitor the location of the user's head and the direction in which the user is looking. Using this information, a computer generates images of the virtual world—a slightly different view for each eye—to match the direction that the user is looking and displays these images on the HMD. Many companies sell or rent virtual-reality interface devices, including Virtual Realities (*www.vrealities.com*), Amusitronix (*www.amusitronix.com*), I-O Display Systems (*www.i-glassesstore.com*), and others.

With current technology, virtual-world scenes must be kept relatively simple so that the computer can update the visual imagery quickly enough (at least ten times per second) to prevent the user's view from appearing jerky and from lagging behind the user's movements.

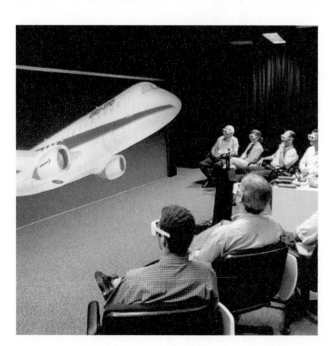

The PowerWall is a virtual reality system that displays large models in accurate dimensions.

(Source: Courtesy of Fakespace Systems, Inc.)

The Electronic Visualization Laboratory at the University of Illinois at Chicago introduced a room constructed of large screens on three walls and the floor on which the graphics are projected. The CAVE, as this room is called, provides the illusion of immersion by projecting stereo images on the walls and floor of a room-sized cube. Several persons wearing lightweight stereo glasses can enter and walk freely inside the CAVE. A head-tracking system continuously adjusts the stereo projection to the current position of the leading viewer.

Military personnel train in an immersive CAVE system.

(Source: Courtesy of Fakespace Systems, Inc.)

Users hear sounds in the virtual world through earphones. The information reported by the position tracker is also used to update audio signals. When a sound source in virtual space is not directly in front of or behind the user, the computer transmits sounds to arrive at one ear a little earlier or later than at the other and to be a little louder or softer and slightly different in pitch.

The *haptic* interface, which relays the sense of touch and other physical sensations in the virtual world, is the least developed and perhaps the most challenging to create. Currently, with the use of a glove and position tracker, the computer locates the user's hand and measures finger movements. The user can reach into the virtual world and handle objects; however, it is difficult to generate the sensations of a person tapping a hard surface, picking up an object, or running a finger across a textured surface. Touch sensations also have to be synchronized with the sights and sounds users experience.

Forms of Virtual Reality

Aside from immersive virtual reality, which we just discussed, virtual reality can also refer to applications that are not fully immersive, such as mouse-controlled navigation through a three-dimensional environment on a graphics monitor, stereo viewing from the monitor via stereo glasses, stereo projection systems, and others.

Some virtual reality applications allow views of real environments with superimposed virtual objects. Motion trackers monitor the movements of dancers or athletes for subsequent studies in immersive virtual reality. Telepresence systems (such as telemedicine and telerobotics) immerse a viewer in a real world that is captured by video cameras at a distant location and allow for the remote manipulation of real objects via robot arms and manipulators. Many believe that virtual reality will reshape the interface between people and information technology by offering new ways to communicate information, visualize processes, and express ideas creatively.

Virtual Reality Applications

You can find hundreds of applications of virtual reality, with more being developed as the cost of hardware and software declines and people's imaginations are opened to the potential of virtual reality. Having been inspired by the 2002 movie *Minority Report*, Pamela Barry of Raytheon is experimenting with a virtual reality system that uses "gesture technology."[31] For example, by pointing an index finger toward a picture on a screen, the computer zooms in

Computer-generated image technology and simulation is used by companies to determine plant capacity, manage bottlenecks, and optimize production rates.

(Source: Courtesy of Flexsim Software Products, Inc.)

on the picture. Moving a hand in one direction causes the computer to scroll down through a video clip, and moving a hand in another direction clears the screen. Raytheon hopes "gesture technology" will have applications in the military and space exploration. The following are additional applications. See Figure 11.15.

Figure 11.15

Virtual Reality Applications

Virtual reality has been used to increase real estate sales in several powerful ways. Cavex, for example, helps real estate firms develop virtual reality tours of their properties.

Medicine

Barbara Rothbaum, the director of the Trauma and Recovery Program at Emory University School of Medicine and cofounder of Virtually Better, uses an immersive virtual reality system to help in the treatment of anxiety disorders.[32] "For most of our applications, we use a head-mounted display that's kind of like a helmet with a television screen in front of each eye and has position trackers and sensors," says Rothbaum. One VR program called SnowWorld helps treat burn patients. Using VR, the patients can navigate through icy terrain and frigid waterfalls. VR helps because it gets a patient's mind off the pain.

In another medical application of virtual reality, researchers are trying to translate biomechanical measurements so that they can be stored in a computer database. Data from actual patients will be gathered from sensors mounted on the fingertips of virtual reality gloves to create a database for examining physicians. When a doctor examines a new patient, a feedback system linked to the virtual reality gloves will capture sensations and compare them with data in the database. This approach could allow surgeons to go right into the operating room without having to obtain a CAT scan, thus saving time—and with many injuries, that's absolutely critical.

Virtual reality technology can also link stroke patients to their physical therapists. Patients put on special gloves and other virtual reality devices at home that are linked to the physical therapist's office. The physical therapist can then see whether the patient is performing the correct exercises without having to travel to the patient's home or hospital room. In this way, using virtual reality can cut travel time and costs.

Education and Training

Virtual environments are used in education to bring exciting new resources into the classroom.[33] Students can stroll among digital bookshelves, learn anatomy on a simulated cadaver, or participate in historical events—all virtually.

Third-grade students at John Cotton Tayloe School in Washington, North Carolina, can take a virtual trip down the Nile for a cross-disciplinary lesson on ancient Egypt. This interactive virtual reality computer lesson integrates social studies, geography, music, art, science, math, and language arts. The software used to design the lesson is 3-D Website Builder (created by the Virtus Corporation), which allowed the designers to create a desert landscape complete with an oasis, camels, and a pyramid. Students can view the scenes from all angles, including front and top views, and can even enter a pyramid and view the sarcophagus holding the mummy in the middle of a room, which contains various Egyptian items. On one wall is a hieroglyphic message that is part of the lesson. On another wall is artwork depicting life in ancient Egypt, and against another are pieces of Egyptian furniture and a harp.

Virtual technology has also been applied by the military. To help with aircraft maintenance, a virtual reality system has been developed to simulate an aircraft and give a user a sense of touch, while computer graphics give the senses of sight and sound. The user sees, touches, and manipulates the various parts of the virtual aircraft during training. The Virtual Aircraft Maintenance System simulates real-world maintenance tasks that are routinely performed on the AV8B vertical takeoff and landing aircraft used by the U.S. Marines. Also, the Pentagon is using a virtual reality training lab to prepare for a military crisis. The virtual reality system simulates various war scenarios.

Real Estate Marketing and Tourism

Virtual reality has been used to increase real estate sales in several powerful ways. From Web publishing to laptop display to a potential buyer, virtual reality provides excellent exposure for properties and attracts potential clients. Clients can take a virtual walk through properties and eliminate wrong choices without wasting valuable time. Virtual walkthroughs can be mailed on disks or posted on the Web as a convenience for nonlocal clients. A CD containing all virtual reality homes can also be sent to clients and other agents. Cavex, located in Madison, Wisconsin (*www.cavex360.com/#VirtualTours*), for example, helps real estate firms develop virtual reality tours of their properties, as shown in Figure 11.15.

In another Web application, virtual reality was used to design a $90 million addition to the Denver Art Museum.[34] According to the contractor, "We're effectively building this with 3-D tools. It's improved quality. It's definitely improved production time." The virtual reality software can be used to view every beam and duct in the large building. The software can also show the picture, length, and diameter of the 50,000 bolts that are being used.

Entertainment

Computer-generated image technology, or CGI, has been around since the 1970s. Many movies use this technology to bring realism to the silver screen, including *Finding*

Nemo, *Spider-Man II*, and *Star Wars Episode II—Attack of the Clones*. A team of artists rendered the roiling seas and crashing waves of *Perfect Storm* almost entirely on computers using weather reports, scientific formulas, and their imagination. Other films include *Dinosaur* with its realistic talking reptiles, *Titan A.E.*'s beautiful 3-D space-scapes, and the casts of computer-generated crowds and battles in *Gladiator* and *The Patriot*. CGI can also be used for sports simulation to enhance the viewers' knowledge and enjoyment of a game. The SimCity (*http://simcity.ea.com/*), a virtual reality game, allows people to experiment with decisions related to urban planning. Natural and man-made disasters test decisions on designing buildings and the surrounding area. Other games can display a 3-D view of the world and allow people to interact with simulated people or avatars in the game. Second Life (*www.secondlife.com*) allows people to play games, interact with avatars, and build structures, such as homes. Many virtual reality entertainment sites charge a monthly fee.

OTHER SPECIALIZED SYSTEMS

In addition to artificial intelligence, expert systems, and virtual reality, other interesting specialized systems have appeared.[35] Segway, for example, is an electric scooter that uses sophisticated software, sensors, and gyro motors to transport people through warehouses, offices, downtown sidewalks, and other spaces. Originally designed to transport people around a factory or around town, more recent versions are being tested by the military for gathering intelligence and transporting wounded soldiers to safety. Experimental computer pens can be used to give commands, such as buying products.[36] By writing the word "buy" and a product description on a piece of paper, these computer pens can search the Internet for online stores for the product.

Some special-purpose devices can help detect crime and bring the criminals to justice. Tracking devices from SC-Integrity were placed inside containers of DVDs being shipped from Minneapolis to Chicago.[37] The tracking devices allowed the FBI to watch what was happening on computer monitors and catch the thieves, two deputy jailers and the truck driver. Increasingly, companies are placing concealed computer-tracking devices in boxes and containers as products are being shipped from one location to another. As mentioned previously, *Radio Frequency Identification (RFID)* tags that contain small chips with information about products or packages can be quickly scanned to perform inventory control or trace a package as it moves from a supplier to a company to its customers. Many companies have used RFID tags to reduce costs, improve customer service, and achieve a competitive advantage. Recall from Chapter 6 that the state of Colorado uses RFID to track elk herds.[38] Farmers are looking into using these tags to track cattle to help identify and control mad cow disease.[39] Recall also how an Italian cheese consortium uses RFID tags in the crust of cheese wheels.[40] The RFID tags contain information about when and where the cheese was made to ensure freshness and avoid spoilage. The technical director of the Italian cheese company that sells cheese under the Virgilio brand believes that the tags will eventually reduce operating costs by about 50 percent. Military planners in Iraq can now carry a black box that allows the planners to view a 3-D hologram of the area, including buildings, streets, and surrounding areas.[41] "You miss an awful lot with a 2-D picture. The hologram puts you right in the place before you've been there," says one U.S. defense official. Two German students have developed a smart beer mat, which uses sensor chips to help determine the weight or amount of beer in a glass or beer mug.[42] When the chips sense that the beer mug is nearly empty, the sensor chip sends an alert to a computer monitor telling the bartender that a customer needs more beer.

Another technology is being used to create "smart containers" for ships, railroads, and trucks. NaviTag (*http://navitag.com/*) and other companies are developing communications systems that allow containers to broadcast the contents, location, and condition of shipments to shipping and cargo managers. A railroad company can use standard radio messages to generate shipment and tracking data for customers and managers.

Navitag, an electronic security device, is attached to a cargo container door and monitors whether the door opens and whether light, radiation, or carbon monoxide enters the container.

(Source: AP Images.)

game theory
The use of information systems to develop competitive strategies for people, organizations, or even countries.

One special application of computer technology is derived from a branch of mathematics called game theory. **Game theory** involves the use of information systems to develop competitive strategies for people, organizations, or even countries. Two competing businesses in the same market can use game theory to determine the best strategy to achieve their goals. The military could also use game theory to determine the best military strategy to win a conflict against another country, and individual investors could use game theory to determine the best strategies when competing against other investors in a government auction of bonds. Groundbreaking work on game theory was pioneered by John Nash, the mathematician whose life was profiled in the book and film *A Beautiful Mind*. Game theory has also been used to develop approaches to deal with terrorism.

informatics
A specialized system that combines traditional disciplines, such as science and medicine, with computer systems and technology.

Informatics, another specialized system, combines traditional disciplines, such as science and medicine, with computer systems and technology. *Bioinformatics*, for example, combines biology and computer science. Also called *computational biology*, bioinformatics has been used to help map the human genome and conduct research on biological organisms. Using sophisticated databases and artificial intelligence, bioinformatics helps unlock the secrets of the human genome, which could eventually prevent diseases and save lives. Stanford University has a course on bioinformatics and offers a bioinformatics certification. Medical informatics combines traditional medical research with computer science. Journals, such as *Healthcare Informatics*, report current research on applying computer systems and technology to reduce medical errors and improve healthcare. The University of Edinburgh even has a School of Informatics. The school has courses on the structure, behavior, and interactions of natural and artificial computational systems. The program combines artificial intelligence, computer science, engineering, and science.

The many other specialized devices used by companies include small radio transceivers, which can be placed in products, such as cell phones. The radio transceivers allow cell phones and other devices to connect to the Internet, cellular phone services, and other devices that use the technology. The radio transceivers could save companies hundreds of thousands of dollars annually. Microsoft's Smart Personal Objects Technology (SPOT) allows small devices to transmit data and messages over the air. SPOT is being used in wrist watches to transmit data and messages over FM radio broadcast bands. The new technology, however, requires a subscription to the Microsoft MSN Direct information service. Automotive software allows cars and trucks to connect to the Internet. The software can track a driver's speed and location, allow gas stations to remotely charge for fuel and related services, and more. Special-purpose bar codes are also being introduced in a variety of settings. For example, to manage office space efficiently, a company gives each employee and office a bar code. Instead of having permanent offices, the employees are assigned offices and supplies as needed, and the bar codes help to make sure that an employee's work, mail, and other materials are routed to the right place. Companies can save millions of dollars by reducing office space and supplies. Manufacturing experiments are also being done with inkjet printers to allow them to "print" 3-D parts. The printer sprays layers of polymers onto circuit boards to form transistors and other electronic components.

SUMMARY

Principle

Knowledge management allows organizations to share knowledge and experience among their managers and employees.

Knowledge is an awareness and understanding of a set of information and the ways that information can be made useful to support a specific task or reach a decision. A knowledge management system (KMS) is an organized collection of people, procedures, software, databases, and devices used to create, store, share, and use the organization's knowledge and experience. Explicit knowledge is objective and can be measured and documented in reports, papers, and rules. Tacit knowledge is hard to measure and document and is typically not objective or formalized.

Knowledge workers are people who create, use, and disseminate knowledge. They are usually professionals in science, engineering, business, and other areas. The chief knowledge officer (CKO) is a top-level executive who helps the organization use a KMS to create, store, and use knowledge to achieve organizational goals. Some organizations and professions use communities of practice (COP) to create, store, and share knowledge. A COP is a group of people or community dedicated to a common discipline or practice, such as open-source software, auditing, medicine, engineering, and other areas.

Obtaining, storing, sharing, and using knowledge is the key to any KMS. The use of a KMS often leads to additional knowledge creation, storage, sharing, and usage. Many tools and techniques can be used to create, store, and use knowledge. These tools and techniques are available from IBM, Microsoft, and other companies and organizations.

Principle

Artificial intelligence systems form a broad and diverse set of systems that can replicate human decision making for certain types of well-defined problems.

The term *artificial intelligence* is used to describe computers with the ability to mimic or duplicate the functions of the human brain. The objective of building AI systems is not to replace human decision making completely but to replicate it for certain types of well-defined problems.

Intelligent behavior encompasses several characteristics, including the abilities to learn from experience and apply this knowledge to new experiences; handle complex situations and solve problems for which pieces of information might be missing; determine relevant information in a given situation, think in a logical and rational manner, and give a quick and correct response; and understand visual images and process symbols. Computers are better than people at transferring information, making a series of calculations rapidly and accurately, and making complex calculations, but human beings are better than computers at all other attributes of intelligence.

Artificial intelligence is a broad field that includes several key components, such as expert systems, robotics, vision systems, natural language processing, learning systems, and neural networks. An expert system consists of the hardware and software used to produce systems that behave as a human expert would in a specialized field or area (e.g., credit analysis). Robotics uses mechanical or computer devices to perform tasks that require a high degree of precision or are tedious or hazardous for humans (e.g., stacking cartons on a pallet). Vision systems include hardware and software that permit computers to capture, store, and manipulate images and pictures (e.g., face-recognition software). Natural language processing allows the computer to understand and react to statements and commands made in a "natural" language, such as English. Learning systems use a combination of software and hardware to allow a computer to change how it functions or reacts to situations based on feedback it receives (e.g., a computerized chess game). A neural network is a computer system that can simulate the functioning of a human brain (e.g., disease diagnostics system). A genetic algorithm is an approach to solving large, complex problems in which a number of related operations or models change and evolve until the best one emerges. The approach is based on the theory of evolution, which requires variation and natural selection. Intelligent agents consist of programs and a knowledge base used to perform a specific task for a person, a process, or another program.

Principle

Expert systems can enable a novice to perform at the level of an expert but must be developed and maintained very carefully.

An expert system consists of a collection of integrated and related components, including a knowledge base, an inference engine, an explanation facility, a knowledge acquisition facility, and a user interface. The knowledge base is an extension of a database, discussed in Chapter 5, and an information and decision support system, discussed in Chapter 10. It contains all the relevant data, rules, and relationships used in the expert system. The rules are often composed of if-then statements, which are used for drawing conclusions. Fuzzy logic allows expert systems to incorporate facts and relationships into expert system knowledge bases that might be imprecise or unknown.

The inference engine processes the rules, data, and relationships stored in the knowledge base to provide answers, predictions, and suggestions the way a human expert would. Two common methods for processing include backward and forward chaining. Backward chaining starts with a conclusion, then searches for facts to support it; forward chaining starts with a fact, then searches for a conclusion to support it.

The explanation facility of an expert system allows the user to understand what rules were used in arriving at a decision. The knowledge acquisition facility helps the user add or update knowledge in the knowledge base. The user interface makes it easier to develop and use the expert system.

The people involved in the development of an expert system include the domain expert, the knowledge engineer, and the knowledge users. The domain expert is the person or group who has the expertise or knowledge being captured for the system. The knowledge engineer is the developer whose job is to extract the expertise from the domain expert. The knowledge user is the person who benefits from the use of the developed system.

The steps involved in the development of an expert system include determining requirements, identifying experts, constructing expert system components, implementing results, and maintaining and reviewing the system.

Expert systems can be implemented in several ways. Previously, traditional high-level languages, including Pascal, FORTRAN, and COBOL, were used. LISP and PROLOG are two languages specifically developed for creating expert systems from scratch. A faster and less-expensive way to acquire an expert system is to purchase an expert system shell or existing package. The shell program is a collection of software packages and tools used to design, develop, implement, and maintain expert systems.

The benefits of using an expert system go beyond the typical reasons for using a computerized processing solution. Expert systems display "intelligent" behavior, manipulate symbolic information and draw conclusions, provide portable knowledge, and can deal with uncertainty. Expert systems can be used to solve problems in many fields or disciplines and can assist in all stages of the problem-solving process. Past successes have shown that expert systems are good at strategic goal setting, planning, design, decision making, quality control and monitoring, and diagnosis.

Applications of expert systems and artificial intelligence include credit granting and loan analysis, catching cheats and terrorists, budgeting, games, information management and retrieval, AI and expert systems embedded in products, plant layout, hospitals and medical facilities, help desks and assistance, employee performance evaluation, virus detection, repair and maintenance, shipping, and warehouse optimization.

Principle

Virtual reality systems can reshape the interface between people and information technology by offering new ways to communicate information, visualize processes, and express ideas creatively.

A virtual reality system enables one or more users to move and react in a computer-simulated environment. Virtual reality simulations require special interface devices that transmit the sights, sounds, and sensations of the simulated world to the user. These devices can also record and send the speech and movements of the participants to the simulation program. Thus, users can sense and manipulate virtual objects much as they would real objects. This natural style of interaction gives the participants the feeling that they are immersed in the simulated world.

Virtual reality can also refer to applications that are not fully immersive, such as mouse-controlled navigation through a three-dimensional environment on a graphics monitor, stereo viewing from the monitor via stereo glasses, stereo projection systems, and others. Some virtual reality applications allow views of real environments with superimposed virtual objects. Virtual reality applications are found in medicine, education and training, real estate and tourism, and entertainment.

Principle

Specialized systems can help organizations and individuals achieve their goals.

A number of specialized systems have recently appeared to assist organizations and individuals in new and exciting ways. Segway, for example, is an electric scooter that uses sophisticated software, sensors, and gyro motors to transport people through warehouses, offices, downtown sidewalks, and other spaces. Originally designed to transport people around a factory or around town, more recent versions are being tested by the military for gathering intelligence and transporting wounded soldiers to safety. Radio Frequency Identification (RFID) tags are used in a variety of settings. Game theory involves the use of information systems to develop competitive strategies for people, organizations, and even countries. Informatics combines traditional disciplines, such as science and medicine, with computer science. Bioinformatics and medical informatics are examples. A number of special-purpose telecommunications systems can be placed in products for varied uses.

CHAPTER 11: SELF-ASSESSMENT TEST

Knowledge management allows organizations to share knowledge and experience among their managers and employees.

1. _____ is a collection of facts organized so that they have additional value beyond the value of the facts themselves.

2. What type of knowledge is objective and can be measured and documented in reports, papers, and rules?
   a. tacit
   b. descriptive
   c. prescriptive
   d. explicit

3. A community of practice (COP) is a group of people or community dedicated to a common discipline or practice, such as open-source software, auditing, medicine, engineering, and other areas. True or False?

Artificial intelligence systems form a broad and diverse set of systems that can replicate human decision making for certain types of well-defined problems.

4. The Turing Test attempts to determine whether the responses from a computer with intelligent behavior are indistinguishable from responses from a human. True or False?

5. _____ are rules of thumb arising from experience or even guesses.

6. What is an important attribute for artificial intelligence?
   a. the ability to use sensors
   b. the ability to learn from experience
   c. the ability to be creative
   d. the ability to make complex calculations

7. _____ involves mechanical or computer devices that can paint cars, make precision welds, and perform other tasks that require a high degree of precision or are tedious or hazardous for human beings.

8. What branch of artificial intelligence involves a computer system that can simulate the functioning of a human brain?
   a. expert systems
   b. neural networks
   c. natural language processing
   d. vision systems

9. A(n) _____ is a combination of software and hardware that allows the computer to change how it functions or reacts to situations based on feedback it receives.

Expert systems can enable a novice to perform at the level of an expert but must be developed and maintained very carefully.

10. What is a disadvantage of an expert system?
    a. the inability to solve complex problems
    b. the inability to deal with uncertainty
    c. limitations to relatively narrow problems
    d. the inability to draw conclusions from complex relationships

11. A(n) _____ is a collection of software packages and tools used to develop expert systems that can be implemented on most popular PC platforms to reduce development time and costs.

12. An expert system heuristic consists of a collection of software and tools used to develop an expert system to reduce development time and costs. True or False?

13. What stores all relevant information, data, rules, cases, and relationships used by the expert system?
    a. the knowledge base
    b. the data interface
    c. the database
    d. the acquisition facility

14. A disadvantage of an expert system is the inability to provide expertise needed at a number of locations at the same time or in a hostile environment that is dangerous to human health. True or False?

15. What allows a user or decision maker to understand how the expert system arrived at a certain conclusion or result?
    a. domain expert
    b. inference engine
    c. knowledge base
    d. explanation facility

16. An important part of an expert system is the _____, which allows a user or decision maker to understand how the expert system arrived at certain conclusions or results.

17. In an expert system, the domain expert is the individual or group who has the expertise or knowledge one is trying to capture in the expert system. True or False?

Virtual reality systems can reshape the interface between people and information technology by offering new ways to communicate information, visualize processes, and express ideas creatively.

18. A(n) _____ enables one or more users to move and react in a computer-simulated environment.

19. What type of virtual reality is used to make human beings feel as though they are in a three-dimensional setting, such as a building, an archaeological excavation site, the human anatomy, a sculpture, or a crime scene reconstruction?
    a. chaining
    b. relative
    c. immersive
    d. visual

Specialized systems can help organizations and individuals achieve their goals.

20. _____ combines traditional disciplines, such as science and medicine, with computer science.

CHAPTER 11: SELF-ASSESSMENT TEST ANSWERS

(1) Information (2) d (3) True (4) True (5) Heuristics (6) d (7) Robotics (8) b (9) learning system (10) c (11) expert system shell (12) False (13) a (14) False (15) d (16) explanation facility (17) True (18) virtual reality system (19) c (20) Informatics

REVIEW QUESTIONS

1. Define the term *artificial intelligence*.
2. What is the difference between knowledge and information?
3. What is a vision system? Discuss two applications of such a system.
4. What is natural language processing? What are the three levels of voice recognition?
5. Describe three examples of the use of robotics. How can a microrobot be used?
6. What is a learning system? Give a practical example of such a system.
7. What is a neural network? Describe two applications of neural networks.
8. What is an expert system shell?
9. Under what conditions is the development of an expert system likely to be worth the effort?
10. Identify the basic components of an expert system and describe the role of each.
11. What is fuzzy logic?
12. What is virtual reality? Give several examples of its use.
13. Expert systems can be built based on rules or cases. What is the difference between the two?
14. Describe the roles of the domain expert, the knowledge engineer, and the knowledge user in expert systems.
15. What are the primary benefits derived from the use of expert systems?
16. What is informatics? Give a few examples.
17. Describe three applications of expert systems or artificial intelligence.
18. Identify three special interface devices developed for use with virtual reality systems.
19. Identify and briefly describe three specific virtual reality applications.
20. Give three examples of other specialized systems.

DISCUSSION QUESTIONS

1. What are the requirements for a computer to exhibit human-level intelligence? How long will it be before we have the technology to design such computers? Do you think we should push to try to accelerate such a development? Why or why not?
2. You work for an insurance company as an entry-level manager. The company contains both explicit and tacit knowledge. Describe the types of explicit and tacit knowledge that might exist in your insurance company. How would you capture each type of knowledge?
3. What are some of the tasks at which robots excel? Which human tasks are difficult for them to master? What fields of AI are required to develop a truly perceptive robot?
4. Describe how natural language processing could be used in a university setting.
5. Discuss how learning systems can be used in a military war simulation to train future officers and field commanders.
6. You have been hired to develop an expert system for a university career placement center. Develop five rules a student could use in selecting a career.

7. What is the purpose of a knowledge base? How is one developed?
8. What is the relationship between a database and a knowledge base?
9. Imagine that you are developing the rules for an expert system to select the strongest candidates for a medical school. What rules or heuristics would you include?
10. Describe how informatics can be used in a business setting.
11. Which interface is the least developed and most challenging to create in a virtual reality system? Why do you think this is so?
12. What application of virtual reality has the most potential to generate increased profits in the future?

PROBLEM-SOLVING EXERCISES

1. You are a senior vice president of a company that manufactures kitchen appliances. You are considering using robots to replace up to ten of your skilled workers on the factory floor. Using a spreadsheet, analyze the costs of acquiring several robots to paint and assemble some of your products versus the cost savings in labor. How many years would it take to pay for the robots from the savings in fewer employees? Assume that the skilled workers make $20 per hour, including benefits.
2. Assume that you have just won a lottery worth $100,000. You have decided to invest half the amount in the stock market. Develop a simple expert system to pick ten stocks to consider. Using your word processing program, create seven or more rules that could be used in such an expert system. Create five cases and use the rules you developed to determine the best stocks to pick.
3. Using a graphics program, develop a set of slides for a presentation to discuss the future of artificial intelligence in a business setting.

TEAM ACTIVITIES

1. With two or three of your classmates, do research to identify three real examples of natural language processing in use. Discuss the problems solved by each of these systems. Which has the greatest potential for cost savings? What are the other advantages of each system?
2. Form a team to debate other teams from your class on the following topic: "Are expert systems superior to human beings when it comes to making objective decisions?" Develop several points supporting either side of the debate.
3. Have your team members explore the use of a special-purpose system in an industry of your choice. Describe the advantages and disadvantages of this special-purpose system.

WEB EXERCISES

1. Use the Internet to find information about the use of robotics. Describe three examples of how this technology is used.
2. This chapter discussed several examples of expert systems. Search the Internet for two examples of the use of expert systems. Which one has the greatest potential to increase profits for the firm? Explain your choice.
3. Use the Internet to get more information about one of the specialized systems discussed at the end of the chapter. Write a report about what you found. Give an example of a new special-purpose system that has great promise in the future.

CAREER EXERCISES

1. Using the Internet or a library, explore how expert systems can be used in a business career. How can expert systems be used in a nonprofit company?

2. Describe the future of artificial intelligence in a career area of your choice.

CASE STUDIES

Case One
BMW Drives Virtual Prototypes

Today's cars consist of about 20,000 parts, and through virtual reality, BMW assembles the parts and examines how they interact in a variety of simulations in virtual space. The analysis takes place in BMW's virtual reality studio, nicknamed the Cave, which features a 175-square-foot PowerWall display and some serious computing power.

Among the many uses of the VR system, the most valuable is simulating test crashes. After an automobile is designed, and prior to building a prototype, technicians wearing 3-D glasses can observe how a crash affects a vehicle down to the smallest details. With this information, they recommend changes to the design to improve driver and passenger safety.

Before a new model is assembled, it has been crashed in a hundred different ways in BMW's Cave. Smashing the virtual car into a virtual wall takes two to four days of computing time. The computer works around the clock, breaking down the moment of impact into stages of milliseconds. The resulting 3-D film is viewed by technicians in superslow motion. At any point in the crash, the vehicle can be dissected to examine the effect of the crash on particular automobile areas and components.

Prior to virtual reality, BMW tested actual prototypes of new models. Such prototypes can cost up to a million dollars to manufacture. Each virtual 3-D simulation costs under $500. After a vehicle's design has been improved based on VR crash tests, actual prototypes are manufactured and also crashed. BMW must prove in real life the theories derived from the virtual reality simulations. Computer simulations are only valid to the extent that they reflect reality. At one time, BMW might have test-crashed six prototypes. Today, they crash only a couple and have the additional benefit of results from the study of over 100 VR crashes. In this way, they can produce safer vehicles for less money.

Safety is only one of several uses of BMW's VR system. Designers use the system to experiment with alternative interior and exterior designs. Surfaces of metal, artificial material, and leather are realistically represented in virtual space to have real textures and properties such as reflection.

Designers can examine a vehicle inside and out in a variety of lighting scenarios, changing colors, and materials as desired to find the perfect combinations.

Another department tests add-ons and pays particular attention to ease of operation. For example, how much energy does it take to close the top of a convertible? What happens when you slam the trunk lid? A technician can repeatedly close a virtual trunk lid, using varying degrees of force, carefully analyzing how the metal reacts each time. How difficult is it to open a door? Recently, BMW engineers tried out the door of a new virtual model and heard a subdued, high-pitched metallic ping when the door was shut. They thought that the sound seemed cheap and tinny. Realizing that a customer's opinion of a car is often shaped when they open and close the door, they set out to find the source of the ping. In a matter of days, they had redesigned the latch in the virtual model to eliminate the ping. Before VR, such an alteration would have taken months of work at a test bench.

BMW is also providing training to its technicians and mechanics using virtual reality. A new VR training facility for BMW technicians in Unterschleissheim, Germany, dramatically increases the number of technicians who can be trained at the same time. Instead of each trainee working one at a time, teams of technicians can now slip on high-tech goggles and gloves to work on a virtual image of a car or car part that's projected onto the inside of the goggles.

Discussion Questions

1. List three ways in which BMW uses VR in the design of new models of vehicles.
2. How does VR save BMW money over the long run?

Critical Thinking Questions

1. How might VR be used in your career area to save money and provide opportunities not available in the real world?
2. When might the results of a simulation in virtual space be considered valid and usable without running the same test in real space?

SOURCES: Staff, "SGI Case Study: Virtual Reality: Crashes Without Dents," accessed July 1, 2006, *www.cgi.com.* "BMW Virtual Reality Room," accessed July 1, 2006, *www.carpages.co.uk/bmw/ bmw_virtual_reality_room_04_09_04.asp.* BMW Group Web site, accessed July1, 2006, *www.bmwgroup.com.*

Case Two

Pacific Gas and Electric Corporation Increases Customer Satisfaction with Speech Recognition

"Our vision is to become the leading utility in the United States by delighting our 15 million customers, energizing our 20,000 employees, and rewarding our shareholders," wrote Peter A. Darbee, chairman of the board, CEO and president of Pacific Gas and Electric (PG&E) Corporation in a recent letter to company stakeholders. One way that PG&E is delighting its customers is through adding customized speech-recognition technology to its existing interactive voice response (IVR) system in its customer service call center.

IVR is a computerized system that allows a caller to interact with the system by speaking commands or other information. Typically used at the front end of call centers, an IVR system might ask you to enter "or say" the extension of the person you want to speak with. If you say "three nine seven two," the IVR system implements speech-recognition techniques to transform the wave pattern of the sound of the spoken numbers to actual numbers that it uses as input. If you are asked a yes or no question, the system might be programmed to accept "yes," "no," "yeah," "yep," "nope," "okay," "no thanks," or even "ahuh," Some of the latest systems use natural language processing techniques to allow callers to ask questions without prompting for specific keywords. If a system asks the customer the purpose of the call, and the caller responds with "the battery in my notebook computer no longer holds a charge," the system interprets the statement and replies, "Hold the line please while I connect you with a notebook power specialist."

PG&E installed the customized speech-recognition system to help automate account identification and provide other customer self-service functionality. The system was developed with ScanSoft Inc. and Nortel Networks Corp. to run on a Windows server. The speech-recognition system is integrated in the company's Nortel IVR system, along with its customer information, outage management, and field order scheduling systems, said Steve Phillips, manager of PG&E Corp's contact center enhancement. The purpose of the implementation was to improve customer satisfaction and the utility's "Technology Take Rates." Technology Take Rates refers to the percentage of customers whose needs are satisfied using the IVR system without speaking directly to a customer service representative.

Based on customer surveys, the speech-recognition system has improved PG&E's customer satisfaction and Technology Take Rates, said Kent Barnes, a senior project manager in PG&E's contact center enhancement group. Over the first year, the percentage of customers rating the IVR system as "excellent" or "very good" rose from 61 percent

to 69 percent. The new speech-recognition technology enabled PG&E Corp to improve its technology take rate from 33 percent to about 38 percent.

Equally important are the savings that PG&E realized with the new system. Based on an average cost increase of $5 to $9 for a PG&E customer service representative to handle a call rather than the IVR system, the utility's $3 million investment in the speech software and associated hardware was paid off in less than a year.

Investments in customer self-service systems will remain one of the most popular areas for investment by power company IT operations, Gartner Inc. analyst Zarko Sumic said in a research note. Utilities are "eager" to reduce customer support costs while improving the quality of service, Sumic said. "To meet higher complexity at the transformational level and increase ROI potential, leading energy companies will complement Web channels with advanced voice technologies."

Discussion Questions

1. What role does speech recognition play in modern-day IVR systems?
2. How did the new speech-recognition system save PG&E Corp. money while increasing customer satisfaction?

Critical Thinking Questions

1. Why do you think PG&E Corp's customers appreciated the ability to speak responses rather than keying them in or waiting for an operator?
2. As IVR systems become more intelligent with broader speech-recognition capabilities, how will the call center operator's duties change?

SOURCES: Hoffman, Thomas, "Speech Recognition Powers Utility's Customer Service," *Computerworld*, September 12, 2005, *www.computerworld.com.* PG&E Corp Web site, accessed July 1, 2006, *www.pgecorp.com.* Nortel Intelligent Call Manager Web site, accessed July 1, 2006, *www.nortel.com.* Microsoft Speech Server Web site, accessed July 1, 2006, *www.microsoft.com/ speech.*

Questions for Web Case

See the Web site for this book to read about the Whitmann Price Consulting case for this chapter. Following are questions concerning this Web case.

Whitmann Price Consulting: Knowledge Management and Specialized Information Systems

Discussion Questions

1. List three forms of AI that are being considered for the AMCI system and how they will be used.
2. List the advantages and disadvantages of implementing the AI systems in the AMCI system.

Critical Thinking Questions

1. What types of considerations might Josh and Sandra take into account when deciding which AI system to include?

2. How might Whitmann Price consultants react when they learn about the Presence system that will track their location? Why?

NOTES

Sources for the opening vignette: "Lafarge, Gensym Success Story," 2005, *www.gensym.com/?p=success_stories&id=15*. Lafarge Web site, accessed June 30, 2006, *www.lafarge.com*.

1   Kimble, Chris, et al., "Dualities, Distributed Communities of Practice and Knowledge Management," *Journal of Knowledge Management*, Vol. 9, 2005, p. 102.

2   Staff, "Ryder Systems: Knowledge Management," *www.accenture.com/xd/xd.asp?it=enweb&xd=services%5Chp% 5Ccase%5Chp_rydersystems.xml*, accessed August 26, 2005.

3   Darroch, Jenny, "Knowledge Management, Innovation, and Firm Performance," *Journal of Knowledge Management*, Vol. 9, 2005, p. 101.

4   Thurm, Scott, "Companies Struggle to Pass On Knowledge That Workers Acquire," *The Wall Street Journal*, January 23, 2006, p. B1.

5   Woods, Ginny Parker, "Sony Sets Its Sights on Digital Books," *The Wall Street Journal*, February 16, 2006, p. B3.

6   Skok, Walter, et al., "Evaluating the Role and Effectiveness of an Intranet in Facilitating Knowledge Management: A Case Study at Surrey County Council," *Information and Management*, July 2005, p. 731.

7   Staff, "CACI Gets National Security Work," *Homeland Security & Defense*, August 3, 2005, p. 4.

8   Staff, "Autonomy Links with Blinkx to Offer Search Facilities in China," *ComputerWire*, July 19, 2005, Issue 5228.

9   Staff, "eArchitect: Share and Enjoy," *Building Design*, June 17, 2005, p. 24.

10 Zolkos, Rodd, "Sharing the Intellectual Wealth," *BI Industry Focus*, March 1, 2005, p. 12.

11 Hsiu-Fen, Lin, et al., "Impact of Organizational Learning and Knowledge Management Factors on E-Business Adoption," *Management Decision*, Vol. 43, 2005, p. 171.

12 Pelz-Sharpe, Alan, "Document Management and Content Management Tucked Away in Several SAP Products," *Computer Weekly*, August 2, 2005, p. 26.

13 McKellar, Hugh, "100 Companies That Matter in Knowledge Management," *KM World*, March 2005, p. 18.

14 Sambamurthy, V., et al., "Special Issue of Information Technologies and Knowledge Management," *MIS Quarterly*, June 2005, p. 193.

15 Kajmo, David, "Knowledge Management in R5," *www-128.ibm.com/ developerworks/lotus/library/ls-Knowledge_Management/ index.html*, accessed August 25, 2005.

16 Staff, "IBM Lotus Domino Document Manager," *www.lotus.com/ lotus/offering4.nsf/wdocs/domdochome*, accessed August 26, 2005.

17 Staff, "Morphy Richards Integrates Its Global Supply Chain with Lotus Domino," *www-306.ibm.com/software/success/cssdb.nsf/cs/ DNSD-6EUNJ7?OpenDocument&Site=lotus*, accessed August 26, 2005.

18 Staff, "Survey Rates Microsoft Number One in Knowledge Management Efforts," *www.microsoft.com/presspass/features/ 1999/11-22award.mspx*, accessed August 26, 2005.

19 Quain, John, "Thinking Machines, Take Two," *PC Magazine*, May 24, 2005, p. 23.

20 Gomes, Lee, "A Back-Cover Brush with a High-Tech Seer and Some of His Pals," *The Wall Street Journal*, October 5, 2005, p. B1.

21 *www.20q.net*, accessed August 23, 2005.

22 Staff, "Send in the Robots," *Fortune*, January 24, 2005, p. 140.

23 *www.cs.cmu.edu/~rll*, accessed July 18, 2006.

24 *www.irobot.com*, accessed August 23, 2005.

25 Freeman, Diane, "RobotDoc," *The Rocky Mountain News*, June 27, 2005, p. 1B.

26 *http://en.wikipedia.org/wiki/DARPA_Grand_Challenge*, accessed August 23, 2005.

27 Gomes, Lee, "Team of Amateurs Cuts Ahead of Experts in Computer-Car Race," *The Wall Street Journal*, October 19, 2005, p. B1.

28 El-Rashidi, Yasime, "Ride'em Robot," *The Wall Street Journal*, October 3, 2005, p. A1.

29 Chamberlain, Ted, "Ultra-Lifelike Robot Debuts in Japan," *National Geographic News*, June 10, 2005.

30 Anthes, Gary, "Self Taught," *Computerworld*, February 6, 2006, p. 28.

31 Karp, Jonathan, "Minority Report Inspires Technology Aimed at Military," *The Wall Street Journal*, April 12, 2005, p. B1.

32 Rosencrance, Linda, "Virtual Therapy," *Computerworld*, March 14, 2005, p. 32.

33 Mearian, Lucas, "Going Virtual Cuts Costs at Palm Beach College," *Computerworld*, January 16, 2006, p. 12.

34 Fillion, Roger, "Easy as 1, 2, 3-D," *Rocky Mountain News*, March 7, 2005, p. 9A.

35 Winkler, Connie, "Data Visualization Software Is Helping Companies Make Decisions," *Computerworld*, January 9, 2006, p. 25.

36 Frostbert, Thomas, "Pen Computer Aims to Be More Than a Fad," *Rocky Mountain News*, February 27, 2006, p. 4B.

37 DeWeese, Chelsea, "New Trackers Help Truckers Foil Hijackings," *The Wall Street Journal*, September 29, 2005, p. B1.

38 Songini, Marc, "Colorado Hopes RFID Can Protect Elk Herds," *Computerworld*, January 16, 2006, p. 8.

39 Brat, Ilan, "New Kind of Cattle Branding," *The Wall Street Journal*, March 1, 2006, p. B1.

40 Kahn, Gabriel, "Who Made My Cheese Tags?" *The Wall Street Journal*, July 7, 2005, p. B1.

41 Fahey, Jonathan, "See It and Touch It," *Forbes*, May 9, 2005, p. 63.

42 Clotheir, Julie, "Beer Mat Knows When It Is Refill Time," *CNN Online*, September 30, 2005.

Kulula.com: The Trials and Tribulations of a South African Online Airline

Anesh Maniraj Singh
University of Durban

Kulula.com was launched in August 2001 as the first online airline in South Africa. Kulula is one of two airlines that are operated by Comair Ltd. British Airways (BA), the other airline that Comair runs, is a full-service franchise operation that serves the South African domestic market. Kulula, unlike BA, is a limited-service operation aimed at providing low fares to a wider domestic market using five aircraft. Since its inception, Kulula has reinvented air travel in South Africa, making it possible for more people to fly than ever before.

Kulula is a true South African e-commerce success. The company boasts as one of its successes the fact that it has been profitable from day one. It is recognized internationally among the top low-cost airlines and participated in a conference attended by other such internationally known low-cost carriers as Virgin Blue, Ryanair, and easy Jet. Kulula also received an award from the South African Department of Trade and Industry for being a Technology Top 100 company.

Kulula's success is based on its clearly defined strategy of being the lowest-cost provider in the South African domestic air travel industry. To this end, Kulula has adopted a no-frills approach. Staff and cabin crew wear simple uniforms, and the company has no airport lounges. There are no business class seats and no frequent-flyer programs. Customers pay for their food and drinks. In addition, Kulula does not issue paper tickets, and very few travel agents book its flights—90 percent of tickets are sold directly to customers. Furthermore, customers have to pay for ticket changes, and the company has a policy of "no fly, no refund." Yet, in its drive to keep costs down, Kulula does not compromise on maintenance and safety, and it employs the best pilots and meets the highest safety standards. Like all B2C companies, Kulula aims to create customer value by reducing overhead costs, including salaries, commissions, rent, and consumables such as paper and paper-based documents. Furthermore, by cutting out the middleman such as travel agents, Kulula is able to keep prices low and save customers the time and inconvenience of having to pick up tickets from travel agents. Instead, customers control the entire shopping experience.

Kulula was the sole provider of low-cost flights in South Africa until early 2004, when One Time launched a no-frills service to compete head-on with Kulula. Due to the high price elasticity of demand within the industry, any lowering of price stimulates a higher demand for flights. The increase in competition in the low-price end of the market has seen Kulula decrease fares by up to 20 percent while increasing passengers by over 40 percent. There, however, has been no brand switching. Kulula has grown in the market at the expense of others.

Apart from its low-cost strategy, Kulula is successful because of its strong B2C business model. As previously mentioned, 90 percent of its revenue is generated from direct sales. However, Kulula has recently ventured into the B2B market by collaborating with Computicket and a few travel agents, who can log on to the Kulula site from their company intranets. Kulula offers fares at substantial reductions to businesses that use it regularly. Furthermore, Kulula bases its success on three simple principles: Any decision taken must bring in additional revenue, save on costs, and/or enhance customer service. Technology contributes substantially to these three principles.

In its first year, Kulula used a locally developed reservation system, which soon ran out of functionality. The second-generation system was AirKiosk, which was developed in Boston for Kulula. The system change resulted in an improvement of functionality for passengers. For example, in 2003, Kulula ran a promotion during which tickets were sold at ridiculously low prices, and the system was overwhelmed. Furthermore, Kulula experienced a system crash that lasted a day and a half, which severely hampered sales and customer service. As a result, year two saw a revamp in all technology: All the hardware was replaced, bandwidth was increased, new servers and database servers were installed, and Web hosting was changed. In short, the entire system was replaced. According to IT Director Carl Scholtz, "Our success depends on infrastructural stability; our current system has an output that is four times better than the best our systems could ever produce." Kulula staff members are conscious of the security needs of customers and have invested in 128-bit encryption, giving customers peace of mind that their transactions and information are safe.

The success behind Kulula's systems lies in its branding—its strong identity in the marketplace, which includes its name and visual appeal. The term kulula means "easy," and Kulula's Web site has been designed with a simple, no-fuss, user-friendly interface. When visiting the Kulula site, you are immediately aware that an airline ticket can be purchased in three easy steps. The first step allows customers to choose destinations and dates. The second step allows customers to choose the most convenient or cheapest flight based on their need. Kulula also allows customers to book cars and accommodations in step two. Step three is the transaction stage, which allows customers to choose the most suitable payment method. The confirmation and ticket can be printed after payment has been settled. Kulula has not embraced mobile commerce yet, because the technology does not support the ability to allow customers to purchase a ticket in three easy steps. Unlike other e-commerce sites, Kulula is uncluttered and simple to understand, enhancing customer service. Kulula is a fun brand—with offbeat advertising campaigns and bright green and blue corporate and aircraft colors—but behind the fun exterior is a group of people who are serious about business.

Kulula's future is extremely promising. Technology changes continually, and Kulula strives to have the best technology in place at all times. B2B e-commerce will continue to be a major focus of the company to develop additional distribution channels with little or no cost. In conjunction with bank partners, Kulula is developing additional methods of payment to replace credit card payments, allowing more people the opportunity to fly. These transactions will be free. Kulula is also involving customers in its marketing efforts by obtaining their permission to promote special offers by e-mail and short message service to customers' cell phones. The Kulula Web site will soon serve as a ticketing portal, where customers can also purchase British Airways tickets, in three easy steps. The company has many other developments in the pipeline that will enhance customer service. According to Scholtz, "We are not an online airline, just an e-tailer that sells airline tickets."

Discussion Questions

1. This case does not mention any backup systems, either electronic or paper-based. What would you recommend to ensure that the business runs 24/7/365?
2. It is clear from this case that Kulula is a low-cost provider. What else could Kulula do with its technology to bring in additional revenue, save on cost, and enhance customer service?
3. Does the approach taken by Kulula in terms of its strategy, its business model, and the three principles of success lend itself to other businesses wanting to engage in e-commerce?
4. Kulula flights are almost always full. Do you think that by partnering with a company such as Lastminute.com the airline could fly to capacity at all times? What are the risks related to such a collaboration?

Critical Thinking Questions

1. Kulula initially developed its systems in-house, which it later outsourced to AirKiosk in Boston. Do you think it is wise for an e-business to outsource its systems development? Is it strategically sound to outsource systems development to a company in a different country?
2. With the current trends in mobile commerce, could Kulula offer its services on mobile devices such as cellular phones? Would the company have to alter its strategic thinking to accommodate such a shift? Is it possible to develop a text-based interface that could facilitate a purchase in three easy steps?

PART

• 4 •

Systems Development

Chapter 12 Systems Development: Investigation and Analysis

Chapter 13 Systems Development: Design, Implementation, Maintenance, and Review

CHAPTER · 12 ·

Systems Development: Investigation and Analysis

PRINCIPLES

LEARNING OBJECTIVES

- Effective systems development requires a team effort from stakeholders, users, managers, systems development specialists, and various support personnel, and it starts with careful planning.

- Identify the key participants in the systems development process and discuss their roles.

- Define the term *information systems planning* and list several reasons for initiating a systems project.

- Discuss three trends that illustrate the impact of enterprise resource planning software packages on systems development.

- Systems development often uses tools to select, implement, and monitor projects, including net present value (NPV), prototyping, rapid application development, CASE tools, and object-oriented development.

- Discuss the key features, advantages, and disadvantages of the traditional, prototyping, rapid application development, and end-user systems development life cycles.

- Identify several factors that influence the success or failure of a systems development project.

- Discuss the use of CASE tools and the object-oriented approach to systems development.

- Systems development starts with investigation and analysis of existing systems.

- State the purpose of systems investigation.

- Discuss the importance of performance and cost objectives.

- State the purpose of systems analysis and discuss some of the tools and techniques used in this phase of systems development.

Information Systems in the Global Economy ⟩⟩
Nordmilch, Germany

German Dairy Products Manufacturer Upgrades Systems to Meet Organizational Goals

Nordmilch produces dairy products such as cheese, cream, long-life milk, dried powdered milk, and yogurt. Based in Bremen, Germany, the company is structured along cooperative lines with ownership of the business in the hands of the region's milk producers. It is a successful global business with annual sales of more than €2.1 billion ($2.65 billion), and around 3,800 employees. Nordmilch has earned its success by constantly evaluating its market position and organizational goals, and then taking action to achieve those goals. New information systems are continuously developed to support changing organizational strategies.

Recently, Nordmilch reached a decisive turning point. Company leaders realized that Nordmilch needed radical strategic and organizational changes to compete with recent entrants into the market. Management set new goals that would allow the company to become more nimble as it responded to changes in the market. The company needed to reduce costs, increase flexibility, and introduce enhanced production quality control. Chairman Stephan Tomat put it this way: "In addition to restructuring our production plants, a company-wide efficiency program is aimed at bringing about a sustained increase in Nordmilch's ability to perform." To accomplish its goals, the company needed to develop new information systems to speed valued information to its decision makers.

Nordmilch was using SAP software as its core suite of business management solutions and was pleased with the software's capabilities and SAP's support of the product. Nordmilch management wanted to implement additional SAP solutions and upgrade their existing SAP applications to provide more detailed production data. By closely monitoring production, Nordmilch could react more quickly to market demands and also provide assurances to customers and government agencies concerned about health and safety issues.

Further systems investigation showed that Nordmilch had yet another reason to upgrade its systems. Its Sun and EMC² servers and storage hardware were nearing the end of their life and could no longer support the needs of the growing organization. A full analysis of Nordmilch systems uncovered that this systems development project would involve implementing new SAP applications, new servers on which those applications would run, new storage hardware for corporate data, and a new database system to manage that data because the current database was not compatible with the SAP applications—in other words, Nordmilch needed a total system overhaul. Karl-Heinz Mansholt, chief information officer at Nordmilch, describes the strategy: "To meet our business objectives, we selected SAP software. To run new SAP applications successfully, we needed a new approach to our IT infrastructure. We did not want to over-invest in servers that would be over-sized in the beginning, and invited several vendors to propose possible solutions."

Nordmilch compared offers for new servers and storage hardware from IBM partners, FSC, and Sun. It also evaluated offers for database systems from IBM and Oracle. Karl-Heinz Mansholt and his team analyzed factors such as ease of administration, maintenance, database performance, scalability, and reliability.

In the end, Nordmilch implemented new SAP applications for the Materials Management, Sales and Distribution, Financials/Controlling, Human Resources, and Production Planning departments. The SAP applications were installed on two IBM p5 model 570 servers running the IBM AIX 5L operating system. The IBM system was chosen because it was scalable, allowing Nordmilch to pay for only the computing power it currently needed with support for growth for at least five years.

Nordmilch chose an IBM TotalStorage DS8100 storage server, providing high-performance storage resources for both p5-570 servers. For a database system, Nordmilch chose IBM's DB2. The interoperability between the SAS applications, and the IBM servers, storage system, and database ideally suited Nordmilch's needs. "The improved technical performance is down to the combination of hardware, AIX operating system, database and storage—it's very hard to split the benefits between these components. In summary, the IBM and SAP solution gives Nordmilch the flexible IT infrastructure and applications required to meet our business goals," concludes Karl-Heinz Mansholt.

As you read this chapter, consider the following:

- What situations can arise within a business to trigger new systems development initiatives?
- What are the best methods for a business to use in approaching new systems development projects?

Why Learn About Systems Development?

Throughout this book, you have seen many examples of the use of information systems in a variety of careers. But where do you start to acquire these systems or have them developed? How can you work with IS personnel, such as systems analysts and computer programmers, to get what you need to succeed on the job? This chapter, the first of two chapters on systems development, gives you the answer. You will see how you can initiate the systems development process and analyze your needs with the help of IS personnel. Systems investigation and systems analysis are the first two steps of the systems development process. This chapter provides specific examples of how new or modified systems are initiated and analyzed in a number of industries. In this chapter, you will learn how your project can be planned, aligned with corporate goals, rapidly developed, and much more. We start with an overview of the systems development process.

When an organization needs to accomplish a new task or change a work process, how does it do it? It develops a new system or modifies an existing one. Systems development is the activity of creating new or modifying existing systems. It refers to all aspects of the process—from identifying problems to solve or opportunities to exploit to implementing and refining the chosen solution.

AN OVERVIEW OF SYSTEMS DEVELOPMENT

In today's businesses, managers and employees in all functional areas work together and use business information systems. As a result, users are helping with development and, in many cases, leading the way. Users might request that a systems development team determine whether they should purchase a few PCs or create an attractive Web site, using the tools discussed in Chapter 7. In other cases, systems development might involve purchasing or leasing an enterprise resource planning (ERP) package discussed in Chapter 9 or building sophisticated programs from scratch to control inventory.

This chapter and the next provide you with a deeper appreciation of the systems development process and help you avoid costly failures. The United Airlines automated baggage systems development project, for example, failed to deliver baggage to airline passengers in good shape or on time.[1] The $250 million systems development project cost United Airlines about $70 million to operate each year. Computers were overwhelmed with data from the cars carrying the baggage. According to Mr. McDonald, the airline's chief operating officer, "We have come to the conclusion that going to a manual approach is best." Participants in systems development, such as Mr. McDonald, determine when a systems development project fails. They are also critical to systems development success.

Participants in Systems Development

Effective systems development requires a team effort. The team usually consists of stakeholders, users, managers, systems development specialists, and various support personnel. This team, called the *development team*, is responsible for determining the objectives of the information system and delivering a system that meets these objectives. Many development teams use a project manager to head the systems development effort and the project management approach to help coordinate the systems development process. A *project* is a planned collection of activities that achieves a goal, such as constructing a new manufacturing plant or developing a new decision support system.[2] All projects have a defined starting point and ending point, normally expressed as dates such as August 4 and November 11. Most have a budget, such as $150,000. A *project manager* is responsible for coordinating all people and resources needed to complete a project on time. In systems development, the project manager can be an IS person inside the organization or an external consultant hired to complete the project. Project managers need technical, business, and people skills. In addition to completing the project on time and within the specified budget, the project manager is usually responsible for controlling project quality, training personnel, facilitating communications, managing risks, and acquiring any necessary equipment, including office supplies and sophisticated computer systems. One study reported that almost 80 percent of responding IS managers believe that it is critical to keep project planning skills in-house instead of outsourcing them.[3] Research studies have shown that project management success factors include good leadership from executives and project managers, a high level of trust in the project and its potential benefits, and the commitment of the project team and organization to successfully complete the project and implement its results.

In the context of systems development, **stakeholders** are people who, either themselves or through the area of the organization they represent, ultimately benefit from the systems development project. **Users** are people who will interact with the system regularly. They can be employees, managers, or suppliers. For large-scale systems development projects, where the investment in and value of a system can be high, it is common for senior-level managers, including functional vice presidents (of finance, marketing, and so on), to be part of the development team.

Depending on the nature of the systems project, the development team might include systems analysts and programmers, among others. A **systems analyst** is a professional who specializes in analyzing and designing business systems. Systems analysts play various roles while interacting with the stakeholders and users, management, vendors and suppliers, external companies, programmers, and other IS support personnel (see Figure 12.1). Like an architect developing blueprints for a new building, a systems analyst develops detailed plans for the new or modified system. The **programmer** is responsible for modifying or developing programs to satisfy user requirements. Like a contractor constructing a new building or renovating an existing one, the programmer takes the plans from the systems analyst and builds or modifies the necessary software.

The other support personnel on the development team are mostly technical specialists, including database and telecommunications experts, hardware engineers, and supplier representatives. One or more of these roles might be outsourced to outside experts or consultants. Depending on the magnitude of the systems development project and the number of IS systems development specialists on the team, one or more IS managers might also belong to the team. The composition of a development team can vary over time and from project to project. For small businesses, the development team might consist of a systems analyst and the business owner as the primary stakeholder. For larger organizations, formal IS staff can include hundreds of people involved in a variety of activities, including systems development. Every development team should have a team leader. This person can be from the IS

stakeholders
People who, either themselves or through the organization they represent, ultimately benefit from the systems development project.

users
People who will interact with the system regularly.

systems analyst
A professional who specializes in analyzing and designing business systems.

programmer
A specialist responsible for modifying or developing programs to satisfy user requirements.

Role of the Systems Analyst

The systems analyst plays an important role in the development team and is often the only person who sees the system in its totality. The one-way arrows in this figure do not mean that there is no direct communication between other team members. These arrows just indicate the pivotal role of the systems analyst—a person who is often called on to be a facilitator, moderator, negotiator, and interpreter for development activities.

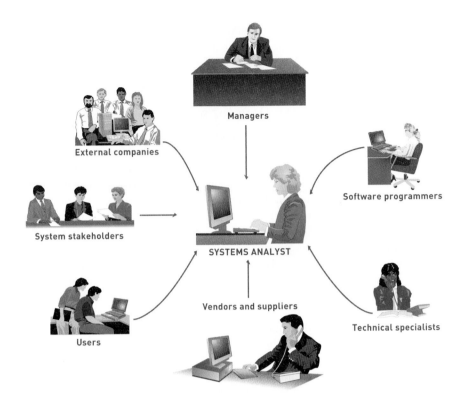

department, a manager from the company, or a consultant from outside the company. The team leader needs both technical and people skills.

Regardless of the specific nature of a project, systems development creates or modifies systems, which ultimately means change. Managing this change effectively requires development team members to communicate well. Because you probably will participate in systems development during your career, you must learn communication skills. You might even be the individual who initiates systems development.

Initiating Systems Development

Systems development initiatives arise from all levels of an organization and are both planned and unplanned. The United States Department of Energy (DOE) and TECSys Development, for example, initiated a systems development project to protect the electrical infrastructure of the United States.[4] According to a consultant for the systems development project, "Our goal here is to develop methods to protect the critical [infrastructure] of the U.S." The new DOE system will include software to monitor servers, routers, firewalls, and other components of the U.S. electrical infrastructure. Solid planning and managerial involvement help ensure that these initiatives support broader organizational goals. Systems development projects are initiated for many reasons, as shown in Figure 12.2.

Mergers and acquisitions can trigger many systems development projects. When MetLife acquired Travelers Life & Annuity for $11 billion, MetLife's vice president and chief information officer Steve Sheinheit had to quickly combine IS operations.[5] It took less than a year to merge the two IS departments. Because the information systems for companies are usually different, a large systems development effort is typically required to unify systems. Even with similar information systems, the procedures, culture, training, and management of the information systems are often different, requiring a realignment of the IS departments. When Harrah's, a casino and entertainment company, merged with Caesars in a $6.8 million deal, the two companies also decided to merge their computerized loyalty programs, resulting in a $130 million systems development effort.[6] The resulting loyalty program collects information from its customers to serve them better and provide customized services. Some believe the systems development project could increase the revenues to the casino and entertainment company by hundreds of millions of dollars annually.

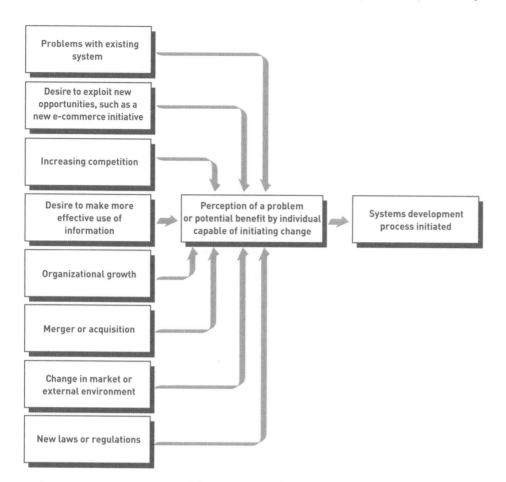

Figure 12.2

Typical Reasons to Initiate a
Systems Development Project

The federal government also fosters new systems development projects in the private sector. In the wake of recent financial scandals, the government has instituted corporate financial reporting rules under the Sarbanes-Oxley Act. These regulations have caused many U.S. companies to initiate systems development efforts. To comply with this law, companies can spend hundreds of thousands or millions of dollars in new systems development efforts. Many U.S. companies are outsourcing their systems development projects that require Sarbanes-Oxley compliance.[7] Some outsourcing companies in India are experiencing a 50 percent growth in business as a result. U.S. public companies are expected to spend almost $6 billion per year to comply with the Sarbanes-Oxley law. New federal legislation also created the Organized Retail Theft Task Force at the FBI.[8] "The establishment of an FBI task force on organized retail theft is a major victory for retailers," said Joseph LaRocca, the National Retail Federation vice president for loss prevention. "This task force and the related database will play a major role in putting these criminals where they belong—behind bars."

Information Systems Planning and Aligning Corporate and IS Goals

Information systems planning and aligning corporate and IS goals are important aspects of any systems development project. Achieving a competitive advantage is often the overall objective of systems development. General Motors, for example, decided to spend about $15 billion to develop a better information system that will allow the various GM subsidiaries and offices around the world to operate as a more unified unit.[9] GM hopes the improved information system will help it increase sales and profits in Africa, Asia, Latin American, the Middle East, and the United States.

Information Systems Planning

The term **information systems planning** refers to translating strategic and organizational goals into systems development initiatives (see Figure 12.3). The Marriott hotel chain, for example, invites its chief information officer (CIO) to board meetings and other top-level management meetings. According to Doug Lewis, former CIO for many *Fortune* 100 companies, "Strategic goals must be finite, measurable, and tangible."[10] Proper IS planning ensures that specific systems development objectives support organizational goals.

Figure 12.3

Information Systems Planning

Information systems planning transforms organizational goals outlined in the strategic plan into specific systems development activities.

Aligning Corporate and IS Goals

Aligning organizational goals and IS goals is critical for any successful systems development effort.[11] Because information systems support other business activities, IS staff and people in other departments need to understand each other's responsibilities and tasks. Determining whether organizational and IS goals are aligned can be difficult, so researchers have increasingly tackled the problem. One measure of alignment uses five levels, ranging from ad hoc processes (Level 1 Alignment) to optimized processes (Level 5 Alignment).

One of the primary benefits of IS planning and alignment of goals is a long-range view of information system's use in the organization. Specific systems development initiatives can spring from the IS plan, but the IS plan must also provide a broad framework for future success. The IS plan should guide development of the IS infrastructure over time. Another benefit of IS planning is that it ensures better use of IS resources—including funds, personnel, and time for scheduling specific projects. The steps of IS planning are shown in Figure 12.4.

Developing a Competitive Advantage

In today's business environment, many companies seek systems development projects that will provide them with a competitive advantage. Thinking competitively usually requires creative and critical analysis. For example, a company might want to achieve a competitive advantage by improving its customer-supplier relationship. Linking customers and suppliers can result in superior products and services. By looking at problems in new or different ways and by introducing innovative methods to solve them, many organizations have gained significant competitive advantage. In some cases, these new solutions are inspired by people and things not directly related to the problem.

Creative analysis involves investigating new approaches to existing problems. By looking at problems in new or different ways and by introducing innovative methods to solve them, many firms have gained a competitive advantage. Typically, these new solutions are inspired by people and events not directly related to the problem.

Critical analysis requires unbiased and careful questioning of whether system elements are related in the most effective ways. It involves considering the establishment of new or different relationships among system elements and perhaps introducing new elements into the system. Critical analysis in systems development involves the following actions:

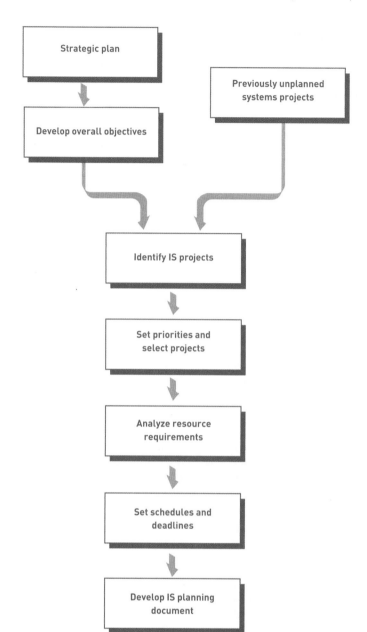

Figure 12.4

The Steps of IS Planning

Some projects are identified through overall IS objectives, whereas additional projects, called *unplanned projects*, are identified from other sources. All identified projects are then evaluated in terms of their organizational priority.

- **Questioning statements and assumptions.** Questioning users about their needs and clarifying their initial responses can result in better systems and more accurate predictions. Too often, stakeholders and users specify certain system requirements because they assume that their needs can only be met that way. Often, an alternative approach would be better. For example, a stakeholder might be concerned because there is always too much of some items in stock and not enough of other items. So, the stakeholder might request a new and improved inventory control system. An alternative approach is to identify the root cause for poor inventory management. This latter approach might determine that sales forecasting is inaccurate and needs improvement or that production cannot meet the set production schedule. All too often, solutions are selected before understanding the complete nature of the problem.
- **Identifying and resolving objectives and orientations that conflict.** Each department in an organization can have different objectives and orientations. The buying department might want to minimize the cost of spare parts by always buying from the lowest-cost supplier, but engineering might want to buy more expensive, higher-quality spare parts to reduce the frequency of replacement. These differences must be identified and resolved before a new purchasing system is developed or an existing one modified.

INFORMATION SYSTEMS @ WORK

Australia's Maroochy Shire Council Gains Control over Information Overload

Maroochy Shire is a region located in Australia's sunshine coast and governed locally by the Maroochy Shire Council. Maroochy Shire's lush green hinterland and over 20 kilometers of beautiful beaches make it a favorite tourist destination as well as an increasingly popular place to live. It is home to Australia's newest university, the University of the Sunshine Coast, and Australia's award-winning regional airport, the Sunshine Coast Airport.

With Australians flocking to the Maroochy Shire in pursuit of the beach lifestyle, the past six years has brought unprecedented growth to the region and extraordinarily turbulent times for the Maroochy Shire Council. Over six years, the council has endured six organizational restructures. It is currently working to replace all of its core information systems to meet the information demands of its ever-expanding population. In so doing, it is also working to transform itself from a traditional local government organization to one that is far more focused on customers and results.

The council hopes that new management information systems will help them dig out from under an overload of information. "We've got information all over the place: There are personal filing systems in hard copy, we have official records file systems, we have e-mails, we have intranet, we have network drives, so there's information all over the place and it's hard to get the complete picture of a particular issue. In a single year between 2003 and 2004, the number of records created rose by 45 percent," explained Wayne Bunker, the council's business process architect. During the same period, the number of e-mails handled grew by 23 percent, the number of files created grew by 32 percent, and the number of electronic documents and images grew by 26 percent.

Wayne Bunker envisioned a new system that would provide two important benefits: electronic document management—the ability to track documents as they entered and moved through the organization—and the capacity to support solid business processes. "We needed to have workflow that could move the information around the organization and encapsulate and code processes, including major exceptions, but we also wanted real-time reporting," Bunker says.

The council started analyzing all its existing processes, classifying the information sources for those processes, documenting employee roles and security, and designing and modeling new processes for the entire organization. After two years of systems development and piloting,

Maroochy recently went live with its first implementation designed for its licensing and compliance area. Bunker says the pilot achieved "terrific benefits," including doubling the number of documents and records captured.

Even with the evident benefits, stakeholders were somewhat reluctant to buy into the new system. Many in the business conceived of technology as a silver bullet that would solve all their problems. Helping them understand that their activities comprised processes that then flowed through to other parts of the business made a major difference. "When they understand that we are all part of the one corporate process, then we start to get some change in behavior and some buying, some understanding," explained Bunker.

"The other point is that the whole business case for doing this was to increase our productivity so that we can manage our growth into the future without having to necessarily put on too many additional bodies. That's why, when they came and said: 'You want to get rid of our jobs,' we said: 'No, we want to be able to do more with less.'"

The council is anticipating significant gains from the new systems, particularly in time saved searching the organization for information. Previous studies indicated that on average, people were spending two hours a week searching for information but not always coming up with the result that they wanted. The new system allows the council to accomplish more in its day-to-day business with less effort. As Maroochy Shire's population grows, the council's staff can remain at the same size and accomplish more in less time, with less frustration.

Discussion Questions

1. What external and internal pressures motivated Maroochy Shire to initiate systems development?
2. What are the primary goals of Maroochy Shire's new system?

Critical Thinking Questions

1. How is Maroochy Shire using prototyping in their approach to systems development?
2. How is the degree of change caused by the new system's pilot being handled by Maroochy Shire employees and managed by the business process architect Wayne Bunker?

SOURCES: Bushell, Sue, "Processes Undergo a Sea Change," CIO Australia, March 2, 2006, *www.cio.com.au*. Maroochy Shire Council Web site, accessed July 7, 2006, *www.maroochy.qld.gov.au*.

Establishing Objectives for Systems Development

The overall objective of systems development is to achieve business goals, not technical goals, by delivering the right information to the right person at the right time. The impact a particular system has on an organization's ability to meet its goals determines the true value of that system to the organization. Although all systems should support business goals, some systems are more pivotal in continued operations and goal attainment than others. These systems are called **mission-critical systems**. An order processing system, for example, is usually considered mission-critical. Without it, few organizations could continue daily activities, and they clearly would not meet set goals.

The goals defined for an organization also define the objectives that are set for a system. A manufacturing plant, for example, might determine that minimizing the total cost of owning and operating its equipment is critical to meet production and profit goals. **Critical success factors (CSFs)** are factors that are essential to the success of certain functional areas of an organization. The CSF for manufacturing—minimizing equipment maintenance and operating costs—would be converted into specific objectives for a proposed system. One specific objective might be to alert maintenance planners when a piece of equipment is due for routine preventative maintenance (e.g., cleaning and lubrication). Another objective might be to alert the maintenance planners when the necessary cleaning materials, lubrication oils, or spare parts inventory levels are below specified limits. These objectives could be accomplished either through automatic stock replenishment via electronic data interchange or through the use of exception reports. One study found that different CSFs might be important during different phases of a systems development project.

Regardless of the particular systems development effort, the development process should define a system with specific performance and cost objectives. The success or failure of the systems development effort will be measured against these objectives.

Performance Objectives

The extent to which a system performs as desired can be measured through its performance objectives. System performance is usually determined by factors such as the following:

- **The quality or usefulness of the output.** Is the system generating the right information for a value-added business process or by a goal-oriented decision maker?
- **The accuracy of the output.** Is the output accurate and does it reflect the true situation? As a result of the accounting scandals of the early 2000s, when some companies overstated revenues or understated expenses, accuracy is becoming more important, and top corporate officers are being held responsible for the accuracy of all corporate reports.
- **The quality or usefulness of the format of the output.** Is the output generated in a form that is usable and easily understood? For example, objectives often concern the legibility of screen displays, the appearance of documents, and the adherence to certain naming conventions.
- **The speed at which output is generated.** Is the system generating output in time to meet organizational goals and operational objectives? Objectives such as customer response time, the time to determine product availability, and throughput time are examples.
- **The scalability of the resulting system.** As mentioned in Chapter 4, *scalability* allows an information system to handle business growth and increased business volume. For example, if a midsized business realizes an annual 10 percent growth in sales for several years, an information system that is scalable will be able to efficiently handle the increase by adding processing, storage, software, database, telecommunications, and other information systems resources to handle the growth.
- **The risk of the system.** One important objective of many systems development projects is to reduce risk.[12] The BRE Bank in Poland, for example, used systems development to create a model-based DSS to analyze and reduce loan risk and a variety of related risks associated with bank transactions. The new project uses a mathematical algorithm, called FIRST (Financial Institutions Risk Scenario Trends), to reduce risk.

In some cases, the achievement of performance objectives can be easily measured (e.g., by tracking the time it takes to determine product availability). In other cases, it is sometimes

mission-critical systems
Systems that play a pivotal role in an organization's continued operations and goal attainment.

critical success factors (CSFs)
Factors that are essential to the success of a functional area of an organization.

more difficult to ascertain in the short term. For example, it might be difficult to determine how many customers are lost because of slow responses to customer inquiries regarding product availability. These outcomes, however, are often closely associated with corporate goals and are vital to the long-term success of the organization. Senior management usually dictates their attainment.

Cost Objectives

Organizations can spend more than is necessary during a systems development project. The U.S. Inspector General, for example, reported that the U.S. Department of Energy spent about $4.1 million more than it should have on acquiring and maintaining desktop computers and software.[13] The benefits of achieving performance goals should be balanced with all costs associated with the system, including the following:

- **Development costs.** All costs required to get the system up and running should be included. Some computer vendors give cash rewards to companies using their systems to reduce costs and as an incentive.[14]
- **Costs related to the uniqueness of the system application.** A system's uniqueness has a profound effect on its cost. An expensive but reusable system might be preferable to a less costly system with limited use.
- **Fixed investments in hardware and related equipment.** Developers should consider costs of such items as computers, network-related equipment, and environmentally controlled data centers in which to operate the equipment.
- **Ongoing operating costs of the system.** Operating costs include costs for personnel, software, supplies, and resources such as the electricity required to run the system.

Balancing performance and cost objectives within the overall framework of organizational goals can be challenging. Setting objectives are important, however, because they allow an organization to allocate resources effectively and measure the success of a systems development effort. For PC manufacturers, for example, parts and components of a typical PC can cost under $500, which includes about $130 for the processor, $100 for a CD or DVD, $100 for memory, $45 for the Windows operating system, and the rest on other hardware parts and components. Some believe these low costs will eventually lead to lower costs for PCs.

SYSTEMS DEVELOPMENT LIFE CYCLES

The systems development process is also called a *systems development life cycle (SDLC)* because the activities associated with it are ongoing. As each system is built, the project has timelines and deadlines, until at last the system is installed and accepted. The life of the system continues as it is maintained and reviewed. If the system needs significant improvement beyond the scope of maintenance, if it needs to be replaced because of a new generation of technology, or if the IS needs of the organization change significantly, a new project will be initiated and the cycle will start over.

A key fact of systems development is that the later in the SDLC an error is detected, the more expensive it is to correct (see Figure 12.5). One reason for the mounting costs is that if an error is found in a later phase of the SDLC, the previous phases must be reworked to some extent. Another reason is that the errors found late in the SDLC affect more people. For example, an error found after a system is installed might require retraining users when a "work-around" to the problem has been found. Thus, experienced systems developers prefer an approach that will catch errors early in the project life cycle.

Several common systems development life cycles exist: traditional, prototyping, rapid application development (RAD), and end-user development. In addition, companies can outsource the systems development process. With some companies, these approaches are formalized and documented so that systems developers have a well-defined process to follow; in other companies, less formalized approaches are used. Keep Figure 12.5 in mind as you are introduced to alternative SDLCs in the next section.

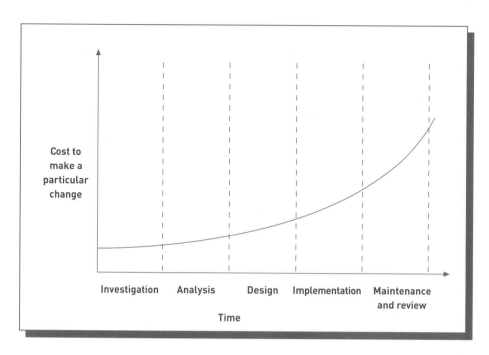

Figure 12.5

Relationship Between Timing of Errors and Costs

The later that system changes are made in the SDLC, the more expensive these changes become.

The Traditional Systems Development Life Cycle

Traditional systems development efforts can range from a small project, such as purchasing an inexpensive computer program, to a major undertaking. The steps of traditional systems development might vary from one company to the next, but most approaches have five common phases: investigation, analysis, design, implementation, and maintenance and review (see Figure 12.6).

Figure 12.6

The Traditional Systems Development Life Cycle

Sometimes, information learned in a particular phase requires cycling back to a previous phase.

systems investigation
The systems development phase during which problems and opportunities are identified and considered in light of the goals of the business.

systems analysis
The systems development phase involving the study of existing systems and work processes to identify strengths, weaknesses, and opportunities for improvement.

systems design
The systems development phase that defines how the information system will do what it must do to obtain the problem solution.

systems implementation
The systems development phase involving the creation or acquiring of various system components detailed in the systems design, assembling them, and placing the new or modified system into operation.

systems maintenance and review
The systems development phase that ensures the system operates as intended and modifies the system so that it continues to meet changing business needs.

In the **systems investigation** phase, potential problems and opportunities are identified and considered in light of the goals of the business. Systems investigation attempts to answer the questions "What is the problem, and is it worth solving?" The primary result of this phase is a defined development project for which business problems or opportunity statements have been created, to which some organizational resources have been committed, and for which systems analysis is recommended. **Systems analysis** attempts to answer the question "What must the information system do to solve the problem?" This phase involves studying existing systems and work processes to identify strengths, weaknesses, and opportunities for improvement. The major outcome of systems analysis is a list of requirements and priorities. **Systems design** seeks to answer the question "How will the information system do what it must do to obtain the problem solution?" The primary result of this phase is a technical design that either describes the new system or describes how existing systems will be modified. The system design details system outputs, inputs, and user interfaces; specifies hardware, software, database, telecommunications, personnel, and procedure components; and shows how these components are related. **Systems implementation** involves creating or acquiring the various system components detailed in the systems design, assembling them, and placing the new or modified system into operation. An important task during this phase is to train the users. Systems implementation results in an installed, operational information system that meets the business needs for which it was developed. The purpose of **systems maintenance and review** is to ensure that the system operates as intended and to modify the system so that it continues to meet changing business needs. As shown in Figure 12.6, a system under development moves from one phase of the traditional SDLC to the next.

The traditional SDLC allows for a large degree of management control. However, a major problem is that the user does not use the solution until the system is nearly complete. Table 12.1 lists advantages and disadvantages of the traditional SDLC.

Advantages	Disadvantages
Formal review at the end of each phase allows maximum management control.	Users get a system that meets the needs as understood by the developers; this might not be what is really needed.
This approach creates considerable system documentation.	Documentation is expensive and time consuming to create. It is also difficult to keep current.
Formal documentation ensures that system requirements can be traced back to stated business needs.	Often, user needs go unstated or are misunderstood.
It produces many intermediate products that can be reviewed to see whether they meet the users' needs and conform to standards.	Users cannot easily review intermediate products and evaluate whether a particular product (e.g., data flow diagram) meets their business requirements.

Table 12.1

Advantages and Disadvantages of Traditional SDLC

Prototyping

Prototyping takes an iterative approach to the systems development process. During each iteration, requirements and alternative solutions to the problem are identified and analyzed, new solutions are designed, and a portion of the system is implemented. Users are then encouraged to try the prototype and provide feedback (see Figure 12.7). Prototyping begins with creating a preliminary model of a major subsystem or a scaled-down version of the entire system. For example, a prototype might show sample report formats and input screens. After they are developed and refined, the prototypical reports and input screens are used as models for the actual system, which can be developed using an end-user programming language such as Visual Basic. The first preliminary model is refined to form the second- and third-generation models, and so on until the complete system is developed (see Figure 12.8).

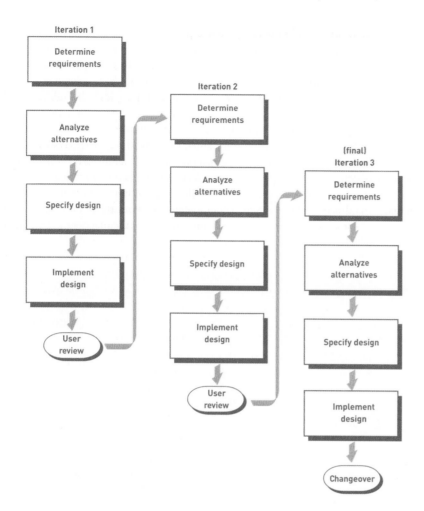

Figure 12.7

Prototyping

Prototyping is an iterative approach to systems development.

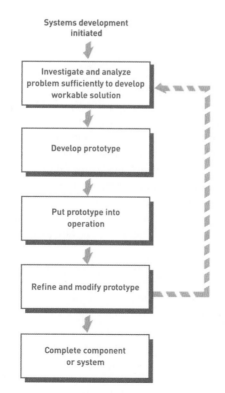

Figure 12.8

Refining During Prototyping

Each generation of prototype is a refinement of the previous generation based on user feedback.

operational prototype
A functioning prototype that accesses real data files, edits input data, makes necessary computations and comparisons, and produces real output.

nonoperational prototype
A mock-up, or model, that includes output and input specifications and formats.

Prototypes can be classified as operational or nonoperational. An **operational prototype** is a prototype that works—accesses real data files, edits input data, makes necessary computations and comparisons, and produces real output. Fully developed financial reports are examples. The operational prototype might access real files but perhaps does not edit input. A **nonoperational prototype** is a mock-up, or model. It typically includes output and input specifications and formats. The outputs include printed reports to managers and the screen layout of reports displayed on personal computers or terminals. The inputs reveal how data is captured, what commands users must enter, and how the system accesses other data files. The primary advantage of a nonoperational prototype is that it can be developed much faster than an operational prototype. Nonoperational prototypes can be discarded, and a fully operational system can be built based on what was learned from the prototypes. Research has shown that systems investigation, design, and implementation are the phases that benefit the most from the prototyping approach. The advantages and disadvantages of prototyping are summarized in Table 12.2.

Table 12.2

Advantages and Disadvantages of Prototyping

Advantages	Disadvantages
Users can try the system and provide constructive feedback during development.	Each iteration builds on the previous one. The final solution might be only incrementally better than the initial solution.
An operational prototype can be produced in weeks.	Formal end-of-phase reviews might not occur. Thus, it is very difficult to contain the scope of the prototype, and the project never seems to end.
As solutions emerge, users become more positive about the process and the results.	System documentation is often absent or incomplete because the primary focus is on development of the prototype.
Prototyping enables early detection of errors and omissions.	System backup and recovery, performance, and security issues can be overlooked in the haste to develop a prototype.

Rapid Application Development, Agile Development, Joint Application Development, and Other Systems Development Approaches

rapid application development (RAD)
A systems development approach that employs tools, techniques, and methodologies designed to speed application development.

Rapid application development (RAD) employs tools, techniques, and methodologies designed to speed application development. Vendors, such as Computer Associates International, IBM, and Oracle, market products targeting the RAD market. Rational Software, a division of IBM, has a RAD tool, called Rational Rapid Developer, to make developing large Java programs and applications easier and faster. Locus Systems, a program developer, used a RAD tool called OptimalJ to generate more than 60 percent of the computer code for three applications it developed.[15] Royal Bank of Canada used OptimalJ to develop some customer-based applications. According to David Hewick, group manager of application architecture for the bank, "It was an opportunity to improve the development life cycle, reduce costs, and bring consistency." Advantage Gen, formerly known as COOL:Gen, is a RAD tool from Computer Associates International. It can be used to rapidly generate computer code from business models and specifications.[16]

RAD reduces paper-based documentation, automatically generates program source code, and facilitates user participation in design and development activities. It makes adapting to changing system requirements easier. Other approaches to rapid development, such as *agile development* or *extreme programming* (*XP*), allow the systems to change as they are being developed. Agile development requires frequent face-to-face meetings with the systems developers and users as they modify, refine, and test how the system meets users' needs and what its capabilities are. XP uses pairs of programmers who work together to design, test, and code parts of the systems they develop. The iterative nature of XP helps companies develop robust systems, with fewer errors. Sabre Airline Solutions, a $2 billion computer company serving the airline travel industry, used XP to eliminate programming errors and shorten program development times.

RAD makes extensive use of the **joint application development (JAD)** process for data collection and requirements analysis. Originally developed by IBM Canada in the 1970s, JAD involves group meetings in which users, stakeholders, and IS professionals work together to analyze existing systems, propose possible solutions, and define the requirements of a new or modified system. JAD groups consist of both problem holders and solution providers. A group normally requires one or more top-level executives who initiate the JAD process, a group leader for the meetings, potential users, and one or more individuals who act as secretaries and clerks to record what is accomplished and to provide general support for the sessions. Many companies have found that groups can develop better requirements than individuals working independently and have assessed JAD as a very successful development technique. Today, JAD often uses *group support systems (GSS)* software to foster positive group interactions, while suppressing negative group behavior. Group support systems were introduced in Chapter 10.

RAD should not be used on every software development project. In general, it is best suited for DSSs and MISs and less well suited for TPSs. During a RAD project, the level of participation of stakeholders and users is much higher than in other approaches. Table 12.3 lists advantages and disadvantages of RAD.

joint application development (JAD)

A process for data collection and requirements analysis in which users, stakeholders, and IS professionals work together to analyze existing systems, propose possible solutions, and define the requirements of a new or modified system.

Advantages	Disadvantages
For appropriate projects, this approach puts an application into production sooner than any other approach.	This intense SDLC can burn out systems developers and other project participants.
Documentation is produced as a by-product of completing project tasks.	This approach requires systems analysts and users to be skilled in RAD systems development tools and RAD techniques.
RAD forces teamwork and lots of interaction between users and stakeholders.	RAD requires a larger percentage of stakeholders' and users' time than other approaches.

Table 12.3

Advantages and Disadvantages of RAD

In addition to the systems development approaches discussed previously, a number of other systems development approaches are available, including adaptive software development, lean software development, Rational Unified Process (RUP), Feature-Driven Development (FDD), and dynamic systems development method. Often created by computer vendors and authors of systems development books, these approaches all attempt to deliver better systems. The Ohio Causality Corporation, for example, uses RUP from IBM and Rational Software. RUP uses an iterative approach to software development that concentrates on software quality as it is changed and updated over time.[17]

The End-User Systems Development Life Cycle

The term **end-user systems development** describes any systems development project in which business managers and users assume the primary effort. Rather than ignoring these initiatives, astute IS professionals encourage them by offering guidance and support. Providing technical assistance, communicating standards, and sharing "best practices" throughout the organization are some ways IS professionals work with motivated managers and employees undertaking their own systems development. In this way, end-user-developed systems can be structured as complementary to, rather than in conflict with, existing and emerging information systems. In addition, this open communication among IS professionals, managers of the affected business area, and users allows the IS professionals to identify specific initiatives so that additional organizational resources, beyond those available to business managers or users, are provided for its development.

end-user systems development

Any systems development project in which the primary effort is undertaken by a combination of business managers and users.

Many end users today are already demonstrating their systems development capability by designing and implementing their own PC-based systems.

(Source: © Tanya Constantine/Getty Images.)

User-developed systems range from the very small (such as a software routine to merge form letters) to those of significant organizational value (such as customer contact databases for the Web). Initially, IS professionals discounted the value of these projects. As the number and magnitude of these projects increased, however, IS professionals began to realize that for the good of the entire organization, their involvement with these projects needed to increase.

End-user systems development does have some disadvantages. Some end users don't have the training to effectively develop and test a system. Multimillion-dollar mistakes, for example, can be made using faulty spreadsheets that were never tested. Some end user systems are also poorly documented. When these systems are updated, problems can be introduced that make the systems error-prone. In addition, some end users spend time and corporate resources developing systems that were already available.

Outsourcing and On-Demand Computing

Many companies hire an outside consulting firm or computer company that specializes in systems development to take over some or all of its development and operations activities.[18] Some companies, such as General Electric, have their own outsourcing subunits or have spun off their outsourcing subunits as separate companies.[19] As mentioned in Chapter 2, *outsourcing* and *on-demand computing* are often used.[20] Table 12.4 describes the circumstances in which outsourcing is a good idea.

The decision to outsource systems development is often a response to downsizing, which reduces the number of employees or managers, equipment and systems, and even functions and departments. Outsourcing allows companies to downsize the IS department and alleviate difficult financial situations by reducing payroll and other expenses.

Organizations can outsource any aspect of their information system, including hardware maintenance and management, software development, database systems, networks and telecommunications, Internet and intranet operations, hiring and staffing, and the development of procedures and rules regarding the information system.[21] Eurostar, a European travel company, for example, hired the outsourcing company Occam to develop a new Web site and back-end database to give its travel customers greater travel information.[22] The new system should be ready in 2007 or 2008. According to Scott Logie, managing director of Occam, "The quality and volume of data that Eurostar possesses is extremely valuable. By working together we will allow the firm to develop real insight into its customers. This can be used to drive a strong customer acquisitions strategy, which will enhance its business and customer relationships."

Reason	Example
When a company believes it can cut costs	PacifiCare outsourced its IS operations to IBM and Keane, Inc. PacifiCare hopes the outsourcing will save it about $400 million over ten years.
When a firm has limited opportunity to distinguish itself competitively through a particular IS operation or application	Kodak outsourced its IS operations, including mainframe processing, telecommunications, and personal computer support, because it had limited opportunity to distinguish itself through these IS operations. Kodak kept application development and support in-house because it thought that these activities had competitive value.
When uninterrupted IS service is not crucial	Airline reservations or catalog shopping systems are mission-critical and should not be trusted to outside firms.
When outsourcing does not strip the company of technical know-how required for future IS innovation	Firms must ensure that their IS staffs remain technically up-to-date and have the expertise to develop future applications.
When the firm's existing IS capabilities are limited, ineffective, or technically inferior	A company might use outsourcing to help it make the transition from a centralized mainframe environment to a distributed client/server environment.
When a firm is downsizing	First Fidelity, a major bank, used outsourcing as part of a program to reduce the number of employees by 1,600 and slash expenses by $85 million.

Table 12.4

When to Use Outsourcing for Systems Development

Reducing costs, obtaining state-of-the-art technology, eliminating staffing and personnel problems, and increasing technological flexibility are reasons that companies have used the outsourcing and on-demand computing approaches.[23] Reducing costs is a primary reason for outsourcing. One American computer company, for example, estimated that a programmer with three to five years of experience in China would cost about $13 per hour, while a programmer with similar experience in the United States would cost about $56 per hour. U.S. companies also provide outsourcing services.[24] Aelera Corporation spent about six months looking for the best outsourcing deal and determined that a company in Savannah, Georgia, was the best.[25] McKesson Corporation saved about $10 million by outsourcing jobs from San Francisco to Dubuque, Iowa. Mattel outsourced to rural Jonesboro, Arkansas. Increasingly, companies are looking to American outsourcing companies to reduce costs and increase services. Individuals, including students, are also outsourcing tasks they have to perform.[26]

A number of companies offer outsourcing and on-demand computing services—from general systems development to specialized services. IBM's Global Services, for example, is one of the largest full-service outsourcing and consulting services.[27] IBM has consultants located in offices around the world. Electronic Data Systems (EDS) is another large company that specializes in consulting and outsourcing.[28] EDS has approximately 140,000 employees in almost 60 countries and more than 9,000 clients worldwide. Accenture, which was once part of Arthur Andersen, is another company that specializes in consulting and outsourcing.[29] The company has more than 75,000 employees in 47 countries. See Figure 12.9.

Organizations can use a number of guidelines to make outsourcing a success, including the following:[30]

- Keep tight controls on the outsourcing project. According to Jerry Bartlett of Ameritrade Holding Company, "The primary area where you've got to make a commitment is project management. You've got to have someone in place who has great communications skills, greater problem-resolution skills, and creativity."
- Treat outsourcing companies as partners. According to the CIO of Keystone Automotive Industries, "We were coming out of the dot-com boom, and we really wanted to take a look at some of these outsourcers to see if they understood the concept of what I call corporate hosting."
- Start with smaller outsourcing jobs.
- Create effective communications channels between the organization and the outsourcing company.
- Carefully review legal outsourcing contracts, including rights and remedies clauses.[31]

Outsourcing has some disadvantages, however. Some companies, such as Sprint Nextel, are starting to reduce their outsourcing and bring systems development back in-house.[32] Because of problems with an outsourcing deal, the British government might sue its outsourcing company to recover part of an estimated $3.5 billion in overpayments to low-income families or families with children in Britain.[33] The overpayment problems were related to procedural errors and technical glitches. Internal expertise can be lost and loyalty can suffer under an outsourcing arrangement. When a company outsources, key IS personnel with expertise in technical and business functions are no longer needed. When these IS employees leave, their experience with the organization and expertise in information systems is lost. For some companies, it can be difficult to achieve a competitive advantage when competitors are using the same computer or consulting company. When the outsourcing or on-demand computing is done offshore or in a foreign country, some people have raised security concerns. How will important data and trade secrets be guarded?

FACTORS AFFECTING SYSTEMS DEVELOPMENT SUCCESS

Successful systems development means delivering a system that meets user and organizational needs—on time and within budget. Getting users and stakeholders involved in systems development is critical for most systems development projects. Having the support of top-level managers is also important. According to Mike McNall, director of application services for Vanguard Car Rental, "If we didn't have the backing of business … no matter what we built, it wouldn't be accepted."[34] Vanguard Car Rental needed to redo its information system when it merged its information system with National and Alamo Car Rental's system to make both systems compatible. A merger often causes information systems to be redone or completely changed. In addition to user involvement and top management support, other factors can contribute to successful systems development efforts—at a reasonable cost. These factors are discussed next.

Degree of Change

A major factor that affects the quality of systems development is the degree of change associated with the project. The scope can vary from implementing minor enhancements to an existing system to major reengineering. The project team needs to recognize where they are on this spectrum of change.

Continuous Improvement Versus Reengineering

As discussed in Chapter 2, continuous improvement projects do not require significant business process or IS changes or retraining of people; thus, they have a high degree of success. Typically, because continuous improvements involve minor improvements, these projects also have relatively modest benefits. On the other hand, reengineering involves fundamental changes in how the organization conducts business and completes tasks. The factors associated with successful reengineering are similar to those of any development effort, including top management support, clearly defined corporate goals and systems development objectives, and careful management of change. Major reengineering projects tend to have a high degree of risk but also a high potential for major business benefits (see Figure 12.10).

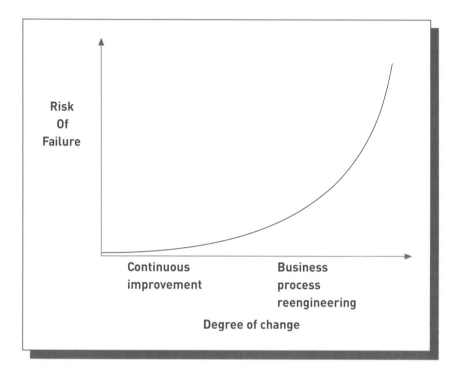

Figure 12.10

The degree of change can greatly affect the probability of a project's success.

Managing Change

The ability to manage change is critical to the success of systems development. New systems inevitably cause change. For example, the work environment and habits of users are invariably affected by the development of a new information system. Unfortunately, not everyone adapts easily, and the increasing complexity of systems can multiply the problems. Managing change requires the ability to recognize existing or potential problems (particularly the concerns of users) and deal with them before they become a serious threat to the success of the new or modified system. Here are several of the most common problems:

- Fear that the employee will lose his job, power, or influence within the organization
- Belief that the proposed system will create more work than it eliminates
- Reluctance to work with "computer people"
- Anxiety that the proposed system will negatively alter the structure of the organization
- Belief that other problems are more pressing than those solved by the proposed system or that the system is being developed by people unfamiliar with "the way things need to get done"
- Unwillingness to learn new procedures or approaches

Preventing or dealing with these types of problems requires a coordinated effort from stakeholders and users, managers, and IS personnel. One remedy is simply to talk with all people concerned and learn what their biggest concerns are. Management can then deal with those concerns and try to eliminate them. After immediate concerns are addressed, people can become part of the project team.

Sentinel: The FBI's New Case Management System

The FBI has hired Lockheed Martin Corporation to build a case management and information-sharing system named Sentinel, designed to replace the bureau's current processes that are predominantly paper-based. This isn't the bureau's first attempt to modernize its systems. Its previous attempt, a system named Virtual Case File, was abandoned after four years of work and a $170 million investment. How does the FBI plan to succeed where it previously failed?

Rather than developing its own proprietary code, as the FBI did with Virtual Case File, Sentinel will be based on commercially available software running on service-oriented architecture. More important, serious changes will be made to project management procedures. According to a 91-page report released by the U.S. Department of Justice's inspector general, the Virtual Case File effort failed partly because the FBI's IT project management office seemed to have a revolving door. Virtual Case File had "15 different key IT managers over the course of its life," the report said. The inspector general recommended that the FBI devote "significant staffing" to Sentinel, saying that doing so "at the outset of the project is key to establishing the stable management staff required to properly oversee the project."

FBI CIO Zalmai Azmi is doing everything possible to improve project management for Sentinel. Previously, the project had only one manager with a professional certification from the Project Management Institute (PMI). Now, the bureau has 47 employees with PMI certifications. Azmi explained that the Sentinel project will be run internally by a dedicated team that will include its own redundant management, with primary and secondary employees assigned to each spot on the team.

If Sentinel is completed successfully, it will radically improve FBI operations. The Sentinel project's primary goal is to provide FBI agents with "one-stop shopping" for data, Azmi said. Currently, agents often file reports on paper and must log on to different databases to search for information.

Because Sentinel will incorporate the FBI's broader goal of seamless information sharing, its benefits will trickle down to all U.S. law enforcement and intelligence agencies. Since the September 11 terrorist attacks, "our mission has changed," said Azmi, who meets regularly with his IT counterparts at agencies such as the CIA. The Service Oriented Architecture (SOA) approach to Sentinel will help ensure that the system works with other current and future systems both inside the FBI and at other agencies.

The time and money invested in Virtual Case File is not a complete waste. Sentinel will use some of the hardware and networking systems installed for that system.

The project is estimated to cost a total of $425 million with $305 million going to Lockheed Martin, who plans to assign 200 developers to the six-year project. The remainder of the funds will cover additional costs, such as independent system validation and verification work.

Sentinel is due to be delivered in four phases. The first phase, expected to be completed in 2007, includes development of a Web portal that will enable FBI workers to access legacy systems via a modern user interface that supports Web services.

Discussion Questions

1. List three organizational goals that the FBI's Sentinel system is being developed to support.
2. What are two perceived problems with the development of Virtual Case File and how are they being addressed in Sentinel?

Critical Thinking Questions

1. Why might the use of 15 different key IT managers over the course of a systems development effort lead to its failure?
2. Why do you think Sentinel is being developed with commercially available software and a SOA rather than proprietary software?

SOURCES: Thibodeau, Patrick, "FBI Shifts into High Gear on $425M Case File System," *Computerworld*, March 20, 2006, *www.computerworld.com*. Dizard, Wilson P. III, "Sentinel System Will Replace FBI's Virtual Case File," *Washington Technology*, May 23, 2005. Arnone, Michael, "Senators Seek to Fast Track FBI's Sentinel," *Federal Computer Week*, July 27, 2005, *www.fwc.com*.

Quality and Standards

Another key success factor is the quality of project planning. The bigger the project, the more likely that poor planning will lead to significant problems. Many companies find that large systems projects fall behind schedule, go over budget, and do not meet expectations. A systems development project for the Child Support Agency of the British government was behind schedule and almost $485 million over budget.[35] The delayed project might have hurt the agencies ability to deliver important services to children. Although proper planning cannot guarantee that these types of problems will be avoided, it can minimize the likelihood of their occurrence. Good systems development is not automatic. Certain factors contribute to the failure of systems development projects. These factors and countermeasures to eliminate or alleviate the problem are summarized in Table 12.5.

Factor	Countermeasure
Solving the wrong problem	Establish a clear connection between the project and organizational goals.
Poor problem definition and analysis	Follow a standard systems development approach.
Poor communication	Communicate, communicate, communicate.
Project is too ambitious	Narrow the project focus to address only the most important business opportunities.
Lack of top management support	Identify the senior manager who has the most to gain from the success of the project, and recruit this person to champion the project.
Lack of management and user involvement	Identify and recruit key stakeholders to be active participants in the project.
Inadequate or improper system design	Follow a standard systems development approach.
Lack of standards	Implement a standards system, such as ISO 9001.
Poor testing and implementation	Plan sufficient time for this activity.
Users cannot use the system effectively	Develop a rigorous user-training program and budget sufficient time in the schedule to execute it.
Lack of concern for maintenance	Include an estimate of employee effort and costs for maintenance in the original project justification.

Table 12.5

Project Planning Issues Frequently Contributing to Project Failure

The development of information systems requires a constant trade-off of schedule and cost versus quality. Historically, the development of application software has put an overemphasis on schedule and cost to the detriment of quality. Techniques, such as use of the ISO 9001 standards, have been developed to improve the quality of information systems. ISO 9001 is a set of international quality standards originally developed in Europe in 1987. These standards address customer satisfaction and are the only standards in the ISO 9000 family where third-party certification can be achieved. Adherence to ISO 9001 is a requirement in many international markets.[36] Many companies in the United States and around the world strive to achieve ISO 9001 certification and various specifications set by computer vendors and other organizations.[37] For example, the Java 2 Enterprise Edition (J2EE) specifications are a set of rules and procedures developed by Sun Microsystems to help reduce costs, minimize development complexities, increase security, and reduce development times.

Organizational experience with the systems development process is also a key factor for systems development success.[38] The *Capability Maturity Model (CMM)* is one way to measure this experience.[39] It is based on research done at Carnegie Mellon University and work by the Software Engineering Institute (SEI).[40] CMM is a measure of the maturity of the software development process in an organization. CMM grades an organization's systems development maturity using five levels: initial, repeatable, defined, managed, and optimized.

ISO 9000 is a set of international quality standards used by IS and other organizations to ensure the quality of products and services.

(Source: *www.iso.org*.)

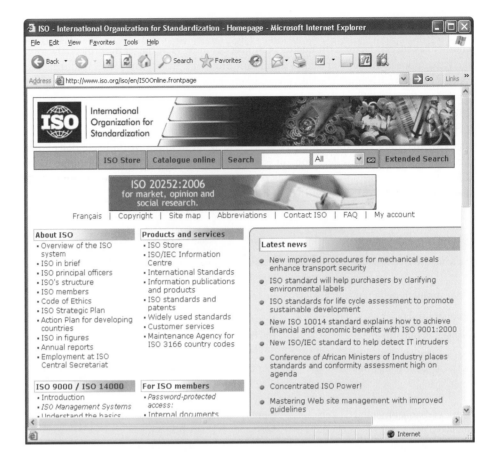

Use of Project Management Tools

Project management involves planning, scheduling, directing, and controlling human, financial, and technological resources for a defined task whose result is achievement of specific goals and objectives. Florida, for example, will spend $15 million on a project to help share data among over 500 law enforcement organizations in the state.[41] Even small systems development projects must employ some type of project management.[42]

A **project schedule** is a detailed description of what is to be done. Each project activity, the use of personnel and other resources, and expected completion dates are described. A **project milestone** is a critical date for the completion of a major part of the project. The completion of program design, coding, testing, and release are examples of milestones for a programming project. The **project deadline** is the date the entire project is to be completed and operational—when the organization can expect to begin to reap the benefits of the project. One company offers a 20 percent refund if it doesn't meet a client's project deadline. In addition, any additional work done after the project deadline is performed free of charge.

In systems development, each activity has an earliest start time, earliest finish time, and slack time, which is the amount of time an activity can be delayed without delaying the entire project. The **critical path** consists of all activities that, if delayed, would delay the entire project. These activities have zero slack time. Any problems with critical-path activities will cause problems for the entire project. To ensure that critical-path activities are completed in a timely fashion, formalized project management approaches have been developed. Tools such as Microsoft Project are available to help compute these critical project attributes.

Although the steps of systems development seem straightforward, larger projects can become complex, requiring hundreds or thousands of separate activities. For these systems development efforts, formal project management methods and tools become essential. A formalized approach called **Program Evaluation and Review Technique (PERT)** creates three time estimates for an activity: shortest possible time, most likely time, and longest possible time. A formula is then applied to determine a single PERT time estimate. A

project schedule

A detailed description of what is to be done.

project milestone

A critical date for the completion of a major part of the project.

project deadline

The date the entire project is to be completed and operational.

critical path

Activities that, if delayed, would delay the entire project.

Program Evaluation and Review Technique (PERT)

A formalized approach for developing a project schedule.

Gantt chart is a graphical tool used for planning, monitoring, and coordinating projects; it is essentially a grid that lists activities and deadlines. Each time a task is completed, a marker such as a darkened line is placed in the proper grid cell to indicate the completion of a task (see Figure 12.11).

Gantt chart
A graphical tool used for planning, monitoring, and coordinating projects.

PROJECT PLANNING DOCUMENTATION — Page 1 of 1

System: **Warehouse Inventory System (Modification)** — Date 12/10

System — Scheduled activity / ▬ Completed activity — Analyst: **Cecil Truman** — Signature

Activity*	Individual assigned	1	2	3	4	5	6	7	8	9	10	11	12	13	14
R — Requirements definition															
R.1 Form project team	VP, Cecil, Bev	▬													
R.2 Define obj. and constraints	Cecil	▬													
R.3 Interview warehouse staff															
for requirements report	Bev		▬	▬											
R.4 Organize requirements	Team			—▬											
R.5 VP review	VP, Team				—▬										
D — Design															
D.1 Revise program specs.	Bev					—▬									
D.2.1 Specify screens	Bev					—▬									
D.2.2 Specify reports	Bev						—▬								
D.2.3 Specify doc. changes	Cecil						▬								
D.4 Management review	Team							—							
I — Implementation															
I.1 Code program changes	Bev								—						
I.2.1 Build test file	Team									—					
I.2.2 Build production file	Bev										—				
I.3 Revise production file	Cecil										—				
I.4.1 Test short file	Bev									—					
I.4.2 Test production file	Cecil											—			
I.5 Management review	Team												—		
I.6 Install warehouse**															
I.6.1 Train new procedures	Bev											—			
I.6.2 Install	Bev											—			
I.6.3 Management review	Team														—

*Weekly team reviews not shown here
**Report for warehouses 2 through 5

Figure 12.11

Sample Gantt Chart

A Gantt chart shows progress through systems development activities by putting a bar through appropriate cells.

Both PERT and Gantt techniques can be automated using project management software. Several project management software packages are identified in Table 12.6. This software monitors all project activities and determines whether activities and the entire project are on time and within budget. Project management software also has workgroup capabilities to handle multiple projects and to allow a team to interact with the same software. Project management software helps managers determine the best way to reduce project completion time at the least cost. Many project managers, however, fear that the quality of a systems development project will suffer with shortened deadlines and think that slack time should be added back to the schedule as a result.

Software	Vendor
AboutTime	NetSQL Partners
Job Order	Management Software
OpenPlan	Welcom
Microsoft Project	Microsoft
Project Scheduler	Scitor
Super Project	Computer Associates

**computer-aided software
engineering (CASE)**

Tools that automate many of the
tasks required in a systems development effort and encourage adherence to the SDLC.

Use of Computer-Aided Software Engineering (CASE) Tools

Computer-aided software engineering (CASE) tools automate many of the tasks required in a systems development effort and encourage adherence to the SDLC, thus instilling a high degree of rigor and standardization to the entire systems development process. VRCASE, for example, is a CASE tool that a team of developers can use when developing applications in C++ and other languages. Prover Technology has developed a CASE tool that searches for programming bugs. The CASE tool searches for all possible design scenarios to make sure that the program is error free. Other CASE tools include Visible Systems (*www.visible.com*) and Popkin Software (*www.popkin.com*). Popkin Software, for example, can generate code in programming languages such as C++, Java, and Visual Basic. Other CASE-related tools include Rational Rose (part of IBM) and Visio, a charting and graphics program from Microsoft. Companies that produce CASE tools include Accenture, Microsoft, and Oracle. Oracle Designer and Developer CASE tools, for example, can help systems analysts automate and simplify the development process for database systems. See Table 12.7 for a list of CASE tools and their providers. The advantages and disadvantages of CASE tools are listed in Table 12.8. CASE tools that focus on activities associated with the early stages of systems development are often called *upper-CASE* tools. These packages provide automated tools to assist with systems investigation, analysis, and design activities. Other CASE packages, called *lower-CASE* tools, focus on the later implementation stage of systems development, and can automatically generate structured program code.

Table 12.7

Typical CASE Tools

CASE Tool	Vendor
Oracle Designer	Oracle Corporation *www.oracle.com*
Visible Analyst	Visible Systems Corporation *www.visible.com*
Rational Rose	Rational Software *www.ibm.com*
Embarcadero Describe	Embarcadero Describe *www.embarcadero.com*

Table 12.8

Advantages and Disadvantages
of CASE Tools

Advantages	Disadvantages
Produce systems with a longer effective operational life	Increase the initial costs of building and maintaining systems
Produce systems that more closely meet user needs and requirements	Require more extensive and accurate definition of user needs and requirements
Produce systems with excellent documentation	Can be difficult to customize
Produce systems that need less systems support	Require more training of maintenance staff
Produce more flexible systems	Can be difficult to use with existing systems

Object-Oriented Systems Development

The success of a systems development effort can depend on the specific programming tools and approaches used. As mentioned in Chapter 4, object-oriented (OO) programming languages allow the interaction of programming objects—that is, an object consists of both data and the actions that can be performed on the data. So, an object could be data about an employee and all the operations (such as payroll, benefits, and tax calculations) that might be performed on the data.

Developing programs and applications using OO programming languages involves constructing modules and parts that can be reused in other programming projects. DTE Energy, a $7 billion Detroit-based energy company, has set up a library of software components that can be reused by its programmers.[43] Systems developers from the company reuse and contribute to software components in the library. DTE's developers meet frequently to discuss ideas, problems, and opportunities of using the library of reusable software components.

Chapter 4 discussed a number of programming languages that use the object-oriented approach, including Visual Basic, C++, and Java. These languages allow systems developers to take the OO approach, making program development faster and more efficient, resulting in lower costs. Modules can be developed internally or obtained from an external source. After a company has the programming modules, programmers and systems analysts can modify them and integrate them with other modules to form new programs.

Object-oriented systems development (OOSD) combines the logic of the systems development life cycle with the power of object-oriented modeling and programming. OOSD follows a defined systems development life cycle, much like the SDLC. The life cycle phases are usually completed with many iterations. Object-oriented systems development typically involves the following tasks:

- **Identifying potential problems and opportunities within the organization that would be appropriate for the OO approach.** This process is similar to traditional systems investigation. Ideally, these problems or opportunities should lend themselves to the development of programs that can be built by modifying existing programming modules.
- **Defining what kind of system users require.** This analysis means defining all the objects that are part of the user's work environment (object-oriented analysis). The OO team must study the business and build a model of the objects that are part of the business (such as a customer, an order, or a payment). Many of the CASE tools discussed in the previous section can be used, starting with this step of OOSD.
- **Designing the system.** This process defines all the objects in the system and the ways they interact (object-oriented design). Design involves developing logical and physical models of the new system by adding details to the object model started in analysis.
- **Programming or modifying modules.** This implementation step takes the object model begun during analysis and completed during design and turns it into a set of interacting objects in a system. Object-oriented programming languages are designed to allow the programmer to create classes of objects in the computer system that correspond to the objects in the actual business process. Objects such as customer, order, and payment are redefined as computer system objects—a customer screen, an order entry menu, or a dollar sign icon. Programmers then write new modules or modify existing ones to produce the desired programs.
- **Evaluation by users.** The initial implementation is evaluated by users and improved. Additional scenarios and objects are added, and the cycle repeats. Finally, a complete, tested, and approved system is available for use.
- **Periodic review and modification.** The completed and operational system is reviewed at regular intervals and modified as necessary.

object-oriented systems development (OOSD)

An approach to systems development that combines the logic of the systems development life cycle with the power of object-oriented modeling and programming.

SYSTEMS INVESTIGATION

As discussed earlier in the chapter, systems investigation is the first phase in the traditional SDLC of a new or modified business information system. The purpose is to identify potential problems and opportunities and consider them in light of the goals of the company. In general, systems investigation attempts to uncover answers to the following questions:

- What primary problems might a new or enhanced system solve?
- What opportunities might a new or enhanced system provide?
- What new hardware, software, databases, telecommunications, personnel, or procedures will improve an existing system or are required in a new system?
- What are the potential costs (variable and fixed)?
- What are the associated risks?

Initiating Systems Investigation

systems request form

A document filled out by someone who wants the IS department to initiate systems investigation.

Because systems development requests can require considerable time and effort to implement, many organizations have adopted a formal procedure for initiating systems development, beginning with systems investigation. The **systems request form** is a document that is filled out by someone who wants the IS department to initiate systems investigation. This form typically includes the following information:

- Problems in or opportunities for the system
- Objectives of systems investigation
- Overview of the proposed system
- Expected costs and benefits of the proposed system

The information in the systems request form helps to rationalize and prioritize the activities of the IS department. Based on the overall IS plan, the organization's needs and goals, and the estimated value and priority of the proposed projects, managers make decisions regarding the initiation of each systems investigation for such projects.

Participants in Systems Investigation

After a decision has been made to initiate systems investigation, the first step is to determine what members of the development team should participate in the investigation phase of the project. Members of the development team change from phase to phase (see Figure 12.12).

Figure 12.12

The Systems Investigation Team

The team consists of upper- and middle-level managers, a project manager, IS personnel, users, and stakeholders.

The Investigation Team

Managers, users, and stakeholders   IS personnel

- Undertakes feasibility analysis
- Establishes systems development goals
- Selects systems development methodology
- Prepares systems investigation report

Ideally, functional managers are heavily involved during the investigation phase. Other members could include users or stakeholders outside management, such as an employee who helped initiate systems development. The technical and financial expertise of others participating in investigation would help the team determine whether the problem is worth solving. The members of the development team who participate in investigation are then responsible for gathering and analyzing data, preparing a report justifying systems development, and presenting the results to top-level managers.

Feasibility Analysis

A key step of the systems investigation phase is **feasibility analysis**, which assesses technical, economic, legal, operational, and schedule feasibility (see Figure 12.13). **Technical feasibility** is concerned with whether the hardware, software, and other system components can be acquired or developed to solve the problem.

feasibility analysis
Assessment of the technical, economic, legal, operational, and schedule feasibility of a project.

technical feasibility
Assessment of whether the hardware, software, and other system components can be acquired or developed to solve the problem.

Figure 12.13

Technical, Economic, Legal, Operational, and Schedule Feasibility

Economic feasibility determines whether the project makes financial sense and whether predicted benefits offset the cost and time needed to obtain them. A securities company, for example, investigated the economic feasibility of sending research reports electronically instead of through the mail. Economic analysis revealed that the new approach could save the company up to $500,000 per year. Economic feasibility can involve cash flow analysis such as that done in net present value or internal rate of return (IRR) calculations.

Net present value is an often-used approach for ranking competing projects and for determining economic feasibility. The net present value represents the net amount by which project savings exceed project expenses, after allowing for the cost of capital and the passage of time. The cost of capital is the average cost of funds used to finance the operations of the business. Net present value takes into account that a dollar returned at a later date is not worth as much as one received today because the dollar in hand can be invested to earn profits or interest in the interim. Spreadsheet programs, such as Lotus and Microsoft Excel, have built-in functions to compute the net present value and internal rate of return.

Legal feasibility determines whether laws or regulations can prevent or limit a systems development project. For example, a Web site that allowed users to share music without paying musicians or music producers was sued. Legal feasibility should have identified this vulnerability during the Web site development project. Legal feasibility involves an analysis of existing and future laws to determine the likelihood of legal action against the systems development project and the possible consequences.

Operational feasibility is a measure of whether the project can be put into action or operation. It can include logistical and motivational (acceptance of change) considerations. Motivational considerations are important because new systems affect people and data flows and can have unintended consequences. As a result, power and politics might come into play, and some people might resist the new system. On the other hand, recall that a new system can help avoid major problems. For example, because of deadly hospital errors, a healthcare consortium looked into the operational feasibility of developing a new computerized physician order-entry system to require that all prescriptions and every order a doctor gives to staff be entered into the computer. The computer then checks for drug allergies and interactions between drugs. If operationally feasible, the new system could save lives and help avoid lawsuits.

economic feasibility
The determination of whether the project makes financial sense and whether predicted benefits offset the cost and time needed to obtain them.

net present value
The preferred approach for ranking competing projects and determining economic feasibility.

legal feasibility
The determination of whether laws or regulations may prevent or limit a systems development project.

operational feasibility
The measure of whether the project can be put into action or operation.

schedule feasibility
The determination of whether the project can be completed in a reasonable amount of time.

Schedule feasibility determines whether the project can be completed in a reasonable amount of time—a process that involves balancing the time and resource requirements of the project with other projects.

Object-Oriented Systems Investigation

The object-oriented approach can be used during all phases of systems development, from investigation to maintenance and review. Consider a kayak rental business in Maui, Hawaii, where the owner wants to computerize its operations, including renting kayaks to customers and adding new kayaks into the rental program (see Figure 12.14). As you can see, the kayak rental clerk rents kayaks to customers and adds new kayaks to the current inventory of kayaks available for rent. The stick figure is an example of an *actor*, and the ovals each represent an event, called a *use case*. In our example, the actor (the kayak rental clerk) interacts with two use cases (rent kayaks to customers and add new kayaks to inventory). The use case diagram is part of the Unified Modeling Language (UML) that is used in object-oriented systems development.

Figure 12.14

Use Case Diagram for a Kayak Rental Application

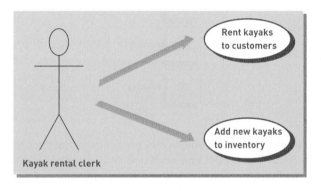

The Systems Investigation Report

systems investigation report
A summary of the results of the systems investigation and the process of feasibility analysis and recommendation of a course of action.

The primary outcome of systems investigation is a **systems investigation report**, also called a *feasibility study*. This report summarizes the results of systems investigation and the process of feasibility analysis and recommends a course of action: Continue on into systems analysis, modify the project in some manner, or drop it. A typical table of contents for the systems investigation report is shown in Figure 12.15.

Figure 12.15

A Typical Table of Contents for a Systems Investigation Report

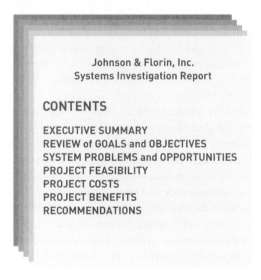

Johnson & Florin, Inc.
Systems Investigation Report

CONTENTS

EXECUTIVE SUMMARY
REVIEW of GOALS and OBJECTIVES
SYSTEM PROBLEMS and OPPORTUNITIES
PROJECT FEASIBILITY
PROJECT COSTS
PROJECT BENEFITS
RECOMMENDATIONS

The systems investigation report is reviewed by senior management, often organized as an advisory committee, or **steering committee**, consisting of senior management and users from the IS department and other functional areas. These people help IS personnel with their decisions about the use of information systems in the business and give authorization to pursue further systems development activities. After review, the steering committee might agree with the recommendation of the systems development team or suggest a change in project focus to concentrate more directly on meeting a specific company objective. Another alternative is that everyone might decide that the project is not feasible and cancel the project.

steering committee
An advisory group consisting of senior management and users from the IS department and other functional areas.

SYSTEMS ANALYSIS

After a project has been approved for further study, the next step is to answer the question "What must the information system do to solve the problem?" The process needs to go beyond mere computerization of existing systems. The entire system, and the business process with which it is associated, should be evaluated. Often, a firm can make great gains if it restructures both business activities and the related information system simultaneously. The overall emphasis of analysis is gathering data on the existing system, determining the requirements for the new system, considering alternatives within these constraints, and investigating the feasibility of the solutions. The primary outcome of systems analysis is a prioritized list of systems requirements.

General Considerations

Systems analysis starts by clarifying the overall goals of the organization and determining how the existing or proposed information system helps meet them. A manufacturing company, for example, might want to reduce the number of equipment breakdowns. This goal can be translated into one or more informational needs. One need might be to create and maintain an accurate list of each piece of equipment and a schedule for preventative maintenance. Another need might be a list of equipment failures and their causes.

Analysis of a small company's information system can be fairly straightforward. On the other hand, evaluating an existing information system for a large company can be a long, tedious process. As a result, large organizations evaluating a major information system normally follow a formalized analysis procedure, involving these steps:

1. Assembling the participants for systems analysis
2. Collecting appropriate data and requirements
3. Analyzing the data and requirements
4. Preparing a report on the existing system, new system requirements, and project priorities

Participants in Systems Analysis

The first step in formal analysis is to assemble a team to study the existing system. This group includes members of the original investigation team—from users and stakeholders to IS personnel and management. Most organizations usually allow key members of the development team not only to analyze the condition of the existing system but also to perform other aspects of systems development, such as design and implementation.

After the participants in systems analysis are assembled, this group develops a list of specific objectives and activities. A schedule for meeting the objectives and completing the specific activities is also devised, along with deadlines for each stage and a statement of the resources required at each stage, such as clerical personnel, supplies, and so forth. Major milestones are normally established to help the team monitor progress and determine whether problems or delays occur in performing systems analysis.

Data Collection

The purpose of data collection is to seek additional information about the problems or needs identified in the systems investigation report. During this process, the strengths and weaknesses of the existing system are emphasized.

Identifying Sources of Data

Data collection begins by identifying and locating the various sources of data, including both internal and external sources (see Figure 12.16).

Figure 12.16

Internal and External Sources of Data for Systems Analysis

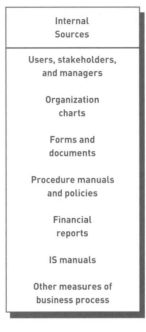

Internal Sources	External Sources
Users, stakeholders, and managers	Customers
Organization charts	Suppliers
Forms and documents	Stockholders
Procedure manuals and policies	Government agencies
Financial reports	Competitors
IS manuals	Outside groups
Other measures of business process	Journals, etc.
	Consultants

Collecting Data

After data sources have been identified, data collection begins. Figure 12.17 shows the steps involved. Data collection might require a number of tools and techniques, such as interviews, direct observation, and questionnaires.

Figure 12.17

The Steps in Data Collection

Identify data sources → Data collection → Follow-up and clarification

structured interview
An interview where the questions are written in advance.

unstructured interview
An interview where the questions are not written in advance.

Interviews can either be structured or unstructured. In a **structured interview**, the questions are written in advance. In an **unstructured interview**, the questions are not written in advance; the interviewer relies on experience in asking the best questions to uncover the inherent problems of the existing system. An advantage of the unstructured interview is that it allows the interviewer to ask follow-up or clarifying questions immediately.

With **direct observation**, one or more members of the analysis team directly observe the existing system in action. One of the best ways to understand how the existing system functions is to work with the users to discover how data flows in certain business tasks. Determining the data flow entails direct observation of users' work procedures, their reports, current screens (if automated already), and so on. From this observation, members of the analysis team determine which forms and procedures are adequate and which are inadequate and need improvement. Direct observation requires a certain amount of skill. The observer must be able to see what is really happening and not be influenced by attitudes or feelings. This approach can reveal important problems and opportunities that would be difficult to obtain using other data collection methods. An example would be observing the work procedures, reports, and computer screens associated with an accounts payable system being considered for replacement.

direct observation
Watching the existing system in action by one or more members of the analysis team.

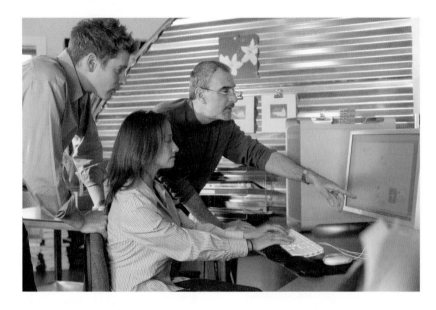

Direct observation is a method of data collection. One or more members of the analysis team directly observe the existing system in action.

(Source: © Marc Romanelli/Getty Images.)

When many data sources are spread over a wide geographic area, **questionnaires** might be the best method. Like interviews, questionnaires can be either structured or unstructured. In most cases, a pilot study is conducted to fine-tune the questionnaire. A follow-up questionnaire can also capture the opinions of those who do not respond to the original questionnaire.

questionnaires
A method of gathering data when the data sources are spread over a wide geographic area.

Other data collection techniques can also be employed. In some cases, telephone calls are an excellent method. Activities can also be simulated to see how the existing system reacts. Thus, fake sales orders, stockouts, customer complaints, and data-flow bottlenecks can be created to see how the existing system responds to these situations. **Statistical sampling**, which involves taking a random sample of data, is another technique. For example, suppose that you want to collect data that describes 10,000 sales orders received over the last few years. Because it is too time consuming to analyze each of the 10,000 sales orders, you can collect a random sample of 100 to 200 sales orders from the entire batch. You can assume that the characteristics of this sample apply to all 10,000 orders.

statistical sampling
Selecting a random sample of data and applying the characteristics of the sample to the whole group.

Data Analysis

The data collected in its raw form is usually not adequate to determine the effectiveness of the existing system or the requirements for the new system. The next step is to manipulate the collected data so that the development team members who are participating in systems analysis can use the data. This manipulation is called **data analysis**. Data and activity modeling and using data-flow diagrams and entity-relationship diagrams are useful during data analysis to show data flows and the relationships among various objects, associations, and activities. Other common tools and techniques for data analysis include application flowcharts, grid charts, CASE tools, and the object-oriented approach.

data analysis
The manipulation of collected data so that the development team members who are participating in systems analysis can use the data.

Data Modeling

Data modeling, first introduced in Chapter 5, is a commonly accepted approach to modeling organizational objects and associations that employ both text and graphics. How data modeling is employed, however, is governed by the specific systems development methodology.

Data modeling is most often accomplished through the use of entity-relationship (ER) diagrams. Recall from Chapter 5 that an entity is a generalized representation of an object type—such as a class of people (employee), events (sales), things (desks), or places (city)—and that entities possess certain attributes. Objects can be related to other objects in many ways. An entity-relationship diagram, such as the one shown in Figure 12.18a, describes a number of objects and the ways they are associated. An ER diagram (or any other modeling tool) cannot by itself fully describe a business problem or solution because it lacks descriptions of the related activities. It is, however, a good place to start because it describes object types and attributes about which data might need to be collected for processing.

Figure 12.18

Data and Activity Modeling

(a) An entity-relationship diagram.
(b) A data-flow diagram.
(c) A semantic description of the business process.

(Source: G. Lawrence Sanders, *Data Modeling*, Boyd & Fraser Publishing, Danvers, MA: 1995.)

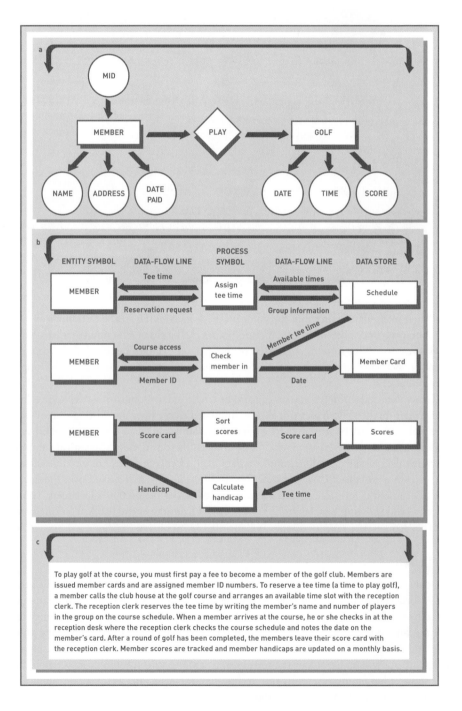

Activity Modeling

To fully describe a business problem or solution, the related objects, associations, and activities must be described. Activities in this sense are events or items that are necessary to fulfill the business relationship or that can be associated with the business relationship in a meaningful way.

Activity modeling is often accomplished through the use of data-flow diagrams. A **data-flow diagram (DFD)** models objects, associations, and activities by describing how data can flow between and around various objects. DFDs work on the premise that every activity involves some communication, transference, or flow that can be described as a data element. DFDs describe the activities that fulfill a business relationship or accomplish a business task, not how these activities are to be performed. That is, DFDs show the logical sequence of associations and activities, not the physical processes. A system modeled with a DFD could operate manually or could be computer based; if computer based, the system could operate with a variety of technologies.

DFDs are easy to develop and easily understood by nontechnical people. Data-flow diagrams use four primary symbols, as illustrated in Figure 12.18b.

- **Data flow.** The **data-flow line** includes arrows that show the direction of data element movement.
- **Process symbol.** The **process symbol** reveals a function that is performed. Computing gross pay, entering a sales order, delivering merchandise, and printing a report are examples of functions that can be represented with a process symbol.
- **Entity symbol.** The **entity symbol** shows either the source or destination of the data element. An entity can be, for example, a customer who initiates a sales order, an employee who receives a paycheck, or a manager who receives a financial report.
- **Data store.** A **data store** reveals a storage location for data. A data store is any computerized or manual data storage location, including magnetic tape, disks, a filing cabinet, or a desk.

Comparing entity-relationship diagrams with data-flow diagrams provides insight into the concept of top-down design. Figure 12.18a and b show an entity-relationship diagram and a data-flow diagram for the same business relationship—namely, a member of a golf club playing golf. Figure 12.18c provides a brief description of the business relationship for clarification.

Application Flowcharts

Application flowcharts show the relationships among applications or systems. Assume that a small business has collected data about its order processing, inventory control, invoicing, and marketing analysis applications. Management is thinking of modifying the inventory control application. The raw facts collected, however, do not help in determining how the applications are related to each other and the databases required for each. These relationships are established through data analysis with an application flowchart (see Figure 12.19). Using this tool for data analysis makes clear the relationships among the order processing functions.

In the simplified application flowchart in Figure 12.19, you can see that the telephone order clerk provides important data to the system about items such as versions, quantities, and prices. The system calculates sales tax and order totals. Any changes made to this order processing system could affect the company's other systems, such as inventory control and marketing.

Grid Charts

A **grid chart** is a table that shows relationships among various aspects of a systems development effort. For example, a grid chart can reveal the databases used by the various applications (see Figure 12.20).

data-flow diagram (DFD)
A model of objects, associations, and activities that describes how data can flow between and around various objects.

data-flow line
Arrows that show the direction of data element movement.

process symbol
Representation of a function that is performed.

entity symbol
Representation of either a source or destination of a data element.

data store
Representation of a storage location for data.

application flowcharts
Diagrams that show relationships among applications or systems.

grid chart
A table that shows relationships among the various aspects of a systems development effort.

Telephone Order Process

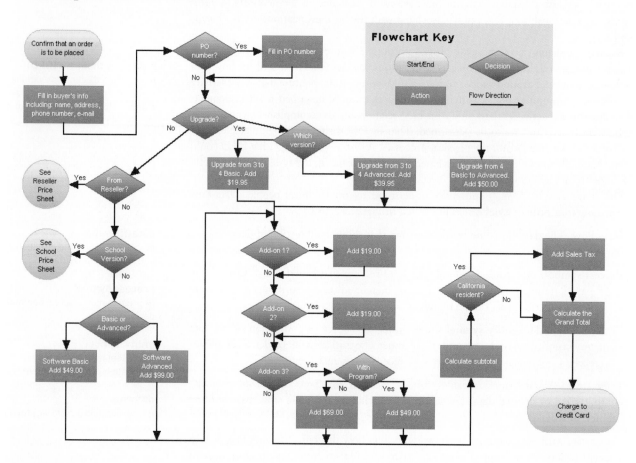

A Telephone Order Process
Application Flowchart

The flowchart shows the
relationships among various
processes.

(Source: Courtesy of
SmartDraw.com.)

A Grid Chart

The chart shows the relationships
among applications and databases.

Databases → Applications	Customer database	Inventory database	Supplier database	Accounts receivable database
Order processing application	X	X		
Inventory control application		X	X	
Marketing analysis application	X	X		
Invoicing application	X			X

The simplified grid chart in Figure 12.20 shows that the customer database is used by the order processing, marketing analysis, and invoicing applications. The inventory database is used by the order processing, inventory control, and marketing analysis applications. The supplier database is used by the inventory control application, and the accounts receivable database is used by the invoicing application. This grid chart shows which applications use common databases and reveals that, for example, any changes to the inventory control application must investigate the inventory and supplier databases.

CASE Tools

As discussed earlier, many systems development projects use CASE tools to complete analysis tasks. Most computer-aided software engineering tools have generalized graphics programs that can generate a variety of diagrams and figures. Entity-relationship diagrams, data-flow diagrams, application flowcharts, and other diagrams can be developed using CASE graphics programs to help describe the existing system. During the analysis phase, a **CASE repository**—a database of system descriptions, parameters, and objectives—will be developed.

Requirements Analysis

The overall purpose of **requirements analysis** is to determine user, stakeholder, and organizational needs. For an accounts payable application, the stakeholders could include suppliers and members of the purchasing department. Questions that should be asked during requirements analysis include the following:

- Are these stakeholders satisfied with the current accounts payable application?
- What improvements could be made to satisfy suppliers and help the purchasing department?

One of the most difficult procedures in systems analysis is confirming user or systems requirements. In some cases, communications problems can interfere with determining these requirements. For example, an accounts payable manager might want a better procedure for tracking the amount owed by customers. Specifically, the manager wants a weekly report that shows all customers who owe more than $1,000 and are more than 90 days past due on their account. A financial manager might need a report that summarizes total amount owed by customers to consider whether to loosen or tighten credit limits. A sales manager might want to review the amount owed by a key customer relative to sales to that same customer. The purpose of requirements analysis is to capture these requests in detail. Numerous tools and techniques can be used to capture systems requirements. Often, various techniques are used in the context of a JAD session.

Asking Directly

One the most basic techniques used in requirements analysis is asking directly. **Asking directly** is an approach that asks users, stakeholders, and other managers about what they want and expect from the new or modified system. This approach works best for stable systems in which stakeholders and users clearly understand the system's functions. The role of the systems analyst during the analysis phase is to critically and creatively evaluate needs and define them clearly so that the systems can best meet them.

Critical Success Factors

Another approach uses critical success factors (CSFs). As discussed earlier, managers and decision makers are asked to list only the factors that are critical to the success of their area of the organization. A CSF for a production manager might be adequate raw materials from suppliers; a CSF for a sales representative could be a list of customers currently buying a certain type of product. Starting from these CSFs, the system inputs, outputs, performance, and other specific requirements can be determined.

The IS Plan

As we have seen, the IS plan translates strategic and organizational goals into systems development initiatives. The IS planning process often generates strategic planning documents that can be used to define system requirements. Working from these documents ensures that requirements analysis will address the goals set by top-level managers and decision makers (see Figure 12.21). There are unique benefits to applying the IS plan to define systems requirements. Because the IS plan takes a long-range approach to using information technology within the organization, the requirements for a system analyzed in terms of the IS plan are more likely to be compatible with future systems development initiatives.

CASE repository
A database of system descriptions, parameters, and objectives.

requirements analysis
The determination of user, stakeholder, and organizational needs.

asking directly
An approach to gather data that asks users, stakeholders, and other managers about what they want and expect from the new or modified system.

Figure 12.21

Converting Organizational
Goals into Systems
Requirements

Screen and Report Layout

Developing formats for printed reports and screens to capture data and display information are some of the common tasks associated with developing systems. Screens and reports relating to systems output are specified first to verify that the desired solution is being delivered. Manual or computerized screen and report layout facilities are used to capture both output and input requirements.

Using a **screen layout**, a designer can quickly and efficiently design the features, layout, and format of a display screen. In general, users who interact with the screen frequently can be presented with more data and less descriptive information; infrequent users should have more descriptive information presented to explain the data that they are viewing (see Figure 12.22).

screen layout

A technique that allows a designer to quickly and efficiently design the features, layout, and format of a display screen.

Figure 12.22

Screen Layouts

(a) A screen layout chart for frequent users who require little descriptive information.

(b) A screen layout chart for infrequent users who require more descriptive information.

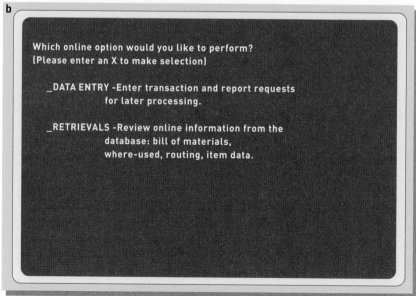

Report layout allows designers to diagram and format printed reports. Reports can contain data, graphs, or both. Graphic presentations allow managers and executives to quickly view trends and take appropriate action, if necessary.

report layout
A technique that allows designers to diagram and format printed reports.

Screen layout diagrams can document the screens users desire for the new or modified application. Report layout charts reveal the format and content of various reports that the application will prepare. Other diagrams and charts can be developed to reveal the relationship between the application and outputs from the application.

Requirements Analysis Tools

A number of tools can be used to document requirements analysis, including CASE tools. As requirements are developed and agreed on, entity-relationship diagrams, data-flow diagrams, screen and report layout forms, and other types of documentation are stored in the CASE repository. These requirements might also be used later as a reference during the rest of systems development or for a different systems development project.

Object-Oriented Systems Analysis

The object-oriented approach can also be used during systems analysis. Like traditional analysis, problems or potential opportunities are identified during object-oriented analysis. Identifying key participants and collecting data is still performed. But instead of analyzing the existing system using data-flow diagrams and flowcharts, an object-oriented approach is used.

In the section "Object-Oriented Systems Investigation," we introduced a kayak rental example. A more detailed analysis of that business reveals that there are two classes of kayaks: single kayaks for one person and tandem kayaks that can accommodate two people. With the OO approach, a class is used to describe different types of objects, such as single and tandem kayaks. The classes of kayaks can be shown in a generalization/specialization hierarchy diagram (see Figure 12.23). KayakItem is an object that will store the kayak identification number (ID) and the date the kayak was purchased (datePurchased).

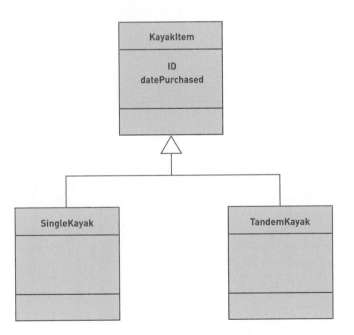

Figure 12.23

Generalization/Specialization Hierarchy Diagram for Single and Tandem Kayak Classes

Of course, there could be subclasses of customers, life vests, paddles, and other items in the system. For example, price discounts for kayak rentals could be given to seniors (people over 65 years) and students. Thus, the Customer class could be divided into regular, senior, and student customer subclasses.

The Systems Analysis Report

Systems analysis concludes with a formal systems analysis report. It should cover the following elements:

- The strengths and weaknesses of the existing system from a stakeholder's perspective
- The user/stakeholder requirements for the new system (also called the *functional requirements*)
- The organizational requirements for the new system
- A description of what the new information system should do to solve the problem

Suppose analysis reveals that a marketing manager thinks a weakness of the existing system is its inability to provide accurate reports on product availability. These requirements and a preliminary list of the corporate objectives for the new system will be in the systems analysis report. Particular attention is placed on areas of the existing system that could be improved to meet user requirements. The table of contents for a typical report is shown in Figure 12.24.

Figure 12.24

A Typical Table of Contents for a Report on an Existing System

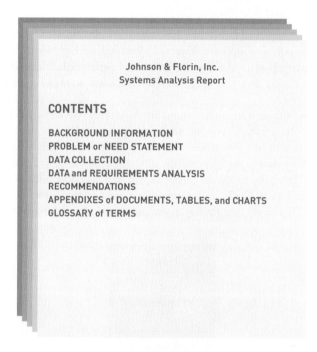

Johnson & Florin, Inc.
Systems Analysis Report

CONTENTS

BACKGROUND INFORMATION
PROBLEM or NEED STATEMENT
DATA COLLECTION
DATA and REQUIREMENTS ANALYSIS
RECOMMENDATIONS
APPENDIXES of DOCUMENTS, TABLES, and CHARTS
GLOSSARY of TERMS

The systems analysis report gives managers a good understanding of the problems and strengths of the existing system. If the existing system is operating better than expected or the necessary changes are too expensive relative to the benefits of a new or modified system, the systems development process can be stopped at this stage. If the report shows that changes to another part of the system might be the best solution, the development process might start over, beginning again with systems investigation. Or, if the systems analysis report shows that it will be beneficial to develop one or more new systems or to make changes to existing ones, systems design, which is discussed in Chapter 13, begins.

SUMMARY

Principle

Effective systems development requires a team effort from stakeholders, users, managers, systems development specialists, and various support personnel, and it starts with careful planning.

The systems development team consists of stakeholders, users, managers, systems development specialists, and various support personnel. The development team determines the objectives of the information system and delivers to the organization a system that meets its objectives.

Stakeholders are people who, either themselves or through the area of the organization they represent, ultimately benefit from the systems development project. Users are people who will interact with the system regularly. They can be employees, managers, customers, or suppliers. Managers on development teams are typically representative of stakeholders or can be stakeholders themselves. In addition, managers are most capable of initiating and maintaining change. For large-scale systems development projects, where the investment in and value of a system can be quite high, it is common to have senior-level managers be part of the development team.

A systems analyst is a professional who specializes in analyzing and designing business systems. The programmer is responsible for modifying or developing programs to satisfy user requirements. Other support personnel on the development team include technical specialists, either IS department employees or outside consultants. Depending on the magnitude of the systems development project and the number of IS development specialists on the team, the team might also include one or more IS managers. At some point in your career, you will likely be a participant in systems development. You could be involved in a systems development team—as a user, as a manager of a business area or project team, as a member of the IS department, or maybe even as a CIO.

Systems development projects are initiated for many reasons, including the need to solve problems with an existing system, to exploit opportunities to gain competitive advantage, to increase competition, to make use of effective information, to create organizational growth, to settle a merger or corporate acquisition, and to address a change in the market or external environment. External pressures, such as potential lawsuits or terrorist attacks, can also prompt an organization to initiate systems development.

Information systems planning refers to the translation of strategic and organizational goals into systems development initiatives. Benefits of IS planning include a long-range view of information technology use and better use of IS resources. Planning requires developing overall IS objectives; identifying IS projects; setting priorities and selecting projects; analyzing resource requirements; setting schedules, milestones, and deadlines; and developing the IS planning document. IS planning can result in a competitive advantage through creative and critical analysis.

Establishing objectives for systems development is a key aspect of any successful development project. Critical success factors (CSFs) can identify important objectives. Systems development objectives can include performance goals (quality and usefulness of the output and the speed at which output is generated) and cost objectives (development costs, fixed costs, and ongoing investment costs).

Principle

Systems development often uses tools to select, implement, and monitor projects, including net present value (NPV), prototyping, rapid application development, CASE tools, and object-oriented development.

The five phases of the traditional SDLC are investigation, analysis, design, implementation, and maintenance and review. Systems investigation identifies potential problems and opportunities and considers them in light of organizational goals. Systems analysis seeks a general understanding of the solution required to solve the problem; the existing system is studied in detail and weaknesses are identified. Systems design creates new or modifies existing system requirements. Systems implementation encompasses programming, testing, training, conversion, and operation of the system. Systems maintenance and review entails monitoring the system and performing enhancements or repairs.

Advantages of the traditional SDLC include the following: It provides for maximum management control, creates considerable system documentation, ensures that system requirements can be traced back to stated business needs, and produces many intermediate products for review. Its disadvantages include the following: Users may get a system that meets the needs as understood by the developers, the documentation is expensive and difficult to maintain, users' needs go unstated or might not be met, and users cannot easily review the many intermediate products produced.

Prototyping is an iterative approach that involves defining the problem, building the initial version, having users work with and evaluate the initial version, providing feedback, and incorporating suggestions into the second version. Prototypes can be fully operational or nonoperational, depending on how critical the system under development is and how much time and money the organization has to spend on prototyping.

Rapid application development (RAD) uses tools and techniques designed to speed application development. Its use reduces paper-based documentation, automates program source code generation, and facilitates user participation in development activities. RAD can use newer programming techniques, such as agile development or extreme programming. RAD makes extensive use of the joint application development (JAD) process to gather data and perform requirements analysis. JAD involves group meetings in which users, stakeholders, and IS professionals work together to analyze existing systems, propose possible solutions, and define the requirements for a new or modified system.

The term *end-user systems development* describes any systems development project in which the primary effort is undertaken by a combination of business managers and users.

Many companies hire an outside consulting firm that specializes in systems development to take over some or all of its systems development activities. This approach is called *outsourcing*. Reasons for outsourcing include companies' belief that they can cut costs, achieve a competitive advantage without having the necessary IS personnel in-house, obtain state-of-the-art technology, increase their technological flexibility, and proceed with development despite downsizing. Many companies offer outsourcing services, including computer vendors and specialized consulting companies.

A number of factors affect systems development success. The degree of change introduced by the project, continuous improvement and reengineering, the use of quality programs and standards, organizational experience with systems development, the use of project management tools, and the use of CASE tools and the object-oriented approach are all factors that affect the success of a project. The greater the amount of change a system will endure, the greater the degree of risk and often the amount of reward. Continuous improvement projects do not require significant business process or IS changes, while reengineering involves fundamental changes in how the organization conducts business and completes tasks. Successful systems development projects often involve such factors as support from top management, strong user involvement, use of a proven methodology, clear project goals and objectives, concentration on key problems and straightforward designs, staying on schedule and within budget, good user training, and solid review and maintenance programs. Quality standards, such as ISO 9001, can also be used during the systems development process.

The use of automated project management tools enables detailed development, tracking, and control of the project schedule. Effective use of a quality assurance process enables the project manager to deliver a high-quality system and to make intelligent trade-offs among cost, schedule, and quality. CASE tools automate many of the systems development tasks, thus reducing an analyst's time and effort while ensuring good documentation. Object-oriented systems development can also be an important success factor. With the object-oriented systems development (OOSD) approach, a project can be broken down into a group of objects that interact. Instead of requiring thousands or millions of lines of detailed computer instructions or code, the systems

development project might require a few dozen or maybe a hundred objects.

Principle

Systems development starts with investigation and analysis of existing systems.

In most organizations, a systems request form initiates the investigation process. Participants in systems investigation can include stakeholders, users, managers, employees, analysts, and programmers. The systems investigation is designed to assess the feasibility of implementing solutions for business problems, including technical, economic, legal, operations, and schedule feasibility. Net present value analysis is often used to help determine a project's economic feasibility. An investigation team follows up on the request and performs a feasibility analysis that addresses technical, economic, legal, operational, and schedule feasibility.

If the project under investigation is feasible, major goals are set for the system's development, including performance, cost, managerial goals, and procedural goals. Many companies choose a popular methodology so that new IS employees, outside specialists, and vendors will be familiar with the systems development tasks set forth in the approach. A systems development methodology must be selected. Object-oriented systems investigation is being used to a greater extent today. The use case diagram is part of the Unified Modeling Language that is used to document object-oriented systems development. As a final step in the investigation process, a systems investigation report should be prepared to document relevant findings.

Systems analysis is the examination of existing systems, which begins after a team receives approval for further study from management. Additional study of a selected system allows those involved to further understand the system's weaknesses and potential areas for improvement. An analysis team is assembled to collect and analyze data on the existing system.

Data collection methods include observation, interviews, questionnaires, and statistical sampling. Data analysis manipulates the collected data to provide information. The analysis includes grid charts, application flowcharts, and CASE tools. The overall purpose of requirements analysis is to determine user and organizational needs.

Data analysis and modeling is used to model organizational objects and associations using text and graphical diagrams. It is most often accomplished through the use of entity-relationship (ER) diagrams. Activity modeling is often accomplished through the use of data-flow diagrams (DFDs), which model objects, associations, and activities by describing how data can flow between and around various objects. DFDs use symbols for data flows, processing, entities, and data stores. Application flowcharts, grid charts, and CASE tools are also used during systems analysis.

Requirements analysis determines the needs of users, stakeholders, and the organization in general. Asking directly, using critical success factors, and determining requirements from the IS plan can be used. Often, screen and report layout charts are used to document requirements during systems analysis.

Like traditional analysis, problems or potential opportunities are identified during object-oriented analysis.

Object-oriented systems analysis can involve using diagramming techniques, such as a generalization/specialization hierarchy diagram.

CHAPTER 12: SELF-ASSESSMENT TEST

Effective systems development requires a team effort from stakeholders, users, managers, systems development specialists, and various support personnel, and it starts with careful planning.

1. _____ is the activity of creating or modifying existing business systems. It refers to all aspects of the process—from identifying problems to be solved or opportunities to be exploited to the implementation and refinement of the chosen solution.

2. Which of the following people ultimately benefit from a systems development project?
   a. computer programmers
   b. systems analysts
   c. stakeholders
   d. senior-level manager

3. _____ requires unbiased and careful questioning of whether systems elements are related in the most effective or efficient ways.

4. Like a contractor constructing a new building or renovating an existing one, the programmer takes the plans from the systems analyst and builds or modifies the necessary software. True or False?

5. The term _____ refers to the translation of strategic and organizational goals into systems development initiatives.

6. What factors are essential to the success of certain functional areas of an organization?
   a. critical success factors
   b. systems analysis factors
   c. creative goal factors
   d. systems development factors

Systems development often uses tools to select, implement, and monitor projects, including net present value (NPV), prototyping, rapid application development, CASE tools, and object-oriented development.

7. What employs tools, techniques, and methodologies designed to speed application development?
   a. rapid application development
   b. joint optimization
   c. prototyping
   d. extended application development

8. System performance is usually determined by factors such as fixed investments in hardware and related equipment. True or False?

9. _____ takes an iterative approach to the systems development process. During each iteration, requirements and alternative solutions to the problem are identified and analyzed, new solutions are designed, and a portion of the system is implemented.

10. Joint application development involves group meetings in which users, stakeholders, and IS professionals work together to analyze existing systems, propose possible solutions, and define the requirements for a new or modified system. True or False?

11. What consists of all activities that, if delayed, would delay the entire project?
    a. deadline activities
    b. slack activities
    c. RAD tasks
    d. the critical path

Systems development starts with investigation and analysis of existing systems.

12. The systems request form is a document that is filled out during systems analysis. True or False?

13. Feasibility analysis is typically done during which systems development stage?
    a. investigation
    b. analysis
    c. design
    d. implementation

14. Data modeling is most often accomplished through the use of _____, whereas activity modeling is often accomplished through the use of _____.

15. The overall purpose of requirements analysis is to determine user, stakeholder, and organizational needs. True or False?

REVIEW QUESTIONS

1. What is an IS stakeholder?
2. What is the goal of IS planning? What steps are involved in IS planning?
3. What are the typical reasons to initiate systems development?
4. What actions can be taken during creative analysis?
5. What is the difference between a programmer and a systems analyst?
6. How is a Gantt chart developed? How is it used?
7. What is the difference between systems investigation and systems analysis? Why is it important to identify and remove errors early in the systems development life cycle?

8. Identify four reasons that a systems development project might be initiated.
9. List factors that have a strong influence on project success.
10. What is the purpose of systems investigation?
11. What are the steps of object-oriented systems development?
12. Define the different types of feasibility that systems development must consider.
13. What is the difference between systems investigation and systems analysis?
14. How does the JAD technique support the RAD systems development life cycle?

DISCUSSION QUESTIONS

1. Why is it important for business managers to have a basic understanding of the systems development process?
2. Briefly describe the role of a system user in the systems investigation and systems analysis stages of a project.
3. How could you use creative analysis to help you develop a better information system for your college or university?
4. Briefly describe when you would use the object-oriented approach to systems development instead of the traditional systems development life cycle.
5. Your company wants to develop or acquire a new sales program to help sales representatives identify new customers. Describe what factors you would consider in deciding whether to develop the application in-house or outsource the application to an outside company.
6. You have been hired as a project manager to develop a new Web site in the next 6 months for a store that sells music and books online. Describe how project management tools, such as a Gantt chart and PERT might be used.
7. For what types of systems development projects might prototyping be especially useful? What are the characteristics of a system developed with a prototyping technique?

8. Assume that you work for an insurance company. Describe three applications that are critical to your business. What tools would you use to develop applications?
9. How important are communications skills to IS personnel? Consider this statement: "IS personnel need a combination of skills—one-third technical skills, one-third business skills, and one-third communications skills." Do you think this is true? How would this affect the training of IS personnel?
10. Discuss three reasons that aligning overall business goals with IS goals is important.
11. Imagine that you are a highly paid consultant who has been retained to evaluate an organization's systems development processes. With whom would you meet? How would you make your assessment?
12. You are a senior manager of a functional area in which a mission-critical system is being developed. How can you safeguard this project from mushrooming out of control?

PROBLEM-SOLVING EXERCISES

1. You are developing a new information system for The Fitness Center, a company that has five fitness centers in your metropolitan area, with about 650 members and 30 employees in each location. This system will be used by both members and fitness consultants to track participation in various fitness activities, such as free weights, volleyball, swimming, stair climbers, and aerobic and yoga classes. One of the performance objectives of the system is that it must help members plan a fitness program to meet their particular needs. The primary purpose of this system, as envisioned by the director of marketing, is to assist The Fitness Center in obtaining a competitive advantage over other fitness clubs. Use word processing software to prepare a brief memo to the required participants on the development team for this systems development project. Be sure to specify what roles these individuals will play and what types of information you hope to obtain from them. Assume that the relational database model will be the basis for building this system. Use a database management system to define the various tables that will make up the database.

2. You have been hired to develop a payroll program for a medium-sized company. At a minimum, the application should have an hours-worked table that contains how many hours each employee worked and an employee table that contains information about each employee, including hourly pay rate. Design and develop these tables that could be used in a database for the payroll program.

TEAM ACTIVITIES

1. Your team should interview people involved in systems development in a local business or at your college or university. Describe the process used. Identify the users, analysts, and stakeholders for a systems development project that has been completed or is currently under development.

2. Your team has been hired to determine the requirements and layout of the Web pages for a company that sells fishing equipment over the Internet. Using the approaches discussed in this chapter, develop a rough sketch of at least five Web pages that you would recommend. Make sure to show the important features and the hyperlinks for each page.

3. Your team has been hired to analyze the potential of developing a database of job openings and descriptions for the companies visiting your campus this year. Describe the tasks your team would perform to complete systems analysis.

WEB EXERCISES

1. A number of tools can be used to develop a new Web-based application. Describe the Web development tools you would use and the steps you would complete to implement a Web site to rent movies and games over the Internet. You might be asked to develop a report or send an e-mail message to your instructor about what you found.

2. Using the Internet, locate an organization that is currently involved in a systems development project. Describe how they are using project management tools. Is project management software being used?

CAREER EXERCISES

1. Pick a career that you are considering. What type of information system would help you on the job? Perform technical, economic, legal, operational, and schedule feasibility for an information system you would like developed.

2. Describe three career areas and the types of systems development projects that could help you be more effective for each career.

CASE STUDIES

Case One

Applebee's Simplifies Life for Managers while Improving Service to Customers

Like many of the large restaurant chains, Applebee's maintains specific performance criteria for its 1,850 restaurants. These include metrics for the speed of getting customers in and out of a restaurant, quality of food, and levels of customer satisfaction. Five hundred of the 1,850 restaurants are company owned, and the rest are franchises. Communication between all Applebee's restaurants is critical for maintaining consistent standards across all restaurants.

To provide the most current corporate information to each Applebee's restaurant, the company provides portal software from BEA Systems' Inc. The portal software provides a browser interface to different enterprise applications. Restaurant managers have access to daily performance records through dashboards in the portal.

Upper-level management at Applebee's in Overland Park, Kansas, recently became concerned over some restaurant manager's complaints about difficulties in keeping up with constantly evolving menus and corporate policies. Managers felt that they were required to spend too much time monitoring performance using the corporate portal software. Applebee's launched a systems investigation to learn more about the complaints and to determine the feasibility of providing a fix.

Systems analysts found that the complexity of using the dashboard systems was too much for most managers. To access performance statistics and metrics, managers were faced with "a sea of information these managers are supposed to weed through every day," explained Patty Cutter, Applebee's IT project manager. What's more, when metrics for performance standards indicated that a restaurant was substandard, the manager would have to phone Applebee's headquarters to learn strategies for improvement—a process that was time consuming for both manager and supervisor.

With a full understanding of the problem, Patty Cutter and fellow systems analysts set out to perform the analysis to define how the problem could be addressed through the development of a new system. They determined that an ideal system should include the following features:

- Provide managers with the latest menu specifications along with instructions on how to execute any changes.
- Provide managers with alerts whenever the restaurant's performance in a specific area falls below the corporate threshold, eliminating the need for managers to dig through a "sea of information."
- Provide suggested action plans with each alert that provides steps the manager should take to rise to corporate standards, eliminating the need to phone headquarters.
- Provide flexibility within the system so that software in all restaurants can be changed and updated from headquarters to reflect changes within the organization.

With a clear view of what the new system should do, Patty Cutter consulted with BEA Systems' to see if the company that provided the original portal software could provide a solution that would meet the new organizational needs and general corporate goals. Developers at BEA Systems recommended their AquaLogic HiPer Workspace for Retail software as an ideal solution. The AquaLogic software uses a Service Oriented Architecture (SOA) that would provide the flexibility that Applebee's wanted. The software is a composite application that can use Web services to quickly notify store managers when a restaurant isn't meeting certain performance criteria, and provide possible solutions to the problem. An added advantage is that the AquaLogic software is designed to work with the portal software already installed in all Applebee's.

Applebee's is rolling out a production version of the new application for its portal to its 500 company-owned restaurants. After being fully tested in those restaurants, the software will be recommended to each of the company's 1,350 franchise restaurants. The company views the software as a big time-saver for its managers. "It is extremely important to us that we minimize the amount of time our managers are spending behind a computer," said Frank Ybarra, associate director of communications at Applebee's. What makes this an ideal solution is that it saves manager's time, while actually improving their ability to maintain Applebee's high standards.

Discussion Questions

1. What situation arose that led to the launch of Applebee's systems development project?
2. What benefits does Applebee's new system provide for restaurant managers and Applebee's headquarters?

Critical Thinking Questions

1. What advantage does a SOA provide for Applebee's? Why do you think SOA and Web services have become so popular in new systems?
2. What benefit(s) did Applebee's gain by choosing BEA Systems to design the new system?

SOURCES: Havenstein, Heather, "Applebee's Taps BEA Tool to Boost Operations," *Computerworld*, June 19, 2006, *www.computerworld.com*. Applebee's International Web site, accessed July 10, 2006, *www.applebees.com*. BEA Systems Web site, accessed July 10, 2006, *www.bea.com*.

Case Two

Waeco Pacific Employees Work Together to Create New ERP System

Waeco Pacific Pty. Ltd. is the Australian branch of Germany-based Waeco International, a worldwide leading supplier of active coolers for all mobile applications—boats and RVs, cars and trucks, buses, and other commercial vehicles. Waeco Pacific's legacy ERP system was frustrating employees throughout the organization from the assemblers in manufacturing plants, to shippers at warehouses, to executives at headquarters. Its antiquated systems were driving Waeco Pacific managers to create separate systems in their departments, limiting interoperability and efficiency.

Problems in the system's warehousing module meant critical operations like stocktaking, stock control, and communication had to be done manually. "We were unable to create bin locations or look at a stocktaking module in the software," Waeco Pacific national logistics manager James Stuart said. "It was all done manually to compensate for having a poor system in place. When staff entered an order they would take the printed order to the old system, type in the docket, and twice a day, an employee would take invoice dockets to the warehouse where staff pick, pack and dispatch the stock."

When Waeco Pacific finally committed to a system overhaul, its first step was to consult the users of the system to find out what they needed. By involving staff from all departments, requirements, procedures, and expectations for the new ERP system were made clear. Stakeholders were briefed on the plans for the new system. Users provided screenshots and instructions to developers to illustrate how they used the current system and how it could be improved. Through their involvement in its design, each user became a subject expert on how the new system would be utilized in her area.

After consultation was complete, the project manager decided the company needed a well-supported, wireless, and unified platform where it could build a system to support growth and improvements. It was hoped that a solution could be developed within three months. With this type of time frame, Waeco Pacific would have to opt for commercially available software; to develop a custom solution would take too much time.

After a bit of research, the company decided to adopt Microsoft's Dynamics NAV software. The primary benefits being its global reach and its ease of integration with Microsoft Word and Excel—software already well established within the organization. Microsoft Dynamics NAV helps growing, midsized companies integrate financial, manufacturing, distribution, customer relationship management, and e-commerce data. Integrating Microsoft Dynamics NAV into Waeco Pacific's existing Microsoft environment would be easier than other solutions because it runs on a Microsoft SQL Server back end.

The rollout of the new system began in late 2005 in the Australian cities of Melbourne, Brisbane, and Perth as well as in New Zealand at a cost of $186,000. Waeco Pacific partnered with Tectura to implement a simple, wireless infrastructure, upgrading its administration complex and warehouse systems.

The chosen solution integrated systems from all departments, which promoted efficiency, flexibility, and communication. Waeco Pacific managing director Andreas Bischof said, "We were starting to run individual solutions for each area of the business but now we have everything under one umbrella. This lets us integrate our business areas such as CRM, warehousing and stock control to make us more productive and efficient in our use of resources."

Today, when an order is placed, the docket automatically prints in the correct warehouse location at the dispatch supervisor's desk and a consignment note is automatically generated. "This has drastically improved performance with less room for human error making the entire warehousing system more reliable and efficient," concludes Waeco Pacific IT manager Mark Maki-Neste.

Discussion Questions

1. What drove Waeco Pacific to implement this systems development initiative?
2. Why did Waeco Pacific decide on a Microsoft solution?

Critical Thinking Questions

1. Waeco Pacific dedicated a considerable amount of time and effort in consulting with employees. What are the costs and benefits of this approach?
2. List what you think are the top benefits that Waeco Pacific will derive from its new system.

SOURCES: Pauli, Darren, "Ausie Manufacturer Puts Outdated ERP System on Ice," July 05, 2006, *Computerworld Australia, www.computerworld.com.au*. Waeco Pacific Web site, accessed July 7, 2006, *www.waeco.com.au*. Microsoft Dynamics NAV Web site, accessed July 7, 2006, *www.microsoft.com/dynamics/nav/default.mspx*.

Questions for Web Case

See the Web site for this book to read about the Whitmann Price Consulting case for this chapter. Following are questions concerning this Web case.

Whitmann Price Consulting: Systems Investigation and Analysis Considerations

Discussion Questions

1. How will the proposed AMCI system help to meet corporate goals and provide Whitmann Price with a competitive advantage?

2. Who are the stakeholders in this systems development project? Who are the primary systems analysts?

Critical Thinking Questions

1. Why did the systems investigation for the proposed AMCI system proceed so quickly and smoothly? What type of proposal might require a more time-consuming, formal investigation?

2. What reasons do you think Josh and Sandra might have for interviewing division managers and not each individual consultant in their system review? What are the pros and cons of both approaches?

NOTES

Sources for the opening vignette: "NORDMILCH Refreshes Five-Year IT Strategy with flexible SAP and IBM Solutions," IBM Case Study, February 15, 2006, www.ibm.com. Nordmilch Web site, accessed July 7, 2006, www.nordmilch.de/nm/web/en/home.html. SAP Web site, accessed July 7, 2006, www.sap.com.

1   McCartney, Scott, "Denver Airport Baggage System Is Canceled by United Airlines," The Wall Street Journal, June 7, 2005, p. D5.

2   Brandel, Mary, "Five Biggest Project Challenges for 2006," Computerworld, January 2, 2006, p. 16.

3   Kolbasuk, Marianne, "Skills That Will Matter," Information Week, January 2, 2006, p. 53.

4   Hoffman, Thomas, "DOE Works with TECSys on Infrastructure Security," Computerworld, January 10, 2005, p. 15.

5   O'Donnell, Anthony, "Nine Months to Merge," Information Week, February 20, 2006, p. 51.

6   Hoffman, Thomas, "Harrah's Bets on Loyalty Program in Caesars Deal," Computerworld, June 27, 2005, p. 10.

7   Bellman, Eric, "One More Cost of Sarbanes-Oxley: Outsourcing to India," The Wall Street Journal, July 14, 2005, p. C1.

8   Staff, "Merchants Welcome Retail Theft Law," National Jeweler, January 11, 2006.

9   Weier, Mary Hayes, "GM's Global Gamble," Information Week, February 6, 2006, p. 22.

10   Lewis, Doug, "The IT Strategic Plan," Computerworld, February 21, 2005, p. 31.

11   Hess, H.M., "Aligning Technology and Business," IBM Systems Journal, Vol. 44, No. 1, 2005, p. 25.

12   Staff, "BRE Bank Subscribes to Risk Management Database," The Asian Banker, January 15, 2006.

13   Rosencrance, Linda, "DOE Overspends on Software, Inspector General Says," Computerworld, February 13, 2006, p. 20.

14   Havenstein, Heather, "Medical Groups Offered Rewards to IT Use," Computerworld, February 6, 2006, p. 14.

15   Havenstein, Heather, "Fast-Moving Development," Computerworld, January 10, 2005, p. 26.

16   Computer Associates Home Page, www.ca.com, accessed September 9, 2005.

17   The Rational Unified Process, www-306.ibm.com/software/rational, accessed September 9, 2005.

18   Engardio, Pete, "The Future of Outsourcing," Business Week, January 30, 2006, p. 50.

19   Kripalani, Manjeet, "Offshoring: Spreading the Gospel," Business Week, March 6, 2006, p. 46.

20   Crawford, C.H. et al., "Toward an On Demand Service-Oriented Architecture," IBM Systems Journal, Vol. 44, No. 1, 2005, p. 81.

21   Shellenbarger, Sue, "Outsourcing Jobs to the Den," The Wall Street Journal, January 12, 2006, p. D1.

22   Staff, "Eurostar Briefs Occam to Boost Traveler Insight," Precision Marketing, January 6, 2006, p. 6.

23   Arsanjani, A., "Empowering the Business Analyst for On Demand Computing," IBM Systems Journal, Vol. 44, No. 1, 2005, p. 67.

24   Fillion, Roger, "All-American Outsourcing Option," Rocky Mountain News, January 14, 2005, p. 1B.

25   King, Julia, "Home Grown," Computerworld, March 28, 2005, p. 45.

26   Gomes, Lee, "Some Students Use Net to Hire Experts to Do Their School Work," The Wall Street Journal, January 18, 2006, p. B1.

27   IBM Web site, www.ibm.com, accessed September 10, 2005.

28   EDS Web site, www.eds.com, accessed September 10, 2005.

29   Accenture Web site, www.accenture.com, accessed September 10, 2005.

30   Vijayan, Jaikumar, "Outsourcing Savvy," Computerworld, January 3, 2005, p. 16.

31   Hoffman, Thomas, "Prenuptials for Outsourcing," Computerworld, January 23, 2006, p. 34.

32   McDougall, Paul, "Sprint Rethinks Outsourcing, Pulls Some Work Back," Information Week, January 30, 2006, p. 34.

33   Songini, Marc, "U.K. Tax Agency Mulls Lawsuit Against EDS," Computerworld, July 4, 2005, p. 9.

34   Thibodeau, Patrick, "Mechanics of a Merger," Computerworld, February 14, 2005, p. 35.

35   Rohde, Laura, "Report Details Flaws in UK Case Management IT System," Computerworld, April 18, 2005, p. 21.

36   ISO – International Standards Organization Web site, www.iso.org, accessed August 1, 2006.

37   The Java 2 Platform, Enterprise Edition, http://java.sun.com/j2ee, accessed September 9, 2005.

38   Capability Maturity Model for Software home page, www.sei.cmu.edu, accessed September 10, 2005.

39   Kay, Russell, "CMMI," Computerworld, January 24, 2005, p. 28.

40 Staff, "Capability Maturity Model for Software," *www.sei.cmu.edu/cmm*, accessed September 11, 2005.

41 Havenstein, Heather, "Florida Begins Linking Its Law Enforcement Agencies," *Computerworld*, February 13, 2006, p. 18.

42 Glen, Paul, "Detecting Disaster Projects," *Computerworld*, February 6, 2006, p. 39.

43 Anthes, Gary, "Software Reuse: Making It Work," *Computerworld*, May 30, 2005, p. 37.

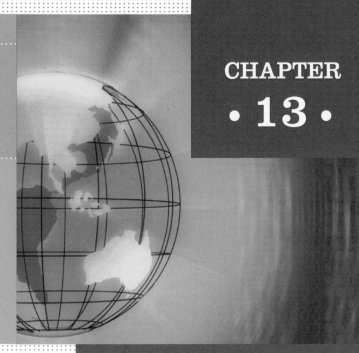

CHAPTER
• 13 •

Systems Development: Design, Implementation, Maintenance, and Review

- Designing new systems or modifying existing ones should always help an organization achieve its goals.

  - State the purpose of systems design and discuss the differences between logical and physical systems design.
  - Describe some considerations in design modeling and the diagrams used during object-oriented design.
  - Outline key considerations in interface design and control and system security and control.
  - Define the term *RFP* and discuss how this document is used to drive the acquisition of hardware and software.
  - Describe the techniques used to make systems selection evaluations.

- The primary emphasis of systems implementation is to make sure that the right information is delivered to the right person in the right format at the right time.

  - State the purpose of systems implementation and discuss the activities associated with this phase of systems development.
  - List the advantages and disadvantages of purchasing versus developing software.
  - Discuss the software development process and some of the tools used in this process, including object-oriented program development tools.

- Maintenance and review add to the useful life of a system but can consume large amounts of resources. These activities can benefit from the same rigorous methods and project management techniques applied to systems development.

  - State the importance of systems and software maintenance and discuss the activities involved.
  - Describe the systems review process.

Information Systems in the Global Economy ⟩⟩
Timex, United States

New Systems Allow Timex to Innovate Faster and with Less Risk

For nearly all of the 152 years that Timex has been manufacturing watches, the watch industry has been more about style than engineering. That trend has changed recently as technology is playing a more active role in all stages of product development, as well as being embedded in products themselves. In the past ten years, innovation has become the industry buzzword as consumers demand new features in their watches. For example, consider watches that include wireless heart rate monitors or can use Earth satellites to compute the wearer's speed and distance.

Timex has 7,500 employees in the United States, South America, Europe, and Asia and sells 30 million watches annually. For years, Timex relied on an active project system to manage its products through their stages of development. The system was limited because it used individual files scattered in various locations that made it difficult for project participants to access information. The inefficient active project system created a "disconnect between R&D and brand managers," as Bernd Becker, vice president of product development, puts it. Becker explains that this is "a typical problem in many companies because engineers and brand managers talk in different languages."

After executing a systems investigation and analysis, Timex systems analysts had a solid understanding of what they needed in a new system. The new system should improve communication and information access for all involved in product development, and help them innovate, speeding new products to market that were sure to sell. The company issued a request for proposals to three software vendors who specialize in product lifecycle management: Sopheon, Centric, and Oracle. Each vendor was invited to present solutions for review. Timex management and systems specialists were most impressed by Sopheon's Accolade software. Accolade employs a stage-gate approach, which uses milestones, or gates, at each step of innovation. It also allows everyone from designers and engineers to product marketing staffers to analyze discussions, reviews, planning, and decisions.

Sopheon software engineers worked with systems specialists at Timex to custom design the software to meet Timex's needs. After creating a detailed design of the system, Sopheon delivered the software to Timex. After a two-month implementation and testing phase, Timex began using the system in early 2006 in its daily operations to support actual projects. The package cost about $250,000 to deploy, including the first year of maintenance and user licenses. The new system is under scrutiny as it is being used to manage 45 projects, including new watch designs and feature innovations. It will take between six and ten months for the first Timex watch project to be completed and sent to market using Accolade.

So far, Timex is impressed with the results of the new system. By using Accolade, Timex expects to shave up to 30 percent off the time needed to develop a watch and get it to market. Furthermore, because Accolade runs on Microsoft Windows and integrates with Microsoft Office applications, with a familiar Windows-style and Web-based interface, it has been easy for employees to learn. Investment in employee training has been minimal.

For the first time, Accolade allows Timex to bring all of its product development data into one place, making it easier and more efficient for Timex employees to offer suggestions in real time. The new system provides decision makers with key information regarding products, market demand, production requirements, and other variables that assist with risk management. New products are introduced to market faster and with less risk, providing Timex with a competitive advantage.

As you read this chapter, consider the following:

- After a company develops specifications for a new system, what steps should it take to implement the system successfully?
- What important factors should be considered when implementing a new system?

Why Learn About Systems Development?

Information systems are designed and implemented for employees and managers every day. A manager at a hotel chain can use an information system to look up client preferences. An accountant at a manufacturing company can use an information system to analyze the costs of a new plant. A sales representative for a music store can use an information system to determine which CDs to order and which to discount because they are not selling. A computer engineer can use an information system to help determine why a computer system is running slowly. Information systems have been designed and implemented for almost every career and industry. This chapter shows how you can be involved in designing and implementing an information system that will directly benefit you on the job. It also shows how to avoid errors or recover from disasters. This chapter starts with how systems are designed.

The way an information system is designed, implemented, and maintained profoundly affects the daily functioning of an organization. Like investigation and analysis covered in Chapter 12, design, implementation, maintenance, and review covered in this chapter strive to achieve organizational goals, such as reducing costs, increasing profits, or improving customer service. The goal is to develop a new or modified system to deliver the right information to the right person at the right time.

SYSTEMS DESIGN

systems design
The stage of systems development that answers the question "How will the information system solve a problem?"

The purpose of **systems design** is to answer the question "How will the information system solve a problem?" The primary result of the systems design phase is a technical design that details system outputs, inputs, and user interfaces; specifies hardware, software, databases, telecommunications, personnel, and procedures; and shows how these components are related.[1] The new system should overcome shortcomings of the existing system and help the organization achieve its goals. The system must also meet certain guidelines, including user and stakeholder requirements and the objectives defined during previous development phases. A California company that lets people meet and interact on a 3-D Web site, for example, spent more than $17 million to design an online simulation that lets people chat, play cards, and flirt with avatars, lifelike characters that appear on Internet sites. The company hopes that the four-year systems development project will generate revenues from subscribers and companies that want to advertise their products on its Web site.

Systems design is typically accomplished using the tools and techniques discussed in Chapter 12.[2] Depending on the specific application, these methods can be used to support and document all aspects of systems design. Two key aspects of systems design are logical and physical design.

Logical and Physical Design

logical design
A description of the functional requirements of a system.

As discussed in Chapter 5, design has two dimensions: logical and physical. The **logical design** refers to what the system will do. Logical design describes the functional requirements of a system. That is, it conceptualizes what the system will do to solve the problems identified through earlier analysis. Without this step, the technical details of the system (such as which hardware devices should be acquired) often obscure the best solution. Logical design involves planning the purpose of each system element, independent of hardware and software considerations. The logical design specifications that are determined and documented include output, input, process, file and database, telecommunications, procedures, controls and security, and personnel and job requirements. Security is always an important design issue for corporations and governments.[3] New rules published in September 2005, for example, require that federal agencies incorporate security procedures in the design of new or modified systems.[4] In addition, the Federal Information Security Management Act, enacted in 2002, requires federal agencies to make sure that security protection measures are incorporated into systems provided by outside vendors and contractors. After the output of a new or modified system has been designed, the other components of logical design can be determined. For example, if the output is a paycheck from a payroll program, we can determine that we need hours worked, pay rate, and various deductions as the input requirements. Multiplying hours worked times pay rate and subtracting any deductions will determine the amount on the paycheck.

The **physical design** refers to how the tasks are accomplished, including how the components work together and what each component does. Physical design specifies the characteristics of the system components necessary to put the logical design into action. In this phase, the characteristics of the hardware, software, database, telecommunications, personnel, and procedure and control specifications must be detailed. These physical design components were discussed in Part 2 on technology.

physical design
The specification of the characteristics of the system components necessary to put the logical design into action.

Object-Oriented Design

Logical and physical design can be accomplished using either the traditional approach or the object-oriented approach to systems development. Both approaches use a variety of design models to document the new system's features and the development team's understandings and agreements. Many organizations today are turning to OO development because of its increased flexibility. This section outlines a few OO design considerations and diagrams.

Using the OO approach, you can design key objects and classes of objects in the new or updated system.[5] This process includes considering the problem domain, the operating environment, and the user interface. The problem domain involves the classes of objects related to solving a problem or realizing an opportunity. In our Maui, Hawaii, kayak rental shop example first introduced in Chapter 12 and referring back to the generalization/specialization hierarchy showing classes we presented there, KayakItem in Figure 12.23 is an example of a problem domain object that will store information on kayaks in the rental program. The operating environment for the rental shop's system includes objects that interact with printers, system software, and other software and hardware devices. The user interface for the system includes objects that users interact with, such as buttons and scroll bars in a Windows program.

Installing Wi-Fi at Southern Wine's Mega Warehouse

Since the late 1970s, Southern Wine and Spirits of America has expanded steadily to more than 20 large distribution centers in 12 state markets. Southern Wine and Spirits of America represents more than 300 wine, spirits, beer, and beverage suppliers from around the world, and markets, promotes, merchandises, and distributes nearly 5,000 individual brands. During a typical working week, Southern Wine's sales, delivery, and support staff collectively calls on or services approximately 140,000 different chain and independent retail and restaurant customers across the country. Southern Wine commands more than 17 percent of total domestic wine and spirits wholesaler revenues, according to a recent survey by Impact Newsletter, a respected drinks industry trade publication.

Recently, Southern Wine and Spirits of America consolidated three of its northern Florida distribution sites into one mega warehouse. The centralized facility is expected to save the company about $4 million per year. The new warehouse, located in Lakeland, Florida, is 650,000 square feet, about the size of eight football fields. Jason Witty, director of information technology at Southern Wine, had seven months to design and implement a wireless network for carrying inventory information to the employees in the new warehouse.

Seven months might seem like a lot of time to those unfamiliar with the demands of such an undertaking. But to Witty and other experienced developers, it would clearly take a miracle to successfully pull off such a feat. Witty said he knew that setting up a Wi-Fi site survey would be a challenge because of all the metal in the warehouse building, including racks to hold bottles of liquor and large metal construction equipment. Metal tends to disrupt wireless signals. He knew that the coverage and placement of Wi-Fi access points would need to be checked, double-checked, and triple-checked to make sure that network coverage is complete.

To design the network properly, Witty and his team needed to consider how the network would be used. The systems analysis report specified the applications for the network, including the following:

- Voice communications between warehouse workers and supervisors
- A voice-based inventory picking system, which allows workers who fill orders by picking bottles from warehouse racks to receive audible commands via a Wi-Fi headset
- Inventory technology that uses four cameras on a main conveyor belt to scan cases as they are loaded onto trucks, and then store the video digitally
- Network terminals mounted on forklifts and connected wirelessly to a warehouse management system, so that workers can access corporate systems to ensure that pallets of cases are stored in the correct location

Witty selected network components from IBM, Cisco Systems, and Symbol Technologies to build the network. The plan for the system included a wireless network for the workers to use for communications and information access, and a wired network for high-bandwidth uses such as video. Witty, his team at Southern Wine, and contractors conducted three site surveys to determine the locations for the Wi-Fi access points. One of the surveys was conducted when the building was nearly built and the physical layout was known. The second took place after some of the goods had arrived. The third survey came after the warehouse was full.

Providing 100 percent coverage was not the only challenge. Southern Wine also needed enough access points to handle large amounts of network traffic. A single access point probably could not support 50 workers in one location. Therefore, the site survey and network design had to include load testing to make sure the warehouse had enough access points to handle heavy network demands. This was crucial considering that workers would be relying on voice commands and communications over the wireless network.

The resulting system includes about 130 wireless endpoints at the facility, including 60 Wi-Fi voice phones, working through 24 access points. Jason Witty estimates the warehouse can process 10,000 cases of liquor per hour using the new system. The resulting system is considered a miraculous success. The new network was installed, set up, and working within seven months. As with most systems development projects, the team endured some failures along the way to its success. "I learned you can never start planning too early," reflects Jason Witty, "There are a lot of gotchas on a building of this size and scale. You have to make sure you talk through every need."

Discussion Questions

1. What unique challenges did this systems development project present to the systems developers?
2. How did Jason Witty and his team confront those challenges, and what lessons did they learn?

Critical Thinking Questions

1. In this systems development project, Jason Witty and his team needed to continuously switch from the logical design to the physical design of the new system. Why do you think this was the case? What can occur during the physical design that might affect changes to the logical design?
2. Witty and his team had a short amount of time to accomplish a tremendous goal. Not only did they work quickly, but the contractors from IBM, Cisco Systems, and Symbol Technologies had to extend an extra effort as well. How do you think Southern Wine motivated everyone to complete the job quickly? In general, describe the costs and benefits of a "rush job" versus one without a time restriction.

SOURCES: Hamblen, Matt, "Southern Wine Opens Huge Warehouse with Wi-Fi," *Computerworld*, July 5, 2006, *www.computerworld.com*. Southern Wine Web site, accessed July 11, 2006, *www.southernwine.com*. Symbol Corporation Web site, accessed July 11, 2006, *www.symbol.com*.

During the design phase, you also need to consider the sequence of events that must happen for the system to function correctly. For example, you might want to design the sequence of events for adding a new kayak to the rental program. A sequence of events is often called a *scenario*, and it can be diagrammed in a sequence diagram (see Figure 13.1).

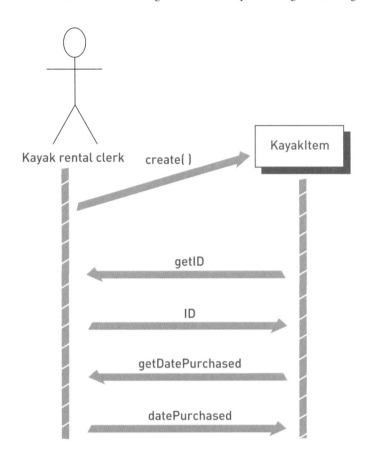

Figure 13.1

A Sequence Diagram to Add a New KayakItem Scenario

You read a sequence diagram starting at the top and moving down.

1. The Create arrow at the top is a message from the kayak rental clerk to the KayakItem object to create information on a new kayak to be placed into the rental program.
2. The KayakItem object knows that it needs the ID for the kayak and sends a message to the clerk requesting the information. See the getID arrow.
3. The clerk then types the ID into the computer. This is shown with the ID arrow. The data is stored in the KayakItem object.
4. Next, KayakItem requests the purchase date. This is shown in the getDatePurchased arrow.
5. Finally, the clerk types the purchase date into the computer. The data is also transferred to KayakItem object. This is shown in the datePurchased arrow at the bottom of Figure 13.1.

This scenario is only one example of a sequence of events. Other scenarios might include entering information about life jackets, paddles, suntan lotion, and other accessories. The same types of use case and generalization/specialization hierarchy diagrams discussed in Chapter 12 can be created for each, and additional sequence diagrams will also be needed.

Interface Design and Controls

Some special system characteristics should be considered during both logical and physical design. These characteristics relate to how users access and interact with the system, including sign-on procedures, interactive processing, and interactive dialogue.

sign-on procedure
Identification numbers, passwords, and other safeguards needed for someone to gain access to computer resources.

- A **sign-on procedure** consists of identification numbers, passwords, and other safeguards needed for someone to gain access to computer resources. See Figure 13.2.

Figure 13.2

The Levels of the Sign-On Procedure

menu-driven system
A system in which users simply pick what they want to do from a list of alternatives.

- With *interactive processing*, people directly interact with the processing component of the system through terminals or networked PCs. With a **menu-driven system** (see Figure 13.3), users simply pick what they want to do from a list of alternatives. Most people can easily operate these types of systems. They select an option or respond to questions (or prompts) from the system, and the system does the rest.

Figure 13.3

Menu-Driven System

A menu-driven system allows you to choose what you want from a list of alternatives.

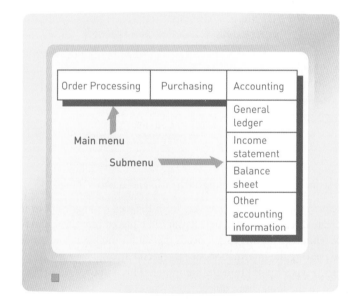

help facility
A program that provides assistance when users want to know more about a program or feature or what type of response is expected.

- Many designers incorporate a **help facility** into the system or applications program. When users want to know more about a program or feature or what type of response is expected, they can activate the help facility.

lookup tables
Tables containing data that computer programs can develop and use to simplify and shorten data entry.

- Computer programs can develop and use **lookup tables** to simplify and shorten data entry. For example, if you are entering a sales order for a company, you can type its abbreviation, such as ABCO. The program will then go to the customer table, normally stored on a disk, and look up all the information pertaining to the company abbreviated ABCO that you need to complete the sales order.

restart procedures
Simplified processes to access an application from where it stopped.

- With a **restart procedure,** users can restart an application where it stopped in case the application crashed or had problems.

Good interactive design refers to the series of messages and prompts communicated between the system and the user. Table 13.1 lists some elements that help designers create good interactive dialogue. These elements should be considered during systems design.

Element	Description
Clarity	The computer system should ask for information using easily understood language. Whenever possible, the users themselves should help select the words and phrases used for dialogue with the computer system.
Response time	Ideally, responses from the computer system should approximate a normal response time from a human being carrying on the same sort of dialogue.
Consistency	The system should use the same commands, phrases, words, and function keys for all applications. After a user learns one application, all others will then be easier to use.
Format	The system should use an attractive format and layout for all screens. The use of color, highlighting, and the position of information on the screen should be considered carefully and applied consistently.
Jargon	All dialogue should be written in easy-to-understand terms. Avoid jargon known only to IS specialists.
Respect	All dialogue should be developed professionally and with respect. Dialogue should not talk down to or insult the user. Avoid statements such as "You have made a fatal error."

Table 13.1

The Elements of Good Interactive Dialogue

Design of System Security and Controls

In addition to considering the system's interface and user interactions, designers must also develop system security and controls for all aspects of the system, including hardware, software, database systems, telecommunications, and Internet operations.[6] These key considerations involve error prevention, detection, and correction; disaster planning and recovery; and systems controls.[7]

Preventing, Detecting, and Correcting Errors

The most cost-effective time to deal with potential errors is early in the design phase. During installation or after the system is operating, it is much more expensive and time consuming to handle errors and related problems. Good systems design attempts to prevent errors before they occur, which involves recognizing what can happen and developing steps and procedures that can prevent, detect, and correct errors. *Enterprise-rights management procedures*, for example, can be used to make sure that sensitive and important documents and data are only distributed to authorized or approved people.[8] Corning Inc., producer of fiber-optic cables and products, uses enterprise-rights management to make sure that sensitive documents on products it sells to the military and government are kept secret. According to the company's director of knowledge and information management, "The government regulations are very explicit." In addition to enterprise-rights management, there are other important security and control measures, including disaster planning and recovery and adequate backup procedures.

Disaster Planning and Recovery

Disaster planning is the process of anticipating and providing for disasters. A disaster can be an act of nature (a flood, fire, or earthquake) or a human act (terrorism, error, labor unrest, or erasure of an important file). An earthquake in Seattle, for example, caused a company to lose its telecommunications capabilities for days, which forced the manufacturing company to shut down some of its operations. Disaster planning often focuses primarily on two issues: maintaining the integrity of corporate information and keeping the information system running until normal operations can be resumed. A California law that forces hospitals to be earthquake-proof has prompted huge investments in information systems. One health company will likely spend about $1 billion to complete its disaster planning and recovery efforts to comply with the California law.

Disaster recovery is the implementation of the disaster plan.[9] When Hurricane Katrina hit New Orleans and the Gulf Coast, Howard Weil Inc. had a plan to keep the firm operating.[10] The investment and trading company would move its 40 employees to

disaster recovery
The implementation of the disaster plan.

Houston, Texas. But when Houston also had to be evacuated, the company had to move its employees to another location—Stamford, Connecticut—according to its disaster plan. The small trading company was able to rapidly re-create its trading desk and IS infrastructure to continue to trade. According to Jefferson Parker, president of Howard Weil, "You don't normally develop a backup plan for the backup plan." See Figure 13.4.

Figure 13.4

Disaster Recovery Efforts

(Source: Getty Images.)

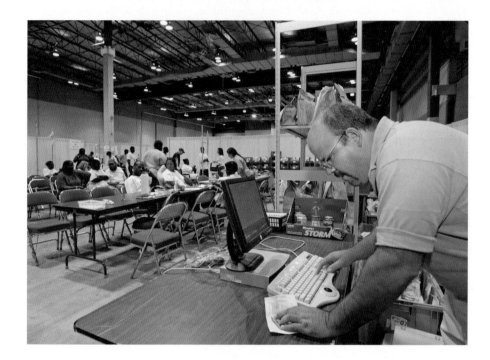

hot site
A duplicate, operational hardware system or immediate access to one through a specialized vendor.

cold site
A computer environment that includes rooms, electrical service, telecommunications links, data storage devices, and the like; also called a *shell*.

incremental backup
Making a backup copy of all files changed during the last few days or the last week.

image log
A separate file that contains only changes to applications.

Although companies have known about the importance of disaster planning and recovery for decades, many do not adequately prepare. The primary tools used in disaster planning and recovery are hardware, software, and database, telecommunications, and personnel backups.[11] Most of these systems were discussed in Part 2 on information technology concepts. For example, hot and cold sites can be used to back up hardware. A duplicate, operational hardware system that is ready to use (or immediate access to one through a specialized vendor) is an example of a **hot site**. If the primary computer has problems, the hot site can be used immediately as a backup. Another approach is to use a **cold site**, also called a *shell*, which is a computer environment that includes rooms, electrical service, telecommunications links, data storage devices, and similar equipment. If a primary computer has a problem, backup computer hardware is brought into the cold site, and the complete system is made operational. Files and databases can be backed up by making a copy of all files and databases changed during the last few days or the last week, a technique called **incremental backup**. This approach to backup uses an **image log**, which is a separate file that contains only changes to applications. Whenever an application is run, an image log is created that contains all changes made to all files. If a problem occurs with a database, an old database with the last full backup of the data, along with the image log, can be used to re-create the current database.

Systems Controls

Security lapses, fraud, and the invasion of privacy can present disastrous problems. For example, because of an inadequate security and control system, a futures and options trader for a British bank lost about $1 billion. A simple systems control might have prevented a problem that caused the 200-year-old bank to collapse. In addition, from time to time, IRS employees have been caught looking at the returns of celebrities and others. Preventing and detecting these problems is an important part of systems design. Prevention includes the following:

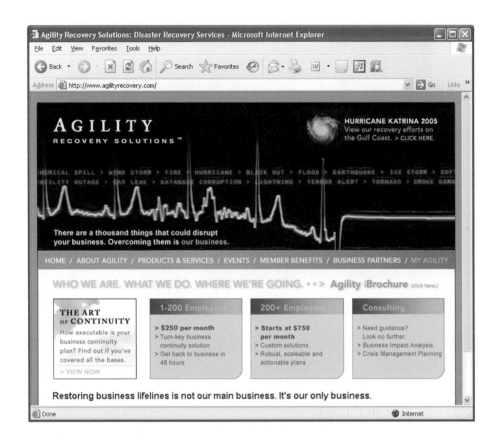

- Determining potential problems
- Ranking the importance of these problems
- Planning the best place and approach to prevent problems
- Deciding the best way to handle problems if they occur

Every effort should be made to prevent problems, but companies must establish procedures to handle problems if they occur, including system controls.

Most IS departments establish tight **systems controls** to maintain data security. Systems controls can help prevent computer misuse, crime, and fraud by managers, employees, and others. The accounting scandals in the early 2000s have caused many IS departments to develop systems controls to make it more difficult for executives to mislead investors and employees. Some of these scandals involved billions of dollars.

Most IS departments have a set of general operating rules that helps protect the system. Some IS departments are **closed shops**, in which only authorized operators can run the computers. Other IS departments are **open shops**, in which other people, such as programmers and systems analysts, are also authorized to run the computers. Other rules specify the conduct of the IS department.

These rules are examples of **deterrence controls**, which involve preventing problems before they occur. Making a computer more secure and less vulnerable to a break-in is another example. Good control techniques should help an organization contain and recover from problems. The objective of containment control is to minimize the impact of a problem while it is occurring, and recovery control involves responding to a problem that has already occurred.

Many types of systems controls can be developed, documented, implemented, and reviewed. These controls touch all aspects of the organization (see Table 13.2).

systems controls
Rules and procedures to maintain data security.

closed shops
IS departments in which only authorized operators can run the computers.

open shops
IS departments in which people, such as programmers and systems analysts, are allowed to run the computers, in addition to authorized operators.

deterrence controls
Rules and procedures to prevent problems before they occur.

The SunGard Data Systems high availability command center oversees services that let companies continuously access and process information—with minimal downtime—even during a disaster.

(Source: Courtesy of SunGard.)

Table 13.2

Using Systems Controls to Enhance Security

Controls	Description
Input controls	Maintain input integrity and security. Their purpose is to reduce errors while protecting the computer system against improper or fraudulent input. Input controls range from using standardized input forms to eliminating data-entry errors and using tight password and identification controls.
Processing controls	Deal with all aspects of processing and storage. The use of passwords and identification numbers, backup copies of data, and storage rooms that have tight security systems are examples of processing and storage controls.
Output controls	Ensure that output is handled correctly. In many cases, output generated from the computer system is recorded in a file that indicates the reports and documents that were generated, the time they were generated, and their final destinations.
Database controls	Deal with ensuring an efficient and effective database system. These controls include the use of identification numbers and passwords, without which a user is denied access to certain data and information. Many of these controls are provided by database management systems.
Telecommunications controls	Provide accurate and reliable data and information transfer among systems. Telecommunications controls include firewalls and encryption to ensure correct communication while eliminating the potential for fraud and crime.
Personnel controls	Make sure that only authorized personnel have access to certain systems to help prevent computer-related mistakes and crime. Personnel controls can involve the use of identification numbers and passwords that allow only certain people access to particular data and information. ID badges and other security devices (such as smart cards) can prevent unauthorized people from entering strategic areas in the information systems facility.

After controls are developed, they should be documented in standards manuals that indicate how the controls are to be implemented. They should then be implemented and frequently reviewed. It is common practice to measure the extent to which control techniques are used and to take action if the controls have not been implemented.

Many companies use ID badges to prevent unauthorized access to sensitive areas in the information systems facility.

(Source: Michael Newman/Photo Edit.)

Generating Systems Design Alternatives

When people or organizations require a system to perform additional functions that an existing system cannot support, they often turn to outside vendors to design and supply their new systems. Such purchases require expertise in both hardware and software. Whether an individual is purchasing a personal computer or a company is acquiring an expensive mainframe computer, the system can be obtained from a single vendor or multiple vendors.[12] In some cases, the vendor simply provides hardware or software. In other cases, the vendor provides additional services. Some of the factors to consider in selecting a vendor are the following:

- The vendor's reliability and financial stability
- The type of service offered after the sale
- The goods and services the vendor offers and keeps in stock
- The vendor's willingness to demonstrate its products
- The vendor's ability to repair hardware
- The vendor's ability to modify its software and its willingness to invest in continued development of its software
- The availability of vendor-offered training of IS personnel and system users
- Evaluations of the vendor by independent organizations

When additional hardware and software are not required, alternative designs are often generated without input from vendors. If the new system is complex, the original development team might want to involve other personnel in generating alternative designs. If new hardware and software are to be acquired from an outside vendor, a formal request for proposal (RFP) can be made.

Request for Proposals

The **request for proposal (RFP)** is an important document for many organizations involved with large, complex systems development efforts. Smaller, less-complex systems often do not require an RFP. A company that is purchasing an inexpensive piece of software that will run on existing hardware, for example, might not need to go through a formal RFP process.

When an RFP is used, it often results in a formal bid that is used to determine who gets a contract for new or modified systems. The RFP specifies in detail the required resources such as hardware and software.[13] Although it can take time and money to develop a high-quality RFP, it can save a company in the long run. Companies that frequently generate RFPs can automate the process. One company, for example, purchased a software package, called The RFP Machine from Pragmatech Software, to improve the quality of its RFPs and to

request for proposal (RFP)
A document that specifies in detail required resources such as hardware and software.

reduce the time it takes to produce them. The RFP Machine stores important data needed to generate RFPs and automates the process of producing RFP documents.

In some cases, separate RFPs are developed for different needs. For example, a company might develop separate RFPs for hardware, software, and database systems. The RFP also communicates these needs to one or more vendors, and it provides a way to evaluate whether the vendor has delivered what was expected. In some cases, the RFP is part of the vendor contract. The Table of Contents for a typical RFP is shown in Figure 13.5.

Figure 13.5

A Typical Table of Contents for a Request for Proposal

Johnson & Florin, Inc.
Systems Investigation Report

Contents

COVER PAGE (with company name and contact person)
BRIEF DESCRIPTION of the COMPANY
OVERVIEW of the EXISTING COMPUTER SYSTEM
SUMMARY of COMPUTER-RELATED NEEDS and/or PROBLEMS
OBJECTIVES of the PROJECT
DESCRIPTION of WHAT IS NEEDED
HARDWARE REQUIREMENTS
PERSONNEL REQUIREMENTS
COMMUNICATIONS REQUIREMENTS
PROCEDURES to BE DEVELOPED
TRAINING REQUIREMENTS
MAINTENANCE REQUIREMENTS
EVALUATION PROCEDURES (how vendors will be judged)
PROPOSAL FORMAT (how vendors should respond)
IMPORTANT DATES (when tasks are to be completed)
SUMMARY

Financial Options

When acquiring computer systems, several choices are available, including purchase, lease, or rent. Cost objectives and constraints set for the system play a significant role in the choice, as do the advantages and disadvantages of each. In addition, traditional financial tools, including net present value and internal rate of return, can be used. Table 13.3 summarizes the advantages and disadvantages of these financial options.

Determining which option is best for a particular company in a given situation can be difficult. Financial considerations, tax laws, the organization's policies, its sales and transaction growth, marketplace dynamics, and the organization's financial resources are all important factors. In some cases, lease or rental fees can amount to more than the original purchase price after a few years. As a result, some companies prefer to purchase their equipment.

On the other hand, constant advances in technology can make purchasing risky. A company would not want to purchase a new multimillion-dollar computer only to have newer and more powerful computers available a few months later at a lower price, unless the computer can be easily and inexpensively upgraded. Some servers, for example, are designed to be *scalable* to allow processors to be added or swapped, memory to be upgraded, and peripheral devices to be installed. Companies often employ several people to determine the best option based on all the factors. This staff can also help negotiate purchase, lease, or rental contracts.

Evaluating and Selecting a Systems Design

The final step in systems design is to evaluate the various alternatives and select the one that will offer the best solution for organizational goals. Depending on their weight, any one of these objectives might result in the selection of one design over another. For example, financial concerns might make a company choose rental over equipment purchase. Specific performance objectives—for example, that the new system must perform online data processing—might result in a complex network design for which control procedures must

Renting (Short-Term Option)	
Advantages	**Disadvantages**
No risk of obsolescence	No ownership of equipment
No long-term financial investment	High monthly costs
No initial investment of funds	Restrictive rental agreements
Maintenance usually included	
Leasing (Longer-Term Option)	
Advantages	**Disadvantages**
No risk of obsolescence	High cost of canceling lease
No long-term financial investment	Longer time commitment than renting
No initial investment of funds	No ownership of equipment
Less expensive than renting	
Purchasing	
Advantages	**Disadvantages**
Total control over equipment	High initial investment
Can sell equipment at any time	Additional cost of maintenance
Can depreciate equipment	Possibility of obsolescence
Low cost if owned for a number of years	Other expenses, including taxes and insurance

Table 13.3

Advantages and Disadvantages
of Acquisition Options

be established. Evaluating and selecting the best design involves achieving a balance of system objectives that will best support organizational goals. Normally, evaluation and selection involves both a preliminary and a final evaluation before a design is selected.

The Preliminary Evaluation

A **preliminary evaluation** begins after all proposals have been submitted. The purpose of this evaluation is to dismiss unwanted proposals. Several vendors can usually be eliminated by investigating their proposals and comparing them with the original criteria. Those that compare favorably are asked to make a formal presentation to the analysis team. The vendors should also be asked to supply a list of companies that use their equipment for a similar purpose. The organization then contacts these references and asks them to evaluate their hardware, their software, and the vendor.

preliminary evaluation
An initial assessment whose purpose is to dismiss the unwanted proposals; begins after all proposals have been submitted.

The Final Evaluation

The **final evaluation** begins with a detailed investigation of the proposals offered by the remaining vendors. The vendors should be asked to make a final presentation and to fully demonstrate the system. The demonstration should be as close to actual operating conditions as possible. Applications such as payroll, inventory control, and billing should be conducted using a large amount of test data. J.D. Powers, the company that gives quality awards to car manufacturers, will start evaluating information systems vendors in terms of their service and support.[14] According to a director at the company, "Typically, every industry evaluated at J.D. Powers has directly or indirectly raised the overall customer satisfaction level of the industry." J.D. Powers is expected to give certifications to those IS vendors that achieve high service and support levels.

final evaluation
A detailed investigation of the proposals offered by the vendors remaining after the preliminary evaluation.

After the final presentations and demonstrations have been given, the organization makes the final evaluation and selection. Cost comparisons, hardware performance, delivery dates, price, flexibility, backup facilities, availability of software training, and maintenance factors are considered. In addition to comparing computer speeds, storage capacities, and other similar characteristics, companies should also carefully analyze whether the characteristics of the proposed systems meet the company's objectives. In most cases, the RFP captures these objectives and goals. Figure 13.6 illustrates the evaluation process.

Figure 13.6

The Stages in Preliminary and Final Evaluations

Note that the number of possible alternatives decreases as the firm gets closer to making a final decision.

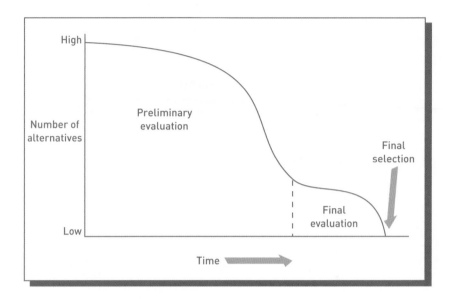

Evaluation Techniques

The exact procedure used to make the final evaluation and selection varies from one organization to the next. Some were first introduced in Chapter 2, including return on investment (ROI), earnings growth, market share, customer satisfaction, and total cost of ownership (TCO). In addition, four other approaches are commonly used: group consensus, cost/benefit analysis, benchmark tests, and point evaluation.

Group Consensus

In **group consensus**, a decision-making group is appointed and given the responsibility of making the final evaluation and selection. Usually, this group includes the members of the development team who participated in either systems analysis or systems design. This approach might be used to evaluate which of several screen layouts or report formats is best.

group consensus
Decision making by a group that is appointed and given the responsibility of making the final evaluation and selection.

Cost/Benefit Analysis

Cost/benefit analysis is an approach that lists the costs and benefits of each proposed system. After they are expressed in monetary terms, all the costs are compared with all the benefits. Table 13.4 lists some of the typical costs and benefits associated with the evaluation and selection procedure. This approach is used to evaluate options whose costs can be quantified, such as which hardware or software vendor to select.

cost/benefit analysis
An approach that lists the costs and benefits of each proposed system. After they are expressed in monetary terms, all the costs are compared with all the benefits.

Benchmark Tests

A **benchmark test** is an examination that compares computer systems operating under the same conditions. Most computer companies publish their own benchmark tests, but some forbid disclosure of benchmark tests without prior written approval. Thus, one of the best approaches is for an organization to develop its own tests, and then use them to compare the equipment it is considering. This approach might be used to compare the end-user system response time on two similar systems. Several independent companies also rate computer systems. *Computerworld*, *PC Week*, and many other publications, for example, not only summarize various systems, but also evaluate and compare computer systems and manufacturers according to a number of criteria.

benchmark test
An examination that compares computer systems operating under the same conditions.

Table 13.4

Cost/Benefit Analysis Table

Costs	Benefits
Development costs	Reduced costs
Personnel	Fewer personnel
Computer resources	Reduced manufacturing costs
	Reduced inventory costs
	More efficient use of equipment
	Faster response time
	Reduced downtime or crash time
	Less spoilage
Fixed costs	**Increased Revenues**
Computer equipment	New products and services
Software	New customers
One-time license fees for software and maintenance	More business from existing customers
	Higher price as a result of better products and services
Operating costs	**Intangible benefits**
Equipment lease and/or rental fees	Better public image for the organization
Computer personnel (including salaries, benefits, etc.)	Higher employee morale
	Better service for new and existing customers
Electric and other utilities	The ability to recruit better employees
Computer paper, tape, and disks	Position as a leader in the industry
Other computer supplies	System easier for programmers and users
Maintenance costs	
Insurance	

Point Evaluation

One of the disadvantages of cost/benefit analysis is the difficulty of determining the monetary values for all the benefits. An approach that does not employ monetary values is a **point evaluation system**. Each evaluation factor is assigned a weight, in percentage points, based on importance. Then each proposed information system is evaluated in terms of this factor and given a score, such as one ranging from 0 to 100, where 0 means that the alternative does not address the feature at all and 100 means that the alternative addresses that feature perfectly. The scores are totaled, and the system with the greatest total score is selected. When using point evaluation, an organization can list and evaluate literally hundreds of factors. Figure 13.7 shows a simplified version of this process. This approach is used when there are many options to be evaluated, such as which software best matches a particular business's needs.

point evaluation system
An evaluation process in which each evaluation factor is assigned a weight, in percentage points, based on importance. Then each proposed system is evaluated in terms of this factor and given a score ranging from 0 to 100. The scores are totaled, and the system with the greatest total score is selected.

Freezing Design Specifications

Near the end of the design stage, some organizations prohibit further changes in the design of the system. Freezing systems design specifications means that the user agrees in writing that the design is acceptable. (See Figure 13.8.) Other organizations, however, allow or even encourage design changes. These organizations often use agile or rapid systems development approaches, introduced in Chapter 12.

Figure 13.7

An Illustration of the Point Evaluation System

In this example, software has been given the most weight (40 percent), compared with hardware (35 percent) and vendor support (25 percent). When system A is evaluated, the total of the three factors amounts to 82.5 percent. System B's rating, on the other hand, totals 86.75 percent, which is closer to 100 percent. Therefore, the firm chooses system B.

		System A			System B		
Factor's importance		Evaluation		Weighted evaluation	Evaluation		Weighted evaluation
Hardware	35%	95	35%	33.25	75	35%	26.25
Software	40%	70	40%	28.00	95	40%	38.00
Vendor support	25%	85	25%	21.25	90	25%	22.50
Totals	100%			82.5			86.75

Figure 13.8

Freezing Design Specifications

The Contract

One of the most important steps in systems design is to develop a good contract if new computer facilities are being acquired. Finding the best terms where everyone makes a profit can be difficult. Most computer vendors provide standard contracts; however, such contracts are designed to protect the vendor, not necessarily the organization buying the computer equipment.

Organizations often use outside consultants and legal firms to help them develop their contracts. Such contracts stipulate exactly what they expect from the system vendor and what interaction will occur between the vendor and the organization. All equipment specifications, software, training, installation, maintenance, and so on are clearly stated. Also, the contract stipulates deadlines for the various stages or milestones of installation and implementation, as well as actions that the vendor will take in case of delays or problems. Some organizations include penalty clauses in the contract, in case the vendor does not meet its obligation by the specified date. Typically, the request for proposal becomes part of the contract. This saves a considerable amount of time in developing the contract, because the RFP specifies in detail what is expected from the vendors. In addition, some companies such as General Motors have shortened the length of the IS contract, giving companies more power to monitor vendors and change contract provisions if needed.[15]

The Design Report

design report
The primary result of systems design, reflecting the decisions made and preparing the way for systems implementation.

System specifications are the final results of systems design. They include a technical description that details system outputs, inputs, and user interfaces, as well as all hardware, software, databases, telecommunications, personnel, and procedure components and the way these components are related. The specifications are contained in a **design report**, which is the primary result of systems design. The design report reflects the decisions made for systems design and prepares the way for systems implementation. The contents of the design report are summarized in Figure 13.9.

```
                 Johnson & Florin, Inc.
                 Systems Design Report

     Contents

     PREFACE
     EXECUTIVE SUMMARY of SYSTEMS
     DESIGN
     REVIEW of SYSTEMS ANALYSIS
     MAJOR DESIGN RECOMMENDATIONS
              Hardware design
              Software design
              Personnel design
              Communications design
              Database design
              Procedures design
              Training design
              Maintenance design
     SUMMARY of DESIGN DECISIONS
     APPENDIXES
     GLOSSARY of TERMS
     INDEX
```

Figure 13.9

A Typical Table of Contents for a Systems Design Report

SYSTEMS IMPLEMENTATION

After the information system has been designed, a number of tasks must be completed before the system is installed and ready to operate.[16] This process, called **systems implementation**, includes hardware acquisition, programming and software acquisition or development, user preparation, hiring and training of personnel, site and data preparation, installation, testing, start-up, and user acceptance. Spending on systems implementation is on the rise.[17] According to one survey, almost 50 percent of IS managers expect spending on systems implementation to increase in the next few years, compared to IS spending in 2005.

Systems implementation can have a profound effect on organizations. The Florida Bankers Association, for example, implemented a fraud database to help detect and prevent check-kiting, skimming, pharming, and phishing.[18] The American Banking Association reports that fraud costs banks more than $5 billion annually. According to Thomas Kerr, senior vice president and CFO for the Florida Bankers Association, "We never intended this to be a national database but that in fact has been created. Now twenty-two states are involved." Melody Shimmel, certified fraud examiner in charge of risk management and fraud at Century Bank in Sarasota, Florida, reports that the implemented fraud system has allowed law enforcement to solve a variety of cases affecting the bank and involving $600 million in bank assets.

DaimlerChrysler AG, the large automotive company, implemented a new Web portal to service its 5,300 dealerships in the United States and Canada and another 1,500 dealerships worldwide.[19] The implementation took about 240 separate applications and unified them into one Web application. The successful implementation resulted in faster information update speeds and a centralized approach to delivering and managing important information to its customers and dealers. The Web portal application is available in ten languages and gets about 300,000 hits, or visits, to the Web site each day.

The typical sequence of systems implementation activities is shown in Figure 13.10. Companies can reap great rewards after implementing new systems. United Parcel Service

systems implementation
A stage of systems development that includes hardware acquisition, software acquisition or development, user preparation, hiring and training of personnel, site and data preparation, installation, testing, start-up, and user acceptance.

(UPS), for example, implemented a project to improve how packages flow and are tracked from pickup through delivery.

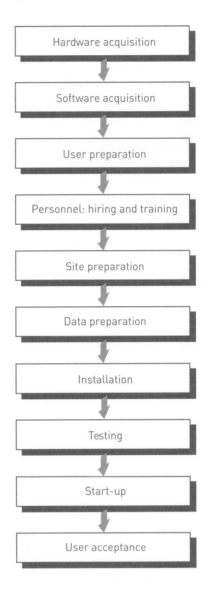

Figure 13.10

Typical Steps in Systems Implementation

Acquiring Hardware from an IS Vendor

To obtain the components for an information system, organizations can purchase, lease, or rent computer hardware and other resources from an IS vendor.[20] An *IS vendor* is a company that offers hardware, software, telecommunications systems, databases, IS personnel, or other computer-related resources. Types of IS vendors include general computer manufacturers (such as IBM and Hewlett-Packard), small computer manufacturers (such as Dell and Gateway), peripheral equipment manufacturers (such as Epson and Cannon), computer dealers and distributors (such as Radio Shack and CompUSA), and leasing companies (such as National Computer Leasing and Paramount Computer Rentals, PLC).

In addition to buying, leasing, or renting computer hardware, companies can pay only for the computing services that it uses. Called "pay-as-you-go," "on-demand," or "utility" computing, this approach requires an organization to pay only for the computer power it uses, as it would pay for a utility such as electricity. A bank, for example, can buy only the computer resources it needs from IBM and other companies. Hewlett-Packard offers its clients a "capacity-on-demand" approach, in which organizations pay according to the computer resources actually used, including processors, storage devices, and network facilities.

Companies can also purchase used computer equipment. This option is especially attractive to firms that are experiencing an economic slowdown. Companies often use traditional Internet auctions to locate used or refurbished equipment. Popular Internet auction sites sometimes sell more than $1 billion of computer-related equipment annually, and companies can purchase equipment for about 20 or 30 cents on the dollar. However, buyers need to beware: Prices are not always low, and equipment selection can be limited on Internet auction sites.

In addition, companies are increasingly turning to service providers to implement some or all of the systems they need. As discussed in Chapter 4, an application service provider (ASP) can help companies implement software and other systems. The ASP can provide both user support and the computers on which to run the software. ASPs often focus on high-end applications, such as database systems and enterprise resource planning packages. As mentioned in Chapter 7, an Internet service provider (ISP) assists a company in gaining access to the Internet. ISPs can also help a company in setting up an Internet site. Some service providers specialize in specific systems or areas, such as marketing, finance, or manufacturing.

Acquiring Software: Make or Buy?

As with hardware, application software can be acquired in several ways. As previously mentioned, it can be purchased from external developers or developed in-house.[21] This decision is often called the **make-or-buy decision**. A comparison of the two approaches is shown in Table 13.5. Today, most software is purchased. SAP, the large international software company headquartered in Germany, produces modular software it sells to a variety of companies. The approach gives its customers using the software more flexibility in what they use and what they pay for SAP's modules.[22] The key is how the purchased systems are integrated into an effective system.

make-or-buy decision
The decision regarding whether to obtain the necessary software from internal or external sources.

Factor	Off the Shelf (Buy)	Developed (Make)
Cost	Low Cost	High Cost
Needs	Might not get what you need	Custom software to satisfy your needs
Quality	Usually high quality	Quality can vary depending on the programming team
Speed	Can acquire it now	Can take years to develop
Competitive advantage	Other organizations can have the same software and same advantage	Can develop a competitive advantage with good software

Table 13.5

Comparison of Off the Shelf and Developed Software

In some cases, companies use a blend of external and internal software development. That is, in-house personnel modify or customize off-the-shelf or proprietary software programs. Software can also be rented. Salesforce.com, for example, rents software online that helps organizations manage their sales force and internal staff. Increasingly, software is being viewed as a utility or service, not a product you purchase.

System software, such as operating systems or utilities, is typically purchased from a software company. Increasingly, however, companies are obtaining open-source systems software, such as the Linux operating system, which can be obtained free or for a low cost.

Externally Acquired Software

A company planning to purchase or lease software from an outside company has many options. Commercial off-the-shelf development is often used. The *commercial off-the-shelf (COTS)* development process involves the use of commonly available products from software vendors. It combines software from various vendors into a finished system. In many cases, it is necessary to write some original software from scratch and combine it with purchased or leased software. For example, a company can purchase or lease software from several software vendors and combine it into a finished software program. COTS can be less expensive than developing an application from scratch. It can streamline and shorten the time needed to develop software. The other steps of the systems development life cycle, such as requirements analysis, testing, and implementation, must still be carefully done. A major challenge with COTS development is integrating all the off-the-shelf components into a unified software package. Other potential problems of the COTS development approach can include no access to the source code, the inability to make changes or updates, and the possibility of quality and security problems concerning the COTS software or components.

Newmarket International, which pioneered hospitality sales and event-management software, created Delphi, a forecasting and tracking tool that gives salespeople instant, updated on-screen room information, including transient commitments, target rates, and room type availability.

(Source: Courtesy of Newmarket International, Inc.)

In-House-Developed Software

Another option is to make or develop software internally. This approach requires the company's IS personnel to be responsible for all aspects of software development. Some advantages inherent with in-house-developed software include meeting user and organizational requirements and having more features and increased flexibility in terms of customization and changes. Software programs developed within a company also have greater potential for providing a competitive advantage because competitors cannot easily duplicate them in the short term. Some in-house software developers, however, often try to incorporate every conceivable feature and option into the software. The resulting software, often called "bloatware" can be slow, unreliable, and insecure with too many lines of software code.[23] If software is to be developed internally, a number of tools and techniques can be used. The following are some examples.

- **Chief programmer teams.** The **chief programmer team** is a group of skilled IS professionals with the task of designing and implementing a set of programs. This team has total responsibility for building the best software possible. Individuals on a chief programmer team often have excellent programming skills.

- **Structured programming.** Some program development projects use a structured approach with three allowable structures. In the **sequence structure**, the program must have definite starting and ending points. After starting the sequence, programming statements are executed one after another until all the statements in the sequence have executed. Then the program either ends or continues on to another structure. The **decision structure** allows the computer to branch, depending on certain conditions. The **loop structure** allows the program to cycle or loop a certain number of times.

- **CASE and object-oriented approaches.** As mentioned in Chapter 12, CASE tools and the object-oriented approach are often used during software development. AXA Financial Services, for example, saved millions of dollars in developing a system by reusing software. JetBlue Airways worked with Visual Studio .NET to implement an inventory tracking system that used the object-oriented approach.

- **Cross-platform development.** One software development technique, called **cross-platform development**, allows programmers to develop programs that can run on computer systems that have different hardware and operating systems, or platforms. Web service tools, such as .NET by Microsoft introduced in Chapter 7, are examples. With cross-platform development, for example, the same program can run on both a personal computer and a mainframe or on two different types of PCs.

- **Integrated development environment. Integrated development environments (IDEs)** combine the tools needed for programming with a programming language in one integrated package. An IDE allows programmers to use simple screens, customized pull-down menus, and graphical user interfaces. Visual Studio 2005 from Microsoft is an example of an IDE. Oracle Designer, which is used with Oracle's database system, is another example of an IDE. The popular Eclipse Workbench supports IDEs that can be used with the C and C++ programming languages. Eclipse Workbench includes a debugger and a compiler, along with other tools.

- **Structured walkthroughs.** As shown in Figure 13.11, a **structured walkthrough** is a planned and preannounced review of the progress of a program or program module. The walkthrough helps team members review and evaluate the progress of components of a project. The structured walkthrough approach is also useful for programming projects that do not use the structured design approach.

- **Documentation.** With internally developed software, documentation is always important. **Technical documentation** is used by computer operators to execute the program and by analysts and programmers to solve problems or modify the program. In technical documentation, the purpose of every major piece of computer code is written out and explained. Key variables are also described. **User documentation** is developed for the people who use the program. This type of documentation shows users, in easy-to-understand terms, how the program can and should be used. Incorporating a description of the benefits of the new application into user documentation can help stakeholders understand the reasons for the program and speed user acceptance.

Acquiring Database and Telecommunications Systems

Because databases are a blend of hardware and software, many of the approaches discussed earlier for acquiring hardware and software also apply to database systems. For example, an upgraded inventory control system might require database capabilities, including more hard disk storage or a new DBMS. If so, additional storage hardware will have to be acquired from an IS vendor. New or upgraded software might also be purchased or developed in-house. MasterCard International, for example, needed to acquire additional storage capacity.[24]

chief programmer team
A group of skilled IS professionals who design and implement a set of programs.

sequence structure
A programming structure in which, after starting the sequence, programming statements are executed one after another until all the statements in the sequence have been executed.

decision structure
A programming structure that allows the computer to branch, depending on certain conditions.

loop structure
A programming structure that allows the program to cycle or loop a certain number of times.

cross-platform development
A development technique that allows programmers to develop programs that can run on computer systems having different hardware and operating systems, or platforms.

integrated development environments (IDEs)
A development approach that combines the tools needed for programming with a programming language into one integrated package.

structured walkthrough
A planned and preannounced review of the progress of a program module.

technical documentation
Written details used by computer operators to execute the program and by analysts and programmers to solve problems or modify the program.

user documentation
Written descriptions developed for people who use a program, showing users, in easy-to-understand terms, how the program can and should be used.

Existing storage capacity was about to run out as the company expanded its business. Instead of adding incremental storage capacity, the company decided to use a large-scale storage area network (SAN). The results were immediate and apparent. According to a company representative, "The company cut time to market, eliminated inefficiencies, lowered storage costs, and improved customer service."

With the increased use of e-commerce, the Internet, intranets, and extranets, telecommunications is one of the fastest-growing applications for today's businesses and people. Over 200 e-commerce Web sites, including www.walmart.com, have implemented a new payment system.[25] The new system, built by I4 systems, asks customers to enter their birth date and the last four numbers of their Social Security number. The software then checks various credit bureaus and databases. In about four seconds, customers are told if their purchase is approved. If purchases are approved, customers are sent bills via traditional mail or e-mail. Like database systems, telecommunications systems require a blend of hardware and software. For personal computer systems, the primary piece of hardware is a modem. For client/server and mainframe systems, the hardware can include multiplexers, concentrators, communications processors, and a variety of network equipment. Communications software will also have to be acquired from a software company or developed in-house. Again, the earlier discussion on acquiring hardware and software also applies to the acquisition of telecommunications hardware and software.

Figure 13.11

Structured Walkthrough

A structured walkthrough is a planned, preannounced review of the progress of a particular project objective.

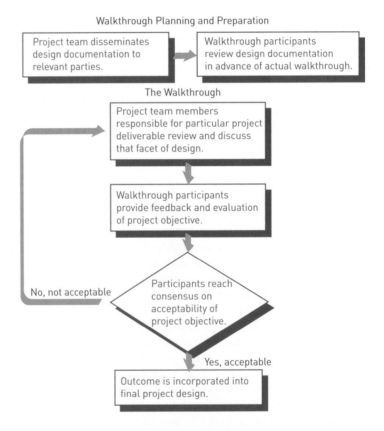

Walkthrough Planning and Preparation

Project team disseminates design documentation to relevant parties.

Walkthrough participants review design documentation in advance of actual walkthrough.

The Walkthrough

Project team members responsible for particular project deliverable review and discuss that facet of design.

Walkthrough participants provide feedback and evaluation of project objective.

No, not acceptable

Participants reach consensus on acceptability of project objective.

Yes, acceptable

Outcome is incorporated into final project design.

User Preparation

user preparation
The process of readying managers, decision makers, employees, other users, and stakeholders for new systems.

User preparation is the process of readying managers, decision makers, employees, other users, and stakeholders for the new systems. This activity is an important but often ignored area of systems implementation. A small airline might not adequately train employees with a new software package. The result could be a grounding of most of its flights and the need to find hotel rooms to accommodate unhappy travelers who are stranded.

Without question, training users is an essential part of user preparation, whether they are trained by internal personnel or by external training firms. In some cases, companies that provide software also train users at no charge or at a reasonable price. The cost of training

Providing users with proper training can help ensure that the information system is used correctly, efficiently, and effectively.

(Source: Ciaran Griffin/Getty Images.)

can be negotiated during the selection of new software. Other companies conduct user training throughout the systems development process. Concerns and apprehensions about the new system must be eliminated through these training programs. Employees should be acquainted with the system's capabilities and limitations by the time they are ready to use it.

IS Personnel: Hiring and Training

Depending on the size of the new system, an organization might have to hire and, in some cases, train new IS personnel. An IS manager, systems analysts, computer programmers, data-entry operators, and similar personnel might be needed for the new system.

As with users, the eventual success of any system depends on how it is used by the IS personnel within the organization. Training programs should be conducted for the IS personnel who will be using the computer system. These programs are similar to those for the users, although they can be more detailed in the technical aspects of the systems. Effective training will help IS personnel use the new system to perform their jobs and support other users in the organization.

Site Preparation

The location of the new system needs to be prepared, a process called **site preparation**. For a small system, site preparation can be as simple as rearranging the furniture in an office to make room for a computer. With a larger system, this process is not so easy because it can require special wiring and air conditioning. One or two rooms might have to be completely renovated, and additional furniture might have to be purchased. A special floor might have to be built, under which the cables connecting the various computer components are placed, and a new security system might be needed to protect the equipment. For larger systems, additional power circuits might also be required.

site preparation
Preparation of the location of a new system.

Data Preparation

Data preparation, or **data conversion**, involves making sure that all files and databases are ready to be used with new computer software and systems. If an organization is installing a new payroll program, the old employee-payroll data might have to be converted into a format that can be used by the new computer software or system. After the data has been prepared or converted, the computerized database system or other software will then be used to maintain and update the computer files.

data preparation, or data conversion
Ensuring all files and databases are ready to be used with new computer software and systems.

Installation

Installation is the process of physically placing the computer equipment on the site and making it operational. Although normally the manufacturer is responsible for installing computer equipment, someone from the organization (usually the IS manager) should oversee the process, making sure that all equipment specified in the contract is installed at the

installation
The process of physically placing the computer equipment on the site and making it operational.

proper location. After the system is installed, the manufacturer performs several tests to ensure that the equipment is operating as it should.

Testing

Good testing procedures are essential to make sure that the new or modified information system operates as intended. Inadequate testing can result in mistakes and problems. A popular tax preparation company, for example, implemented a Web-based tax preparation system, but people could see one another's tax returns. The president of the tax preparation company called it "our worst-case scenario." Better testing can prevent these types of problems.

Several forms of testing should be used, including testing each program (**unit testing**), testing the entire system of programs (**system testing**), testing the application with a large amount of data (**volume testing**), and testing all related systems together (**integration testing**), as well as conducting any tests required by the user (**acceptance testing**). Figure 13.12 lists the types of testing. In addition to these forms of testing, there are different types of testing.

unit testing
Testing of individual programs.

system testing
Testing the entire system of programs.

volume testing
Testing the application with a large amount of data.

integration testing
Testing all related systems together.

acceptance testing
Conducting any tests required by the user.

Figure 13.12

Types of Testing

alpha testing
Testing an incomplete or early version of the system.

beta testing
Testing a complete and stable system by end users.

Alpha testing involves testing an incomplete or early version of the system, while **beta testing** involves testing a complete and stable system by end users. Alpha-unit testing, for example, is testing an individual program before it is completely finished. Beta-unit testing, on the other hand, is performed after alpha testing, when the individual program is complete and ready for use by end users.

Unit testing is accomplished by developing test data that will force the computer to execute every statement in the program. In addition, each program is tested with abnormal data to determine how it will handle problems.

System testing requires the testing of all the programs together. It is not uncommon for the output from one program to become the input for another. So, system testing ensures that the output from one program can be used as input for another program within the system. Volume testing ensures that the entire system can handle a large amount of data under normal operating conditions. Integration testing ensures that the new programs can interact with other major applications. It also ensures that data flows efficiently and without error to other applications. For example, a new inventory control application might require data input from an older order processing application. Integration testing would be done to ensure smooth data flow between the new and existing applications. Integration testing is typically done after unit and system testing. Metaserver, a software company for the insurance industry, has developed a tool called iConnect to perform integration testing for different insurance applications and databases.

Finally, acceptance testing makes sure that the new or modified system is operating as intended. Run times, the amount of memory required, disk access methods, and more can be tested during this phase. Acceptance testing ensures that all performance objectives defined for the system or application are satisfied. Insurance company AON, for example, uses acceptance testing to make sure that new software is ready for use. Involving users in acceptance testing can help them understand and effectively interact with the new system. Acceptance testing is the final check of the system before start-up.

Start-Up

Start-up, also called *cutover*, begins with the final tested information system. When start-up is finished, the system is fully operational. Start-up can be critical to the success of the organization. If not done properly, the results can be disastrous. One of the authors is aware of a small manufacturing company that decided to stop an accounting service used to send out bills on the same day they were going to start their own program to send out bills to customers. The manufacturing company wanted to save money by using their own billing program developed by an employee of the company. The new program didn't work, the accounting service wouldn't help because they were upset about being terminated, and the manufacturing company wasn't able to send out any bills to customers for more than 3 months. The manufacturing company almost went bankrupt.

Various start-up approaches are available (see Figure 13.13). **Direct conversion** (also called *plunge* or *direct cutover*) involves stopping the old system and starting the new system on a given date. Direct conversion is usually the least desirable approach because of the potential for problems and errors when the old system is shut off and the new system is turned on at the same instant.

start-up
The process of making the final tested information system fully operational.

direct conversion (also called *plunge* or *direct cutover*)
Stopping the old system and starting the new system on a given date.

Figure 13.13

Start-Up Approaches

Direct conversion

Phase-in approach

Pilot start-up

Parallel start-up

phase-in approach
Slowly replacing components of the old system with those of the new one. This process is repeated for each application until the new system is running every application and performing as expected; also called a *piecemeal approach*.

pilot start-up
Running the new system for one group of users rather than all users.

parallel start-up
Running both the old and new systems for a period of time and comparing the output of the new system closely with the output of the old system; any differences are reconciled. When users are comfortable that the new system is working correctly, the old system is eliminated.

user acceptance document
A formal agreement signed by the user that states that a phase of the installation or the complete system is approved.

The **phase-in approach** is a popular technique preferred by many organizations. In this approach, sometimes called a *piecemeal approach*, components of the new system are slowly phased in while components of the old one are slowly phased out. When everyone is confident that the new system is performing as expected, the old system is completely phased out. This gradual replacement is repeated for each application until the new system is running every application. In some cases, the phase-in approach can take months or years.

Pilot start-up involves running the new system for one group of users rather than all users. For example, a manufacturing company with many retail outlets throughout the country could use the pilot start-up approach and install a new inventory control system at one of the retail outlets. When this pilot retail outlet runs without problems, the new inventory control system can be implemented at other retail outlets. Carnival Cruise Lines, for example, used a pilot start-up for a systems development project to remotely manage PCs. England's National Health Service Cancer Registry, for example, used a pilot start-up approach to implement and test a new system to manage and integrate hundreds of cancer-related data sources.[26]

Parallel start-up involves running both the old and new systems for a period of time. The output of the new system is compared closely with the output of the old system, and any differences are reconciled. When users are comfortable that the new system is working correctly, the old system is eliminated.

User Acceptance

Most mainframe computer manufacturers use a formal **user acceptance document**—a formal agreement the user signs stating that a phase of the installation or the complete system is approved. This is a legal document that usually removes or reduces the IS vendor's liability for problems that occur after the user acceptance document has been signed. Because this document is so important, many companies get legal assistance before they sign the acceptance document. Stakeholders can also be involved in acceptance testing to make sure that the benefits to them are indeed realized.

SYSTEMS OPERATION AND MAINTENANCE

systems operation
Use of a new or modified system.

Systems operation involves all aspects of using the new or modified system in all kinds of operating conditions. Getting the most out of a new or modified system during its operation is the most important aspect of systems operations for many organizations. Throughout this book, we have seen many examples of information systems operating in a variety of settings and industries. Thus, we will not cover the operation of an information system in detail in this section. The operation of any information system, however, does require adequate training and support before the system is used and continual support while the system is being operated. This training and support is required for all stakeholders, including employees, customers, and others. Companies typically provide training through seminars, manuals, and online documentation. To provide adequate support, many companies use a formal help desk. A *help desk* consists of people with technical expertise, computer systems, manuals, and other resources needed to solve problems and give accurate answers to questions. Wal-Mart, for example, set up a help desk for its employees after Hurricane Katrina.[27] Wal-Mart's IS department set up a system to allow employees to get $250 on the spot and start work if they wanted after the storm. With today's advances in telecommunications, help desks can be located around the world. If you are having trouble with your PC and call a toll-free number for assistance, you might reach a help desk in India, China, or another country. For most organizations, operations costs over the life of a system are much greater than the development costs.

systems maintenance
A stage of systems development that involves checking, changing, and enhancing the system to make it more useful in achieving user and organizational goals.

Systems maintenance involves checking, changing, and enhancing the system to make it more useful in achieving user and organizational goals.[28] Maintenance is important for individuals, groups, and organizations.[29] Individuals, for example, can use the Internet, computer vendors, and independent maintenance companies, including YourTechOnline.com, Geek Squad, PC Pinpoint, Geek Help, and others.[30] Organizations often have personnel dedicated to maintenance.

This maintenance process can be especially difficult for older software. A *legacy system* is an old system that might have been patched or modified repeatedly over time. An old payroll program in COBOL developed decades ago and frequently changed is an example of a legacy system. Legacy systems can be very expensive to maintain. At some point, it becomes less expensive to switch to new programs and applications than to repair and maintain the legacy system. Maintenance costs for older legacy systems can be 50 percent of total operating costs in some cases.

Software maintenance is a major concern for organizations. In some cases, organizations encounter major problems that require recycling the entire systems development process. In other situations, minor modifications are sufficient to remedy problems. Hardware maintenance is also important. Companies such as IBM are investigating *autonomic computing*, in which computers will be able to manage and maintain themselves.[31] The goal is for computers to be self-configuring, self-protecting, self-healing, and self-optimizing. Being self-configuring allows a computer to handle new hardware, software, or other changes to its operating environment. Being self-protecting means a computer can identify potential attacks, prevent them when possible, and recover from attacks if they occur. Attacks can include viruses, worms, identity theft, and industrial espionage. Being self-healing means a computer can fix problems when they occur, and being self-optimizing allows a computer to run faster and get more done in less time. Exempla Health is using an autonomic approach to maintaining its 4,000 workstations to prevent virus attacks and save money.[32] The autonomic approach to maintenance is expected to save Exempla Health about $800,000 per month.

Getting rid of old equipment is an important part of maintenance. The options include selling it on Web auction sites such as eBay, recycling the equipment at a computer-recycling center, and donating it to a charitable organization, such as a school, library, or religious organization. When discarding old computer systems, it is always a good idea to permanently remove sensitive files and programs. Companies, including McAfee and Blancco, have software to help people remove data and programs from old computers and transfer them to new ones. [33]

Reasons for Maintenance

After a program is written, it will need ongoing maintenance. To some extent, a program is similar to a car that needs oil changes, tune-ups, and repairs at certain times. Experience shows that frequent, minor maintenance to a program, if properly done, can prevent major system failures later. Some of the reasons for program maintenance are the following:

- Changes in business processes
- New requests from stakeholders, users, and managers
- Bugs or errors in the program
- Technical and hardware problems
- Corporate mergers and acquisitions
- Government regulations
- Change in the operating system or hardware on which the application runs
- Unexpected events, such as severe weather or the terrorist attacks of September 11, 2001

Most companies modify their existing programs instead of developing new ones because existing software performs many important functions, and companies can have millions of dollars invested in their old legacy systems. So, as new systems needs are identified, the burden of fulfilling the needs most often falls on the existing system. Old programs are repeatedly modified to meet ever-changing needs. Yet, over time, repeated modifications tend to interfere with the system's overall structure, reducing its efficiency and making further modifications more burdensome.

San Francisco's Bay Area Rapid Transit System Upgrade

Today's technologically advanced societies are dependant on properly maintained information systems for many of our day-to-day activities. We are not aware of most of the information systems we depend on unless they fail. The most common time for an information system to fail is shortly after its initial installation, or when it is being upgraded or altered in some way. Commuters in San Francisco were made painfully aware of these facts during the spring of 2006.

San Francisco's Bay Area Rapid Transit System, commonly called BART, began passenger service in the Bay Area in 1972 and now has 670 rail cars in its fleet, which cover 104 route miles through 43 stations. A large portion of the BART system carries commuters through tubes under the San Francisco Bay between Oakland and San Francisco. The system carries about 300,000 riders daily.

The Integrated Control System, which controls the automated trains, has been undergoing a major renovation. About 15 of BART's approximately 100 information system workers are performing the system upgrade, which is taking place in stages over 19 months. Software testing is done in advance on a separate "virtual environment" so that it won't interfere with everyday operations.

Implementing changes to a system that controls transportation 24 hours a day, 7 days a week, without shutting down the system is tricky. Typically, BART implements system upgrades during weekends so that if anything goes wrong, the inconvenience to commuters is minimized. After 14 months of upgrading the system without incident, perhaps BART systems workers were feeling more confident than they should have. They implemented an upgrade that worked flawlessly in the simulated environment but brought the system to a standstill on a Monday, leaving thousands of commuters stranded for up to two hours.

The system was repaired and brought back online, only to fail again on Tuesday, again leaving thousands stranded.

Again, BART systems specialists repaired the problem within hours. In an effort to avoid another system failure, BART developed a redundant backup system that would take over if the primary system failed again—eliminating delays to train schedules. Unfortunately, when bringing the untested backup system online, the new system interfered with network switches causing the primary system to fail and stranding passengers for the third day in a row.

BART systems developers were experiencing the worst nightmare of every systems developer. Linton Johnson, a BART spokesman, stated that "We're taking measures to make sure that this never happens again. This is essentially a mistake on BART's part."

Discussion Questions

1. What sound practices do BART systems developers typically apply to system upgrades prior to start-up?
2. How might BART systems developers have minimized the risk of implementing the upgrade described in this article? What lessons were learned?

Critical Thinking Questions

1. The results of a failure of a system that controls trains could be much, much worse than just inconvenience to commuters. What safeguards would you assume are built in to the BART system to ensure that there is no loss of life in the event of system failure?
2. Do you think the idea of a redundant system is a good one? Why? How might BART implement a redundant system in a more thoughtful manner?

SOURCES: Weiss, Todd R., "IT Upgrades Slow BART Trains in San Francisco," *Computerworld*, March 31, 2006, *www.computerworld.com*. Sebastian, Simone, et al., "35,000 Evening Commuters Left in Lurch—System Takes Blame for Software Snafu," San Francisco Chronicle, March 30, 2006, *www.sfgate.com*. BART Web site, accessed July 10, 2006, *www.bart.gov*.

Types of Maintenance

Software companies and many other organizations use four generally accepted categories to signify the amount of change involved in maintenance. A **slipstream upgrade** is a minor upgrade—typically a code adjustment or minor bug fix. Many companies don't announce to users that a slipstream upgrade has been made. A slipstream upgrade usually requires recompiling all the code, so it can create entirely new bugs. This maintenance practice can explain why the same computers sometimes work differently with what is supposedly the same software. A **patch** is a minor change to correct a problem or make a small enhancement. It is usually an addition to an existing program. That is, the programming code representing the system enhancement is usually "patched into," or added to, the existing code. Although slipstream upgrades and patches are minor changes, they can cause users and support personnel big problems if the programs do not run as before. A new **release** is a significant program change that often requires changes in the documentation of the software. Finally, a new **version** is a major program change, typically encompassing many new features.

The Request for Maintenance Form

Because of the amount of effort that can be spent on maintenance, many organizations require a **request for maintenance form** to authorize modification of programs. This form is usually signed by a business manager, who documents the need for the change and identifies the priority of the change relative to other work that has been requested. The IS group reviews the form and identifies the programs to be changed, determines the programmer who will be assigned to the project, estimates the expected completion date, and develops a technical description of the change. A cost/benefit analysis might be required if the change requires substantial resources.

Performing Maintenance

Depending on organizational policies, the people who perform systems maintenance vary. In some cases, the team who designs and builds the system also performs maintenance. This ongoing responsibility gives the designers and programmers an incentive to build systems well from the outset: If there are problems, they will have to fix them. In other cases, organizations have a separate **maintenance team**. This team is responsible for modifying, fixing, and updating existing software.

In the past, companies had to maintain each computer system or server separately. With hundreds or thousands of computers scattered throughout an organization, this task could be very costly and time consuming. Today, the maintenance function is becoming more automated. Some companies, for example, use maintenance tools and software that will allow them to maintain and upgrade software centrally.

A number of vendors have developed tools to ease the software maintenance burden. Relativity Technologies unveiled RescueWare, a product that converts third-generation code such as COBOL to highly maintainable C++, Java, or Visual Basic object-oriented code. Using RescueWare, maintenance personnel download mainframe code to Windows NT or Windows 2000 workstations. They then use the product's graphical tools to analyze the original system's inner workings. RescueWare lets a programmer see the original system as a set of object views, which visually illustrate module functioning and program structures. IS personnel can choose one of three levels of transformation: revamping the user interface, converting the database access, and transforming procedure logic.

The Financial Implications of Maintenance

The cost of maintenance is staggering. For older programs, the total cost of maintenance can be up to five times greater than the total cost of development. In other words, a program that originally cost $25,000 to develop might cost $125,000 to maintain over its lifetime. The average programmers can spend more than half their time on maintaining existing programs instead of developing new ones. In addition, as programs get older, total maintenance expenditures in time and money increase, as illustrated in Figure 13.14. With the use of

slipstream upgrade
A minor upgrade—typically a code adjustment or minor bug fix—not worth announcing. It usually requires recompiling all the code and, in so doing, it can create entirely new bugs.

patch
A minor change to correct a problem or make a small enhancement. It is usually an addition to an existing program.

release
A significant program change that often requires changes in the documentation of the software.

version
A major program change, typically encompassing many new features.

request for maintenance form
A form authorizing modification of programs.

maintenance team
A special IS team responsible for modifying, fixing, and updating existing software.

Relativity Technologies' Modernization Workbench is a PC-based software solution that enables companies to consolidate legacy or redundant systems into one, more maintainable and modern application.

(Source: Courtesy of Relativity Technologies.)

Figure 13.14

Maintenance Costs as a Function of Age

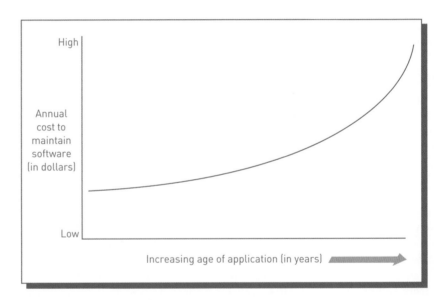

newer programming languages and approaches, including object-oriented programming, maintenance costs are expected to decline. Even so, many organizations have literally millions of dollars invested in applications written in older languages (such as COBOL), which are both expensive and time consuming to maintain. According to Robert Rosen, chief information officer for the National Institute of Arthritis and Skin Diseases in Bethesda, Maryland, "Maintenance is a huge black hole. I have not seen anyone taking a step back and saying, 'OK we have all these systems; what does it really cost to keep these things running?'"[34]

The financial implications of maintenance mean companies must keep track of why systems are maintained, instead of simply keeping cost figures. This is another reason that documentation of maintenance tasks is so crucial. A determining factor in the decision to replace a system is the point at which it is costing more to fix it than to replace it.

The Relationship Between Maintenance and Design

Programs are expensive to develop, but they are even more expensive to maintain. Programs that are well designed and documented to be efficient, structured, and flexible are less expensive to maintain in later years. Thus, there is a direct relationship between design and maintenance. More time spent on design up front can mean less time spent on maintenance later.

In most cases, it is worth the extra time and expense to design a good system. Consider a system that costs $250,000 to develop. Spending 10 percent more on design would cost an additional $25,000, bringing the total design cost to $275,000. Maintenance costs over the life of the program could be $1,000,000. If this additional design expense can reduce maintenance costs by 10 percent, the savings in maintenance costs would be $100,000. Over the life of the program, the net savings would be $75,000 ($100,000 − $25,000). This relationship between investment in design and long-term maintenance savings is graphically displayed in Figure 13.15.

Figure 13.15

The Value of Investment in Design

The need for good design goes beyond mere costs. Companies risk ignoring small system problems when they arise, but these small problems can become large in the future. As mentioned earlier, because maintenance programmers spend an estimated 50 percent or more of their time deciphering poorly written, undocumented program code, they have little time to spend on developing new, more effective systems. If put to good use, the tools and techniques discussed in this chapter will allow organizations to build longer-lasting, more reliable systems.

SYSTEMS REVIEW

Systems review, the final step of systems development, is the process of analyzing systems to make sure that they are operating as intended. This process often compares the performance and benefits of the system as it was designed with the actual performance and benefits of the system in operation.[35] A payroll application being developed for the Irish Health Service, for example, was almost $170 million over budget.[36] As a result, work on the application that serves about 37,000 workers was halted so the entire project could be reviewed in detail. The purpose of the systems review is to make sure that any additional work will result in a program that will work as intended.

Problems and opportunities uncovered during systems review trigger systems development and begin the process anew. For example, as the number of users of an interactive system increases, it is not unusual for system response time to increase. If the increase in response time is too great, it might be necessary to redesign some of the system, modify databases, or increase the power of the computer hardware. Wayne State University, for

systems review
The final step of systems development, involving the analysis of systems to make sure that they are operating as intended.

example, uses the Vantage review tool to make sure that its systems are available and running correctly for faculty and students.[37] When faced with a possible patent infringement problem, RIM, the maker of the popular BlackBerry phone and e-mail service, developed backup software that could be used in case the courts ruled against the company.[38] Even though RIM was able to settle the suit out-of-court, BlackBerry users were happy that the company had a backup plan.

Internal employees, external consultants, or both can perform systems review. When the problems or opportunities are industry-wide, people from several firms can get together. In some cases, they collaborate at an IS conference or in a private meeting involving several firms.

Types of Review Procedures

There are two types of review procedures: event driven and time driven (see Table 13.6). An **event-driven review** is triggered by a problem or opportunity such as an error, a corporate merger, or a new market for products.[39] UPS, for example, used an event-driven review for a new package-flow system that used bar-coded labels and geographic information systems (GISs).[40] The systems development project was expected to save UPS about $700 million per year, but the actual system only saved the company about $100 million. As a result of the review, UPS is retraining many of its employees to take better advantage of the new package-flow system. Natural disasters often revealed flaws in older systems, causing many companies and organizations to review their existing systems.[41] After an event-driven review caused by Hurricane Katrina, the city of Houston decided to revamp its software that analyzes storm-preparation costs.

event-driven review

A review triggered by a problem or opportunity such as an error, a corporate merger, or a new market for products.

Table 13.6

Examples of Review Types

Event Driven	Time Driven
Problem with an existing system	Monthly review
Merger	Yearly review
New accounting system	Review every few years
Executive decision that an upgraded Internet site is needed to stay competitive	Five-year review

In some cases, companies wait until a large problem or opportunity occurs before a change is made, ignoring minor problems. In contrast, some companies use a continuous improvement approach to systems development. With this approach, an organization makes changes to a system even when small problems or opportunities occur. Although continuous improvement can keep the system current and responsive, repeatedly designing and implementing changes can be both time consuming and expensive.

A **time-driven review** is performed after a specified amount of time. Many application programs are reviewed every six months to one year. With this approach, an existing system is monitored on a schedule. If problems or opportunities are uncovered, a new systems development cycle can be initiated. A payroll application, for example, can be reviewed once a year to make sure that it is still operating as expected. If it is not, changes are made.

Many companies use both approaches. A billing application, for example, might be reviewed once a year for errors, inefficiencies, and opportunities to reduce operating costs. This is a time-driven approach. In addition, the billing application might be redone after a corporate merger, if one or more new managers require different information or reports, or if federal laws on bill collecting and privacy change. This is an event-driven approach.

time-driven review

Review performed after a specified amount of time.

Factors to Consider During Systems Review

Systems review should investigate a number of important factors, such as the following:

- **Mission.** Is the computer system helping the organization achieve its overall mission? Are stakeholder needs and desires satisfied or exceeded with the new or modified system?
- **Organizational goals.** Does the computer system support the specific goals of the various areas and departments of the organization?
- **Hardware and software.** Are hardware and software up to date and adequate to handle current and future processing needs?
- **Database.** Is the current database up to date and accurate? Is database storage space adequate to handle current and future needs?
- **Telecommunications.** Is the current telecommunications system fast enough, and does it allow managers and workers to send and receive timely messages? Does it allow for fast order processing and effective customer service?
- **Information systems personnel.** Are there sufficient IS personnel to perform current and projected processing tasks?
- **Control.** Are rules and procedures for system use and access acceptable? Are the existing control procedures adequate to protect against errors, invasion of privacy, fraud, and other potential problems?
- **Training.** Are there adequate training programs and provisions for both users and IS personnel?
- **Costs.** Are development and operating costs in line with what is expected? Is there an adequate IS budget to support the organization?
- **Complexity.** Is the system overly complex and difficult to operate and maintain?
- **Reliability.** Is the system reliable? What is the mean time between failures (MTBF)?
- **Efficiency.** Is the computer system efficient? Are system outputs generated by the right amount of inputs, including personnel, hardware, software, budget, and others?
- **Response time.** How long does it take the system to respond to users during peak processing times?
- **Documentation.** Is the documentation still valid? Are changes in documentation needed to reflect the current situation?

System Performance Measurement

Systems review often involves monitoring the system, called **system performance measurement**. The number of errors encountered, the amount of memory required, the amount of processing or CPU time needed, and other problems should be closely observed.[42] If a particular system is not performing as expected, it should be modified, or a new system should be developed or acquired.

Setting up benchmarks for performance measurement can be critical. DuPont, for example, developed benchmarks for computer system availability and network speed.[43] The chemical manufacturer also developed benchmarks for its human resources department. The benchmarks are tied to performance goals and compares DuPont to competitors in its industry.

System performance products have been developed to measure all components of the information system, including hardware, software, database, telecommunications, and network systems. When properly used, system performance products can quickly and efficiently locate actual or potential problems.

A number of products have been developed to assist in assessing system performance. OMEGAMON from IBM can monitor system performance in real time. Precise Software Solutions has system performance products that provide around-the-clock performance monitoring for Oracle database applications. Mercury Interactive offers a software tool called Diagnostic to help companies analyze the performance of their computer systems, diagnose potential problems, and take corrective action if needed.[44]

Measuring a system is, in effect, the final task of systems development. The results of this process can bring the development team back to the beginning of the development life cycle, where the process begins again.

system performance measurement
Monitoring the system—the number of errors encountered, the amount of memory required, the amount of processing or CPU time needed, and other problems.

system performance products
Software that measures all components of the computer-based information system, including hardware, software, database, telecommunications, and network systems.

SUMMARY

Principle

Designing new systems or modifying existing ones should always help an organization achieve its goals.

The purpose of systems design is to prepare the detailed design needs for a new system or modifications to the existing system. Logical systems design refers to the way that the various components of an information system will work together. The logical design includes data requirements for output and input, processing, files and databases, telecommunications, procedures, personnel and job design, and controls and security design. Physical systems design refers to the specification of the actual physical components. The physical design must specify characteristics for hardware and software design, database and telecommunications, and personnel and procedures design.

Logical and physical design can be accomplished using the traditional systems development life cycle or the object-oriented approach. Using the OO approach, analysts design key objects and classes of objects in the new or updated system. The sequence of events that a new or modified system requires is often called a *scenario*, which can be diagrammed in a sequence diagram.

A number of special design considerations should be taken into account during both logical and physical system design. Interface design and control relates to how users access and interact with the system. A sign-on procedure consists of identification numbers, passwords, and other safeguards needed for individuals to gain access to computer resources. If the system under development is interactive, the design must consider menus, help facilities, table lookup facilities, and restart procedures. A good interactive dialogue will ask for information in a clear manner, respond rapidly, be consistent among applications, and use an attractive format. Also, it will avoid use of computer jargon and treat the user with respect.

System security and control involves many aspects. Error prevention, detection, and correction should be part of the system design process. Causes of errors include human activities, natural phenomena, and technical problems. Designers should be alert to prevention of fraud and invasion of privacy.

Disaster recovery is an important aspect of systems design. Disaster planning is the process of anticipating and providing for disasters. A disaster can be an act of nature (a flood, fire, or earthquake) or a human act (terrorism, error, labor unrest, or erasure of an important file). The primary tools used in disaster planning and recovery are hardware, software, database, telecommunications, and personnel backup.

Security, fraud, and the invasion of privacy are also important design considerations. Most IS departments establish tight systems controls to maintain data security. Systems controls can help prevent computer misuse, crime, and fraud by employees and others. Systems controls include input, output, processing, database, telecommunications, and personnel controls.

Whether an individual is purchasing a personal computer or an experienced company is acquiring an expensive mainframe computer, the system could be obtained from one or more vendors. Some of the factors to consider in selecting a vendor are the vendor's reliability and financial stability, the type of service offered after the sale, the goods and services the vendor offers and keeps in stock, the vendor's willingness to demonstrate its products, the vendor's ability to repair hardware, the vendor's ability to modify its software, the availability of vendor-offered training of IS personnel and system users, and evaluations of the vendor by independent organizations.

If new hardware or software will be purchased from a vendor, a formal request for proposal (RFP) is needed. The RFP outlines the company's needs; in response, the vendor provides a written reply. In addition to responding to the company's stated needs, the vendor provides data on its operations. This data might include the vendor's reliability and stability, the type of postsale service offered, the vendor's ability to perform repairs and fix problems, the available vendor training, and the vendor's reputation. Financial options to consider include purchase, lease, and rent.

RFPs from various vendors are reviewed and narrowed down to the few most likely candidates. In the final evaluation, a variety of techniques—including group consensus, cost/benefit analysis, point evaluation, and benchmark tests—can be used. In group consensus, a decision-making group is appointed and given responsibility for making the final evaluation and selection. With cost/benefit analysis, all costs and benefits of the alternatives are expressed in monetary terms. Benchmarking involves comparing computer systems operating under the same condition. Point evaluation assigns weights to evaluation factors, and each alternative is evaluated in terms of each factor and given a score from 0 to 100. After the vendor is chosen, contract negotiations can begin.

At the end of the systems design step, the final specifications are frozen and no changes are allowed so that implementation can proceed. One of the most important steps in systems design is to develop a good contract if new computer facilities are being acquired. A final design report is developed at the end of the systems design phase.

Principle

The primary emphasis of systems implementation is to make sure that the right information is delivered to the right person in the right format at the right time.

The purpose of systems implementation is to install the system and make everything, including users, ready for its operation. Systems implementation includes hardware acquisition, software acquisition or development, user preparation, hiring and training of personnel, site and data preparation, installation, testing, start-up, and user acceptance. Hardware acquisition requires purchasing, leasing, or renting computer resources from an IS vendor. Hardware is typically obtained from a computer hardware vendor.

Software can be purchased from vendors or developed in-house—a decision termed the *make-or-buy decision*. A purchased software package usually has a lower cost, less risk regarding the features and performance, and easy installation. The amount of development effort is also less when software is purchased. Developing software can result in a system that more closely meets the business needs and has increased flexibility in terms of customization and changes. Developing software also has greater potential for providing a competitive advantage. Increasingly, companies are using service providers to acquire software, Internet access, and other IS resources.

Software development is often performed by a chief programmer team—a group of IS professionals who design, develop, and implement a software program. Structured design is a philosophy of designing and developing application software. Other tools, such as cross-platform development and integrated development environments (IDEs), make software development easier and more thorough. CASE tools are often used to automate some of these techniques.

Database and telecommunications software development involves acquiring the necessary databases, networks, telecommunications, and Internet facilities. Companies have a wide array of choices, including newer object-oriented database systems.

Implementation must address personnel requirements. User preparation involves readying managers, employees, and other users for the new system. New IS personnel might need to be hired, and users must be well trained in the system's functions. Preparation of the physical site of the system must be done, and any existing data to be used in the new system will require conversion to the new format. Hardware installation is done during the implementation step, as is testing. Testing includes program (unit) testing, systems testing, volume testing, integration testing, and acceptance testing.

Start-up begins with the final tested information system. When start-up is finished, the system is fully operational. There are a number of different start-up approaches. Direct conversion (also called *plunge* or *direct cutover*) involves stopping the old system and starting the new system on a given date. With the phase-in approach, sometimes called a *piecemeal approach*, components of the new system are slowly phased in while components of the old one are slowly phased out. When everyone is confident that the new system is performing as expected, the old system is completely phased out. Pilot start-up involves running the new system for one group of users rather than all users. Parallel start-up involves running both the old and new systems for a period of time. The output of the new system is compared closely with the output of the old system, and any differences are reconciled. When users are comfortable that the new system is working correctly, the old system is eliminated. Many IS vendors ask the user to sign a formal user acceptance document that releases the IS vendor from liability for problems that occur after the document is signed.

Principle

Maintenance and review add to the useful life of a system but can consume large amounts of resources. These activities can benefit from the same rigorous methods and project management techniques applied to systems development.

Systems operation is the use of a new or modified system. Systems maintenance involves checking, changing, and enhancing the system to make it more useful in obtaining user and organizational goals. Maintenance is critical for the continued smooth operation of the system. The costs of performing maintenance can well exceed the original cost of acquiring the system. Some major causes of maintenance are new requests from stakeholders and managers, enhancement requests from users, bugs or errors, technical or hardware problems, newly added equipment, changes in organizational structure, and government regulations.

Maintenance can be as simple as a program patch to correct a small problem to the more complex upgrading of software with a new release from a vendor. For older programs, the total cost of maintenance can be greater than the total cost of development. Increased emphasis on design can often reduce maintenance costs. Requests for maintenance should be documented with a request for maintenance form, a document that formally authorizes modification of programs. The development team or a specialized maintenance team can then make approved changes. Maintenance can be greatly simplified with the object-oriented approach.

Systems review is the process of analyzing and monitoring systems to make sure that they are operating as intended. The two types of review procedures are the event-driven review and the time-driven review. An event-driven review is triggered by a problem or opportunity. A time-driven review is started after a specified amount of time.

Systems review involves measuring how well the system is supporting the mission and goals of the organization. System performance measurement monitors the system for number of errors, amount of memory and processing time required, and so on.

CHAPTER 13: SELF-ASSESSMENT TEST

Designing new systems or modifying existing ones should always help an organization achieve its goals.

1. _____ details system outputs, inputs, and user interfaces; specifies hardware, software, databases, telecommunications, personnel, and procedures; and shows how these components are related.

2. Determining the needed hardware and software for a new system is an example of _____.
   a. logical design
   b. physical design
   c. interactive design
   d. object-oriented design

3. Disaster planning is an important part of designing security and control systems. True or False?

4. The _____ often results in a formal bid that is used to determine who gets a contract for designing new or modifying existing systems. It specifies in detail the required resources such as hardware and software.

5. Scenarios and sequence diagrams are used with _____.
   a. object-oriented design
   b. point evaluation
   c. incremental design
   d. nominal evaluation

6. Near the end of the design stage, an organization prohibits further changes in the design of the system. This is called _____.

7. In object-oriented systems design, a sequence of events is called a *scenario* and can be diagrammed in a sequence diagram. True or False?

The primary emphasis of systems implementation is to make sure that the right information is delivered to the right person in the right format at the right time.

8. ASP is an example of an IS vendor that offers hardware and software solutions. True or False?

9. Software can be purchased from external developers or developed in-house. This decision is often called the _____ decision.

10. What type of documentation is used by computer operators to execute a program and by analysts and programmers?
    a. unit documentation
    b. integrated documentation
    c. technical documentation
    d. user documentation

11. _____ testing involves testing the entire system of programs.

12. The phase-in approach to conversion involves running both the old system and the new system for three months or longer. True or False?

Maintenance and review add to the useful life of a system but can consume large amounts of resources. These activities can benefit from the same rigorous methods and project management techniques applied to systems development.

13. A(n) _____ is a minor change to correct a problem or make a small enhancement to a program or system.

14. Many organizations require a request for maintenance form to authorize modification of programs. True or False?

15. A systems review that is caused by a problem with an existing system is called _____.
    a. object review
    b. structured review
    c. event-driven review
    d. critical factors review

16. Corporate mergers and acquisitions can be a reason for systems maintenance. True or False?

17. Monitoring a system after it has been implemented is called _____.

CHAPTER 13: SELF-ASSESSMENT TEST ANSWERS

(1) Systems design (2) b (3) True (4) request for proposal (RFP) (5) a (6) freezing design specifications (7) True (8) False (9) make-or-buy (10) c (11) System (12) False (13) patch (14) True (15) c (16) True (17) system performance measurement

REVIEW QUESTIONS

1. What is the purpose of systems design?
2. What are procedures requirements?

3. What is interactive processing? What design factors should be taken into account for this type of processing?

4. How can the object-oriented approach be used during systems design?
5. What is the difference between logical and physical design?
6. What are the different types of software and database backup? Describe the procedure you use to back up your homework files.
7. Identify specific controls that are used to maintain input integrity and security.
8. What is an RFP? What is typically included in one? How is it used?
9. What activities go on during the user preparation phase of systems implementation?

10. What is systems operation?
11. What are the major steps of systems implementation?
12. What are some tools and techniques for software development?
13. Give three examples of an IS vendor.
14. Describe how you back up the files you use at school.
15. What are the steps involved in testing the information system?
16. What are some of the reasons for program maintenance? Explain the types of maintenance.
17. How is system performance measurement related to the systems review?

DISCUSSION QUESTIONS

1. Describe the participants in the systems design stage. How do these participants compare with the participants of systems investigation?
2. Assume that you are the owner of a company that is about to start marketing and selling bicycles over the Internet. Describe your top three objectives in developing a new Web site for this systems development project.
3. Assume that you want to start a new video rental business for students at your college or university. Go through logical design for a new information system to help you keep track of the videos in your inventory.
4. Assume that you are the owner of an online stock-trading company. Describe how you could design the trading system to recover from a disaster.
5. Identify some of the advantages and disadvantages of purchasing versus developing software.
6. Discuss the relationship between maintenance and systems design.
7. Is it equally important for all systems to have a disaster recovery plan? Why or why not?
8. Several approaches were discussed to evaluate a number of systems acquisition alternatives. No one approach is always the best. How would you decide which approach to use for evaluation when selecting a new personal computer and printer?
9. What are the advantages and disadvantages of the object-oriented approach to systems implementation?

10. You have been hired to oversee a major systems development effort to purchase a new accounting software package. Describe what is important to include in the contract with the software vendor.
11. Assume that you are starting an Internet site to sell clothing. Describe how you would design the interactive processing system for this site. Draw a diagram showing the home Web page for the site. Describe the important features of this home page.
12. Identify the various forms of testing. Why are there so many different types of tests?
13. What is the goal of conducting a systems review? What factors need to be considered during systems review?
14. Describe how you would select the best admissions software for your college or university. What features would be most important for school administrators? What features would be most important for students?
15. What issues might you expect to arise if you initiate the use of a request for maintenance form when none had been required previously? How would you deal with these issues?
16. Assume that you have a personal computer that is several years old. Describe the steps you would use to perform systems review to determine whether you should acquire a new PC.

PROBLEM-SOLVING EXERCISES

1. You have been hired to develop a new computer system for a video rental business using the object-oriented approach. Using the information presented in this chapter and in Chapter 12, describe in a report the approach you would use. Using a graphics program, use the object-oriented approach to document the parts of the new computer system.

2. A project team has estimated the costs associated with the development and maintenance of a new system. One approach requires a more complete design and will result in a slightly higher design and implementation cost but a lower maintenance cost over the life of the system. The second approach cuts the design effort, saving some dollars but with a likely increase in maintenance cost.

   a. Enter the following data in the spreadsheet. Print the result.

The Benefits of Good Design

	Good Design	Poor Design
Design costs	$14,000	$10,000
Implementation cost	$42,000	$35,000
Annual maintenance cost	$32,000	$40,000

   b. Create a stacked bar graph that shows the total cost, including the design, implementation, and maintenance costs. Be sure that the chart has a title and that the costs are labeled on the chart.

   c. Use your word processing software to write a paragraph that recommends an approach to take and why.

3. You have been hired to design a new sales ordering program. The program needs a database that contains a customer table containing important customer information, an inventory table that contains current inventory levels, and an order table that contains customer number, inventory number, and order quantity. Develop a database that shows the fields in each table. Include ten sample records in each table.

TEAM ACTIVITIES

1. Assume that your project team has been working for three months to complete the systems design of a new Web-based customer ordering system. Two possible options seem to meet all users' needs. The project team must make a final decision on which option to implement. The following table summarizes some of the key facts about each option.

   a. What process would you follow to make this important decision?
   b. Who needs to be involved?
   c. What additional questions need to be answered to make a good decision?
   d. Based on the data, which option would you recommend and why?
   e. How would you account for project risk in your decision making?

2. Your team must select one PC for your team to use for classroom assignments. The cost of the PC should be $2,000 or less. Only using the Internet to communicate without using any face-to-face meetings, determine the best PC for the team.

Factor	Option 1	Option 2
Annual gross savings	$1.5 million	$3.0 million
Total development cost	$1.5 million	$2.2 million
Annual operating cost	$0.5 million	$1.0 million
Time required to implement	9 months	15 months
Risk associated with project (expressed in probabilities)		
Benefits will be 50% less than expected	20%	35%
Cost will be 50% greater than expected	25%	30%
Organization will not/cannot make changes necessary for system to operate as expected	20%	25%
Does system meet all mandatory requirements?	Yes	Yes

3. Your team has been asked to purchase and install a network system that includes five PCs, two printers, and a wireless network for a small business. Develop an RFP that is to be sent to four PC vendors that specifies all the equipment and software that is needed.

WEB EXERCISES

1. Use the Internet to find two different systems development projects that failed to meet cost or performance objectives. Summarize the problems and what should have been done. You might be asked to develop a report or send an e-mail message to your instructor about what you found.

2. Using the Web, search for information on structured design and programming. Also search the Web for information about the object-oriented approach to systems design and implementation. Write a report on what you found. Under what conditions would you use these approaches to systems development and implementation?

CAREER EXERCISES

1. Describe what type of information system you would need in your chosen job. Your description should include logical and physical design. What specific steps would you include to be able to recover from a natural or man-made disaster, such as a hurricane or terrorist attack?

2. What is the most important software product that could help you succeed in a career of your choice? Describe how you would select a vendor for the software.

CASE STUDIES

Case One

Port of Brisbane Develops System to Tighten Security

Since the terrorist attacks of September 11, security has become a major concern for countries around the world, especially security at borders, airports, and seaports. In 2005, the Australian government passed a law that mandates Australian airports and ports allow the government to run background security checks on all employees. After they are cleared, airport and seaport employees should be issued security ID cards: Aviation Security Identification Cards (ASICs) and Maritime Security Identification Cards (MSICs).

According to government documentation, a maritime security identification card system should be in place by the end of 2006, with all people unmonitored within a maritime security zone required to display a pass starting January 1, 2007. The Australian government established a new department called AusCheck that should work with the Australian Security Intelligence Organization (ASIO) and the Australian Federal Police (AFP) to perform the background checks and issue ID cards. It is left to the seaports to provide security training for employees, make sure employees fill out the application forms for ID cards, submit applications to AusCheck, distribute ID cards to cleared employees, and work with government officials to deal with employees who don't clear the security check.

The government-owned Port of Brisbane Corporation is responsible for the operation and management of Australia's third busiest container port and the first to develop a system to process their 250 employees through the security check process. Believing strongly in the power of computer-based information systems, the administrators at the Port of Brisbane provided the requirements of the system to their system analysts as the basis for developing a system to automate the process.

Systems analysts first considered alternatives to training Port employees. They decided that the ideal method of delivery for the required 15-page security awareness course was over the Web. Rather than creating the training system themselves, they turned to the experts: TodayCorp, the Asia-Pacific's largest online learning, development, and recruitment company, and one of the top ten eLearning solutions companies globally.

TodayCorp worked with Port of Brisbane systems experts to design an all-in-one solution that includes a Web-based eLearning program and ID card registration system. Providing the training online saves the Port of Brisbane from having to hire trainers to provide face-to-face classes. An online system allows employees to work through the training at times convenient for them rather than attending a class at a required time. It also saves the Port the expense and inconvenience of taking employees off duty for training.

TodayCorp developed a training system that provided an enjoyable and thorough method of training. Quizzes are provided throughout the training to make sure employees are obtaining and retaining the necessary information. At the conclusion of the lesson, employees fill out an electronic registration form for a security ID card. When submitted, the system collects the employee information gathered from the online form and transfers it directly to the AusCheck database over a secure VPN connection.

After AusCheck has completed the security check, data is returned to the Port, which then notifies the employee. Employees report to the Port office where a photo is taken and the ID card is issued.

It took two months for systems developers at the Port of Brisbane and TodayCorp to develop and deliver the system. The system was reviewed and tested by Port administrators and given approval for use in February 2006.

TodayCorp developed the training and registration system using Macromedia Flash and XML. The system was developed in a manner that will allow TodayCorp to custom-fit the system for other Australian ports with only a few minor changes.

Discussion Questions

1. What motivated the Port of Brisbane to develop its new system?
2. What were the requirements of the system design?

Critical Thinking Questions

1. What were the benefits and costs of the new system to the Port of Brisbane Corporation, its employees, the Australian government, and Australian citizens?
2. What benefits were obtained by developing this system using commonly accepted standards such as Flash and XML in a manner that allows for customization?

SOURCES: Crawford, Michael, "Aussie Ports Float Online Training to Meet Security Mandate," *Computerworld*, April 10, 2006, *www.computerworld.com.* Ferguson, Iain, "Government Turns to IT for Airport, Port Security," ZDNet Australia, *www.zdnet.com.au.* TodayCorp Web site, accessed July 12, 2006, *www.todaycorp.com.* Port of Brisbane Web site, accessed July 12, 2006, *www.portbris.com.au.* Australian Government Web site, accessed July 12, 2006, *www.tisn.gov.au/agd/WWW/agdHome.nsf/Page/RWP5C5D02388E1D3F35CA25719300234577.*

Case Two

The Christian Science Monitor Moves to Open Source for Content Management System

Selecting a software development platform, which usually takes place during the implementation stage of systems development, can dramatically affect the system in the long term. The platform chosen initially dictates how the software will be maintained and updated over its lifetime. To change the platform for an established system usually requires major changes to the system and to the development team. System changes recently made at the *Christian Science Monitor* are a case in point.

The *Christian Science Monitor* is a popular international daily newspaper published Monday through Friday. It is not a religious periodical but rather a newspaper, in the traditional sense of the word, owned by The First Church of Christ, Scientist in Boston, Massachusetts. Distribution channels for the *Monitor* include paper, e-mail, and the Web.

For years, the *Christian Science Monitor* has used a content management system that many at the *Monitor* consider inflexible. A side effect of the problems with the system made it difficult to modify the newspaper's Web site or deliver content to new devices, such as smart phones. When investigating possibilities for a new content management system, CIO Curt Edge wanted to avoid a single vendor system. Although Curt Edge believes that commercial software is fine for established business practices, such as Oracle Financials for accounting, he does not want to be tied to a specific vendor's approach for an emerging, core technology such as content management.

To build a system based on the collaboration of many experts, Curt Edge turned to the open-source community. Recall that open-source software is freely available to anyone in a form that can be easily modified. Because it is easily modified, communities of developers work together to create what they consider to be the best software.

"I like the open source community because anyone at any level can talk," Curt Edge explains. "I want my developers to talk with other developers, rather than with a help desk or a support organization. There's a lot of benefit from a grassroots community figuring out what's good for everyone." Curt Edge adopted an open-source content management system called Alfresco for the *Monitor*.

Moving to an open-source development platform led to major changes in the culture and makeup of the development team. "I need to hire inventors, and I want to be partnered with groups that want to do the same thing. We need a culture that does that," Edge says. He has implemented changes in his IT staff to ensure that the church's in-house IT culture fits the open-source culture. In the past 18 months, 60 percent of his IT staff has turned as he changes the culture to a more flexible, entrepreneurial—even creative—approach to technology development.

The new development philosophy of Curt Edge and his team is a radical departure from traditional philosophies. Curt Edge's team is not only developing a content management system for the *Monitor*, but for the entire open-source community. Likewise, those in the community are assisting Curt Edge's team in developing the content management system for the *Monitor*. Within months of adopting Alfresco, the *Monitor*'s development team had already submitted two code modules to Alfresco, which Curt Edge hopes will be improved and extended by other members of the community.

Joining the open-source community has helped Curt Edge find other organizations with similar needs. He hopes this will lead to faster development of the Alfresco platform to support the needs of publishers managing content across multiple media. Curt Edge sees open source as a way to bring more resources to bear on the development of content management systems, but he also likes that the open-source model allows him to decide what technology to adopt and when. "You choose your path," he says. "We want to build in a modular manner."

Discussion Questions

1. Why did Curt Edge want to avoid single vendor, commercial software for the Monitor's content management system?
2. What advantages does Curt Edge find in using an open-source platform?

Critical Thinking Questions

1. Why did the change of development platforms lead to a 60 percent turnover in the Monitor's IT staff?
2. What concerns might a large competitive corporation like GM have over using the open-source development platform?

SOURCES: Gruman, Galen, "Christian Science Monitor Seeks Closer Technology Relationships," *InfoWorld*, April 3, 2006, *www.infoworld.com*. Alfresco Web site, accessed July 12, 2006, *http://dev.alfresco.com*. Christian Science Monitor Web site, accessed July 12, 2006, *www.csmonitor.com/aboutus/about_the_monitor.html*.

Questions for Web Case

See the Web site for this book to read about the Whitmann Price Consulting case for this chapter. Following are questions concerning this Web case.

Whitmann Price Consulting: Systems Design, Implementation, and Review Considerations

Discussion Questions

1. What function(s) did the systems analysis report and the design report play in the creation of the AMCI system?
2. What precautions did Josh and Sandra take to make sure the AMCI system was stable and customer-pleasing prior to mass production?

Critical Thinking Questions

1. What did Josh and Sandra do that was contrary to "textbook execution" of the systems development life cycle? Why?
2. Why do you think Whitmann Price executives shelved the suggestions made by the systems analysts regarding extensions to the system?

NOTES

Sources for the opening vignette: Weiss, Todd, "Timex Ticking with New Product Innovation Software," *Computerworld*, June 30, 2006, *www.computerworld.com*. Timex Web site, accessed July 10, 2006, *www.timex.com*. Sopheon Web site, accessed July 10, 2006, *www.sopheon.com*.

1. Arnott, David, "Cognitive Biases and Decision Support Systems Development: A Design Science Approach," *Information Systems Journal*, January, 2006, p. 55.
2. Huisman, Magda et al., "Deployment of Systems Development Methodologies," *Information & Management*, January, 2006, p. 29.
3. Ambrosio, Johanna, "Is Linux Next?" *Information Week*, February 6, 2006, p. 70.
4. Vijayan, Jaikumar, "Feds Make Security a Priority in IT Purchases," *Computerworld*, October 10, 2005, p. 7.
5. Hung, Wing Yan, "Object Oriented Design Implementation," *European Journal of Operational Research*, March 16, 2006, p. 1064.
6. Greenemeier, Larry, "Wanted: Up-Front Security," *Information Week*, January 2, 2006, p. 35.
7. Cavusoglu, Huseyin et al., "The Value of Intrusion Detection Systems in Information Technology Security Architecture," *Information Systems Research*, March 2005, p. 28.
8. Mitchell, Robert, "Rights of Passage," *Computerworld*, July 4, 2005, p. 17.
9. Mearian, Lucas, "CNL Financial Updates Disaster Recovery Plan," *Computerworld*, January 9, 2006, p. 8.
10. Hadi, Mohammed, "New Orleans Firm Fled Houston for Stamford," *The Wall Street Journal*, September 27, 2005, p. C3.
11. Mearian, Lucas, "CNL Financial Updates Disaster Recovery Plan," *Computerworld*, January 9, 2006, p. 8.
12. Thibodeau, Patrick, "GM Splits IT Service Work," *Computerworld*, February 6, 2006, p. 1.
13. Brandel, Mary, "Getting to Know You," *Computerworld*, February 21, 2005, p. 36.
14. Hall, Mark, "J.D. Powers to Bestow IT Service and Support Certifications," *Computerworld*, March 28, 2005, p. 8.
15. Thibodeau, Patrick, "GM Splits IT Service Work," *Computerworld*, February 6, 2006, p. 1.
16. Zha, Xuan, "Knowledge-Intensive Collaborative Design Modeling and Support: System Implementation and Application," *Computers in Industry*, January 2006, p. 56.
17. McGee, Marianne, "Outlook 2006," *Information Week*, January 2, 2006, p. 28.
18. Entzminger, Angela, "Fraud Database Allows Bankers to Cut Down on Losses," *ABA Banking Journal*, January 6, 2005, p. 7.
19. Ulfelder, Steve, "Great Works," *Computerworld*, January 3, 2005, p. 22.
20. Thibodeau, Patrick, "HP Gives Reprieve on Support to e300 Users," *Computerworld*, January 2, 2006, p. 8.
21. Hoffman, Thomas, "Return on Software," *Computerworld*, January 31, 2005, p. 39.
22. Reinhardt, Andy, "SAP: A Sea of Change in Software," *Business Week*, July 11, 2005, p. 46.
23. McCracken, Harry, "How to Build Better Software," *PC World*, February 2005, p. 17.
24. Ulfelder, Steve, "Great Works," *Computerworld*, January 3, 2005, p. 22.
25. Burrows, Peter, "Bill Me Later," *Business Week*, January 16, 2006, p. 38.
26. Havenstein, Heather, "Pilot Project Aims to Improve Analysis and Delivery of Cancer Treatment," *Computerworld*, January 30, 2006, p. 24.
27. Sullivan, Laurie, "Wal-Mart: Lessons of the Storm," *Information Week*, September 26, 2005, p. 24.
28. Koten, C.; Gray A. R., "An Application of Bayesian Network for Predicting Object-Oriented Software Maintainability," *Information and Software Technology*, January 2006, p. 59.
29. Pratt, Mark, "Shining a Light on Maintenance," *Computerworld*, February 13, 2006, p. 41.
30. Karp, David, "Who You Gonna Call," *PC Magazine*, September 20, 2005, p. 95.
31. Thibodeau, Patrick, "IBM Adds Autonomic Tools to Speed Up Error Detection," *Computerworld*, July 4, 2005, p. 7.
32. Sullivan, Laurie, "A Patch for the Money Pit," *Information Week*, February 13, 2006, p. 64.
33. Nuzu, Christine M., "Before You Throw It Out," *The Wall Street Journal*, September 12, 2005, p. R9.
34. Pratt, Mark, "Shining a Light on Maintenance," *Computerworld*, February 13, 2006, p. 41.
35. Songini, Marc, "Buggy App Causes Tax Problems in Wisconsin," *Computerworld*, January 9, 2006, p. 12.

36 Songini, Marc, "Irish Agency Halts Work on Two SAP Application Projects," *Computerworld*, October 17, 2005, p. 12.

37 Hildreth, Sue, "Application Monitoring Software," *Computerworld*, January 31, 2005, p. 26.

38 Malykhina, Elena, "BlackBerry Backup Plan," *Information Week*, February 13, 2006, p. 28.

39 Hayashi, Yuka, "Tokyo Exchange to Retool Trading System," *The Wall Street Journal*, January 21, 2006, p. B1.

40 Rosencrance, Linda, "Planning System Isn't Fully Delivering at UPS," *Computerworld*, February 28, 2005, p. 12.

41 Babcock, Charles, "Hurricanes Expose Obsolete Software," *Information Week*, January 9, 2006, p. 57.

42 Yang, Ching-Chien et al., "A Study of the Factors Impacting ERP System Performance – From the User's Perspective," *Journal of American Academy of Business*, March 2006, p. 161.

43 Violino, Bob, "Benchmarking Takes Roots," *Information Week*, October 3, 2005, p. 60.

44 Havenstein, Heather, "Tools Bridge IT, Operations," *Computerworld*, April 4, 2005, p. 12.

Brandon Trust Develops MIS Capability for Improved Operations and Services

Andy Igonor
University of the West of England

Organizations typically exist to make money and to save money. But what about those organizations registered as charities, whose goals are something other than profit making? Brandon Trust, a major player in the healthcare sector, with an annual turnover exceeding £20 million, is a registered charity in England. It does not exist to make money; rather, in the words of its chief executive officer, Steve Bennett, Brandon Trust is a "people supporting people" organization. As a visionary, Brandon Trust's CEO believes that making a difference is essential to sustaining the organization's aim and purpose.

In the quest to improve service operations, Steve Bennett met with Charles Harvey, a business strategist and expert in leadership and management practices, who also currently serves as dean of the UK's Bristol Business School. The decision to invest in information systems followed from this preliminary study of the Brandon Trust's strategy and business systems. The study revealed a lack of consistency and some redundancy in existing processes stemming from a lack of integration. Brandon Trust's situation is similar to other enterprises and was of both theoretical and practical interest to researchers at the Bristol Business School who were studying business systems and innovation. The outcome of this consultation led to a partnership agreement to help in the strategic streamlining of service operations and delivery. Sponsored by the Knowledge Transfer Partnership (KTP)—a government-funded program—the project involves the automation, rationalization, and reengineering of part of Brandon Trust's activities and procedures, with the goal of eliminating bottlenecks in information production, distribution, and use.

Brandon Trust operates community teams, employment and training units, and day-care centers spread across various areas of Bristol and the South West region of England (of which Bristol is a part). Brandon Trust also provides intensive support and supported living so that people with disabilities can live in a place of their preference. In meeting its obligations, the Trust runs a coordinated operational network, where people, both office and field staff, have to effectively communicate in meeting clients' needs. Apart from its full-time employees, the Trust also recruits part-time, hourly workers to facilitate the efficient flow of services. Brandon Trust exists in a large market for learning disability care in which purchasers demand greater efficiency and effectiveness. Brandon Trust is a market leader in providing learning disability services, and it has grown by taking over contracts. Its service area is mainly confined to Bristol, South Gloucestershire, Bath, Northeast Somerset, and North Somerset. Its competitive advantage lies in service innovation, user-centered provision, and quality management. Its main competitors are local authorities and other charitable trusts, which are typically quite small and less developed in terms of management and systems. The Trust is now trying to improve the quality of its own management systems by using innovative approaches to maximize its effectiveness.

Brandon Trust's finance and business systems director, Hilary Pearce, believes that information systems can, if efficiently applied and deployed, be used to improve facets of the organization's business. In her words, "I believe that by working with our academic partner and our KTP Associate that the Brandon Trust will derive real benefit from the project. The pooling of ideas and expertise will be vital as we analyze our information needs and data sources and identify key result areas that we need to focus on to improve the management of the Trust. I envisage an innovative and integrated business system solution that will continue to evolve and grow along with the Trust's operational needs." The role of information systems in strengthening the Trust's service level cannot be overemphasized. Its chief executive officer, Steve Bennett, had this to say: "As an organization that has grown 500% in the ten years of its existence, it is important [for Brandon Trust] to take stock and look for continual improvements in our performance. Information and the systems that provide it are fundamental to performance management, and the importance of having systems that utilize available technology in a pragmatic and useable way cannot be overstated. I therefore am excited to address this with our academic partners and KTP associates in a two-year project, the outcome of which will be an integrated business system solution that will continue to evolve and grow with the continuing development of Brandon Trust."

The Trust's services are distributed via 50 sites, and it is essential to improve information flows for efficiency and the management of risk. This development project was, therefore, focused on implementing new management systems and integrating them with other core business systems, including finance, following a detailed analysis of information requirements. Lack of systems integration currently makes provision of sound information for management controls, budgeting, and costing of contract proposals difficult to achieve. Improving information and business systems across the Trust will enhance its capacity for further growth by ensuring the best possible quality and design in its services. It is anticipated that the program will significantly reduce costs through business process reengineering, resulting in savings in the region of £250,000 per annum.

One KTP associate, Abid Mohammed, was recruited to serve as industrial manager and to work with Andy Igonor (academic advisor and project manager) and Hilary Pearce (director of finance and business systems). Abid's initial task included a review and evaluation of all existing business systems, particularly computer systems at the Trust. Although a number of systems existed in the Trust, the systems did not talk to one another; notably, the crucial financial system needed to manage all finance-related issues from payroll to payment for contract services. The hoped-for result of the development project is a system with the following capabilities:

An effective system capable of delivering relevant information when needed

An integrated system with simplified information presentation and elimination of redundancy

The development of an IS/IT audit report indicating current operating resources, which also serves as a guide for future work, including identification of appropriate software and hardware for improved operations, improved access to information, improved management information systems reporting, and improved performance

Discussion Questions

1. What are the key issues to be looked into during systems investigation? How might they affect the overall success of the project?
2. Brandon Trust does not exist to make money. How relevant will this systems development project be to the organization?
3. Why is it necessary for Brandon Trust's systems to be integrated? Of what benefit would this be to the organization?

Critical Thinking Questions

1. In what ways can a management information system benefit an organization?
2. How often should reports be generated within such systems to fulfill managerial needs?

Note: All information provided in this case is courtesy of the Brandon Trust. Used with permission.

Strategic Enterprise Management at International Manufacturing Corporation (IMC)

Bernd Heesen
University of Applied Sciences Furtwangen, Germany

The International Manufacturing Corporation (IMC), with 3,000 employees worldwide, is a supplier of automotive parts and is headquartered in the United States. Given the increasing market pressure and the need to operate internationally, IMC has recently made three acquisitions. The acquisition of one company in Great Britain and two companies in Germany is expected to leverage the companies' complementary manufacturing and sales capabilities to gain a stronger market presence in North America and Europe. The subsidiaries are still managed by their former management teams, who are now reporting to the U.S. headquarters. Each of the subsidiaries still operates its own management information system because the system fulfilled the requirements in the past and at the time of the acquisition there was no visible benefit to change what was working. A couple of interfaces were created to facilitate the monthly financial reporting to the headquarters.

Recently, the CIO of IMC realized some disadvantages of the diversity of hardware and software platforms. When one IT expert from one German subsidiary terminated her employment, no other IT expert within IMC had detailed knowledge of the local system, which threatened regular operations in that plant. The company needed to seek external consulting support to maintain the system until a new IT expert could be hired. This dependency on individual experts in each location was expensive and caused problems during periods when the experts were on leave. In addition, the interfaces between the subsidiaries' systems and the headquarters' system that were created right after the acquisition had to be modified whenever a change or upgrade was made to one of the subsidiary systems. Even minor changes, such as the reorganization of product codes or sales organizations, required reprogramming, maintenance of conversion tables, and subsequent testing of the related business processes in the system. The IT departments worked overtime and still did not find the time to invest in new initiatives such as enabling mobile computing for the sales force and developing a strategic enterprise management system that would allow consolidated planning and monitoring of all key performance indicators for the organization, a project long requested by the CEO. The two German subsidiaries had recently upgraded their systems, and problems identified during the testing of the interfaces could not be corrected in time for the submission of the monthly reports to headquarters. The data was finally corrected, but the monthly reporting was several days late. This was not the first time that the monthly financial reporting had been delayed or that the data needed to be corrected because of system-related problems, which caused other problems at headquarters.

The CEO finally requested consolidation of the systems in the coming year to allow for an integrated strategic enterprise management system that would provide current information from all legal entities and support a consolidated budgeting and planning process. He compared the company's current system to the cockpit of an airplane and stated that "no one would expect a pilot to fly an airplane with malfunctioning instruments; hence nobody can expect management to run a company based on incomplete or incorrect information." The CIO was asked to develop a business case for the implementation of such a system, considering the savings from personnel, hardware, and software, as well as process improvements.

After IMC's board approved the business case to implement enterprise resource planning (ERP) software from SAP, the CIO established a project in January 1999. The project schedule called for completion of the financial, human resource, supply chain management, and customer relationship management functions for the complete organization—called the Big Bang—in January 2000. A prototype of the application was planned to be ready by July 1999 so that the departments could test all functionality with a set of converted data extracted from the current systems. The plan also called for complete conversion of live data by the end of November 1999 so that in December both the current system and the new system could be used in parallel before making the decision to switch systems. Consultants from SAP were hired to support the implementation—for project management and configuration of the system—and an independent training organization was charged with planning the end-user training and developing computer-based training (CBT) programs for ongoing use. The project plan allowed for hiring temporary staff to help with the cutover and redundant data entry in December 1999.

The project started with a kickoff meeting in February, after which nearly all team members began their project work. The complete prototype of the application could only be made available in September because some of the team members had to fulfill their regular jobs while providing their expertise to the project. Some of the early project decisions had to be revised to accommodate the best-practice business model from IMC, after the IMC team members better understood how the SAP system worked and the consultants had gained a better understanding of how IMC wanted to operate. The CIO still wanted to maintain the timeline and called weekly meetings of the project's team leaders for finance, human resources, supply chain management, and customer relationship management. Those meetings led to the decision to delay rollout of some of the functionality and to reduce the functionality to what was really needed (eliminating nice-to-have functionality). The prototype testing produced additional feedback from the departments that needed to be incorporated in the design of the final solution prior to the conversion. Many of the department heads, who had not been part of the core project team, only then became aware of the changes and started to talk about the new software, saying that it was not working properly yet and some of the team members were scared they would not make the deadline. In November, the data conversion revealed some additional problems (e.g., with data-entry errors made by the temporary staff who were keying data into the new system in instances in which an automated conversion was not cost-effective). As a result, the parallel use of the current and new system, which depended on all data being available, was delayed. To add to the pressure, the CIO believed that delaying the project's completion could be perceived as his failure. The CIO and steering committee of the project needed to make a decision on how to move forward: (1) Continue with the project's completion in January 2000 as planned or (2) delay conversion to the new system until a proper parallel test could be completed.

Discussion Questions

1. How would you develop a business case to justify an investment in an integrated strategic enterprise management system?
2. What are the essential requirements for a strategic management system?

Critical Thinking Questions

1. Which tasks would you consider when developing the original project plan for the implementation?
2. Given IMC's current situation, what would you decide if you were the CIO?

Note: The name International Manufacturing Corporation was selected to protect the identity of the real company featured in the case.

PART · 5 ·

Information Systems in Business and Society

Chapter 14 The Personal and Social Impact of Computers

CHAPTER
· 14 ·

The Personal and Social Impact of Computers

PRINCIPLES	LEARNING OBJECTIVES
▪ Policies and procedures must be established to avoid computer waste and mistakes.	▪ Describe some examples of waste and mistakes in an IS environment, their causes, and possible solutions. ▪ Identify policies and procedures useful in eliminating waste and mistakes. ▪ Discuss the principles and limits of an individual's right to privacy.
▪ Computer crime is a serious and rapidly growing area of concern requiring management attention.	▪ Explain the types and effects of computer crime. ▪ Identify specific measures to prevent computer crime.
▪ Jobs, equipment, and working conditions must be designed to avoid negative health effects.	▪ List the important effects of computers on the work environment. ▪ Identify specific actions that must be taken to ensure the health and safety of employees. ▪ Outline criteria for the ethical use of information systems.

Information Systems in the Global Economy ⟩⟩
Johnson & Johnson, United States
Maintaining Secure Global Connections

Johnson & Johnson (J&J) is a healthcare industry giant with more than 200 separate companies operating in 54 countries. Many J&J companies and their partners conduct e-commerce business with J&J through extranet connections to the corporate network. This presented a problem to J&J network security specialists because partners' networks sometimes introduced worms and viruses into J&J's network. Not only was security an issue, but also the process of reviewing business requests for J&J network access had become burdensome, often delaying e-commerce transactions.

To remedy the problem, the J&J systems group set out to define policies and processes to streamline the process of connecting to the J&J intranet in a secure manner. J&J information system and security professionals worked with the legal department to design standard procedures for requests and evaluations. They also designed a contract or memo of understanding regarding the network connection to be established.

Under the new system, when a business manager at J&J wants to provide counterparts in outside firms with access to internal applications for e-commerce, the IT department is summoned to assess risk. First, the J&J unit and the outside firm complete a detailed questionnaire about the nature of the connection request, says Denise Medd, information security senior analyst. In addition, J&J expects the intended e-commerce partner to submit to a security assessment and evaluation. A neutral third party typically carries out the vulnerability assessment. The goal is to ensure that doing business via the network connection, which is typically opened up through the J&J firewall, presents no unnecessary risks. The outside firm must maintain systems free of viruses and worms, protected with a firewall, with up-to-date security patches. The J&J operating company, officially known as "the sponsor," is held to the same standards.

Occasionally, a request for network access is turned down, especially if the servers lack proper patch-update mechanisms or have other shortcomings. Patches are released by software vendors to fix bugs in software that often leave the system vulnerable to hackers.

After they are connected, the partner company's system is subject to an inspection process every six months to ascertain the security of the network connection. The risk management procedure has resulted in a dramatic drop in virus and worm outbreaks. Sometimes, business project managers grumble about the assessment process, but J&J management's solid backing of it has made it a uniformly enforced process that is in effect with hundreds of outside firms, says Thomas Bunt, director of worldwide information security at J&J. Also, companies are typically more willing to undergo the security assessment after J&J explains why they need to do it and how they will benefit.

As you read this chapter, consider the following:

- What are the primary concerns of corporations regarding security, privacy, and ethics?
- What strategies can assist a company with issues of security and privacy, and at what cost?

Why Learn about Security, Privacy, and Ethical Issues in Information Systems and the Internet?

A wide range of nontechnical issues associated with the use of information systems and the Internet provide both opportunities and threats to modern organizations. The issues span the full spectrum—from preventing computer waste and mistakes, to avoiding violations of privacy, to complying with laws on collecting data about customers, to monitoring employees. If you become a member of a human resources, information systems, or legal department within an organization, you will likely be charged with leading the rest of the organization in dealing with these and other issues covered in this chapter. Also, as a user of information systems and the Internet, it is in your own self-interest to become well versed on these issues. You need to know about the topics in this chapter to help avoid or recover from crime, fraud, privacy invasion, and other potential problems. This chapter begins with a discussion of preventing computer waste and mistakes.

Earlier chapters detailed the amazing benefits of computer-based information systems in business, including increased profits, superior goods and services, and higher quality of work life. Computers have become such valuable tools that today's businesspeople have difficulty imagining work without them. Yet the information age has also brought the following potential problems for workers, companies, and society in general:

- Computer waste and mistakes
- Work environment
- Computer crime
- Ethical issues
- Privacy

This chapter discusses some of the social and ethical issues as a reminder of these important considerations underlying the design, building, and use of computer-based information systems. No business organization, and, hence, no information system, operates in a vacuum. All IS professionals, business managers, and users have a responsibility to see that the potential consequences of IS use are fully considered.

Managers and users at all levels play a major role in helping organizations achieve the positive benefits of IS. These people must also take the lead in helping to minimize or eliminate the negative consequences of poorly designed and improperly utilized information systems. For managers and users to have such an influence, they must be properly educated. Many of the issues presented in this chapter, for example, should cause you to think back to some of the systems design and systems control issues discussed previously. They should also help you look forward to how these issues and your choices might affect your future use of information systems.

COMPUTER WASTE AND MISTAKES

Computer-related waste and mistakes are major causes of computer problems, contributing as they do to unnecessarily high costs and lost profits. Computer waste involves the inappropriate use of computer technology and resources. Computer-related mistakes refer to errors, failures, and other computer problems that make computer output incorrect or not useful, caused mostly by human error. This section explores the damage that can be done as a result of computer waste and mistakes.

Computer Waste

The United States government is the largest single user of information systems in the world. It should come as no surprise, then, that it is also perhaps the largest abuser. The government is not unique in this regard—the same type of waste and misuse found in the public sector also exists in the private sector. Some companies discard old software and even complete computer systems when they still have value. Others waste corporate resources to build and maintain complex systems never used to their fullest extent.

A less-dramatic, yet still relevant, example of waste is the amount of company time and money employees can waste playing computer games, sending unimportant e-mail, or accessing the Internet. Junk e-mail, also called *spam*, and junk faxes also cause waste. People receive hundreds of e-mail messages and faxes advertising products and services not wanted or requested. Not only does this waste time, but it also wastes paper and computer resources. Worse yet, spam messages often carry attached files with embedded viruses that can cause networks and computers to crash or allow hackers to gain unauthorized access to systems and data.[1] Image-based spam is a new tactic spammers use to circumvent spam-filtering software that rejects e-mail based on the content of messages and the use of keywords. The message is presented in a graphic form that can be read by people but not computers. This form of spam can be quite offensive and can contain photos of naked people and extremely graphic language.[2]

When waste is identified, it typically points to one common cause: the improper management of information systems and resources.

Computer-Related Mistakes

Despite many people's distrust, computers themselves rarely make mistakes. Even the most sophisticated hardware cannot produce meaningful output if users do not follow proper procedures. Mistakes can be caused by unclear expectations and a lack of feedback. Or a programmer might develop a program that contains errors. In other cases, a data-entry clerk might enter the wrong data. Unless errors are caught early and prevented, the speed of computers can intensify mistakes. As information technology becomes faster, more complex, and more powerful, organizations and computer users face increased risks of experiencing the results of computer-related mistakes. Take, for example, these cases from recent news.

The U.S. Postal Service relies on centralized systems to process credit card payments and automate the weighing of letters and packages. System failures in the computer network that links over half the 15,000 post offices caused delays and limited services across the United States.[3]

Worldspan is a technology company that specializes in travel-related software and systems. Its global distribution system is software used by travel agents, corporations, and travel-related Web sites to book airline tickets, hotel rooms, and rental cars. When this system temporarily lost its ability to connect to Northwest Airlines airport computer systems, more than 200 flights were delayed or canceled.[4]

PREVENTING COMPUTER-RELATED WASTE AND MISTAKES

To remain profitable in a competitive environment, organizations must use all resources wisely. Preventing computer-related waste and mistakes like those just described should therefore be a goal. Today, nearly all organizations use some type of Computer Based Information System (CBIS). To employ IS resources efficiently and effectively, employees and managers alike should strive to minimize waste and mistakes. Preventing waste and mistakes involves (1) establishing, (2) implementing, (3) monitoring, and (4) reviewing effective policies and procedures.

Establishing Policies and Procedures

The first step to prevent computer-related waste is to establish policies and procedures regarding efficient acquisition, use, and disposal of systems and devices. Computers permeate organizations today, and it is critical for organizations to ensure that systems are used to their full potential. As a result, most companies have implemented stringent policies on the acquisition of computer systems and equipment, including requiring a formal justification statement before computer equipment is purchased, definition of standard computing platforms (operating system, type of computer chip, minimum amount of RAM, etc.), and the use of preferred vendors for all acquisitions.

Prevention of computer-related mistakes begins by identifying the most common types of errors, of which there are surprisingly few. Types of computer-related mistakes include the following:

- Data-entry or data-capture errors
- Errors in computer programs
- Errors in handling files, including formatting a disk by mistake, copying an old file over a newer one, and deleting a file by mistake
- Mishandling of computer output
- Inadequate planning for and control of equipment malfunctions
- Inadequate planning for and control of environmental difficulties (electrical problems, humidity problems, etc.)
- Installing computing capacity inadequate for the level of activity on corporate Web sites
- Failure to provide access to the most current information by not adding new and not deleting old URL links

To control and prevent potential problems caused by computer-related mistakes, companies have developed policies and procedures that cover the acquisition and use of computers, with a goal of avoiding waste and mistakes. Training programs for individuals and workgroups and manuals and documents on how computer systems are to be maintained and used also help prevent problems. Other preventative measures include approval of certain systems and applications before they are implemented and used to ensure compatibility and cost-effectiveness and a requirement that documentation and descriptions of certain applications be filed or submitted to a central office, including all cell formulas for spreadsheets and a description of all data elements and relationships in a database system; such standardization can ease access and use for all personnel.

After companies have planned and developed policies and procedures, they must consider how best to implement them.

Sometimes, computer error combines with human procedural errors to lead to the loss of human life. In March 2003, a Patriot missile battery on the Kuwait border accidentally shot down a British Royal Air Force Tornado GR-4 aircraft that was returning from a mission over Iraq. Two British pilots were killed in the incident. Many defense industry experts think the accident was caused by problems with the Patriot's radar combined with human error to result in friendly fire.

Implementing Policies and Procedures

Implementing policies and procedures to minimize waste and mistakes varies according to the business conducted. Most companies develop such policies and procedures with advice from the firm's internal auditing group or its external auditing firm. The policies often focus on the implementation of source data automation and the use of data editing to ensure data accuracy and completeness, and the assignment of clear responsibility for data accuracy within each information system. Some useful policies to minimize waste and mistakes include the following:

- Changes to critical tables, HTML, and URLs should be tightly controlled, with all changes authorized by responsible owners and documented.
- A user manual should be available that covers operating procedures and documents the management and control of the application.
- Each system report should indicate its general content in its title and specify the time period it covers.
- The system should have controls to prevent invalid and unreasonable data entry.

- Controls should exist to ensure that data input, HTML, and URLs are valid, applicable, and posted in the right time frame.
- Users should implement proper procedures to ensure correct input data.

Training is another key aspect of implementation. Many users are not properly trained in using applications, and their mistakes can be very costly. One home in the small town of Valparaiso, Indiana, fairly valued at $121,900, was incorrectly recorded in the county's computer system as being worth over $400 million. The erroneous figure was used to forecast future income from property taxes. When the error was uncovered, the local school district and government agencies were forced to slash their budgets by $3 million when they found they wouldn't be getting the tax dollars after all.[5]

Because more and more people use computers in their daily work, it is important that they understand how to use them. Training is often the key to acceptance and implementation of policies and procedures. Because of the importance of maintaining accurate data and of people understanding their responsibilities, companies converting to ERP and e-commerce systems invest weeks of training for key users of the system's various modules.

Monitoring Policies and Procedures

To ensure that users throughout an organization are following established procedures, the next step is to monitor routine practices and take corrective action if necessary. By understanding what is happening in day-to-day activities, organizations can make adjustments or develop new procedures. Many organizations implement internal audits to measure actual results against established goals, such as percentage of end-user reports produced on time, percentage of data-input errors detected, number of input transactions entered per eight-hour shift, and so on.

The passage of the Sarbanes-Oxley Act has caused many companies to monitor their policies and procedures and to plan changes in financial information systems. These changes have affected many business activities. The act requires public companies to implement procedures to ensure that their audit committees can document underlying financial data to validate earnings reports. Companies that fail to comply could find their top executives behind bars. Jefferson Wells International provides professional services to clients in the areas of internal audit, technology risk management, tax, finance, and accounting. The firm decided to unplug its instant messaging systems as part of the unit's evaluation of whether its IT controls met the provisions of Sarbanes-Oxley. The primary concern was that the company wouldn't be able to detect software viruses embedded in messages. Because many of the company's employees work at client locations, executives from Jefferson Wells did not want to run the risk of having a virus infect a customer's network.[6]

Reviewing Policies and Procedures

The final step is to review existing policies and procedures and determine whether they are adequate. During review, people should ask the following questions:

- Do current policies cover existing practices adequately? Were any problems or opportunities uncovered during monitoring?
- Does the organization plan any new activities in the future? If so, does it need new policies or procedures on who will handle them and what must be done?
- Are contingencies and disasters covered?

This review and planning allows companies to take a proactive approach to problem solving, which can enhance a company's performance, such as by increasing productivity and improving customer service. During such a review, companies are alerted to upcoming changes in information systems that could have a profound effect on many business activities. An example is the need for healthcare organizations to meet the requirements of the Health Insurance Portability and Accountability Act of 1996 (HIPAA). The goal of this act is to require healthcare organizations to implement cost-effective procedures for exchanging medical data. Healthcare organizations must employ standard electronic transactions, codes, and identifiers designed to enable them to fully "digitize" medical records and make it possible to use the Internet rather than expensive private networks for electronic data interchange. The regulations affect 1.5 million healthcare providers, 7,000 hospitals, and 2,000 healthcare plans. Now that the full details of HIPAA are becoming clear, many experts are concerned. Some fear that the HIPAA provisions are too complicated and will not meet the original objective of reducing medical industry costs and instead increase costs and paperwork for doctors without improving medical care.

Information systems professionals and users still need to be aware of the misuse of resources throughout an organization. Preventing errors and mistakes is one way to do so. Another is implementing in-house security measures and legal protections to detect and prevent a dangerous type of misuse: computer crime.

COMPUTER CRIME

Even good IS policies might not be able to predict or prevent computer crime. A computer's ability to process millions of pieces of data in less than one second can help a thief steal data worth millions of dollars. Compared with the physical dangers of robbing a bank or retail store with a gun, a computer criminal with the right equipment and know-how can steal large amounts of money from the privacy of a home. Computer crime often defies detection, the amount stolen or diverted can be substantial, and the crime is "clean" and nonviolent. The largest consumer fraud in the United States was committed by the Gambino crime family involving two different computer-related ploys and resulted in a loss to the public of over $250 million. One of the schemes offered "free" tours of adult Internet sites but required the victim to provide a credit card supposedly for age-verification purposes. Victims took the free tours and then their credit cards were hit for charges over and over again. The second prong to this scheme involved the use of a third-party billing provider to add charges on peoples' telephone bills for services not provided.[7]

The following is a sample of recent computer crimes:

- A 20-year-old man was sentenced to 57 months in prison for hijacking more than 400,000 PCs over the Internet and turning them into a network of zombie computers that would follow the commands of their master. (A zombie is a personal computer used to perform a task or tasks without the owner's knowledge.) He then would rent the zombie network out to spyware distributors, hackers, and spammers to use in performing work.[8]
- NETGEAR is a seller of computer networking products and two of its key customers are Marvell and Broadcom. A NETGEAR employee, pursuant to a nondisclosure agreement and other restrictions, was permitted controlled access to Marvell trade secrets. It is alleged that when the NETGEAR employee accepted a position at Broadcom (a Marvell competitor), he downloaded dozens of files from Marvell's extranet that contained proprietary and trade secret information. He then allegedly loaded the files onto a laptop issued to him by Broadcom and later e-mailed certain Marvell trade secrets to other Broadcom employees. The man now faces a nine-count indictment alleging computer fraud, theft, and unauthorized downloading of trade secrets. The maximum penalty for each of the trade secret counts is 10 years imprisonment and a $250,000 fine.[9]
- Two men were convicted of sabotaging the computer network of American Flood Research (a national provider of flood zone determinations and real estate hazard insurance tracking to mortgage lenders). The firm's computers stopped functioning and deleted critical information, thereby preventing the company from conducting business. They also programmed the computers to erase evidence of their crime. Damage to the firm is estimated to exceed $600,000.[10]
- A computer security specialist working at the Department of Education placed software on a supervisor's computer that enabled him to access the computer's storage at will. He used that access on numerous occasions to view his supervisor's e-mail and Internet activity and then shared this information with others in the office.[11]

- Russian organized crime extorted untold thousands of dollars from firms doing business on the Internet by demanding $10,000 or more for protection from being hit by a denial-of-service attack on their Web site. Some firms bought the "protection"; some of those that did not were attacked.[12]

Although no one really knows how pervasive cybercrime is, here is what Andrew G. Arena, special agent in charge of the FBI's criminal division in New York, said in a recent interview: "Cybercrime really overlaps every other program in the FBI. It's not just some 18-year-old kid with no social life trying to hack into the system. It's organized groups, it's state sponsored organizations, it's terrorist organizations, for whatever purpose, trying to infiltrate our country. It's economic espionage targeting our infrastructure, trying to damage us financially. There's a lot of different reasons and a lot of different groups involved in this."[13]

The Computer Emergency Response Team Coordination Center (CERT/CC) is located at the Software Engineering Institute (SEI), a federally funded research and development center at Carnegie Mellon University in Pittsburgh, Pennsylvania. It is charged with coordinating communication among experts during computer security emergencies and helping to prevent future incidents. CERT employees study Internet security vulnerabilities, handle computer security incidents, publish security alerts, research long-term changes in networked systems, develop information and training to help organizations improve security at their sites, and conduct an ongoing public awareness campaign. Most attacks go undetected—as many as 60 percent, according to security experts. What's more, of the attacks that are exposed, only an estimated 15 percent are reported to law enforcement agencies. Why? Companies don't want the bad press. Such publicity makes the job even tougher for law enforcement. Most companies that have been electronically attacked won't talk to the press. A big concern is loss of public trust and image—not to mention the fear of encouraging copycat hackers.

The Computer Security Institute, with the participation of the San Francisco Federal Bureau of Investigation (FBI) Computer Intrusion Squad, conducts an annual survey of computer crime and security. The aim of the survey is to raise awareness of security, as well as to determine the scope of computer crime in the United States. The following are a few of the highlights of the 11th annual Computer Crime and Security Survey based on responses from 616 companies and government agencies that are members of the Computer Security Institute:

- The leading causes of financial loss cited in the survey were: virus attacks, unauthorized access, theft of laptop computers or PDAs, and theft of proprietary information with 68 percent of these losses resulting from the actions of insiders within the organization.
- Only 29 percent of the respondents use any form of cyber insurance to offset potential losses.
- Companies with annual revenue in excess of $1 billion report spending less than $20 per year on end-user security awareness training.
- Over 80 percent of the organizations conduct computer security audits.[14]

Today, computer criminals are a new breed—bolder and more creative than ever. With the increased use of the Internet, computer crime is now global. It's not just on U.S. shores that law enforcement has to battle cybercriminals. Regardless of its nonviolent image, computer crime is different only because a computer is used. It is still a crime. Part of what makes computer crime so unique and hard to combat is its dual nature—the computer can be both the tool used to commit a crime and the object of that crime.

THE COMPUTER AS A TOOL TO COMMIT CRIME

A computer can be used as a tool to gain access to valuable information and as the means to steal thousands or millions of dollars. It is, perhaps, a question of motivation—many people who commit computer-related crime claim they do it for the challenge, not for the money. Credit card fraud—whereby a criminal illegally gains access to another's line of credit with

social engineering
Using one's social skills to get computer users to provide you with information to access an information system or its data.

dumpster diving
Going through the trash cans of an organization to find secret or confidential information, including information needed to access an information system and/or its data.

stolen credit card numbers—is a major concern for today's banks and financial institutions. In general, criminals need two capabilities to commit most computer crimes. First, the criminal needs to know how to gain access to the computer system. Sometimes, obtaining access requires knowledge of an identification number and a password. Second, the criminal must know how to manipulate the system to produce the desired result. Frequently, a critical computer password has been talked out of a person, a practice called **social engineering**. Or, the attackers simply go through the garbage—**dumpster diving**—for important pieces of information that can help crack the computers or convince someone at the company to give them more access. In addition, over 2,000 Web sites offer the digital tools—for free—that will let people snoop, crash computers, hijack control of a machine, or retrieve a copy of every keystroke.

Also, with today's sophisticated desktop publishing programs and high-quality printers, crimes involving counterfeit money, bank checks, traveler's checks, and stock and bond certificates are on the rise. As a result, the United States Treasury Department redesigned and printed new currency that is much more difficult to counterfeit.

Cyberterrorism

Government officials and IS security specialists have documented a significant increase in Internet probes and server scans since early 2001. A growing concern among federal officials is that such intrusions are part of an organized effort by cyberterrorists, foreign intelligence services, or other groups to map potential security holes in critical systems. A **cyberterrorist** is someone who intimidates or coerces a government or organization to advance his political or social objectives by launching computer-based attacks against computers, networks, and the information stored on them.

cyberterrorist
Someone who intimidates or coerces a government or organization to advance his political or social objectives by launching computer-based attacks against computers, networks, and the information stored on them.

Even before the September 11, 2001, terrorist attacks, the U.S. government considered the potential threat of cyberterrorism serious enough that it established the National Infrastructure Protection Center in February 1998. This function was transferred to the Homeland Security Department's Information Analysis and Infrastructure Protection Directorate to serve as a focal point for threat assessment, warning, investigation, and response for threats or attacks against our country's critical infrastructure, which provides telecommunications, energy, banking and finance, water systems, government operations, and emergency services. Successful cyberattacks against the facilities that provide these services could cause widespread and massive disruptions to the normal function of our society. A 20-year-old hacker created a zombie network that disrupted critical care systems used by Seattle's Northwest hospital—disrupting operating room doors, physicians' pagers, and computers in the intensive care unit. This person faces 10 years in prison and a $250,000 fine.[15] An unemployed United Kingdom systems administrator was charged with breaking into and attempting to damage almost 100 computer systems belonging to the Air Force, Army, Defense Department, Navy, and NASA. He was accused of causing $700,000 in damages over a 13-month period in the largest-ever known crime involving U.S. military computers. He faces 70 years in prison and fines of $2 million.[16]

Identity Theft

Identity theft is a crime in which an imposter obtains key pieces of personal identification information, such as Social Security or driver's license numbers, to impersonate someone else. The information is then used to obtain credit, merchandise, and/or services in the name of the victim or to provide the thief with false credentials. Nearly 9 million U.S. adults were victims of identity fraud in 2005, according to Javelin Strategy & Research, which compiles a widely accepted survey.[17] Such a wide range of methods are used by the perpetrators of these crimes that it makes investigating them difficult.

In some cases, the identity thief uses personal information to open new credit accounts, establish cellular phone service, or open a new checking account to obtain blank checks. In other cases, the identity thief uses personal information to gain access to the person's existing accounts. Typically, the thief changes the mailing address on an account and runs up a huge bill before the person whose identity has been stolen realizes there is a problem. The Internet has made it easier for an identity thief to use the stolen information because transactions can be made without any personal interaction.

Another popular method to get information is "shoulder surfing"—the identity thief simply stands next to someone at a public office, such as the Bureau of Motor Vehicles, and watches as the person fills out personal information on a form.

Consumers can help protect themselves by regularly checking their credit reports with major credit bureaus, following up with creditors if their bills do not arrive on time, not revealing any personal information in response to unsolicited e-mail or phone calls (especially Social Security numbers and credit card account numbers), and shredding bills and other documents that contain sensitive information.[18]

Research shows that identities were stolen most often from healthcare-related institutions and then from financial institutions. Indeed a stolen name and Social Security number can enable a thief to receive medical treatments, obtain prescription drugs, and submit fraudulent Medicare claims. A man in Pennsylvania discovered an imposter had used his identity at five different hospitals to receive more than $100,000 in medical treatment.[19] A Tennessee doctor's Medicare provider number was used to bill false claims in his name for more than $1 million in payments from an insurance company.[20] All this creates not only unpaid bills in your name but also generates incorrect entries into your health records at hospitals, doctors' offices, pharmacies, and insurance companies. Unfortunately, the process to correct all this misinformation is extremely difficult, lengthy, and time consuming.[21]

The U.S. Congress passed the Identity Theft and Assumption Deterrence Act of 1998 to fight identity theft. Under this act, the Federal Trade Commission (FTC) is assigned responsibility to help victims restore their credit and erase the impact of the imposter. It also makes identity theft a federal felony punishable by a prison term ranging from 3 to 25 years.

THE COMPUTER AS THE OBJECT OF CRIME

A computer can also be the object of the crime, rather than the tool for committing it. Tens of millions of dollars of computer time and resources are stolen every year. Each time system access is illegally obtained, data or computer equipment is stolen or destroyed, or software is illegally copied, the computer becomes the object of crime. These crimes fall into several categories: illegal access and use, data alteration and destruction, information and equipment theft, software and Internet piracy, computer-related scams, and international computer crime.

Illegal Access and Use

Crimes involving illegal system access and use of computer services are a concern to both government and business. Since the outset of information technology, computers have been plagued by criminal hackers. Originally, a **hacker** was a person who enjoys computer technology and spends time learning and using computer systems. A **criminal hacker**, also called a **cracker**, is a computer-savvy person who attempts to gain unauthorized or illegal access to computer systems to steal passwords, corrupt files and programs, or even transfer money. In many cases, criminal hackers are people who are looking for fun and excitement—the challenge of beating the system. Today, many people use the term hacker and cracker interchangeably. **Script bunnies** admire crackers, but have little technical savvy. They are crackers who download programs called *scripts* that automate the job of breaking into computers. **Insiders** are employees, disgruntled or otherwise, working solo or in concert with outsiders to compromise corporate systems.

Catching and convicting criminal hackers remains a difficult task. The method behind these crimes is often hard to determine. Even if the method behind the crime is known, tracking down the criminals can take a lot of time. It took years for the FBI to arrest one criminal hacker for the alleged "theft" of almost 20,000 credit card numbers that had been sent over the Internet.

hacker
A person who enjoys computer technology and spends time learning and using computer systems.

criminal hacker (cracker)
A computer-savvy person who attempts to gain unauthorized or illegal access to computer systems to steal passwords, corrupt files and programs, or even transfer money.

script bunny
A cracker with little technical savvy who downloads programs called scripts, which automate the job of breaking into computers.

insider
An employee, disgruntled or otherwise, working solo or in concert with outsiders to compromise corporate systems.

Data and information are valuable corporate assets. The intentional use of illegal and destructive programs to alter or destroy data is as much a crime as destroying tangible goods. The most common of these programs are viruses and worms, which are software programs that, when loaded into a computer system, will destroy, interrupt, or cause errors in processing. Such programs are also called *malware*. Internet security firm McAfee released virus threat definition number 200,000 in July 2006 and predicts that there will be twice that many viruses within two years.[22] McAfee also predicts the increased connectivity of smartphones will lead to serious and widespread attacks on these devices. "… a mobile threat targeting several (smartphone) operating systems could infect up to 200 million connected smartphones simultaneously because the majority of these devices do not currently have mobile security protection installed."[23]

virus

A computer program file capable of attaching to disks or other files and replicating itself repeatedly, typically without the user's knowledge or permission.

A **virus** is a computer program file capable of attaching to disks or other files and replicating itself repeatedly, typically without the user's knowledge or permission. Some viruses attach to files, so when the infected file executes, the virus also executes. Other viruses sit in a computer's memory and infect files as the computer opens, modifies, or creates the files. They are often disguised as games or images with clever or attention-grabbing titles such as "Boss, nude." Some viruses display symptoms, and some viruses damage files and computer systems. Computer viruses are written for several operating systems, including Windows, Macintosh, UNIX, and others. The m00p virus gang conspired to infect computers with a virus that would turn each infected machine into a zombie machine under their control. The zombie network could then be used to spread viruses and other malware across the Internet, without the owners of the compromised computers even being aware.[24]

worm

A parasitic computer program that can create copies of itself on the infected computer or send copies to other computers via a network.

Worms are parasitic computer programs that replicate but, unlike viruses, do not infect other computer program files. Worms can create copies on the same computer or can send the copies to other computers via a network. Worms often spread via Internet Relay Chat (IRC). For example, the MyDoom worm, also known as Shimgapi and Novarg, started spreading in January 2004 and quickly became the most virulent e-mail worm ever. The worm arrived as an e-mail with an attachment that has various names and extensions, including .exe, .scr, .zip, and .pif. When the attachment executed, the worm sent copies of itself to other e-mail addresses stored in the infected computer. The first version of the virus, MyDoom.A, was designed to attack The SCO Group Inc.'s Web site. A later variant, dubbed MyDoom.B, was designed to enable similar denial-of-service attacks against the Microsoft Web site. The B variant also included a particularly nasty feature in that it blocks infected computers from accessing sites belonging to vendors of antivirus products. Infected e-mail messages carrying the MyDoom worm have been intercepted from over 142 countries and at one time accounted for 1 in every 12 e-mail messages. The Hewlett Packard Active Countermeasures is a security service designed to locate and eliminate critical security vulnerabilities for organizations that subscribe to this service. As part of the service, HP releases a worm that seeks out and then fixes weaknesses in customers' networks and computers.[25]

Trojan horse

A malicious program that disguises itself as a useful application and purposefully does something the user does not expect.

A **Trojan horse** program is a malicious program that disguises itself as a useful application and purposefully does something the user does not expect. Trojans are not viruses because they do not replicate, but they can be just as destructive. Many people use the term to refer only to nonreplicating malicious programs, thus making a distinction between Trojans and viruses. With nearly 7 million registered players worldwide, World of Warcraft is an extremely popular multiplayer, online, role-playing game. A password-stealing Trojan horse program targets players of the game and steals their password. After the attacker has a player's password, he can transfer virtual, in-game goods that the player had accumulated to his own account to other players for cash.[26] A German language e-mail is being used to spread a Trojan horse that steals passwords and logon details of customers' online bank accounts and then relays them back to a remote server. The malware tries to get users to install the Trojan horse by disguising itself as a software patch for a new flaw in Microsoft software.[27]

A *logic bomb* is a type of Trojan horse that executes when specific conditions occur. Triggers for logic bombs can include a change in a file by a particular series of keystrokes or at a specific time or date.

A *variant* is a modified version of a virus that is produced by the virus's author or another person who amends the original virus code. If changes are small, most antivirus products will also detect variants. However, if the changes are significant, the variant might go undetected by antivirus software.

In some cases, a virus or a worm can completely halt the operation of a computer system or network for days or longer until the problem is found and repaired. In other cases, a virus or a worm can destroy important data and programs. If backups are inadequate, the data and programs might never be fully functional again. The costs include the effort required to identify and neutralize the virus or worm and to restore computer files and data, as well as the value of business lost because of unscheduled computer downtime.

Investec is an international, specialist banking group that provides a diverse range of financial products and services to a select client base. The bank can't allow computer viruses to disrupt the information systems that are crucial to the operation of its financial systems that run on 4,000 workstations in eight countries, so it installed Spectator Professional software from Promisec. The software allows Investec to monitor the use of external devices, such as memory sticks, wireless cards, and software configurations on all its computers. The system can lock down computers and prevent staff from plugging in unauthorized devices or downloading unauthorized software—a frequent source of allowing viruses or worms to enter a system.[28]

The F-Secure Corporation provides centrally managed security solutions, and its products include antivirus, file encryption, and network security solutions for all major platforms—from desktops to servers and from laptops to handhelds. F-Secure is headquartered in Helsinki, Finland. F-Secure provides real-time virus statistics on the most active viruses in the world at its Web site, *www.f-secure.com/virus-info/statistics.*[29] Table 14.1 lists the five most active viruses on August 24, 2006 as identified by F-Secure volunteer reports located primarily in Scandinavia.

Place	Virus Name	Number of Incidences Reported August 24, 2006
1	E-mail-Worm Win32.MyDoom.m	500
2	Win32/MyDoom.M@mm	120
3	E-mail-worm.Win32.Nyxem.e	90
4	Net-worm.Win32.Nytob.bi	80
5	Net-worm.Win32.Nytob.gen	60

Table 14.1

Most Active Viruses in the World—August 24, 2006

(Source: "F-Secure Virus Statistics," *www.f-secure.com/virus-info/statistics*, accessed August 26, 2006.)

McAfee Security for Consumers is a division of Network Associates Inc. that delivers retail and online solutions designed to secure, protect, and optimize the computers of consumers and home office users. McAfee's retail desktop products include premier antivirus, security, encryption, and desktop optimization software. McAfee delivers software through an Internet browser to provide these services to users online through its Web site *www.mcafee.com*, one of the largest paid subscription sites on the Internet with over 2 million active paid subscribers. McAfee provides a real-time map of where the latest viruses are infecting computers worldwide at *http://us.mcafee.com/virusInfo/default.asp*. See Figure 14.1. The site also provides software for scanning your computer for viruses and tips on how to remove a virus.[30]

Using Antivirus Programs

As a result of the increasing threat of viruses and worms, most computer users and organizations have installed **antivirus programs** on their computers. Such software runs in the background to protect your computer from dangers lurking on the Internet and other possible sources of infected files. Some antivirus software is even capable of repairing common virus infections automatically, without interrupting your work. The latest virus definitions are downloaded automatically when you connect to the Internet, ensuring that your PC's

antivirus program
Software that runs in the background to protect your computer from dangers lurking on the Internet and other possible sources of infected files.

Figure 14.1

Global Virus Infections—
Number of Infected Computers
per Million Citizens

(Source: Courtesy of McAfee, Inc.)

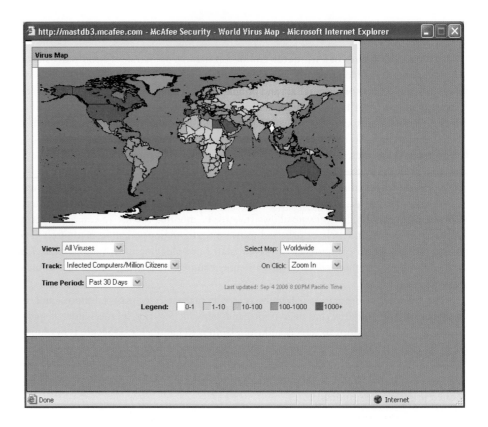

protection is current. To safeguard your PC and prevent it from spreading viruses to your friends and coworkers, some antivirus software scans and cleans both incoming and outgoing e-mail messages. Table 14.2 lists some of the most popular antivirus software.

Table 14.2

Antivirus Software

Antivirus Software	Software Manufacturer	Web Site
Symantec's Norton AntiVirus 2005	Symantec	www.symantec.com
McAfee Virus Scan	McAfee	www.mcafee.com
Panda Antivirus Platinum	Panda Software	www.pandasoftware.com
Vexira Antivirus	Central Command	www.centralcommand.com
Sophos Antivirus	Sophos	www.sophos.com
PC-cillin	Trend Micro	www.trendmicro.com

Proper use of antivirus software requires the following steps:

1. **Install antivirus software and run it often.** Many of these programs automatically check for viruses each time you boot up your computer or insert a disk or CD, and some even monitor all e-mail and file transmissions and copying operations.
2. **Update antivirus software often.** New viruses are created all the time, and antivirus software suppliers are constantly updating their software to detect and take action against these new viruses.
3. **Scan all removable media, including CDs, before copying or running programs from them.** Hiding on disks or CDs, viruses often move between systems. If you carry

document or program files on removable media between computers at school or work and your home system, always scan them.

4. **Install software only from a sealed package or secure Web site of a known software company.** Even software publishers can unknowingly distribute viruses on their program disks or software downloads. Most scan their own systems, but viruses might still remain.

5. **Follow careful downloading practices.** If you download software from the Internet or a bulletin board, check your computer for viruses immediately after completing the transmission.

6. **If you detect a virus, take immediate action.** Early detection often allows you to remove a virus before it does any serious damage.

Antivirus software should be used and updated often.

Despite careful precautions, viruses can still cause problems. They can elude virus-scanning software by lurking almost anywhere in a system. Future antivirus programs might incorporate "nature-based models" that check for unusual or unfamiliar computer code. The advantage of this type of virus program is the ability to detect new viruses that are not part of an antivirus database.

Hoax, or false, viruses are another problem. Criminal hackers sometimes warn the public of a new and devastating virus that doesn't exist to create fear. Companies sometimes spend hundreds of hours warning employees and taking preventative action against a nonexistent virus. Security specialists recommend that IS personnel establish a formal paranoia policy to thwart virus panic among gullible end users. Such policies should stress that before users forward an e-mail alert to colleagues and higher-ups, they should send it to the help desk or the security team. The corporate intranet can be used to explain the difference between real viruses and fakes, and it can provide links to Web sites to set the record straight.

Be aware that virus writers also use known hoaxes to their advantage. For example, AOL4FREE began as a hoax virus warning. Then, a hacker distributed a destructive Trojan attached to the original hoax virus warning. Always remain vigilant and never open a suspicious attachment.[31] Read the "Information Systems @ Work" article to learn how one company reacted to a virus infestation.

Westinghouse Finds New Tools to Guard Against Viruses

In 2004, Westinghouse suffered serious damage to information systems at the hands of the most dangerous variant of the mass-mailer MyDoom worm. "It was diabolical and self-propagating. It infected 100 PCs, 42 servers and deleted 9.3 million files," says Tom Moser, Westinghouse's manager of IT services. The day it hit, "people were saying, 'Where's the file? It's gone. I went to lunch, and it's gone.'" Ever since that day, the company has been searching for the most effective tools to combat viruses on user PCs.

Westinghouse employs McAfee security software on all corporate PCs, which includes virus protection, but Moser and his associates feel that such virus protection falls short of total protection. "There is a period when you're totally vulnerable," Moser says, noting that antivirus vendors take two or three hours to release a signature for a new threat. It's not realistic to presume that patching vulnerable software can be done in minutes because the process requires careful testing and a scheduled time it can be applied.

So, Moser has turned to behavior-based security software from Cisco, Cisco Security Agent (CSA), to complement the McAfee software. Behavior-based security runs on PCs and watches for suspicious code activity and blocks it based on behavior. The software proved effective in 2005 when it stopped the Zotob worm in its tracks, in advance of antivirus vendors identifying the threat and producing signatures to detect it, says Tom Moser.

The implementation of the new security system has not been without challenges for Westinghouse. Westinghouse began piloting the software with a group of 150 users. Users found that CSA was trigger-happy in its alerts when placed in monitoring mode. The software generated a lot of false alerts that distract both end users and help desk staff. Other desktop applications that Westinghouse was running, including McAfee antivirus and programming debugging tools, induced CSA to wrongly tell the user in a screen-display message that the machine faced an unknown security threat.

"One thing we ran into early on was it generated 30,000 alerts per day for McAfee anti-virus," Moser says. With Cisco consultants called in to help, Westinghouse was able to reduce that number to 50.

After initial testing, systems engineers spent six months installing the software for use by almost all its 7,500 employees. Helping thousands of employees adapt to the concept of behavior-based software was a challenge, requiring considerable training to help them interact with CSA. The software requires users to make a decision about whether their machine is being attacked. CSA gives the user the chance to override any blocking of a suspected threat. But in training its employees, Westinghouse has encouraged them to always let CSA block the code activity it detects, just to be on the safe side. Users are instructed to check with the help desk if CSA seems to be blocking something legitimate.

Westinghouse feels that its $50,000 software investment and the additional cost of training is paying off. When the Zotob worm appeared, the number of alerts jumped from around 1,000 per day to more than 8,000, and Westinghouse was spared damage. CSA caught the worm before McAfee knew of its existence. With the desktop CSA now in place, Westinghouse plans to start testing CSA on its servers later this year, with the intent of a large-scale deployment.

Discussion Questions

1. What makes behavior-based security software different from traditional virus protection? Why is it able to catch viruses sooner than traditional methods?
2. What is the downside of deploying behavior-based security software in a large organization such as Westinghouse?

Critical Thinking Questions

1. Why do you think Westinghouse runs both McAfee and CSA?
2. Which professionals are responsible for keeping viruses off PCs at Westinghouse? Do you agree with placing the responsibility in their hands? Why or why not?

Sources: Messmer, Ellen, "Westinghouse Tightens Security," *Network World*, October 3, 2005, *www.networkworld.com*. Westinghouse Web site, accessed July 14, 2006, *www.westinghouse.com*. Cisco Security Agent Web site, accessed July 14, 2006, *www.cisco.com/en/US/products/sw/secursw/ps5057/index.html*.

Information and Equipment Theft

Data and information are assets or goods that can also be stolen. People who illegally access systems often do so to steal data and information. To obtain illegal access, criminal hackers require identification numbers and passwords. Some criminals try different identification numbers and passwords until they find ones that work. Using password sniffers is another approach. A **password sniffer** is a small program hidden in a network or a computer system that records identification numbers and passwords. In a few days, a password sniffer can record hundreds or thousands of identification numbers and passwords. Using a password sniffer, a criminal hacker can gain access to computers and networks to steal data and information, invade privacy, plant viruses, and disrupt computer operations.

In addition to theft of data and software, all types of computer systems and equipment have been stolen from offices. Portable computers such as laptops (and the data and information stored in them) are especially easy for thieves to take. In many cases, the data and information stored in these systems are more valuable than the equipment. A laptop and external hard drive containing the names and personal data of 26.5 million military personal was stolen during a home burglary of a Veterans Affairs analyst. Because of the sensitivity of the data and the number of people involved, the FBI and Department of Veteran Affairs launched an intensive investigation. The computer was eventually turned in by an unidentified person who claimed a $50,000 reward offered by authorities. The well-publicized theft led to hearings in the Senate and House, the dismissal of at least one manager in Veterans Affairs, and a memo from the White House's Office of Management and Budget with new security recommendations for all federal agencies.[32]

A break-in at one of the offices of American International Group, one of the world's largest insurance companies, resulted in the loss of two laptops and a server containing the personal information on nearly 1 million employees at client companies. Police and insurance officials hope that the physical hardware was the target, not the data.[33]

password sniffer
A small program hidden in a network or a computer system that records identification numbers and passwords.

To fight computer crime, many companies use devices that disable the disk drive and/or lock the computer to the desk.

(Source: Courtesy of Kensington Technology Group.)

Software and Internet Software Piracy

Each time you use a word processing program or access software on a network, you are taking advantage of someone else's intellectual property. Like books and movies—other intellectual properties—software is protected by copyright laws. Often, people who would never think of plagiarizing another author's written work have no qualms about using and copying software programs they have not paid for. Such illegal duplicators are called *pirates*; the act of illegally duplicating software is called **software piracy**.

Technically, software purchasers are granted the right only to use the software under certain conditions; they don't really own the software. Licenses vary from program to program and can authorize as few as one computer or one person to use the software or as many as several hundred network users to share the application across the system. Making additional copies, or loading the software onto more than one machine, might violate copyright law and be considered piracy.

The Business Software Alliance estimates that the software industry loses over $11 billion per year in revenue to software piracy annually. Half the loss comes from Asia, where China and Indonesia are the biggest offenders. In Western Europe, annual piracy losses range between $2.5 and $3 billion dollars. Although the rate of software piracy is quite high in Latin America and Central Europe, those software markets are so small that the dollar losses are considerably lower. About $2 billion in annual piracy losses come from North America. Overall, it is estimated that 35 percent of the world's software is pirated.[34]

software piracy
The act of illegally duplicating software.

Internet-based software piracy occurs when software is illegally downloaded from the Internet. It is the most rapidly expanding type of software piracy and the most difficult form to combat. The same purchasing rules apply to online software purchases as for traditional purchases. Internet piracy can take several forms, including the following:

- Pirate Web sites that make software available for free or in exchange for uploaded programs
- Internet auction sites that offer counterfeit software, which infringes copyrights
- Peer-to-peer networks, which enable unauthorized transfer of copyrighted programs

Penalties for software piracy can be severe. If the copyright owner brings a civil action against someone, the owner can seek to stop the person from using its software immediately and can also request monetary damages. The copyright owner can then choose between compensation for actual damages—which includes the amount it has lost because of the person's infringement, as well as any profits attributable to the infringement—and statutory damages, which can be as much as $150,000 for each program copied. In addition, the government can prosecute software pirates in criminal court for copyright infringement. If convicted, they could be fined up to $250,000 or sentenced to jail for up to five years, or both.[35]

Operation Copycat is an ongoing undercover investigation into warez groups, which are online organizations engaged in the illegal uploading, copying, and distribution of copyrighted works such as music, movies, games, and software, often even before they are released to the public. The investigation is led by the Computer Hacking and Intellectual Property (CHIP) Unit of the United States Attorney's Office and the Federal Bureau of Investigation. From July 2005 to August 2006, 37 people were criminally charged as part of this ongoing investigation and 32 of them, including two film critics, were convicted.[36]

Computer-Related Scams

People have lost hundreds of thousands of dollars on real estate, travel, stock, and other business scams. Today, many of these scams are being perpetrated with computers. Using the Internet, scam artists offer get-rich-quick schemes involving bogus real estate deals, tout "free" vacations with huge hidden costs, commit bank fraud, offer fake telephone lotteries, sell worthless penny stocks, and promote illegal tax-avoidance schemes.

Over the past few years, credit card customers of various banks have been targeted by scam artists trying to get personal information needed to use their credit cards. The scam works by sending customers an e-mail including a link that seems to direct users to their bank's Web site. At the site, they are greeted with a pop-up box asking them for their full debit card numbers, their personal identification numbers, and their credit card expiration dates. The problem is that the Web site customers are directed to is a fake site operated by someone trying to gain access to that information. As discussed previously, this form of scam is called *phishing*. During November 2005, the Anti-Phishing Working Group received 16,882 unique reports of phishing attacks aimed at the consumers of 93 different brands—this was double the number of reports received the previous November.[37]

In the weeks following Hurricane Katrina that hit New Orleans and parts of the Gulf Coast, the FBI warned that over half the Hurricane Katrina aid sites it checked were registered to people outside the United States and likely to be fraudulent. A 20-year-old man was charged with setting up Web sites designed to look like those of the American Red Cross and other organizations accepting donations to help the victims. He then sold these to "would-be scammers" for $150 each. For his trouble, this person is facing 50 years in prison and a fine of $1 million.[38]

The following is a list of tips to help you avoid becoming a scam victim:

- Don't agree to anything in a high-pressure meeting or seminar. Insist on having time to think it over and to discuss things with your spouse, your partner, or even your lawyer. If a company won't give you the time you need to check it out and think things over, you don't want to do business with it. A good deal now will be a good deal tomorrow; the only reason for rushing you is if the company has something to hide.

- Don't judge a company based on appearances. Flashy Web sites can be created and published in a matter of days. After a few weeks of taking money, a site can vanish without a trace in just a few minutes. You might find that the perfect money-making opportunity offered on a Web site was a money maker for the crook and a money loser for you.

- Avoid any plan that pays commissions simply for recruiting additional distributors. Your primary source of income should be your own product sales. If the earnings are not made primarily by sales of goods or services to consumers or sales by distributors under you, you might be dealing with an illegal pyramid.

- Beware of shills, people paid by a company to lie about how much they've earned and how easy the plan was to operate. Check with an independent source to make sure that you aren't having the wool pulled over your eyes.

- Beware of a company's claim that it can set you up in a profitable home-based business but that you must first pay up front to attend a seminar and buy expensive materials. Frequently, seminars are high-pressure sales pitches, and the material is so general that it is worthless.

- If you are interested in starting a home-based business, get a complete description of the work involved before you send any money. You might find that what you are asked to do after you pay is far different from what was stated in the ad. You should never have to pay for a job description or for needed materials.

- Get in writing the refund, buy-back, and cancellation policies of any company you deal with. Do not depend on oral promises.

- Do your homework. Check with your state attorney general and the National Fraud Information Center before getting involved, especially when the claims about a product or potential earnings seem too good to be true.

If you need advice about an Internet or online solicitation, or if you want to report a possible scam, use the Online Reporting Form or Online Question & Suggestion Form features on the Web site for the National Fraud Information Center at *http://fraud.org*, or call the NFIC hotline at 1-800-876-7060.

International Computer Crime

Computer crime is also an international issue, and it becomes more complex when it crosses borders. As already mentioned, the software industry loses about $11 to $12 billion in revenue to software piracy annually, with about $9 billion of that occurring outside the United States.[39]

With the increase in electronic cash and funds transfer, some are concerned that terrorists, international drug dealers, and other criminals are using information systems to launder illegally obtained funds. Computer Associates International developed software called CleverPath for Global Compliance for customers in the finance, banking, and insurance industries to eliminate money laundering and fraud. Companies that are required to comply with legislation such as the USA Patriot Act and Sarbanes-Oxley Act might lack the resources and processes to do so. The software automates manual tracking and auditing processes that are required by regulatory agencies and helps companies handle frequently changing reporting regulations. The application can drill into a company's transactions and detect transaction patterns that suggest fraud or other illegal activities based on built-in business rules and predictive analysis. Suspected fraud cases are identified and passed on to the appropriate personnel for action to thwart criminals and help companies avoid paying fines.

PREVENTING COMPUTER-RELATED CRIME

Because of increased computer use today, greater emphasis is placed on the prevention and detection of computer crime. Although all states have passed computer crime legislation, some believe that these laws are not effective because companies do not always actively detect and pursue computer crime, security is inadequate, and convicted criminals are not severely punished. However, all over the United States, private users, companies, employees, and public officials are making individual and group efforts to curb computer crime, and recent efforts have met with some success.

Crime Prevention by State and Federal Agencies

State and federal agencies have begun aggressive attacks on computer criminals, including criminal hackers of all ages. In 1986, Congress enacted the Computer Fraud and Abuse Act, which mandates punishment based on the victim's dollar loss. The Department of Defense also supports the Computer Emergency Response Team (CERT), which responds to network security breaches and monitors systems for emerging threats. Law enforcement agencies are also increasing their efforts to stop criminal hackers, and many states are now passing new, comprehensive bills to help eliminate computer crimes. A complete listing of computer-related legislation by state can be found at *www.onlinesecurity.com/forum/article46.php*. Recent court cases and police reports involving computer crime show that lawmakers are ready to introduce newer and tougher computer crime legislation.

Crime Prevention by Corporations

Companies are also taking crime-fighting efforts seriously. Many businesses have designed procedures and specialized hardware and software to protect their corporate data and systems. Specialized hardware and software, such as encryption devices, can be used to encode data and information to help prevent unauthorized use. As discussed in Chapter 7, encryption is the process of converting an original electronic message into a form that can be understood only by the intended recipients. A key is a variable value that is applied using an algorithm to a string or block of unencrypted text to produce encrypted text or to decrypt encrypted text. Encryption methods rely on the limitations of computing power for their effectiveness—if breaking a code requires too much computing power, even the most determined code crackers will not be successful. The length of the key used to encode and decode messages determines the strength of the encryption algorithm.

Public-key Infrastructure (PKI) enables users of an unsecured public network such as the Internet to securely and privately exchange data through the use of a public and a private cryptographic key pair that is obtained and shared through a trusted authority. PKI is the most common method on the Internet for authenticating a message sender or encrypting a message. PKI uses two keys to encode and decode messages. One key of the pair, the message receiver's public key, is readily available to the public and is used by anyone to send that individual encrypted messages. The second key, the message receiver's private key, is kept secret and is known only by the message receiver. Its owner uses the private key to *decrypt* messages—convert encoded messages back into the original message. Knowing a person's public key does not enable you to decrypt an encoded message to that person.

Using biometrics is another way to protect important data and information systems. **Biometrics** involves the measurement of one of a person's traits, whether physical or behavioral. Biometric techniques compare a person's unique characteristics against a stored set to detect differences between them. Biometric systems can scan fingerprints, faces, handprints, irises, and retinal images to prevent unauthorized access to important data and computer resources. Most of the interest among corporate users is in fingerprint technology, followed by face recognition. Fingerprint scans hit the middle ground between price and effectiveness. Iris and retina scans are more accurate, but they are more expensive and involve more equipment.

Co-op Mid Counties is the first United Kingdom retailer to implement a payment by biometrics system with fingerprint readers supplied by the U.S. company Pay By Touch. The system is installed in just three of its stores in Oxford, but if successful, the system will be expanded to all of its 150 stores. To use the system, customers must register with Co-op Mid Counties by providing a photo ID and submit to fingerprinting. In addition to providing improved security, the system takes less time to process a payment—3 seconds compared to 7 seconds for traditional payment approval methods.[40]

Public-key Infrastructure (PKI)
A means to enable users of an unsecured public network such as the Internet to securely and privately exchange data through the use of a public and a private cryptographic key pair that is obtained and shared through a trusted authority.

biometrics
The measurement of one of a person's traits, whether physical or behavioral.

As employees move from one position to another at a company, they can build up access to multiple systems if inadequate security procedures fail to revoke access privileges. It is clearly not appropriate for people who have changed positions and responsibilities to still have access to systems they no longer use. To avoid this problem, many organizations create role-based system access lists so that only people filling a particular role (e.g., invoice approver) can access a specific system.

Fingerprint authentication devices provide security in the PC environment by using fingerprint information instead of passwords.

(Source: Permission granted by Pay By Touch.)

Crime-fighting procedures usually require additional controls on the information system. Before designing and implementing controls, organizations must consider the types of computer-related crime that might occur, the consequences of these crimes, and the cost and complexity of needed controls. In most cases, organizations conclude that the trade-off between crime and the additional cost and complexity weighs in favor of better system controls. Having knowledge of some of the methods used to commit crime is also helpful in preventing, detecting, and developing systems resistant to computer crime (see Table 14.3). Some companies actually hire former criminals to thwart other criminals.

Table 14.3

Common Methods Used to Commit Computer Crimes

Methods	Examples
Add, delete, or change inputs to the computer system.	Delete records of absences from class in a student's school records.
Modify or develop computer programs that commit the crime.	Change a bank's program for calculating interest to make it deposit rounded amounts in the criminal's account.
Alter or modify the data files used by the computer system.	Change a student's grade from C to A.
Operate the computer system in such a way as to commit computer crime.	Access a restricted government computer system.
Divert or misuse valid output from the computer system.	Steal discarded printouts of customer records from a company trash bin.
Steal computer resources, including hardware, software, and time on computer equipment.	Make illegal copies of a software program without paying for its use.
Offer worthless products for sale over the Internet.	Send e-mail requesting money for worthless hair growth product.
Blackmail executives to prevent release of harmful information.	Eavesdrop on organization's wireless network to capture competitive data or scandalous information.
Blackmail company to prevent loss of computer-based information.	Plant logic bomb and send letter threatening to set it off unless paid considerable sum.

Although the number of potential computer crimes appears to be limitless, the actual methods used to commit crime are limited. The following list provides a set of useful guidelines to protect your computer from criminal hackers.

- Install strong user authentication and encryption capabilities on your firewall.
- Install the latest security patches, which are often available at the vendor's Internet site.
- Disable guest accounts and null user accounts that let intruders access the network without a password.
- Do not provide overfriendly logon procedures for remote users (e.g., an organization that used the word *welcome* on their initial logon screen found they had difficulty prosecuting a criminal hacker).
- Restrict physical access to the server and configure it so that breaking into one server won't compromise the whole network.
- Give each application (e-mail, File Transfer Protocol, and domain name server) its own dedicated server.
- Turn audit trails on.
- Consider installing caller ID.
- Install a corporate firewall between your corporate network and the Internet.
- Install antivirus software on all computers and regularly download vendor updates.
- Conduct regular IS security audits.
- Verify and exercise frequent data backups for critical data.

Companies are also joining together to fight crime. The Software and Information Industry Alliance (SIIA) was the original antipiracy organization, formed and financed by many of the large software publishers. Microsoft financed the formation of a second antipiracy organization, the Business Software Alliance (BSA). The BSA, through intense publicity, has become the more prominent organization. Other software companies, including Apple, Adobe, Hewlett-Packard, and IBM, now contribute to the BSA.

The former owner of BuysUSA, a large, for-profit software piracy Web site was sentenced to six years in federal prison and ordered to pay restitution of $4.1 million in a case brought to the attention of the FBI by the BSA. The software products purchased on the Web site (software from Adobe, Macromedia, and Autodesk) included a serial number that allowed the consumer to activate and use the product. Some expert witnesses estimated that his activities caused copyright owner losses of nearly $20 million.[41]

Using Intrusion Detection Software

intrusion detection system (IDS)

Software that monitors system and network resources and notifies network security personnel when it senses a possible intrusion.

An **intrusion detection system (IDS)** monitors system and network resources and notifies network security personnel when it senses a possible intrusion. Examples of suspicious activities include repeated failed logon attempts, attempts to download a program to a server, and access to a system at unusual hours. Such activities generate alarms that are captured on log files. Intrusion detection systems send an alarm, often by e-mail or pager, to network security personnel when they detect an apparent attack. Unfortunately, many IDSs frequently provide false alarms that result in wasted effort. If the attack is real, network security personnel must make a decision about what to do to resist the attack. Any delay in response increases the probability of damage from a criminal hacker attack. Use of an IDS provides another layer of protection in the event that an intruder gets past the outer security layers—passwords, security procedures, and corporate firewall.

The following story is true, but the company's name has been changed to protect its identity. The ABCXYZ company employs more than 25 IDS sensors across its worldwide network, enabling it to monitor 90 percent of the company's internal network traffic. The remaining 10 percent comes from its engineering labs and remote sales offices, which are not monitored because of a lack of resources. The company's IDS worked very well in providing an early warning of an impending SQL Slammer attack. The Slammer worm had entered the network via a server in one of the engineering labs. The person monitoring the IDS noticed outbound traffic consistent with SQL Slammer at about 7:30 a.m. He contacted the network operations group by e-mail and followed up with a phone call and a voice mail message. Unfortunately, the operations group gets so many e-mails that if a message is not highlighted as URGENT, the message might be missed. That is exactly what happened—the e-mail alert wasn't read, and the voice message wasn't retrieved in time to block the attack. A few hours later, the ABCXYZ company found itself dealing with a massive number of reports of network and server problems.

A firm called Internet Security Systems (ISS) manages security for other organizations through its Managed Protection Services. The company's IDSs are designed to recognize 30 of the most-critical threats, including worms that go after Microsoft software and those that exploit Apache Web servers and other programs. When an attack is detected, the service automatically blocks it without requiring human intervention. Taking the manual intervention step out of the process enables a faster response and minimizes damage from a criminal hacker. To encourage customers to adopt its service, ISS guaranteed up to $50,000 in cash if the prevention service failed.

Using Managed Security Service Providers (MSSPs)

Keeping up with computer criminals—and with new regulations—can be daunting for organizations. Criminal hackers are constantly poking and prodding, trying to breach the security defenses of companies. Also, such recent legislation as HIPAA, the Sarbanes-Oxley, and the USA Patriot Act requires businesses to prove that they are securing their data. For most small and midsized organizations, the level of in-house network security expertise needed to protect their business operations can be quite costly to acquire and maintain. As a result, many are outsourcing their network security operations to managed security service providers (MSSPs) such as Counterpane, Guardent, Internet Security Services, Riptech, and Symantec. MSSPs monitor, manage, and maintain network security for both hardware and software. These companies provide a valuable service for IS departments drowning in reams of alerts and false alarms coming from virtual private networks (VPNs); antivirus, firewall, and intrusion detection systems; and other security monitoring systems. In addition, some provide vulnerability scanning and Web blocking/filtering capabilities.

Internet Laws for Libel and Protection of Decency

To help parents control what their children see on the Internet, some companies provide software called *filtering software* to help screen Internet content. Many of these screening programs also prevent children from sending personal information over e-mail or through chat groups. This stops children from broadcasting their name, address, phone number, or other personal information over the Internet. The two approaches used are filtering, which blocks certain Web sites, and rating, which places a rating on Web sites. According to the 2004 Internet Filter Review, the five top-rated filtering software packages are, in order: ContentProtect, Cybersitter, Net Nanny, CyberPatrol, and FilterPack.[42]

The Internet Content Rating Association (ICRA) is a nonprofit organization whose members include Internet industry leaders such as America Online, Bell South, British Telecom, IBM, Microsoft, UUNet, and Verizon. Its specific goals are to protect children from potentially harmful material, while also safeguarding free speech on the Internet. Using the ICRA rating system, Web authors fill out an online questionnaire describing the content of their site—what is and isn't present. The broad topics covered are the following: chat capabilities, the language used on the site, the nudity and sexual content of a site, the violence depicted on the site, and other areas such as alcohol, drugs, gambling, and suicide. Based on the authors' responses, ICRA then generates a content label (a short piece of computer code) that the authors add to their site. Internet users (and parents) can then set their browser to allow or disallow access to Web sites based on the objective rating information declared in the content label and their own subjective preferences. Reliance on Web site authors to do their own rating has its weaknesses, though. Web site authors can lie when completing the ICRA questionnaire so that their site receives a content label that doesn't accurately reflect the site's content. In addition, many hate groups and sexually explicit sites don't have an ICRA rating, so they will not be blocked unless a browser is set to block all unrated sites. Also, this option would block out so many acceptable sites that it could make Web surfing useless. For these reasons, at this time, site labeling is at best a complement to other filtering techniques.

ContentProtect is a filtering software program that helps block unwanted Internet content from children and young adults.

(Source: Courtesy of ContentWatch Inc.)

The Children's Internet Protection Act (CIPA) is a federal law passed in December 2000 that required federally funded libraries to use some form of prevention measure (such as Internet filters) to block access to obscene material and other material considered harmful to minors. Opponents of the law feared that it transferred power over the education process to private software companies that develop the Internet filters and define which sites are to be blocked.

With the increased popularity of networks and the Internet, libel becomes an important legal issue. A publisher, such as a newspaper, can be sued for libel, which involves publishing an intentionally false written statement that is damaging to a person's reputation. Generally, a bookstore cannot be held liable for statements made in newspapers or other publications it sells. Online services, such as CompuServe and America Online, might exercise some control over who puts information on their service but might not have direct control over the content of what is published by others on their service. So, can online services be sued for libel for content that someone else publishes on their service? Do online services more closely resemble a newspaper or a bookstore? This legal issue has not been completely resolved, but some court cases have been decided. The *Cubby, Inc. v. CompuServe* case ruled that CompuServe was similar to a bookstore and not liable for content put on its service by others. In this case, the judge stated, "While CompuServe can decline to carry a given publication altogether, in reality, after it does decide to carry a given publication, it will have little or no editorial control over that publication's content." This case set a legal precedent that has been applied in similar, subsequent cases. Companies should be aware that publishing Internet content to the world can subject them to different countries' laws in the same way that exporting physical products does.

Geolocation tools match the user's IP address with outside information to determine the actual geographic location of the online user where the customer's computer signal enters the Internet. This enables someone to identify the user's actual location within approximately 50 miles. Internet publishers can now limit the reach of their published speech to avoid potential legal risks. Use of such technology is also dividing the global Internet into separate content regions, with readers in Brazil, Japan, and the United States all receiving variations of the same information from the same publisher.

Individuals, too, must be careful what they post on the Internet to avoid libel charges. In many cases, disgruntled former employees are being sued by their former employers for material posted on the Internet.

Preventing Crime on the Internet

As mentioned in Chapter 7, Internet security can include firewalls and many methods to secure financial transactions. A firewall can include both hardware and software that act as a barrier between an organization's information system and the outside world. Some systems have been developed to safeguard financial transactions on the Internet.

To help prevent crime on the Internet, the following steps can be taken:

1. Develop effective Internet usage and security policies for all employees.
2. Use a stand-alone firewall (hardware and software) with network monitoring capabilities.
3. Deploy intrusion detection systems, monitor them, and follow up on their alarms.
4. Monitor managers and employees to make sure that they are using the Internet for business purposes.
5. Use Internet security specialists to perform audits of all Internet and network activities.

Even with these precautions, computers and networks can never be completely protected against crime. One of the biggest threats is from employees. Although firewalls provide good perimeter control to prevent crime from the outside, procedures and protection measures are needed to protect against computer crime by employees. Passwords, identification numbers, and tighter control of employees and managers also help prevent Internet-related crime.

PRIVACY ISSUES

Another important social issue in information systems involves privacy. In 1890, U.S. Supreme Court Justice Louis Brandeis stated that the "right to be left alone" is one of the most "comprehensive of rights and the most valued by civilized man." Basically, the issue of privacy deals with this right to be left alone or to be withdrawn from public view. With information systems, privacy deals with the collection and use or misuse of data. Data is constantly being collected and stored on each of us. This data is often distributed over easily accessed networks and without our knowledge or consent. Concerns of privacy regarding this data must be addressed.

With today's computers, the right to privacy is an especially challenging problem. More data and information are produced and used today than ever before. When someone is born, takes certain high school exams, starts working, enrolls in a college course, applies for a driver's license, purchases a car, serves in the military, gets married, buys insurance, gets a library card, applies for a charge card or loan, buys a house, or merely purchases certain products, data is collected and stored somewhere in computer databases. A difficult question to answer is, "Who owns this information and knowledge?" If a public or private organization spends time and resources to obtain data on you, does the organization own the data, and can it use the data in any way it desires? Government legislation answers these questions to some extent for federal agencies, but the questions remain unanswered for private organizations.

Privacy and the Federal Government

The U.S. federal government is perhaps the largest collector of data. Over 4 billion records exist on citizens, collected by about 100 federal agencies, ranging from the Bureau of Alcohol, Tobacco, and Firearms to the Veterans Administration. Other data collectors include state and local governments and profit and nonprofit organizations of all types and sizes.

In May 2006, it became widely known through published newspaper reports that the U.S. National Security Agency (NSA) had secretly collected the phone call records of tens of millions of U.S. citizens starting shortly after the September 11, 2001, terrorist attacks on the United States. The phone call records were then used to create a data warehouse for data-mining analysis to determine who might be a terrorist.[43] President Bush defended this data-collection program saying the government is not monitoring the individual calls of innocent Americans but is monitoring records to detect calling patterns that suggest terrorist activity.[44] Some Americans objected to what they saw as the collection and scrutiny of phone records based on suspicion and without obtaining court-ordered warrants. Other Americans defended the program saying they had nothing to hide from the government and that if the program would help track down terrorists it was justified. In August 2006, a federal judge ruled that the NSA program to wiretap the telephone and Internet traffic of U.S. residents is unconstitutional, illegal, and must be stopped.[45] As of this writing, the U.S. Department of Justice has appealed the federal judge's order and the final resolution is not known.

The European Union has a data-protection directive that requires firms transporting data across national boundaries to have certain privacy procedures in place. This directive affects virtually any company doing business in Europe, and it is driving much of the attention being given to privacy in the United States.

Privacy at Work

The right to privacy at work is also an important issue. Currently, the rights of workers who want their privacy and the interests of companies that demand to know more about their employees are in conflict. Recently, companies that have been monitoring have raised their employee's concerns. For example, workers might find that they are being closely monitored via computer technology. These computer-monitoring systems tie directly into workstations; specialized computer programs can track every keystroke made by a user. This type of system can determine what workers are doing while at the keyboard. The system also knows when the worker is not using the keyboard or computer system. These systems can estimate what people are doing and how many breaks they are taking. Needless to say, many workers consider this close supervision very dehumanizing.

E-Mail Privacy

E-mail also raises some interesting issues about work privacy. Federal law permits employers to monitor e-mail sent and received by employees. Furthermore, e-mail messages that have been erased from hard disks can be retrieved and used in lawsuits because the laws of discovery demand that companies produce all relevant business documents. On the other hand, the use of e-mail among public officials might violate "open meeting" laws. These laws, which apply to many local, state, and federal agencies, prevent public officials from meeting in private about matters that affect the state or local area.

E-mail has changed how workers and managers communicate in the same building or around the world. E-mail, however, can be monitored and intercepted. As with other services—such as cellular phones—the convenience of e-mail must be balanced with the potential of privacy invasion.

(Source: © Gary Conner/Photo Edit.)

Privacy and the Internet

Some people assume that there is no privacy on the Internet and that you use it at your own risk. Others believe that companies with Web sites should have strict privacy procedures and be accountable for privacy invasion. Regardless of your view, the potential for privacy invasion on the Internet is huge. People wanting to invade your privacy could be anyone from criminal hackers to marketing companies to corporate bosses. Your personal and professional information can be seized on the Internet without your knowledge or consent. E-mail is a prime target, as discussed previously. Sending an e-mail message is like having an open conversation in a large room—people can listen to your messages. When you visit a Web site on the Internet, information about you and your computer can be captured. When this information is combined with other information, companies can know what you read, what products you buy, and what your interests are. According to an executive of an Internet software monitoring company, "It's a marketing person's dream."

Most people who buy products on the Web say it's very important for a site to have a policy explaining how personal information is used, and the policy statement must make people feel comfortable and be extremely clear about what information is collected and what will and will not be done with it. However, many Web sites still do not prominently display their privacy policy or implement practices completely consistent with that policy. The real issue that Internet users need to be concerned with is—what do content providers want with their personal information? If a site requests that you provide your name and address, you have every right to know why and what will be done with it. If you buy something and provide a shipping address, will it be sold to other retailers? Will your e-mail address be sold on a list of active Internet shoppers? And if so, you should realize that it's no different than the lists compiled from the orders you place with catalog retailers. You have the right to be taken off any mailing list.

A potential solution to some consumer privacy concerns is the screening technology called the **Platform for Privacy Preferences (P3P)** being proposed to shield users from sites that don't provide the level of privacy protection they desire. Instead of forcing users to find and read through the privacy policy for each site they visit, P3P software in a computer's browser will download the privacy policy from each site, scan it, and notify the user if the policy does not match his preferences. (Of course, unethical marketers can post a privacy policy that does not accurately reflect the manner in which the data is treated.) The World Wide Web Consortium, an international industry group whose members include Apple, Commerce One, Ericsson, and Microsoft, is supporting the development of P3P. Version 1.1 of the P3P was released in February 2006 and can be found at *www.w3.org/TR/2006/WD-P3P11-20060210/Overview.html.*

The Children's Online Privacy Protection Act (COPPA) was passed by Congress in October 1998. This act was directed at Web sites catering to children, requiring them to post comprehensive privacy policies on their sites and to obtain parental consent before they collect any personal information from children under 13 years of age. Web site operators who violate the rule could be liable for civil penalties of up to $11,000 per violation. The Act has made an impact in the design and operations of Web sites that cater to children.

A social network service employs the Web and software to connect people for whatever purpose. There are thousands of such networks, which have become popular among teenagers. Some of the more popular social networking Web sites include Bebo, Classmates.com, Facebook, Hi5, Imbee, MySpace, Namesdatabase.com, Tagged, and XuQa. Most of these Web sites allow one to easily create a user profile that provides personal details, photos, even videos that can be viewed by other visitors to the Web site. Some of the Web sites have age restrictions or require that a parent register their preteen by providing a credit card to validate the parent's identity. Teens can provide information about where they live, go to school, their favorite music, and interests in hopes of meeting new friends. Unfortunately, they can also meet ill-intentioned strangers at these sites. Many documented encounters involve adults masquerading as teens attempting to meet young people for illicit purposes. Parents are advised to discuss potential dangers, check their children's profiles, and monitor their activities at such Web sites.

Fairness in Information Use

Selling information to other companies can be so lucrative that many companies will continue to store and sell the data they collect on customers, employees, and others. When is this information storage and use fair and reasonable to the people whose data is stored and sold? Do people have a right to know about data stored about them and to decide what data is stored and used? As shown in Table 14.4, these questions can be broken down into four issues that should be addressed: knowledge, control, notice, and consent.

In the past few decades, significant laws have been passed regarding a person's right to privacy. Others relate to business privacy rights and the fair use of data and information.

Platform for Privacy Preferences (P3P)

A screening technology that shields users from Web sites that don't provide the level of privacy protection they desire.

Fairness Issues	Database Storage	Database Usage
The right to know	Knowledge	Notice
The ability to decide	Control	Consent

Knowledge. Should people know what data is stored about them? In some cases, people are informed that information about them is stored in a corporate database. In others, they do not know that their personal information is stored in corporate databases.

Control. Should people be able to correct errors in corporate database systems? This is possible with most organizations, although it can be difficult in some cases.

Notice. Should an organization that uses personal data for a purpose other than the original purpose notify individuals in advance? Most companies don't do this.

Consent. If information on people is to be used for other purposes, should these people be asked to give their consent before data on them is used? Many companies do not give people the ability to decide if information on them will be sold or used for other purposes.

Table 14.4

The Right to Know and the Ability to Decide Federal Privacy Laws and Regulations

The Privacy Act of 1974

The major piece of legislation on privacy is the Privacy Act of 1974 (PA74). PA74 applies only to certain federal agencies. The act, which is about 15 pages long, is straightforward and easy to understand. The purpose of this act is to provide certain safeguards for people against an invasion of personal privacy by requiring federal agencies (except as otherwise provided by law) to do the following:

- Permit people to determine what records pertaining to them are collected, maintained, used, or disseminated by such agencies
- Permit people to prevent records pertaining to them from being used or made available for another purpose without their consent
- Permit people to gain access to information pertaining to them in federal agency records, to have a copy of all or any portion thereof, and to correct or amend such records
- Ensure that they collect, maintain, use, or disseminate any record of identifiable personal information in a manner that ensures that such action is for a necessary and lawful purpose, that the information is current and accurate for its intended use, and that adequate safeguards are provided to prevent misuse of such information
- Permit exemptions from this act only in cases of an important public need for such exemption, as determined by specific law-making authority
- Be subject to civil suit for any damages that occur as a result of willful or intentional action that violates anyone's rights under this act

PA74, which applies to all federal agencies except the CIA and law enforcement agencies, also established a Privacy Study Commission to study existing databases and to recommend rules and legislation for consideration by Congress. PA74 also requires training for all federal employees who interact with a "system of records" under the act. Most of the training is conducted by the Civil Service Commission and the Department of Defense. Another interesting aspect of PA74 concerns the use of Social Security numbers—federal, state, and local governments and agencies cannot discriminate against people for not disclosing or reporting their Social Security number.

Gramm-Leach-Bliley Act

This act was passed in 1999 and required all financial institutions to protect and secure customers' nonpublic data from unauthorized access or use. Under terms of this act, it was assumed that all customers approve of the financial institutions' collecting and storing their personal information. The institutions were required to contact their customers and inform them of this fact. Customers were required to write separate letters to each of their individual financial institutions and state in writing that they wanted to opt out of the data collection and storage process. Most people were overwhelmed with the mass mailings they received from their financial institutions and simply discarded them without ever understanding their importance.

USA Patriot Act

As discussed previously, the 2001 Uniting and Strengthening America by Providing Appropriate Tools Required to Intercept and Obstruct Terrorism Act (USA Patriot Act) was passed in response to the September 11 terrorism acts. Proponents argue that it gives necessary new powers to both domestic law enforcement and international intelligence agencies. Critics argue that the law removes many of the checks and balances that previously allowed the courts to ensure law enforcement agencies did not abuse their powers. For example, under this act, Internet service providers and telephone companies must turn over customer information, including numbers called, without a court order if the FBI claims that the records are relevant to a terrorism investigation. Also, the company is forbidden to disclose that the FBI is conducting an investigation.

Other Federal Privacy Laws

In addition to PA74, other pieces of federal legislation relate to privacy. A federal law that was passed in 1992 bans unsolicited fax advertisements. This law was upheld in a 1995 ruling by the Ninth U.S. Circuit Court of Appeals, which concluded that the law is a reasonable way to prevent the shifting of advertising costs to customers. Table 14.5 lists additional laws related to privacy.

Table 14.5

Federal Privacy Laws and Their Provisions

Law	Provisions
Fair Credit Reporting Act of 1970 (FCRA)	Regulates operations of credit-reporting bureaus, including how they collect, store, and use credit information
Tax Reform Act of 1976	Restricts collection and use of certain information by the Internal Revenue Service
Electronic Funds Transfer Act of 1979	Outlines the responsibilities of companies that use electronic funds transfer systems, including consumer rights and liability for bank debit cards
Right to Financial Privacy Act of 1978	Restricts government access to certain records held by financial institutions
Freedom of Information Act of 1970	Guarantees access for individuals to personal data collected about them and about government activities in federal agency files
Education Privacy Act	Restricts collection and use of data by federally funded educational institutions, including specifications for the type of data collected, access by parents and students to the data, and limitations on disclosure
Computer Matching and Privacy Act of 1988	Regulates cross-references between federal agencies' computer files (e.g.,to verify eligibility for federal programs)
Video Privacy Act of 1988	Prevents retail stores from disclosing video rental records without a court order
Telephone Consumer Protection Act of 1991	Limits telemarketers' practices
Cable Act of 1992	Regulates companies and organizations that provide wireless communications services, including cellular phones
Computer Abuse Amendments Act of 1994	Prohibits transmissions of harmful computer programs and code, including viruses
Gramm-Leach-Bliley Act of 1999	Requires all financial institutions to protect and secure customers' nonpublic data from unauthorized access or use
USA Patriot Act of 2001	Requires Internet service providers and telephone companies to turn over customer information, including numbers called, without a court order, if the FBI claims that the records are relevant to a terrorism investigation

Corporate Privacy Policies

Even though privacy laws for private organizations are not very restrictive, most organizations are very sensitive to privacy issues and fairness. They realize that invasions of privacy can hurt their business, turn away customers, and dramatically reduce revenues and profits. Consider a major international credit card company. If the company sold confidential financial information on millions of customers to other companies, the results could be disastrous. In a matter of days, the firm's business and revenues could be reduced dramatically. Therefore, most organizations maintain privacy policies, even though they are not required by law. Some companies even have a privacy bill of rights that specifies how the privacy of employees, clients, and customers will be protected. Corporate privacy policies should address a customer's knowledge, control, notice, and consent over the storage and use of information. They can also cover who has access to private data and when it can be used.

Multinational companies face an extremely difficult challenge in implementing data-collection and dissemination processes and policies because of the multitude of differing country or regional statutes. For example, Australia requires companies to destroy customer data (including backup files) or make it anonymous after it's no longer needed. Firms that transfer customer and personnel data out of Europe must comply with European privacy laws that allow customers and employees to access data about them and let them determine how that information can be used.

A good database design practice is to assign a single unique identifier to each customer—so that each has a single record describing all relationships with the company across all its business units. That way, the organization can apply customer privacy preferences consistently throughout all databases. Failure to do so can expose the organization to legal risks—aside from upsetting customers who opted out of some collection practices. Again, the 1999 Gramm-Leach-Bliley Financial Services Modernization Act required all financial service institutions to communicate their data privacy rules and honor customer preferences.

LexisNexis Gains Public Trust Through Disclosure

LexisNexis is a leading global provider of comprehensive information and business solutions to professionals in a variety of areas, including legal, risk management, corporate, government, law enforcement, accounting, and academic. Like most corporations, LexisNexis understands that keeping customer information and records private and secure is important for many reasons. The company maintains a strong and well-advertised privacy statement (*www.lexisnexis.com/terms/privacy/*) that informs customers of how their information is used and protected and allows customers to opt out of information-sharing practices.

Recently, LexisNexis learned that even with the best intentions and efforts, it is difficult if not impossible to protect private information completely. In late 2004, LexisNexis acquired Seisint Inc., a data broker that collects personal information and provides it to law enforcement and private companies for services such as debt recovery and fraud detection. In 2005, Seisint was the victim of a sophisticated hack attack. Leo Cronin, LexisNexis senior director for information security, described the attack as a social engineering ploy that targeted Seisint's "less sophisticated customers."

Some Seisint's customers received an e-mail with a pornographic lure, Cronin explained. The mail also contained a worm and a keystroke logger, which stole LexisNexis credentials, specifically for its risk-management services. Using the stolen usernames and passwords, criminal hackers logged on to the LexisNexis network for a total of 59 incidents of unauthorized access to information, LexisNexis said in a statement. The security breach exposed the personal data of hundreds of thousands of LexisNexis customers.

When the damage became clear, LexisNexis made an immediate decision to be forthcoming and transparent about the breach, Lee Cronin said. "We tried to do the best job we could." The company contacted all those who were affected by the attack using the framework of a California data security disclosure law passed in 2003 as a guide.

Since the breach, LexisNexis has reviewed the security of all its Web applications and created new procedures for verifying customers with access to sensitive data. The company has published a Web site, *http://privacyfacts.lexisnexis.com* in an effort to educate the public on matters of privacy.

LexisNexis encouraged certain customers to sign up for antivirus software. It revamped online security access, looking at password complexity and expiration times. The company also implemented measures to automatically detect anomalies in use of its products to identify potential security problems, Cronin said.

LexisNexis learned other lessons as well. Although passwords do not provide adequate security against sophisticated hack attacks, such front-door perimeter attacks are less likely than the persistent weak link: people. "Attackers are effective at going after low hanging fruit," Cronin said.

By being forthcoming with the public and victims, the company survived with minimal impact. "I think that's why we were so successful in dealing with this," Cronin said of the decision to be open and direct about the breach.

Discussion Questions

1. What security hole was exploited by criminal hackers to illegally access the LexisNexis network? What technique was employed?
2. How did LexisNexis turn a bad situation into an opportunity to improve its reputation?

Critical Thinking Questions

1. What type of protection can be applied against attacks such as the one described here? Was LexisNexis really at fault? Why or why not?
2. What pressures must LexisNexis have felt that motivated it to respond quickly and openly to the problem?

Sources: Kirk, Jeremy, "Data Theft Disclosure Meant Less Pain for LexisNexis," IDG News Service, April 25, 2006, *www.idg.com*. Roberts, Paul, "LexisNexis: 280,000 More Possible Data Theft Victims," *InfoWorld*, April 25, 2006, *www.infoworld.com*. LexisNexis Web site, accessed July 14, 2006, *www.lexisnexis.com*.

Individual Efforts to Protect Privacy

Although numerous state and federal laws deal with privacy, privacy laws do not completely protect individual privacy. In addition, not all companies have privacy policies. As a result, many people are taking steps to increase their own privacy protection. Some of the steps that you can take to protect personal privacy include the following:

- **Find out what is stored about you in existing databases.** Call the major credit bureaus to get a copy of your credit report. You are entitled to a free credit report every 12 months (see *freecreditreport.com*). You can also obtain a free report if you have been denied credit in the last 60 days. The major companies are Equifax (800-685-1111, *www.equifax.com*), TransUnion (800-916-8800, *www.transunion.com*), and Experian (888-397-3742, *www.experian.com*). You can also submit a Freedom of Information Act request to a federal agency that you suspect might have information stored on you.

- **Be careful when you share information about yourself.** Don't share information unless it is absolutely necessary. Every time you give information about yourself through an 800, 888, or 900 call, your privacy is at risk. Be vigilant in insisting that your doctor, bank, or financial institution not share information about you with others without your written consent.

- **Be proactive to protect your privacy.** You can get an unlisted phone number and ask the phone company to block caller ID systems from reading your phone number. If you change your address, don't fill out a change-of-address form with the U.S. Postal Service; you can notify the people and companies that you want to have your new address. Destroy copies of your charge card bills and shred monthly statements before disposing of them in the garbage. Be careful about sending personal e-mail messages over a corporate e-mail system. You can also get help in avoiding junk mail and telemarketing calls by visiting the Direct Marketing Association Web site at *www.the-dma.org*. Go to the Web site and look under Consumer Help-Remove Name from Lists.

- **When purchasing anything from a Web site, make sure that you safeguard your credit card numbers, passwords, and personal information.** Do not do business with a site unless you know that it handles credit card information securely (with FireFox look for a solid blue key in a small blue rectangle; with Microsoft Internet Explorer, look for the words "Secure Web Site"). Do not provide personal information without reviewing the site's data privacy policy. Many credit card companies issue single-use credit card numbers on request. Charges appear on your usual bill, but the number is destroyed after a single use, eliminating the risk of stolen credit card numbers.

THE WORK ENVIRONMENT

The use of computer-based information systems has changed the makeup of the workforce. Jobs that require IS literacy have increased, and many less-skilled positions have been eliminated. Corporate programs, such as reengineering and continuous improvement, bring with them the concern that, as business processes are restructured and information systems are integrated within them, the people involved in these processes will be removed.

However, the growing field of computer technology and information systems has opened up numerous avenues to professionals and nonprofessionals of all backgrounds. Enhanced telecommunications has been the impetus for new types of business and has created global markets in industries once limited to domestic markets. Even the simplest tasks have been aided by computers, making cash registers faster, smoothing order processing, and allowing people with disabilities to participate more actively in the workforce. As computers and other IS components drop in cost and become easier to use, more workers will benefit from the increased productivity and efficiency provided by computers. Yet, despite these increases in productivity and efficiency, information systems can raise other concerns.

Health Concerns

Organizations can increase employee effectiveness by paying attention to the health concerns in today's work environment. For some people, working with computers can cause occupational stress. Anxieties about job insecurity, loss of control, incompetence, and demotion are just a few of the fears workers might experience. In some cases, the stress can become so severe that workers might sabotage computer systems and equipment. Monitoring employee stress can alert companies to potential problems. Training and counseling can often help the employee and deter problems.

Computer use can affect physical health as well. Strains, sprains, tendonitis, and other problems account for more than 60 percent of all occupational illnesses and about a third of workers' compensation claims, according to the Joyce Institute in Seattle. The cost to U.S. corporations for these types of health problems is as high as $27 billion annually. Claims relating to *repetitive motion disorder,* which can be caused by working with computer keyboards and other equipment, have increased greatly. Also called *repetitive stress injury (RSI),* the problems can include tendonitis, tennis elbow, the inability to hold objects, and sharp pain in the fingers. Also common is *carpal tunnel syndrome (CTS),* which is the aggravation of the pathway for nerves that travel through the wrist (the carpal tunnel). CTS involves wrist pain, a feeling of tingling and numbness, and difficulty grasping and holding objects. It can be caused by many factors, such as stress, lack of exercise, and the repetitive motion of typing on a computer keyboard. Decisions on workers' compensation related to repetitive stress injuries have been made both for and against employees.

Other work-related health hazards involve emissions from improperly maintained and used equipment. Some studies show that poorly maintained laser printers can release ozone into the air; others dispute the claim. Numerous studies on the impact of emissions from display screens have also resulted in conflicting theories. Although some medical authorities believe that long-term exposure can cause cancer, studies are not conclusive at this time. In any case, many organizations are developing conservative and cautious policies.

Most computer manufacturers publish technical information on radiation emissions from their CRT monitors, and many companies pay close attention to this information. San Francisco was one of the first cities to propose a video display terminal (VDT) bill. The bill requires companies with 15 or more employees who spend at least four hours a day working with computer screens to give 15-minute breaks every two hours. In addition, adjustable chairs and workstations are required if employees request them.

In addition to the possible health risks from radio-frequency exposure, cell phone use has raised a safety issue—an increased risk of traffic accidents as vehicle operators become distracted by talking on their cell phones (or operating their laptop computers, car navigation systems, or other computer devices) while driving. As a result, some states have made it illegal to operate a cell phone while driving.

Avoiding Health and Environmental Problems

Many computer-related health problems are caused by a poorly designed work environment. The computer screen can be hard to read, with glare and poor contrast. Desks and chairs can also be uncomfortable. Keyboards and computer screens might be fixed in place or difficult to move. The hazardous activities associated with these unfavorable conditions are collectively referred to as *work stressors.* Although these problems might not be of major concern to casual users of computer systems, continued stressors such as repetitive motion, awkward posture, and eyestrain can cause more serious and long-term injuries. If nothing else, these problems can severely limit productivity and performance.

Research has shown that developing certain ergonomically correct habits can reduce the risk of RSI when using a computer.

(Source: Courtesy of Balt, Inc.)

ergonomics
The science of designing machines, products, and systems to maximize the safety, comfort, and efficiency of the people who use them.

The science of designing machines, products, and systems to maximize the safety, comfort, and efficiency of the people who use them, called **ergonomics**, has suggested some approaches to reducing these health problems. The slope of the keyboard, the positioning and design of display screens, and the placement and design of computer tables and chairs have been carefully studied. Flexibility is a major component of ergonomics and an important feature of computer devices. People come in many sizes, have differing preferences, and require different positioning of equipment for best results. Some people, for example, want to place the keyboard in their laps; others prefer it on a solid table. Because of these individual differences, computer designers are attempting to develop systems that provide a great deal of flexibility. In fact, the revolutionary design of Apple's iMac computer came about through concerns for users' comfort. After using basically the same keyboard design for over a decade, Microsoft introduced a new split keyboard called the Natural Ergonomic Keyboard 4000. The keyboard provides improved ergonomic features such as improved angles that reduce motion and how much you must stretch your fingers when you type. The design of the keyboard also provides more convenient wrist and arm postures which make typing more convenient for users.[46]

Computer users who work at their machines for more than an hour per day should consider using LCD screens, which are much easier on your eyes than CRT screens. If you stare at a CRT screen all day long, your eye muscles can get fatigued from all the screen flicker and bright backlighting of the monitor. LCD screens provide a much better viewing experience for your eyes by virtually eliminating flicker and while still being bright without harsh incandescence.[47]

In addition to steps taken by hardware manufacturing companies, computer users must also take action to reduce RSI and develop a better work environment. For example, when working at a workstation, the top of the monitor should be at or just below eye level. Your wrists and hands should be in line with your forearms, with your elbows close to your body and supported. Your lower back needs to be well supported. Your feet should be flat on the floor. Take an occasional break to get away from the keyboard and screen. Stand up and stretch while at your workplace. Do not ignore pain or discomfort. Many workers ignore early signs of RSI, and as a result, the problem becomes much worse and more difficult to treat.

ETHICAL ISSUES IN INFORMATION SYSTEMS

code of ethics
A code that states the principles and core values that are essential to a set of people and, therefore, govern their behavior.

As you've seen throughout the book in our "Ethical and Societal Issues" boxes, ethical issues deal with what is generally considered right or wrong. As we have seen, laws do not provide a complete guide to ethical behavior. Just because an activity is defined as legal does not mean that it is ethical. As a result, practitioners in many professions subscribe to a **code of ethics** that states the principles and core values that are essential to their work and, therefore, govern their behavior. The code can become a reference point for weighing what is legal and what is ethical. For example, doctors adhere to varying versions of the 2000-year-old Hippocratic Oath, which medical schools offer as an affirmation to their graduating classes.

Some IS professionals believe that their field offers many opportunities for unethical behavior. They also believe that unethical behavior can be reduced by top-level managers developing, discussing, and enforcing codes of ethics. Various IS-related organizations and associations promote ethically responsible use of information systems and have developed useful codes of ethics. The Association for Computing Machinery (ACM) is the oldest computing society founded in 1947 and boasts more than 80,000 members in more than 100 countries. The ACM has a code of ethics and professional conduct that includes eight general moral imperatives that can be used to help guide the actions of IS professionals. These guidelines can also be used for those who employ or hire IS professionals to monitor and guide their work. These imperatives are outlined in the following list:

As an ACM member I will ...

1. Contribute to society and human well-being.
2. Avoid harm to others.
3. Be honest and trustworthy.
4. Be fair and take action not to discriminate.
5. Honor property rights including copyrights and patents.
6. Give proper credit for intellectual property.
7. Respect the privacy of others.
8. Honor confidentiality.

(Source: ACM Code of Ethics and Professional Conduct, *http://www.acm.org/constitution/code.html*.)

The mishandling of the social issues discussed in this chapter—including waste and mistakes, crime, privacy, health, and ethics—can devastate an organization. The prevention of these problems and recovery from them are important aspects of managing information and information systems as critical corporate assets. Increasingly, organizations are recognizing that people are the most important component of a computer-based information system and that long-term competitive advantage can be found in a well-trained, motivated, and knowledgeable workforce.

SUMMARY

Principle

Policies and procedures must be established to avoid computer waste and mistakes.

Computer waste is the inappropriate use of computer technology and resources in both the public and private sectors. Computer mistakes relate to errors, failures, and other problems that result in output that is incorrect and without value. Waste and mistakes occur in government agencies as well as corporations. At the corporate level, computer waste and mistakes impose unnecessarily high costs for an information system and drag down profits. Waste often results from poor integration of IS components, leading to duplication of efforts and overcapacity. Inefficient procedures also waste IS resources, as do thoughtless disposal of useful resources and misuse of computer time for games and personal processing jobs. Inappropriate processing instructions, inaccurate data entry, mishandling of IS output, and poor systems design all cause computer mistakes.

A less-dramatic, yet still relevant, example of waste is the amount of company time and money employees can waste playing computer games, sending unimportant e-mail, or accessing the Internet. Junk e-mail, also called *spam*, and junk faxes also cause waste.

Preventing waste and mistakes involves establishing, implementing, monitoring, and reviewing effective policies and procedures. Careful programming practices, thorough testing, flexible network interconnections, and rigorous backup procedures can help an information system prevent and recover from many kinds of mistakes. Companies should develop manuals and training programs to avoid waste and mistakes. Company policies should specify criteria for new resource purchases and user-developed processing tools to help guard against waste and mistakes.

Principle

Computer crime is a serious and rapidly growing area of concern requiring management attention.

Some crimes use computers as tools (e.g., to manipulate records, counterfeit money and documents, commit fraud via telecommunications links, and make unauthorized electronic transfers of money). Identity theft is a crime in which an imposter obtains key pieces of personal identification information to impersonate someone else. The information is then used to obtain credit, merchandise, and services in the name of the victim, or to provide the thief with false credentials.

A cyberterrorist is someone who intimidates or coerces a government or organization to advance his political or social objectives by launching computer-based attacks against computers, networks, and the information stored on them. A criminal hacker, also called a *cracker*, is a computer-savvy person who attempts to gain unauthorized or illegal access to computer systems to steal passwords, corrupt files and programs, and even transfer money. Script bunnies are crackers with little technical savvy. Insiders are employees, disgruntled or otherwise, working solo or in concert with outsiders to compromise corporate systems.

Computer crimes target computer systems and include illegal access to computer systems by criminal hackers, alteration and destruction of data and programs by viruses (system, application, and document), and simple theft of computer resources. A virus is a program that attaches itself to other programs. A worm functions as an independent program, replicating its own program files until it destroys other systems and programs or interrupts the operation of computer systems and networks. Malware is a general term for software that is harmful or destructive. A Trojan horse program is a malicious program that disguises itself as a useful application and purposefully does something the user does not expect. A logic bomb is designed to "explode" or execute at a specified time and date. A variant is a modified version of a virus that is produced by the virus's author or another person by amending the original virus code. A password sniffer is a small program hidden in a network or computer system that records identification numbers and passwords.

Because of increased computer use, greater emphasis is placed on the prevention and detection of computer crime. Antivirus software is used to detect the presence of viruses, worms, and logic bombs. Use of an intrusion detection system (IDS) provides another layer of protection in the event that an intruder gets past the outer security layers—passwords, security procedures, and corporate firewall. It monitors system and network resources and notifies network security personnel when it senses a possible intrusion. Many small and midsized organizations are outsourcing their network security operations to managed security service providers (MSSPs), which monitor, manage, and maintain network security hardware and software.

Software and Internet piracy might represent the most common computer crime. It is estimated that the software industry loses nearly $12 billion in revenue each year to software piracy. Computer scams have cost people and companies thousands of dollars. Computer crime is also an international issue.

Many organizations and people help prevent computer crime, among them state and federal agencies, corporations, and people. Security measures, such as using passwords, identification numbers, and data encryption, help to guard against illegal computer access, especially when supported by effective control procedures. Public-key Infrastructure (PKI) enables users of an unsecured public network such as the Internet to securely and privately exchange data through the use of a public and a private cryptographic key pair that is obtained and shared through a trusted authority. The use of biometrics, involving the measurement of a person's unique characteristics, such as the iris, retina, or voice pattern, is another way to protect important data and information systems. Virus-scanning software identifies and removes damaging computer programs. Law enforcement agencies armed with new legal tools enacted by Congress now actively pursue computer criminals.

Although most companies use data files for legitimate, justifiable purposes, opportunities for invasion of privacy abound. Privacy issues are a concern with government agencies, e-mail use, corporations, and the Internet. The Children's Internet Protection Act was enacted to protect minors using the Internet. The Privacy Act of 1974, with the support of other federal laws, establishes straightforward and easily understandable requirements for data collection, use, and distribution by federal agencies; federal law also serves as a nationwide moral guideline for privacy rights and activities by private organizations. The USA Patriot Act, passed just five weeks after the September 11 terrorist attacks, requires Internet service providers and telephone companies to turn over customer information, including numbers called, without a court order, if the FBI claims that the records are relevant to a terrorism investigation. Also, the company is forbidden to disclose that the FBI is conducting an investigation. Only time will tell how this act will be applied in the future. The Gramm-Leach-Bliley Act requires all financial institutions to protect and secure customers' nonpublic data from unauthorized access or use. Under terms of this act, it is assumed that all customers approve of the financial institutions collecting and storing their personal information.

A business should develop a clear and thorough policy about privacy rights for customers, including database access. That policy should also address the rights of employees, including electronic monitoring systems and e-mail. Fairness in information use for privacy rights emphasizes knowledge, control, notice, and consent for people profiled in databases. People should know about the data that is stored about them and be able to correct errors in corporate database systems. If information on people is to be used for other purposes, they should be asked to give their consent beforehand. Each person has the right to know and the ability

to decide. Platform for Privacy Preferences (P3P) is a screening technology that shields users from Web sites that don't provide the level of privacy protection they desire.

Principle

Jobs, equipment, and working conditions must be designed to avoid negative health effects.

Computers have changed the makeup of the workforce and even eliminated some jobs, but they have also expanded and enriched employment opportunities in many ways. Computers and related devices affect employees' emotional and physical health, especially by causing repetitive stress injury (RSI). Some critics blame computer systems for emissions of ozone and electromagnetic radiation. No conclusive data connects cell phone use and cancer; however, heavy cell phone users might want to use "hands-free" phone sets. Use of cell phones while driving has been linked to increased car accidents.

The study of designing and positioning computer equipment, called *ergonomics*, has suggested some approaches to reducing these health problems. Ergonomic design principles help to reduce harmful effects and increase the efficiency of an information system. The slope of the keyboard, the positioning and design of display screens, and the placement and design of computer tables and chairs are essential for good health. RSI prevention includes keeping good posture, not ignoring pain or problems, performing stretching and strengthening exercises, and seeking proper treatment. Although they can cause negative health consequences, information systems can also be used to provide a wealth of information on health topics through the Internet and other sources.

Ethics determine generally accepted and discouraged activities within a company and society at large. Ethical computer users define acceptable practices more strictly than just refraining from committing crimes; they also consider the effects of their IS activities, including Internet usage, on other people and organizations. The Association for Computing Machinery developed guidelines and a code of ethics. Many IS professionals join computer-related associations and agree to abide by detailed ethical codes.

CHAPTER 14: SELF-ASSESSMENT TEST

Policies and procedures must be established to avoid computer waste and mistakes.

1. It is solely up to IS professionals to implement and follow proper IS usages policies to ensure effective use of company resources. True or False?

2. Computer-related waste and mistakes are major causes of computer problems, contributing to unnecessarily high ____ and lost ____.

3. Preventing waste and mistakes involves establishing, implementing, _____, and reviewing effective policies and procedures.

Computer crime is a serious and rapidly growing area of concern requiring management attention.

4. Computer crime is frequently easily detected and the amount of money involved is often quite small. True or False?

5. _____ is a federally funded research and development center at Carnegie Mellon University charged with coordinating communication among experts during computer security emergencies.

6. The vast majority of organizations conduct some form of computer security audit. True or False?

7. _____ is a crime in which an imposter obtains key pieces of personal identification information, such as Social Security or driver's license numbers, to impersonate someone else.

8. Someone who intimidates or coerces a government to advance his political objectives by launching computer-based attacks against computers, networks, and the information stored on them is called a _____.
   a. cyberterrorist
   b. hacker
   c. criminal hacker or cracker
   d. social engineer

9. A logic bomb is a type of Trojan horse that executes when specific conditions occur. True or False?

10. Malware capable of spreading itself from one computer to another is called a _____:
    a. logic bomb
    b. Trojan horse
    c. virus
    d. worm

11. A(n) _____ is a modified version of a virus that is produced by the virus's author or another person amending the original virus code.

12. Half the loss from software piracy comes from countries in the European Union. True or False?

13. Phishing is a computer scam that seems to direct users to a bank's Web site but actually captures key personal information about its victims. True or False?

Jobs, equipment, and working conditions must be designed to avoid negative health effects.

14. CTS, or _____, is the aggravation of the pathway of nerves that travel through the wrist.

15. Positive evidence suggests that excessive use of cell phones increases the risk of brain cancer. True or False?

16. The study of designing and positioning computer equipment to improve worker productivity and minimize worker injuries is called _____.

CHAPTER 14: SELF-ASSESSMENT TEST ANSWERS

(1) False (2) costs, profits (3) monitoring (4) False (5) CERT/CC (6) True (7) Identity theft (8) a (9) True (10) d (11) variant (12) False (13) True (14) carpal tunnel syndrome (15) False (16) ergonomics

REVIEW QUESTIONS

1. What special issues are associated with the prevention of image-based spam?
2. Give two recent examples of computer mistakes causing serious repercussions.
3. Identify three types of common computer-related mistakes.
4. According to the Computer Security Institute, what are the leading causes of financial loss associated with computer crime?
5. What is a variant? What dangers are associated with such malware?
6. What is phishing? What actions can you take to reduce the likelihood that you will be a victim of this crime?
7. What is a virus? What is a worm? How are they different?
8. Outline measures you should take to protect yourself against viruses and worms.
9. What role does the Business Software Alliance play in the protection of software?
10. Identify at least five tips to follow to avoid becoming a victim of a computer scam.
11. What is biometrics, and how can it be used to protect sensitive data?
12. What is the difference between antivirus software and an intrusion detection system?
13. What is the Children's Online Privacy Protection Act?
14. What is ergonomics? How can it be applied to office workers?
15. What specific actions can you take to avoid RSI?
16. What is a code of ethics? Give an example.

DISCUSSION QUESTIONS

1. If your company were guilty of flagrant violation of software copyright protection and illegally pirated dozens of different software packages, should you report it to the Business Software Alliance? Would you? Why or why not?

2. Outline an approach, including specific techniques (e.g., dumpster diving, phishing, social engineering) that you could employ to gain personal data about the members of your class.

3. Your 12-year-old niece shows you a profile of her math teacher posted on MySpace that includes a list of dozens of students as the instructor's friends and a quote: "I hope to make lots of new friends and, who knows, maybe find Miss Right." What would you do?

4. Imagine that you are a hacker and have developed a Trojan horse program. What tactics might you use to get unsuspecting victims to load the program onto their computer?

5. You are the director of public relations for a large credit card firm that just had its customer database hacked and data on more than 1 million customers stolen. What course of action would you recommend to the CEO? Should the firm be forthright in disclosing the hack and its potential impact or keep quiet and hope to minimize negative publicity?

6. Briefly discuss the potential for cyberterrorism to cause a major disruption in our daily life. What are some likely targets of a cyberterrorist? What sort of action could a cyberterrorist take against these targets?

7. You travel a lot in your role of vice president of sales and carry a laptop containing customer data, budget information, product development plans, and promotion information. What measures should you take to ensure against potential theft of your laptop and its critical data?

8. Do you believe that the National Security Agency should be able to collect the telephone call records of U.S. citizens without the use of search warrants? Why or why not?

9. Using information presented in this chapter on federal privacy legislation, identify which federal law regulates the following areas and situations: cross-checking IRS and Social Security files to verify the accuracy of information, customer liability for debit cards, your right to access data contained in federal agency files, the IRS obtaining personal information, the government obtaining financial records, and employers' access to university transcripts.

10. Briefly discuss the difference between acting morally and acting legally. Give an example of acting legally and yet immorally.

PROBLEM-SOLVING EXERCISES

1. Access the CSI/FBI Annual Survey results for the past five years. Choose one of the types of computer crime incidents reported there and determine the number of such incidents and the associated cost. Use a graphics package to show the variation in number of incidents and cost over time.

2. Using spreadsheet software and appropriate forecasting routines, develop a forecast for next year. Document any assumptions you make in developing your forecast.

3. Using your word processing software, write a few brief paragraphs summarizing the trends you see from reviewing the data for the past few years. Then cut and paste the graph from exercise 1 and your forecast from exercise 2 into your report.

TEAM ACTIVITIES

1. Imagine that your team has been hired as a group of consultants to improve the computer security at your school. How would you approach this challenging task? Outline some of your initial steps and identify specific organizations and/or people who need to be involved.

2. Have each member of your team access ten different Web sites and summarize their findings in terms of the existence of data privacy policy statements: Did the site have such a policy? Was it easy to find? Was it complete and easy to understand? Did you find any sites using the P3P standard or ICRA rating method?

WEB EXERCISES

1. Search the Web for a site that provides software to detect and remove spyware. Write a short report for your instructor summarizing your findings.
2. Do research on the Web to find evidence of an increase or decrease in the number of viruses being developed and released. To what is the change attributed? Write a brief memo to your instructor identifying your sources and summarizing your findings.
3. Do research to find out the current state of the NSA program to wiretap the telephone and Internet traffic of U.S. residents. Write a paragraph or two summarizing your findings.

CAREER EXERCISES

1. Computer forensics is a relatively new but growing field, which involves the discovery of computer-related evidence and data. It relies on formal computer evidence-processing protocols. Its findings can be presented in a court of law. Cases involving trade secrets, commercial disputes, employment discrimination, misdemeanor and felony crimes, and personal injury can be won or lost solely with the introduction of recovered e-mail messages and other electronic files and records. Computer forensics tools and methods are used extensively by law enforcement, military, intelligence agencies, and businesses. Do research to identify the experience and training necessary to become a certified computer forensics specialist. Would you consider this as a possible career field? Why, or why not?
2. You have just begun a new position in customer relations for a midsized bank. Within your first week on the job, several customers have expressed concern about potential theft of customer data from the bank's computer databases and identity theft. Who would you talk with to develop a satisfactory response to address your customers' concerns? What key points would you need to verify with bank employees?

CASE STUDIES

Case One

Indigo Books and Music Align Security Strategies with Corporate Goals

Indigo Books & Music Inc. is the largest bookstore chain in Canada. It was created in 1996, and their first store, called Indigo Books, Music & More, was opened in Burlington, Ontario, on September 4, 1997. Indigo merged with its largest competitor, Chapters Inc., on August 14, 2001. Together, the companies created a hugely successful e-commerce Web site, chapters.indigo.ca, which is similar in style to amazon.com.

Indigo is a large business, with headquarters in Toronto. It has around 6,000 employees and annual revenue approaching a billion Canadian dollars. The corporation has an approach toward information security that security experts think should act as an example to other businesses. Indigo's security spending is based on its impact to the overall business model, assessing risks from a business perspective, instead of a technology perspective, explained Ricky Mehra, director of IT security and internal controls for Indigo.

"Our security investments are strategic in nature," he said, adding that Indigo looks at its security investments and determines how it will fit the company's business model five years down the line.

Indigo invests its substantial security budget in what Mehra refers to as a "defence in depth" security framework. Such a framework includes technical controls such as firewalls, password control mechanisms, and intrusion detection devices. It also includes the establishment of internal policies such as who gets access to what information, and user awareness and training to better secure its online business and corporate network.

Stephen Lawson, vice president of technology with Fox Group Consulting in Mount Albert, Ontario, thinks that Mehra and Indigo have the right idea. Lawson learned about Indigo security practices at an online security round table hosted by Microsoft Canada. Although many organizations like Indigo use a "defence in depth" framework, some companies still believe that installing a piece of technology is all they need for IT security. Technology is only one piece of the puzzle, Lawson said. It is important for the business side to first create security policies, and then look into technology to enforce those policies.

Steve Lloyd, Microsoft Canada's chief security advisor, observed that implementing IT security policy as part of the corporate agenda is becoming more important in view of increasing government regulations. With compliance becoming more of an issue, security executives are citing compliance as a way of providing a return on investment for security initiatives. "To convince the higher ranks that security is worth investing in, talk about compliance and the penalties that will be levied if you don't comply," said Lloyd. Like many countries, Canada has increasing amounts of laws regulating the manner in which businesses keep private customer information safe and secure.

Discussion Questions

1. What is unusual about Indigo's philosophy regarding information security?
2. What are the components of a "defence in depth" security framework?

Critical Thinking Questions

1. How can information security support an organization's overall goals and play a role in a business strategy?
2. Why do you think governments are stepping in and regulating business's information security?

Sources: Ho, Vanessa, "Strategy Pays Off for Bookseller," *ComputerWorld Canada*, July 11, 2006, *www.itworldcanada.com*. Indigo Web site, accessed July 15, 2006, *www.chapters.indigo.ca*. Yahoo! Finance Web site, accessed July 15, 2006, *http://biz.yahoo.com/ic/57/57232.html*.

Case Two

Disgruntled Employee Plants Time Bomb in Get-Rich-Quick Scheme

Recently, another hacker went to trial to determine if he will be spending the next several years in jail or as a free man. Unlike many hacker trials, the defendant in this case is not an adolescent, but a 63-year-old systems administrator earning $160,000 per year with a big name financial company.

After working for a financial company for many years, the systems administrator came to expect a $25,000 bonus at the end of each year. One year, the company suffered financial losses and the employee received only a $10,000 bonus. The employee had been counting on $25,000 for his son's college tuition. Feeling cheated, the employee began building the code that would punish his employer while creating a windfall for him and his family.

According to the prosecution, the systems administrator developed malicious code to delete files and cause a major disruption on his company's network. The time bomb was ingenious in design. Working remotely on the corporate

system from his home, the employee allegedly built four separate components of the time bomb:

- Component 1, the Payload: This destructive portion of the code told the servers to delete all files.
- Component 2, Distribution: This code pushed the bomb from the central server in the company's data center out to the 370 branch offices scattered across the country.
- Component 3, Persistence: This code kept the bomb running despite reboots and any loss of power.
- Component 4, Triggers: To avoid mistakes, he built not one, but two triggers for the bomb. If one trigger was accidentally discovered and deleted off the system, another one would be silently waiting to go off, setting a destructive chain of events into motion.

The employee set the malicious code, or bomb, to run, or detonate, on April 4.

With the bomb in place, the employee went to his supervisor and demanded the bonus that he felt he was due, and threatened to quit if he didn't get it. Then, he packed his things and left. Prosecutors said that "within an hour or so" of walking out the door, he was at a securities office buying "puts" against his company. "Puts" is a high-risk, high-payoff type of trade where the buyer profits if the company stock goes down. Over the three weeks that followed, the employee spent nearly $25,000 to purchase a total of 330 puts, almost all of them against his company. He had not bought one before that month, and he never bought another one afterward. He purchased more than half of the "puts" the day before the disaster struck.

The damage caused by the malicious code impaired trading at the firm that day, hampering more than 1,000 servers and 17,000 individual workstations. The attack cost the company about $3 million to assess and repair. The prosecution claimed "It took hundreds of people, thousands of man hours and millions of dollars to correct."

Whether the employee is found guilty is yet to be determined. Certainly, someone planted a time bomb in the financial company's computer systems. The unusual purchase of "puts" is the primary incriminating evidence against the employee. Investigators also determined that the bomb was planted by someone logged on with the employee's username and password. The employee's primary defense is that other company users could have accessed the system using his password and that the systems were vulnerable to outside attackers. The employee faces one count each of securities fraud and computer sabotage, and two counts of mail fraud in U.S. District Court.

Discussion Questions

1. What were the four components of this time bomb and why might they be considered ingenious?
2. Name two pieces of evidence that are most damaging to the employee. Explain why.

Critical Thinking Questions

1. Based on the information presented here, do you think the employee is guilty beyond a reasonable doubt? Why or why not? Is guilt more difficult to prove in cases of cybercrime as opposed to ordinary crimes?
2. What steps could the company have taken to avoid this type of destruction?

Sources: Solheim, Shelley, "UBS Employee Stands Trial for Detonating 'Computer Bomb,'" IDG News Service, June 08, 2006, *www.infoworld.com*. Gaudin, Sharon, "Prosecutors: UBS Sysadmin Believed 'He Had Created The Perfect Crime,'" *Information Week*, July 10, 2006. "Disgruntled UBS PaineWebber Employee Charge with Allegedly Unleashing 'Logic Bomb' on Company Computers," U.S. Department of Justice Web site, accessed July 15, 2006, *www.usdoj.gov/criminal/cybercrime/duronioIndict*.

Questions for Web Case

See the Web site for this book to read about the Whitmann Price Consulting case for this chapter. Following are questions concerning this Web case.

Whitmann Price Consulting: Security, Privacy, and Ethical Considerations

Discussion Questions

1. Why do you think extending access to the Whitmann Price network beyond the business's walls dramatically elevated the risk to information security?
2. What was the primary tool used to minimize that risk, and how does it work?

Critical Thinking Questions

1. Why is it that information security usually comes at the cost of user convenience?
2. List the security policies put in place for the AMCI system and the rational that you think is behind them.

NOTES

Sources for the opening vignette: Messmer, Ellen, "Johnson & Johnson Tackles Security Pain," *Network World*, March 30, 2005, *www.networkworld.com*. Johnson & Johnson Web site, accessed July 14, 2006, *www.jnj.com*.

1   McGillicuddy, Shamus, "Thwarting Spam from the Inside and the Outside," *ComputerWeekly.com*, July 11, 2006.
2   McGillicuddy, Shamus, "Image-Based Spam on the Rise," *ComputerWeekly.com*, August 3, 2006.
3   Reuters, "Computer Woes Briefly Hobble Postal Services," *Computerworld*, March 3, 2006.
4   Rosencrance, Linda, "Brief: Computer Glitch Grounds, Delays Northwest Flights," *Computerworld*, July 26, 2006.
5   Whiting, Rick, "Hamstrung by Defective Data," *Information Week*, May 8, 2006.
6   Hoffman, Thomas, "Sarbanes-Oxley Trumps IM at Some Firms," *Computerworld*, August 5, 2006.
7   Mitchell, Robert, "Q&A: Making A Federal Case – How the FBI Collars Cybercriminals," *Computerworld*, July 28, 2006.
8   Koprowski, Gene J., "Study: Nearly a Quarter Million PCs Turned into 'Zombies' Daily", *Ecommerce Times*, January 14, 2006.
9   Staff, "Silicon Valley Engineer Indicted for Stealing Trade Secrets and Computer Fraud," U.S. Department of Justice Web site, December 22, 2005, *www.usdoj.gov/criminal/cybercrime*, accessed August 10, 2006.
10 Staff, "Plano Man Convicted of Computer Sabotage," U.S. Department of Justice Web site, November 21, 2005, *www.usdoj.gov/criminal/cybercrime*, accessed August 16, 2006.
11 Staff, "Former Federal Computer Security Specialist Sentenced for Hacking Department of Education Computer," U.S. Department of Justice Web site, May 12, 2006, *www.usdoj.gov/criminal/cybercrime*, accessed August 16, 2006.
12 McMillian, Robert, "Internet Sieges Can Cost Businesses a Bundle," *Computerworld*, August 25, 2005.
13 Mitchell, Robert, "Q&A: Making A Federal Case – How the FBI Collars Cybercriminals," *Computerworld*, July 28, 2006.

14 2006 CSI/FBI Computer Crime and Security Survey, Computer Security Institute Web site, *www.gocsi.com/press*, accessed August 17, 2006.
15 Greenemeier, Larry, "Judges and Prosecutors Throw the Book at Hackers," *Information Week*, May 15, 2006.
16 Greenemeier, Larry, "Judges and Prosecutors Throw the Book at Hackers," *Information Week*, May 15, 2006.
17 Hatlestad, Luc, "Brief: Hacks Decline, Worries Don't," *Information Week*, August 4, 2006.
18 Keizer, Gregg, "U.S. Consumers Taking Steps to Stymie ID Theft," *Information Week*, May 19, 2006.
19 Hatlestad, Luc, "Brief: Hacks Decline, Worries Don't," *Information Week*, August 4, 2006.
20 Hatlestad, Luc, "Brief: Hacks Decline, Worries Don't," *Information Week*, August 4, 2006.
21 Heun, Christopher, "E-Health Initiatives Could Lead to New Forms of ID Theft," *Information Week*, June 16, 2006.
22 Savaas, Antony, "McAfee: 400,000 Virus Definitions on Users' Machines by 2008," *ComputerWeekly.com*, July 6, 2006.
23 Lyman, Jay, "Study: Mobile Malware Threat to Grow in '06," *Ecommerce Times*, December 10, 2005.
24 Savvas, Antony, "Police Arrest m00p Gang Suspects," *ComputerWeekly.com*, June 28, 2006.
25 Saran, Cliff, "HP Unleashes a Worm to Find and Fix Server Holes," *ComputerWeekly.com*, July 4, 2006.
26 Keizer, Gregg, "Trojan Snags World of Warcraft Passwords to Cash Out Accounts," *Information Week*, May 2, 2006.
27 Savvas, Antony, "Trojan Steals Bank Details After Pretending to Be Microsoft Patch," *ComputerWeekly.com*, May 31, 2006.
28 Goodwin, Bill, "Investec Rolls Out System to Monitor Desktop Security," *ComputerWeekly.com*, July 4, 2006.
29 F-Secure Web site, *www.f-secure.com/corporate/intro.shtml*, accessed August 26, 2006.
30 McAfee Web site, *www.mcafee.com/us*, accessed August 26, 2006.
31 "Virus Hoaxes," McAfee Web site, *http://vil.mcafee.com/hoax.asp*, August 24, 2006.

32 Keizer, Gregg, "FBI Recovers Stolen Veterans Affairs Laptop," *Information Week*, June 29, 2006.

33 Babcock, Charles; Greenemeier, Larry, "A Million Identities Stolen from Two Financial Services Firms," *Information Week*, June 20, 2006.

34 The Business Software Alliance Web site home page, *www.bsa.org/usa*, accessed August 26, 2006.

35 The Business Software Alliance Web site, "Piracy and the Law," *www.bsa.org/usa/antipiracy/Piracy-and-the-Law.cfm*, accessed August 28, 2006.

36 Staff, "Four Men Sentenced and Another Film Critic Pleads Guilty in Operation Copycat," U.S. Department of Justice Web site, *www.usdoj.gov/criminal/cybercrime*, accessed August 14, 2006.

37 Garretson, Cara, "Stats Show Phishing Attacks Doubled," *Computerworld*, January 13, 2006.

38 McMillan, Robert, "Man Charged in Hurricane Katrina Phishing Scams," *Computerworld*, August 18, 2006.

39 Software and Information Industry Association, "Anti-Piracy," *www.siia.net/piracy/whatis.asp*, accessed August 26, 2006.

40 Hadfield, Will, "Co-op Goes Live with First Payment by Biometrics System," *ComputerWeekly.com*, March 10, 2006.

41 "Former Owner of Massive For-Profit Software Piracy Website Sentenced to 6 Years," BSA Web site, *www.bsa.org/usa/press/newsreleases/Ferrer-sentenced.cfm*, accessed August 28, 2006.

42 2006 Internet Filter Report, *www.internetfilterreview.com*, accessed August 28, 2006.

43 Lai, Eric; Fisher, Sharon, "NSA's Alleged Phone-Records Program Puts Spotlight on Data Mining," *Computerworld*, May 25, 2006.

44 Gross, Grant, "Qwest Praised for Declining NSA Phone Records Request," *IT World*, May 12, 2006.

45 Gross, Grant, "NSA Wiretap Program Ruled Unconstitutional," *Computerworld*, August 17, 2006.

46 Shah, Agam, "Microsoft Revamps Keyboards and Mice," *Computerworld*, September 6, 2005.

47 Merrin, John, "Review: Six 19-inch LCD Monitors," *Information Week*, June 8, 2005.

acceptance testing Conducting any tests required by the user.

accounting MIS An information system that provides aggregate information on accounts payable, accounts receivable, payroll, and many other applications.

accounting systems Systems that include budget, accounts receivable, payroll, asset management, and general ledger.

ad hoc DSS A DSS concerned with situations or decisions that come up only a few times during the life of the organization.

alpha testing Testing an incomplete or early version of the system.

analog signal A variable signal continuous in both time and amplitude so that any small fluctuations in the signal are meaningful.

antivirus program Software that runs in the background to protect your computer from dangers lurking on the Internet and other possible sources of infected files.

applet A small program embedded in Web pages.

application flowcharts Diagrams that show relationships among applications or systems.

application program interface (API) An interface that allows applications to make use of the operating system.

application service provider (ASP) A company that provides software, support, and the computer hardware on which to run the software from the user's facilities.

arithmetic/logic unit (ALU) The part of the CPU that performs mathematical calculations and makes logical comparisons.

ARPANET A project started by the U.S. Department of Defense (DoD) in 1969 as both an experiment in reliable networking and a means to link DoD and military research contractors, including many universities doing military-funded research.

artificial intelligence (AI) A field in which the computer system takes on the characteristics of human intelligence. Also the ability of computers to mimic or duplicate the functions of the human brain.

artificial intelligence systems People, procedures, hardware, software, data, and knowledge needed to develop computer systems and machines that demonstrate the characteristics of intelligence.

asking directly An approach to gather data that asks users, stakeholders, and other managers about what they want and expect from the new or modified system.

asynchronous communications A form of communications where the receiver gets the message after some delay—sometimes hours or days after the message is sent.

attribute A characteristic of an entity.

auditing Analyzing the financial condition of an organization and determining whether financial statements and reports produced by the financial MIS are accurate.

automated clearing house (ACH) A secure, private network that connects all U.S. financial institutions to one another by way of the Federal Reserve Board or other ACH operators.

backbone One of the Internet's high-speed, long-distance communications links.

backward chaining The process of starting with conclusions and working backward to the supporting facts.

batch processing system A form of data processing where business transactions are accumulated over a period of time and prepared for processing as a single unit or batch.

benchmark test An examination that compares computer systems operating under the same conditions.

best practices The most efficient and effective ways to complete a business process.

beta testing Testing a complete and stable system by end users.

biometrics The measurement of one of a person's traits, whether physical or behavioral.

blade server A server that houses many individual computer motherboards that include one or more processors, computer memory, computer storage, and computer network connections.

bot A software tool that searches the Web for information, products, or prices.

brainstorming A decision-making approach that often consists of members offering ideas "off the top of their heads."

bridge A telecommunications device that connects one LAN to another LAN that uses the same telecommunications protocol.

broadband communications A telecommunications system in which a very high rate of data exchange is possible.

business intelligence The process of gathering enough of the right information in a timely manner and usable form and analyzing it to have a positive impact on business strategy, tactics, or operations.

business-to-business (B2B) e-commerce A subset of e-commerce where all the participants are organizations.

business-to-consumer (B2C) e-commerce A form of e-commerce in which customers deal directly with an organization and avoid intermediaries.

byte (B) Eight bits that together represent a single character of data.

cache memory A type of high-speed memory that a processor can access more rapidly than main memory.

CASE repository A database of system descriptions, parameters, and objectives.

central processing unit (CPU) The part of the computer that consists of three associated elements: the arithmetic/logic unit, the control unit, and the register areas.

centralized processing Processing alternative in which all processing occurs at a single location or facility.

certificate authority (CA) A trusted third-party organization or company that issues digital certificates.

certification A process for testing skills and knowledge, which results in a statement by the certifying authority that states an individual is capable of performing a particular kind of job.

change model A representation of change theories that identifies the phases of change and the best way to implement them.

channel bandwidth The rate at which data is exchanged over a communications channel, usually measured in bits per second (bps).

character A basic building block of information, consisting of uppercase letters, lowercase letters, numeric digits, or special symbols.

chat room A facility that enables two or more people to engage in interactive "conversations" over the Internet.

chief knowledge officer (CKO) A top-level executive who helps the organization use a KMS to create, store, and use knowledge to achieve organizational goals.

chief programmer team A group of skilled IS professionals who design and implement a set of programs.

choice stage The third stage of decision making, which requires selecting a course of action.

click fraud A problem arising in a pay-per-click online advertising environment where additional clicks are generated beyond those that come from actual, legitimate users.

clickstream data The data gathered based on the Web sites you visit and the items you click.

client/server An architecture in which multiple computer platforms are dedicated to special functions such as database management, printing, communications, and program execution.

clock speed A series of electronic pulses produced at a predetermined rate that affects machine cycle time.

closed shops IS departments in which only authorized operators can run the computers.

code of ethics A code that states the principles and core values that are essential to a set of people and, therefore, govern their behavior.

cold site A computer environment that includes rooms, electrical service, telecommunications links, data storage devices, and the like; also called a *shell*.

command-based user interface A user interface that requires you to give text commands to the computer to perform basic activities.

communications protocol A set of rules that governs the exchange of information over a communications channel.

compact disc read-only memory (CD-ROM) A common form of optical disc on which data, once it has been recorded, cannot be modified.

competitive advantage A significant and (ideally) long-term benefit to a company over its competition.

competitive intelligence One aspect of business intelligence limited to information about competitors and the ways that knowledge affects strategy, tactics, and operations.

compiler A special software program that converts the programmer's source code into the machine-language instructions consisting of binary digits.

computer literacy Knowledge of computer systems and equipment and the ways they function; it stresses equipment and devices (hardware), programs and instructions (software), databases, and telecommunications.

computer network The communications media, devices, and software needed to connect two or more computer systems and/or devices.

computer programs Sequences of instructions for the computer.

computer-aided software engineering (CASE) Tools that automate many of the tasks required in a systems development effort and encourage adherence to the SDLC.

computer-assisted manufacturing (CAM) A system that directly controls manufacturing equipment.

computer-based information system (CBIS) A single set of hardware, software, databases, telecommunications, people, and procedures that are configured to collect, manipulate, store, and process data into information.

computer-integrated manufacturing (CIM) Using computers to link the components of the production process into an effective system.

concurrency control A method of dealing with a situation in which two or more people need to access the same record in a database at the same time.

consumer-to-consumer (C2C) e-commerce A subset of e-commerce that involves consumers selling directly to other consumers.

content streaming A method for transferring multimedia files over the Internet so that the data stream of voice and pictures plays more or less continuously without a break, or very few breaks; enables users to browse large files in real time.

continuous improvement Constantly seeking ways to improve business processes to add value to products and services.

control unit The part of the CPU that sequentially accesses program instructions, decodes them, and coordinates the flow of data in and out of the ALU, registers, primary storage, and even secondary storage and various output devices.

coprocessor The part of the computer that speeds processing by executing specific types of instructions while the CPU works on another processing activity.

cost center A division within a company that does not directly generate revenue.

cost/benefit analysis An approach that lists the costs and benefits of each proposed system. After they are expressed in monetary terms, all the costs are compared with all the benefits.

counterintelligence The steps an organization takes to protect information sought by "hostile" intelligence gatherers.

creative analysis The investigation of new approaches to existing problems.

criminal hacker (cracker) A computer-savvy person who attempts to gain unauthorized or illegal access to computer systems to steal passwords, corrupt files and programs, or even transfer money.

critical analysis The unbiased and careful questioning of whether system elements are related in the most effective ways.

critical path Activities that, if delayed, would delay the entire project.

critical success factors (CSFs) Factors that are essential to the success of a functional area of an organization.

cross-platform development A development technique that allows programmers to develop programs that can run on computer systems having different hardware and operating systems, or platforms.

culture A set of major understandings and assumptions shared by a group.

customer relationship management (CRM) system A system that helps a company manage all aspects of customer encounters, including marketing and advertising, sales, customer service after the sale, and programs to retain loyal customers.

cybermall A single Web site that offers many products and services at one Internet location.

cyberterrorist Someone who intimidates or coerces a government or organization to advance his political or social objectives by launching computer-based attacks against computers, networks, and the information stored on them.

data Raw facts, such as an employee number, number of hours worked in a week, inventory part numbers, or sales orders.

data administrator A nontechnical position responsible for defining and implementing consistent principles for a variety of data issues.

data analysis The manipulation of collected data so that the development team members who are participating in systems analysis can use the data.

data cleanup The process of looking for and fixing inconsistencies to ensure that data is accurate and complete.

data collection Capturing and gathering all data necessary to complete the processing of transactions.

data correction The process of reentering data that was not typed or scanned properly.

data definition language (DDL) A collection of instructions and commands used to define and describe data and relationships in a specific database.

data dictionary A detailed description of all the data used in the database.

data editing The process of checking data for validity and completeness.

data entry Converting human-readable data into a machine-readable form.

data input Transferring machine-readable data into the system.

data item The specific value of an attribute.

data manipulation The process of performing calculations and other data transformations related to business transactions.

data manipulation language (DML) The commands that are used to manipulate the data in a database.

data mart A subset of a data warehouse.

data mining An information-analysis tool that involves the automated discovery of patterns and relationships in a data warehouse.

data model A diagram of data entities and their relationships.

data preparation, or data conversion Ensuring all files and databases are ready to be used with new computer software and systems.

data storage The process of updating one or more databases with new transactions.

data store Representation of a storage location for data.

data warehouse A database that collects business information from many sources in the enterprise, covering all aspects of the company's processes, products, and customers.

database An organized collection of facts and information.

database administrator (DBA) A skilled IS professional who directs all activities related to an organization's database.

database approach to data management An approach whereby a pool of related data is shared by multiple application programs.

database management system (DBMS) A group of programs that manipulates the database and provides an interface between the database and the user of the database and other application programs.

data-flow diagram (DFD) A model of objects, associations, and activities that describes how data can flow between and around various objects.

data-flow line Arrows that show the direction of data element movement.

decentralized processing Processing alternative in which processing devices are placed at various remote locations.

decision room A room that supports decision making, with the decision makers in the same building, combining face-to-face verbal interaction with technology to make the meeting more effective and efficient.

decision structure A programming structure that allows the computer to branch, depending on certain conditions.

decision support system (DSS) An organized collection of people, procedures, software, databases, and devices used to support problem-specific decision making.

decision-making phase The first part of problem solving, including three stages: intelligence, design, and choice.

delphi approach A decision-making approach in which group decision makers are geographically dispersed; this approach encourages diversity among group members and fosters creativity and original thinking in decision making.

demand report A report developed to give certain information at someone's request.

design report The primary result of systems design, reflecting the decisions made and preparing the way for systems implementation.

design stage The second stage of decision making, in which alternative solutions to the problem are developed.

desktop computer A relatively small, inexpensive, single-user computer that is highly versatile.

deterrence controls Rules and procedures to prevent problems before they occur.

dialogue manager A user interface that allows decision makers to easily access and manipulate the DSS and to use common business terms and phrases.

digital audio player A device that can store, organize, and play digital music files.

digital camera An input device used with a PC to record and store images and video in digital form.

digital certificate An attachment to an e-mail message or data embedded in a

Web site that verifies the identity of a sender or Web site.

Digital Rights Management (DRM) Refers to the use of any of several technologies to enforce policies for controlling access to digital media such as movies, music, and software.

digital signal A signal that represents bits.

digital subscriber line (DSL) A telecommunications service that delivers high-speed Internet access to homes and small businesses over the existing phone lines of the local telephone network.

digital video disc (DVD) A storage medium used to store digital video or computer data.

direct access A retrieval method in which data can be retrieved without the need to read and discard other data.

direct access storage device (DASD) A device used for direct access of secondary storage data.

direct conversion (also called *plunge* or *direct cutover*) Stopping the old system and starting the new system on a given date.

direct observation Watching the existing system in action by one or more members of the analysis team.

disaster recovery The implementation of the disaster plan.

disaster recovery plan (DRP) A formal plan describing the actions that must be taken to restore computer operations and services in the event of a disaster.

disk mirroring A process of storing data that provides an exact copy that protects users fully in the event of data loss.

distance learning The use of telecommunications to extend the classroom.

distributed database A database in which the data can be spread across several smaller databases connected via telecommunications devices.

distributed processing Processing alternative in which computers are

placed at remote locations but are connected to each other via a network.

document production The process of generating output records and reports.

documentation The text that describes the program functions to help the user operate the computer system.

domain The allowable values for data attributes. Also the area of knowledge addressed by the expert system.

domain expert The individual or group who has the expertise or knowledge one is trying to capture in the expert system.

downsizing Reducing the number of employees to cut costs.

drill-down report A report providing increasingly detailed data about a situation.

dumpster diving Going through the trash cans of an organization to find secret or confidential information, including information needed to access an information system and/or its data.

dynamic Web pages Web pages containing variable information that are built to respond to a specific Web visitor's request.

e-commerce Any business transaction executed electronically between companies (business-to-business), companies and consumers (business-to-consumer), consumers and other consumers (consumer-to-consumer), business and the public sector, and consumers and the public sector.

economic feasibility The determination of whether the project makes financial sense and whether predicted benefits offset the cost and time needed to obtain them.

economic order quantity (EOQ) The quantity that should be reordered to minimize total inventory costs.

effectiveness A measure of the extent to which a system achieves its goals; it can be computed by dividing the goals actually achieved by the total of the stated goals.

efficiency A measure of what is produced divided by what is consumed.

eGovernment The use of information and communications technology to simplify the sharing of information, speed formerly paper-based processes, and improve the relationship between citizen and government.

electronic bill presentment A method of billing whereby a vendor posts an image of your statement on the Internet and alerts you by e-mail that your bill has arrived.

electronic business (e-business) Using information systems and the Internet to perform all business-related tasks and functions.

electronic cash An amount of money that is computerized, stored, and used as cash for e-commerce transactions.

electronic commerce Conducting business activities (e.g., distribution, buying, selling, marketing, and servicing of products or services) electronically over computer networks such as the Internet, extranets, and corporate networks.

electronic data interchange (EDI) An intercompany, application-to-application communication of data in a standard format, permitting the recipient to perform a standard business transaction, such as processing purchase orders.

electronic document distribution A process that enables the sending and receiving of documents in a digital form, without being printed (although printing is possible).

electronic exchange An electronic forum where manufacturers, suppliers, and competitors buy and sell goods, trade market information, and run back-office operations.

electronic funds transfer (EFT) A system of transferring money from one bank account directly to another without any paper money changing hands.

electronic retailing (e-tailing) The direct sale from business to consumer through electronic storefronts, typically designed around an electronic catalog and shopping cart model.

empowerment Giving employees and their managers more responsibility and authority to make decisions, take certain actions, and have more control over their jobs.

end-user systems development Any systems development project in which the primary effort is undertaken by a combination of business managers and users.

enterprise data modeling Data modeling done at the level of the entire enterprise.

enterprise resource planning (ERP) system A set of integrated programs capable of managing a company's vital business operations for an entire multisite, global organization.

enterprise sphere of influence The sphere of influence that serves the needs of the firm in its interaction with its environment.

enterprise system A system central to the organization that ensures information can be shared across all business functions and all levels of management to support the running and managing of a business.

entity A generalized class of people, places, or things for which data is collected, stored, and maintained.

entity symbol Representation of either a source or destination of a data element.

entity-relationship (ER) diagrams Data models that use basic graphical symbols to show the organization of and relationships between data.

ergonomics The science of designing machines, products, and systems to maximize the safety, comfort, and efficiency of the people who use them.

event-driven review A review triggered by a problem or opportunity such as an error, a corporate merger, or a new market for products.

exception report A report automatically produced when a situation is unusual or requires management action.

execution time (E-time) The time it takes to execute an instruction and store the results.

executive support system (ESS) Specialized DSS that includes all hardware, software, data, procedures, and people used to assist senior-level executives within the organization.

expert system Hardware and software that stores knowledge and makes suggestions and inferences, similar to a human expert.

explanation facility Component of an expert system that allows a user or decision maker to understand how the expert system arrived at certain conclusions or results.

Extensible Markup Language (XML) The markup language for Web documents containing structured information, including words, pictures, and other elements.

external auditing Auditing performed by an outside group.

extranet A network based on Web technologies that allows selected outsiders, such as business partners and customers, to access authorized resources of a company's intranet.

feasibility analysis Assessment of the technical, economic, legal, operational, and schedule feasibility of a project.

feedback Output that is used to make changes to input or processing activities.

field Typically a name, number, or combination of characters that describes an aspect of a business object or activity.

file A collection of related records.

File Transfer Protocol (FTP) A protocol that describes a file transfer process between a host and a remote computer and allows users to copy files from one computer to another.

final evaluation A detailed investigation of the proposals offered by the vendors remaining after the preliminary evaluation.

financial MIS An information system that provides financial information not only for executives but also for a broader set of people who need to make better decisions on a daily basis.

five-forces model A widely accepted model that identifies five key factors that can lead to attainment of competitive advantage, including (1) the rivalry among existing competitors, (2) the threat of new entrants, (3) the threat of substitute products and services, (4) the bargaining power of buyers, and (5) the bargaining power of suppliers.

flash memory A silicon computer chip that, unlike RAM, is nonvolatile and keeps its memory when the power is shut off.

flat organizational structure An organizational structure with a reduced number of management layers.

flexible manufacturing system (FMS) An approach that allows manufacturing facilities to rapidly and efficiently change from making one product to making another.

forecasting Predicting future events to avoid problems.

forward chaining The process of starting with the facts and working forward to the conclusions.

front-end processor A special-purpose computer that manages communications to and from a computer system serving hundreds or even thousands of users.

full-duplex channel A communications channel that permits data transmission in both directions at the same time, so a full-duplex channel is like two simplex channels.

game theory The use of information systems to develop competitive strategies for people, organizations, or even countries.

Gantt chart A graphical tool used for planning, monitoring, and coordinating projects.

gateway A telecommunications device that serves as an entrance to another network.

genetic algorithm An approach to solving large, complex problems in which a number of related operations or models change and evolve until the best one emerges.

geographic information system (GIS) A computer system capable of assembling, storing, manipulating, and displaying geographic information, that is, data identified according to its location.

gigahertz (GHz) Billions of cycles per second.

goal-seeking analysis The process of determining the problem data required for a given result.

graphical user interface (GUI) An interface that uses icons and menus displayed on screen to send commands to the computer system.

grid chart A table that shows relationships among the various aspects of a systems development effort.

grid computing The use of a collection of computers, often owned by multiple individuals or organizations, to work in a coordinated manner to solve a common problem.

group consensus Decision making by a group that is appointed and given the responsibility of making the final evaluation and selection.

group consensus approach A decision-making approach that forces members in the group to reach a unanimous decision.

group support system (GSS) Software application that consists of most elements in a DSS, plus software to provide effective support in group decision making; also called *group support system* or *computerized collaborative work system.*

hacker A person who enjoys computer technology and spends time learning and using computer systems.

half-duplex channel A communications channel that can transmit data in either direction, but not simultaneously.

handheld computer A single-user computer that provides ease of portability because of its small size.

hardware Computer equipment used to perform input, processing, storage, and output activities in an information system.

help facility A program that provides assistance when users want to know more about a program or feature or what type of response is expected.

heuristics Commonly accepted guidelines or procedures that usually find a good solution.

hierarchy of data Bits, characters, fields, records, files, and databases.

highly structured problems Problems that are straightforward and require known facts and relationships.

home page A cover page for a Web site that has graphics, titles, and text.

hot site A duplicate, operational hardware system or immediate access to one through a specialized vendor.

HTML tags Codes that let the Web browser know how to format text—as a heading, as a list, or as body text—and whether images, sound, and other elements should be inserted.

human resource MIS An information system that is concerned with activities related to employees and potential employees of an organization, also called a personnel MIS.

hypermedia Tools that connect the data on Web pages, allowing users to access topics in whatever order they want.

Hypertext Markup Language (HTML) The standard page description language for Web pages.

IF-THEN statements Rules that suggest certain conclusions.

image log A separate file that contains only changes to applications.

implementation stage A stage of problem solving in which a solution is put into effect.

incremental backup Making a backup copy of all files changed during the last few days or the last week.

inference engine Part of the expert system that seeks information and relationships from the knowledge base and provides answers, predictions, and suggestions the way a human expert would.

informatics A specialized system that combines traditional disciplines, such as science and medicine, with computer systems and technology.

information A collection of facts organized in such a way that they have additional value beyond the value of the facts themselves.

information center A support function that provides users with assistance, training, application development, documentation, equipment selection and setup, standards, technical assistance, and troubleshooting.

information service unit A miniature IS department.

information system (IS) A set of interrelated components that collect, manipulate, store, and disseminate data and information and provide a feedback mechanism to meet an objective.

information systems literacy Knowledge of how data and information are used by individuals, groups, and organizations.

information systems planning Translating strategic and organizational goals into systems development initiatives.

input The activity of gathering and capturing raw data.

insider An employee, disgruntled or otherwise, working solo or in concert with outsiders to compromise corporate systems.

installation The process of physically placing the computer equipment on the site and making it operational.

instant messaging A method that allows two or more people to communicate online using the Internet.

institutional DSS A DSS that handles situations or decisions that occur more than once, usually several times per year or more. An institutional DSS is used repeatedly and refined over the years.

instruction time (I-time) The time it takes to perform the fetch-instruction and decode-instruction steps of the instruction phase.

integrated development environments (IDEs) A development approach that combines the tools needed for programming with a programming language into one integrated package.

integration testing Testing all related systems together.

intellectual property Includes works of the mind such as books, films, music, processes, and software, which are distinct somehow and are owned and/ or created by a single entity.

intelligence stage The first stage of decision making, in which potential problems or opportunities are identified and defined.

intelligent agent Programs and a knowledge base used to perform a specific task for a person, a process, or another program; also called *intelligent robot* or *bot*.

intelligent behavior The ability to learn from experiences and apply knowledge acquired from experience, handle complex situations, solve problems when important information is missing, determine what is important, react quickly and correctly to a new situation, understand visual images, process and manipulate symbols, be creative and imaginative, and use heuristics.

internal auditing Auditing performed by individuals within the organization.

international network A network that links users and systems in more than one country.

Internet A collection of interconnected networks, all freely exchanging information.

Internet Protocol (IP) A communication standard that enables traffic to be routed from one network to another as needed.

Internet service provider (ISP) Any company that provides people or organizations with access to the Internet.

intranet An internal network based on Web technologies that allows people within an organization to exchange information and work on projects.

intrusion detection system (IDS) Software that monitors system and network resources and notifies network security personnel when it senses a possible intrusion.

Java An object-oriented programming language from Sun Microsystems based on C++ that allows small programs (applets) to be embedded within an HTML document.

joining Manipulating data to combine two or more tables.

joint application development (JAD) A process for data collection and requirements analysis in which users, stakeholders, and IS professionals work together to analyze existing systems, propose possible solutions, and define the requirements of a new or modified system.

just-in-time (JIT) inventory A philosophy of inventory management in which inventory and materials are delivered just before they are used in manufacturing a product.

kernel The heart of the operating system, which controls the most critical processes.

key A field or set of fields in a record that is used to identify the record.

key-indicator report A summary of the previous day's critical activities; typically available at the beginning of each workday.

knowledge The awareness and understanding of a set of information and ways that information can be made useful to support a specific task or reach a decision.

knowledge acquisition facility Part of the expert system that provides convenient and efficient means of capturing and storing all the components of the knowledge base.

knowledge base The collection of data, rules, procedures, and relationships that must be followed to achieve value or the proper outcome. Also a component of an expert system that stores all relevant information, data, rules, cases, and relationships used by the expert system.

knowledge engineer A person who has training or experience in the design, development, implementation, and maintenance of an expert system.

knowledge user The person or group who uses and benefits from the expert system.

LCD displays Flat displays that use liquid crystals—organic, oil-like material placed between two polarizers—to form characters and graphic images on a backlit screen.

learning systems A combination of software and hardware that allows the computer to change how it functions or reacts to situations based on feedback it receives.

legal feasibility The determination of whether laws or regulations may prevent or limit a systems development project.

linking Manipulating two or more tables that share at least one common data attribute to provide useful information and reports.

local area network (LAN) A network that connects computer systems and devices within a small area, such as an office, home, or several floors in a building.

logical design A description of the functional requirements of a system.

lookup tables Tables containing data that computer programs can develop and use to simplify and shorten data entry.

loop structure A programming structure that allows the program to cycle or loop a certain number of times.

machine cycle The instruction phase followed by the execution phase.

magnetic disk A common secondary storage medium, with bits represented by magnetized areas.

magnetic stripe card A type of card that stores limited amounts of data by modifying the magnetism of tiny iron-based particles contained in a band on the card.

magnetic tape A secondary storage medium; Mylar film coated with iron oxide with portions of the tape magnetized to represent bits.

mainframe computer A large, powerful computer often shared by hundreds of concurrent users connected to the machine via terminals.

maintenance team A special IS team responsible for modifying, fixing, and updating existing software.

make-or-buy decision The decision regarding whether to obtain the necessary software from internal or external sources.

management information system (MIS) An organized collection of people, procedures, software, databases, and devices that provides routine information to managers and decision makers.

market segmentation The identification of specific markets to target them with advertising messages.

marketing MIS An information system that supports managerial activities in product development, distribution, pricing decisions, and promotional effectiveness.

massively parallel processing systems A form of multiprocessing that speeds processing by linking hundreds or thousands of processors to operate at the same time, or in parallel, with each processor having its own bus, memory, disks, copy of the operating system, and applications.

material requirements planning (MRP) A set of inventory-control techniques that help coordinate thousands of inventory items when the demand of one item is dependent on the demand for another.

megahertz (MHz) Millions of cycles per second.

menu-driven system A system in which users simply pick what they want to do from a list of alternatives.

mesh networking A way to route communications between network nodes (computers or other device) by allowing for continuous connections and reconfiguration around blocked paths by "hopping" from node to node until a connection can be established.

meta tag A special HTML tag, not visible on the displayed Web page, that contains keywords representing your site's content, which search engines to use to build indexes pointing to your Web site.

meta-search engine A tool that submits keywords to several search engines and returns the results from all search engines queried.

metropolitan area network (MAN) A telecommunications network that connects users and their devices in a geographical area that spans a campus or city.

microcode Predefined, elementary circuits and logical operations that the processor performs when it executes an instruction.

middleware Software that allows different systems to communicate and exchange data.

MIPS Millions of instructions per second.

mission-critical systems Systems that play a pivotal role in an organization's continued operations and goal attainment.

mobile commerce (m-commerce) Transactions conducted anywhere, anytime using wireless communications.

model base Part of a DSS that provides decision makers access to a variety of models and assists them in decision making.

model management software Software that coordinates the use of models in a DSS.

modem A telecommunications hardware device that converts (modulates and demodulates) communications signals so they can be transmitted over the communication media.

monitoring stage The final stage of the problem-solving process, in which decision makers evaluate the implementation.

Moore's Law A hypothesis that states that transistor densities on a single chip double every 18 months.

MP3 A standard format for compressing a sound sequence into a small file.

multicore microprocessor A microprocessor that combines two or more independent processors into a single computer so they can share the workload and deliver a big boost in processing capacity.

multiple instruction/multiple data (MIMD) A form of parallel computing in which the processors all execute different instructions.

multiplexer A device that encodes data from two or more data sources onto a single communications channel, thus reducing the number of communications channels needed and therefore, lowering telecommunications costs.

multiprocessing The simultaneous execution of two or more instructions at the same time.

narrowband communications A telecommunications system that supports a much lower rate of data exchange than broadband.

natural language processing Processing that allows the computer to understand and react to statements and commands made in a "natural" language, such as English.

net present value The preferred approach for ranking competing projects and determining economic feasibility.

network operating system (NOS) Systems software that controls the computer systems and devices on a network and allows them to communicate with each other.

network-attached storage (NAS) Storage devices that attach to a network instead of to a single computer.

network-management software Software that enables a manager on a networked desktop to monitor the use of individual computers and shared hardware (such as printers), scan for viruses, and ensure compliance with software licenses.

networks Computers and equipment that are connected in a building, around the country, or around the world to enable electronic communications.

neural network A computer system that can simulate the functioning of a human brain.

newsgroups Online discussion groups that focus on specific topics.

nominal group technique A decision-making approach that encourages feedback from individual group members, and the final decision is made by voting, similar to the way public officials are elected.

nonoperational prototype A mock-up, or model, that includes output and input specifications and formats.

nonprogrammed decision A decision that deals with unusual or exceptional situations that can be difficult to quantify.

object-oriented database A database that stores both data and its processing instructions.

object-oriented database management system (OODBMS) A group of programs that manipulate an object-oriented database and provide a user interface and connections to other application programs.

object-oriented systems development (OOSD) An approach to systems development that combines the logic of the systems development life cycle with the power of object-oriented modeling and programming.

object-relational database management system (ORDBMS) A DBMS capable of manipulating audio, video, and graphical data.

on-demand computing Contracting for computer resources to rapidly respond to an organization's varying workflow. Also called on-demand business and utility computing.

online analytical processing (OLAP) Software that allows users to explore data from a number of perspectives.

online transaction processing (OLTP) A form of data processing where each transaction is processed immediately, without the delay of accumulating transactions into a batch.

open shops IS departments in which people, such as programmers and systems analysts, are allowed to run the computers, in addition to authorized operators.

open-source software Software that is freely available to anyone in a form that can be easily modified.

operating system (OS) A set of computer programs that controls the computer hardware and acts as an interface with application programs.

operational feasibility The measure of whether the project can be put into action or operation.

operational prototype A functioning prototype that accesses real data files, edits input data, makes necessary computations and comparisons, and produces real output.

optical disc A rigid disc of plastic onto which data is recorded by special lasers that physically burn pits in the disc.

optimization model A process to find the best solution, usually the one that will best help the organization meet its goals.

organization A formal collection of people and other resources established to accomplish a set of goals.

organizational change The responses that are necessary so that for-profit and nonprofit organizations can plan for, implement, and handle change.

organizational culture The major understandings and assumptions for a business, corporation, or other organization.

organizational learning The adaptations to new conditions or alterations of organizational practices over time.

organizational structure Organizational subunits and the way they relate to the overall organization.

output Production of useful information, usually in the form of documents and reports.

outsourcing Contracting with outside professional services to meet specific business needs.

parallel computing The simultaneous execution of the same task on multiple processors to obtain results faster.

parallel start-up Running both the old and new systems for a period of time and comparing the output of the new system closely with the output of the old system; any differences are reconciled. When users are comfortable that the new system is working correctly, the old system is eliminated.

password sniffer A small program hidden in a network or a computer system that records identification numbers and passwords.

patch A minor change to correct a problem or make a small enhancement. It is usually an addition to an existing program.

perceptive system A system that approximates the way a person sees, hears, and feels objects.

personal area network (PAN) A network that supports the interconnection of information technology within a range of 33 feet or so.

personal productivity software The software that enables users to improve their personal effectiveness, increasing the amount of work they can do and its quality.

personal sphere of influence The sphere of influence that serves the needs of an individual user.

personalization The process of tailoring Web pages to specifically target individual consumers.

phase-in approach Slowly replacing components of the old system with those of the new one. This process is repeated for each application until the new system is running every application and performing as expected; also called a *piecemeal approach*.

phishing A practice that entails sending bogus messages purportedly from a legitimate institution to pry personal information from customers by convincing them to go to a "spoofed" Web site.

physical design The specification of the characteristics of the system components necessary to put the logical design into action.

pilot start-up Running the new system for one group of users rather than all users.

pipelining A form of CPU operation in which multiple execution phases are performed in a single machine cycle.

pixel A dot of color on a photo image or a point of light on a display screen.

planned data redundancy A way of organizing data in which the logical database design is altered so that certain data entities are combined, summary totals are carried in the data records rather than calculated from elemental data, and some data attributes are repeated in more than one data entity to improve database performance.

Platform for Privacy Preferences (P3P) A screening technology that shields users from Web sites that don't provide the level of privacy protection they desire.

point evaluation system An evaluation process in which each evaluation factor is assigned a weight, in percentage points, based on importance. Then each proposed system is evaluated in terms of this factor and given a score ranging from 0 to 100. The scores are totaled, and the system with the greatest total score is selected.

point-of-sale (POS) device A terminal used in retail operations to enter sales information into the computer system.

Point-to-Point Protocol (PPP) A communications protocol that transmits packets over telephone lines.

policy-based storage management Automation of storage using previously defined policies.

portable computer A computer small enough to be carried easily.

predictive analysis A form of data mining that combines historical data with assumptions about future conditions to predict outcomes of events such as future product sales or the probability that a customer will default on a loan.

preliminary evaluation An initial assessment whose purpose is to dismiss the unwanted proposals; begins after all proposals have been submitted.

primary key A field or set of fields that uniquely identifies the record.

primary storage (main memory; memory) The part of the computer that holds program instructions and data.

problem solving A process that goes beyond decision making to include the implementation and monitoring stages.

procedures The strategies, policies, methods, and rules for using a CBIS.

process A set of logically related tasks performed to achieve a defined outcome.

process symbol Representation of a function that is performed.

processing Converting or transforming data into useful outputs.

productivity A measure of the output achieved divided by the input required.

profit center A department within an organization that focuses on generating profits.

Program Evaluation and Review Technique (PERT) A formalized approach for developing a project schedule.

programmed decision A decision made using a rule, procedure, or quantitative method.

programmer A specialist responsible for modifying or developing programs to satisfy user requirements.

programming languages Sets of keywords, symbols, and a system of rules for constructing statements by which humans can communicate instructions to be executed by a computer.

project deadline The date the entire project is to be completed and operational.

project milestone A critical date for the completion of a major part of the project.

project organizational structure A structure centered on major products or services.

project schedule A detailed description of what is to be done.

projecting Manipulating data to eliminate columns in a table.

public network services Systems that give personal computer users access to vast databases and other services, usually for an initial fee plus usage fees.

Public-key Infrastructure (PKI) A means to enable users of an unsecured public network such as the Internet to securely and privately exchange data through the use of a public and a private cryptographic key pair that is obtained and shared through a trusted authority.

push technology The automatic transmission of information over the Internet rather than making users search for it with their browsers.

quality The ability of a product (including services) to meet or exceed customer expectations.

quality control A process that ensures that the finished product meets the customers' needs.

questionnaires A method of gathering data when the data sources are spread over a wide geographic area.

Radio Frequency Identification (RFID) A technology that employs a microchip with an antenna that broadcasts its unique identifier and location to receivers.

random access memory (RAM) A form of memory in which instructions or data can be temporarily stored.

rapid application development (RAD) A systems development approach that employs tools, techniques, and methodologies designed to speed application development.

read-only memory (ROM) A nonvolatile form of memory.

record A collection of related data fields.

redundant array of independent/ inexpensive disks (RAID) A method of storing data that generates extra bits of data from existing data, allowing the system to create a "reconstruction map" so that if a hard drive fails, the system can rebuild lost data.

reengineering (process redesign) The radical redesign of business processes, organizational structures, information systems, and values of the organization to achieve a breakthrough in business results.

register A high-speed storage area in the CPU used to temporarily hold small units of program instructions and data immediately before, during, and after execution by the CPU.

relational model A database model that describes data in which all data elements are placed in two-dimensional tables, called *relations,* which are the logical equivalent of files.

release A significant program change that often requires changes in the documentation of the software.

reorder point (ROP) A critical inventory quantity level.

replicated database A database that holds a duplicate set of frequently used data.

report layout A technique that allows designers to diagram and format printed reports.

request for maintenance form A form authorizing modification of programs.

request for proposal (RFP) A document that specifies in detail required resources such as hardware and software.

requirements analysis The determination of user, stakeholder, and organizational needs.

restart procedures Simplified processes to access an application from where it stopped.

return on investment (ROI) One measure of IS value that investigates the additional profits or benefits that are generated as a percentage of the investment in IS technology.

revenue center A division within a company that generates sales or revenues.

robotics Mechanical or computer devices that perform tasks requiring a high degree of precision or that are tedious or hazardous for humans.

router A telecommunications device that forwards data packets across two or more distinct networks toward their destinations, through a process known as routing.

satisficing model A model that will find a good—but not necessarily the best— problem solution.

scalability The ability to increase the capability of a computer system to process more transactions in a given period by adding more, or more powerful, processors.

schedule feasibility The determination of whether the project can be completed in a reasonable amount of time.

scheduled report A report produced periodically, or on a schedule, such as daily, weekly, or monthly.

schema A description of the entire database.

screen layout A technique that allows a designer to quickly and efficiently design the features, layout, and format of a display screen.

script bunny A cracker with little technical savvy who downloads programs called scripts, which automate the job of breaking into computers.

search engine A Web search tool.

secondary storage (permanent storage) Devices that store larger amounts of data, instructions, and information more permanently than allowed with main memory.

Secure Sockets Layer (SSL) A communications protocol is used to secure sensitive data during e-commerce.

selecting Manipulating data to eliminate rows according to certain criteria.

semistructured or unstructured problems More complex problems in which the relationships among the pieces of data are not always clear, the data might be in a variety of formats, and the data is often difficult to manipulate or obtain.

sequence structure A programming structure in which, after starting the sequence, programming statements are executed one after another until all the statements in the sequence have been executed.

sequential access A retrieval method in which data must be accessed in the order in which it is stored.

sequential access storage device (SASD) A device used to sequentially access secondary storage data.

Serial Line Internet Protocol (SLIP) A communications protocol that transmits packets over telephone lines.

server A computer designed for a specific task, such as network or Internet applications.

shareware and freeware Software that is very inexpensive or free, but whose source code cannot be modified.

sign-on procedure Identification numbers, passwords, and other safeguards needed for someone to gain access to computer resources.

simplex channel A communications channel that can transmit data in only one direction.

simulation The ability of the DSS to duplicate the features of a real system.

single instruction/multiple data (SIMD) A form of parallel computing in which the processors all execute the same instruction on many data values simultaneously.

site preparation Preparation of the location of a new system.

slipstream upgrade A minor upgrade—typically a code adjustment or minor bug fix—not worth announcing. It usually requires recompiling all the code and, in so doing, it can create entirely new bugs.

smart card A credit card–sized device with an embedded microchip to provide electronic memory and processing capability.

smartphone A phone that combines the functionality of a mobile phone, personal digital assistant, camera, Web browser, e-mail tool, and other devices into a single handheld device.

social engineering Using one's social skills to get computer users to provide you with information to access an information system or its data.

software The computer programs that govern the operation of the computer.

software piracy The act of illegally duplicating software.

software suite A collection of single application programs packaged in a bundle.

source data automation Capturing and editing data where the data is initially created and in a form that can be directly input to a computer, thus ensuring accuracy and timeliness.

speech-recognition technology Input devices that recognize human speech.

split-case distribution A distribution system that requires cases of goods to be opened on the receiving dock and the individual items from the cases to be stored in the manufacturer's warehouse.

stakeholders People who, either themselves or through the organization they represent, ultimately benefit from the systems development project.

start-up The process of making the final tested information system fully operational.

static Web pages Web pages that always contain the same information.

statistical sampling Selecting a random sample of data and applying the characteristics of the sample to the whole group.

steering committee An advisory group consisting of senior management and users from the IS department and other functional areas.

storage area network (SAN) The technology that provides high-speed connections between data-storage devices and computers over a network.

storefront broker Companies that act as intermediaries between your Web site and online merchants who have the products and retail expertise.

strategic alliance (strategic partnership) An agreement between two or more companies that involves the joint production and distribution of goods and services.

strategic planning Determining long-term objectives by analyzing the strengths and weaknesses of the organization, predicting future trends, and projecting the development of new product lines.

structured interview An interview where the questions are written in advance.

structured walkthrough A planned and preannounced review of the progress of a program module.

supercomputers The most powerful computer systems with the fastest processing speeds.

switch A telecommunications device that uses the physical device address in each incoming message on the network to determine to which output port it should forward the message to reach another device on the same network.

synchronous communications A form of communications where the receiver gets the message instantaneously, when it is sent.

syntax A set of rules associated with a programming language.

system A set of elements or components that interact to accomplish goals.

system performance measurement Monitoring the system—the number of errors encountered, the amount of memory required, the amount of processing or CPU time needed, and other problems.

system performance products Software that measures all components of the computer-based information system, including hardware, software, database, telecommunications, and network systems.

system performance standard A specific objective of the system.

system testing Testing the entire system of programs.

systems analysis The systems development phase involving the study of existing systems and work processes to identify strengths, weaknesses, and opportunities for improvement.

systems analyst A professional who specializes in analyzing and designing business systems.

systems controls Rules and procedures to maintain data security.

systems design The systems development phase that defines how the information system will do what it must do to obtain the problem solution.

systems development The activity of creating or modifying existing business systems.

systems implementation A stage of systems development that includes hardware acquisition, software acquisition or development, user preparation, hiring and training of personnel, site and data preparation, installation, testing, start-up, and user acceptance.

systems investigation The systems development phase during which problems and opportunities are identified and considered in light of the goals of the business.

systems investigation report A summary of the results of the systems investigation and the process of feasibility analysis and recommendation of a course of action.

systems maintenance A stage of systems development that involves checking, changing, and enhancing the system to make it more useful in achieving user and organizational goals.

systems maintenance and review The systems development phase that ensures the system operates as intended and

modifies the system so that it continues to meet changing business needs.

systems operation Use of a new or modified system.

systems request form A document filled out by someone who wants the IS department to initiate systems investigation.

systems review The final step of systems development, involving the analysis of systems to make sure that they are operating as intended.

team organizational structure A structure centered on work teams or groups.

technical documentation Written details used by computer operators to execute the program and by analysts and programmers to solve problems or modify the program.

technical feasibility Assessment of whether the hardware, software, and other system components can be acquired or developed to solve the problem.

technology acceptance model (TAM) A model that describes the factors that lead to higher levels of acceptance and usage of technology.

technology diffusion A measure of how widely technology is spread throughout the organization.

technology infrastructure All the hardware, software, databases, telecommunications, people, and procedures that are configured to collect, manipulate, store, and process data into information.

technology infusion The extent to which technology is deeply integrated into an area or department.

technology-enabled relationship management Occurs when a firm obtains detailed information about a customer's behavior, preferences, needs, and buying patterns and uses that information to set prices, negotiate terms, tailor promotions, add product features, and otherwise customize its entire relationship with that customer.

telecommunications The electronic transmission of signals for

communications; enables organizations to carry out their processes and tasks through effective computer networks.

telecommunications medium Any material substance that carries an electronic signal and serves as an interface between a sending device and a receiving device.

telecommuting A work arrangement whereby employees work away from the office using personal computers and networks to communicate via e-mail with other workers and to pick up and deliver results.

Telnet A terminal emulation protocol that enables users to log on to other computers on the Internet to gain access to public files.

thin client A low-cost, centrally managed computer with essential but limited capabilities and no extra drives, such as a CD or DVD drive, or expansion slots.

time-driven review Review performed after a specified amount of time.

total cost of ownership (TCO) The measurement of the total cost of owning computer equipment, including desktop computers, networks, and large computers.

traditional approach to data management An approach whereby separate data files are created and stored for each application program.

traditional organizational structure An organizational structure similar to a managerial pyramid, where the hierarchy of decision making and authority flows from strategic management at the top down to operational management and nonmanagement employees. Also called a hierarchical structure.

transaction Any business-related exchange, such as payments to employees, sales to customers, and payments to suppliers.

transaction processing cycle The process of data collection, data editing, data correction, data manipulation, data storage, and document production.

transaction processing system (TPS) An organized collection of people, procedures, software, databases, and devices used to record completed business transactions.

transaction processing system audit A check of a firm's TPS systems to prevent accounting irregularities and/or loss of data privacy.

Transmission Control Protocol (TCP) The widely used Transport-layer protocol that most Internet applications use with IP.

Trojan horse A malicious program that disguises itself as a useful application and purposefully does something the user does not expect.

tunneling The process by which VPNs transfer information by encapsulating traffic in IP packets over the Internet.

Uniform Resource Locator (URL) An assigned address on the Internet for each computer.

unit testing Testing of individual programs.

unstructured interview An interview where the questions are not written in advance.

Usenet A system closely allied with the Internet that uses e-mail to provide a centralized news service; a protocol that describes how groups of messages can be stored on and sent between computers.

user acceptance document A formal agreement signed by the user that states that a phase of the installation or the complete system is approved.

user documentation Written descriptions developed for people who use a program, showing users, in easy-to-understand terms, how the program can and should be used.

user interface The element of the operating system that allows you to access and command the computer system.

user preparation The process of readying managers, decision makers, employees, other users, and stakeholders for new systems.

users People who will interact with the system regularly.

utility programs Programs that help to perform maintenance or correct problems with a computer system.

value chain A series (chain) of activities that includes inbound logistics, warehouse and storage, production, finished product storage, outbound logistics, marketing and sales, and customer service.

version A major program change, typically encompassing many new features.

videoconferencing A telecommunications system that combines video and phone call capabilities with data or document conferencing.

virtual organizational structure A structure that employs individuals, groups, or complete business units in geographically dispersed areas that can last for a few weeks or years, often requiring telecommunications or the Internet.

virtual private network (VPN) A secure connection between two points on the Internet.

virtual reality The simulation of a real or imagined environment that can be experienced visually in three dimensions.

virtual reality system A system that enables one or more users to move and react in a computer-simulated environment.

virtual tape A storage device that manages less frequently needed data so that it appears to be stored entirely on tape cartridges, although some parts of it might actually be located on faster hard disks.

virtual workgroups Teams of people located around the world working on common problems.

virus A computer program file capable of attaching to disks or other files and replicating itself repeatedly, typically without the user's knowledge or permission.

vision systems The hardware and software that permit computers to capture, store, and manipulate visual images.

voice mail Technology that enables users to send, receive, and store verbal messages for and from other people around the world.

volume testing Testing the application with a large amount of data.

Web auction An Internet site that matches buyers and sellers.

Web browser Software that creates a unique, hypermedia-based menu on a computer screen, providing a graphical interface to the Web.

Web log (blog) A Web site that people can create and use to write about their observations, experiences, and feelings on a wide range of topics.

Web page construction software Software that uses Web editors and extensions to produce both static and dynamic Web pages.

Web services Standards and tools that streamline and simplify communication among Web sites for business and personal purposes. Also software modules supporting specific business processes that users can interact with over a network (such as the Internet) on an as-needed basis.

Web site development tools Tools used to develop a Web site, including HTML or visual Web page editor, software development kits, and Web page upload support.

what-if analysis The process of making hypothetical changes to problem data and observing the impact on the results.

wide area network (WAN) A telecommunications network that ties together large geographic regions.

wire transfer An extremely fast, reliable means to move funds from one account to another.

workgroup Two or more people who work together to achieve a common goal.

workgroup application software
Software that supports teamwork,
whether in one location or around the
world.

workgroup sphere of influence The
sphere of influence that serves the needs
of a workgroup.

workstation A more powerful personal
computer that is used for technical
computing, such as engineering, but
still fits on a desktop.

World Wide Web (WWW or W3) A
collection of tens of thousands of
independently owned computers that
work together as one in an Internet
service.

worm A parasitic computer
program that can create copies of
itself on the infected computer or send
copies to other computers via a
network.

INDEX

Subject

A boldface page number indicates a key term and the location of its definition in the text.

A

acceptance testing, **548–549**

access methods, **93–97**

Access (Microsoft), 153, 182, 184, 187, 194, 436–437

accounting systems, 30, 48, **359–361**, 370–371, 405–406

accuracy, of information, 7, 321, 348

ACD (automatic call distributor), 242–243

ACE (Automated Commercial Environment), 245, 455

ACH (automated clearing house), **247**

ActiveX (Microsoft), 271

activity modeling, **509**

ADSL (asymmetric DSL), 227, 239

advertising, 263, 316–317, 402

after-sales service, 310

agile development, 490

AI (artificial intelligence), 23–25, 431–432, 436–446, 453–355

airline industry, 30, 127–128, 218–219, 400, 403, 472–474, 478–479, 545

algorithms, 445

alpha testing, **548–549**

alphanumeric data, **5**

alt text, 269

ALUs (arithmetic logic units), **83**

analog signals, **225**

anomalies, 186

anonymous surfing, 288

antennae, 239

Apache Web server, 160, 162, 328

API (application program interface), **134**, 170

applets, **268–269**, 271. *See also* Java

application software. *See also* software
    described, **129**, 130, 146–156
    personal, 147–148

ARPANET, 259–260

ASCII (American Standard Code for Information Interchange), 266

asking directly, use of the term, **511**

ASPs (application service providers), **147**, 543

assembly language, 157

asynchronous communications, **219**

ATM (Asynchronous Transfer Mode), **237**

ATMs (automated teller machines), **103**, 108, 333, 338–339

attacks, cost of, 28

attributes, 178–181

auctions, 281–282, 307, **320**, 323, 332

audio. *See also* music
    data, **5**
    database systems, 206–207
    players, 69, **108–109**

auditing, 355, 397

automobile industry, 3–4, 13, 24, 41–43, 58, 82–83, 248, 468, 493

autonomic computing, 64, 409, 551

B

B2B (business-to-business) e-commerce, 305, 324, 329–331, 474
    described, **17–18**, 306–307
    multistage model for, 308–310

B2C (business-to-consumer) e-commerce, 305, 324, 472–473
    described, **17–18**, 307
    multistage model for, 308–310
    software, 329–331

backbones, **261**

back-end applications, 196

backups, 92–99, **532**, 585. *See also* storage

backward chaining, **450**

bandwidth, **220**

banking industry, 122–123, 177, 234, 247, 287, 313, 317–323, 348, 453, 490, 541–542. *See also* financial services

bar-code scanners, **105**

batch processing systems, **346–347**

benchmark tests, **538–539**

best practices, **364**

beta testing, **548–549**

BI (business intelligence), **202**, 435

bioinformatics, 462

biometric devices, 212–213, 322, **592–593**

bits, **177**

blade servers, **113**

blocks, **97**

bloggers, 278

blogosphere, 278

blogs (Web logs), 15, **278–279**

Bluetooth, 224, 238

BOM (bill of materials), 367

bots, **281**

BPR (business process reengineering), **52**. *See also* reengineering

brainstorming, **414**

branding, 202

bridges, **239–240**

broadband communications, 238, 242, 263
    described, **220**
    over power lines (BPL), **222**

brokerages, 443–444

browsers. *See also* Web pages; Web sites
    ActiveX and, 271
    anonymity and, 288
    content development and, 272–273
    described, 268–269
    e-commerce and, 314–315
    open-source software and, 162
    privacy issues and, 599

buddy lists, 276

bugs, 160. *See also* errors

business information systems, **17–25**

bytes, **87**, 88

C

C++ (high-level language), 158–159, 191, 192, 271, 500, 501, 553
C2C (consumer-to-consumer) e-commerce, 17–18, **307**
cable modems, 226, 227, 242
cache memory, **89**, 116
CAD (computer-assisted design), 30, 113, 153, 398
call
    centers, **242–243**, 403, 456
    management systems, 241
CAM (computer-assisted manufacturing), 30, **399–400**
career opportunities, 63–69, 258, 277
CAs (certificate authorities), **331**
CASE (computer-aided software engineering)
    described, **500–501**
    repositories, **511**
case management systems, 495
catalog management, 330
CAVE system, 458
CBIS (Computer Based Information System), 12, 16, 577
CDs (compact discs), 92, 93, **96**, 116
cell phones, 141–142, 224–225, 276–277, 313. *See also* wireless networks
centralized processing, **235**
CEOs (chief executive officers), 23, 29, 53
certification, **68**
CFOs (chief financial officers), 66
CGI (computer-generated image) technology, 460–461
change, organizational
    described, **49–52**, 435
    degrees of, 494–495
    implementing, 364
    managing, 495
    models, **50**
    sustaining, 49
channel(s)
    bandwidth, **220**
    described, **220**
    full-duplex, **220**
    half-duplex, **220**
    simplex, **220**
characters, 177
chat rooms, **279–280**
chief programmer teams, **545**
choice stage, **385**
CIM (computer-integrated manufacturing), 30, **399**
CIOs (chief information officers), **66–67**, 75, 139, 482
CIPA (Children's Internet Protection Act), 596
CKDs (chief knowledge officers), **434**
cladding, 222
click fraud, **323**
clickstream data, **324**
client/server systems, **235–236**, 266. *See also* servers
Clinger-Cohen Act, 66
clock speed, **85–86**
closed shops, **533**
CMDA (Code-Division Multiple Access), 239
CMM (Capability Maturity Model), 497
CMS (content management systems), 273
coaxial cable, **222**
COBOL, 158, 192, 453, 551, 553, 554

cold sites, **532**
collaborative
    software, 154
    work, 48
command-based user interface, **133**
competitive advantage, **56–59**, 348, 482–483
compilers, **159**
computer(s)
    downsizing, 114
    general-purpose, 110
    literacy, **28**
    mistakes, 576–580
    recycling, 115
    selecting, 114–116
    types of, 110–116
    upgrading, 114–116
    waste, 576–580
Computer Fraud and Abuse Act, 592
concurrency control, **190**
containers, 187
content streaming, **281**
continuous improvement, 50–51, **52**, 495
contracts, 540
control(s)
    described, **532–534**
    design of, 531–535
    units, **83**
COP (communities of practice), **433–434**
COPPA (Children's Online Privacy Protection Act), 599
coprocessors, **90**
copyright laws, 139, 160, 163, 170, 283, 589–590. *See also* intellectual property cost(s)
    benefit analysis, **538**
    centers, **397**
    ERP and, 364, 365
    leadership strategy, **58**
    MISs and, 397
    objectives, 486
    of software, 544–545
    systems development and, 486, 487, 492–494, 544–545
COTS (commercial off-the-shelf) software, 544
counterintelligence, 202
covert intelligence gathering, 288
CPUs (central processing units). *See also* processors
    clock speed and, 85–86
    described, 83, 85–91
    memory and, 89, 134
    multiprocessing and, 90
    physical characteristics of, 86–87
    pipelining and, 84
    programming languages and, 159
    requirements, 115
    telecommunications and, 240
    workstations and, 113
crackers, **593**
creative analysis, **482–483**
credit card transaction(s), 322, 333. *See also* e-commerce; transactions
    cell phones and, 313
    expert systems and, 447
    fraud, 323, 581–582

information, protecting, 604
scams and, 590–591
verification, 322
credit granting, 453
credit reports, 58, 604
crime. *See also* security
the computer as an object of, 583–591
disgruntled employees and, 613–614
international, 591
overview, 580–591
preventing, 591–597
crisis management, 419–420
critical
analysis, 482–483
path, **497**
CRM (customer relationship management), 49, 81–82, 143, 294, 400
described, **45**, 367–370
DSSs and, 403
enterprise systems and, 343, 367–370
Salesforce.com and, 161
Select Comfort and, 170–171
cross-platform development, **545**
CRTs (cathode ray tubes), 106, 107, 115
CSFs (critical-success factors), **485**, 511
CSMA/CD (Carrier Sense Multiple Access with Collision Detection), 237
CSS (Cascading Style Sheets), 267–268
CTOs (chief technology officers), 67
CTS (carpal tunnel syndrome), 605–606
culture, organizational, 31, 49–52, 372
currencies, 373
customer(s). *See also* CRM (customer relationship management)
awareness, 61
bargaining power of, 58
churn, 202
loyalty, 348, 349, 401
satisfaction, 61
service, 322, 348
customs rules, 372–373
cybermall, **314–315**
cyberterrorism, **582**. *See also* terrorism
cycle time, 85

D

DASD (direct access storage devices), **93**
dashboards, 390–393, 436–437
data. *See also* files
administrator, **194**
automation, **100**
cleanup, **185–186**
clickstream, **324**
collection, 350–351, 353, 506–507
described, **5–6**
dictionaries, **189–190**
editing, **351**
entry, **100**
flow, transborder, 232
flow-/flow lines, **509**

havens, 232
hierarchies, **177–178**
identifying sources of, 506
input, **100**
items, **178–181**
management, **174–215**
manipulating, 183–185, 191–193, **353**
marts, **198–199**
mining, **199–202**, 204, 435
modeling, **181–196**, 508–509
nature of, 100
organizing, 174–215
preparation/conversion, **547**
redundancy, **181**, 189
reliability, 189
retrieving, 190
sorting, 190
store, **509**
synchronization, 203–204
transforming, into information, 6
types of, 5–6
warehouses, **197–199**
workers, **433–434**
database(s). *See also* DBMSs (database management systems); relational databases; tables
acquiring, 545–547
applications, 157, 196–207
approach, **179**
cost of, 195
creating, 188–190
described, 14, 176, 177
DSSs and, 410–412
linking, to the Internet, 196–197
modifying, 188–190
normalization, 186
output, 193
overview, 174–215
privacy issues and, 604
size of, 195
software, **148**, 149–150
special-purpose, 195
systems development and, 534, 545–547, 557
types of, 187–188
unauthorized use of, 14
used with other software, 196
DBAs (database administrators), 177, 193–194
DBMSs (database management systems), 23, **179–181**, **186–196**. *See also* databases
DDL (data definition language), **188–190**
decency, protection of, 595–597
decentralized processing, **235**
decision(s)
nonprogrammed, **386–387**
programmed, **386–387**
room, **417**
structure, **545**
decision making
application software and, 155
EDSSs and, 418
ERP and, 363–364
levels, support for, 410

overview, 384–286
phase, **385**
decode instructions, 84
delphi approach, **414**
demand management, 367
design stage, **385**
desktop
computers, **111–112**
search tools, **145**
detailed scheduling, 367
deterrence controls, **533**
DFDs (data-flow diagrams), **509**
DHTML (Dynamic HTML), 267, 268
dialogue manager, **410**, 412–413
differentiation, **58**
digital
cameras, **101–102**
certificates, **331**
signals, **225**
Digital Dashboard, 436–437
direct
access, **93**
conversion (cutover), **549–550**
observation, **507**
disaster recovery, 176, 235, 243, **354–356**, 412, 531–532,
550, 556, 590–591
disintermediation, 307
disk mirroring, **94**
display monitors, **106**
distance learning, **247**
distributed
databases, **202–204**
processing, **228–241**
DML (data manipulation language), **191**
document production, **353**
documentation, **128**, 545, 557
domain(s)
described, **183**, 451
experts, **451–452**
names, top-level, 261, 312
dot pitch, 106
downsizing, **54**, 114
downstream management, 44
dpi (dots per inch), 108, 116
drill-down analysis, **197**, 407, 418
DRM (Digital Rights Management), 323
DRP (disaster recovery plan), **354–355**. *See also* disaster
recovery
DSL (Digital Subscriber Line), 226, **227–228**, 237, 242,
263, 264
DSSs (decision support systems), **20–23**, 382–429, 403,
406–413, 451
dumpster diving, **582**
DVDs (digital video discs), 14, 92–93, 96–97, 116

E

e-business, **18–19**
e-commerce. *See also* payment systems; transactions
advantages of, 321–322
applications, 314–322

challenges, 310–311
data mining and, 201
described, **17–18**, 306–311
multistage model for, 308–310
overview, 304–341
software, 329–331
strategies for successful, 325–327
technology infrastructure, 327–331
threats to, 322–325
economic feasibility, **503**
EDI (electronic data interchange), 74, 245–246, 348
EDRs (event data recorders), 13
EDS (Electronic Data Systems), 69, 493
EDSSs (executive support systems), **417–420**
EFF (Electronic Frontier Foundation), 200
effectiveness, **9**
efficiency, **8**, 348, 557
EFT (electronic funds transfer), **246–247**
eGovernment, **307–308**
electronic
bill presentment, **318**
cash, **331–332**
document distribution, **242**
exchange, **315–316**
wallets, 333
e-mail
abbreviations, 275
junk (spam), 29, 144
open-source software and, 160
overview, 275–276
privacy, 598
e-marketplaces, 4
embedded computer(s)
information systems, **5**
system software and, 141–142
empowerment, **47**
EMS (environmental management systems), 74–75
encryption, 95, 333, 339
end-user systems development, **491–492**
enterprise(s). *See also* enterprise systems
application software, **154–156**
data modeling, **181–182**
-rights management procedures, 531
sphere of influence, **130**
storage, 98–99
system software and, 136, 140
utilities, 144
enterprise systems. *See also* enterprises
control issues, 354–356
described, **344**
hosted software model for, 371–372
international issues associated with, 372–374
management issues for, 354–356
overview, 342–381
TPS and, 19–20
entities, **178–181**
entity symbols, **509**
EOQ (economic order quantity), **399**
e-procurement, **18**
EPROM (erasable programmable read-only memory),
88, 105

equipment theft, 589
ER (entity-relationship) diagrams, **182**, 184, 508
ergonomics, 28, 100, **606**
ERP (enterprise resource planning), 170–171, 257, 343–349, 373–374, 400, 416
    described, **20**, 154–155, 362–366
    overview, 345–349
    systems development and, 478, 521, 570–571
errors, 531, 576–580. *See also* bugs
e-tailing, **314–315**
Ethernet, 221, 237, 242
ethics, 27–29, 200, 356, 576, 606–607
e-time (execution time), **84**
EULAs (End-User License Agreements), **128**, 131
evaluations, **537–539**
event-driven reviews, **556**
exabyte, **88**
Excel (Microsoft), 153, 383, 387, 412, 503
Exchange (Microsoft), 416
exchange rates, 373
Exchange Server (Microsoft), 154
execute instructions, 84
executive
    dashboards, 390–393
    information system, 23
expert systems, **23–25**, 440, 446–456
explanatory facility, **450**
extranets, **15**, 30–31, 256–302, 285–287
eyebud screens, 109–110

F

fair use, issue of, 599–602
fault detection, 240
feasibility analysis, **503–504**
Federal Information Security Management Act, 527
feedback mechanisms, **8**, 11–12
fetch instructions, 84
fiber-optic cable, **222**
fields, **177**
file(s). *See also* data
    -compression utilities, 144
    described, 177
    management, 135
    remote access to, 284
    servers, **235–236**
filtering software, 595–596
final evaluations, **537–538**
financial MISs, **396–397**
financial services, 47, 61, 86, 113, 122–123. *See also* banking industry; investment services
fingerprint readers, 592–593
firewalls, 28, 242
five-forces model, **57**
flash memory, 92, **97**
flat
    file databases, **187**
    organizational structures, **47**
FLEXPERT, 159, 454
floppy disks, 92
flowcharts, **509–510**

FMS (flexible manufacturing system), **399**
Food Allergen and Consumer Protection Act, 191
forecasting, **12**
forward chaining, **450**
frame relay, **237**
fraud, 202, 287, 323–324, 348, 581–582
Freedom of Information Act, 604
freeware, **162–163**
front-end
    applications, 196
    processors, **226**
FTP (File Transfer Protocol), 272, **278**
full mesh topology, 233
functional requirements, 514
fuzzy logic, 449

G

game theory, **462**
games, 26–27, 68–69, 88, 454, 462
Gantt charts, **499**
gateways, **239–240**
genetic algorithms, **445**
gigabyte, **88**
gigahertz, **86**
GIS (geographic information system), **406**
goal(s)
    aligning, 481–483
    -seeking analysis, **408**
    supporting, 129–130
    systems reviews and, 557
goods, flow of, speed of, 321
GPS (Global Positioning Systems), 41, 109–110, 147, 228, 248, 313, 441–442
Gramm-Leach-Bliley Act, 355, 600, 601, 602
graphics software, **148**, 150–151
grid
    charts, **509–510**
    computing, **90**, 142, 212
group(s)
    consensus approach, **414**, 538
    goals, supporting, 129–130
    support systems, 23
groupware, 23, 154, 415–416, 435
GSM (Global System for Mobile) communications, 239, 312
GSS (group support system), **413–417**, 491
GUI (graphical user interface), **133**
guided transmission media, **220–221**

H

hackers, 29, 242, 575, **593**, 613
handheld computers, 10, **110–114**, 121, 276–277
hard copy, **108**
hard disks, **94–95**, 115. *See also* storage
hardware. *See also specific devices*
    acquiring, 542–543
    components, 83–85
    costs, 486
    described, **13–14**, 82
    e-commerce and, 328

functions, common, 132–133
  independence, **134**
  system software and, 132–133
  telecommunications, 225–226
  understanding, importance of, 82
  utilities, 142
health care, 20, 30, 98, 112, 146, 196, 231, 380, 454–455, 462, 551
  DSSs and, 384–386
  issues related to computers, 605–606
  virtual reality and, 459–460
help desks, 455, **550**
heuristic approaches, **387–388**, 408, 438, 439
hierarchical structure, 46–47
high-level application software, 157, 159
HIPAA (Health Insurance Portability and Accountability Act), 231, 355, 579–580
hiring, 59, 547
HMDs (head-mounted displays), 25, 457–458
home
  networks, 242
  pages, **266**
hosted software model, 371–372
hosts, **261**
hot sites, **532**
HTML (HyperText Markup Language), 271–273, 578–579
  alt text and, 269
  described, **266–269**, 326
  e-commerce and, 314, 326, 329
human resource(s). *See also* personnel
  EDSSs and, 419
  MISs, **404–405**
  planning, 305
Hurricane Katrina, 176, 235, 243, 356, 412, 531–532, 550, 556, 590–591
HVAC equipment, 113
HVD (holographic versatile disc), **97**
hypermedia, **266**

I

identification services, 329
identity theft, 95, **582–583**
Identity Theft and Assumption Deterrence Act, 583
IDEs (integrated development environments), **545**
IDS (intrusion detection systems), **594–595**
iDTV (interactive digital television), 326–327
IF-THEN statements, **449**
image
  data, 5
  logs, 532
implementation stage, **386**
inference engine, **450**
informatics, **462**
information. *See also* IS (information systems)
  center, **66**
  flow of, speed of, 321
  organizing, 174–215
  service unit, **66**
  turning data into, 6

  use, fairness in, 599–602
  use of the term, 5–6
  valuable, characteristics of, 6–7
infrared transmission, 225
input
  described, **8**, 11
  devices, **99–110**
  controls, 534
  /output management, **133**
insiders, 593
installation, use of the term, 547–548
instant messaging, **275–276**
integrated
  application packages, 153
  information systems, 346–347
integration testing, **548–549**
intellectual property, 322–323. *See also* copyrights
intelligence
  nature of, 437–440
  stage, **385**
intelligent
  agents, **445**
  application software, 157
  behavior, **437**
interactive
  design, 530–531
  processing, 530
Internet. *See also* networks; World Wide Web
  accessing, 262
  applications, 274–285
  careers, 67–68
  described, **14–16**, 258
  general security issues related to, 27–29
  growth of, 259
  how it works, 261–263
  lack of access to, 324
  linking databases to, 196–197
  overview, 256–302
  phone service, 280–281
  privacy issues and, 598–599
  use and functioning of, 259–274
  utilities, 144
Internet Explorer browser (Microsoft), 135, 269. *See also* browsers
Internet2 (I2), 260–261
interpreters, **159**
intranets, **15**, 30–31, 285–287
investment services, 15, 27, 30, 317–319. *See also* financial services
IP (Internet Protocol), **260–261**, 265, 283, 288, 596
iPod, 108–109, 278, 323. *See also* MP3 players
IPTV (Internet Protocol Television) standard, 135–136, 283
IRC (Internet Relay Chat), 280, 584
IRR (internal rate of return), 503, 532
IS (information systems). *See also* information
  components of, 11
  concepts, 5–9
  described, **3**, 11–16
  global challenges facing, 31–32
  literacy, **28**, 30
  overview, 382–429

specialized, 430–474
value of, 60–61
ISPs (Internet Service Providers), 230, 239–240, 263–264, 287, 288
Israeli Holocaust Database, 195
i-time (instruction time), **84**
iTunes, 108, 109, 144, 323, 324

J

JAD (joint application development), **491**
Java, 271, 294, 326, 329
described, **157–158**
systems development and, 500, 501, 553
JIT (just-in-time inventory), 43, **399**
jobs. *See* career opportunities
joining tables, **184**

K

kernel, **132**
keyboards, 100–101
keys, **178–181**
keywords, 326
kilobyte, **88**
kiosks, 103, 217–218
knowledge
acquisition facility, **450–451**
base, **25**, 448–450
described, 6, 432
engineers, **452**
explicit/tacit, **433**
management systems (KMSs), **23–25**, 432–437
maps, 435
obtaining, 434–435
repository, 434
sharing, 434–435
storing, 434–435
system concepts and, 8
users, 452
workers (KWs), **6**, 63, 433–434

L

language differences, 31, 372
LANs (local area networks), 238–239, 241, 258, 312–314. *See also* networks
administrators, **67**
described, 229–230
servers and, 262–263
sharing workstations and, 231
wireless, 238
laptop computers, **111–112**, 581
LCDs (liquid crystal displays), 106, **107**
learning
organizational, 50, 435
systems, 24–25, 444
legacy systems, 551, 364. *See also* mainframe computers
legal
feasibility, **503**

information systems, 30
jurisdictions, 324–325
liability issues, 595–597
license agreements, 160, 161
line positions, **46**
linking tables, **184**, 185
links, sponsored, 269. *See also* URLs (Uniform Resource Locators)
Linux, 194, 257, 269, 298–299
described, **139**, 140
software and, 133, 136, 139, 140, 152, 160, 162
logic bombs, 584
logical
access path (LAP), 190
design, 181, **526–527**
lookup tables, **530**
loop structure, **545**

M

machine
cycles, **84–85**
language, 157
Macintosh, 136, 144–145, 269, 272
GUI, 133
operating system, **137–139**
programming languages and, 159, 271
magnetic
disks, 93–94
stripe card, **103**
tape, **93**
main memory, **83**. *See also* memory
mainframe computers, **111**, 113–114, 122, 144–145, 241. *See also* legacy systems
maintenance
systems, **550–555**
teams, **552**
make-or-buy decision, **543–545**
management consulting firms, 30–31
MANs (metropolitan area networks), **230–231**
manufacturing, 30, 314–316, 397–400, 454
market. *See also* marketing
segmentation, 202, **317**
share, 61
marketing, 30, 202, 316–317, 455, 460. *See also* market
MISs, **400–403**
research, 400–401
mashing, 15
mash-up, 272–273
massively parallel processing systems, **90**
m-commerce, **17–18**, 304–341
megabyte, **88**
megahertz, **86**
memory, 134, 139, 181
devices, **85–91**
requirements, 115–116
secondary storage, **92–99**
MEMS (micro-electromechanical systems), 442
menu-driven systems, **530**
mergers and acquisitions, 480

mesh networking, **233–234**
meta tags, 269, **326**
meta-search engines, **270**
methods, **158–159**, 206
metropolitan services, 248
MICR (magnetic ink character recognition), **103**
microcode, **85**
microwave transmission, **222–224**
middleware, **145**, 170
military systems, 15, 82–83, 99, 365, 408, 415, 582
MIMD (multiple instruction/multiple data), **90**
MIS (management information system), 345, 383,
    410–411, 491
        described, **21–22**, 388–394
        functional aspects of, 395–406
        inputs/outputs, 390–393
mission-critical systems, **485**
mistakes, computer-related, 576–580. *See also* errors
MMS (model management system), 23, **412**
model base, **412–413**
modeling data, **508–509**
modems, **225–227**, 242
monitoring, **386**, 598, 579
monitors, **106**, 115
Moore's Law, **86**
mouse, **100**
moving, use of the term, **50**
MP3 format, **108**, 144
MP3 players, 69, **108–109**. *See also* iPod
MPE/iX, **140**
MPIS (millions of instructions per second), **85**
MRO (manufacturing, repair, and operations), 314–316
MRP (material requirements planning), 367, **399**
MSN (Microsoft Network), 145, 263, 462
MSSPs (managed security service providers), 595
MTSO (mobile telephone subscriber office), 225
multicore microprocessors, **90**
multiplexers, **226–227**, 228
multiprocessing, **90**
multitasking, **134–135**
music, 272, 282–284, 322–324. *See also* audio; MP3 format;
    MP3 players
        e-commerce and, 339–340
        files, compression of, 144

N

nanoseconds, 85
narrowband communications, **220**
NAS (network-attached storage), **98**
natural
        application software, 157, 159
        language processing, 24–25, **443–444**
NDIS (Network Driver Interface Specification), 262
net present value, 503
.NET platform (Microsoft), 273, 545
NetMeeting (Microsoft), 281, 416
Netscape Navigator browser, 269. *See also* browsers
network(s). *See also specific types*
        applications, 241–248
        described, **14–16**, 228

distributed processing and, 228–241
        home, 242
        international, **232–233**
        linking PCs to, 241
        -management software, 240–241
        mesh, **233–234**
        overview, 216–255
        small business, 242
        system software and, 135
        types, **229**
        utilities, 144
neural networks, 24–25, **444–445**
newsgroups, 279–280
NGI (Next Generation Internet), 260–261
niche strategies, **58**
nodes, **228**
nominal group technique, **414**
nonoperational prototypes, **490**
nonprofit organizations, **42**, 49
NOS (network operating system), **240–241**
notebook computers, 10, 107, **111–112**
Novell NetWare, **139**, 164, 240

O

object(s). *See also* object-oriented systems
        described, 157
        query language (OQL), 206
object-oriented systems. *See also* objects
        analysis, 513
        design, **527–529**
        development, 501
        investigation, 504–505
        programming languages, 156–157
objectives, establishing, 485–486
OCR (optical character recognition), 102
off-the-shelf software, **146–147**
Office (Microsoft), 14, 145, 152, 164, 247–248, 415–416,
    436
        described, **153**
        enterprise software and, 155
offices, on the Web, 284
offshoring, 81, 243
OLAP (online application processing), **204**, 213
OLED (organic light-emitting diode), 107–108
OLTP (online transaction processing), 197, 198, 346–347
on-demand computing, 54, **492–494**
OODBMSs (object-oriented DBMSs), **206**
open shops, **533**
open-source software, **160–162**, 169–170
operating systems, 132–135. *See also specific operating systems*
operational
        feasibility, **503**
        prototypes, **490**
operations component, 64–65
optical data readers, **102**
optical discs, 96. *See also* CDs (compact discs); DVDs (digital
    video discs)
optimization, 149, 387–388
ORDBMSs (object-relational DBMSs), 206
ordering systems, 257–258, 357–358, 370, 386

organizations
  competitive advantage and, 56–59
  described, **42–43**
  information systems in, overview of, 40–77
  structure of, **46–49**
Outlook (Microsoft), 151, 374
outplacement, 405
output
  controls, 534
  described, **8**, 11
  devices, **99–110**
outsourcing, 26, 54, 492–494

P

P3P (Platform for Privacy Preferences), 599
pairing sequences, 127
PANs (personal area networks), **229**
parallel
  computing, **90–91**
  start-up, 550
partial mesh topology, 233
password(s)
  e-commerce and, 322–323, 332
  protecting, 604
  sniffers, **589**
patches, 552, 575
patents, 162
Patriot Act, 200
payment systems, 331–333. *See also* transactions
PBX networks, 224, 265
PDAs (personal digital assistants), **110–111**, 141–143, 367, 581
pen input devices, **105**
performance
  -based information systems, **59–63**
  management, 240
  objectives, 485–486
permanent storage. *See* secondary storage
personal
  productivity software, **130**
  sphere of influence, **130**
personalization, **327**
personnel. *See also* human resources
  privacy issues and, 598
  staff positions, 46–47
  systems development and, 534, 547
  titles, 66–67
perspective system, **439**
PERT (Program Evaluation and Review Technique) charts, **497–498**
petabytes, **88**
phase-in approach, **550**
phishing, 323, 590–591
PHP (hypertext preprocessor), 271
physical
  access path (PAP), 190
  design, 181, **526–527**
picoseconds, 85
pilot start-up, 550
PIMs (personal information managers), **151**

pipelining, **84**
piracy, **589–590**
pixels, **106**
PKI (public-key infrastructure), **592**
place shifting, 15
planning. *See also* ERP (enterprise resource planning)
  described, **481–483**, 511–512
  steps, 483
  strategic, 58–59, **419**
plotters, **108**
plug-ins, 266, 320
Pocket PCs, 141–142, 358
podcasting, 15, **278–279**
point evaluation system, **539–540**
point-of-sale transactions, 350–351
policies, 99, 577–580, 602
policy-based storage management, **99**
pop-up ads, 144, 263
portable computers, **111–112**
portals, 266, 293
POS (point-of-sale) devices, **103**, 121
PowerPoint (Microsoft), 105, 151, 153
PPP (Point-to-Point Protocol), **263**
predictive analysis, **201**
preliminary evaluations, **537–538**
primary storage, **83**. *See also* memory
printers, 108, 109, 116
Privacy Act of 1974, 600
privacy issues, 27–28, 231, 287, 324, 576, 597–602, 604
problem solving. *See also* problems
  decision making as a component of, 384–385
  described, **386**
  phases, support for, 409
problems. *See also* problem solving
  highly structured, **410**
  semistructured/unstructured, **410**
procedures, **16**, 577–580
process(es)
  described, **6**
  symbols, **509**
processing
  controls, 534
  described, **11**
  mechanisms, **8**
processors. *See* CPUs (central processing units)
product(s)
  configuration software, 330
  creating new, 58
  customized, 59
  development, 401–402
  and service delivery stage, 308, 309–310
  substitute, threat of, 57
productivity, 59–60, 82, 148
profit centers, **397**
program(s)
  code, 156
  described, **128**
  trading, 12
programmed decision, **386**
programmers, 479
programming languages, **156–159**, 271–272

project(s)
    deadlines, **497**
    described, **479**
    managers, **479**
    milestones, **497**
    organizational structure, 47–48
    schedules, **497**
Project (Microsoft), 498
projecting, use of the term, **183**
PROM (programmable read-only memory), 88
proprietary software, 146–147
protocols. *See also specific protocols*
    described, **236–237**
    wireless, 237–240
prototyping, **488–490**
public
    domain software, **162–163**
    network services, **246**
purchasing systems, 308, 309, 358–359, 367
push technology, **275–276**

Q

quality
    control, **400**
    described, **53**
    development and, **497–498**
queries, 157, 191–192
questionnaires, **507**

R

RAD (rapid application development), 486, **490–491**
radio, 224, 282–284
RAID (redundant array of independent disks), **94**, 98
RAM (random access memory), 87–88, 115–116, 134.
    See also memory
read/write heads, 94
records, **177**
recovery utilities, **142**
recruiting, 405
recycling, 115
Red Hat Linux, 133, **139**, 140, 162
reengineering, **50–52**, 495
refreezing, 50
registers, **83**
relational databases, **181–196**. *See also* databases
releases, **552**
remote
    access, to files, 284
    logon, 278
replicated databases, **203–204**
report(s)
    demand, **391–392**
    developing effective, 392
    drill-down, **392**
    DSSs and, 407
    enterprise systems and, 348, 353–354, 366
    exception, **392**
    generating, 191–193
    key-indicator, **391**, 392

layouts, **513**
    MIS, 390–393, 402
    scheduled, **391**
    systems analysis, 513, 514
    systems design, **540–541**
    systems investigation, **504–505**
request for maintenance form, **552**
requirements analysis, **511–513**
resource-based view, 57
restart procedures, **530**
revenue centers, **397**
reviews, 579–580
RFIDs (Radio Frequency Identification) tags, 3, 14, 62, 88,
    197, 234, 254, 461
    described, **105–106**
    direct-order companies and, 59
    enterprise systems and, 350, 362, 367
    PANs and, 229
    radio transmissions and, 224
RFP (request for proposal), **535–536**, 540
robotics, 24–25, **440–442**
robots. *See* bots
ROI (return on investment), **60–61**, 324, 469, 538
roll up, 197
ROM (read-only memory), **88**, 116
ROP (reorder point), **399**
routers, **239–240**, 242, 258
rows, deleting, 183
RSI (repetitive stress injuries), 605–606
RSS (Really Simple Syndication), 278–279
rules, **449**

S

salaries, 405
sales
    analysis, 402
    forecasting, 366
    general role of information systems in, 30
    growth in, 59
    and operations plan, 366–367
    tax, 325
SANs (storage area networks), **98–99**, 356, 546
SaR (Sense and Respond), 388
Sarbanes-Oxley Act, 481, 354, 355, 579, 591
SASD (sequential access storage), **93**
satellites, 222–223, 263
satisficing, **387–388**, 408
scalability, **113**, 135, 485
scaling out/up, 113
scams, 590–591
scanners, 59, **102**
scenarios, **529**
schedules, 367, 391, 497, 504
schema, **188–190**
SCM (supply chain management), **43–44**, 62, 218, 294,
    321, 366–367
screen layout, **512**
script bunnies, **593**
SDRAM (synchronous RAM), **88**, 92
search and identification stage, 308, 309

search engines, **269–271**, 307

secondary storage, **92–99**

security. *See also* passwords; privacy issues; terrorism; viruses
 e-commerce and, 322–323, 329
 encryption, 95, 333, 339
 extranets and, 286
 firewalls, **28**, 242
 importance of, 576
 Internet and, 287
 intranets and, 286
 officers, 68
 overview, 27–28, 531–535
 spyware and, 162–163
 storage media and, 95
 system software and, 135
 systems development and, 531–535
 Trojan horses and, **584**

selecting, use of the term, **183**

semantic Web, 197

Sendmail, 160, 162

sequential access, **93**

servers, 122, 262–263, 271. *See also* client/server systems
 described, **111**, 113, 236
 farms, 113
 utilities, 144–145
 Web, **266**

service(s)
 creating new, 58
 customized, 59
 differentiation, 58
 improving, 59
 substitute, threat of, 57
 telecommunications and, 226–227
 Web, 273–274, 331

shareware, **162–163**

shells, **532**

shopping carts, **330**

sign-on procedures, **530**

SIMD (single instruction/multiple data), **90**

simulation, **408**

single user databases, **187**

site preparation, **547**

skill, use of the term, 8

SLAs (service license agreements), 161

SLIP (Serial Line Internet Protocol), **263**

slipstream upgrades, **553**

small business networks, 242

smart antenna technology, 239

smart cards, **333**, 352

smartphones, **111**, 141–142

SMS (Short Message Service), 276, 313

SNMP (Simple Network Management Protocol), 240

SOA (Service Oriented Architecture), 495, 520

SOAP (Simple Object Access Protocol), 274

social engineering, **582**

software. *See also specific types*
 acquiring, 543–545
 described, **14**
 e-commerce and, 328–329
 externally acquired, 544
 in-house-developed, 544–545
 installing, 130–132

 issues, 160–164
 overview, 126–172
 packages, integrated, 152–154
 piracy, 589–590
 removing, 130–132
 suites, **152–154**
 support, global, 164
 trends, 160–164
 upgrades, 163

sorting data, 190

source data automation, **100**

spam (junk mail), 29, 144

spatial data technology, 207

speech-recognition technology, **101**

split-case distribution, 311

sponsored links, 269

spreadsheet software, **148–150**

spyware, 162–163

SQL (Structured Query Language), 191–192, 594–595

SSL (Secure Sockets Layer), **331**, 332

stakeholders, **479**, 511

standards, **8–9**, 497–498

start-up, **549–550**

statistical sampling, **507**

steering committees, **505**

storage. *See also* memory
 described, **353**
 devices, 93–94
 secondary, 92–99

storefront brokers, **326**

strategic
 alliances/partnerships, **58**
 control, 419
 planning, 58–59, **419**

striping, 94

structured
 interview, **506**
 programming, **545**
 walkthrough, 545, 546

subnotebook computers, **111–112**

supercomputers, **111**, 114

support services, **65–66**, 164

switches, **239–240**

synchronous communications, **219**

syntax, **156**

system(s)
 analysis, 26–27, **488**, 505–514
 analyst, **479–480**
 components of, 8–9
 concepts, 8–9
 controls, **532–534**
 described, **8**
 design, 26–27, **488**, 526–541, 555
 development, 26–27, **486–501**
 implementation, 26–27, **488**, 541–550
 investigation, 26–27, **488**, 502–505
 maintenance, 26–27, **488**, 550–555
 operation, 550–555
 operators, **64–65**
 performance, **8–9**, 557
 request form, **502**

resources, 135
reviews, 26–27, **555–557**
security, 531–535
software, **130**, 128–145
testing, **548–549**

T

T-1/T-3 connections, 263–264, 265
tables, linking/joining, **184**, 185
Tablet PCs, 10, 104–105, 112, 231
TAM (technology acceptance model), **52–53**
tape storage, 92
task management, 134–135
taxation, 325
T-carrier systems, **237**
TCO (total cost of ownership), **61**
TCP (Transmission Control Protocol), **261**, 538
TCP/IP (Transmission Control Protocol/Internet Protocol), 263, 264
team organizational structure, **47–48**
technology
    diffusion, 53
    -enabled relationship management, **317**
    infrastructure, 13
    infusion, **53**
    transfer issues, 32
telecommunications, 534, 545–547. *See also* wireless
    networks
        described, **14–16**, 216–255
        medium, **219**
        network applications and, 241–248
        services, 226–227
        systems, elements of, **219**
telecommuting, **243–244**
teleconferencing, 417
television, on the Internet, 282–284
Telnet, **278**
terabyte, **88**
terminals, **102**
terrorism, 200, 235, 284
    cyber-, 582
    exception reports and, 392
    knowledge management and, 454
    MISs and, 394
    privacy issues and, 598, 601
    systems development and, 551, 563
testing, 400, 548–549
text
    alt (alternate), 269
    messaging, 276–277
thin clients, **111–112**
time-driven reviews, 556
TPS (transaction processing systems), **19–22**, 343–354, 491. *See also* payment systems; transactions
        auditing, 355
    traditional applications for, 357–361
trade laws/agreements, 32, 372–373
trademarks, 323–324

traditional approach, to data management, 179
training, 66, 405, 547, 557, 571
transaction(s). *See also* credit-card transactions; e-commerce;
    TPS (transaction processing systems)
        described, **19–20**
        processing cycle, **350**
trigger points, 392
Trojan horses, **584**
tunneling, **287**
Turing Test, 438–439
twisted-pair wire, **221–225**

U

UML (Unified Modeling Language), 504
unfreezing, 50
unit testing, 548–549
UNIX, 136, **139**, 158, 271, 298
unstructured interview, **506**
UPC codes, 350, 353, 367
upstream management, 44
URLs (Uniform Resource Locators), **261**, 266, 326, 329, 578–579
USA Patriot Act, 591, 601
usenet newsgroups, **279–280**
user(s)
    acceptance documents, **550**
    described, **479**
    documentation, **545**
    interface, **133**, 450–451, 529–531
    knowledge, **452**
    preparation, **546–547**
    satisfaction, **52–53**
    systems development and, 479, 501, 546–547, 550
    views, providing, 188
utility programs, **142–145**
UWB (Ultra Wideband), **239**

V

value chain, **43**, 44
versions, **552**
video
    -conferencing, **244–245**, 280–281, 416
    data, **5**
    games, 26–27, 68–69, 88
    on the Internet, 282–284
    logs (vlogs), **278–279**
virtual
    database systems, 207
    machines, 144–145
    memory, 134
    organizational structure, **48–49**
    reality, **23–25**, 457–461, 468
    tape, **94–95**
    workers, **243–244**, 248
    workgroups, **243–244**, 417
Virtual PC, 144–145
virtualization software, 144–145

virus(es)
    -detection software, **142**, 144, 585–588
    described, **584**, 585
    neural networks and, 455
visa programs, 64
vision systems, 24–25, **443**
visual application software, 157
Visual Basic (Microsoft), **157**, 329, 488, 500, 501, 553
visual database systems, 206–207
visual programming languages, 156–157
voice
    mail, **241**
    recognition, 443
VoIP (Voice over IP), 217, 222, 224, 233, **280–281**, 320
volume testing, **548–549**
VPNs (virtual private networks), **287**, 288, 595

W

WANs (wide area networks), **232**
WAP (wireless application protocol), 263, 314
warehouses, data, **197–199**
Web browsers. *See* browsers
Web page(s). *See also* Web sites; World Wide Web
    construction software, **329**
    dynamic/static, **329**
Web site(s). *See also* Web pages; World Wide Web
    building traffic to, **326–327**
    development tools, **329**
    establishing, 326
    functions, defining, **325–326**
    maintaining/improving, 327
    technology infrastructure, 327–331
    tracking, 329
what-if analysis, **408**
Wi-Fi networks, 224, 234, 237–238, 271, 528.
    See also wireless networks
Windows (Microsoft). *See also specific versions*
    described, **135–136**
    GUI, 133

programming languages and, 158
    running software on older versions of, 145
Windows Advanced Server (Microsoft), 136, **139**
Windows Embedded (Microsoft), **141**
Windows Mobile (Microsoft), 136, **141–142**, 253
Windows Server (Microsoft), **139**
Windows Vista (Microsoft), **136**, 137
Windows XP (Microsoft), 14, 133, **135–136**, 242
wire transfers, **246–247**
wireless networks, 222–223, 228, 237–240, 253.
    See also Wi-Fi networks
WML (Wireless Markup Language), 267, 314
Word (Microsoft), 105, **148–149**, 153
word processing software, 105, **148–150**, 153
World Wide Web. *See also* browsers; Web pages; Web sites
    described, **15**, 266–274
    developing content for, 272–373
    programming languages, 271–272
    research, **269–271**
    services, **273–274**, 331
workgroup(s), 136, 139–140, 144, 415–416
    application software, **154**
    described, 130
    sphere of influence, **130**
    virtual, **243–244**, 417
workstations, **111**, 113, 231, 553
worms, 584

X

XML (Extensible Markup Language), 212, **267–268**, 273–274, 338, 563
XP (extreme programming), 490

Z

Zip disks, 94
z/OS (IBM), **140**

Company

A

ABC News, 278
Accenture, 23, 56, 69, 500
Accolade, 525–526
Acxiom Corporation, 29, 91
Adobe Systems, 14, 594
Aelera Corporation, 493
Alaska Airlines, 403
Allstate, 412
Allulent, 348
Altra Energy Technologies, 16
Amazon.com, 53, 55, 68, 267, 272–274, 307
AMD (Advanced Micro Devices), 90
American Airlines, 82–83
American Century Investments, 63
American Electric Power (AEP), 353–354
American Express, 333
American Fidelity Assurance Company, 63
American Home Products, 164
American Red Cross, 590
Ameritrade, 318, 493
AMR Research, 327
Anheuser-Busch, 400
ANSI (American National Standards Institute), 191
AOL (America Online), 246, 248, 263, 264, 276, 280, 595, 596
AON, Inc., 549
Apple Computer, 90, 59, 108, 269
Applebee's, 520–521
AppRiver, 354
Arch Chemicals, 354
Archuleta, James, 403
Arena, Andrew G., 581
Army (United States), 409, 445, 582
Arthur Andersen, Inc., 493
Ask Jeeves, 145, 270
AT&T (American Telephone & Telegraph), 228, 263–264, 280, 312, 354, 455
Atlanta Veterans Administration, 392
Australia National University, 298–299
Autodesk, 415
Autonomy Corporation, 435
Autosketch, 153
Avaya, 353
Avvenu, 284
AXA Financial Services, 545
Azmi, Zalmai, 495

B

Bandyopadhyay, Pulak, 234
Bank of America, 177, 318
Bank of New York, 122–123
Bank One Corporation, 247
Barilla Group, 343–344
Barnes, Ken, 469
Barnes & Noble, 37
Bartlett, Jerry, 493
Basto, Jorge, 213, 214

BEA Systems, 520
Beaulieu Group, 358
Bedford Industries, 365
Bell, Clifford, 161
BellSouth Corporation, 264, 394
Benioff, Marc, 161
Bennett, Steve, 568
Berners-Lee, Tim, 266
Best Buy, 179, 366
Bezos, Jeff, 272
Bigstep, 326
BillMeLater, 332
BillPoint, 247
BIP (Beauté Prestige International), 383–384
Bischof, Andreas, 521
Bloomingdale's, 364
Blue Cross and Blue Shield of Louisiana, 356
BMW, 468
Boeing, 15, 218–219
Boston Celtics, 205
Boyes, Scott, 254
Bradley, Todd, 75
Brandeis, Louis, 597
Brandon Trust, 568–569
BRE Bank, 485
Briggs & Stratton, 403
British Petroleum, 370
Broadcom, 580
BroadVision, 273
Brookes, James, 456
Buckman Laboratories, 16
Bunker, Wayne, 484
Bunt, Thomas, 575
Bureau of Labor Statistics (United States), 63–64
Business Software Alliance, 589
Butler Group, 60
Buttasi, Carlo, 3
Byrd, Geoff, 282

C

C2Net, 162
Cable & Wireless PLC, 233
CACI International, 435
Caesars, 480
Carefree Technologies, 38
Carnegie Mellon University, 160, 441, 444, 581
Carnival Cruise Lines, 550
Carter, Steve, 446
Catholic Healthcare West, 380
CDC (Centers for Disease Control), 207, 267
Centra Software, 48
CERN, 266
CERT/CC (Computer Emergency Response Team/Coordination Center), 581, 592
Chabot, Christian, 195
Chamberlain, D. D., 191
Charles Schwab & Company, 443
Chase, Rory, 436
CHED (Centre for Higher Education Development), 300
Chicago Mercantile Exchange, 96

Child Support Agency, 497
Children's Memorial Research Center, 201
China Netcom Corporation, 435
CIA (Central Intelligence Agency), 110, 496, 600
Cinergy Corporation, 409
Cingular Wireless, 228, 312
Circuit City, 56
Cisco, 217, 265, 588
Citibank, 322, 323
Citigroup, 177
Citizens Property Insurance, 243
Citrix Online, 284
CIX (Commercial Internet Exchange) Association, 274
Clear Channel Communications, 278
Clearwire, 237
Cloudmark SpamNet, 144
Collins, Jim, 56
Comcast, 280
CompuServe, 248, 596
Computer Associates International, 243, 490, 591
Computer Security Institute, 581
ComScore Networks, 317
Comshare, 204
Congress (United States), 64, 95
Consolidated Edison, 318
Cottrell, Lance, 288
Council of Supply Chain Management Professionals, 321
Coventry University, 300–302
Cox, John, 10
Craigslist, 307
Cronin, Leo, 603
Customs Service (United States), 58, 242, 245–246
Cutter, Patty, 520

D

DaimlerChrysler, 154, 541
Darbee, Peter, 469
Davy, Christophe, 383
Dawson, John, 154
DCI (Discovery Communications, Inc.), 38
DeArment, Gary, 43
Deere & Company, 387
Defense Department (United States), 259, 582, 600
Del Monte, 61
Dell Computer, 59, 107, 109, 115, 243, 307
Deloitte & Touche, 397
Department of Education (United States), 580
Department of Homeland Security (United States), 178
Department of Veteran Affairs (United States), 589
Despres, Jonathan, 403
Destra, Frank, 319
DHL, 309
DigitalGlobe, 229
DLJdirect, 443
Dobrin, David, 171
DoCoMo, 313, 314
DOE (Department of Energy), 480, 486
Dogpile, 270
DoubleClick, 317
Dragon Systems, 443

DreamWorks SKG, 169
DTE Energy, 501
Duke University, 196
Dunkin Brands, 154
DuPont, 371–372
Dupuis, Marcel, 175

E

EarthLink, 233, 263, 264
EasyReach, 284
EasyStreet, 230
eBay, 68, 267, 273, 281–282, 307, 320, 323–325
EBT, 273
Edge, Curt, 564
Edwards, Karen, 435
eHarmony, 446–447
El-Al Airlines, 127
Electronic Arts, 312
Ellison, Larry, 161
E-Loan, 284
EMC, 99, 477
Emory University, 459
Environmental Defense Fund, 16
EPA (Environmental Protection Agency), 115
Epicor Vantage, 366
Equifax, 604
Erickson, Craig, 407
Erickson Retirement Communities, 407
Ericsson, 312
E*Trade, 318
Eurostar, 492
Everest Software, 371
Excelon, 206
Exempla Health, 551
Expedia, 37, 184, 325
ExxonMobile, 370

F

Fairchild Semiconductor, 306
FBI (Federal Bureau of Investigation), 29, 481, 495, 581, 589, 590
FCC (Federal Communications Commission), 280
FDA (Food and Drug Administration), 106
Federal Credit Union, 356
Federal Express, 15, 309–310, 345–346, 349
Federal Reserve Bank, 246
Federal Reserve Board, 247
Feikin, Jennifer, 283
Feilden, Clegg, Bradley, and Aesas, 435
Felsing, Grant, 403
Fidelity Investments, 443
Fiducia AG, 86
Fiorina, Carly, 75
First National Bank (FNB) of South Africa, 313
Firwood Paints, 257–258
Florida Bankers Association, 541
Florida Chamber of Commerce, 354
Focumentum, 273
Ford Motor Company, 224

Forrester Research, 213, 307, 319, 403
FoxNews, 270
Frayne, David, 196
Friedman, Thomas, 31
Froggle, 307
FSC, Inc., 477
F-Secure Corporation, 585
FTC (Federal Trade Commission), 243, 323, 583
FUJI Market, 294
Fujitsu, 101, 112, 121, 352

G

Garage-Band.com, 282
Gartner Group, 61, 177, 213, 287, 338–339
Gas Natural, 143
Gates, Bill, 161
Gaultier, Jean Paul, 383
Geffen, David, 169
Genalytics, 201
General Electric, 74, 196, 244, 492
General Motors, 41–42, 234, 400, 481, 540
Georgia State Courts, 213
Geospatial, 308
Ghosen, Carlos, 3
Gile, Keith, 213, 403
Gillette, 56, 398
Gilsbar, Inc., 356
Glamorgan University Business School, 272
GlaxoSmithKline, 331
Global Trade Services, 373
Godiva Chocolatier, 16
Goldman Industrial Group, 16
Google, 145, 200, 269, 270, 278, 273, 283, 285, 307, 323
Gore Mutual Insurance, 10
Gracenote, 339
Granite Rock Company, 202
Greencine, 400
Grokster, 283
Groove Networks, 415
GT Nexus, 242

H

Hackensack University Medical Center, 384
Halifax Bank, 323
Handspring, 110
Hannaford Brothers, 139
Harrah's, 401, 480
Harris Corporation, 253
Harrisdirect, 318
Hartford Life Insurance Company, 199
Harvard University, 43, 385, 454
Hawking, Steven, 437
Health Record Network Foundation, 196
Heesen, Bernd, 570–571
Hewlett-Packard, 75, 140, 164, 243, 309–310, 542, 584, 594
    disposal program, 115

DSSs and, 409
    enterprise systems and, 354
    MIS reporting and, 390
High Tech Computer Corporation, 253
Hitachi Data Systems Corporation, 99
H.J. Heinz, 61
HMRC (Her Majesty's Revenue & Customs), 26
Home Depot, 176, 358
HotBot, 270
HotJobs, 277
Howard Weil, Inc., 531–532
Huber, Chet, 41
Huber, George, 385
Hyperion Solutions, 204

I

i2 Technologies, 366
I4 Commerce, 332
IATA (International Air Transport Association), 127
IBM (International Business Machines), 69, 81, 90, 127, 154, 161, 164, 217, 257, 388, 455, 477–478, 490–494, 500, 542, 551, 557, 594–595
    databases and, 182, 188, 191, 194, 197, 206, 207
    diskette format, 92
    disposal program, 115
    DSSs and, 409
    global support operations and, 243
    GSSs and, 415–416
    job listings, 277
    mainframes, 114, 122
    MISs and, 405
    Nissan and, 3
    PalmOne and, 110
    processors and, 13
    SANs and, 99
    software patents and, 162
    system software and, 135–137, 139–140
    utilities, 144–145
    Web portals, 38, 293
ICANN (Internet Corporation for Assigned Names and Numbers), 262, 312
ICRA (Internet Content Rating Association), 595
IEA (International Energy Agency), 176
IFS, Inc., 373
iHealthRecord, 412
IMC (International Manufacturing Corporation), 570–571
Indigo Books & Music, 612–613
Information Builders Incorporated, 175
Informix, 272, 329
ING Direct, 409
Inline Internet Systems, 197
Institute for Electrical and Electronics Engineers (IEEE), 237, 238
Intel, 90
Intermountain Health Care, 196
Invensys PLC, 235
Iomega, 109
IRI (Information Resources, Inc.), 351
Irish Health Service, 555

Iron Mountain, 355
IRS (Internal Revenue Service), 20, 323, 371
ISM, Inc., 368–369
isMobile, 143
ISO (International Organization for Standardization), 74–75, 497, 498
Itochu Kenzai Corporation, 74
ITU (International Telecommunications Union), 239
Ixquick, 270

J

Jackson, Shirley, 276
Jade Stadium, 400
Javelin Strategy & Research, 582
J.D. Edwards, 365
J.D. Powers, 537
Jeffereson Wells International, 579
JetBlue Airways, 545
Jobs, Steve, 90
JobSpot, 273
John Cotton Taylor School, 460
Johnson, Linton, 552
Johnson & Johnson, 575–576
JP Morgan, 177, 318, 247
Justice Department (United States), 29, 200, 371, 406, 598

K

Kafner, Katie, 260
Kahn, Robert, 260
Kaiser Permanente, 455
Karlson, Maakon, 397
Katzenberg, Jeffrey, 169
Kay, Alan, 133
KDDI, 339–340
Keesler Federal Credit Union, 356
Kemmons Wilson, 104
Kenexa, 331
Kennedy Space Center, 414
Kerr, Thomas, 541
Keystone Automotive Industries, 64, 493
Klepper, Ken, 245
Kodak, 55, 101, 107
Kovac, John, 265
Kravchenko, Sergev, 54
Kulula.com, 472–474
Kurzweil, Ray, 438

L

Labor Department (United States), 63
Ladisernia, Anthony, 265
Lafarge, 431–432
Lamson, George, 121
LandAmerica Financial Group, 235
Lanier, Jaron, 457
LAPD (Los Angeles Police Department), 22, 103, 407
Laplink, 284

LaRocca, Joseph, 481
Laurel Pub Company, 391–392
Lawrence Livermore National Laboratory, 115
Lawson, Stephen, 612
Lee, Sharon, 305
LendingTree, 284, 319
LetsGoDigital, 196
Lewin, Kurt, 50
Lewis, Doug, 482
Lexis-Nexis, 435, 603
Library of Congress (United States), 87
Liquid Films, 298–299
Litan, Avivah, 338–339
Livermore National Laboratories, 13
Liz Claiborne, Inc., 265
Lloyd, Steve, 613
Lloyds TSB Bank, 323
LMG (Liberty Mutual Group), 348
Lockheed Martin Corporation, 495
Logie, Scott, 492
Look-Look, 305
Lowe's Companies, Inc., 358, 366
LowerMyBills.com, 319
Lucasfilms, 193
LUCIE, 431–432
Lufthansa, 455
Lynx Services, 243
Lyons Bakeries, 154

M

M2 Systems, 294
MAA Bozell, 53
Macromedia, 197
Macworld, 90
Macy's, 364
Magnify, Inc., 201
Magnolia Hi-Fi, 179
Mansholt, Karl-Heinz, 477
MapInfo, 207
Marks & Spencer PLC, 233
Maroochy Sire Council, 484
Marr, Brian, 29
Marriott, 482
Marvell, 580
MasterCard, 333, 339, 545–546
Maytag Corporation, 228
Mazda, 248
McAfee, 130, 142, 144, 584, 585, 586
McCarthy, John, 437
McClellan, Mark, 176
MCI, 263
McNall, Mike, 27, 494
Medco Health Solutions, 245
Medd, Denise, 575
Medicaid, 176
Medicare, 176, 583
Mental Images, 13
Mercury Interactive, 144
Mercy Health System, 231

MetLife, 480
Metro AG, 229
Michigan, State of, 293–294
MineShare, 204
Minnesota Life Insurance, 63
Misterovich, Pat, 69
Mitchell, Tom, 444
Mitre Corporation, 63
Mitsubishi Motors, 273
MobileDataforce, 254
Mohammed, Abid, 569
Monster.com, 68, 258, 267, 277
Morey, Darly, 205
Moser, Tom, 588
Motorola, 109, 137, 212–213, 225, 312
Mott, Randy, 75
Mulligan, Edward, 122
Munich Re Group, 435
Murphy, Mick, 231
Music Unlimited, 272

N

NACHA (National Automated Clearing House Association), 247
Napster, 283
NASA (National Aeronautics and Space Administration), 110, 196, 207, 267, 441, 582
Nash, John, 462
National Car Rental, 494
National Do Not Call Registry, 243
National Foods, 315–316
National Infrastructure Protection Center, 582
National Insurance Crime Bureau, 68–69
National Recycling Coalition, 115
NatWest, 323
Navajo Nation, 247
NaviTag, 461, 462
NCR Teradata, 201
Neff, Dianah, 67
Net2Phone, 280
NETGEAR, 580
Network Appliance, 63
Network Associates, 585
New York City Police Department, 203
New York Metropolitan Transport Authority, 102
Newmarket International, 544
NexTag.com, 319
NFIC (National Fraud Information Center), 591
Nicolas, Peter, 48
Nike, 305
Nissan Motor Company, 3–4, 24
Nokia, 111, 312
Nordmilch, 477–478
Nordstrom, 364
Nortel, 469
NorthernLight, 270
Northrop Grumman Corporation, 356
NPR (National Public Radio), 278
NSA (National Security Agency), 394, 597–598
NSI (Network Solutions), 262

O

Occam, 492
Ochsner Clinic Foundation, 356
Office Depot, 92, 314
OnFiber, 230
Opportunity International, 412
Oracle, 49, 81–82, 182, 188, 194, 197–198, 201, 204, 272, 329, 373, 490, 500, 525, 557
    global support operations and, 243
    Salesforce.com and, 161
    Select Comfort and, 170–171
Orbitz, 184, 325
Oversight Systems, 353

P

Pacific Gas and Electric (PG&E), 101, 389, 469
Panasonic, 139, 366–367
Parker, Jefferson, 532
Parker, Robert, 234
Partnership America, 16
PayPal, 313, 332
PCMCIA (Personal Computer Memory Card International Association), 97
Pearce, Hilary, 568
Penril/Bay Networks, 225
PeopleSoft, 365
Perot Systems Corporation, 380
Pfizer, 146
Pharmascience, 403
Phillips, Dan, 196
Phillips, Steve, 469
Phoenix Technologies, 161
Pico Electronics, 379
Pienaar, Len, 313
Pike, Linda, 380
Pixar, 13
Polygon, 193
Popkin Software, 500
Porsche, 58
Port of Brisbane, 563–564
Porter, Michael, 43, 57, 59
Porter Adventist Hospital, 441
Pragmatech Software, 535–536
Precise Software, 557
PriceGrabber, 307
Priceline, 37, 320, 325
PriceWaterhouseCoopers, 48, 397
Procter & Gamble, 241, 398, 412
Prodigy, 246, 248
ProFlowers, 361
ProFusion, 270
Prover Technology, 500
Providence Washington Insurance Company, 390

Q

QAD, Inc., 373
Quadrem, 315–316
Quadstone, 201

Qualcomm, 63, 312
Quotron, 248

R

RailCrewXpress, 254
Red Brick Systems, 201
Relativity Technologies, 554
Renault Motors, 3
Restoration Hardware, 443
RIAA (Recording Industry Association of America), 283
Richards, Murphy, 435–436
Rideout, Jan, 356
RIM, Inc., 556
Robot Learning Laboratory, 441
Rodriguez, Narcisco, 383
Rogge, Eric, 213
Rolex, 58
Rosen, Robert, 554
RosettaNet, 306–307
Royal Bank of Canada, 490
Royal Bank of Scotland, 47
Ryder Systems, 23, 433

S

Sable Holdings Corporation, 403
Sakamoto, Yoshihiro, 74
Salesforce.com, 161, 371
Salvation Army, 368–369
Sam Goody, 179
Samsung, 109
San Diego Gas & Electric, 389
Sanderson, Tom, 53
SAP AG, 21, 154, 331, 338, 343, 363–365, 370–372, 374, 379–380, 435, 477, 543
SAS, 201, 478
SBC Communications, 228
SC-Integrity, 461
Schein, Edgar, 50
Schlegel, Kurl, 213
SCO Group, 584
Scripter, David, 356
Seagate Technology, 455
Sears, 55, 366
SEC (Securities and Exchange Commission), 371
Secret Service (United States), 29
Securities Industry Association, 51
SEI (Software Engineering Institute), 160, 497, 581
Seibel Systems, 365
Select Comfort, 170–171
services, 263
7-Eleven, 198
Severino, Victoria, 199
Sharp Corporation, 107
Shimmel, Melody, 541
Shopzilla, 307
Shroff, Arjun, 37
Siebel Systems, 49, 81–82, 170–171
Siemens, 16
Sierra Pacific, 241

Sightward, 201
Simon, Herbert, 385
Singh, Anesh Maniraj, 472–473
SITCI (Shroff International Travel Care), 37–38
Skype, 320
Sling Media, 283–284
Smalltalk, 158
Snodgrass, Steve, 202
Softsmart, 164
Software AG, 199
Sony, 109
Southern California Edison, 389
Southern Wine and Spirits of America, 527
Southwest Airlines, 288
Spencer Gifts, 232
Spielberg, Steven, 169
Sprint Nextel, 253, 264, 312, 494
Stanford University, 454
Staples, 121–122, 406
Stark, Philip, 200
Starwood Hotels & Resorts, 114, 206, 324
State Department (United States), 178
State Farm, 412
Strothkamp, Brad, 319
Sumic, Zarko, 469
Sumitomo Mitsui Banking Corporation, 113
Sun Microsystems, 61, 82, 86, 139, 152, 157, 271, 477, 497
Suncoast, 179
Supreme Court (United States), 325
Surrey County Council, 434
SWIFT (Society for Worldwide Interbank Financial Telecommunication), 246
Sybase, 164, 182, 188, 199, 201
Symantec, 142, 586

T

Tactician Corporation, 406
Takahashi, Makoto, 339–340
Tandem, 201
Target, 398
TD Waterhouse, 318, 443
TeamWare, 416
TechConnect, 228
TECSys Development, 480
Telos, 436
Temple, William C., 29
Tesco, 45
Tew, Alex, 402
TicketMaster, 205
Tiffany & Company, 323–324
Time, Inc., 55
Timex, 525–526
TodayCorp, 563
Torvalds, Linux, 139
TowerGroup, 339
Toyota, 105, 248
Travelers Life & Annuity, 480
Travelocity, 37, 184, 267, 325
TSA (Transportation Security Administration), 394
Turkish Army, 408

Turning, Alan, 438
Tyndall, Gene, 23, 433

U

uBid, 307
UCAID (University Corporation for Advanced Internet
    Development), 260
Union Bank, 61
United Airlines, 478–479
United Distillers, 455
United Nations, 176
Universal Health Services, 63
UPS (United Parcel Service), 44–45, 245–246, 309, 345,
    541–542, 556
URS Corporation, 356
U.S. DataTrust, 95
U.S. Robotics, 225
USPS (United States Postal Service), 53, 577, 604

V

Valero Energy, 174–175, 235, 351
Vanguard Car Rental, 27, 494
Vecchio, Dale, 122
Velline, Jonathan, 338
Ventana Research, 213
VeriSign, 320
Verizon Communications, 280, 312, 394, 595
Versant, 206
Vetco International, 370
Vignette, 273
Virgilio, 461
Virgin Entertainment Group, 217–218
Virginia, State of, 409
Visa International, 322, 323, 333
Visible Systems, 500
Visio, 500
Vonage, 280
VPL Research, 457

W

W3C (World Wide Web Consortium), 267, 599
Waeco Pacific, 521
Walgreens, 56, 412
Wallace, Catriona, 456
Wallen, Martin, 257
Wal-Mart, 4–5, 58, 62, 75, 121, 154, 176, 195–196, 313,
    314, 316, 398, 550
Walther, Arne, 176
Warren, Neil Clark, 446
Wells Fargo Bank, 56, 247, 338–339
Westpac Bank, 323
Western Union, 55, 332
Westinghouse, 588
Whirlpool Corporation, 20, 366
WhiteLight, 204
Whitmann Price Consulting, 39, 76
Widgit Software, 145
Wikipedia, 271
Wild Oats Market, 54
Witty, Jason, 528
Wong, Abe, 170
Wong, Nicole, 200

X

Xerox PARC (Palo Alto Research Center), 133
Xinhua News Agency, 259
Xiotech, 99

Y

Yahoo!, 68, 145, 196, 258, 266–267, 269–270, 276–277,
    307, 323
Yamanaka, Toshihiko, 294
Yankee Group, 312